AF705727

FORTSCHRITTE DER CHEMIE ORGANISCHER NATURSTOFFE

PROGRESS IN THE CHEMISTRY OF ORGANIC NATURAL PRODUCTS

BEGRÜNDET VON · FOUNDED BY

L. ZECHMEISTER

HERAUSGEGEBEN VON · EDITED BY

W. HERZ
TALLAHASSEE, FLA.

H. GRISEBACH
FREIBURG i. BR.

G. W. KIRBY
GLASGOW

VOL. 35

VERFASSER · AUTHORS

O. R. GOTTLIEB · K. HERRMANN · R. D. H. MURRAY
G. OHLOFF · G. PATTENDEN

1978

WIEN · SPRINGER-VERLAG · NEW YORK

This work is subject to copyright
All rights are reserved, whether the whole or part of the material is concerned, specifically those of translation, reprinting, re-use of illustrations, broadcasting, reproduction by photocopying machine or similar means, and storage in data banks

© 1978 by Springer-Verlag/Wien

Library of Congress Catalog Card Number AC 39-1015

ISBN-13: 978-3-7091-8507-0 e-ISBN-13: 978-3-7091-8505-6
DOI: 10.1007/978-3-7091-8505-6

Inhaltsverzeichnis. Contents

I. Introduction

According to HAWORTH (*73, 74*), lignans are plant products based on carbon skeleta having two n-propylbenzene residues linked by the β-carbon atoms of the side chains. Around 1940, when this definition was published, practically all known bis-phenylpropanoids were derivatives of structural variants of only **1** and **2** (Table 1). Since then, the additional skeleta **3**—**15** were recognized and the restriction imposed on the mode of union of the two units seems now unprofitable. It was therefore proposed that the term lignan be extended to cover all natural products of low molecular weight that arise primarily from the oxidative coupling of p-hydroxyphenylpropene units (*106*), a concept which implicitly also refers to variants of skeleta such as **15** in which the two units are linked by an oxygen bridge. This is, of course, fundamentally a biogenetic definition and implies knowledge of the monomeric precursor units.

Four such units seem to be involved: cinnamic acid (and exceptionally aldehyde), cinnamyl alcohol, propenylbenzene and allylbenzene. These are conveniently distributed among groups, one encompassing the

Neolignans

By O. R. GOTTLIEB, University of São Paulo, Brasil

Contents

Mitarbeiterverzeichnis. List of Contributors

Gottlieb, Prof. Dr. O. R., Instituto de Quimica, Universidade de São Paulo, Caixa Postal 20780, 01000 São Paulo, Brazil.

Herrmann, Prof. Dr. rer. nat. K., Lehrstuhl für Lebensmittelchemie, Technische Universität, Wunstorferstraße 4, D-3000 Hannover, Bundesrepublik Deutschland.

Murray, Dr. R. D. H., Chemistry Department, The University, Glasgow G12 8QQ, Scotland.

Ohloff, Dr. G., Laboratoire de Recherches, Firmenich SA, Case Postale 239, CH-1211 Genève 8, Switzerland.

Pattenden, Prof. Dr. G., Department of Chemistry, The University of Nottingham, Nottingham, England.

former two and one the latter two of the precursors, on account of two facts. 1. Structural analysis of the coupling products reveals involvement either of cinnamic acid + cinnamic acid, cinnamic acid + cinnamyl alcohol, cinnamyl alcohol + cinnamyl alcohol, or of propenylbenzene + propenylbenzene, propenylbenzene + allylbenzene, allylbenzene + allylbenzene. Furoguaiacidin (**1.10**) (Table 2), the sole known mixed oxidative dimer (one oxygenated and one unoxygenated γ-carbon) co-occurs in *Guaiacum officinale* with furoguaiacin (**1.9a**), its probable biosynthetic precursor. 2. Distributional analysis of the coupling products shows that acid and/or alcohol derivatives are of widespread occurrence, in contradistinction to propenyl and/or allyl derivatives which have so far been located only in relatively few and, apparently, related taxa.

All this testifies to a certain degree of biosynthetic independence of the two groups of coupling products, which are thus conveniently designated by different terms. Since most of the "Haworth lignans" are derived by coupling of acid and/or alcohol, the term lignan is retained for this group. The propenyl and/or allyl derivatives are termed neolignans (*65*).

The application of systematic nomenclature to neolignans (see subtitles of Sections in Contents) detracts from the fact that, in spite of the diversity of skeleta and functions, they form a singularly homogeneous group of natural compounds. The fact is honoured in this review in the following way: If present, the $Ar \cdot C_3$ group is written towards the left, as in Table 1. Its C-atoms are numbered 1—9, while the C-atoms of the additional $C_6 \cdot C_3$ unit are numbered 1′—9′, *e.g.* **16** according to FREUDENBERG and WEINGES (*57*). Direct or oxygen links between the units are indicated by identification of the bridgehead positions through the smallest possible numerals, respectively, *e.g.* 7.2′ (and not 6′), 8.8′ (Table 1, type **2**) and 8.O.4′-neolignans (Table 1, type **15**). With these simple rules in mind, designations for neolignan skeleta (Table 1) and structures (Table 2) are easily grasped.

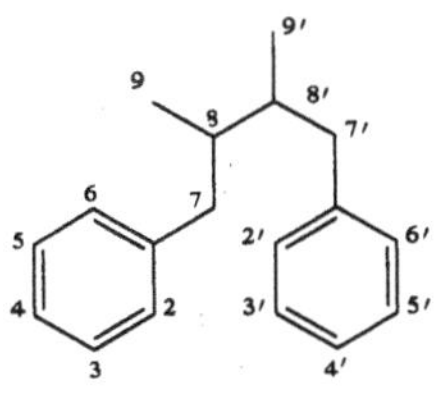

(16)

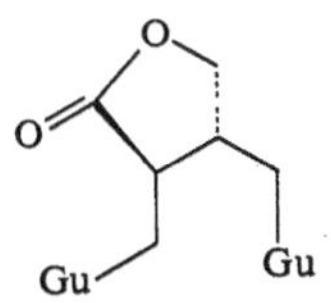

(17)

The neolignans are listed in Table 2 by two digit arabic numerals and a letter which represent, in order, skeletal type (Table 1), configurational and constitutional differences. The same symbols are also used in the

Table 1. *Numbers of Derivatives of Bis-Phenylpropanoid Carbon-Skeleta, Classified According to the Apparent Nature of the Precursors (data on lignans based on ref. 103)*

Type	Structure of skeleta	Designation	cinnamoyl / cinnamoyl (LIGNANS)	cinnamoyl / cinnamyl (LIGNANS)	cinnamyl / cinnamyl (LIGNANS)	propenyl / propenyl (NEOLIGNANS)	propenyl / allyl (NEOLIGNANS)	allyl / allyl (NEOLIGNANS)
1	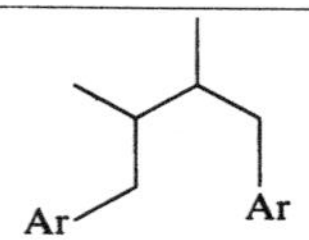	8.8′		27	55	22		
2	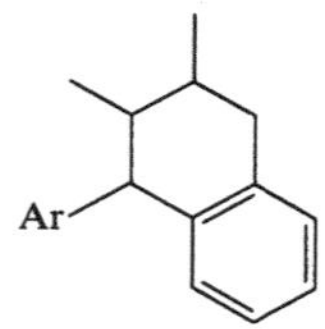	8,8′,7.2′		31	11	15		
3	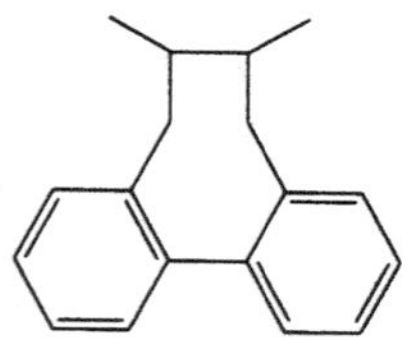	8.8′,2.2′		4		7		
4	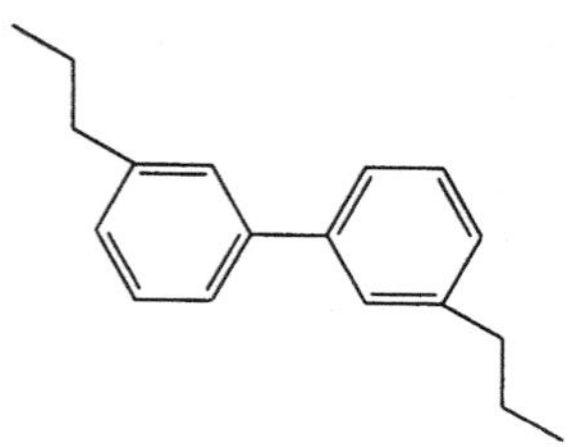	3.3′		1		1		2

5	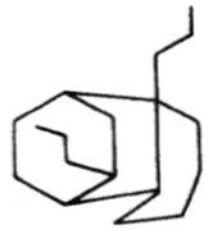	5.1′,2.2′				1
6	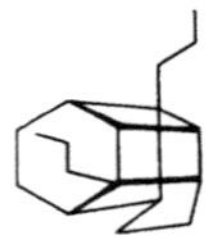	5.1′,2.2′,1.5′,6.6′				1
7	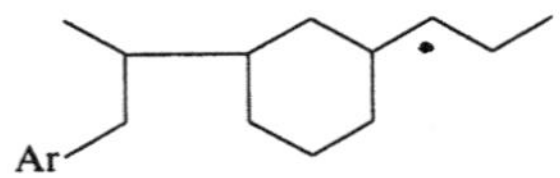	8.3′	2	16	4	
8	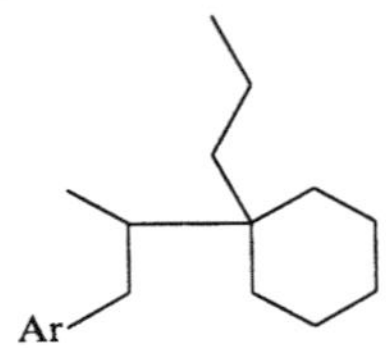	8.1′			11	
9	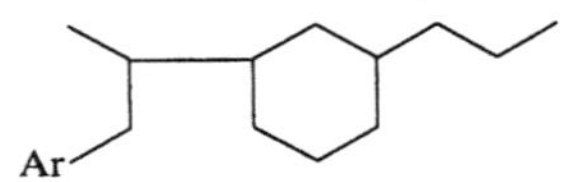	8.3′*			8	
10	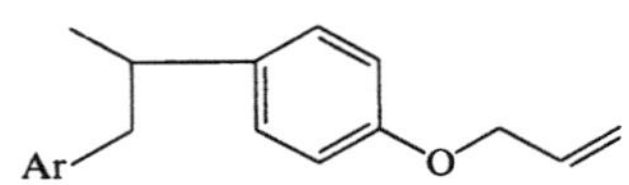	8.4′,1.O.7′*			6	

* Possibly formed by rearrangement of type 8 (see Section VIII).

Table 1 (continued)

Type	Structure of skeleta	Designation	LIGNANS: cinnamoyl | cinnamoyl	LIGNANS: cinnamoyl | cinnamyl	LIGNANS: cinnamyl | cinnamyl	NEOLIGNANS: propenyl | propenyl	NEOLIGNANS: propenyl | allyl	NEOLIGNANS: allyl | allyl
11	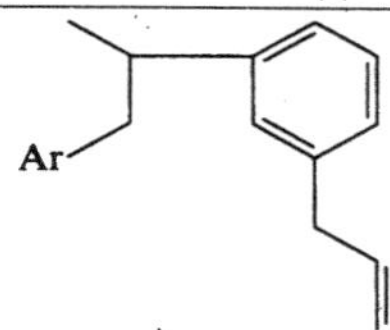	8.3′*					7	
12	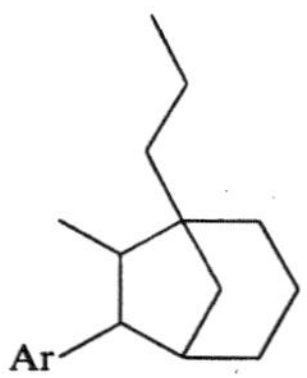	8.1′,7.3′					13	
13	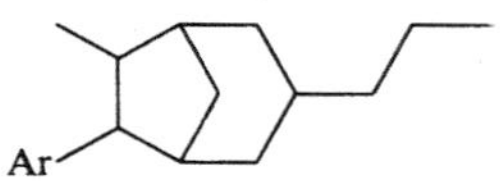	8.3′,7.5′					2	
14	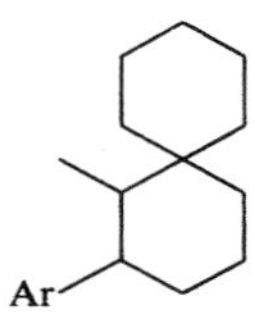	8.1′,7.9′					1	
15	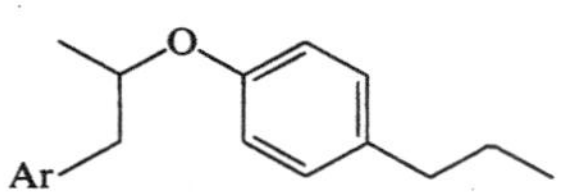	8.O.4′				1	2	

Table 2. *Structure and Occurrence of Neolignans*
(symbols for Ar-groups: Table 3, symbols for source species: Table 4)

Structure		Ar¹	Ar²	Name	Source
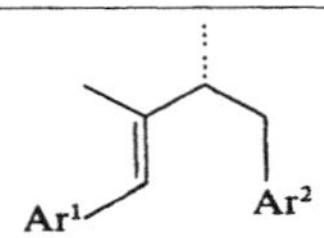	**1.1**			(8′R)-Δ^7-8.8′-lignan	
	1.1a	Gu	Gu	(−)-guaiaretic acid	ZYGO 1 (*85*)
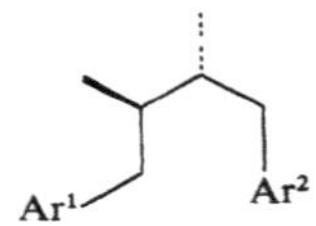	**1.2**			(8R,8′R)-8.8′-lignan	
	1.2a	Gu	Gu	(−)-dihydroguaiaretic acid	ZYGO 1 (*85*)
	1.2b	Pi	Gu	(−)-austrobailignan-6	AUST 1 (*112*)
	1.2c	Pi	Pi	(−)-austrobailignan-5	AUST 1 (*112*)
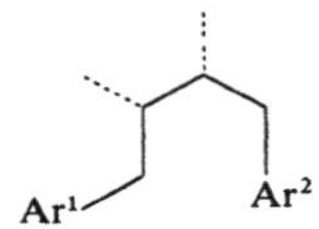	**1.3**			(8S,8′R)-8.8′-lignan	
	1.3a	Hy	Hy	conocarpol (mixt. with **1.4a**)	COMB 1 (*76*)
	1.3b	Ca	Ca	*meso-nor*-dihydroguaiaretic acid (NDGA)	ZYGO 2 (*126*), 3 (*143*), 4 (*96*);
	1.3c	Ca	Gu		ZYGO 3 (*63*)
	1.3d	Gu	Gu	*meso*-dihydroguaiaretic acid	ZYGO 1 (*85, 133*), 3 (*63*); LAUR 18 (x) (*101*)
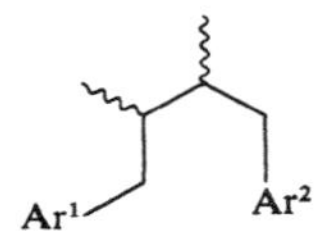	**1.4**			8.8′-lignan	
	1.4a	Hy	Hy	(±)-conocarpol (mixt. with **1.3c**)	COMB 1 (*76*)
	1.4b	Hy	Mr	2′-methoxyconocarpol	COMB 1 (*76*)
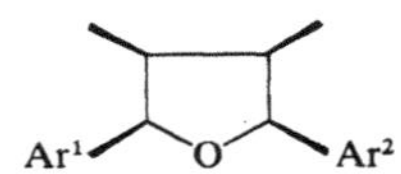	**1.5**			*rel*-(7R,8S,7′S,8′R)-7.O.7′,8.8′-lignan	
	1.5a	Gu	Gu	tetrahydrofuroguaiacin B	ZYGO 1 (*85*)
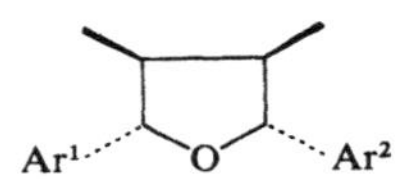	**1.6**			*rel*-(7S,8S,7′R,8′R)-7.O.7′,8.8′-lignan	
	1.6a	Gu	Gu	tetrahydrofuroguaiacin A	ZYGO 1 (*85*)
	1.6b	Ve	Ve	galgravin	HIMA 1 (*83*); MAGN 1 (*51*)

Table 2 (continued)

		Ar^1	Ar^2	Name	Source
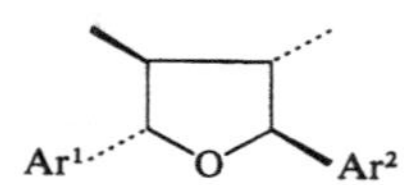	**1.7**			*rel*-(7S,8S,7′S,8′S)-7.O.7′,8.8′-lignan	
	1.7a	Pi	Pi	(−)-galbacin	HIMA 2 (*83*); MYRI 4 (*24*)
	1.7b	Ve	Ve	(−)-galbelgin	HIMA 1, 2 (*83*)
	1.7c	Mp	Mp	(−)-grandinsin	LAUR 15, 16 (*82*)
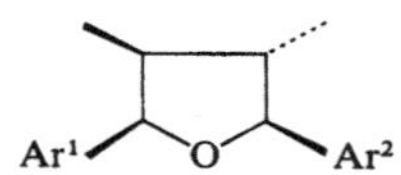	**1.8**			(7R,8S,7′S,8′S)-7.O.7′,8.8′-lignan	
	1.8a	Gu	Pi	(+)-austrobailignan-7	AUST 1 (*112*)
	1.8b	Ve	Pi	(+)-calopiptin	MAGN 1 (*51*); TRIM 1 (*105, 124*)
	1.8c	Ve	Ve	(+)-veraguensin	MAGN 1 (*51*); LAUR 22 (*47*); MYRI 4 (*24*); TRIM 2 (*105*)
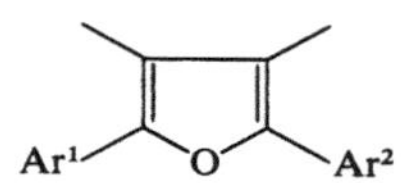	**1.9**			$\Delta^{7,7'}$-7.O.7′,8.8′-lignan	
	1.9a	Gu	Gu	furoguaiacin	ZYGO 1 (*85*)
	1.9b	Gu	Ve	methylfuroguaiacin	ZYGO 1 (*85*)
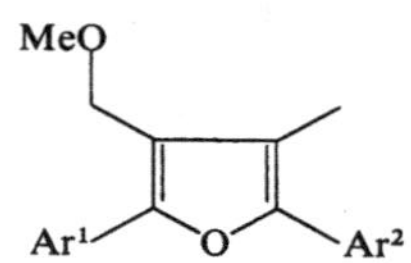	**1.10**			9-methoxy-$\Delta^{7,7'}$-7.O.7′,8.8′-lignan	
	1.10a	Gu	Gu	furoguaiacidin	ZYGO 1 (*100*)

Structure		Ar	R	R¹	Name	Source
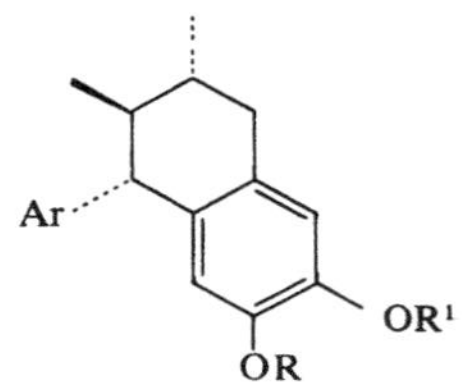	**2.1**				(7S,8S,8′R)-7.2′,8.8′-lignan	
	2.1a	Gu	H	Me	(+)-guaiacin	ZYGO 1 (*85*); LAUR 18 (*101*)
	2.1b	Pi	H	Me	(+)-otobaphenol	MYRI 1 (*88*)
	2.1c	Gu	$-CH_2-$		(−)-iso-otobaphenol	MYRI 2 (*67*)
	2.1d	Ve	$-CH_2-$		(−)-galcatin	HIMA 2 (*80*); MYRI 2 (*67*)
	2.1e	Pi	Me	Me	(+)-isogalcatin (=iso-otobain) (y)	MYRI 1 (*142*)
	2.1f	Ve	Me	Me	(−)-galbulin	HIMA 2 (*83*)
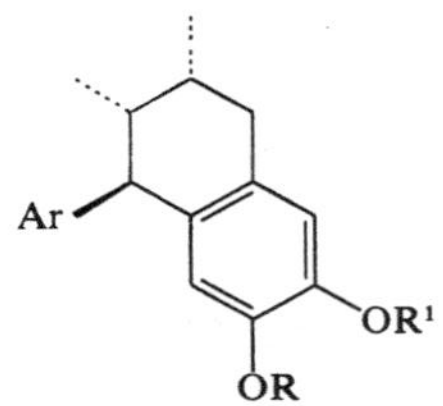	**2.2**				(7R,8R,8′R)-7.2′,8.8′-lignan	
	2.2a	Hy	H	Me	3′-demethoxyisoguaiacin	ZYGO 3 (*63*)
	2.2b	Gu	H	H	(−)-*nor*-isoguaiacin	ZYGO 3 (*63*)
	2.2c	Gu	H	Me	(−)-isoguaiacin	ZYGO 1 (*85*)
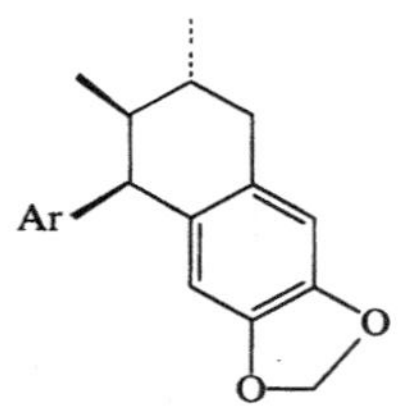	**2.3**				(7R,8S,8′R)-7.2′,8.8′-lignan	
	2.3a	Mp			(−)-austrobailignan-3	AUST 1 (*112*)
	2.3b	Tp			(−)-austrobailignan-4	AUST 1 (*112*)

(x) Insufficiently characterized. (y) Isolated from a methylated extract.

Table 2 (continued)

Structure		Ar	R	Name	Source
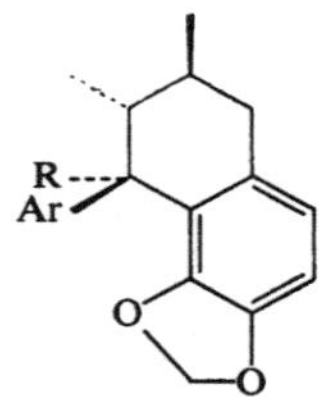	2.4			(7R,8R,8′S)-7.2′,8.8′-lignan	
	2.4a	Pi	H	(−)-otobain	MYRI 1 (*62*)
	2.4b	Pi	OH	(−)-hydroxyotobain	MYRI 1 (*142*), 3 (*29*)
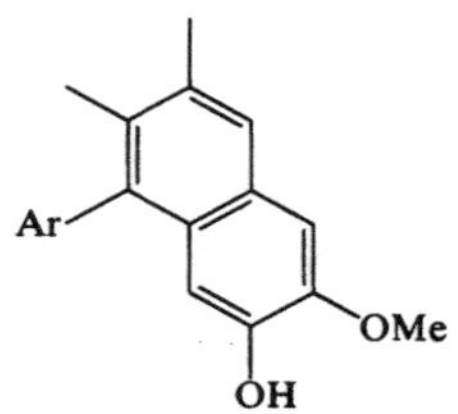	2.5			$\Delta^{7,7'}$-7.2′,8.8′-lignan	
	2.5a	Gu		dehydroguaiaretic acid	ZYGO 1 (*85*)
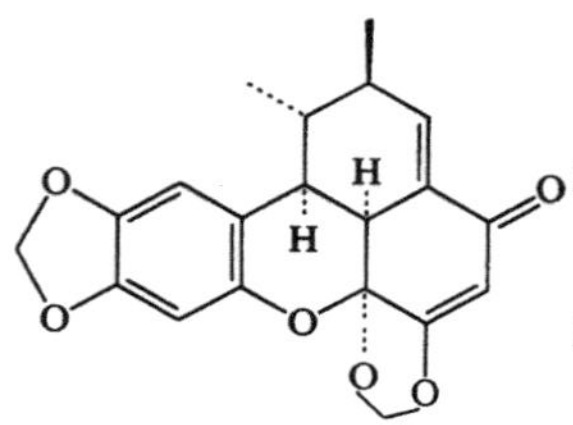	2.6			*rel*-(7S,8R,2′S,3′S,8′R)-$\Delta^{1'(7')}$-1′,2′,3′,6′-tetrahydro-6′-oxo-2.O.3′,7.2′,8.8′-lignan	
	2.6a			carpanone	LAUR 7 (*34*)

Structure	No.	R	R	Name	Source
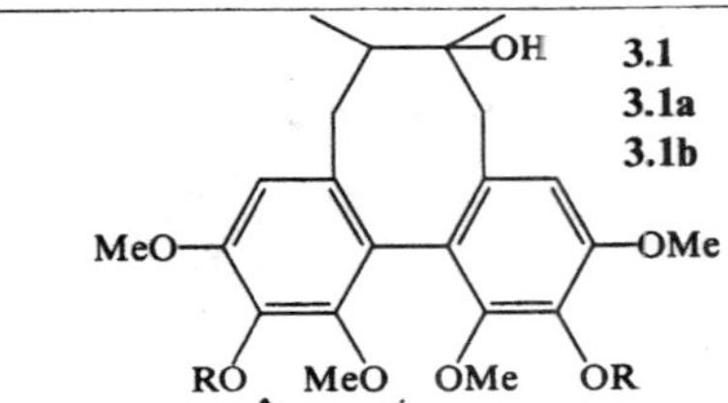	**3.1**			8′-hydroxy-2.2′,8.8′-lignan	
	3.1a	H	H	schizandrol (x)	SCHI 1 (*89*)
	3.1b	Me	Me	schizandrin	SCHI 1 (*89, 90, 94*)
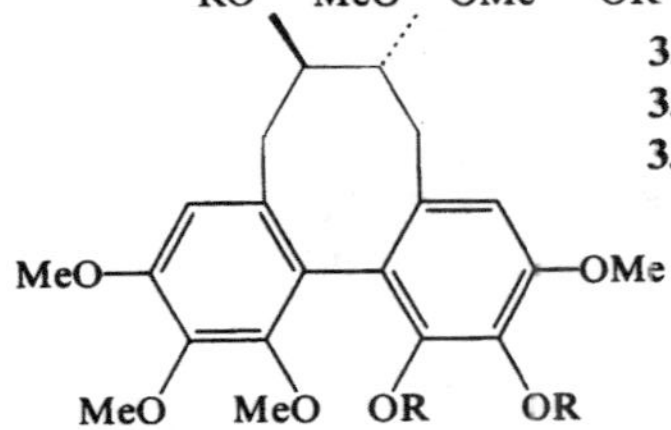	**3.2**			*rel*-(8R,8′S)-2.2′,8.8′-lignan	
	3.2a	Me	Me	deoxyschizandrin	SCHI 1 (*92, 93*)
	3.2b	$-CH_2-$		pseudo-γ-schizandrin	SCHI 1 (*92*)
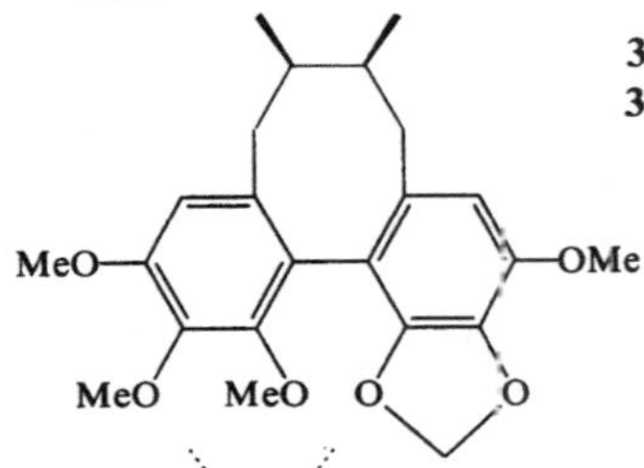	**3.3**			*rel*-(8R,8′R)-2.2′,8.8′-lignan	
	3.3a			γ-schizandrin	SCHI 1 (*92*)
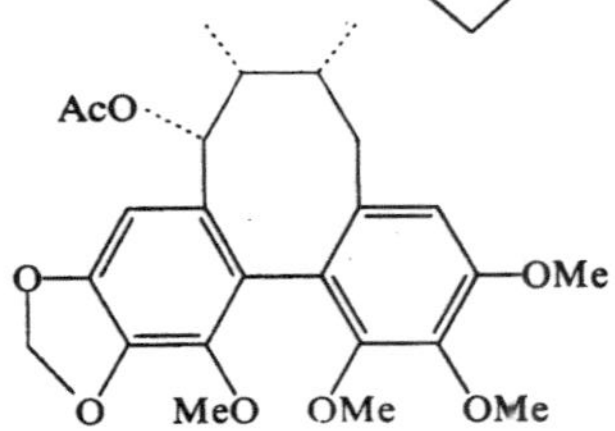	**3.4**			*rel*-(7R,8R,8′S)-7-acetoxy-2.2′,8.8′-lignan	
	3.4a			kadsurin	MAGN 3 (*43*)

(x) Insufficiently characterized.

Table 2 (continued)

		R	R	Name	Source
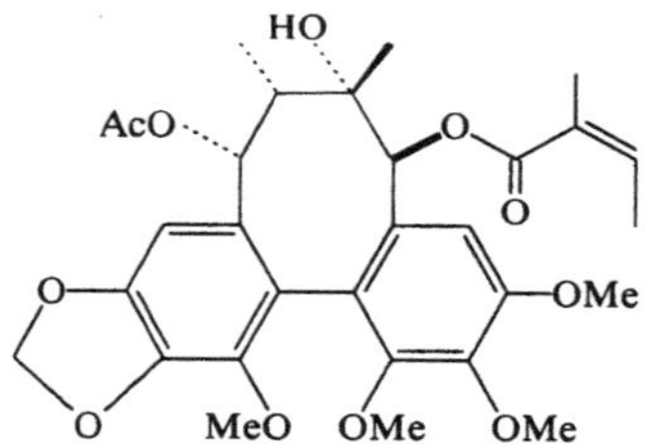	**3.5**			*rel*-(7R,8S,7′R,8′S)-7-acetoxy-7′-angeloxy-8′-hydroxy-2.2′,8.8′-lignan	
	3.5a			kadsurarin	MAGN 3 (*43*)
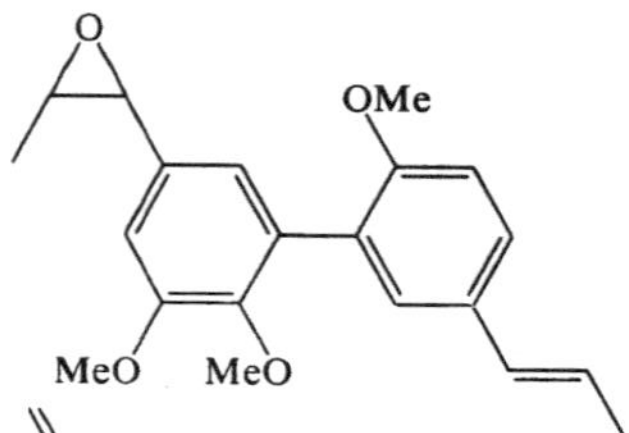	**4.1**			7,8-epoxy-$\Delta^{7'(E)}$-3.3′lignan	
	4.1a			acuminatin	MAGN 1 (*51*)
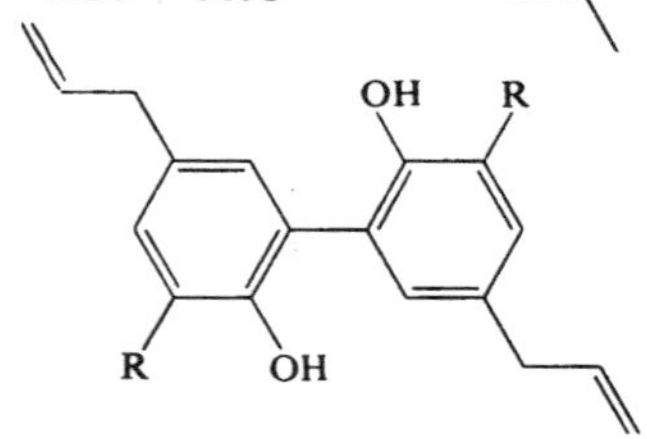	**4.2**			$\Delta^{8,8'}$-3.3′-lignan	
	4.2a	H	H	magnolol	MAGN 2 (*138*)
	4.2b	OMe	OMe	dehydrodieugenol	LAUR 17 (*81*)
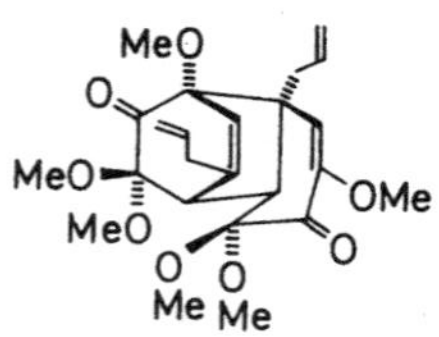	**5.1**			$\Delta^{8,8'}$-2,3,4,5,1′,2′,3′,4′-octahydro-4,4′-dioxo-5.1′,2.2′-lignan	
	5.1a			(±)-asatone	ARIS 1 (*42, 147*)

		Ar	R	Name	Source
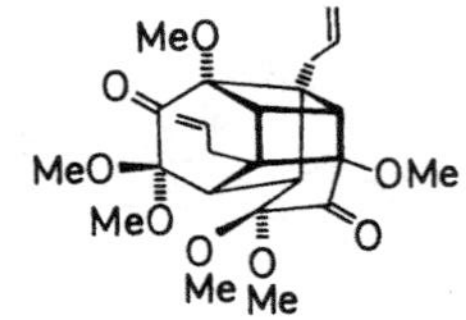	**6.1**			$\Delta^{8,8'}$-1,2,3,4,5,6,1′,2′,3′,4′,5′,6′-dodecahydro-4,4′-dioxo-5.1′,2.2′,1.5′,6.6′-lignan	
	6.1a			(±)-isoasatone	ARIS 1 (*131*)
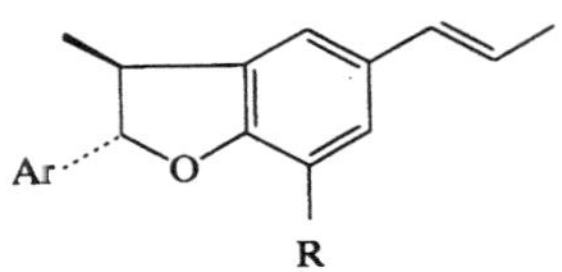	**7.1**			(7S,8S)-$\Delta^{7'(E)}$-7.O.4′,8.3′-lignan	
	7.1a	Gu	OMe	licarin-A	LAUR 9 (*8*)
	7.1b	Pi	OMe	licarin-B, (+)-eupomatenoid-8	LAUR 9 (*8*); EUPO 1 (*31*)
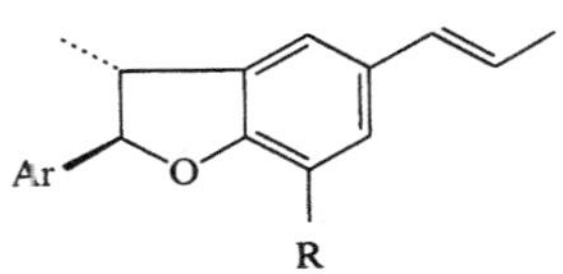	**7.2**			(7R,8R)-$\Delta^{7'(E)}$-7.O.4′,8.3′-lignan	
	7.2a	Hy	H	conocarpin	COMB 1 (*76*)
	7.2b	Ve	OMe		LAUR 1 (*55*)
	7.2c	Tp	OMe	licarin-C	LAUR 19 (*16*)

Table 2 (continued)

		Ar	R	Name	Source
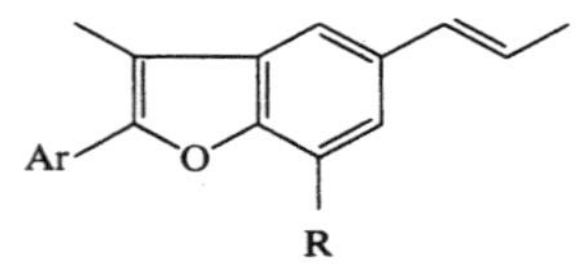	**7.3**			$\Delta^{7,7'(E)}$-7.O.4',8.3'-lignan	
	7.3a	Pi	OMe	eupomatenoid-1 (= eupomatene)	EUPO 1 (*106*)
	7.3b	Rh	H	eupomatenoid-2	EUPO 1 (*31*)
	7.3c	Pi	H	eupomatenoid-3	EUPO 1 (*31*)
	7.3d	Ve	H	eupomatenoid-4	EUPO 1 (*31*)
	7.3e	Gu	H	eupomatenoid-5	EUPO 1 (*31*)
	7.3f	Hy	H	eupomatenoid-6	EUPO 1 (*31*)
	7.3g	Gu	OMe	eupomatenoid-7	EUPO 1 (*31*)
	7.3h	Ve	OMe	eupomatenoid-12	EUPO 1 (*122*)
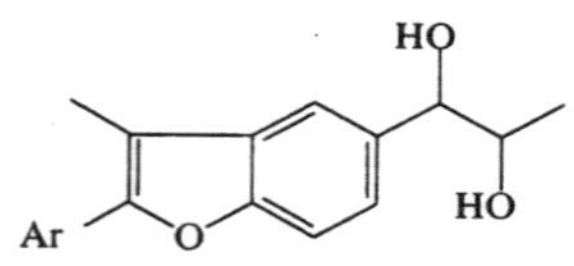	**7.4**			7',8'-dihydroxy-Δ^{7}-7.O.4',8.3'-lignan	
	7.4a	Gu		eupomatenoid-9	EUPO 1 (*122*)
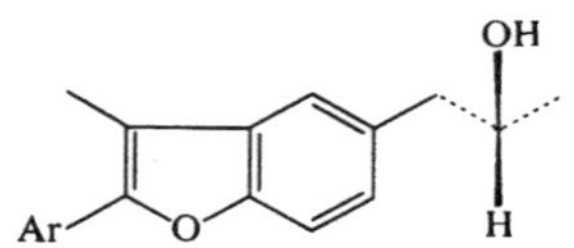	**7.5**			(8'R)-8'-hydroxy-Δ^{7}-7.O.4',8.3'-lignan	
	7.5a	Pi		eupomatenoid-11	EUPO 1 (*122*)
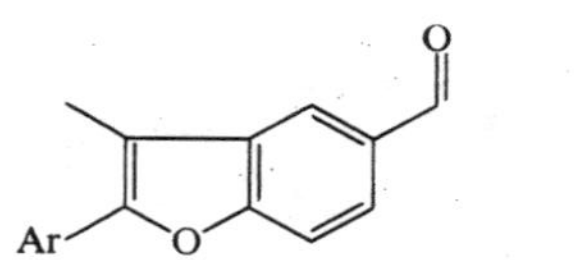	**7.6**			Δ^{7}-7'-oxo-*nor*-8',9'-7.O.4',8.3'-lignan	
	7.6a	Gu		eupomatenoid-10	EUPO 1 (*122*)
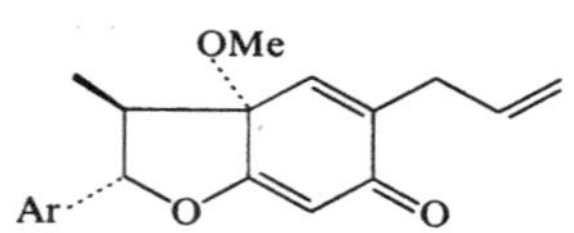	**7.7**			(7S,8S,3'S)-$\Delta^{8'}$-3',6'-dihydro-6'-oxo-7.O.4',8.3'-lignan	
	7.7a	Tp		mirandin-A	LAUR 19 (*17*), 20 (*32*)

Structure	No.	Ar	R	Name	Source
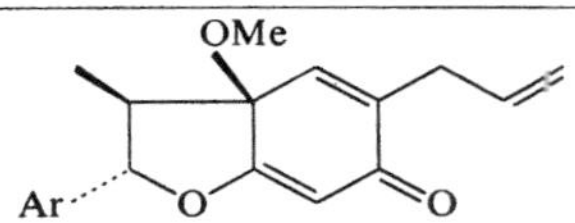	**7.8**			(7S,8S,3′R)-$\Delta^{8'}$-3′,6′-dihydro-6-oxo-7.O.4′,8.3′-lignan	
	7.8a	Tp		mirandin-B	LAUR 19 (*17*)
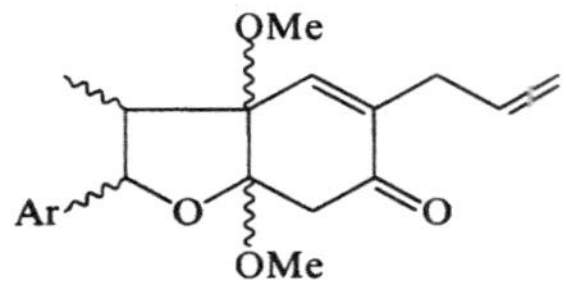	**7.9**			$\Delta^{8'}$-3′,4′,5′,6′-tetrahydro-6′-oxo-7.O.4′,8.3′-lignan	
	7.9a	Ve		piperenone	PIPE 1 (*104*)
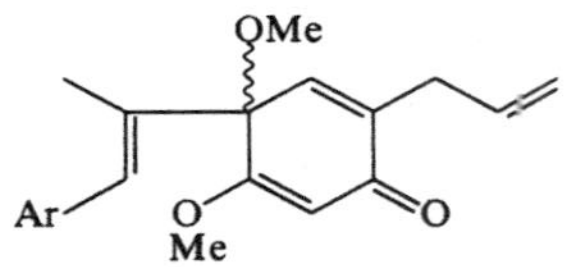	**7.10**			$\Delta^{7,8'}$-3′,6′-tetrahydro-6′-oxo-8.3′-lignan	
	7.10a	Pi		futoquinol	PIPE 1 (*139*)
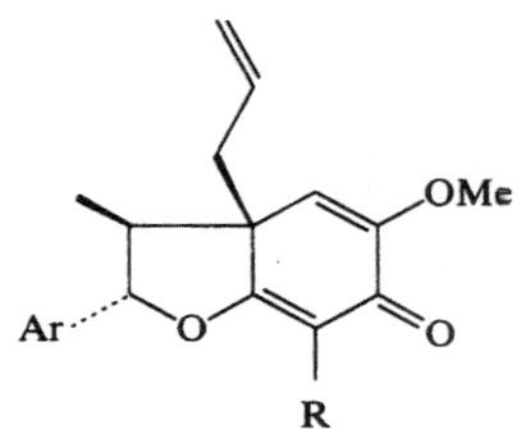	**8.1**			(7S,8S,1′R)-$\Delta^{8'}$-1′,4′-dihydro-4′-oxo-7.O.2′,8.1′-lignan	
	8.1a	Pi	H	burchellin	LAUR 1 (*55*), 2 (*21*), 5 (*69*)
	8.1b	Ve	H		LAUR 1 (*55*), 2 (*13*)
	8.1c	Ve	OMe		LAUR 1 (*55*)

Table 2 (continued)

		Ar	R	Name	Source
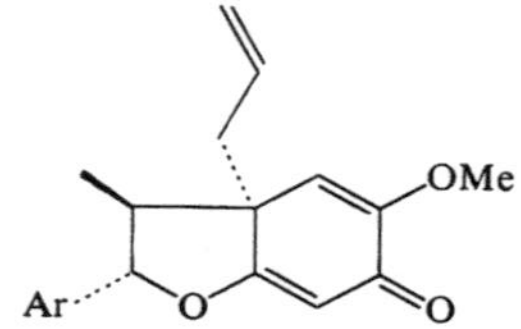	**8.2**			(7S,8S,1′S)-$\Delta^{8'}$-1′,4′-dihydro-4′-oxo-7.O.2′,8.1′-lignan	
	8.2a	Pi			LAUR 5 (*69*)
	8.2b	Tp			LAUR 4 (*9, 10*), 19 (*16*)
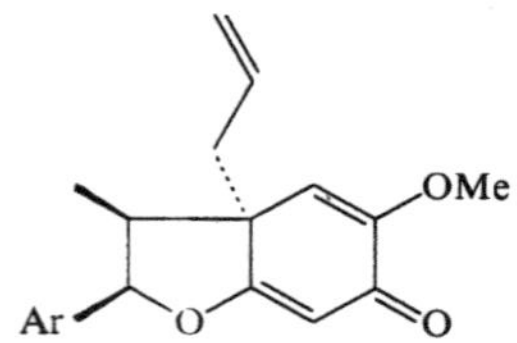	**8.3**			(7R,8S,1′S)-$\Delta^{8'}$-1',4′-dihydro-4′-oxo-7.O.2′,8.1′-lignan	
	8.3a	Pi			LAUR 5 (*69, 70*); ANNO 1 (*15*)
	8.3b	Mp			LAUR 4 (*9, 10, 19, 70*)
	8.3c	Tp			LAUR 4 (*9, 10, 70*)
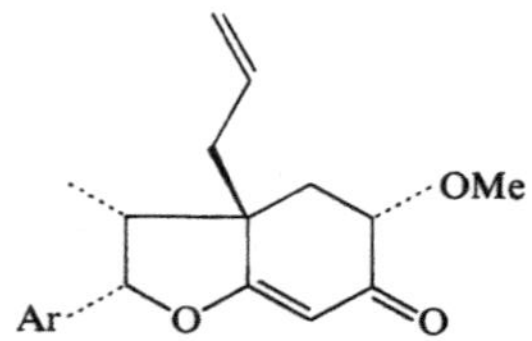	**8.4**			(7S,8R,1′S,5′S)-$\Delta^{8'}$-1′,4′,5′,6′-tetrahydro-4′-oxo-7.O.2′,8.1′-lignan	
	8.4a	Ve		porosin	LAUR 21 (*7, 18*)
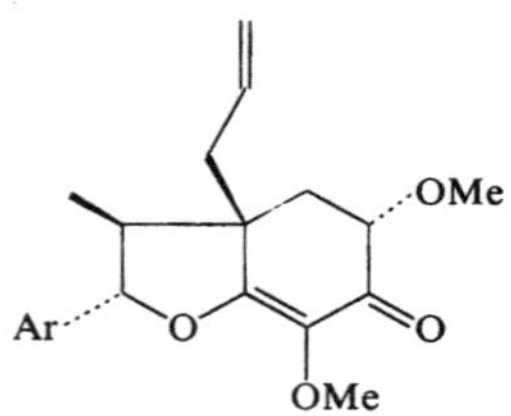	**8.5**			(7S,8S,1′S,5′S)-$\Delta^{8'}$-1′,4′,5′,6′-tetrahydro-4′-oxo-7. O.2′,8.1′-lignan	
	8.5a	Pi		canellin-B	LAUR 11 (*11, 61*)

Structure	No.	Ar	R	Name	Occurrence
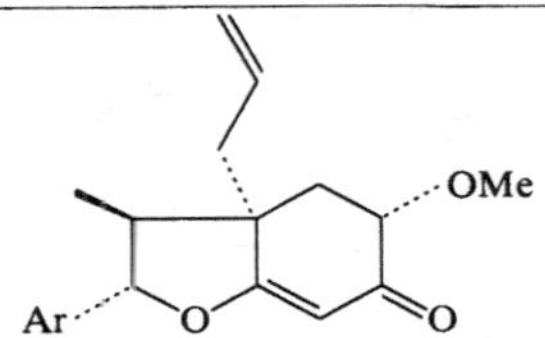	**8.6**			(7S,8S,1′R,5′S)-$\Delta^{8'}$-1′,4′,5′,6′-tetrahydro-4′-oxo-7.O.2′,8.1′-lignan	
	8.6a	Pi			LAUR 10 (*10*)
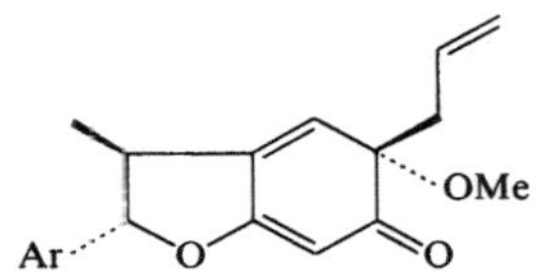	**9.1**			(7S,8S,1′S)-$\Delta^{8'}$-1′,6′-dihydro-6′-oxo-7.O.4′,8.3′-lignan	
	9.1a	Pi			LAUR 2 (*13*), 5 (*69*)
	9.1b	Ve			LAUR 1 (*55*), 2 (*13*); ANNO 1 (*15*)
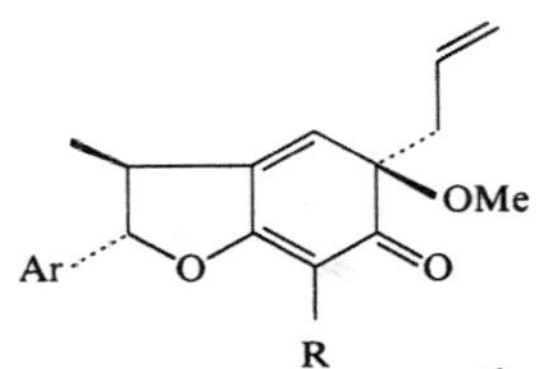	**9.2**			(7S,8S,1′R)-$\Delta^{8'}$-1′,6′-dihydro-6′-oxo-7.O.4′,8.3′-lignan	
	9.2a	Mp	H		LAUR 4 (*9, 10, 19*)
	9.2b	Mp	OMe		LAUR 4 (*10*)
	9.2c	Tp	H		LAUR 4 (*9, 10, 32*), 19 (*16*), 20 (*32*)
	9.2d	Tp	OMe		LAUR 4 (*10*)
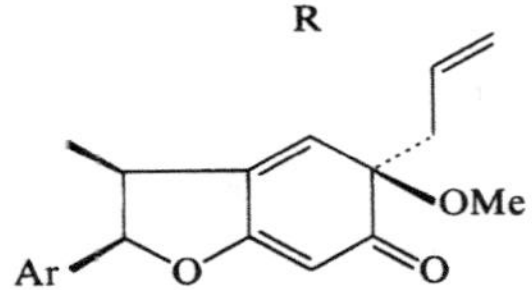	**9.3**			(7R,8S,1′R)-$\Delta^{8'}$-1′,6′-dihydro-6′-oxo-7.O.4′,8.3′-lignan	
	9.3a	Pi			LAUR 5 (*69, 70*)
	9.3b	Mp			LAUR 4 (*10, 19, 70*)

Table 2 (continued)

		Ar	R	Name	Source
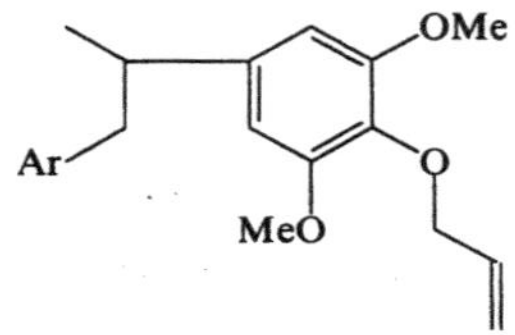	**10.1**			$\Delta^{8'}$-8.4′,1′.O.7′-lignan	
	10.1a	Tp	H	aurein	LAUR 9a (*20*), 13 (*66*)
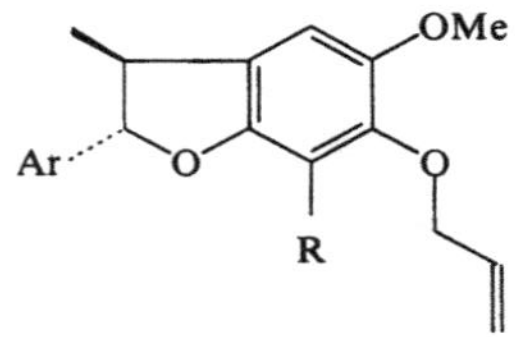	**10.2**			(7S,8S)-$\Delta^{8'}$-7.O.3′,8.4′,1′.O.7′-lignan	
	10.2a	Pi	H		LAUR 2 (*13*), 5 (*69*)
	10.2b	Mp	H		LAUR 4 (*19*)
	10.2c	Tp	H		LAUR 4 (*9*), 19 (*16*)
	10.2d	Tp	OMe		LAUR 4 (*9*)
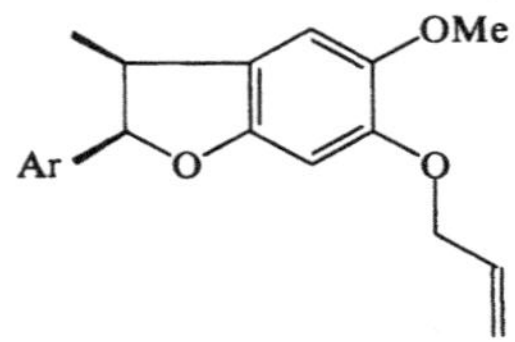	**10.3**			(7R,8S)-$\Delta^{8'}$-7.O.3′,8.4′,1′.O.7′-lignan	
	10.3a	Mp			LAUR 4 (*19*)
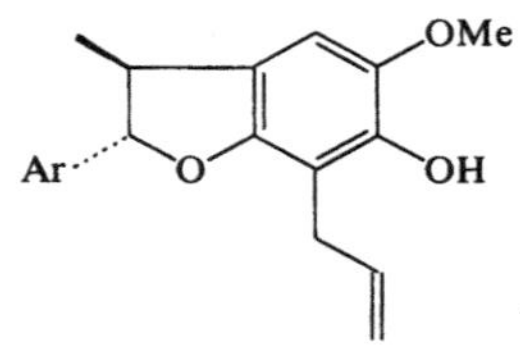	**11.1**			(7S,8S)-$\Delta^{8'}$-7.O.2′,8.3′-lignan	
	11.1a	Pi			LAUR 5 (*69*)
	11.1b	Mp			LAUR 4 (*9*)
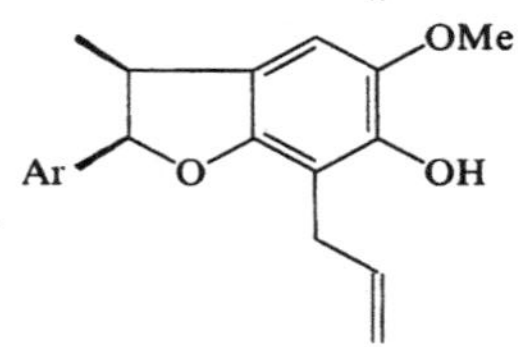	**11.2**			(7R,8S)-$\Delta^{8'}$-7.O.2′,8.3′-lignan	
	11.2a	Pi			LAUR 5 (*69*)
	11.2b	Mp			LAUR 4 (*9*)

Structure	No.	Ar	R¹	R²	R³	Name	Source
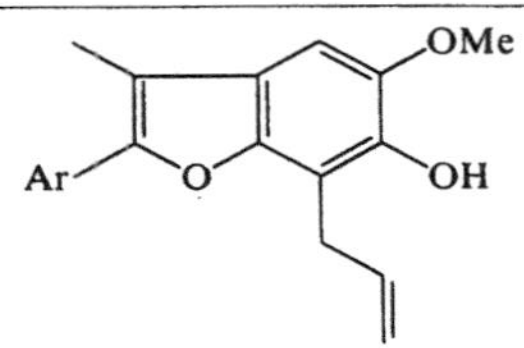	**11.3**					$\Delta^{7,8'}$-7.O.2′,8.3′-lignan	
	11.3a	Pi					LAUR 5 (*69*)
	11.3b	Mp					LAUR 4 (*19*)
	11.3c	Tp					LAUR 19 (*16*)

Structure	No.	Ar	R¹	R²	R³	Name	Source
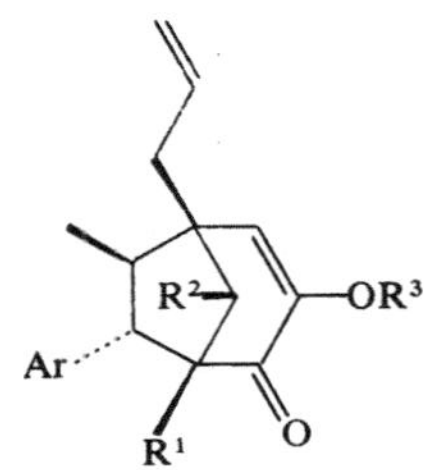	**12.1**					*rel*-(7R,8R,1′S,2′S,3′R)-$\Delta^{8'}$-1′,2′,3′,4′-tetrahydro-4′-oxo-7.3′,8.1′-lignan	
	12.1a	Pi	H	OH	Me	guianin	LAUR 1 (*55*), 3 (*36*)
	12.1b	Pi	OMe	OH	Me		LAUR 1 (*55*)
	12.1c	Pi	OMe	OAc	Me		LAUR 1 (*55*)
	12.1d	Pi	OMe	OAc	H		LAUR 1 (*55*)
	12.1e	Mp	H	OH	H		LAUR 4 (*19*)
	12.1f	Mp	OMe	OH	H		LAUR 4 (*19*)
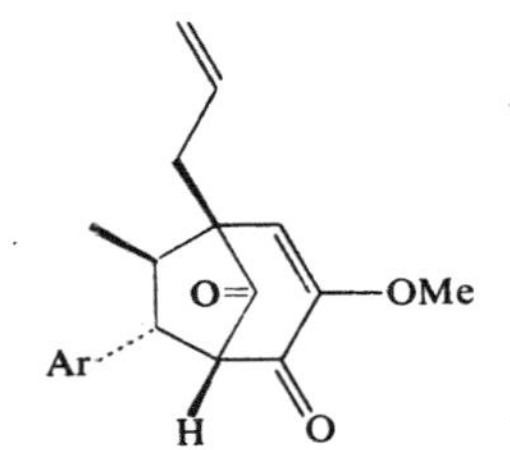	**12.2**					*rel*-(7R,8R,1′S,3′R)-$\Delta^{8'}$-1′,2′,3′,4′-tetrahydro-2′,4′-dioxo-7.3′,8.1′-lignan	
	12.2a	Pi					LAUR 1 (*55*)
	12.2b	Gu					LAUR 2 (*13*)

Table 2 (continued)

		Ar	R	Name	Source
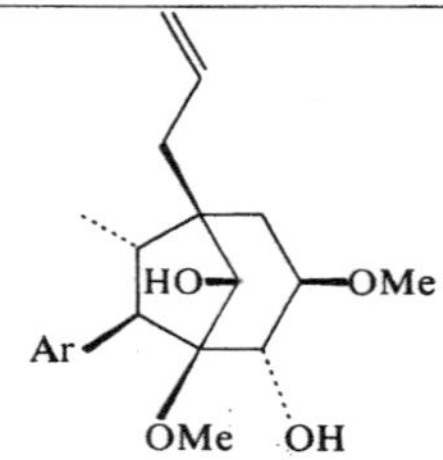	**12.3**			*rel*-(7R,8S,1′S,2′R,3′S,4′S,5′R)-$\Delta^{8'}$-1′,2′,3′,4′,5′,6′-hexahydro-7.3′,8.1′-lignan	
	12.3a	Pi		canellin-A	LAUR 11 (*61*)
	12.3b	Mp			LAUR 6 (*49*)
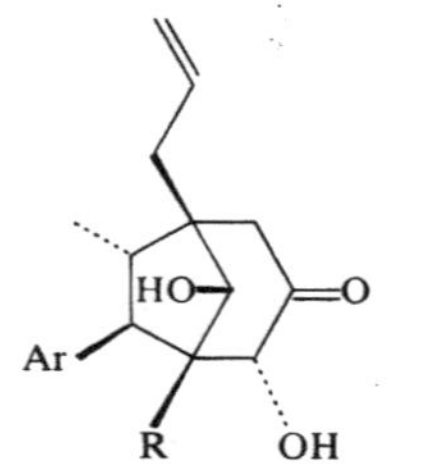	**12.4**			*rel*-(7R,8S,1′S,2′R,3′S,4′R)-$\Delta^{8'}$-1′,2′,3′,4′,5′,6′-hexahydro-5′-oxo-7.3′,8.1′-lignan	
	12.4a	Pi	OMe	canellin-C (x)	LAUR 11 (*61*)
	12.4b	Mp	H		LAUR 4 (*9, 19*)
	12.4c	Mp	OMe		LAUR 4 (*19*)
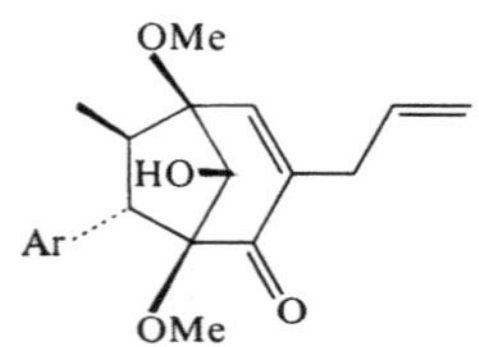	**13.1**			*rel*-(7S,8R,3′S,4′R,5′S)-$\Delta^{8'}$-2′,3′,4′,5′-tetrahydro-2′-oxo-7.3′,8.5′-lignan	
	13.1a	Mp		macrophyllin (x)	LAUR 12 (*56*)
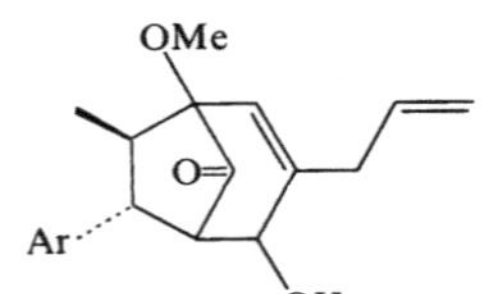	**13.2**			7,8-*trans*-$\Delta^{8'}$-2′,3′,4′,5′-tetrahydro-4′-oxo-7.3′,8.5′-lignan	
	13.2a	Tp		(x)	LAUR 20 (*32*)

Structure	No.		Name	Occurrence
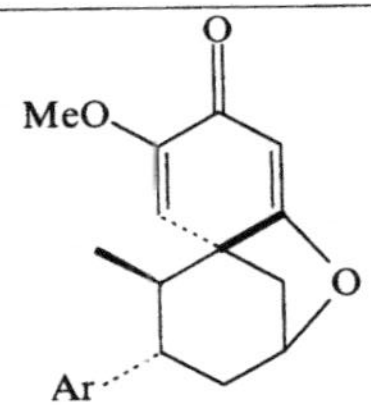	**14.1**		*rel*-(7S,8R,1′S,8′S)-2′,8′-epoxy-1′,4′-dihydro-4′-oxo-7.9′,8.1′-lignan	
	14.1a	Pi	futoenone	PIPE 1 (*114*)
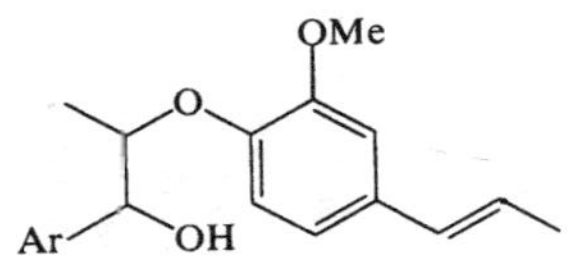	**15.1**		$\Delta^{7'(E)}$-8.O.4′-lignan	
	15.1a	Tp	surinamensin	MYRI 4 (*24*)
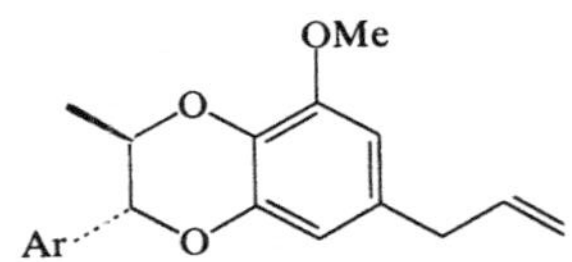	**15.2**		*rel*-(7S,8R)-$\Delta^{8'}$-7.O.3′,8.O.4′-lignan	
	15.2a	Pi	eusiderin-B	LAUR 14 (*33*)
	15.2b	Tp	eusiderin	LAUR 6 (*49*), 8 (*80*), 13 (*66*), 9a (*20*), 14 (*33*)

(x) Insufficiently characterized.

text to refer to these natural neolignans. Systematic designations in Table 2 refer only to skeleta, configurations and C_3-substitution. Full names of compounds are completed by citation of substitution of the C_6-rings by hydroxy, methoxy and methylenedioxy groups. Thus, *e.g.*, **1.2a** is (8R,8′R)-4,4′-dihydroxy-3,3′-dimethoxy-8.8′-lignan. There is no need to add the prefix neo in such complete designations, which are in themselves sufficient to discriminate between propenyl-/allyl-benzene and cinnamoyl-/cinnamyl derived lignans. Indeed, according to the system, **17,** a lignan structurally related to the neolignan **1.2a,** is (8R,8′R)-4,4′-dihydroxy-3,3′-dimethoxy-9-oxo-8.8′,9.O.9′-lignan. Extension of the proposal to all classes of oligophenylpropanoids is an obvious possibility.

Table 3. *Symbols for Aryl Groups Used in Table 2 and in Formulae*

Ar	
An	Anisyl (4-methoxyphenyl)
Ca	Catechyl (3,4-dihydroxyphenyl)
Hy	4-Hydroxyphenyl
Gu	Guaiacyl (4-hydroxy-3-methoxyphenyl)
Mp	O-Methyl-O,O-methylenylpyrogallyl (3-methoxy-4,5-methylenedioxyphenyl)
Mr	2-O-Methylresorcinyl (4-hydroxy-2-methoxyphenyl)
Pi	Piperonyl (3,4-methylenedioxyphenyl)
Re	Resorcinyl (2,4-dihydroxyphenyl)
Rh	O-L-Rhamnosyl-4-hydroxyphenyl
Tp	Tri-O-methylpyrogallyl (3,4,5-trimethoxyphenyl)
Ve	Veratryl (3,4-dimethoxyphenyl)

Table 4. *Plant Sources of Neolignans*
(Classification of the numbered *Aniba* and *Licaria* spp. will be possible upon completion of the revision of Lauraceae by Dr. K. KUBITZKI, Hamburg)

Subclass, Order	Family	Species name	Species No.
Magnoliidae			
Magnoliales	Austrobaileyaceae	*Austrobaileya scandens* C. T. White	1
	Magnoliaceae	*Magnolia acuminata* L.	1
		M. officinalis Rhed. et Wils.	2
		Kadsura japonica Dunal	3
	Himantandraceae	*Galbulimina belgraveana* (F. Muell.) Sprague (= *Himantandra belgraveana* F. Muell.)	1
		H. baccata Bail.	2

Table 4 (continued)

Subclass, Order	Family	Species name	Species No.
	Annonaceae	*Duguetia surinamensis* R. E. Fries	1
	Myristicaceae	*Dialyanthera otoba* (H. et B.) Warb. (= *Myristica otoba* H. et B.)	1
		Virola carinata (Benth.) Warb.	2
		V. cuspidata (Benth.) Warb.	3
		V. surinamensis (Rol.) Warb.	4
	Schisandraceae	*Schisandra chinensis* Bail.	1
	Eupomatiaceae	*Eupomatia laurina* R. Br.	1
	Trimeniaceae	*Piptocalyx moorei* Oliv.	1
		Trimenia papuana	2
	Lauraceae	*Aniba affinis* (Meissn.) Mez.	1
		A. burchellii Kosterm.	2
		A. guianensis Aubl.	3
		A. simulans Allen	4
		A. terminalis Ducke	5
		A. sp. 41	6
		Cinnamomum sp.	7
		Eusideroxylon zwageri T. et B.	8
		Licaria aritu Ducke	9
		L. aritu Ducke (2nd sample)	9a
		L. armeniaca (Nees) Kosterm.	10
		L. canella (Meissn.) Kosterm.	11
		L. macrophylla (A. C. Smith) Kosterm.	12
		L. sp. 04	13
		L. sp. 63	14
		Litsea gracilipes	15
		L. grandis (Wall.) Hook f.	16
		L. turfosa Hook f.	17
		Machilus edulis King	18
		Nectandra miranda Sandw.	19
		N. sp.	20
		Ocotea porosa (Nees) L. Barr.	21
		O. veraguensis Mez	22
Piperales	Piperaceae	*Piper futokadzura* Sieb. et Zucc.	1
Aristolochiales	Aristolochiaceae	*Asarum taitonense* Hayata	1
Rosidae			
Myrtales	Combretaceae	*Conocarpus erectus*	1
Sapindales	Zygophyllaceae	*Guaiacum officinale* L.	1
		Larrea cuneifolia Cav.	2
		L. divaricata Cav. (= Covillea *tridentata* (D.C.) Vail.)	3
		L. nitida Cav.	4

II. 8.8′-Neolignans

1. 1.1—1.4

Chemical knowledge on guaiaretic acid (**1.1a**), as well as on (−)- (**1.2a**), *meso*- (**1.3d**) and *meso-nor*- (**1.3b**) dihydroguaiaretic acids was reviewed up to 1955 (*78*).

The similar shape of the isomorphous molecules **1.1a** and **1.3d** made it appear likely that the hydrogen at C-7 and the methyl group at C-8 of **1.1a** are *trans*. The configuration at C-8′, originally derived from rotational values (*134*), was confirmed through correlation with L-dihydroxyphenylalanine (**18,** Chart 1) (*135*).

The configuration of **1.2a** was established by synthesis of its dimethyl ether through defunctionalization of the lactone moiety of (−)-matairesinol (**17**) dimethyl ether (*133*) and by stepwise degradation to (2S,3S)-2,3-dimethylsuccinic acid and (2S)-2-methylbutanol (*39, 40*). While the stereochemistry of some compounds (**1.4a, b**) of the dihydroguaiaretic acid series is unknown (*76*), the absolute configuration of others (**1.2b, c**) was inferred by comparison of optical rotation with **1.2a** (*112*). Optically inactive compounds of this series were generally assumed to possess the *meso*-configuration (**1.3**), rather than to be racemates. Confirmatory proof of this postulate was accomplished by reduction of (−)-guaiaretic acid to an optically inactive **1.3d** (*78*); a stereospecific synthesis of the methyl ether **19,** which led inclusively to the racemate **20** (Chart 2) (*132*); and an X-ray analysis (*107*).

Totally synthetic *nor*-dihydroguaiaretic acid (**1.3b**) became available in 1934 (*75*). When NDGA began to be required in quantity (see Section XIII), this and subsequently published syntheses (*30, 60, 78,*

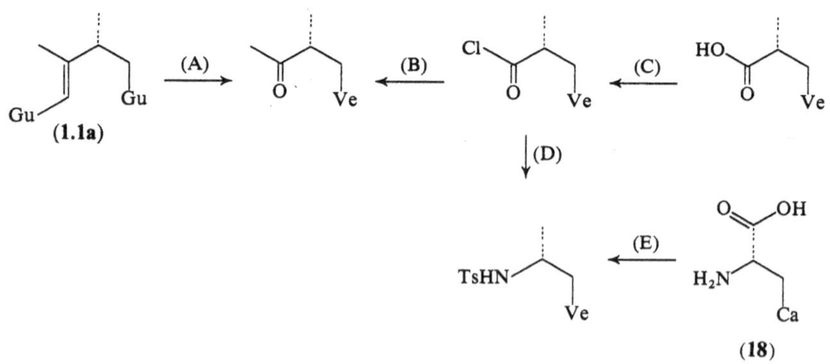

(A) ... 1. Me_2SO_4; 2. OsO_4, then alk. mannitol; 3. HIO_4. (B) ... MeMgBr, low temp. (C) ... $(COCl)_2$. (D) ... 1. Curtius; 2. TsCl, C_5H_5N. (E) ... 1. HCO_2H, Ac_2O; 2. Me_2SO_4; 3. MeOH, HCl; 4. $LiAlH_4$; 5. TsCl, C_5H_5N. 7. $LiAlH_4$.

Chart 1. Determination of the absolute configuration of (−)-guaiaretic acid (**1.1a**)

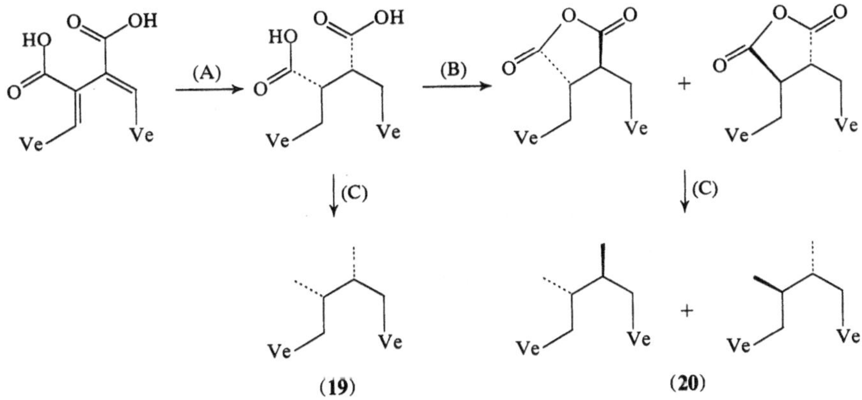

(A) ... RaNi, 10% NaOH, 90°. (B) ... Ac_2O, Δ. (C) ... 1. $LiAlH_4$, THF; 2. TsCl, C_5H_5N; 3. $LiAlH_4$, THF.

Chart 2. Synthesis of the dimethyl ethers of *meso*- (**19**) and of (±)-dihydroguaiaretic (**20**) acids

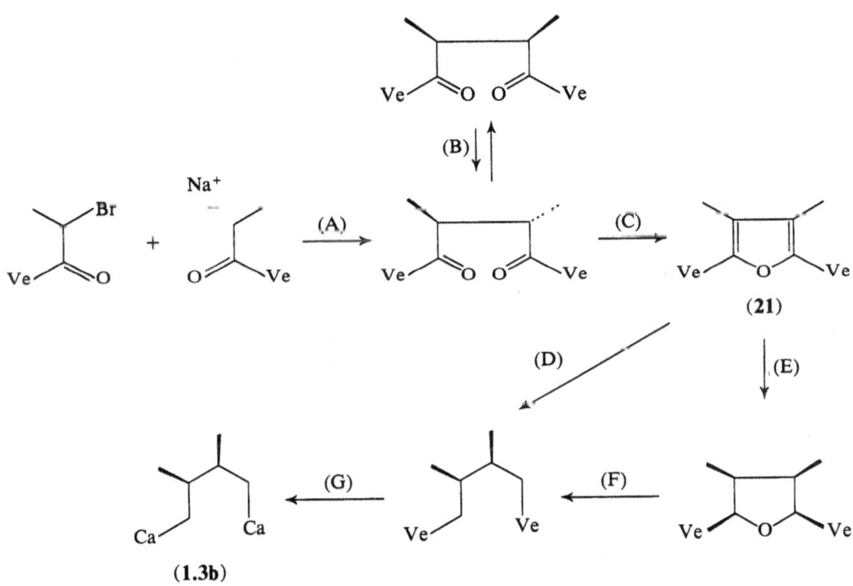

(A) ... liq. NH_3, −33°. (B) ... MeONa, MeOH, EtOH, C_6H_6, refl. (C) ... 1% HCl in MeOH, CH_2Cl_2, refl. (D) ... 1 500 psig H_2, PdO, THF, 50°; or 50 psig H_2, $PdCl_2$, NaOAc, THF, 75°. (E) ... 1 500 psig H_2, 10% Pd/$CaCO_3$, EtOH, 125°. (F) ... 1 atm. H_2, PdO, THF, 25°. (G) ... HBr conc., N_2, refl.

Chart 3. Synthesis of *nor*-dihydroguaiaretic acid (**1.3b**)

119, 137) could not compete with natural material because of low yields, lengthy reaction sequences or expensive starting materials. A recently published process (Chart 3) (*121*), however, is stated to be able to produce NDGA industrially at a time when natural **1.3b** is no longer available (*117*).

Subsequent to the isolation of partially O-demethylated dihydroguaiaretic acid (**1.3c**), it was proposed to call this compound *nor*-dihydroguaiaretic acid, and **1.3b** *bisnor*-dihydroguaiaretic acid (*63*). Adoption of this proposal would lead to confusion, in view of the voluminous literature on NDGA (**1.3b**) (see Section XIII).

2. 1.5—1.9

The early literature on galgravin (**1.6b**), galbacin (**1.7a**) and galbelgin (**1.7b**) was reviewed up to 1955 (*78*) and 1963 (*3, 48*).

The preparation of **21** (Chart 3) by selenium dehydrogenation of the dimethyl ethers of pinoresinol (**22**) and epipinoresinol (**23**) was reported (*22*) prior to the isolation of the furoguaiacins (**1.9a, b**) from a natural

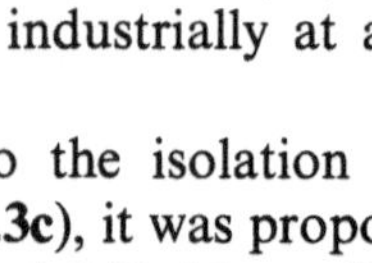

(**22**)

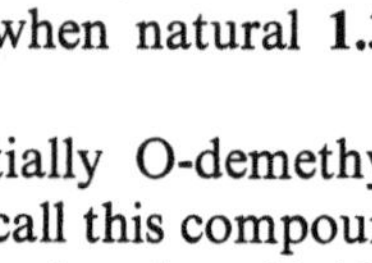

(**23**)

source. The report included a discussion of total synthetic methods which, in modified form, proved indeed useful in the preparation of the dimethyl [Chart 4 (B_1)] (*5*) and diethyl [Chart 4 (B_2)] (*86*) ethers not only of **1.9a**, but also of the ethyl ether of asymetrical **1.9b** (Chart 5) (*86*), and even of *nor*-dihydroguaiaretic acid (**1.3b**) (Chart 3) (*121*).

Catalytic addition of hydrogen to **21** (see Chart 3) gave di-O-methyl **1.5a**. The all-*cis* structure which, accordingly, was assigned to tetrahydrofuroguaiacin-B (**1.5a**), would be expected to isomerize to the more stable *trans*-arrangement of aryl and methyl groups upon acid

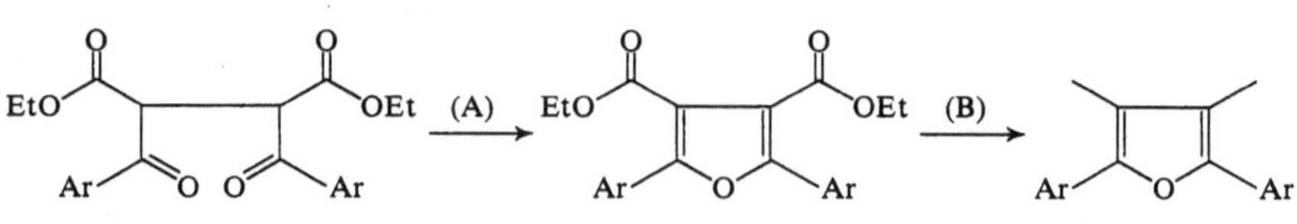

(A) ... Ac_2O, H_2SO_4. (B_1) ... 1. $LiAlH_4$, THF, refl.; 2. H_2, 30 atm., AcOH, MeOH, Pd/C; or (B_2) ... 1. $LiAlH_4$; 2. CrO_3, C_5H_5N; 3. EtSH, $ZnCl_2$, Na_2SO_4; 4. RaNi.

Chart 4. Syntheses of 7.O.7,8.8′-$\Delta^{7,7'}$-neolignans

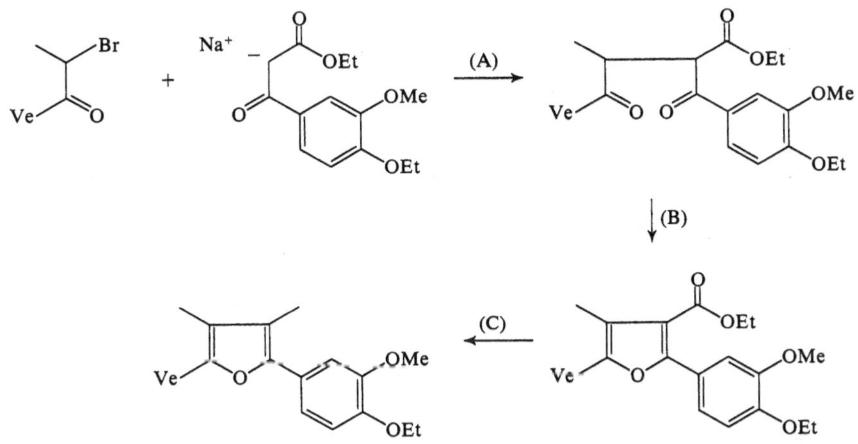

(A) ... C_6H_6, refl. (B) ... HCl, MeOH. (C) ... 1. $LiAlH_4$; 2. CrO_3, C_5H_5N; 3. EtSH, $ZnCl_2$, Na_2SO_4; 4. RaNi.

Chart 5. Synthesis of the ethyl ether of **1.9b**

catalyzed ring opening and closure. Indeed, heating with perchloric acid of di-O-methyl **1.5a** led to an isomer which proved to be identical with galgravin (**1.6b**). This is also optically inactive and thus must possess

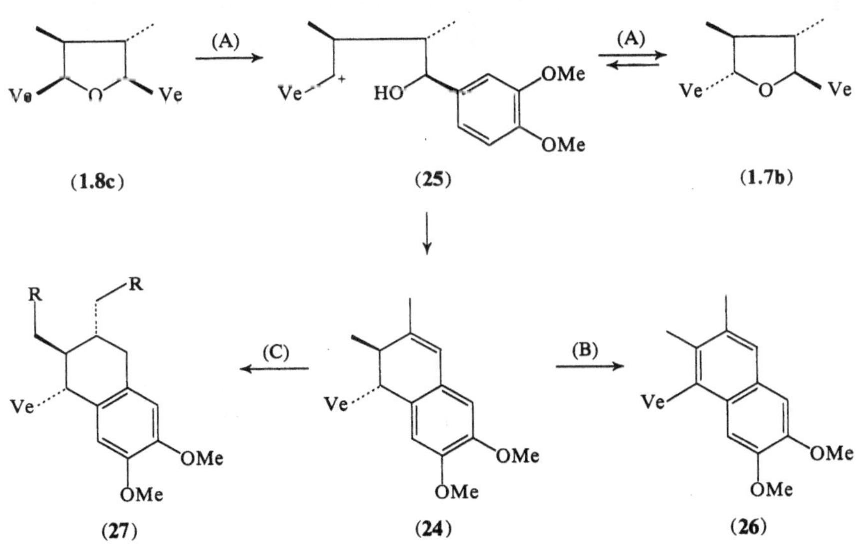

(A) ... $HClO_4$, AcOH, room temp. (B) ... 10% Pd/C, 200°, 4 min. (C) ... H_2, Pd/C.

Chart 6. Typical reactions of 7.O.7′,8.8′-neolignans

the *meso*-7,8-*trans*-7′,8′-*trans*-configuration (*85*). The stability of **1.6b** to $HClO_4$ is paralleled by **1.7b** which must, in view of its optical activity, be written in the alternative 7,8-*trans*-7′,8′-*trans*-configuration (*28*). Although cyclogalbelgin (**24**), instead of the expected galbelgin (**1.7b**), was isolated from the isomerization product of veraguensin, its structure is nevertheless represented by **1.8c** (*47*). Indeed, more vigorous or prolonged treatment with acid of all tetrahydrofuran neolignans, including galbelgin (**1.7b**), leads to such 1-aryl-2,3-dimethyl-3,4-dihydronapthalenes (Chart 6), which also proved useful in stereochemical assignments (*28*). The product ratio (diaryltetrahydrofuran isomer *vs.* dihydronaphthalene) depends on the relative nucleophilicity of the hydroxyl and the aryl group towards the carbonium ion in **25** (*129*).

Compounds such as **24** are transformed into the corresponding arylnaphthalenes (**26**) by dehydrogenation or aryltetralins (**27**, R=H) by hydrogenation (*28*). Such aryltetralins can be correlated with lignans (**27**, R=OH) of known configuration.

An additional technique which yielded structural and stereochemical information in this series involves reduction with sodium in liquid ammonia. This also leads, according to the conditions used, to two types of product, **28**, a dihydroguaiaretic acid, and **29**, again an aryltetralin. Chart 7 exemplifies the reaction sequence for veraguensin (*47*).

Tetrahydrofuranoid neolignans of the galbelgin (**1.7**) and veraguensin (**1.8**) types were synthesized by a biomimetic oxidative coupling procedure which revealed a remarkable stereospecificity in the 8,8′-coupling

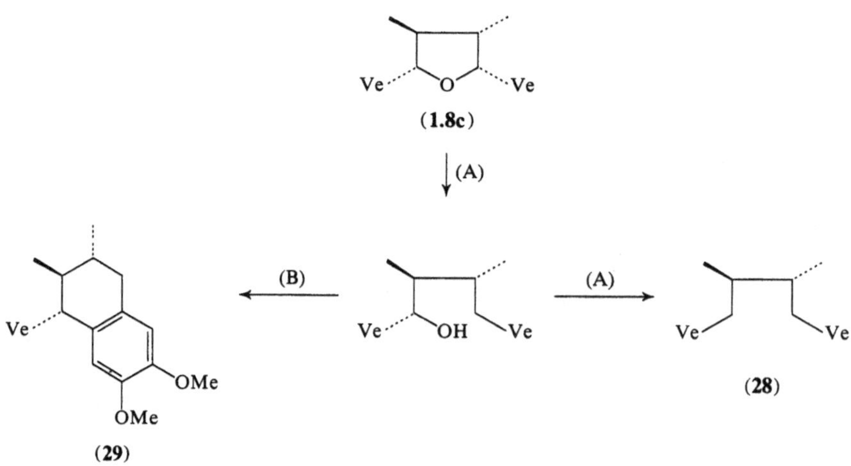

(A) ... Na, NH_3 liq., $(CH_2OMe)_2$. (B) ... SiO_2 chromatography.

Chart 7. Typical reactions of 7.O.7′,8.8′-neolignans

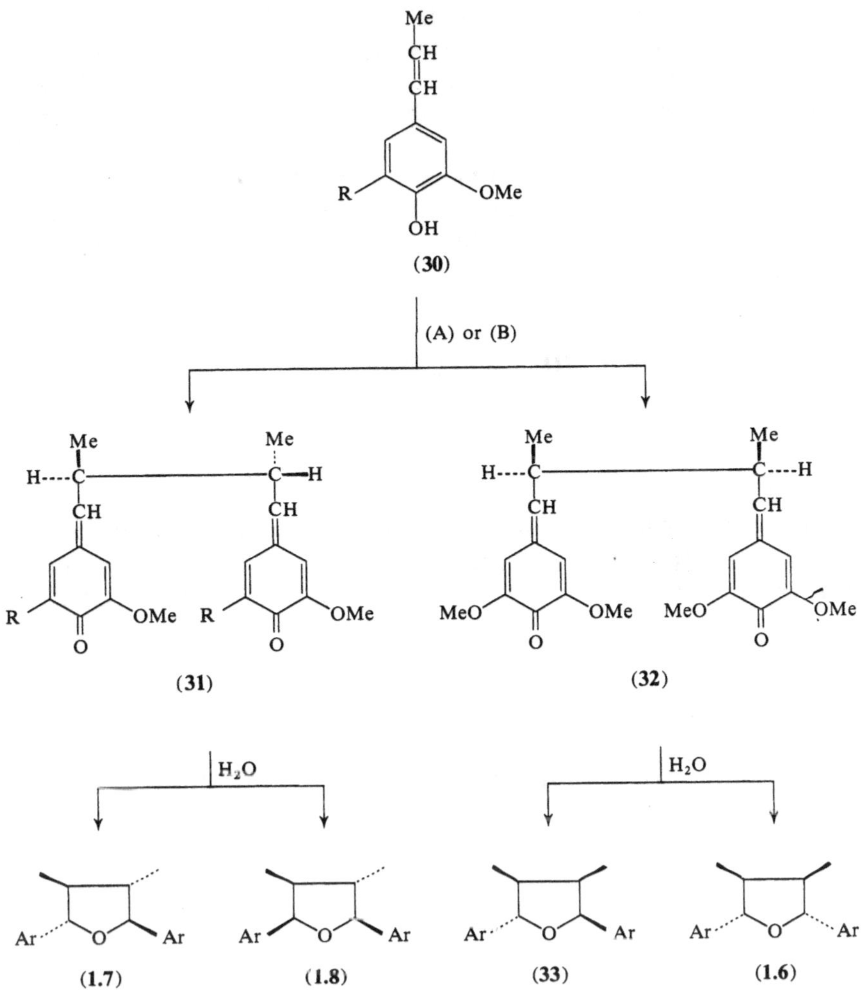

(A) ... H_2O_2, peroxidase. (B) ... $K_3[Fe(CN)_6]$.

Chart 8. Biomimetic syntheses of 7.O.7′,8.8′-neolignans

of (*E*) and (*Z*)-4-propenylphenols (**30**; R=H, *E* or *Z*; R=OMe, *E*) to *threo*-bisquinone methides (**31**). Only in one case (**30**; R=OMe, *Z*) did these intermediates appear in the *threo*- (**31**) and *erythro*- (**32**) forms, hydration thus leading also to a galgravin type (**1.6**) and a tetrahydrofuranoid analogue **33** not yet isolated from a natural source (Chart 8) (*129*).

In contradistinction to ^1H NMR spectra (*130*), mass spectra (*120*) are insensitive to the stereochemistry of the substituents of tetrahydrofuranoid neolignans. The absolute configurations of galbacin (**1.7a**) and calopiptin (**1.8b**) were determined [basic evidence: ozonolysis leads to (2*S*,3*S*)-2,3-dimethylsuccinic acid (*105*)]. Conversions of **1.8b** into **1.7b** and **1.8c** (*105*), as well as of **1.8a** into **1.8b** (*112*) proved, consequently, of stereochemical significance. An X-ray analysis of grandinsin (**1.7c**) and ORD/CD correlations with other neolignans of the series were planned (*82*).

III. 8.8′,7.2′-Neolignans

1. 2.1—2.5

The literature on galcatin (**2.1d**) and galbulin (**2.1f**) was reviewed with respect to structural elucidation up to 1955 (*78*) and to synthesis up to 1963 (*1*—*4*). Both have identical configurations since **2.1d** can be converted into **2.1f** by alkaline demethylation and methylation (*28*).

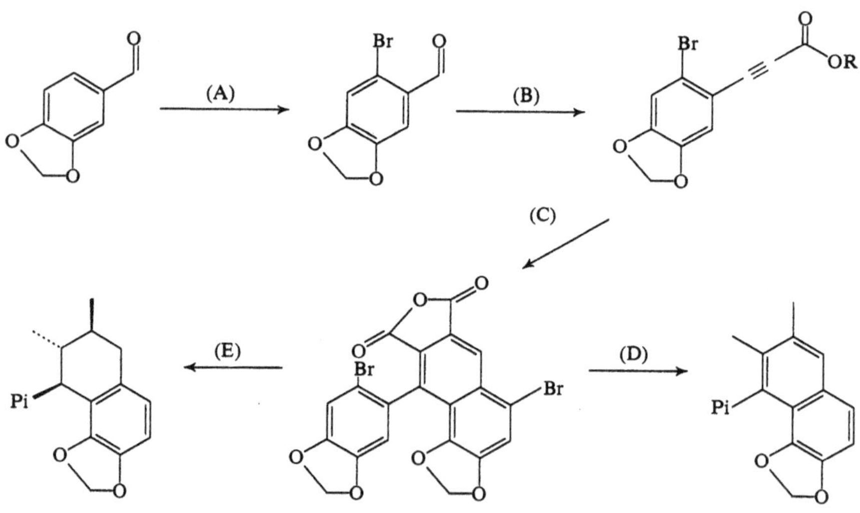

(A) ... Br_2, I_2, CS_2, room temp. (B) ... triethyl phosphonoacetate, NaH, $(CH_2OMe)_2$. (C) ... N,N-dicyclohexylcarbodiimide, $(CH_2OMe)_2$, $-12°$. (D) ... 1. KOH, MeOH; 2. CH_2N_2; 3. $LiAlH_4$, Et_2O; 4. H_2, 10% Pd/C. (E) ... 1. 4% Na/Hg, aq. NaOH; 2. CH_2N_2; 3. $LiAlH_4$, Et_2O; 4. TsCl, C_5H_5N; 5. $LiAlH_4$, THF, refl.

Chart 9. Synthesis of tetrahydro-otobain (**35**) and of (±)-otobain (**2.4a**)

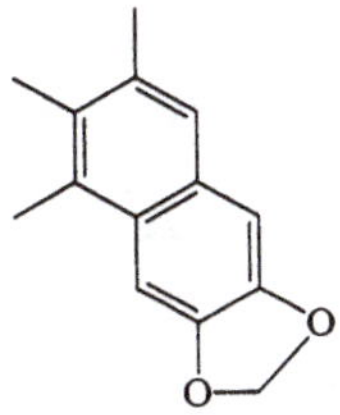

(**34**)

The initial postulate for otobain was based on the galcatin (**2.1d**) model. The corresponding tetralin, however, prepared either by oxidative cyclization of dipiperonylidenesuccinic anhydride (*62*) or by acid isomerization of galbacin (**1.7a**) followed by dehydrogenation to "dehydro-epigalbacin" (**34**) (*26*), was not identical with tetrahydrootobain (**35**)

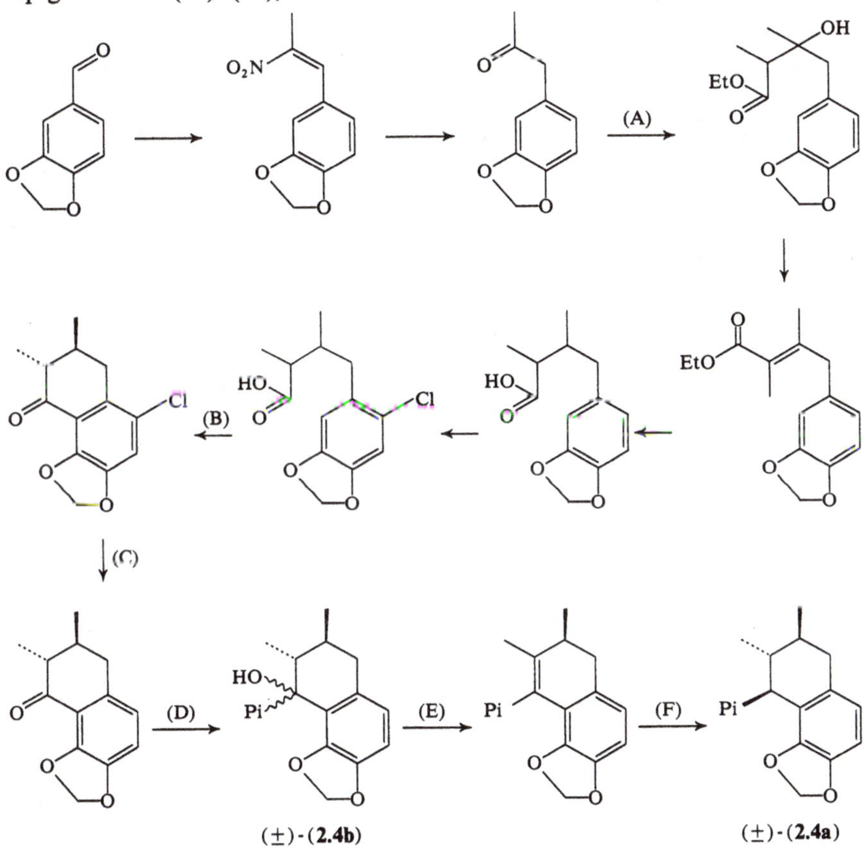

(A) ... Reformatsky. (B) ... *via* acid chloride, $SnCl_4$, $-70°$. (C) ... H_2, Pd/$SrCO_3$, NaOAc. (D) ... PiLi, LiBr, $-70°$. (E) ... $POCl_3$, C_5H_5N. (F) ... Na, isopropanol.

Chart 10. Synthesis of (±)-hydroxyobtain (**2.4b**) and of (±)-otobain (**2.4a**)

(Chart 9), obtained from otobain by heating with DDQ or Pd/C. This suggested that otobain might be represented by **2.4a,** a fact which was substantiated by ^{1}H NMR (*26, 62*). Significantly, the CH_2O_2-protons of the piperonyl moieties of **34** and **35** give rise, respectively, to sharp and broadened NMR singlets (*71*). Synthesis, initially of tetrahydro-otobain (**35**) (*35*) and subsequently of (±)-otobain, either by the same propiolic acid condensation (Chart 9) (*99*) or a tetralone route (Chart 10) (*108*), provided the ultimate solution for the problem.

Attempted acetylation of hydroxyotobain gave didehydrootobain, whose hydrogenation furnished otobain (**2.4a**). Hence both must contain the same basic skeleton (*62*). The absence of ^{1}H NMR absorption in the τ 6.6 region, characteristic of a doubly benzylic proton, showed that the compound is 4-hydroxyotobain (**2.4b**) (*142*), a postulate which was proved by synthesis (Chart 10) (*108*). The absolute configuration indicated in the formulae **2.4a** and **2.4b** was assigned on the basis of ORD measurements (*87*).

Isogalcatin would be a preferable name (*3*) for the compound **2.1e**, originally designated iso-otobain (*142*). The similarities in the structures of **2.1e** and galcatin (**2.1d**) are reflected in their ^{1}H NMR spectra. On such evidence, the relative configurations at C-7, 8 and 8′ of **2.1d, 2.1e, 2.4a** and **2.4b** are identical (*142*), thereby confirming the stereochemical relations derived for galcatin (**2.1d**) by other means (*28, 133*).

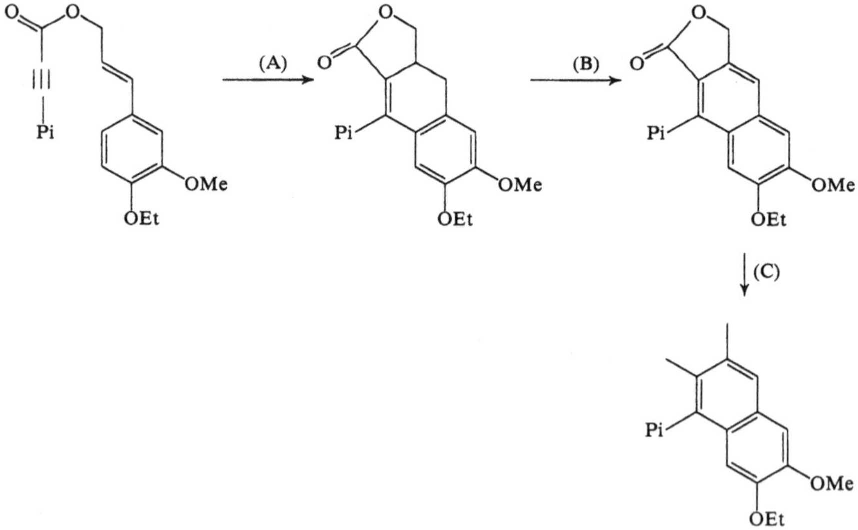

(**36**)

(A) ... Ac_2O, refl. (B) ... Pd/C, Δ. (C) ... 1. $LiAlH_4$, Et_2O; 2. H_2, 10% Pd/C, AcOEt.

Chart 11. Synthesis of dehydro-otobaphenol ethyl ether (**36**)

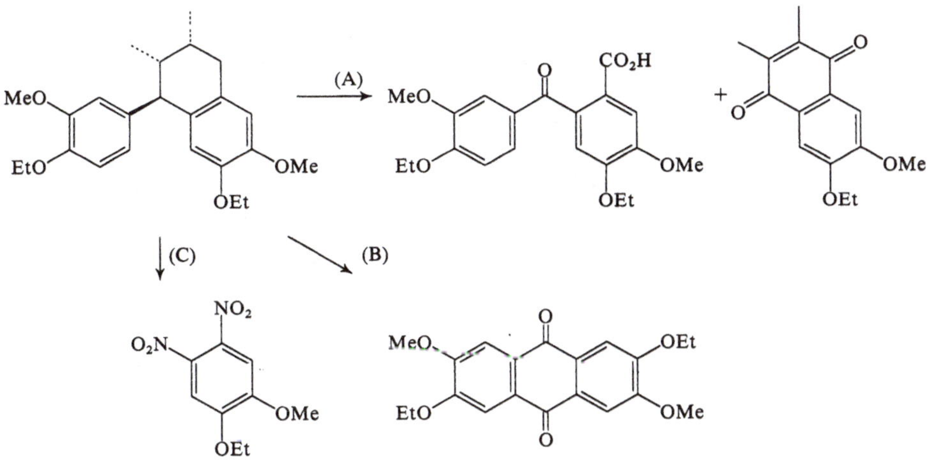

(A) ... $K_2Cr_2O_7$, AcOH, 100°. (B) ... CrO_3, AcOH$-H_2O$, refl. (C) ... HNO_3 conc., refl.

Chart 12. Oxidative degradation of diethylisoguaiacin

Isogalcatin (**2.1e**) was obtained after methylation of a mixture of natural phenols which may have included otobaphenol (**2.1b**). This, isolated in later years, had its structure confirmed by synthesis of the dehydrogenation product of the derived ethyl ether **36** (Chart 11) (*88*).

Guaiacin (**2.1a**), isoguaiacin (**2.2c**) and dehydroguaiaretic acid (**2.5a**) were equally isolated only as ethers after methylation or ethylation of a phenolic resin. Guaiacin (**2.1a**) and isoguaiacin (**2.2c**), recognized as phenyltetralins by oxidations (Chart 12), are stereoisomers: the diethylethers gave dehydrodiethylguaiaretic acid upon dehydrogenation with tetrachloro-*o*-benzoquinone. Their structures were formulated, respectively, **2.1a** and **2.2c** (*85*) upon comparisons of (+)-diethylguaiacin with (−)-galbulin (**2.1f**) and of (−)-dimethylisoguaiacin with (+)-isogalbulin, a product obtained by defunctionalization of the lactone moiety of dimethyl-β-conidendrin (**37**) (*134*). The analogous defunctionalization of deoxypodophyllotoxin (**38**) gave austrobailignan-4 (**2.3b**),

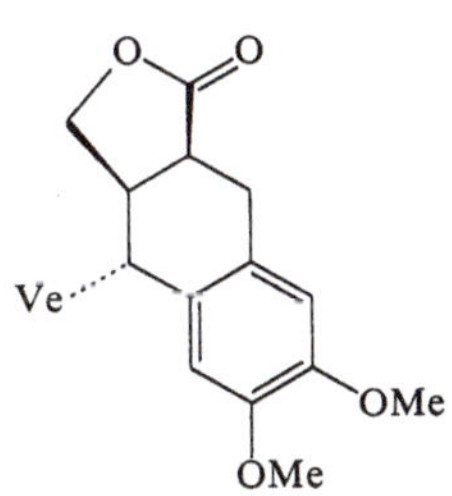

(**37**)

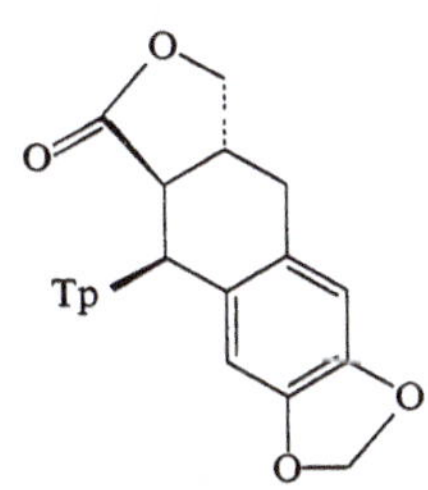

(**38**)

settling the absolute configuration of this neolignan and, by force of ^1H NMR spectra and optical rotation, that of its companion substance **2.3a** (*112*).

In the report on 3′-demethoxyisoguaiacin and nor-isoguaiacin (**2.2b**) (*63*), which was based on ^1H NMR spectra and methylation of **2.2b** to isoguaiacin (**2.2c**), the structural representation of the first named compound features the vicinal OH and OMe groups at exchanged positions. This is probably due to a mistake, and not reproduced in **2.2a** (Table 2).

It is interesting to note that dehydroguaiaretic acid (**2.5a**) is among the $FeCl_3$-dehydrogenation products of isoeugenol (*98*).

2. 2.6

Treatment of carpanone with H_2/Pt in AcOH gave dihydrocarpanone (**39**), which with H_2SO_4 in Ac_2O gave a diacetate (**40**). These trans-

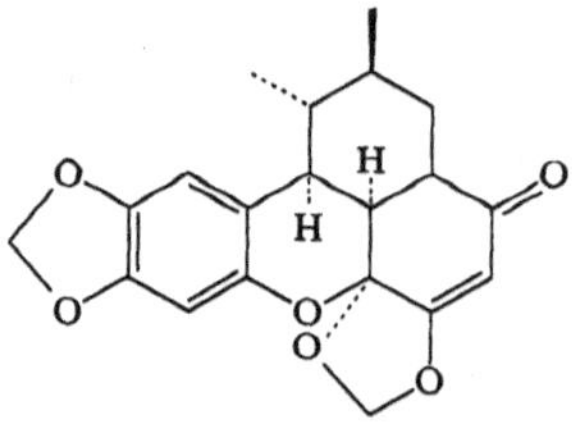

(**39**)

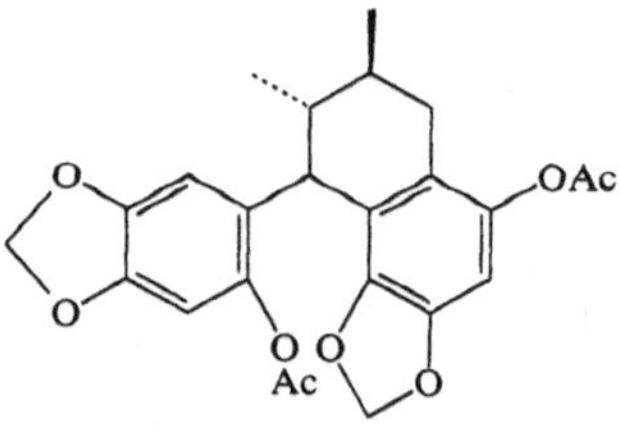

(**40**)

formations, together with MS and ^1H NMR data, including the results of spin decoupling experiments, led to structure **2.6a** for carpanone. Configurational isomers (designated carpanonones) are also natural products (*34*).

The synthesis of carpanone (**2.6a**) was achieved by isomerization of the naturally occurring **41** (*14*) and treatment of **42** by a divalent cation which, besides bringing the two phenolic units together (**43**), effected the required net two-electron oxidation. Chart 13 rationalizes, but does not diminish the marvel, how this single 46% yield reaction introduced all five contiguous chiral centers with the correct stereochemistry (*41*).

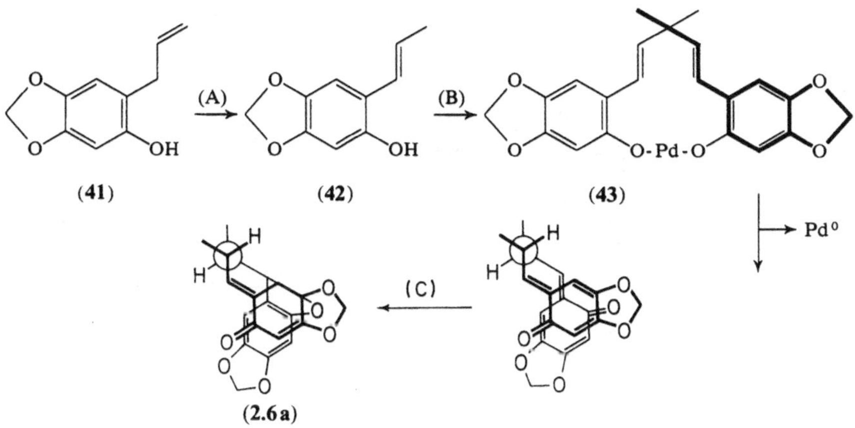

(A) ... t-BuOK, DMSO. (B) ... $PdCl_2$, AcONa, MeOH, H_2O, 38°. (C) ... Δ.

Chart 13. Synthesis of carpanone (**2.6a**)

IV. 8.8′,2.2′-Neolignans

1. 3.1—3.5

Schizandrol (**3.1a**) and its di-O-methyl derivative schizandrin (**3.1b**) were the first representatives of this neolignan type to be reported (*89*). $KMnO_4$-oxidation of **3.1b** to **44a** (*90*) pointed to the structure of the biphenyl moiety and revealed the site of attachment of the aliphatic ring. This was recognized to be eight-membered by UV comparison with model biphenyl derivatives (*e. g.* **45a**) containing bridges of varying number of members between the 2,2′-positions. These facts and the existence of 4 benzylic protons (^1H NMR evidence) pointed to structure **3.1b** for schizandrin, a proposal which is consistent with the inertness of the carbinol function towards acetylation and hydrogenolysis (*94*).

Structure **3.2a** was assigned to deoxyschizandrin on basis of its UV (identical to the UV of **3.1b**) and ^1H NMR spectra, as well as its incapability of racemization. Optically active 6,7-*cis*-substituted 1,2,3,4-dibenzocycloöcta-1,3-dienes racemize readily (*93*). Indeed, synthetic dimethyl *trans*-2′,3′,4′,1″,2″,3″-hexamethoxy-1,2,3,4-dibenzocycloöcta-1,3-diene-6,7-dicarboxylate (**45a**) (*91*) proved to be a key intermediate in the synthesis of deoxyschizandrin (**3.2a**) (Chart 14) (*93*).

Examination of IR and ^1H NMR spectra, as well as racemization experiments, showed γ- and *pseudo*-γ-schizandrin to be a pair of *cis*- and *trans*-isomers, represented probably by **3.3a** and **3.2b** (*92*).

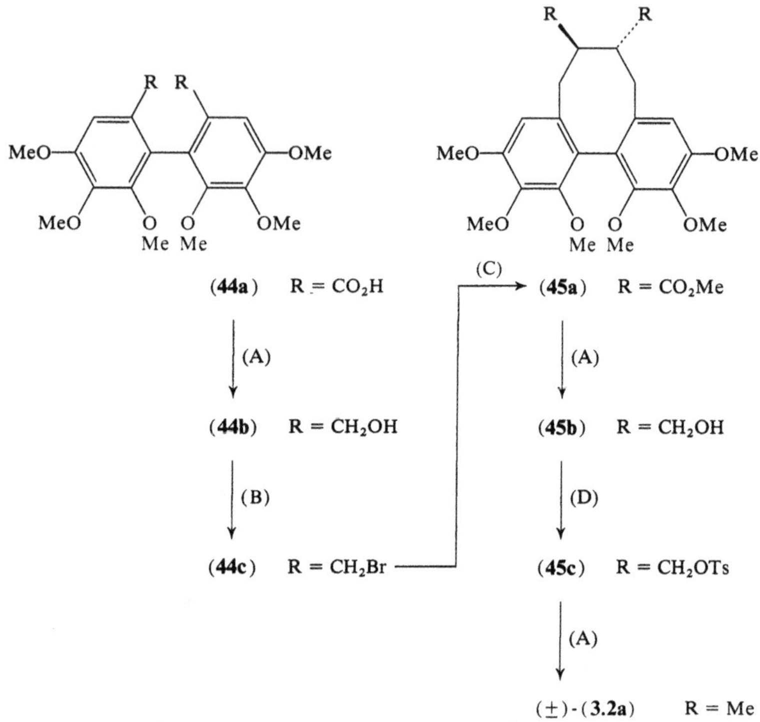

(A) ... $LiAlH_4$, THF, 50°. (B) ... PBr_3, $CHCl_3$, C_5H_5N, 20°. (C) ... $H_2C(CO_2Me)_2$, EtONa, THF, 50°. (D) ... TsCl, C_5H_5N, 0°.

Chart 14. Synthesis of (±)-deoxyschizandrin (**3.2a**)

Table 5. *Nuclear Overhauser Enhancement of Selected* ^1H *NMR Signals of Kadsurin* (**3.4a**) *and Kadsurarin* (**3.5a**)

Low intens. irrad. of	Signal observed of	Enhancement of intensity % (**3.4a**)	% (**3.5a**)
H-7	H-6	21	23
Me-8	H-6	0	
OMe-3	H-6	0	
OMe-5′	H-6	0	
OMe-4′	H-6	0	
OMe-3′	H-6	0	
Me-8	H-7	12	11
Me-8′	H-7′		17
H-7	H-6′	0	
H-7′	H-6′	10	24
Me-8′	H-6′	13	0
OMe-5′	H-6′	11	18

The UV spectrum of kadsurin is almost superimposable on that of schizandrin (**3.1b**), a fact which, in addition to other spectral evidence, led to structural proposal **3.4a**. The position of each functional group and the conformation of kadsurin and the related kadsurarin (**3.5a**) were elucidated by measurements of nuclear Overhauser enhancement of signal intensities (Table 5) (*43*).

V. 3.3′-, 5.1′,2.2′- and 5.1′,2.2′,1.5′,6.6′-Neolignans

1. 4.1

Structure **4.1a** for acuminatin resulted by consideration of spectral data. *Trans*-isoeugenol and the *cis*-epoxide prepared from *cis*-isoeugenol benzoate with *m*-chloroperbenzoic acid served as model compounds in the structural assignments of the two side chains. The synthetic *cis*-epoxide is characterized by a larger vicinal coupling ($J = 4.2$ Hz) than the equally synthetic *trans*-analogue ($J = 2.0$ Hz). Acuminatin, however, gives rise to a still larger value, due to influence of the biphenyl system. Evidence for placing the two methoxyls on the phenyl ring bearing the propylene oxide group was obtained from MS studies, while the exact location of all three methoxyls was ascertained by Lemieux-von Rudloff oxidation of acuminatin to the epoxy-aldehyde **46** and the biphenyl dicarboxylic acid **47.** The dimethyl ester of this acid was identical with a synthetic sample (*51*).

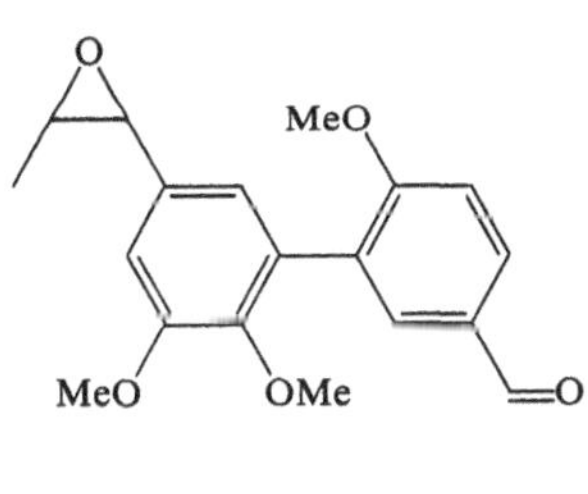

(**46**)

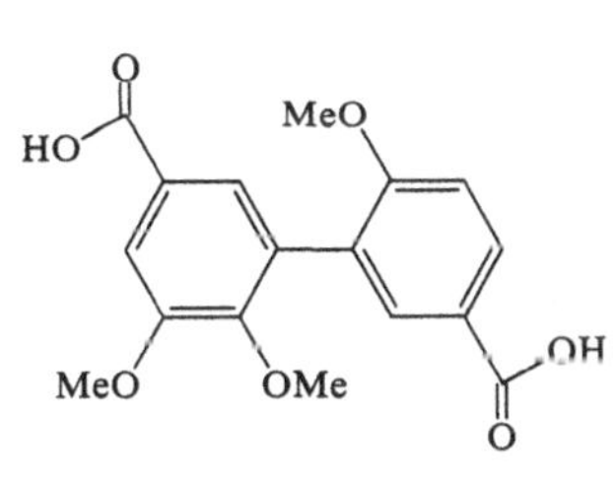

(**47**)

2. 4.2

Catalytic hydrogenation of magnolol (**4.2a**) led to tetrahydromagnolol, for which structure **49b** was confirmed by synthesis (Chart 15). The nature of the side chain of **4.2a** was ascertained by treatment with alkali in AmOH, which gave isomagnolol, and $KMnO_4$-oxidation of the diacetate to diacetoxybiphenyldiacetic acid and diacetoxybiphenyldicarboxylic acid. Saponification and decarboxylation (HCl at 270°)

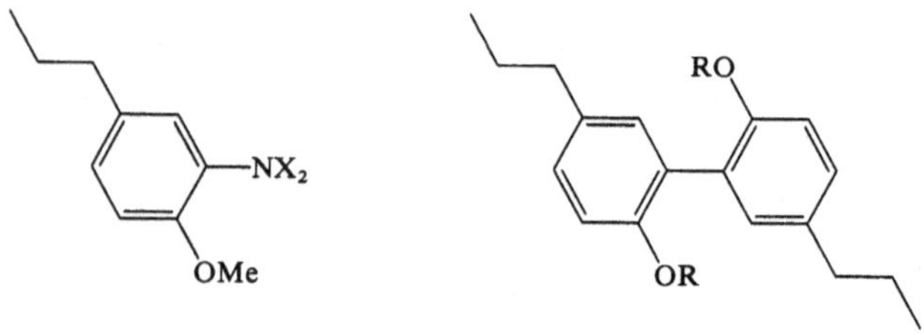

(**48a**) X = O → (B) → (**49a**) R = Me

(**48a**) —(A)→ (**48b**) X = H

(**48b**) —(B)→ (**49a**)

(**49a**) —(C)→ (**49b**) R = H

(A) ... red.; (B) ... Cu, 270°; (C) ... HI

Chart 15. Synthesis of tetrahydromagnolol (**49b**)

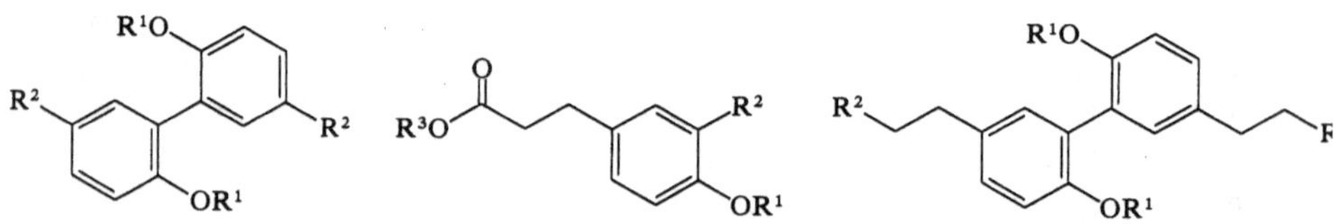

(**50a**) $R^1 = R^2 = H$

(A)

(**50b**) $R^1 = H$, $R^2 = Br$

(B)

(**50c**) $R^1 = Me$, $R^2 = Br$

(C)

(**50d**) $R^1 = Me$,
$R^2 = CH_2CH = CH_2$

(D)

(**4.2a**)

(**51a**) $R^1 = R^2 = R^3 = H$

(E)

(**51b**) $R^1 = R^3 = H$, $R^2 = I$

(F)

(**51c**) $R^1 = R^3 = Me$, $R^2 = I$

(G) → (**52a**)

(**52a**) $R^1 = Me$, $R^2 = CO_2Me$

(H)

(**52b**) $R^1 = Me$, $R^2 = CH_2OH$

(I)

(**52c**) $R^1 = Me$, $R^2 = CH_2OS_2N$

(J) → (**50d**)

(A) ... Br_2, $CHCl_3$. (B) ... Me_2SO_4, 10% NaOH. (C) ... 1. EtMgBr, THF; 2. $BrCH_2CH{=}CH$
(D) ... 1. MeMgI, Et_2O; 2. 180°, 1 hr. (E) ... I_2, KI, aq., NH_3 conc. (F) ... CH_2N_2, Et_2O.
(G) ... copper bronze, 225°. (H) ... $LiAlH_4$. (I) ... Chugaev. (J) ... 250°.

Chart 16. Synthetic pathways to magnolol (**4.2a**)

gave diphenylene oxide and 2,2′-dihydroxybiphenyl (*138*). While tetrahydromagnolol (**49b**) was obtained additionally by dehydrogenation of p-(n-propyl-)phenol with aqueous H_2O_2, $FeSO_4$, 0 · 2N H_2SO_4 (*116*) or aqueous K_3 [$Fe(CN)_6$], 0 · 2N $NaHCO_3$ (*77*), magnolol itself (**4.2a**) was prepared by dehydrogenation of chavicol with $FeCl_3$ (*53*) and by the pathways outlined in Chart 16 (*125*).

Dehydrodieugenol (**4.2b**) was long known as a synthetic product (*44, 58*) before its isolation from nature (*81*).

3. 5.1, 6.1

In addition to the biphenyls **4.2a** and **b** only two further allyl- + allylphenol derived neolignans, asatone and isoasatone, are known.

The stereostructure of isoasatone (**6.1a**) was established through the X-ray crystallographic analysis of its $LiAlH_4$-reduction product, "dihydroxyisoasatone" (*131*), and rationalizes the fact that the compound yields a mixture of 5-allyl-2-hydroxy-1,3-dimethoxybenzene (46%) and 6,6′-diallyl-3,3′-dihydroxy-2,4,2′,4′-tetramethoxybiphenyl (12%) upon reduction with zinc in acetic acid (*146*). Since asatone is cleaved by this reagent into the same aromatic compounds and can be converted photochemically into isoasatone (see Section XIV, Chart 26, **5.1a**→**6.1a**), its previously proposed structure (*42*) must be revised to **5.1a** (*147*).

VI. 8.3′-Neolignans

1. 7.1, 7.2

Dehydrodiisoeugenol was first synthesized in 1908 by oxidative dimerization of isoeugenol with $FeCl_3$ as well as with H_2O_2-peroxidase (*45*). Its constitution was elucidated in 1933 by Erdtman, who immediately grasped that one "könnte die Möglichkeit ins Auge fassen, daß das Lignin durch oxydative Polymerisation eines in der Seitenkette oxydierten Propylguajakols entsteht" (*52*). This view was subsequently widely accepted, dehydrodiisoeugenol being considered a lignin model compound (*127*) 40 years before its natural occurrence in optically active form (**7.1a**) was recognized (*8*).

The *trans*-relation of the substituents in synthetic (±)-dehydrodiisoeugenol was elucidated by oxidative degradation and ^1H NMR (*23*). In order to explain the lack of formation of *cis*-arylmethyl-dihydrobenzofurans, even by the action of H_2O_2-peroxidase on (Z)-isoeugenol, relieve of steric compression of adjacent *cis*-methyl and quinonemethide moieties by rotation about the 7,8-bond prior to cyclization of the initially formed intermediate **53** was invoked (*128*). ^{1}H NMR, in conjunction with MS, was the major tool in the elucidation of all natural compounds of this series. The absolute stereochemistries of **7.1a** and **7.1b** (*8*), as well as of the antipodal **7.2a** (*76*), **7.2b** (*55*) and **7.2c** (*16*) were deduced from ORD curves which are comparable to the analogous curves respectively of (2S,3S)- (*50*) and (2R,3R)- (*72, 123*) 2-aryl-3-methyl-2,3-dihydrobenzofurans.

Structure **7.1b** was also assigned to (+)-eupomatenoid-8, described as a gum (*31*), and thus possibly a partly racemized product (*12*). The racemate, also a gum, was obtained (as intermediate in an eupomatene synthesis), even before either product had been isolated, using dehydrodiisoeugenol as starting material. Selective demethylation with periodate gave an *o*-quinone which was reduced to the catechol derivative

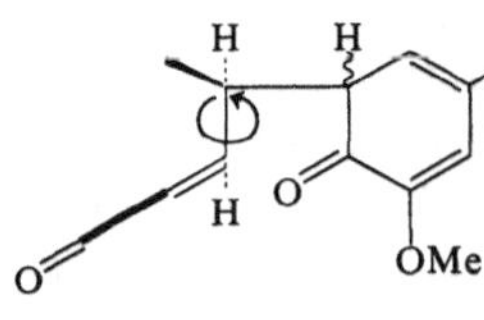

(**53**)

whose methylenation gave the desired product (*106*). Since in this process both aromatic rings of the final product stem from the same phenol which, by necessity, contains a propenyl side chain, it is useful only for the synthesis of 2-aryl-5-*propenyl*-coumarans with a limited number of oxygenation patterns. A more generally applicable synthetic approach to these structures involves the stereospecific "abnormal" Claisen rearrangement outlined in Chart 17 (*12*).

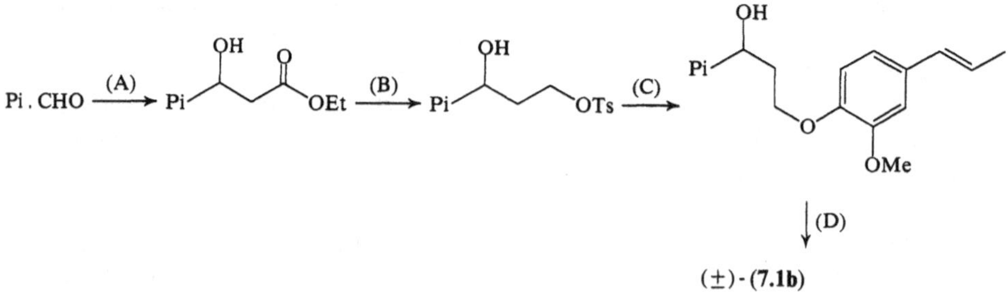

(A) ... $BrCH_2CO_2Et$, Zn, C_6H_6. (B) ... 1. $LiAlH_4$, Et_2O; 2. TsCl, C_5H_5N. (C) ... sodium salt of isoeugenol, DMF. (D) ... $C_6H_5NEt_2$, 235°, 10 h.

Chart 17. Synthesis of (±)-licarin-B (**7.1b**)

The racemate of conocarpin (**7.2a**), the simplest known member of this small group of neolignans, was again synthesized by oxidative dimerization (with $FeCl_3$) of *p*-propenylphenol (*76*).

2. 7.3—7.6

Eupomatene [later designated eupomatenoid-1 (*31*)] was the first representative of the series to be described. The ^1H NMR spectrum suggested the correct structure **7.3a,** confirmed by ozonolysis of the dihydro-derivative **54** to the ester **55,** equally available by synthesis (Chart 18).

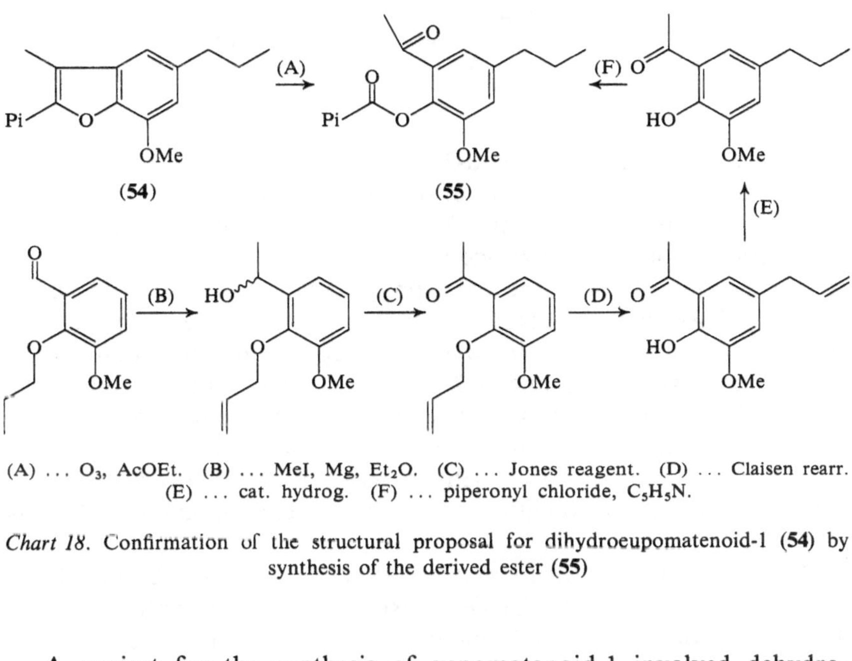

(A) ... O_3, AcOEt. (B) ... MeI, Mg, Et_2O. (C) ... Jones reagent. (D) ... Claisen rearr. (E) ... cat. hydrog. (F) ... piperonyl chloride, C_5H_5N.

Chart 18. Confirmation of the structural proposal for dihydroeupomatenoid-1 (**54**) by synthesis of the derived ester (**55**)

A project for the synthesis of eupomatenoid-1 involved dehydrogenation of (±)-eupomatenoid-8 (**7.1b**), available *via* dehydrodiisoeugenol (**7.1a**) (see Section VI.1). It turned out, however, that the double bond in the side chain has an adverse effect on this reaction attempted under a variety of conditions (*106*), and the preparation of 5-propenylbenzofurans by this sequence is satisfactory only if the conjugated ethylenic function is protected, *e.g.* by bromine addition (*6*).

A second and more practicable entry to the 2-arylbenzofuran series used as a key step the oxidation of flavylium salts (Chart 19) (*106*).

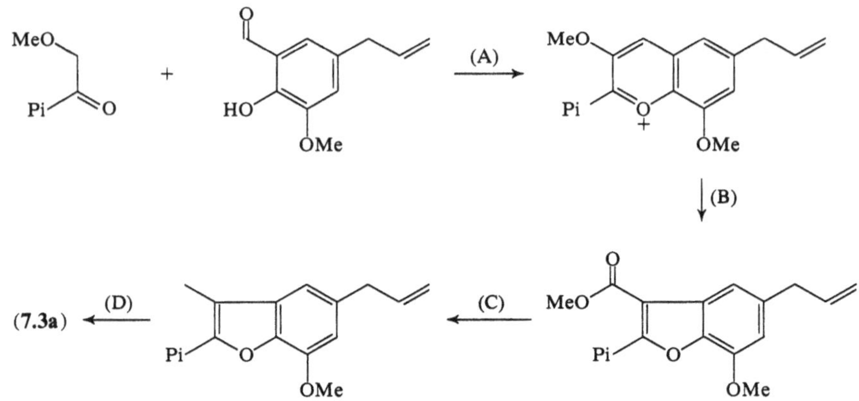

(A) ... HCl, AcOH. (B) ... pH 5.8, H_2O_2, MeOH. (C) ... 1. $LiAlH_4$, Et_2O; 2. MnO_2, C_6H_6; 3. $HS(CH_2)_2SH$, AcOH, BF_3/Et_2O; 4. RaNi, EtOH. (D) ... t-BuOK, DMSO, N_2.

Chart 19. Synthesis of eupomatenoid-1 (**7.3a**)

Hydrolysis of eupomatenoid-2 (**7.3b**) gave L-rhamnose and eupomatenoid-6 (**7.3f**). The structure of the methyl ether of eupomatenoid-6, as well as of the eupomatenoids-3 (**7.3c**) and -4 (**7.3d**), were confirmed by ozonolysis of the dihydroderivatives (*cf.* **54**→**55**) and hydrolysis of the resulting esters to the corresponding aromatic acids and acetophenones. These experiments also defined the structure of eupomatenoid-5 (**7.3e**), since it gives eupomatenoid-4 upon O-methylation (*106*). Eupomatenoids-7 (**7.3g**) and -12 (**7.3h**) are linked by an analogous relationship (*122*). Again, the structures of both compounds became known when eupomatenoid-7 proved to be identical with a synthetic sample of **7.3g** prepared by S-dehydrogenation of dehydrodiisoeugenol (**7.1a**) (*31*). Reaction conditions were critical and optimum yield was only 10% (*106*).

The constitution of eupomatenoid-9 (**7.4a**) was established by a partial synthesis (m-$ClC_6H_4CO_3H$ oxidation of **7.3e** acetate). Since the ethylenic double bond of eupomatenoid-5 (**7.3e**) has the (*E*)-configuration, the resulting glycol, and consequently also eupomatenoid-9, are in the *erythro* form. Periodate oxidation of the synthetic product [(±)-**7.4a**] gave the aldehyde **7.6a** which proved to be identical with eupomatenoid-10 (*122*).

The nature of the side chain in eupomatenoid-11 (**7.5a**) was determined by spectroscopy and confirmed by oxidation of the compound to a substituted propanone. Conversion into eupomatenoid-3 (**7.3c**) by substitution of the alcoholic OH by Cl and dehydrohalogenation established the additional structural features. The absolute configuration was determined as R by Horeau's method (*122*).

3. 7.7—7.9

The structural proposals for the mirandins A (**7.7a**) and B (**7.8a**) were based on MS and ^{1}H NMR evidence (*17*). This is different from the analogous evidence for **8.2b** in two relevant points, suggesting the existence of methoxyl at a sp^3-C (τ **7.7a** 7.00, **7.8a** 6.88) and not at a sp^2-C (τ **8.2b** 6.38) and of allyl at a sp^2-C (τ_{CH_2} **7.7a** 6.89, **7.8a** 6.84, both d, $J = 7.0$ Hz) and not at a sp^3-C (τ_{CH_2} **8.2b** 7.72 and 7.83, both dd, $J = 6.3$, 12.5 Hz). Since both mirandins maintain the Ar and Me (τ **7.7a** 8.89, **7.8a** 8.85) substituents in a *trans*-relationship, they must differ in the orientation of the aliphatic methoxyl. This is with respect to H-8 *cis*-related, thus deshielding this proton (τ 7.35), in **7.7a** and *trans*-related in **7.8a** (τ 7.84).

Constitution and relative stereochemistry of the mirandins were confirmed by X-ray crystallography (*17*) and by ^{13}C NMR (see Section IX) (*144*). The formulae **7.7a** and **7.8a** represent absolute configurations, since photolysis and pyrolysis of **7.7a** leads to **56,** a compound of established

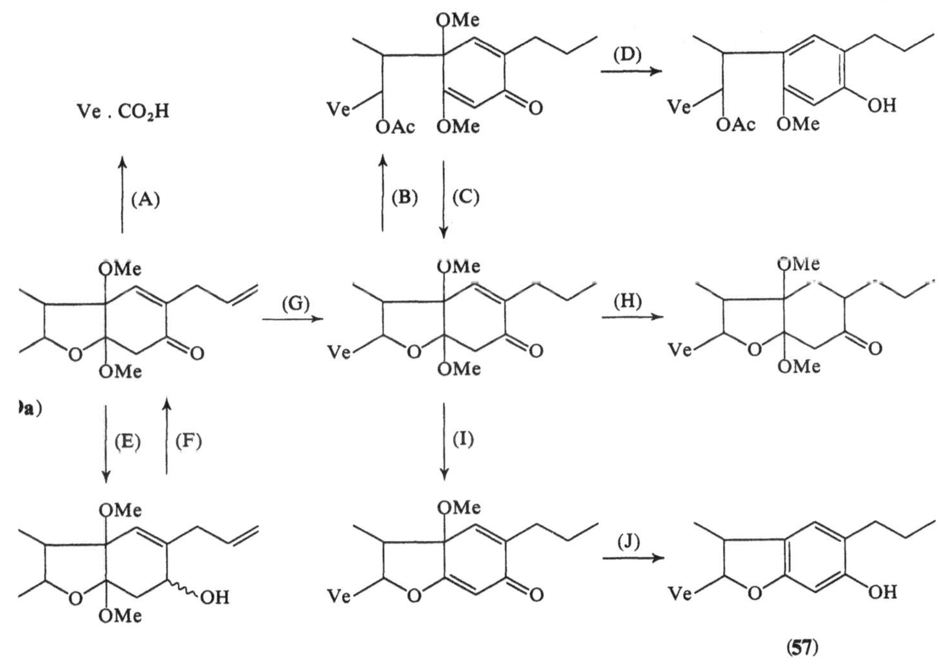

(A) ... $KMnO_4$. (B) ... AcONa, Ac_2O, refl. (C) ... KOH, aq., EtOH. (D) ... Zn, AcOH. (E) ... $LiAlH_4$. (F) ... MnO_2. (G) ... H_2, Pd/C, EtOH. (H) ... H_2, Pd/C, MeOH. (I) ... KOH, EtOH, refl. (J) ... Zn, AcOH, room temp.

Chart 20. Chemical reactions of piperenone (**7.9a**)

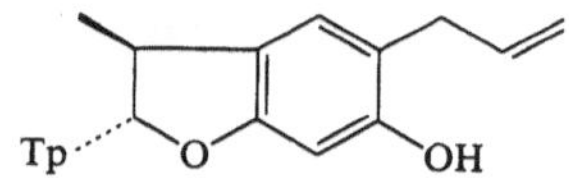

(56)

stereochemistry (*9*), and the mirandins were correlated through ORD curves.

The constitutional proposal **7.9a** for piperenone was deduced by reasoning based on the reactions summarized in Chart 20, the diagnostically most significant one leading to a coumaran (**57**) (*104*).

4. 7.10

Analysis of ^1H NMR data on futoquinol and the derived mixture of epimeric alcohols (**58**) led to structural proposal **7.10a** which justifies the dienone-phenol and dienone-benzene rearrangements formulated in Chart 21 (*139*).

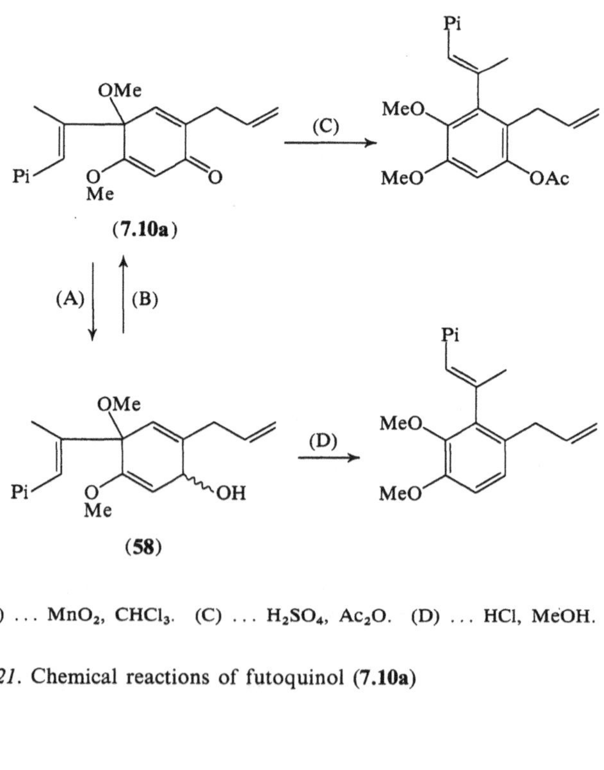

(A) ... $NaBH_4$, EtOH. (B) ... MnO_2, $CHCl_3$. (C) ... H_2SO_4, Ac_2O. (D) ... HCl, MeOH.

Chart 21. Chemical reactions of futoquinol (**7.10a**)

VII. 8.1′-Neolignans

1. 8.1—8.3

The first 8.1′-neolignan to be isolated was burchellin to which structure **8.1a** was accorded by consideration of IR, UV and 220 MHz ^{1}H NMR spectral data (*21*). The MS was of little help in the analysis. The rearrangements which seem to account for some of the major peaks could only be formulated with confidence *a posteriori* (Chart 22).

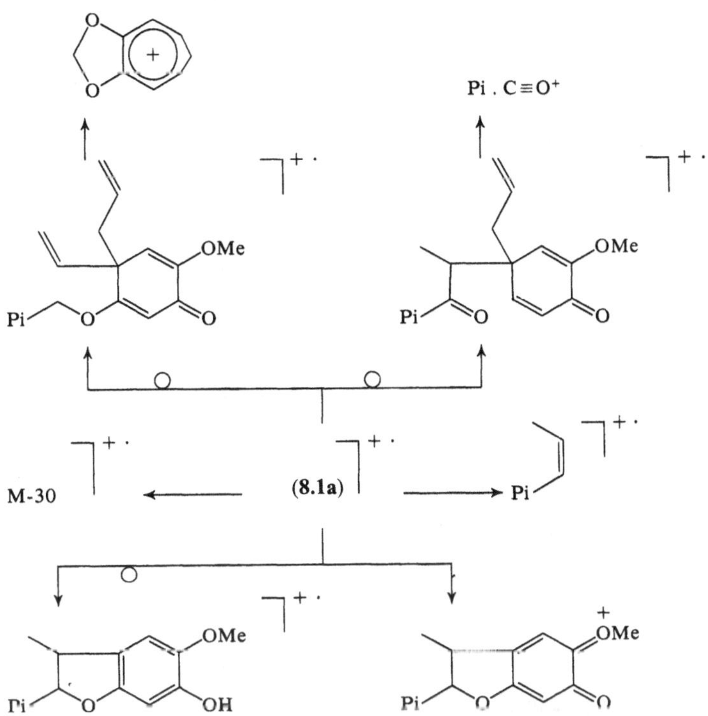

Chart 22. Interpreted mass spectrum of burchellin (**8.1a**)

Interpretation of the MS of a hydrolytic product (**59**), however, was exempt from such difficulties. The stereochemistry of the 7,8-bond in **8.1a** was established upon comparing ORD curves of the hydrogenolysis product **60** and appropriate 2,3-*trans*-2-aryl-3-methyl-2,3-dihydrobenzofurans of known absolute configuration.

Subsequently to the isolation of a further 7S, 8S derivative (**8.2b**), the stereochemistry at C-1′ was revealed also. Indeed, the two ^{1}H NMR signals due to the CH_2-group existent at this centre occur at higher field in the spectrum of **8.2b** than in the spectrum of **8.1a** (Table 6). The

Table 6. ^1H *NMR* (τ) *and* ^{13}C *NMR* (δ) *Values Relevant in the Assignment of Configuration to Compounds* **8.1, 8.2** *and* **8.3** [^1H NMR: $CDCl_3$ **8.1a, 8.2b** (220 MHz); **8.2c, 8.3a, 8.3b** (60 MHz). CCl_4 **8.3c** (220 MHz); **8.1b, 8.1c** (60 MHz). ^{13}C NMR: $CDCl_3$ (25.20 MHz)]

	2H-7′	3H-9	H-8	H-7	$J_{H\text{-}7,H\text{-}8}$ (Hz)	C-7′	C-9	C-8	C-7
8.1a	7.45, 7.66	8.84	7.12	4.83	9.5	36.6	8.3	49.5	90.0
8.1b	7.5—7.8	8.89	7.5—7.8	4.97	9.0	36.7	8.5	49.3	91.0
8.1c	7.5—7.8	8.89	7.5—7.8	4.94	9.0	36.7	8.5	49.6	91.5
8.2c	>7.6	8.87							
8.2b	7.72, 7.83	8.82	7.30	4.71	indet.				
8.3a	7.15—7.45	9.47	7.15—7.45	4.04	5.0	43.9	12.0	44.6	87.2
8.3b	7.07—7.60	9.50	7.07—7.60	4.0—4.5		43.8	11.9	44.5	87.1
8.3c	7.46—7.69	9.53	7.28—7.46	4.16	5.0	43.9	12.0	44.5	87.2

Table 7. *Pr(fod)*$_3$ *Induced* ^1H *NMR Shifts* (Δδ) *Obtained by Extrapolation of Observed Shifts (in* $CDCl_3$*) to 1 : 1 Shift Reagent Substrate Ratio*

	H-3′	OMe-5′	H-6′
8.1a	9.1	27.0	11.7
8.3b	9.5	23.0	11.4
8.3c	9.3	24.8	11.5
8.4a	2.4	13.4	ax. 1.4; eq. 4.3

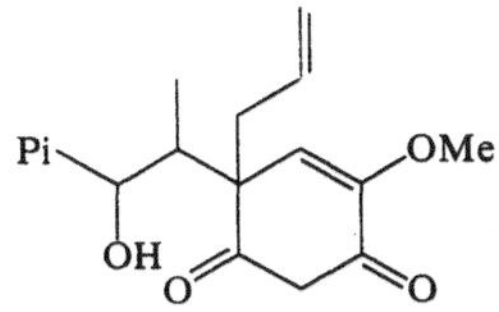

(59)

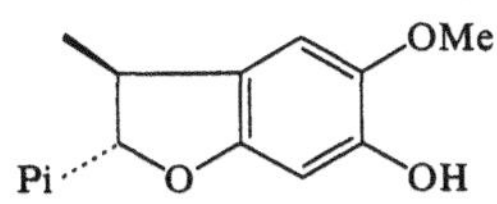

(60)

allyl group of **8.2b** must thus protrude into the relatively protected zone above the aromatic ring (*9*). The structures of the additional 7,8-*trans*-compounds **8.1b, c** (*55*) and **8.2a** (*69*) were deduced in a similar manner.

The relative shielding (with respect to **8.1** and **8.2**) of the 3H-9 and relative deshielding (with respect to **8.2**) of the 2H-7′ observed in three further $\Delta^{8'}$-1′,4′-dihydro-4′-oxo-7.O.2′,8.1′-lignans (Table 6) was rationalized by formulae **8.3a** (*69*), **8.3b** and **8.3c** (*9*). The absolute stereochemistries were suggested by the analogous Cotton effects of the compounds and of 7-*epi*-mirandin-A, the product of isomerization by acid of **7.7a** (*17*), as well as by the sequence **8.3b**-(pyrolysis) → **10.3a**-(acid) → **10.2b.** The 7S,8S-configuration of **10.2b** was deduced by ORD comparison with dihydrobenzofurans of known configuration (see above). ^{13}C NMR data confirmed these structural assignments (Table 6) (*144*) and merit further comment in Section IX.

2. 8.4—8.6

The original structural proposal **61** for porosin was based on ^{1}H NMR and MS evidence (*7*). All ^{1}H NMR data are, however, equally interpretable on grounds of structure **8.4a** in which the oxymethine hydrogen, assigned to C-2′ in **61,** occupies C-5′. Indeed Pr(fod)$_3$ ^{1}H NMR shifts are relatively feeble in the case of porosin (Table 7), and, thus, incompatible with the existence of the planar CH=C·OMe·CO system shown in **61** (*18*).

Further evidence for the validity of structure **8.4a** for porosin was obtained by direct and indirect UV evidence (*18*). The indirect method relied on the argument that elimination of a methoxyl from a double bond of an α,β-unsaturated carbonyl system (such as in *61*) would modify the UV absorption, while removal of the methoxyl from a saturated C should not modify the chromophore of **8.4a;** and, indeed, the UV spectra of porosin (**8.4a**) and of its photoproduct **62** were found to be superimposable.

The revision of the porosin structure was necessitated initially by the ^{13}C NMR analysis of the natural product which inclusively confirmed ^{1}H NMR evidence concerning its stereochemistry (see Section IX) (*144*).

Table 8. *Postulated Biosynthetic Relationship and Occurrence in Lauraceae Species (Numbered acc. to Table 4) of Type* **8—11** *Neolignans*

	Cope		retro-Claisen		Claisen		dehydrogenation			
Ar	Cmpd.	LAUR	Cmpd.	LAUR	Cmpd.	LAUR	Cmpd.	LAUR	Cmpd.	LAUR
Pi	**8.1a**	1,2,5	**9.1a**	2,5	**10.2a**	2,5	**11.1a**	5	**11.3a**	5
Pi	**8.2a**	5								
Pi	**8.3a**	5	**9.3a**	5			**11.2a**	5		
Ve	**8.1b**	1,2	**9.1b**	1,2						
Ve	**8.1c**	1								
Mp			**9.2a**	4	**10.2b**	4	**11.1b**	4	**11.3b**	4
Mp	**8.3b**	4	**9.3b**	4	**10.3a**	4	**11.2b**	4		
Tp	**8.2b**	4	**9.2b**	4	**10.2c**	4,19			**11.3c**	19
Tp	**8.3c**	4								

References, pp. 65—72

The tentative formulation **63** for canellin-B was based on spectral comparison with porosin, then formulated as **61** (*61*). If the analogy is correct, the structure of canellin-B must be revised to **8.5a,** a postulate which is also more reasonable from the biosynthetic point of view, and in conformity with the fact that additional representatives of this type, such as **8.6a**, are presently being discovered (*10*).

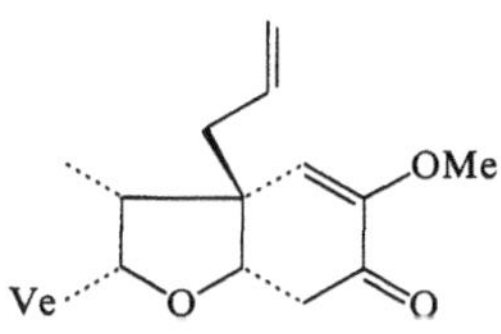

(61)

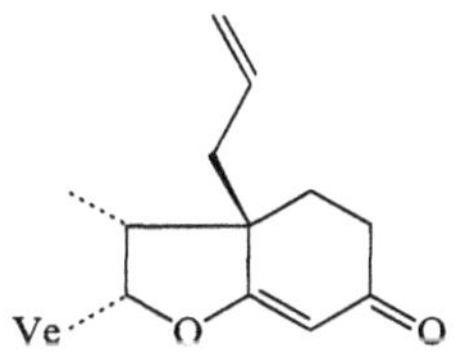

(62)

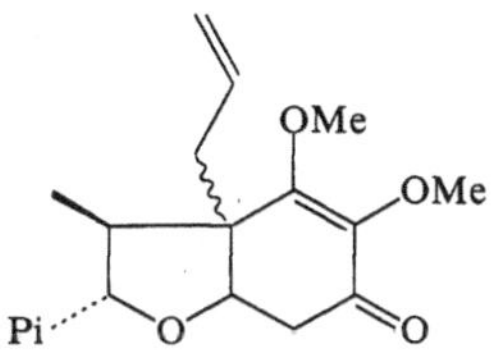

(63)

VIII. Rearranged 8.1′-Neolignans

1. 9.1—9.4, 10.2, 10.3, 11.1—11.3

The structures of the derivatives belonging to types **9—11** were again determined by spectral means. Structural comparison and co-occurrence led to the postulate that compounds belonging to type **8** are responsible for the biosynthesis of the representatives of types **9, 10** and **11** by successive Cope, retro-Claisen and Claisen rearrangements (Section XIV) (*9, 55, 69*). Stereochemical assignments for the tetrasubstituted carbon of the dienone ring of **9.1, 9.2** and **9.3** were based on consideration of the chiralities at C-1′ of the co-occurring precursors of type **8** (Table 8) and confirmed for **9.2c** by X-ray crystallography (*70*). Additional representatives of the **9.1, 9.2** or **9.4** group can be classified with ease

Table 9. *Pyrolysis of Neolignans (in $C_6H_5NEt_2$ under N_2)*

Substrate	Temp.	Products
7.8a	200°	**64c**
8.1a	115°	**9.1a**
8.1a	225°	**11.1a** + **64a**
8.3b	115°	**9.3b** + **10.3a** + **11.2b**
8.3b	225°	**11.2b** + **65a**
8.3c	225°	**65b** + **65c**
9.2a	225°	**11.1b** + **64b**
9.2c	155°	**10.2c** + **64c**
9.2c	225°	**64c** + **64d**
10.2c	225°	**64c** + **64d** + **66**

by their ORD curves (*9, 13, 55*). *In vitro*, pyrolysis of neolignans of types **8.1, 8.3, 9.2** and **10.2** led to the expected derivatives of types **9.1, 9.3, 10.2, 10.3, 11.1** and **11.2** (Table 9), thus corroborating their structural proposals (**9, 13, 19**).

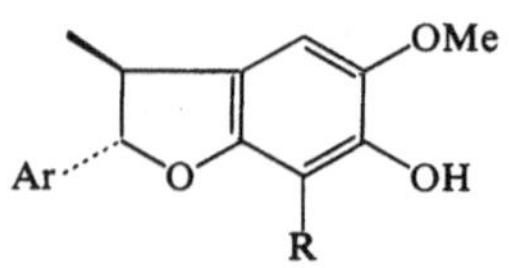

(**64a**) Ar = Pi, R = H
(**64b**) Ar = Mp, R = H
(**64c**) Ar = Tp, R = H
(**64d**) Ar = Tp, R = allyl

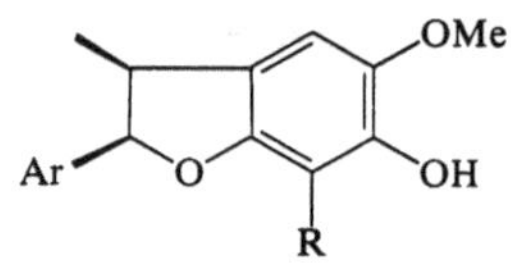

(**65a**) Ar = Mp, R = H
(**65b**) Ar = Tp, R = H
(**65c**) Ar = Tp, R = allyl

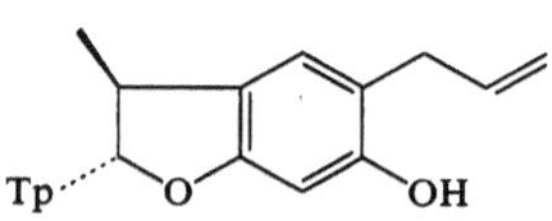

(**66**)

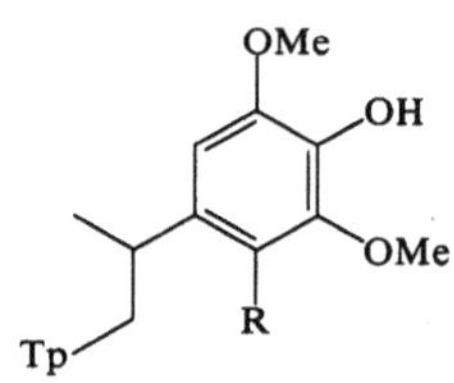

(**67a**) R = H
(**67b**) R = allyl

2. 10.1

The structure of aurein (**10.1a**) was deduced from MS, ^{1}H NMR (*66*) and ^{13}C NMR (*144*) data. It shows an O-allyl which is easily cleaved by pyrolysis (to **67a**) and acid treatment (to **67b**). The former reaction is relevant, since loss of the allyl group demonstrates that the *ortho*- and *para*-positions relative to the O-allyl group are substituted (*66*).

IX. ^{13}C NMR Spectroscopy of Neolignans

The ^{13}C NMR spectra of neolignans pertaining to structural types **7.1, 7.7, 7.8, 8.1, 8.3, 8.4, 9.1, 9.2** and **16.2** were analyzed and their carbon shifts assigned. Clearly, CMR constitutes a powerful analytical method, of use in the recognition not only of specific neolignan types, but also of configuration and conformation (*144*).

In this connection, the C-methyl (C-9) shift of 8.1′- and 8.3′-neolignans is of diagnostic importance (Table 10). Its similarity in the licarins

References, pp. 65—72

Table 10. Selected ^{13}C NMR Shifts (δ Values in ppm Downfield from TMS) of 8.1′- and 8.3′-Neolignans in $CDCl_3$

(68a) **(68b)**

$J_{H\text{-}7,H\text{-}8}$	9 Hz
	C-9
(7.1a)	17.2
(7.1b)	17.6

$J_{H\text{-}7,H\text{-}8}$	3 Hz
	C-9
(9.1a)	16.1
(9.2b)	16.3

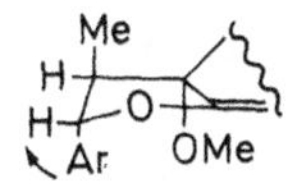

(69)

$J_{H\text{-}7,H\text{-}8}$	<0.5 Hz
	C-9
(7.7a)	16.1

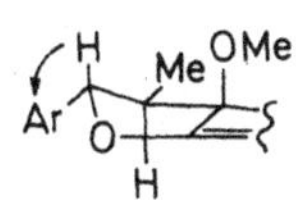

(70)

$J_{H\text{-}7,H\text{-}8}$	9.5 Hz
	C-9
(7.8a)	6.9

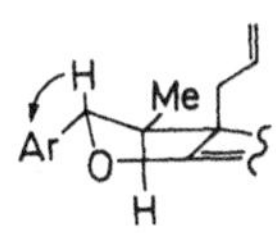

(71)

$J_{H\text{-}7,H\text{-}8}$	9.5 Hz	
	C-9	C-7′
(8.1a)	8.3	36.6
(8.1b)	8.5	36.7
(8.1c)	8.5	36.7

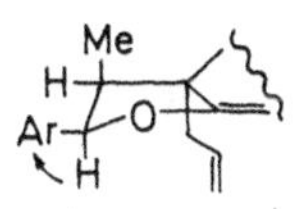

(72)

$J_{H\text{-}7,H\text{-}8}$	≤5 Hz	
	C-9	C-7′
(8.3a)	12.0	43.9
(8.3b)	11.9	43.8
(8.3c)	12.0	43.9

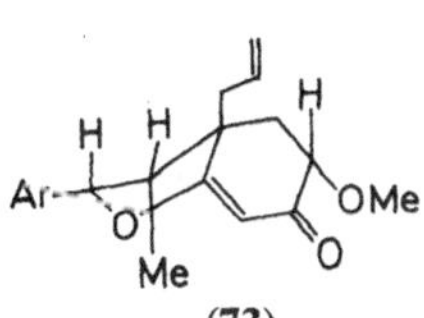

(73)

$J_{H\text{-}7,H\text{-}8}$	5 Hz	
	C-9	C-7′
(8.4a)	11.6	39.0

(7.1a, b) and the cyclohexadienones **(9.1a. 9.2b)**, both classes containing nearly planar benzofuranoid rings in which only C-7 can be non-coplanar, shows the aryl groups to contribute no significant γ-effect on C-9. While the H-7, H-8 coupling constants reflect the adoption of predominantly the conformation **68a** and **68b** by **7.1a, 7.1b** and **9.1a, 9.2b** respectively, the *pseudo*-equatorial methyl group of compounds **7.1** must be mostly beyond non-bonded interaction range of the neighbouring aryl side chain. The identity of the C-9 shift of mirandin-A **(7.7a)** with that of compounds **9.1a** and **9.2b** shows this angularly methoxylated substance to possess a conformation related to **68b,** i.e. **69** in which the 1,3-diaxial interaction of the methoxy and the aryl groups is relieved (see arrow on **69**), in conformity with the small value of $J_{H\text{-}7,H\text{-}8}$. Contrastingly, mirandin-B (**7.8a**) reveals its C-9 signal over 9 ppm upfield of that of its isomer A (**7.7a**) and its $J_{H\text{-}7,H\text{-}8}$ to be large. A relaxed version of conformer **70** (see arrow) can account for these facts, especially in view of its support of the presence of at least two γ-effects on C-9 exerted by its neighbours. Thus, the ^{13}C NMR analysis of the mirandins identifies them clearly as 3′-epimers.

Burchellin (**8.1a**) and its relatives (**8.1b, c**) possess the same relative configurations of the substituents of the furanoid ring as mirandin-B (**7.8a**), and hence would be expected to prefer a conformation as in **71.** In accord therewith the three substances were shown to exhibit large values of $J_{H\text{-}7,H\text{-}8}$, and an upfield C-9 signal. The *ca.* 1.5 ppm decrease of shielding of C-9 of compounds **8.1** vs. that of mirandin-B (**7.8a**) may be due to the known higher γ-shift by alkoxy over allyl groups. In view of the reciprocity of the γ-effect, the α-methylene of the allyl group of substances **8.1** is shielded by *ca.* 7 ppm, contrasted with C-7′ in the 8-epimers (**8.3a, b, c**). The cyclohexadienones **8.3,** confined to conformation **72** with a flattened five-membered ring (see arrow), cause C-9 to feel one less γ-effect than in substances **8.1** and hence to be deshielded (*ca.* 3.5 ppm). Having the furanoid substituents in the same configuration as compounds **8.3,** porosin (**8.4a**) is restricted to conformation **72.** In view of this constraint and the axiality of H-5′ ($J = 5$ and 12 Hz), the C-5′ stereochemistry is as depicted in **73.** This feature is confirmed by the shielding of C-7′ of porosin (**8.4a**) with respect to this subunit in cyclohexadienones **8.3,** induced by a γ-effect from H-5′.

X. 8.1′,7.3′- and 8.5′,7.3′-Neolignans

1. 12.1, 12.2

The existence of bicyclo[3,2,1]octane neolignans was recognized through the isolation of guianin* (**12.1a**) (*36*), whose gross structural features were ascertained by MS. Guianin is cleaved cleanly into two fragments $[C_{10}H_{10}O_2]^{+\cdot}$ and $[C_{10}H_{12}O_3]^{+\cdot}$. The structural moieties responsible for the generation of these ions were formulated respectively as **74** and **75** on the grounds of IR and ^1H NMR spectra. In partial structure **75** two alternatives for the definition of the tertiary C-atom indicated in **74** are offered. The proton at this centre, however,

* Guianine and macrophylline A and B are designations given, previously, to alkaloids from *Strychnos* species (*95*).

gives rise to a ^{1}H NMR doublet ($J = 7.0$ Hz) at such low field (τ 6.81), that it must be located in the anisotropically deshielded region vicinal to the carbonyl system, as shown in **12.1a.**

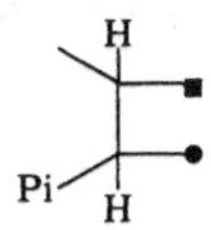

(**74**)

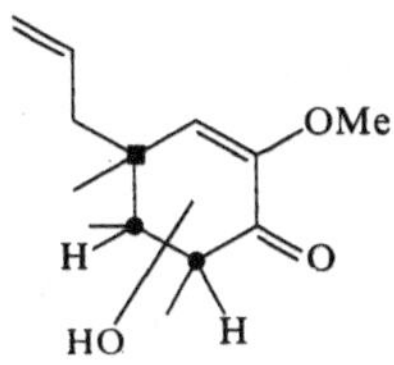

(**75**)

■ tetrasubstituted carbon

● trisubstituted carbon

The orientation of the hydroxyl towards the pentacycle became evident upon consideration of ^{1}H NMR coupling and dihedral angle involving the CHOH–CH system. Since this hydroxyl is easily acetylated at room temperature and addition of hydrogen on the β-side of the molecule is favoured, the bulky aryl group should be located on the endo-face of this molecule.

Jones oxidation of guianin led to the naturally occurring diketone (ν_{max} 1748, 1683 cm^{-1}) **12.2a.** Three additional compounds (**12.1b, c, d**) belong to the guianin type. Oxidation of **12.1b** gave again a diketone (ν_{max} 1758, 1696 cm^{-1}), confirming the existence of a secondary carbinol in a 5-membered ring. Methylation of the enol **12.1d** gave **12.1c.** The ORD curves of the compounds of series **12.1** are sufficiently similar to warrant the conclusion that all possess the stereochemistry assigned to guianin (**12.1a**) (*55*).

2. 12.3, 12.4

Spectral evidence as well as the biogenetic requirement of a 1,4-relationship between the allyl side chain and an oxy-function led to partial structures **76** and **77** for canellin-A (**12.3a**). Since linkage of the two units here occurs through two tetra-substituted centres, their structural relation was not as obvious as in the case of guianin. Room temperature acetylation of canellin-A affected only one of the hydroxyls, evidence for the *cis*-relation between the other OH and the aryl groups. The monoacetate was oxidized to a 5-membered ring ketone (ν_{max} 1760 cm^{-1}). This transformation was accompanied by a ^{1}H NMR shift of one methoxy signal only, indicating its correlation to the bridgehead methoxyl. Since it is this same signal which suffers solvent shifts, the bridgehead methoxyl (and not the allyl) is located in the vicinity of the differentially solvatable aryl, as in **12.3a.** The configurational assignments were based on interpretation of proton coupling constants (*61*).

The structural proposals for **12.3b** (*49*), **12.4a** (*61*), **12.4b** (*9*, revised in *19*) and **12.4c** (*19*) were based on spectral comparisons with **12.3a,** and Bi_2O_3-oxidations of **12.4b** to **12.1e** and of **12.1c** to **12.1f** (*19*).

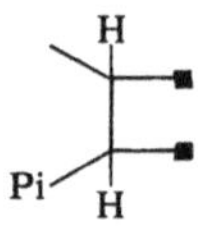

(76)

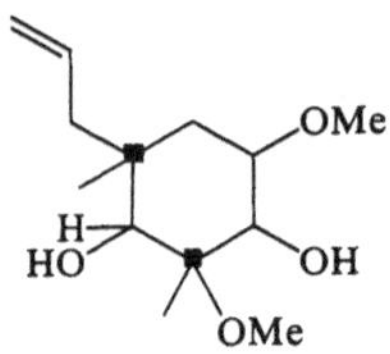

(**77**)

3. 13.1, 13.2

Formulation of a structural proposal for macrophyllin* based on the partial structures **78** and **79** met with difficulties. There was no doubt that the aliphatic methoxyls are situated on tetrasubstituted carbons. Since, however, no differential shielding of these groups occurred, due to the *endo*-arrangement of the aryl group, it is impossible to know in which way **78** and **79** are connected, and **13.1a** is only one of two alternative representations of macrophyllin. Evidence for the *endo*-arrangement, which requires an *exo*-methyl ($J_{H\text{-}7,H\text{-}8}$ 8.7 Hz), was seen in the fact that the hydroxyl is as completely acetylated at room temperature as that of guianin (**12.1a**). This hydroxyl must again be directed towards the pentacycle, since the ^{1}H NMR frequency of the carbinolic proton is strongly affected by hydrogenation of the cyclic double bond (*56*).

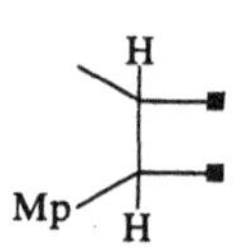

(78)

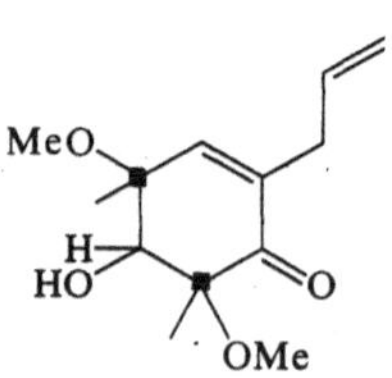

(79)

Compound **13.2a** diverges from macrophyllin not only in aromatic substitution and the lack of an aliphatic methyl, but also by the presence of the carbonyl and carbinol functions at interchanged positions, facts which were deduced by the comparison of IR and ^{1}H NMR spectra. At the time of writing, the type of doubt mentioned in the case of **13.1a** also applied here (*32*).

References, pp. 65—72

XI. 8.1′,7.9′-Neolignan

1. 14.1

The structure of futoenone (**14.1a**) was deduced through the careful interpretation of ^{1}H NMR spectra (*114*) aided by partial decoupling and observation of Overhauser effects (*145*) of the compound and such derivatives as the Lemieux oxidation products **80** and **81**,

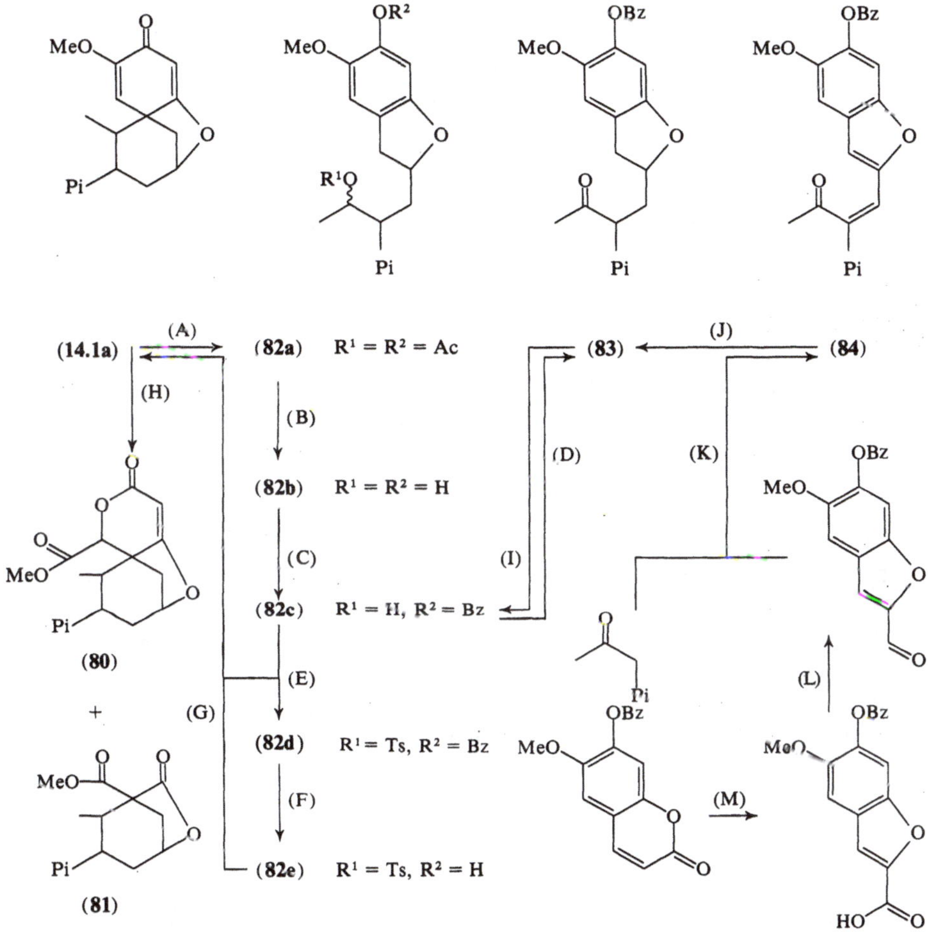

(A) ... Ac_2O, BF_3. (B) ... KOH, EtOH. (C) ... BzCl, K_2CO_3, DMF. (D) ... CrO_3, C_5H_5N. (E) ... TsCl, C_5H_5N. (F) ... H_2, Pd/C, AcOH. (G) ... Al_2O_3, AcOEt. (H) ... 1. $NaIO_4$, $KMnO_4$, Na_2CO_3, dioxane; 2. CH_2N_2. (I) ... $NaBH_4$, EtOH. (J) ... H_2, Pt, AcOH. (K) ... piperidine, C_6H_6, refl. (L) ... 1. $LiAlH_4$, THF; 2. MnO_2, $CHCl_3$. (M) ... 1. Br_2, $CaCO_3$, $CHCl_3$; 2. KOH, refl.

Chart 23. Reactions and synthesis of futoenone (**14.1a**)

as well as the ring opened products **82a, b, c, d, e.** Since it was possible to reconvert these into futoenone, they not only supported the spiro-dienone structure for the compound, but were also useful in the accomplishment of the total synthesis which employed the Knoevenagel adduct **84** as key intermediate (Chart 23) (*115*). Assignment of the relative stereochemistry **85** to futoenone was based on interpretation of the coupling constants of the cyclohexane protons and the appearance of the secondary methyl signal at high field, indicating its orientation with respect to the dienone ring.

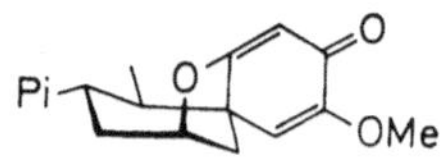

(85) (14.1 a)

XII. 8.O.4′-Neolignans

1. 15.1

The structural determination of **15.1a** relied on ^{1}H NMR evidence. The doublet due to the benzylic H-7 suffered a 1.25 ppm paramagnetic shift upon acetylation, a fact which established the relative position of the ether and alcohol functions. The interrelationships of the corresponding methine, as well as of all other side chain protons, was confirmed by double irradiation experiments and ^{13}C NMR (*24*). The reaction of sodium isoeugenolate with α-bromo-3,4,5-trimethoxy-propiophenone and $NaBH_4$ reduction of the resulting product led to a mixture of *threo*- and *erythro*-compounds, the former predominating in natural **15.1a** (*25*). Mixtures of both diastereomeric forms of structural analogues were obtained, prior to the isolation of **15.1a** from a natural source, through the oxidation of isoeugenol by oxygen in a dye sensitized photoreaction (*54*), by tri-t-butylphenoxyl radical (*111*) and by H_2O_2-peroxidase (*129*).

2. 15.2

Structural proposals for eusiderin were based mainly on recognition that no phenolic degradation products were observed other than pyrogallol derivatives. Thus under demethylation conditions the dihydro-derivative gave 5-n-propylpyrogallol, indicating that eusiderin arises from the oxidative coupling through oxygen of two C_6-C_3 phenols. Since, however, a series of attempted additional classical reactions were either without effect on eusiderin or yielded impurifiable products, no decision was possible among two benzodioxane alternatives (**15.2b** and **86b**) and even the isomeric structure **87** was stated to possess all the observed attributes of eusiderin (*80*).

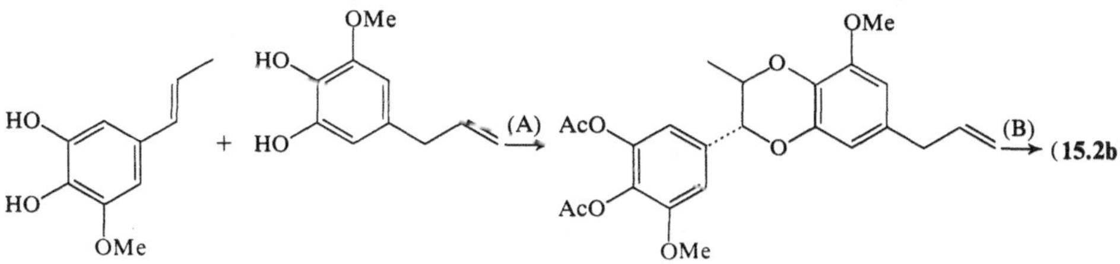

(A) ... 1. Ag_2O, C_6H_6, room temp. 2. acetylation, TLC. (B) ... CH_2N_2, Et_2O, MeOH, 2 days.

Chart 24. Synthesis of (±)-eusiderin

Reexamination of this problem by modern spectrometric techniques invalidated any but the benzodioxane structures (*66*), **15.2b** constituting the favoured alternative upon consideration of biosynthetic mechanism (*64, 109*) and a biogenetically patterned, but not unambiguous, synthesis of the racemate (Chart 24) (*109*). This postulate was confirmed by a study of the ^{1}H NMR spectral behaviour of eusiderin (**15.2b**) and eusiderin-B (**15.2a**) in the presence of $Pr(fod)_3$ (*33*).

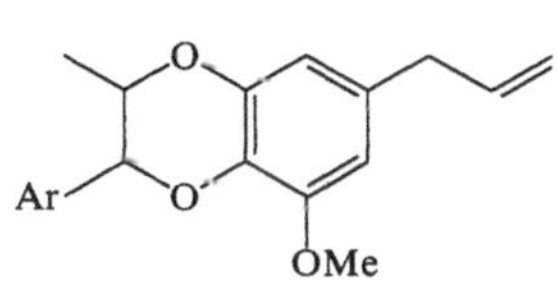

(**86a**) Ar = Pi
(**86b**) Ar = Tp

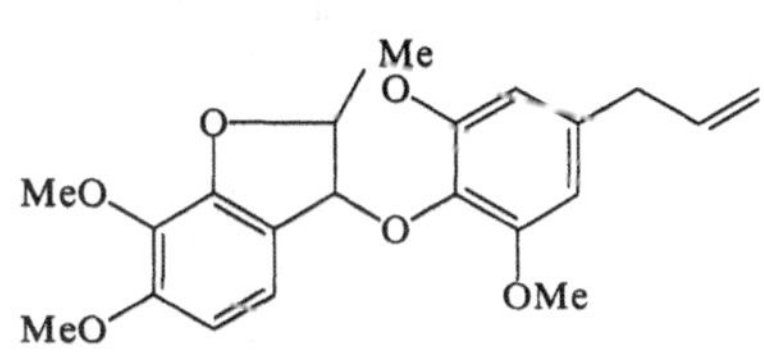

(**87**)

Isolated aromatic ethers associate only weakly with this NMR shift reagent. The same occurs with *ortho*-diethers in which the oxyfunctions are part of an additional ring (e.g. in methylenedioxybenzene and benzodioxane), unless an OR group substitutes at least one of the vicinal positions. *Ortho*-di- and triethers associate strongly with the reagent.

These facts lend significance to lanthanide-induced shift data concerning eusiderin-B (**15.2a**). Coordination with Pr affects H-7 (Δ 9.2 ppm) less strongly than H-8 (Δ 19.2 ppm), and the methoxyl must thus, be located closer to the latter site (as in **15.2a**) and not to the former (as in **86a**). Structure **15.2a** for eusiderin-B was corroborated by examining lanthanide-induced shift data for licarin-B (**7.1b**). Here the Δ values for H-7 (12.0 ppm) and H-8 (6.2 ppm) show, as expected, that the methoxyl is closer to the former site. While absolute Δ values are not directly useful in such correlations, they indicate that the complexation ratio in **15.2a** (Δ_{OMe} 16.2 ppm) is higher than in **7.1b** (Δ_{OMe} 10.9 ppm) by a factor of 1.5. Turning to the case of eusiderin (**15.2b**), the complexation ratio at the oxy-methoxy site (Δ_{OMe} 4.6 ppm) is also higher than in licarin-C (**7.2c**) (Δ_{OMe} 3.0 ppm), and again by a factor of 1.5. Since this phenomenon is here as clearly linked to differential steric hindrance at the coordination site as it was in the case of **15.2a** vs. **7.1b,** only **15.2b** can represent correctly the structure of eusiderin. This conclusion is consistent with other Δ values which indicate that H-2′ is additionally under the influence of the neighbouring coordination site at the aryl group.

The *trans*-arrangement of the substituents at C-7 and C-8 of eusiderin was consistent with ^{1}H NMR data (*66*) and inferred through the ^{13}C NMR spectrum. This showed, in comparison with model neolignan spectra, that Me-8 must be beyond non-bonded interaction range of the Ar-7 substituent (*144*).

XIII. Function and Use of Neolignans

Investigations concerning the function of neolignans in plants would appear profitable in view of the insect antifeeding property of piperenone (**7.9a**) (*104*).

Investigations concerning the pharmacologic activity of neolignans should derive motivation by their relation to those lignans which attained such importance, especially as antitumor agents (*84*), that *Podophyllum peltatum* L. is much in demand in the drug trade (*110*). Indeed, also *nor*-dihydroguaiaretic acid (**1.3b**) has been implicated in cancer therapy (*118*), and was described as the most potent cancer antimetabolite *in vitro* (*37*). A review of these and other astoundingly varied pharmacologic, antimicrobial and biochemical activities mentions also industrial applications, and considers the literature on the use of NDGA as antioxidant for food material (*78, 79*) up to 1970 (*117*). Among all other neolignans, so far only schizandrin (**3.1b**) has been studied properly. This is one of the active principles of *Schizandra chinensis*, long used in the Orient as a stimulant (*89*). Reports on the pharmacologic properties of crude drugs (*136, 141*) as well as of schizandrin and its relatives (*97, 113, 140*), are available. Otobain (**2.4a**) and co-occurring neolignans are the fungicide or fungistatic ingredients of otoba butter, used in Colombia in veterinary practice (*3*).

XIV. Biogenesis of Neolignans

The biogenesis of 116 out of 120 known neolignans can be explained by oxidative coupling of a propenylphenol derived starter with either a propenylphenol or an allylphenol derived termination unit. Of the remaining 4 cases, 2 should involve the coupling of allylphenol derived units (Chart 25), while the biogenesis of asatone (**5.1a**) and isoasatone (**6.1a**) may follow a rather different route, involving an oxidized 4-allyl-2,6-dimethoxyphenol as precursor (Chart 26) (*131, 147*). The relative proneness to oxidation of propenylphenols is to be expected in view of the higher stabilization of the derived radical. Indeed, stability of

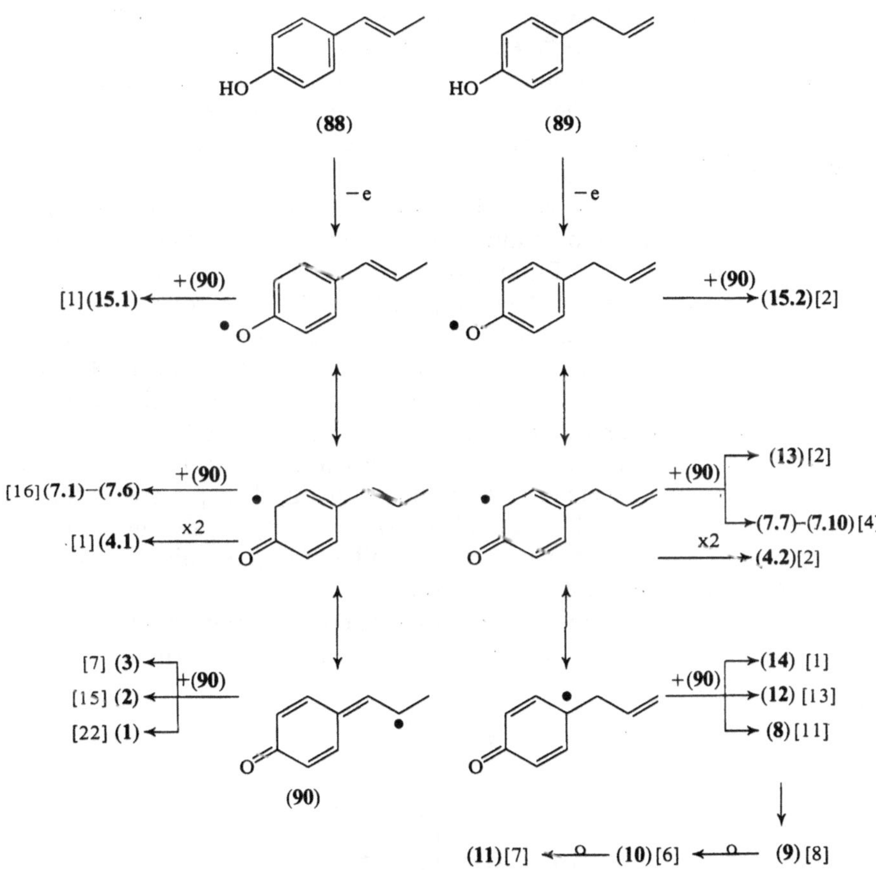

Chart 25. Schematic representation of coupling processes which lead from propenylphenol (**88**) and allylphenol (**89**) derived radicals, either by dimerization (× 2) or by addition of radical **90** [+(**90**)] to the indicated (in brackets) numbers of neolignans belonging to the specified skeletal types

reactive species and number of neolignans it generates seem to correlate quite well. It is, nevertheless, not implied that all reactions represented in Chart 25 proceed *in vivo* by radical pairing (see below).

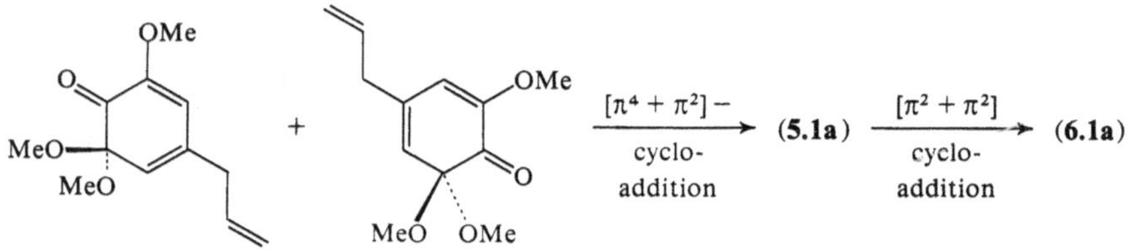

Chart 26. Postulated biosynthesis of asatone (**5.1a**) and isoasatone (**6.1a**)

As originally postulated by ERDTMAN (*52, 64*), the coupling step should produce a quinone methide intermediate which may add water, hydroxyl, hydride or carbanion. *E.g.*, two such follow-up reactions are postulated to explain the biogenesis of futoenone (**91→14.1a**) (*64*). In the *in vitro* oxidative dimerization of propenylphenols arise mixtures of products whose structures reveal the co-occurrence of 8.8′-, 8.3′- and 8.O.4′-coupling (*129*). *In vivo,* the coupling step appears to be fairly selective for a family, *e.g.* leading to 8.8′-neolignans in Zygophyllaceae and to 8.3′-neolignans in Eupomatiaceae, and for a genus, *e.g.* leading to **92** in *Aniba* and **93** in *Nectandra* (Chart 27). The follow-up steps are of general occurrence, *e.g.* attack of hydroxyl leads to hydrocoumarans in *Aniba* (**8.1—8.3**) and *Nectandra* (**7.7, 7.8**) and to bicycloöctanes in *Aniba* (**12.1—12.4**) and *Nectandra* (**13.2**). The pathway **93→7.8a→9.2c→8.2b** seems, thus, to be a more plausible route than **92→8.2b** to this constituent (**8.2b**) of *Nectandra miranda* (LAUR 19). The allylic rearrangement of the methoxyl involved in the step **7.8a→9.2c,** however, was not achieved *in vitro* by pyrolysis, in contradistinction to the transposition of the allyl group required in **8→9** which proceeds with ease (see Section VIII. 1.).

In vitro, the coupling and follow-up steps are remarkably stereospecific (*41, 129*), a fact which was explained by the formation of a

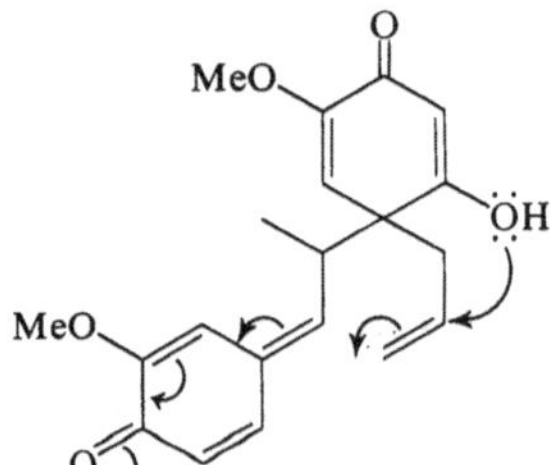

(**91**)

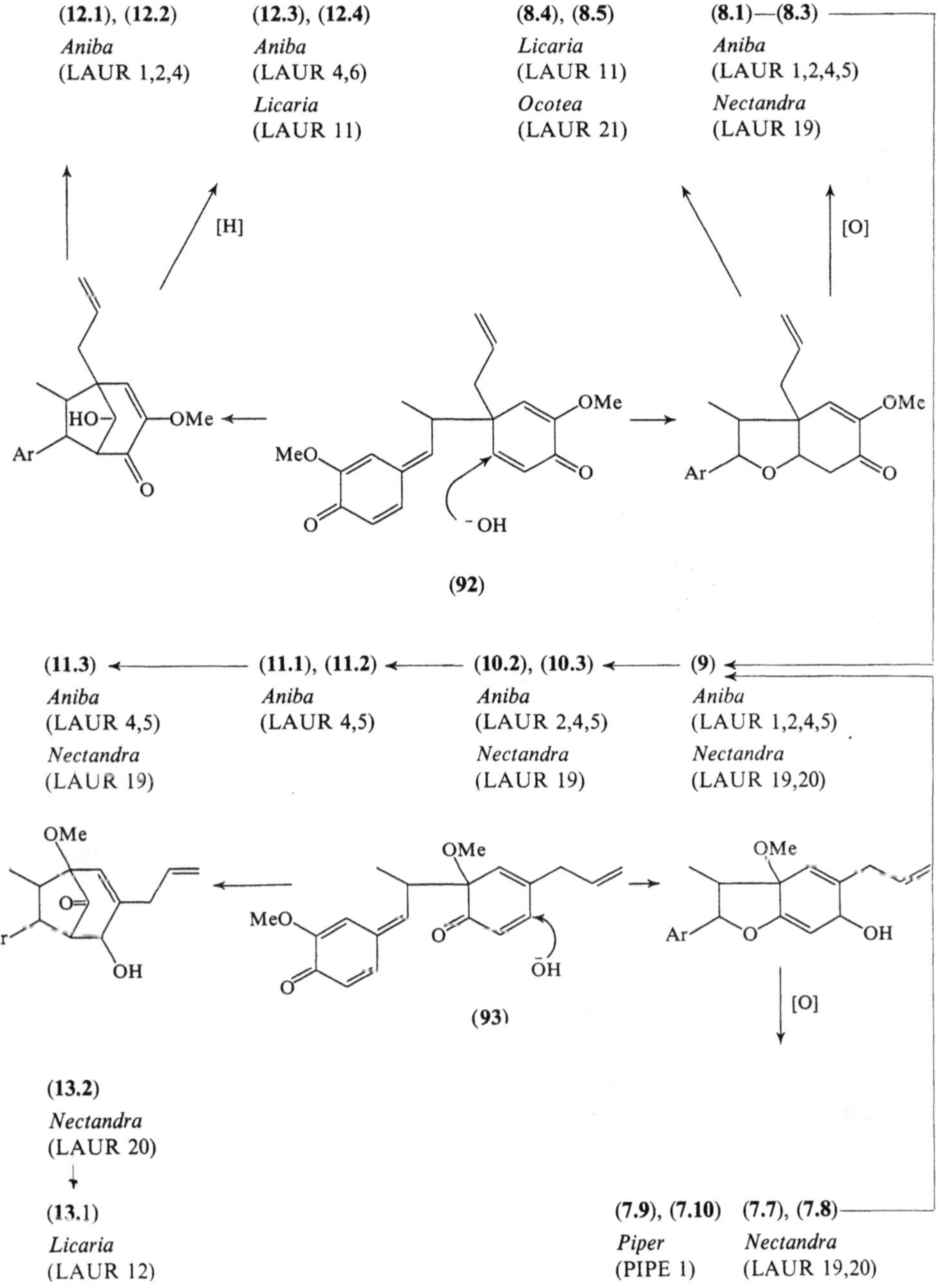

Chart 27. Postulated biosynthetic pathways to neolignan types of Lauraceae and Piperaceae

short lived dimeric intermediate with the aromatic rings hold in close association by charge transfer (**94**) (*129*) or ionic binding to cation (**43**) (*41*). Although *in vivo* in certain cases (*e.g.* **7.7a/7.8a** in LAUR 19) the all-*trans* isomer predominates, this is not always the case (*e.g.* **8.1a/8.2a** in LAUR 5) and the neolignan fraction of any Lauraceae is always a complex mixture of optically active stereoisomers. The C_6-substitution pattern of neolignans is again rigourously species specific.

The coupling of two units originated from an identical precursor is easily rationalized by the radical pairing mechanism (*59*), and experimental evidence testifies in favour of this postulate (*111*). The gene-

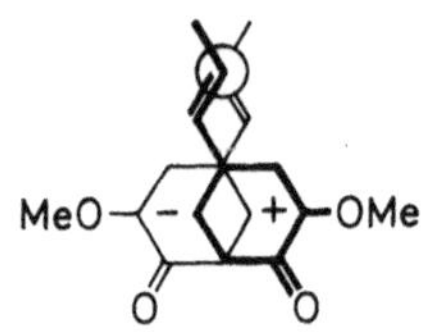

(**94**)

sis of propenyl- + allylphenol neolignans, however, presents matter for thought. Coupling of units, endowed with different reactivities, suggests: 1. that the reaction is under strict enzymatic control; 2. that the coupling mechanism involves different reactive species such as propenylphenoxonium and allylphenolate ions rather than radicals; 3. that propenyl- and allylphenol molecules exist in close spacial proximity and, hence, bifurcation of the biosynthetic pathway leading to both occurs at an immediate precursor. This point merited attention (*68*) in view of the experimental observation that propenyl- and allylphenols are formed by considerably divergent pathways (*38, 102*).

A study of the natural distribution of propenyl- vs. allylphenols (*103*) is summarized in Table 11. In a total of 66 cases, exclusive occurrence of allylphenols is registered 35 times. The predominance of allyl- over propenylphenols in nature, which has been noted (*27*), could be due to their relative stability (*64*), the accumulation of propenylphenols being hindered by oxidative polymerization or scavenging of other metabolites, such as allylphenols. Exclusive occurrence of propenylphenols is restricted to three cases, and the compound asarone, but this hardly affects the general impression that both types are close biosynthetic relatives. Furthermore, it is certainly true that propenyl- and allylphenols attained climactic natural distribution and structural diversity along CRONQUIST's (*46*) evolutionary line Magnoliidae-Rosidae-Asteridae, which nicely rationalizes the presently known distribution pattern of neolignans (Table 11, column B). Propenyl- and allylphenols of identical oxygenation pattern are, nevertheless, formed throughout the Embryobionta, and it is quite possible that neolignans may be much more widespread than the presently available findings seem to suggest.

Table 11. *Number of Families of the Indicated Orders* [division acc. to CRONQUIST (*46*)] *Containing A. Propenylphenols/Allylphenols of the Indicated Aromatic Oxygenation Pattern, and B. Neolignans of Propenyl- + Propenylphenol/Propenyl- + Allylphenol/Allyl- + Allylphenol Derivation*

	A									B
Oxygenation pattern	4	3,4	2,4	3,4,5	2,4,5	2,3,4	2,4,6	2,3,4,5	2,3,4,5,6	
POLYPODIOPSIDA										
Polypodiales	1/1	1/2								
PINOPSIDA										
Coniferales	1/1	–/4		–/3						
MAGNOLIOPSIDA										
Hamamelidae										
Hamamelidales		–/1								
Urticales		–/1								
Fagales	–/1	–/1								
Dilleniidae										
Taxales		–/1								
Violales		–/1								
Ericales		–/1								
Magnoliidae										
Magnoliales	2/4	5/7		1/2	3/1			–/2		8/2/2
Piperales	1/1	–/2		–/1	1/–			–/1		–/1/–
Aristolochiales	–/1	1/1		–/1	1/1	–/1	–/1	–/1		1/–/1
Rosidae										
Geraniales		–/1								
Umbellales	1/1	1/1	1/1	1/1	1/–			–/1	–/1	
Sapindales	2/1	1/1		–/2						1/–/–
Rosales	–/1	1/3								
Myrtales	1/1	1/1		1/1						1/–/–
Asteridae										
Polemoniales		–/1								
Lamiales	1/1	1/2		–/1	1/–			–/1		
Scrophulariales		–/2		–/1						
Rubiales		1/1								
Asterales	1/1	–/1			–/1					
LILIOPSIDA										
Commelinidae										
Cyperales		1/1		–/1						
Bromeliales	–/1									
Zingiberales	–/1									
Arecidae										
Arales	–/1	1/1		1/1	1/1					
Liliidae										
Liliales		–/1								

XV. Conclusion

Structural variation makes the neolignans a most interesting group of natural products from the chemical (Sections II—XII), practical (XIII) and biochemical (XIV) standpoint. Is the list of skeletal variants and their derivatives nearing completion?

Many neolignans are difficult to crystallize and their isolation from plant extracts is very tedious. It is possible, consequently, that they have been frequently overlooked in fractionation work and that the compounds described in the present review are merely representative of a vast class of natural products. It is noteworthy in this sense, that nearly all neolignans discovered in recent years do not bear free phenolic hydroxyls. Since, however, such groups are a prerequisite for the oxidative coupling reaction which initiates their biosynthetic process, etherification must, of necessity, be the last step of this process. This implies the, even if transitory, existence of phenolic neolignans which may be inactivated against oxidation additionally by continuation of the polymerization process. Rate of formation (*129*) and structure (*111*) of dimeric and polymeric material seem to be different and a search for naturally occurring oligomers of propenyl- and allylphenols in the viscous masses which form large parts of the extractives from tropical woods may prove to be of interest.

The neolignans constitute, thus, at present a rapidly expanding domain of natural products chemistry whose delimitations are not yet in sight.

Addendum

For "Unpublished work" mentioned in (*10*) see (*148*), for (*13*) see (*150*), for (*16*) and (*17*) see (*149*), for (*19*) see (*151*), for (*70*) see (*154*). New information on compounds in Table 2 is as follows:

3. The structures of four new representatives from SCHI 1 (*156, 157*) and a synthesis of **3.2a** (*154*) were reported.

5.1a, 6.1a. Experimental details of structural work (*157*) and additional plant sources (*156*) were disclosed.

7—11. The absolute configurations of the benzofuranoid neolignans were correlated by chemical interconversions and ORD comparisons (*154*).

7.2d, Ar=Gu, R=OMe and **12.1g,** Ar=Pi, R^1=OMe, R^2=OH, R^3=H, absolute configuration as indicated, were isolated besides several known neolignans from LAUR 1 (*152*).

7.1, 15.1 (allyl for propenyl). The structures of seven new representatives from *Myristica fragrans* Houtt. Were reported (*153*).

References, pp. 65—72

12. Chemical correlation of **12.1e** with **8.3b** (*151*) and ORD comparisons (*150*) led to the absolute configurational proposals shown below:

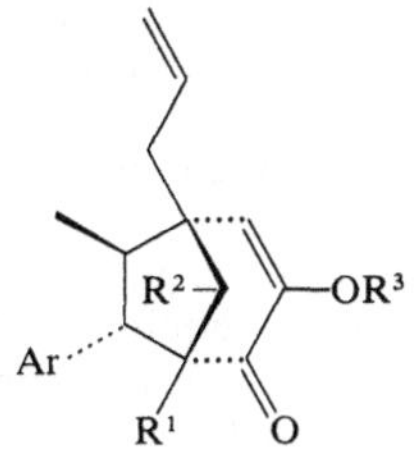

(12.1a—d)

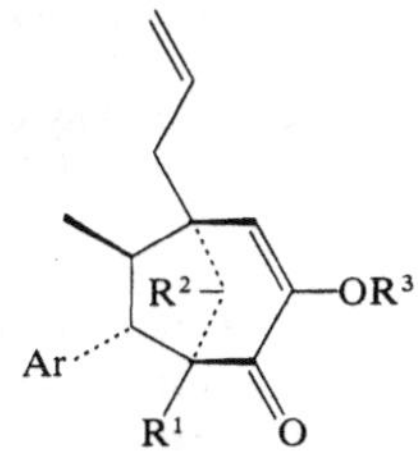

(12.1e, f)

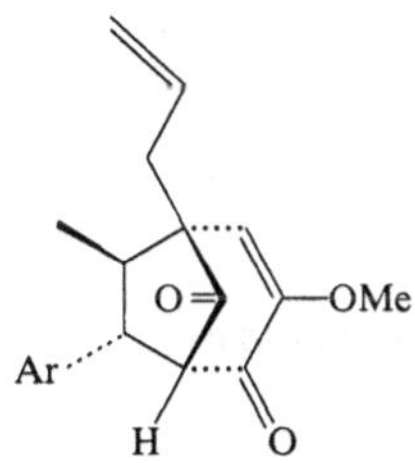

(12.2a, b)

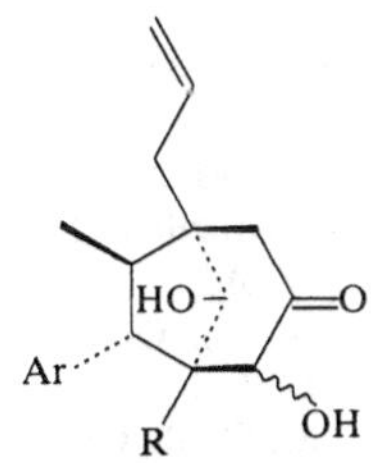

(12.4b, c)

References

1. ADJANGBA, M. S.: Recherches sur les Lignanes. IV. Condensation d'un Ether Phenolique avec une Lactone en présence de Chlorure Stannique. Bull. soc. chim. France **1963,** 1007.

2. — Recherches sur les Lignanes. V. Synthèses Totales de quelques Lignanes Naturels et de leurs Derivés. Bull. soc. chim. France **1963,** 1942.

3. — Progrès Récents dans la Chimie des Lignanes. Bull. soc. chim. France **1963,** 2344.

4. ADJANGBA, M. S., and D. BILLET: Recherches sur les Lignanes. III. Confirmation par Synthèse Totale de la Structure de la Galcatine. Bull. soc. chim. France **1962,** 1970.

5. ADJANGBA, M. S., D. BILLET, and C. MENTZER: Recherches sur les Lignanes. II. Sur une Nouvelle Synthèse des Diaryl-2,5-diméthyl-3,4-furanes. Bull. soc. chim. France **1962,** 132.

6. AHMED, R., and R. STEVENSON: Synthesis of Eupomatenoid-7 and Eupomatenoid-12. Phytochem. **14,** 2710 (1975).

7. AIBA, C. J., R. BRAZ F°., and O. R. GOTTLIEB: The Chemistry of Brazilian Lauraceae. XX. Porosin: a Neolignan from *Ocotea porosa*. Phytochem. **12,** 413 (1973).

8. AIBA, C. J., R. G. C. CORRÊA, and O. R. GOTTLIEB: The Chemistry of Brazilian Lauraceae. XXII. Natural Occurrence of Erdtman's Dehydrodiisoeugenol. Phytochem. **12,** 1163 (1973).

9. AIBA, C. J., J. B. FERNANDES, O. R. GOTTLIEB, and J. G. S. MAIA: The Chemistry of Brazilian Lauraceae. XXX. Neolignans from an *Aniba* Species (later classified *A. simulans*). Phytochem. **14,** 1597 (1975).

10. AIBA, C. J., A. M. GIESBRECHT, O. R. GOTTLIEB, and F. M. PAGLIOSA: Unpublished work (1976).

11. AIBA, C. J., A. M. GIESBRECHT, O. R. GOTTLIEB, and M. YOSHIDA: A Revision of the Structure of the Neolignan Canellin-B. Communication presented at the XXVIII Annual Meeting of Sociedade Brasileira para o Progresso da Ciência, Brasília (1976).
12. AIBA, C. J., and O. R. GOTTLIEB: The Chemistry of Brazilian Lauraceae. XXVIII. Synthesis of (±)-Licarin-B. Phytochem. **14,** 253 (1975).
13. AIBA, C. J., O. R. GOTTLIEB, U. KAMPHAUSEN, M. T. MAGALHÃES, and M. YOSHIDA: Unpublished work (1976).
14. AIBA, C. J., O. R. GOTTLIEB, and M. T. MAGALHÃES: Unpublished work (1976).
15. AIBA, C. J., O. R. GOTTLIEB, and J. G. S. MAIA: Neolignans from *Duguetia surinamensis.* Communication presented at the XXVII Annual Meeting of Sociedade Brasileira para o Progresso da Ciência, Belo Horizonte (1975).
16. AIBA, C. J., O. R. GOTTLIEB, F. M. PAGLIOSA, and M. T. MAGALHÃES: Unpublished work (1976).
17. AIBA, C. J., O. R. GOTTLIEB, F. M. PAGLIOSA, and M. YOSHIDA: Unpublished work (1976).
18. AIBA, C. J., O. R. GOTTLIEB, M. YOSHIDA, J. C. MOURÃO, and H. E. GOTTLIEB: The Chemistry of Brazilian Lauraceae. XXXV. The Structure of Porosin. Phytochem. **15,** 1031 (1976).
19. ALVARENGA, M. A. DE, O. CASTRO C., O. R. GOTTLIEB, and J. G. S. MAIA: Unpublished work (1976).
20. ANDRADE, C. H. S., O. R. GOTTLIEB, and J. G. S. MAIA: Unpublished work (1974).
21. ARAÚJO LIMA, O., O. R. GOTTLIEB, and M. T. MAGALHÃES: The Chemistry of Brazilian Lauraceae. XX. Burchellin, a Neolignan from *Aniba burchellii.* Phytochem. **11,** 2031 (1972).
22. ATKINSON, J. R., and R. D. HAWORTH: The Constituents of Natural Phenolic Resins. XII. The Action of Selenium on Lignans. J. Chem. Soc. (London) **1938,** 1681.
23. AULIN-ERDTMAN, G., Y. TOMITA, and S. FORSÉN: Studies on the Degradation of Lignin and Model Compounds. I. The Configuration of Dehydrodi-isoeugenol. Acta Chem. Scand. **17,** 535 (1963).
24. BAKER, P. M., and L. E. S. BARATA: Chemical Study of Amazonian Plants with Biological Action: The Structure of a New Lignan. Communication presented at the XXVI Annual Meeting of Sociedade Brasileira para o Progresso da Ciência, Recife (1974).
25. BARATA, L. E. S., P. M. BAKER, O. R. GOTTLIEB, and E. RUVEDA: Syntheses of Neolignans. Communication presented at the XXVIII Annual Meeting of Sociedade Brasileira para o Progresso da Ciência, Brasília (1976).
26. BHACCA, N. S., and R. STEVENSON: The Constitution of Otobain. J. Organ. Chem. (USA) **28,** 1638 (1963).
27. BIRCH, A. J.: Biosynthetic Pathways. In Chemical Plant Taxonomy (Ed. T. SWAIN), p. 141. London: Academic Press. 1963.
28. BIRCH, A. J., B. MILLIGAN, E. SMITH, and R. N. SPEAKE: Some Stereochemical Studies of Lignans. J. Chem. Soc. (London) **1958,** 4471.
29. BLAIR, G. E., J. M. CASSADY, J. E. ROBBERS, V. E. TYLER, and R. F. RAFFAUF: Chemical Investigation of *Virola* Species. I. Isolation of 3,4′,5-Trimethoxy-*trans*-stilbene, Otobaene and Hydroxyotobain from *Virola cuspidata.* Phytochem. **8,** 497 (1969).
30. BLEARS, J. G., and R. D. HAWORTH: The Constituents of Natural Phenolic Resines. XXIV. A Synthesis of Galgravin. J. Chem. Soc. (London) **1958,** 1985.
31. BOWDEN, B. F., E. RITCHIE, and W. C. TAYLOR: Constituents of *Eupomatia* Species. II. Isolation and Structure Determination of Further Eupomatenoid Lignans from the Bark of *Eupomatia laurina.* Austral. J. Chem. **25,** 2659 (1972).

32. BRAZ F°., R., R. FIGLIUOLO, O. R. GOTTLIEB, and J. G. S. MAIA: Neolignans from *Endlicheria verticillata* (later classified as a *Nectandra* sp). Communication presented at the XXVIII Annual Meeting of Sociedade Brasileira para o Progresso da Ciência, Brasília (1976).

33. BRAZ F°., R., J. C. MOURÃO, O. R. GOTTLIEB, and J. G. S. MAIA: Lanthanide Induced Shifts as Aid in the Structural Determination of Eusiderins. Tetrahedron Letters **1976,** 1157.

34. BROPHY, G. C., J. MOHANDAS, M. SLAYTOR, S. STERNHELL, T. R. WATSON, and L. A. WILSON: Novel Lignans from a *Cinnamomum* Sp. from Bougainville. Tetrahedron Letters **1969,** 5159.

35. BROWN, D., and R. STEVENSON: Synthesis of Dehydrootobain. J. Organ. Chem. (USA) **30,** 1759 (1965).

36. BÜLOW, M. V. VON, N. C. FRANCA, O. R. GOTTLIEB, and A. M. PUENTES SUAREZ: The Chemistry of Brazilian Lauraceae. XXIV. Guianin: a Neolignan from *Aniba guianensis*. Phytochem. **12,** 1805 (1973).

37. BURK, D., and M. WOODS: Hydrogen Peroxide, Catalase, Glutathione Peroxidase, Quinones, Nor-dihydroguaiaretic Acid, and Phosphopyridine Nucleotides in Relation to X-Ray Action on Cancer Cells. Radiation Res. Suppl. **3,** 212 (1963); *apud* Chem. Abstr. **59,** 1934 (1963).

38. CANONICA, L., P. M. MANITTO, P. MONTI, and M. SANCHEZ A.: Biosynthesis of Allylphenols in *Ocimum basilicum* L. Chem. Commun. **1971,** 1108.

39. CARNMALM, B.: The Absolute Configuration of Dihydroguaiaretic Acid. Chem. and Ind. **1956,** 1093.

40. — Absolute Configuration in the Lignan Series. Ark. Kemi **15,** 215 (1960); *apud* Chem. Abstr. **55,** 5367 (1961).

41. CHAPMAN, O. L., M. R. ENGEL, J. P. SPRINGER, and J. C. CLARDY: The Total Synthesis of Carpanone. J. Amer. Chem. Soc. **93,** 6696 (1971).

42. CHEN, Y.-P., M. HONG, H.-Y. HSU, S. YAMAMURA, and Y. HIRATA: Isolation and Structure of Asatone. Tetrahedron Letters **1972,** 1607.

43. CHEN, Y.-P., R. LIU, and H.-Y. HSU: The Structures and Conformations of Two New Lignans, Kadsurin and Kadsurarin. Tetrahedron Letters **1973,** 4257.

44. COUSIN, H., and H. HÉRISSEY: Oxidation of Eugenol by Oxidizing Enzyme of Mushrooms and by Ferric Chloride; Obtention of Dehydrodi-eugenol. Bull. soc. chim. France Ser. 4, **3,** 1066 (1908).

45. — — Oxidation of Isoeugenol. On Dehydrodi-isoeugenol. Bull. soc. chim. France Ser. 4, **3,** 1070 (1908).

46. CRONQUIST, A.: The Evolution and Classification of Flowering Plants. London: Nelson. 1968.

47. CROSLEY, N. S., and C. DJERASSI: Naturally Occurring Oxygen Heterocyclics. XI. Veraguensin. J. Chem. Soc. (London) **1962,** 1459.

48. DEAN, F. M.: Naturally Occurring Oxygen Ring Compounds, p. 41. London: Butterworths. 1963.

49. DIAS, S. M., J. B. FERNANDES, O. R. GOTTLIEB, and J. G. S. MAIA: Neolignans from a Species of the Genus *Aniba*. Communication presented at the XXVII Annual Meeting of Sociedade Brasileira para o Progresso da Ciência, Belo Horizonte (1975).

50. DONNELLY, B. J., D. M. X. DONNELLY, A. M. O'SULLIVAN, and J. P. PRENDERGAST: *Dalbergia* Species. VII. The Isolation and Structure of Melanoxin, a New Dihydrobenzofuran from *Dalbergia melanoxylon* Guill. and Perr. (Leguminosae). Tetrahedron **25,** 4409 (1969).

51. DOSKOTCH, R. W., and M. S. FLOM: Acuminatin, a New Bis-Phenylpropide from *Magnolia acuminata* L. Tetrahedron **28,** 4711 (1972).

52. ERDTMAN, H.: Dehydrierungen in der Coniferylreihe. II. Dehydro-di-isoeugenol. Liebigs Ann. Chem. **503**, 283 (1933).

53. ERDTMAN, H., and J. RUNEBERG: Phenol Dehydrogenations. VII. Dehydrogenation of Chavicol to Magnolol. Acta Chem. Scand. **11**, 1060 (1957).

54. ESKINS, K., C. GLASS, W. ROHWEDDER, R. KLEIMAN, and J. SLONEKER: Dimers of Isoeugenol by Dye-Sensitized Photooxidation. Tetrahedron Letters **1972**, 861.

55. FERNANDES, J. B., O. R. GOTTLIEB, and J. G. S. MAIA: The Chemistry of Brazilian Lauraceae. XXXIV. Neolignans from an *Aniba* Species (later classified *A. affinis*). Phytochem. **15**, 1033 (1976).

56. FRANCA, N. C., O. R. GOTTLIEB, and J. G. S. MAIA: The Chemistry of Brazilian Lauraceae. XXVII. Macrophyllin, a Neolignan from *Licaria macrophylla*. Phytochem. **13**, 2839 (1974).

57. FREUDENBERG, K., and K. WEINGES: Systematik und Nomenklatur der Lignane. Tetrahedron **15**, 115 (1961).

58. FUJITA, Y., and J. SHIGENOI: Oxidation of Eugenol with Hydrogen Peroxide. Nippon Kagaku Zasshi **87**, 1002 (1966); *apud* Chem. Abstr. **65**, 18450 (1966).

59. GEISSMAN, T. A., and D. H. CROUT: Organic Chemistry of Secondary Plant Metabolism, p. 378. San Francisco: Freeman, Cooper & Co. 1969.

60. GERCHUK, M. P., and V. M. IVANOVA: Synthesis of Nordihydroguaiaretic Acid and its Use as a Preservative of Edible Fats. Khim. Nauka i Prom. **3**, 685 (1958); *apud* Chem. Abstr. **53**, 7094 (1958).

61. GIESBRECHT, A. M., N. C. FRANCA, O. R. GOTTLIEB, and A. I. DA ROCHA: The Chemistry of Brazilian Lauraceae. XXVI. The Neolignans of *Licaria canella*. Phytochem. **13**, 2285 (1974).

62. GILCHRIST, T., R. HODGES, and A. L. PORTE: The Structure of Otobain. J. Chem. Soc. (London) **1962**, 1780.

63. GISVOLD, O., and E. THAKER: Lignans from *Larrea divaricata*. J. Pharm. Sci. **63**, 1905 (1974).

64. GOTTLIEB, O. R.: Chemosystematics of the Lauraceae. Phytochem. **11**, 1537 (1972).

65. — Lignans and Neolignans. Revista Latinoamericana de Química **5**, 1 (1974).

66. GOTTLIEB, O. R., J. G. S. MAIA, and J. C. MOURÃO: The Chemistry of Brazilian Lauraceae. XXXVI. Neolignans from a *Licaria* species. Phytochem. **15**, 1289 (1976).

67. GOTTLIEB, O. R., J. G. S. MAIA, and M. N. DE S. RIBEIRO: The Chemistry of Brazilian Myristicaceae. VII. Neolignans from *Virola carinata*. Phytochem. **15**, 773 (1976).

68. GOTTLIEB, O. R., M. C. MARX, and C. M. A. DA M. REZENDE: Proposals concerning the Biosynthesis of Phenylpropanoids. Communication presented at the XXVII Annual Meeting of Sociedade Brasileira para o Progresso da Ciência, Belo Horizonte (1975).

69. GOTTLIEB, O. R., M. L. DA SILVA, and Z. S. FERREIRA: The Chemistry of Brazilian Lauraceae. XXXI. Neolignans from *Aniba terminalis*. Phytochem. **14**, 1825 (1975).

70. GOTTLIEB, O. R., J. C. MOURÃO, M. YOSHIDA, R. D. ROSENSTEIN, and K. TOMITA: Unpublished work (1976).

71. GOVINDACHARI, T. R., K. NAGARAJAN, N. VISWANATHAN, and H. FUHRER: Nonequivalence of Methylenedioxy Protons in Cyclolignans. Indian J. Chem. **9**, 546 (1971).

72. GREGSON, M., W. D. OLLIS, B. T. REDMAN, I. O. SUTHERLAND, and H. H. DIETRICHS: The Constitution and Stereochemistry of Obtusafuran. Chem. Commun. **1968**, 1394.

73. HAWORTH, R. D.: Natural Resins. Annu. Rep. Progr. Chem. **33**, 266 (1937).

74. — The Chemistry of the Lignan Group of Natural Products. J. Chem. Soc. (London) **1942**, 448.

75. HAWORTH, R. D., C. R. MAVIN, and G. SHELDRICK: The Constituents of Guaiacum

Resin. II. Synthesis of dl-Guaiaretic Acid Dimethyl Ether. J. Chem. Soc. (London) **1934,** 1423.

76. Hayashi, T., and R. H. Thomson: New Lignans in *Conocarpus erectus*. Phytochem. **14,** 1085 (1975).

77. Haynes, C. G., A. H. Turner, and W. A. Waters: The Oxidation of Monohydric Phenols by Alkaline Ferricyanide. J. Chem. Soc. (London) **1956,** 2823.

78. Hearon, W. M., and W. S. MacGregor: The Naturally Occurring Lignans. Chem. Rev. **55,** 958 (1955).

79. Herrmann, K.: The Lignans and their Practical Significance. Pharmazie **12,** 147 (1957).

80. Hobbs, J. J., and F. E. King: The Chemistry of Extractives from Hardwoods. XXIX. Eusiderin, a Possible By-product of Lignin Synthesis in *Eusideroxylon zwageri*. J. Chem. Soc. (London) **1960,** 4732.

81. Holloway, D. M., and F. Scheinmann: Co occurrence of Aporphine and Biphenyl Constituents in *Litsea turfosa*. Phytochem. **12,** 1503 (1973).

82. — — Two Lignans from *Litsea grandis* and *L. gracilipes*. Phytochem. **13,** 1233 (1974).

83. Hughes, G. K., and E. Ritchie: The Chemical Constituents of *Himantandra* Species. I. The Lignans of *Himantandra baccata* and *H. belgraveana*. Austral. J. Chem. **7,** 104 (1954).

84. Jewers, K., A. H. Manchanda, and H. M. Rose: Naturally-occurring Antitumour Agents. In Progress in Medicinal Chemistry (Edited by G. P. Ellis and G. B. West), Vol. 9, p. 33. London: Butterworths. 1971.

85. King, F. E., and J. G. Wilson: The Chemistry of Extractives from Hardwoods. XXXVI. The Lignans of *Guaiacum officinale* L. J. Chem. Soc. (London) **1964,** 4011.

86. — — The Chemistry of Extractives from Hardwoods. XXXVII. The Synthesis of Furoguaiacin Ethers. J. Chem. Soc. (London) **1965,** 1572.

87. Klyne, W., R. Stevenson, and R. J. Swan: Optical Rotatory Dispersion. XXVIII. Absolute Configuration of Otobain and Derivatives. J. Chem. Soc. (C) (London) **1966,** 893.

88. Kohen, F., I. Maclean, and R. Stevenson: The Constitution of Otobaphenol. J. Chem. Soc. (C) (London) **1966,** 1775.

89. Kotchetkov, N. K., A. Ya. Khorlin, and O. S. Chizhov: Chemical Study of *Schizandra chinensis*. I. Schizandrin and Related Compounds. Zhurn. Obshchei Khimii (USSR) **31,** 3454 (1961); *apud* Chem. Abstr. **57,** 3334 (1962).

90. Kotchetkov, N. K., A. Ya. Khorlin, O. S. Chizhov, and V. I. Sheichenko: Chemical Study of *Schizandra chinensis*. II. The Structure of Schizandrine. Izv. Akad. Nauk SSSR, Otd. Khim. Nauk 850 (1962); *apud* Chem. Abstr. **57,** 13704 (1962).

91. Kotchetkov, N. K., A. Ya. Khorlin, and O. S. Chizhov: Chemical Study of *Schizandra chinensis*. III. Synthesis and Ultraviolet Spectra of Some Derivatives of 2,3,4,2′,3′,4′-hexamethoxybiphenyl. Izv. Akad. Nauk SSSR, Otd. Khim. Nauk 856 (1962); *apud* Chem. Abstr. **57,** 13704 (1962).

92. — — — Chemical Study of *Schizandra chinensis*. IV. Isolation, Structure and Synthesis of Deoxyschizandrine and the Structure of γ-Schizandrine. Izv. Akad. Nauk SSSR, Ser. Khim. [6], 1036 (1964); *apud* Chem. Abstr. **61,** 8250 (1964).

93. — — — Deoxyschizandrin — Structure and Total Synthesis. Tetrahedron Letters **1962,** 361.

94. Kotchetkov, N. K., A. Ya. Khorlin, O. S. Chizhov, and V. I. Sheichenko: Schizandrin — Lignan of Unusual Structure. Tetrahedron Letters **1961,** 730.

95. Krukoff, B. A.: American Species of *Strychnos*. Lloydia **35,** 197 (1972).

96. Labán, E. A.: Extraction and Assay of Nordihydroguaretic Acid in *Larrea nitida*. Tesis quím. Univ. Chile **2,** 17 (1950); *apud* Chem. Abstr. **47,** 7126 (1953).

97. LEBEDEV, A. A., and I. K. KAMILOV: Schizandrin Pharmacology. Med. Zh. Uzb. **1966** (9), 60; *apud* Chem. Abstr. **66,** 74744 (1967).
98. LINDBERG, B.: Dehydrogenation of Phenols. VII. Dehydroguaiaretic Acid by Dehydrogenation of Isoeugenol. Svensk Papperstidn. **56,** 6 (1953); *apud* Chem. Abstr. **49,** 6866 (1955).
99. MACLEAN, I., and R. STEVENSON: Synthesis of (±)-Otobain. J. Chem. Soc. (C) (London) **1966,** 1717.
100. MAJUMDER, P. L., and M. BHATTACHARYYA: Structure of Furoguaiacidin: a New Furanoid Lignan of the Heartwood of *Guaiacum officinale* L. Chem. and Ind. **1974,** 77.
101. MAJUMDER, P. L., A. CHATTERJEE, and G. C. SENGUPTA: Lignans from *Machilus edulis*. Phytochem. **11,** 811 (1972).
102. MANITTO, P., D. MONTI, and P. GRAMATICA: Biosynthesis of Anethole in *Pimpinella anisum* L. Tetrahedron Letters **1974,** 1567.
103. MARX, M. C.: Arylpropanoids in Embryobionta. M. S. thesis presented to Universidade Federal Rural do Rio de Janeiro (1975).
104. MATSUI, K., and K. MUNAKATA: The Structure of Piperenone, a New Antifeeding Substance from *Piper futokadzura*. Tetrahedron Letters **1975,** 1905.
105. MCALPINE, J. B., N. V. RIGGS, and P. G. GORDON: Absolute Stereochemistry of Calopiptin. Austral. J. Chem. **21,** 2095 (1968).
106. MCCREDIE, R. S., E. RITCHIE, and W. C. TAYLOR: Constituents of *Eupomatia* Species. The Structure and Synthesis of Eupomatene. A Lignan of Novel Type from *Eupomatia laurina* R. Br. Austral. J. Chem. **22,** 1011 (1969).
107. MCKECHNIE, J. S., and I. C. PAUL: Molecular Structure of Nordihydroguaiaretic Acid. J. Chem. Soc. (B) (London) **1969,** 699.
108. MCMURRY, T. B. H., and H. K. KENNEDY-SKIPTON: A Synthesis of (±)-Otobain. Tetrahedron Letters **1966,** 975.
109. MERLINI, L., and A. ZANAROTTI: A Biogenetically Patterned Synthesis of (±)-Eusiderin. Tetrahedron Letters **1975,** 3621.
110. MEYER, W.: *Podophyllum peltatum* — May Apple. A Potential New Cash-Crop Plant of Eastern North America. Economic Botany **28,** 68 (1974).
111. MILLER, I. J.: The Dimerization of Isoeugenol by Free Radicals. Tetrahedron Letters **1972,** 4955.
112. MURPHY, S. T., E. RITCHIE, and W. C. TAYLOR: Some Constituents of *Austrobaileya scandens* (Austrobaileyaceae): Structures of Seven New Lignanes. Austral. J. Chem. **28,** 81 (1975).
113. NASIROV, S. KH., A. A. LEBEDEV, N. A. ISHMUKHAMEDOV, and I. K. KAMILOV: Electrophysiological Analysis of the Schizandrin Effect on the Central Nervous System. Med. Zh. Uzb. **1967** (5), 34; *apud* Chem. Abstr. **68,** 20792 (1968).
114. OGISO, A., M. KURABAYASHI, H. MISHIMA, and M. C. WOODS: The Structure of Futoenone, a Novel Spiro-cyclohexadienone Derivative. Tetrahedron Letters **1968,** 2003.
115. OGISO, A., M. KURABAYASHI, S. TAKAHASHI, H. MISHIMA, and M. C. WOODS: The Structure and Total Synthesis of Futoenone, a Constituent of *Piper futokadzura* Sieb. et Zucc. Chem. Pharm. Bull. (Japan) **18,** 105 (1970).
116. OHTA, T., and Y. MORI: Oxidation of *p*-Propylphenol with Hydrogen Peroxide. Ann. Rept. Tokyo Coll. Pharm. **3,** 206 (1953); *apud* Chem. Abstr. **49,** 7513 (1955).
117. OLIVETO, E. P.: Nordihydroguaiaretic Acid. A Naturally Occurring Antioxidant. Chem. and Ind. **1972,** 677.
118. PARDINI, R. S., J. C. HEIDKER, and D. C. FLETCHER: Inhibition of Mitochondrial Electron Transport by Nor-Dihydroguaiaretic Acid (NDGA). Biochem. Pharmac. **19,** 2695 (1970).

119. PEARL, I. A.: 2,3-Bis(hydroxybenzyl)butane Derivatives. US Pat. 2,644,822 (1953); *apud* Chem. Abstr. **48,** 6468 (1954).

120. PELTER, A., A. P. STAINTON, and M. BARBER: The Mass Spectra of Oxygen Heterocycles. III. An Examination of Simple Lignans. J. Heterocycl. Chem. **3,** 191 (1966).

121. PERRY, C. W., M. V. KALNINS, and K. H. DEITCHER: Synthesis of Lignans. Nordihydroguaiaretic Acid. J. Organ. Chem. (USA) **37,** 4371 (1972).

122. PICKER, K., E. RITCHIE, and W. C. TAYLOR: Constituents of *Eupomatia* Species. III. New Eupomatenoid Lignans from the Leaves and Wood of *Eupomatia laurina*. Austral. J. Chem. **26,** 1111 (1973).

123. REDMAN, B. T.: Phytochemical Studies in the Leguminosae and the Guttiferae Families. Ph. D. thesis, p. 86, University of Sheffield (1968).

124. RIGGS, N. V., and J. D. STEVENS: Some Constituents of *Piptocalyx moorei* Oliv. Austral. J. Chem. **15,** 305 (1962).

125. RUNEBERG, J.: Phenol Dehydrogenations. VIII. Synthesis of Magnolol. Acta Chem. Scand. **12,** 188 (1958).

126. RUTH, E. F.: Isolation of nor-Dihydroguaiaretic Acid from Argentinian *Larrea* Species. Anales asoc. quím. Argentina **34,** 163 (1946).

127. SARKANEN, K. V.: In: Lignins, Occurrence, Formation, Structure and Reactions (Eds. K. V. SARKANEN and C. H. LUDWIG), p. 95. New York: Wiley-Interscience. 1971.

128. SARKANEN, K. V., and A. F. A. WALLIS: Stereospecific Formation of *trans*-Phenylcoumarans by Dehydrogenative Dimerization of *trans*- and *cis*-Isoeugenol. Chem. Commun. 298 (1969).

129. — — Oxidative Dimerizations of (E)- and (Z)-Isoeugenol (2-Methoxy-4-propenylphenol) and (E)- and (Z)-2,6-Dimethoxy-4-propenylphenol. J. Chem. Soc. Perkin I (London) **1973,** 1869.

130. — — PMR Analysis and Conformation of 2,5-bis-(3,4,5-Trimethoxyphenyl)-3,4-dimethyltetrahydrofuran Isomers. I. J. Heterocycl. Chem. **10,** 1025 (1973).

131. SASAKI, K., Y. HIRATA, S. YAMAMURA, Y.-P. CHEN, M. HONG, and H.-Y. HSU: The Isolation and Structure of Isoasatone. Tetrahedron Letters **1973,** 4881.

132. SCHRECKER, A. W.: Meso-Dihydroguaiaretic Acid and its Derivatives. J. Amer. Chem. Soc. **79,** 3823 (1957).

133. SCHRECKER, A. W., and J. L. HARTWELL: Application of Tosylate Reductions and Molecular Rotations to the Stereochemistry of Lignans. J. Amer. Chem. Soc. **77,** 432 (1955).

134. — — Components of Podophyllin. XX. The Absolute Configuration of Podophyllotoxin and Related Lignans. J. Organ. Chem. (USA) **21,** 381 (1956).

135. — — The Absolute Configuration of Lignans. J. Amer. Chem. Soc. **79,** 3827 (1957).

136. SHIPOCHLIEV, T., and ST. ILIEVA: Pharmacologic Study of Bulgarian *Schizandra chinensis*. Farmatsiya (Sofia) **17,** 56 (1967); *apud* Chem. Abstr. **68,** 1870 (1968).

137. SUGIMOTO, N., and K. OKUMURA: Nordihydroguaiaretic Resin Acid. Japan. Pat. 4626 (1954); *apud* Chem. Abstr. **50,** 4225 (1956).

138. SUGII, Y.: Constituents of the Bark of *Magnolia officinalis* Rhed. et Wils. and *Magnolia obovata* Thumb. J. Pharmac. Soc. Japan **50,** 183 (1930); *apud* Chem. Abstr. **24,** 3505 (1930).

139. TAKAHASHI, S., and A. OGISO: The Structure of Futoquinol, a Constituent of *Piper futokadzura* Sieb. et Zucc. Chem. Pharm. Bull. (Japan) **18,** 100 (1970).

140. TITLYANOV, A. A.: Biologically Active Substances of Chinese Magnolia Vine Fruit. Tr. Vses. Semin. Biol. Aktiv. (Lech.) Veshchestvam Plodov Yagod (Edited by L. I. VIGOROV, Ural. Lesotekh. Inst., Sverdlovsk, USSR), 3rd ed., p. 363 (1966, publ. 1968); *apud* Chem. Abstr. **73,** 91186 (1970).

141. VOLICER, L., M. SRAMKA, I. JANKU, R. CAPEK, R. SMETANA, and V. DITTEOVA:

Some Pharmacological Effects of *Schizandra chinensis*. Arch. Intern. Pharmacodyn. Ther. **163,** 249 (1966); *apud* Chem. Abstr. **66,** 1413 (1967).

142. WALLACE, R., A. L. PORTE, and R. HODGES: Lignans from *Myristica otoba*. The Structures of Hydroxyotobain and Iso-otobain. J. Chem. Soc. (London) **1963,** 1445.

143. WALLER, C. W., and O. GISVOLD: A Phytochemical Investigation of *Larrea divaricata* Cav. J. Am. Pharm. Assoc. **34,** 78 (1945).

144. WENKERT, E., H. E. GOTTLIEB, O. R. GOTTLIEB, M. O. DA S. PEREIRA, and M. D. FORMIGA: Carbon-13 Nuclear Magnetic Spectroscopy of Naturally Occurring Substances. XLV. ^{13}C NMR Spectroscopy of Neolignans. Phytochem. **15,** 1547 (1976).

145. WOODS, M. C., I. MIURA, A. OGISO, M. KURABAYASHI, and M. MISHIMA: The NMR Spectra of Futoenone and Derivatives: Partial Decoupling as an Aid in the Assignment of Complex Spectra, and Further Observations of Overhauser Effects. Tetrahedron Letters **1968,** 2009.

146. YAMAMURA, S., K. SASAKI, Y. HIRATA, Y.-P. CHEN, and H.-Y. HSU: Further Studies on Isoasatone. Tetrahedron Letters **1973,** 4877.

147. YAMAMURA, S., Y. TERADA, Y.-P. CHEN, H.-Y. HSU, and Y. HIRATA: Photochemical Conversion of Asatone to Isoasatone. The Revised Structure of Asatone. Tetrahedron Letters **1975,** 1903.

148. AIBA, C. J., M. A. DE ALVARENGA, O. CASTRO C., A. M. GIESBRECHT, O. R. GOTTLIEB, and F. M. PAGLIOSA: The Chemistry of Brazilian Lauraceae. XXXIX. Benzofuranoid Neolignans from *Aniba simulans*. Phytochem. **16,** 741 (1977).

149. AIBA, C. J., O. R. GOTTLIEB, F. M. PAGLIOSA, M. YOSHIDA, and M. T. MAGALHÃES: The Chemistry of Brazilian Lauraceae. XXXVII. Neolignans from *Nectandra miranda*. Phytochem. **16,** 745 (1977).

150. ALVARENGA, M. A. DE, U. BROCKSOM, O. CASTRO C., O. R. GOTTLIEB, and M. T. MAGALHÃES: The Chemistry of Brazilian Lauraceae. XLIV. Neolignans from *Aniba burchellii*. Phytochem. **16** (1977).

151. ALVARENGA, M. A. DE, O. CASTRO C., A. M. GIESBRECHT, and O. R. GOTTLIEB: The Chemistry of Brazilian Lauraceae. XLV. Bicyclo [3,2,1]octanoid Neolignans from *Aniba simulans*. Phytochem. **16** (1977).

152. DIAZ, A. M. P. DE, O. R. GOTTLIEB, A. F. MAGALHÃES, E. G. MAGALHÃES, and J. G. S. MAIA: The Chemistry of Brazilian Lauraceae. XLVI. Notes on *Aniba* Species. Acta Amazonica **7,** 41 (1977).

153. FORREST J. E., R. A. HEACOCK, and J. P. FORREST: Diarylpropanoids from Nutmeg and Mace (*Myristica fragrans* Houtt.). J. Chem. Soc. Perkin I (London) **1974,** 205.

154. GHERA, E., Y. BEN-DAVID, and D. BECKER: Desoxyschizandrin, Stereochemistry and Total Synthesis. Tetrahedron Letters **1977,** 463.

155. GOTTLIEB, O. R., J. C. MOURÃO, M. YOSHIDA, Y. P. MASCARENHAS, M. RODRIGUES, R. D. ROSENSTEIN, and K. TOMITA: The Chemistry of Brazilian Lauraceae. XLIII. Absolute Configuration of the Benzofuranoid Neolignans. Phytochem. **16,** 1003 (1977).

156. IKEYA, Y., H. TAGUCHI, and Y. IITAKA: The Constituents of *Schizandra chinensis* Baill. The Structure of a New Lignan, Gomisin D. Tetrahedron Letters **1976,** 1359.

157. TAGUCHI, H., and Y. IKEYA: The Constituents of *Schizandra chinensis* Baill. I. The Structures of Gomisin A, B and C. Chem. Pharm. Bull. (Japan) **23,** 3296 (1975).

158. YAMAMURA, S., Y.-P. CHEN, H.-Y. HSU, and Y. HIRATA: Asatone in Plants of the Aristolochiaceae. Phytochem. **15,** 426 (1976).

159. YAMAMURA, S., Y. TERADA, Y.-P. CHEN, M. HONG, H.-Y. HSU, K. SASAKI, and Y. HIRATA: The Structures of Two Novel Neolignans, Asatone and Isoasatone. Bull. Chem. Soc. Japan **49,** 1940 (1976).

(Received August, 1976)

Hydroxyzimtsäuren und Hydroxybenzoesäuren enthaltende Naturstoffe in Pflanzen

Von K. Herrmann, Lehrstuhl für Lebensmittelchemie, Technische Universität, Hannover, Bundesrepublik Deutschland

Inhaltsübersicht

I. Einleitung

Hydroxyzimtsäuren, aber auch Hydroxybenzoesäuren, sind in Blütenpflanzen (Angiospermen wie Gymnospermen) ubiquitär verbreitet und auch in Farnpflanzen (Pterophyten) (*31, 91*) und anderen niederen Pflanzen, z. B. Lycopodiales (*282*), bekannt. Das Vorkommen beschränkt sich im wesentlichen auf einige bestimmte Grundtypen. Auch treten diese in der Regel nicht frei, sondern in gebundener Form auf, womit sich eine Vielzahl von Verbindungen ergibt.

Hydroxyzimtsäure-Verbindungen sind dank der chromatographischen Methoden in einer sehr großen Zahl von Pflanzen erkannt worden. Nach Hydrolyse der Pflanzenextrakte findet man freie Hydroxyzimtsäuren fast in jeder Blütenpflanze, wenn gezielt nach ihnen gesucht wird.

Literaturverzeichnis: SS. 118—132

Von jeder Hydroxyzimtsäure wiederum kann eine Reihe von Verbindungen in der einzelnen Pflanze vorliegen, womit eine beträchtliche Zahl von Inhaltsstoffen pro Species möglich ist und gelegentlich auch aufgefunden wurde, z. B. Hafer (*65—69*) und Kaffee (*53, 55, 57, 251* bis *253*).

Dagegen ist die Zahl der Pflanzen, in denen Hydroxybenzoesäuren und besonders Verbindungen der Hydroxybenzoesäuren bisher nachgewiesen wurden, geringer. Dies mag vielleicht daran liegen, daß Hydroxybenzoesäuren weniger chromatographisch auffallen bzw. ihre Flecke auf dem Chromatogramm von entsprechenden Hydroxyzimtsäuren überlagert sein können. Auch scheinen die Gehalte an Hydroxybenzoesäuren häufig geringer als die an Hydroxyzimtsäuren zu sein (*234, 257—259*). Indessen sind bei gezielten Untersuchungen von Familien oder Gattungen einzelne Hydroxybenzoesäuren fast immer aufgefunden worden (vgl. *126*), selbst in Farnen (*31, 32, 91*). Bei Untersuchung der in Deutschland angebauten Obst- und Gemüsearten haben wir neben Hydroxyzimtsäuren fast in jeder Species Hydroxybenzoesäuren (oft mehrere) nachweisen können, wenn auch meist in geringer Konzentration, häufig nur in Spurenmengen (um 1 mg/kg) (*234, 256—259*).

Phenolsäuren haben COOH- und OH-Gruppen. Daraus resultieren die zahlreichen Möglichkeiten von Verbindungen. Die Carboxylgruppe gibt mit Alkoholen und Phenolen Ester und mit Amino-Verbindungen Amide. Damit sind Reaktionen mit Kohlenhydraten oder Glykosiden wie mit Aminosäuren oder Proteinen möglich und auch bekannt. Die phenolische Hydroxylgruppe andererseits ist ebenfalls sehr reaktionsfähig. Mit Zuckern entstehen Glykoside.

Phenolische Verbindungen, vor allem 1,2-Di-, 1,4-Di- und 1,2,3-Trihydroxyl-Verbindungen, werden leicht oxydiert. Dabei können sehr unterschiedliche Stoffe auftreten, z. B. das Benzotropolon Purpurogallincarbonsäure durch enzymatische Oxydation der Gallussäure (*121*) oder unterschiedlich gebaute Dimere durch phenolische Kupplung der Grundsubstanzen (*179*), wie Polyhydroxydiphensäuren aus Hydroxybenzoesäuren, oder Lignane, wie die Dehydrodiferulasäure und das Dehydrodisinapinsäuredilacton. Zum Teil werden die Dimere, z. B. der Kaffeesäure, leicht zu höher molekularen Verbindungen kondensiert, wie wir es von der enzymatischen Bräunung des Obstes her kennen. Auf das umfangreiche Gebiet der Oxydation der Phenolsäuren soll indessen nicht näher eingegangen werden.

Ebenfalls sollen Gallotannine und Ellagitannine, die bekannte Typen der hydrolysierbaren Gerbstoffe darstellen, hier nicht behandelt werden. Diese Verbindungen der Gallussäure oder der Hexahydroxydiphensäure, die aus 2 Molekülen Gallussäure durch eine C–C-Brücke entsteht (stabiles Dilacton = Ellagsäure), werden seit jeher in Beiträgen über Gerbstoffe oder Tannine zusammengefaßt. Hier sei u. a. auf HASLAM (*115*) verwiesen.

Desgleichen kann auf die Beziehungen der Hydroxyzimtsäuren zum Lignin und deren Bausteinen, wie Hydroxyzimtalkohole, nur hingewiesen werden. Doch sei vermerkt, daß das Lignin von Gramineen und einigen anderen Pflanzen Phenolsäuren als Ester aufweist; vgl. S. 98.

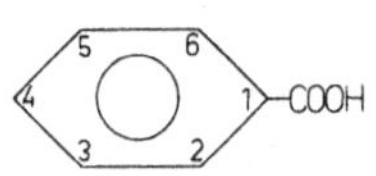

Hydroxybenzoesäure

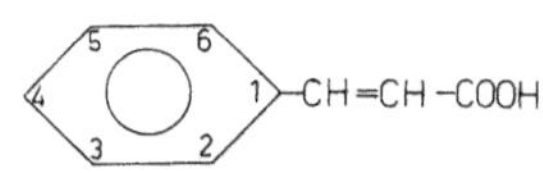

Hydroxyzimtsäure

Substituenten	Formel-Nr.	Trivialname	UV-Maxima (nm)	Eich-faktor
		Hydroxybenzoesäuren		
2-OH	**(1)**	Salicylsäure	207, 236, 305	1,74
3-OH	**(2)**	*m*-Hydroxybenzoesäure	210, 234, 298	2,60
4-OH	**(3)**	*p*-Hydroxybenzoesäure	207, 253	0,49
2,3-di-OH	**(4)**	*o*-Pyrocatechusäure	213, 246, 318	2,43
2,4-di-OH	**(5)**	β-Resorcylsäure	210, 256, 295	1,34
2,5-di-OH	**(6)**	Gentisinsäure	212, 230 sh, 330	1,78
2,6-di-OH	**(7)**	γ-Resorcylsäure	216, 249, 317	2,47
3,4-di-OH	**(8)**	Protocatechusäure	217 sh, 258, 295	0,82
3,5-di-OH	**(9)**	α-Resorcylsäure	215, 250, 310	2,88
3,4,5-tri-OH	**(10)**	Gallussäure	215, 271	0,89
2,4,6-tri-OH	**(11)**	Phloroglucinsäure	220, 261, 297	0,61
3-OCH_3, 4-OH	**(12)**	Vanillinsäure	207, 216, 258, 290	0,81
3-OH, 4-OCH_3	**(13)**	Isovanillinsäure		
3,5-di-OCH_3, 4-OH	**(14)**	Syringasäure	219, 272	0,88
		Hydroxyzimtsäuren		
2-OH	**(15)**	*o*-Cumarsäure	214, 273, 325	0,98
3-OH	**(16)**	*m*-Cumarsäure	214, 232 ,276 ,312 sh	0,44
4-OH	**(17)**	*p*-Cumarsäure	210, 223, 293 sh, 302 sh, 310	0,40
3,4-di-OH	**(18)**	Kaffeesäure	217, 240, 297 sh, 325	0,56
3-OCH_3, 4-OH	**(19)**	Ferulasäure	217, 233, 297 sh, 320	0,58
3-OH, 4-OCH_3	**(20)**	Isoferulasäure	217, 240, 292, 322	0,64
3,5-di-OCH_3, 4-OH	**(21)**	Sinapinsäure	225 sh, 235, 322	0,61

Anmerkung: sh = Schulter; Hauptmaxima sind unterstrichen.
Eichfaktor = einer Extinktion von 0,500 entsprechenden Menge Substanz in mg/100 ml Methanol

Schema 1. Wichtige Hydroxybenzoesäuren und Hydroxyzimtsäuren

Auch soll auf die Biochemie dieser Naturstoffe nicht eingegangen werden; darüber stehen Übersichtsarbeiten (z. B. *96*, *111*) zur Verfügung.

II. Hydroxyzimtsäuren und Hydroxybenzoesäuren

Die wesentlichen Hydroxyzimt- und Hydroxybenzoesäuren sind in Schema 1 angegeben.

Die *Hydroxyzimtsäuren* stehen als phenolische Phenylpropane (C_6C_3) in enger Beziehung zur großen Gruppe der Flavonoide (*111*), die das Grundgerüst des Diphenylpropans ($C_6C_3C_6$) besitzen, z. B. Flavon(ol)e (*111, 290*)., Anthocyanidine (*111*), Catechine (*294*) etc. Daher werden bisweilen die Hydroxyzimtsäuren mit zu den Flavonoiden gerechnet, was aufgrund ihrer biochemischen Zugehörigkeit und ihrer ähnlichen Verbreitung in Blütenpflanzen einige Berechtigung hat. Auch in der wirtschaftlichen Bedeutung ergeben sich Parallelen.

Die Hydroxyzimtsäuren und ihre Ester weisen *cis-trans*-Isomerie auf. Die *trans*-Verbindungen sind die stabileren und dürften in der Regel in der Pflanze vorliegen. Die *cis*-Verbindungen sind energiereicher und entstehen z. B. beim Belichten mit UV-Licht. Die beiden Isomere zeigen bei der Chromatographie in wasserhaltigen Fließmitteln wie verdünnter Essigsäure unterschiedliche Rf-Werte; auch unterscheiden sich ihre Spektren etwas voneinander (*73*).

In der Natur kommen *p*-Cumarsäure (**17**), Ferulasäure (**19**) und vor allem Kaffeesäure (**18**), weniger häufig Sinapinsäure (**21**) als Grundkörper vor. Die ersten Hinweise über ihre weite Verbreitung im Pflanzenreich verdanken wir u. a. Bate-Smith (*10*, vgl. auch *12*), der in papierchromatographischen Untersuchungen fand, daß von 467 Dicotylen-Species 66% Kaffeesäure, 48% *p*-Cumarsäure, 33% Ferulasäure und 26% Sinapinsäure enthielten. Für 94 Monocotylen-Species betrug der Anteil der genannten Säuren 50, 55, 67 und 57%.

Andere Grundkörper sind bisher nur vereinzelt nachgewiesen worden, wie *o*-Cumarsäure (**15**) z. B. im Echten Steinklee (*Melilotus altissimus*) (*140*) und Isoferulasäure (**20**) z. B. in Wurzel und Rhizom des Baldrian (*Valeriana officinalis*) (*263*), vgl. auch (*4, 92*).

Im Moos (*Sphagnum magellanicum*) wurde über das Vorkommen von *p*-Hydroxy-(β-carboxy-methyl)-zimtsäure (Sphagnumsäure) (**22**), Schmp. 178—180° C, berichtet (*285*). Auch sei erwähnt, daß 4-Methoxyzimtsäure und 3,4-Dimethoxyzimtsäure in einigen Pflanzen vorkommen. Papierchromatographische Hinweise auf 3,4-Methylendioxy-zimtsäure wurden in *Piper tuberculatum* gefunden (*243*).

Von der *p*-Cumarsäure (**17**) ist das Cumarin Umbelliferon (**23**), von der Ferulasäure (**19**) Scopoletin (**24**) und von der Kaffeesäure (**18**) Aesculetin (**25**) abgeleitet. Die beiden letzteren, weniger Umbelliferon, sind im Pflanzenreich recht verbreitet. Besonders leicht erfolgt die Umlagerung der Kaffeesäure in Aesculetin, bereits beim Aufbewahren in Lösung. Indessen soll auf Cumarine in diesem Beitrag nicht eingegangen werden.

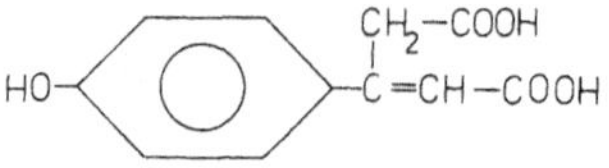

(**22**) Sphagnumsäure

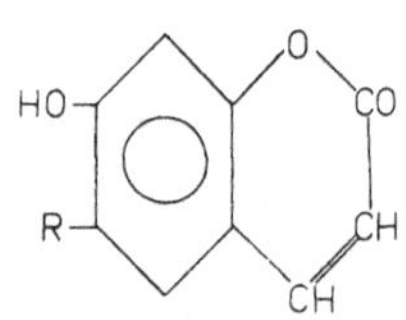

(**23**) R = H Umbelliferon
(**24**) R = OCH_3 Scopoletin
(**25**) R = OH Aesculetin

Besonders auffällig ist, daß die 3,4,5-Trihydroxyzimtsäure in der Natur noch nicht aufgefunden werden konnte, obwohl unabhängig voneinander BATE-SMITH in England und HERRMANN intensiv nach dieser Verbindung im Pflanzenreich gesucht haben (vgl. *123*). Es ist dies bei der weiten Verbreitung der Mono- und Dihydroxyzimtsäuren überraschend. Die 3,4,5-Trimethoxyzimtsäure ist z. B. im Molekül des Alkaloids Rescinnamin und in Alkaloiden von *Alstonia lanceolifera* (*162*) enthalten. Auch das chemisch eng verwandte 3,4,5-Trihydroxycumarin scheint in der Natur noch nicht aufgefunden zu sein. Bei den meisten Flavonoiden sind dagegen Verbindungen mit drei benachbarten OH-Gruppen nach den 1,2-Dihydroxy-Verbindungen die nächsthäufigen.

Von den *Hydroxybenzoesäuren* sind bisher Salicylsäure (**1**), *p*-Hydroxybenzoesäure (**3**), Gentisinsäure (**6**), Protocatechusäure (**8**), Gallussäure (**10**), Vanillinsäure (**12**) und Syringasäure (**14**) als Grundkörper häufiger aufgefunden worden (*126, 234, 256—259*). Meist ist der Gehalt der Pflanzen an Hydroxybenzoesäure-Verbindungen gering; es können aber gelegentlich auch Konzentrationen im Prozentbereich auftreten, z. B. 1—2% Protocatechusäure in den trockenen Schalen gefärbter Zwiebeln *(Allium cepa)* (*122, 234*) oder 1 und mehr Prozent an Gallussäure in der Trockensubstanz tanninhaltiger Pflanzen. Phloroglucinsäure

(**11**) entsteht beim Abbau von Flavonoiden, z. B. in Weintrestern beim Befall durch Saprophyten (*13*).

Wegen der von der Gallussäure (**10**) abgeleiteten Stoffe, die in Verbindung mit Tanninen stehen: *m*-Digallussäure = 4-Gallylgallussäure, *m*-Trigallussäure, die beide Depside darstellen, sowie Hexahydroxydiphensäure und deren Dilacton Ellagsäure (**26**), vgl. HASLAM (*115*). Ellagsäure (**26**) bzw. Hexahydroxydiphensäure soll in der Natur verbreitet vorkommen (*11, 12*) und wurde häufig zusammen mit Gallussäure aufgefunden, z. B. in Erdbeeren (*Fragaria ananassa*) (*256*).

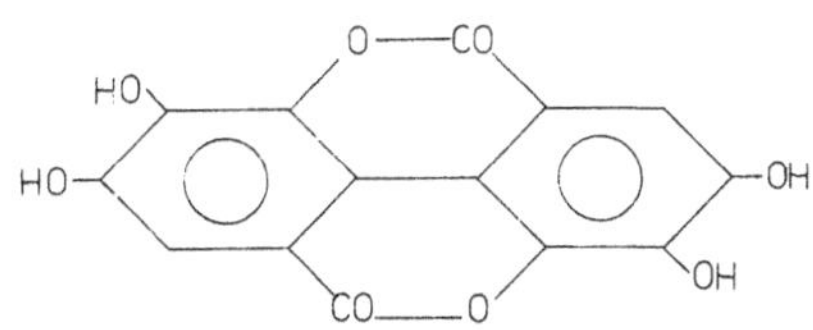

(**26**) Ellagsäure

Auch soll in diesem Beitrag auf spezifische Flechteninhaltsstoffe, die sich von Hydroxy-alkyl-benzoesäuren, z. B. Orsellinsäure (2,4-Dihydroxy-6-methyl-benzoesäure) ableiten, nicht näher eingegangen werden. Sie werden in der Regel in speziellen Arbeiten über Flechteninhaltsstoffe (*60, 134*) abgehandelt.

Ebenfalls werden Hydroxyphenylessigsäuren, deren Trivialnamen aus denen der Hydroxybenzoesäuren durch das Vorsetzen des Wortes „Homo“ gebildet wurden, wie Homovanillinsäure, Homoprotocatechusäure, Homogallussäure, hier nicht besprochen. Sie sind in Pflanzen gelegentlich meist in recht geringen Mengen aufgefunden worden; doch ist ihr Auftreten vermutlich auf eine kleinere Zahl von Species beschränkt. Bei Untersuchung der in Deutschland angebauten Obst- und Gemüsearten fanden wir nur sehr selten einen Hinweis auf eine dieser Verbindungen.

Weiterhin werden Hydroxyphenylpropionsäuren (Hydroxy-dihydrozimtsäuren), deren Trivialnamen aus denen der Hydroxyzimtsäuren durch Vorsetzen des Wortes „Hydro“ gebildet wurden, wie Hydroferulasäure, Hydrokaffeesäure, und die in der Natur auch nur gelegentlich aufgefunden worden sind, in diesem Beitrag nicht behandelt.

Wegen der zahlreichen synthetisch hergestellten Phenolsäuren und deren Verbindungen sei auf BEILSTEINS Handbuch der Organischen Chemie (Springer-Verlag, Berlin-Heidelberg-New York), Band X, verwiesen.

Schließlich seien hier das gelbe Curcumin = Diferulylmethan sowie Di-*p*-cumarylmethan und *p*-Cumaryl-ferulylmethan in *Curcuma*-Species noch erwähnt (*145*).

III. Ester der Hydroxyzimtsäuren und der Gallussäure mit anderen Hydroxycarbonsäuren

Nach den bisher vorliegenden Untersuchungen scheinen in der Natur die Hydroxyzimtsäuren am häufigsten mit Hydroxycarbonsäuren verestert zu sein. Dabei ist die COOH-Gruppe der Hydroxyzimtsäure mit der oder einer OH-Gruppe der anderen Säure verbunden, womit sogenannte Depside entstehen. Im Vordergrund steht offensichtlich die D-Chinasäure. Dies gilt vor allem für die als Chlorogensäuren bezeichneten Kaffeesäureester. Auch sind Gallyl-chinasäuren in der Natur bekannt. Als Esterkomponenten sind außerdem bisher (−)-Äpfelsäure, (−)- und *meso*-Weinsäure, Shikimisäure sowie α-Hydroxyhydrokaffeesäure aufgefunden worden.

1. Caffeyl-D-chinasäuren (Chlorogensäuren)*

Theoretisch sind vier Monocaffeyl-mono-D-chinasäuren möglich; sie sind — wenn auch recht unterschiedlich — im Pflanzenreich aufgefunden worden. Am verbreitetsten ist die Chlorogensäure = 3-O-Caffeyl-D-chinasäure (**28**), wenn auch ihr Vorkommen in einer beträchtlichen Zahl von Pflanzen, in denen sie mit einfachen chromatographischen Mitteln angegeben wurde, der Nachprüfung bedarf. Sie wurde bereits vor 130 Jahren von Payen in Kaffeebohnen (*Coffea*-Species) nachgewiesen und Chlorogensäure genannt, weil sich ihre alkalische Lösung an der Luft grün färbte. Allen weiteren Autoren des 19. Jahrhunderts mißglückte ihre Gewinnung, und erst zu Beginn dieses Jahrhunderts wurde sie wieder in kristallisierter Form aus Kaffee erhalten.

Nach 1950 wurden Neochlorogensäure (**30**) erstmals in Pfirsichen (*Prunus persica*) (*52*) und Cryptochlorogensäure (**29**) (ursprünglich als „Band 510" bezeichnet) in Kaffeebohnen und verschiedenen Früchten (*251*) aufgefunden. Später wurden sie durch Perjodat-Oxydation der

* Die Numerierung am D-Chinasäure-Ring ist in dieser Übersicht in der bisher üblichen Weise vorgenommen worden, um einen Vergleich mit den bisher erschienenen Arbeiten zu erleichtern. Neuerdings wird nach dem IUPAC-System die 3-O-Acyl-chinasäure als 5′-O-Acyl-chinasäure und entsprechend die 5-O-Acyl-chinasäure als 3′-O-Acyl-chinasäure bezeichnet, z. B. Chlorogensäure bisher als 3-O-Caffeyl-chinasäure und jetzt als 5′-O-Caffeyl-chinasäure.

Literaturverzeichnis: SS. 118—132

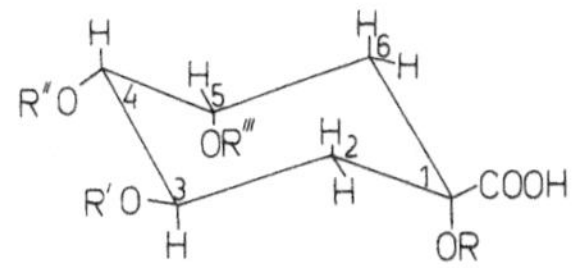

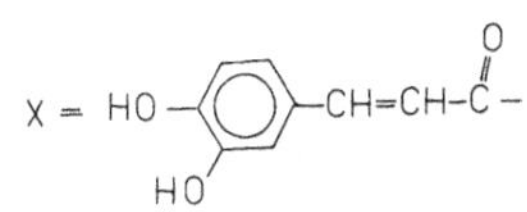

Caffeyl-

(Wegen der Numerierung der Chinasäure vgl. Fußnote S. 80.)

	R	R′	R″	R‴	
	H	H	H	H	D-(−)-Chinasäure, Schmp. 157—160°, $\alpha_D = -45°$ (Wasser) (*54*)
(27)	X	H	H	H	1-O-Caffeyl-D-chinasäure
(28)	H	X	H	H	Chlorogensäure, Schmp. 205—208°, $\alpha_D = -33°$ (Wasser); λ_{max} 322 nm, ε 17 200 (*252, 253*)
(29)	H	H	X	H	Cryptochlorogensäure („Band 510"), Schmp. 182—184°, $\alpha_D = -75°$ (Wasser) (*228*); $\alpha_D = -54°$ (Wasser), λ_{max} 322 nm, ε 15 400 (Wasser) (*54*)
(30)	H	H	H	X	Neochlorogensäure, Schmp. 210°, $\alpha_D = +2°$ (Methanol) (*119*), λ_{max} 322 nm, ε 16 600 (*253*)
(32)	X	X	H	H	1,3-Di-O-caffeyl-D-chinasäure, $\alpha_D = +168°$ (Methanol) (*228*)
(33)	X	H	X	H	1,4-Di-O-caffeyl-D-chinasäure, $\alpha_D = -30°$ (Methanol) (*191*)
(34)	X	H	H	X	1,5-Di-O-caffeyl-D-chinasäure = Cynarin, Schmp. 226—228°, $\alpha_D = -59°$ (Methanol), λ_{max} 325 nm, log ε 4,50 (*190, 191*)
(35)	H	X	X	H	3,4-Di-O-caffeyl-D-chinasäure, $\alpha_D = -288°$, λ_{max} 329 nm, log ε 4,51 (*55*)
(36)	H	X	H	X	3,5-Di-O-caffeyl-D-chinasäure, Schmp. 170—172°, $\alpha_D = -197°$, λ_{max} 329 nm, log ε 4,55 (*55*)
(37)	H	H	X	X	4,5-Di-O-caffeyl-D-chinasäure, $\alpha_D = -166°$, λ_{max} 329 nm, log ε 4,51 (*55*)

Schema 2. Mono- und Di-caffeyl-D-chinasäuren

mit Diazomethan behandelten Verbindungen als 5- bzw. 4-O-Caffeyl-D-chinasäuren erkannt (*221*); durch Aufnahme der NMR-Spektren wurde dies bestätigt (*291*). Zuvor waren andere Bindungsverhältnisse angenommen worden, die als überholt anzusehen sind. Neochlorogensäure herrscht im Stein- und Beerenobst weitgehend vor (*104*) und scheint von den Kaffeesäureestern nach der Chlorogensäure im Pflanzenreich am verbreitetsten zu sein. Ob auch die Cryptochlorogensäure natürlich verbreitet vorkommt, kann noch nicht gesagt werden, da bei den Estern mit Chinasäure Umesterungsreaktionen bei der Aufarbeitung auftreten können. So werden Chlorogensäure oder Neochlorogensäure oder Cryptochlorogensäure durch 30 min langes Erhitzen in Pufferlösung pH 7 jeweils in eine Mischung der drei Chlorogensäure (in ähnlichen Mengen) umgewandelt (*222*). Häufig wird die Umesterung durch Erhitzen mit gesättigter wäßriger $NaHCO_3$-Lösung vorgenommen. Die Auftrennung der Isomere kann über eine Silicagelsäule erfolgen (*313*).

Schließlich wurde das Vorkommen von 1-Caffeyl-chinasäure (**27**) vom Arbeitskreis LITVINENKO (*164*) neben anderen Chlorogensäuren und Rosmarinsäure (vgl. S. 87) in einer Reihe von Labiaten angegeben.

Die Chlorogensäuren sind leicht löslich in Alkoholen und Aceton, schwer löslich in kaltem Wasser, sehr schwer löslich in Essigester und Äther und praktisch unlöslich in Chloroform und Benzol.

Weiterhin wurde wiederholt über das Vorkommen einer „Isochlorogensäure" berichtet, die ursprünglich (1950) als 5-Caffeyl-chinasäure angesehen und später als ein Gemisch mehrerer Dicaffeyl-chinasäuren erkannt wurde (*55, 223*); vgl. S. 83). Auch in neueren Arbeiten wird sie noch häufig als Pflanzeninhaltsstoff angegeben; günstiger erscheint die Bezeichnung „Isochlorogensäure-Fraktion", wenn eine Isolierung und Aufklärung der einzelnen Komponenten nicht erfolgte, was offensichtlich auf beträchtliche Schwierigkeiten stößt.

Es sei noch erwähnt, daß vor 20 Jahren über eine weitere Chlorogensäure, die „Pseudochlorogensäure" $\alpha_D = -155°$ in Süßkartoffeln *(Ipomoea batatas)* (*286*) und in Viktoria-Pflaumen *(Prunus domestica)* (*71*) berichtet wurde. Sie ist seitdem nie wieder aufgefunden worden. Vermutlich hat es sich dabei um ein Gemisch aus verschiedenen Caffeyl- und/oder Dicaffeylchinasäuren gehandelt.

Chlorogensäuren können in einigen Pflanzen in beträchtlicher Konzentration vorliegen, z. B. in rohem Kaffee zu 5—8%. In der Regel dürften die Konzentrationen in der Pflanze unter 1000 mg/kg, oft unter 100 mg/kg Frischgewicht betragen (*104*).

Aus Kaffee wurde ein kristallisierter Komplex aus 1 Mol Kaliumchlorogenat und 1 Mol Coffein isoliert (*252*), der als ein hydrophobisch gebundener π-Molekularkomplex bezeichnet werden kann und dessen Konformation diskutiert wurde (*132*). Auch konnten aus geröstetem Kaffee nach Gegenstromverteilung Chlorogensäure, Neochlorogensäure und Cryptochlorogensäure isoliert werden (*53*). Sonnenblumensamen *(Helianthus annuus)* enthalten Äthylchlorogenat = 3-O-Caffeyl-D-chinasäureäthylester (**31**), Schmp. 91—92° (*1*).

Die einzelnen Mono-caffeyl-D-chinasäuren sind bisher wiederholt durch spezifische Kondensation von Kaffeesäurechlorid, dessen beide OH-Gruppen in geeigneter Weise geschützt wurden, mit Chinasäureester bzw. Chinid nach gezielter Substitution von OH-Gruppen und anschließender saurer Hydrolyse der Kondensationsprodukte synthetisch hergestellt worden (vgl. *119, 228* und Seite 84). Auch wurden die NMR-Spektren beschrieben und diskutiert (*54, 291*).

2. Dicaffeyl-D-chinasäuren*

Als erste Dicaffeyl-chinasäure wurde das Cynarin (**34**) aus Artischocken *(Cynara scolymus)* gewonnen (*190*) und später auch in anderen Compositen aufgefunden. Es wirkt wie einige andere Hydroxyzimtsäure-Verbindungen anregend auf die Gallesekretion (vgl. S. 116). Daher werden Artischockenextrakte therapeutisch verwendet.

Später wurde im gleichen Arbeitskreis gefunden, daß in Artischockenblättern ursprünglich 1,3-Di-O-caffeyl-D-chinasäure (**32**) vorliegt, die bei der Aufarbeitung in der Wärme in 1,5-Di-O-caffeyl-D-chinasäure = Cynarin (**34**) umgeestert wird. Der Beweis erfolgte durch oxydative Spaltung von Cynarinderivaten (*191*). Auch wurden die wirkliche 1,4-Di-O-caffeyl-D-chinasäure (**33**), ausgehend von 5-Tetrahydropyranylchinid (*191*), und die 1,3-Di-O-caffeyl-D-chinasäure (**32**) aus 3,4-Carbonyl-dioxy-cinnamylchlorid und 4,5-Isopropylidenchinasäuremethylester synthetisch hergestellt (*228*). Ursprünglich war für Cynarin die Struktur einer 1,4-Di-O-caffeyl-chinasäure angenommen worden (*190*). Eine ähnliche Verbindung, Schmp. 228—231°, Acetylderivat Schmp. 103—105° (als 1,4-Di-O-caffeyl-chinasäure bezeichnet), wurde in Blättern von *Sempervivum ruthenicum* angegeben (*101*).

Im Komplex der „Isochlorogensäure" (vgl. S. 82), Schmp. 170 bis 172°, des Kaffees wurden 3,4-Di-O-caffeyl-D-chinasäure (**35**), 3,5-Di-O-caffeyl-D-chinasäure (**36**), 4,5-Di-O-caffeyl-D-chinasäure (**37**) nachgewiesen (*55*, vgl. auch *223*). Eine vierte Fraktion scheint ein Gemisch der 3'-Methyläther der 3,5-Di-O-caffeyl-chinasäure zu sein. Die drei Verbindungen konnten durch Gegenstromverteilung mit Äthylacetat und 1m-Phosphatpuffer pH 5,5 bzw. 5,7 aufgetrennt werden. Auch wurden die NMR-Spektren diskutiert (*54, 55*). Schließlich unterliegen die drei Isomere ebenso leicht der Umesterung wie die Chlorogensäuren. Aus jeder der einzelnen Verbindungen entsteht in Phosphatpuffer pH 7—7,2 ein Gemisch der drei Isomere in ungefähr gleichem Verhältnis. So sind die drei Stoffe durch Umesterung der leicht synthetisch zu erhaltenen 3,4-Di-O-caffeyl-chinasäure zugänglich (*223*).

3. Ester der *p*-Cumarsäure, Ferulasäure, Sinapinsäure und Gallussäure mit Chinasäure*

Obwohl *p*-Cumarsäure (**17**), Ferulasäure (**19**) und Sinapinsäure (**21**) im Pflanzenreich weit verbreitet vorkommen, haben sich bisher nur relativ wenige Arbeiten mit dem einwandfreien Nachweis der einzelnen

* Siehe Fußnote S. 80.

Acyl-chinasäuren in Pflanzen befaßt, wobei meist 3-O-Ferulyl-D-chinasäure (**43**) aufgefunden wurde. So sind die meisten der nachstehenden Acyl-chinasäuren bisher nur durch Synthese bekannt:

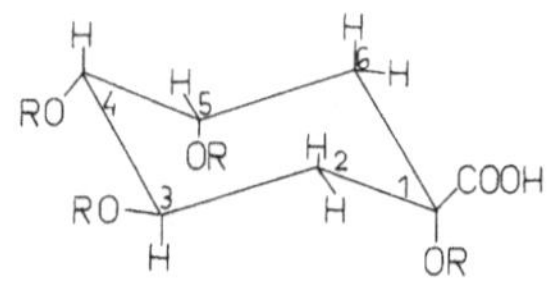

D-Chinasäure R = H

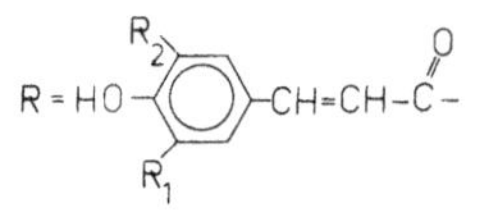

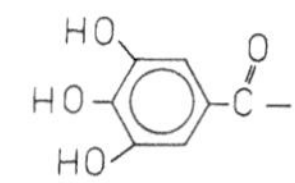

R_1	R_2		
H	H	*p*-Cumaryl-	Gallyl-
OCH_3	H	Ferulyl-	
OCH_3	OCH_3	Sinapyl-	

Schema 3. Weitere Acyl-D-chinasäuren

(**38**) 1-O-*p*-Cumaryl-D-chinasäure, amorph, $\alpha_D = -5°$ (Methanol) (*119*)

(**39**) 3-O-*p*-Cumaryl-D-chinasäure, Schmp. 247—248°, $\alpha_D = -54°$ (Methanol), λ_{max} 250 nm, ε 14 000 (95%ig. Äthanol) (*118*)

(**40**) 4-O-*p*-Cumaryl-D-chinasäure, Schmp. 192—193°, $\alpha_D = -47°$ (Methanol) (*119*)

(**41**) 5-O-*p*-Cumaryl-D-chinasäure, Schmp. 193—194°, $\alpha_D = -6°$ (Methanol) (*119*)

(**42**) 1-O-Ferulyl-D-chinasäure, amorph, $\alpha_D = +12°$ (Methanol) (*313*)

(**43**) 3-O-Ferulyl-D-chinasäure, Schmp. 196—197°, $\alpha_D = -43°$ (Äthanol), λ_{max} 325 nm, ε 19 200 (Äthanol) (*57*)

(**44**) 4-O-Ferulyl-D-chinasäure, Schmp. 195—196° (*313*)

(**45**) 5-O-Ferulyl-D-chinasäure, Schmp. 199—200° (*313*)

(**46**) 3-O-Sinapyl-D-chinasäure, Schmp. 209—210°, $\alpha_D = -43°$ (Äthanol), λ_{max} 327 nm; ε 19 400 (*56*)

(**47**) 3-O-Isoferulyl-D-chinasäure, Schmp. 201—202°, $\alpha_D = -40°$ (Methanol) (*206, 54*), λ_{max} 326 nm, ε 21 000 (Äthanol) (*54*)

(**48**) 1-O-Gallyl-D-chinasäure, amorph, $\alpha_D = -14°$ (Wasser) (*117*)

(**49**) Theogallin = 3-O-Gallyl-D-chinasäure, amorph, $\alpha_D = -41°$ (Wasser) (*117*)

(**50**) 4-O-Gallyl-D-chinasäure, amorph, $\alpha_D = -37°$ (Aceton) (*117*)

(**51**) 5-O-Gallyl-D-chinasäure, amorph, $\alpha_D = -15°$ (Wasser) (*117*)

Die Synthesen erfolgten ähnlich denen der Chlorogensäuren durch Kondensation von Acetoxy-cinnamyl-chloriden bzw. Tri-O-Benzylgallylchlorid mit Chinasäureestern oder Chiniden mit leicht abspaltbaren

Schutzgruppen, die in der gewünschten Position freie OH-Gruppen besaßen.

Außerdem wurden in amorpher Form 4,5-Di-O-gallyl-D-chinasäure, $\alpha_D = -58°$ (Aceton), 3,4,5-Tri-O-gallyl-D-chinasäure, $\alpha_D = -130°$ (Wasser) und Tetra-O-gallyl-D-chinasäure, $\alpha_D = -177°$ (Wasser) hergestellt (*117*). Die Polygallyl-chinasäuren leiten zu natürlichen Gallotanninen über. So enthält Tara-Gallotannin aus den Fruchthülsen von *Caesalpinia spinosa* Chinasäure, die pro Molekül mit 4 bis 5 Molekülen Gallussäure — wie in anderen Gallotanninen zum Teil depsidartig — verbunden ist (*117*).

3-O-Ferulyl-D-chinasäure (**43**) wurde aus Kaffeebohnen (*Coffea* Species) isoliert (*57*) und u. a. in Tabakblättern (*Nicotiana tabacum*) (*311*), Sonnenblumenblättern (*Helianthus annuus*) (*283, 314*) und Tomatenpflanzen (*Lycopersicon esculentum*) (*34*) aufgefunden; vgl. auch (*238*). Auch scheinen 4- und 5-O-Ferulyl-D-chinasäuren in Kaffeebohnen vorzukommen (*55*).

4-O-*p*-Cumaryl-D-chinasäure (**40**) wurde in kristallisierter Form, Schmp. 190—192°, $\alpha_D = —77°$ (Methanol), aus (unreifen) Äpfeln besonders der Sorte „Yarlington Mill" gewonnen (*299*), wobei sich eine Diskrepanz zu $\alpha_D = —47°$ der synthetischen Verbindung ergibt. Ursprünglich war das Vorliegen von 3-O-*p*-Cumaryl-D-chinasäure angenommen worden. 4-O-*p*-Cumaryl-D-chinasäure wurde bei erneuter, sehr schonender Aufarbeitung, bei der eine Umesterung ausgeschlossen werden konnte, isoliert und durch Schmelzpunkt, Kristallform, optische Rotation, NMR-Spektrum, fehlende Reaktion mit Perjodat und Bildung der 3- und 5-Ester durch $NaHCO_3$-Behandlung identifiziert.

Darüber hinaus sind Chinasäureester der beiden Hydroxyzimtsäuren in einer größeren Zahl von Pflanzen durch Hydrolyse in beide Komponenten nachgewiesen worden, ohne daß die Stellung der Acylgruppe geklärt wurde.

3-O-Sinapyl-D-chinasäure (**46**), $\alpha_D = —35°$ (Äthanol), wurde in der Asclepiadacee *Antitoxicum sibiricum* angegeben (*238*). *o*-Cumaryl-D-chinasäure (**52**) wurde chromatographisch nach Fütterungsversuchen von *Cestrum poepigii* mit *trans*-Zimtsäure-[2-^{14}C] nachgewiesen (*181*). Theogallin (**49**) wurde in Teeblättern aufgefunden (*210*); später konnte seine Identität mit 3-O-Gallyl-D-chinasäure bewiesen werden (*255*).

Wie die Caffeyl-D-chinasäuren unterliegen auch diese Hydroxycinnamyl-D-chinasäuren leicht einer Umesterung. Beim Erhitzen in gesättigter $NaHCO_3$-Lösung (30 Minuten 90° C) erfolgt eine Wanderung der *p*-Cumaryl-Gruppe von der Position 1 der Chinasäure zu 3 bzw. $3 \leftrightarrow 4 \leftrightarrow 5$. Unterhalb von 50° C ist die Geschwindigkeit der Umesterung bei den 3-, 4- und 5-O-Acyl-chinasäuren gering (*119*), stärker bei der 1-O-Acyl-chinasäure. Auch 1-O-Ferulyl-chinasäure wird in

30 Minuten bei 100° C in ein Gemisch aus 3-, 4- und 5-O-Ferulyl-chinasäure umgewandelt (*313*). Diese leichte Umesterung erschwert verständlicherweise den Nachweis der ursprünglich in der Pflanze vorliegenden Verbindung, wofür Äpfel (*299*) ein schönes Beispiel sind.

4. Ester der Hydroxyzimtsäuren mit Weinsäure

Nach Acyl-chinasäuren wurde eine Reihe von Acyl-weinsäuren im Pflanzenreich nachgewiesen:

(**53**) Mono-caffeyl-S,S-(−)-weinsäure, Schmp. 146—147°, $\alpha_D = +37°$ (Wasser). Vorkommen: Cichorienblätter (*Cichorium intybus*) (*220*).

(**54**) Cichoriumsäure = Dicaffeyl-S,S-(−)-weinsäure, Schmp. 204—206°, $\alpha_D = +383°$ (Methanol) (*224*). Vorkommen: Cichorienblätter (*Cichorium intybus*) (*224*), Endivien (*Cichorium endivia*) (*306*) und Salat (*Lactuca sativa*) (*79*). SCARPATI und ORIENTE (*224*) synthetisierten neben den beiden optisch aktiven Dicaffeyl-weinsäuren auch die Dicaffeyl-*meso*-weinsäure (**55**), Schmp. 225°.

(**56**) Mono-*p*-cumaryl-R,S-*meso*-weinsäure, Schmp. 198—200°. Vorkommen: Spinatblätter (*Spinacia oleracea*) als Mono-Ammoniumsalz (*188, 266*); die Konstitution der Säure wurde auf Grund der UV-, IR- und NMR-Spektren und durch Synthese aufgeklärt (*267*). Auch wurde Mono-*p*-cumaryl-D-weinsäure (**57**), Schmp. 194—195°, hergestellt (*267*).

Weiterhin wurde über das Vorliegen von Depsiden der 3-Hydroxyzimtsäuren *p*-Cumar-, Ferula- und Kaffeesäure mit Weinsäure in den Blättern und den Beerenhäutchen des Weins (*Vitis vinifera*) als Monohydroxycinnamyl-weinsäuren auf Grund papierchromatographischer Arbeiten (*208*) und in Bohnenblättern (*Phaseolus vulgaris*) mit *meso*-Weinsäure (?) (*298*) berichtet.

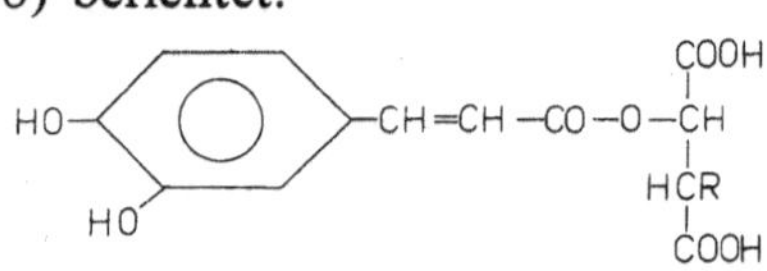

	R	
(**53**)	OH	Caffeyl-(−)-weinsäure
(**54**)	O-Caffeyl	Dicaffeyl-(−)-weinsäure
(**58**)	H	Caffeyl-(−)-äpfelsäure

5. Ester der Hydroxyzimtsäuren mit anderen Hydroxycarbonsäuren

Neben den verschiedenen Acyl-chinasäuren und Acyl-weinsäuren wurden vor allem durch den Arbeitskreis um SCARPATI in Rom einige weitere Ester der Kaffeesäure aufgefunden:

(**58**) Phaseolussäure = Caffeyl-S-(−)-äpfelsäure, $\alpha_D = +28°$ (Wasser). Vorkommen: Bohnenblätter (*Phaseolus vulgaris*) (*226*) und Rotklee (*Trifolium pratense*) (*309*). — In Keimblättern junger Bohnenpflanzen wurden auch die entsprechenden Verbindungen der *p*-Cumarsäure und Ferulasäure vermutet (*274*).

Literaturverzeichnis: SS. 118—132

(**59**) Dattelsäure = 5-O-Caffeyl-shikimisäure*, Schmp. 224—225°; $\alpha_D = -124°$ (62,5%ig. wäßr. Äthanol) aus frischen grünen Datteln *(Phoenix dactylifera)* (*169*). Dattelsäure ist in Palmen verbreitet (*112*) und wurde auch synthetisch dargestellt (*169*). Weiterhin wurde das Vorkommen von Isodattelsäure (4-O-Caffeyl-shikimisäure) und Neodattelsäure (3-O-Caffeyl-shikimisäure) erwähnt (*169*).

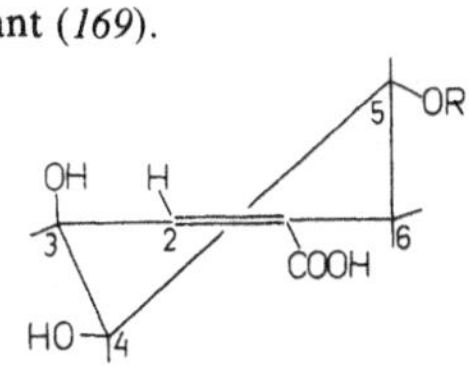

R = H Shikimisäure

R = Caffeyl Dattelsäure

Im Splintholz von *Tsuga canadensis* wurden Caffeyl-, Ferulyl- und *p*-Cumaryl-shikimisäure auf Grund papierchromatographischer Untersuchung der durch saure Hydrolyse gewonnenen Spaltprodukte angegeben (*93*).

(**60**) Rosmarinsäure = 2-O-Caffeyl-3-(3,4-dihydroxyphenyl)-D-milchsäure (Depsid aus Kaffeesäure und α-Hydroxy-hydrokaffeesäure), Schmp. 203—204° (*124, 225, 227, 244*), $\alpha_D = +145°$ (Äthanol) (*225*)

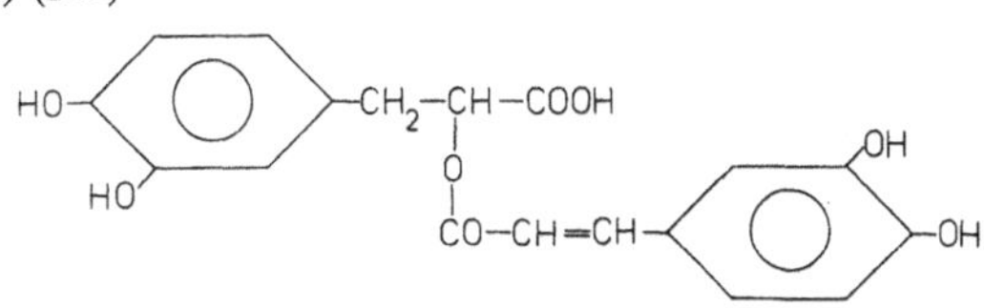

(**60**) Rosmarinsäure

Rosmarinsäure wurde ursprünglich aus Blättern von *Rosmarinus officinalis* (*225*) isoliert und unabhängig davon in weiteren Labiaten aufgefunden (*124*). Sie ist in einigen Pflanzenfamilien recht verbreitet, häufig zusammen mit Chlorogensäure, z. B. in Labiaten, Genus *Thymus* 1—1,5% (*244, 245*) und zahlreichen anderen Labiaten-Species (*124*), in Saniculoideen <0,05—3% (*129*) und in Umbelliferen, Borraginaceen, Hydrophyllaceen und Acanthaceen (*108*).

IV. Verbindungen der Phenolsäuren mit Kohlenhydraten

1. Ester der Hydroxyzimtsäuren mit Mono- und Disacchariden

Zu den weitverbreiteten Pflanzeninhaltsstoffen gehören nach den Estern der Hydroxyzimtsäuren mit Hydroxycarbonsäuren (vor allem D-Chinasäure) auch die Ester mit Zuckern. Sie wurden praktisch erst Anfang der 60er Jahre bekannt. In der Pflanze treten sie frei als

* Die Numerierung entspricht den neuen IUPAC-Regeln.

1-O-Acyl-glykosen auf. In letzter Zeit wurden allerdings auch 6-O-Caffeyl-glucose und 1,6-Di-O-caffeyl-glucose neben 1-O-Caffeyl-glucose in *Thymus*-Species (Labiaten) nachgewiesen; siehe unten.

Dagegen nimmt der Säurerest in acylierten Flavonoiden eine andere Position am Zucker ein, da dessen C_1-Atom glykosidisch verknüpft ist. Diese werden später unter „IX. Mit Hydroxyzimtsäuren oder Hydroxybenzoesäuren acylierte Flavonoidglykoside" behandelt (S. 107).

Am verbreitetsten erscheinen in der Natur die Ester der bekannten Hydroxyzimtsäuren mit Glucose. So wurden in chromatographischen Untersuchungen an 204 willkürlich ausgewählten Pflanzenspecies in 31% 1-O-Caffeyl-glucose und in 5% 1-O-*p*-Cumaryl-glucose aufgefunden; zum Vergleich: 43% enthielten Chlorogensäure und 13% *p*-Cumarylchinasäure (*109*). Auch kommt 1-O-Ferulyl-glucose häufiger vor.

Isoliert bzw. chromatographisch identifiziert wurden bisher folgende Verbindungen:

(61) 1-O-*p*-Cumaryl-β-D-glucose, Schmp. 212°; $\alpha_D = -8°$ (Methanol) (*25, 279;* vgl. auch *195*); Acetylderivat: Schmp. 166—167°, $\alpha_D = -16°$ (Chloroform) (*279*); bei Synthese wurden Schmp. 191—193°, $\alpha_D = +4{,}5°$ (90%ig. wäßr. Dioxan) gefunden (*24*)

(62) 1-O-Ferulyl-β-D-glucose, Schmp. 123—126°, $\alpha_D = -14°$ (Wasser) (*25*)

(63) 1-O-Caffeyl-β-D-glucose, Schmp. 96—98°, $\alpha_D = -47°$ (Äthanol) (*245*)

(64) 1-O-Sinapyl-β-D-glucose

(65) 6-O-Caffeyl-β-D-glucose, amorph, λ_{max} 332 nm, ε 24 500 (Äthanol) (*293*)

(66) 1,6-Di-O-caffeyl-β-D-glucose (Harashengin), Schmp. 127—129°, $\alpha_D = -78°$ (Methanol) (*2, 245*)

1-O-*p*-Cumaryl, 1-O-Ferulyl- und 1-O-Caffeylglucosen wurden u. a. in Blüten von Petunien *(Petunia hybrida)*, Löwenmaul *(Antirrhinum majus)* und *Brunfelsia calycina,* in jungen Rettichblättern *(Raphanus sativus)*, in Beeren, Blättern und Blüten knollentragender *Solanum*-Species (*109*) und in Keimblättern von Bohnenpflänzchen *(Phaseolus vulgaris)* (*274*) chromatographisch aufgefunden. In Blattstielen des Rhabarbers *(Rheum rhaponticum)* wurden 1-O-*p*-Cumaryl-, 1-O-Ferulyl- und 1-O-Sinapyl-glucosen (*27*), in Äpfeln (*167*) und Erdbeeren (*25*) 1-O-*p*-Cumaryl-glucose, in Blättern der Kohlarten *(Brassica oleracea)* (*109*) 1-O-Sinapylglucose, in Tabakblättern 1-O-*p*-Cumaryl- und 1-O-Ferulylglucosen (*311*), in Blüten von Levkojen *(Matthiola incana)* und Stechapfel *(Datura stramonium)* (*25*) 1-O-*p*-Cumaryl- und 1-O-Ferulyl-glucosen angegeben. In Blättern und Rinden von *Populus*-Species ist zumindest 1-O-*p*-Cumarylglucose verbreitet (*195, 199, 279*). In *Thymus*-Arten (Labiaten) wurden 1- und 6-O-Caffeyl- und 1,6-Di-O-caffeylglucosen (*245*) und in einigen Farn-Species 1-O-Caffeylglucose (*31*) nachgewiesen. 1,6-Di-O-caffeylglucose wurde erstmals in *Scrophularia grosshelmii* gefunden (*2*). Kürzlich wurde auch 1-(3-β-D-Glucosyloxy-*p*-cumaryl)-glucose in den Blüten von *Lantana hybrida* aufgefunden (*135a*).

Literaturverzeichnis: SS. 118—132

Über Ester mit anderen Mono- und Disacchariden als Glucose ist bisher nur vereinzelt berichtet worden. So wurden durch chromatographische Arbeiten bisher folgende aufgezeigt:

(67) 1-O-*p*-Cumaryl-rutinose in Blüten von *Lathyrus odoratus* (*109*)

(68) 1-O-*p*-Cumaryl-galactosylglucose in Blättern von *Vaccinium arctostaphylos* (*180*)

(69) 1-O-Caffeyl-gentiobiose in Blüten von *Petunia hybrida* (*109*)

(70) 1-O-Caffeyl-rhamnose in Blüten von *Lantana hybrida* (*136*)
(Rutinose = α-L-Rhamnopyranosyl (1→6)-D-glucose; Gentiobiose = β-D-Glucopyranosyl (1→6)-D-glucose)

Schließlich wurde Pajaneelin = *p*-Cumaryl-D-fructose (**71**), Schmp. 237—239°, α_D = —173° (Äthanol), als bitter schmeckende Substanz aus der Rinde von *Pajaneelia rheedii* (Bignoniaceae) isoliert (*141*). Allerdings ist das vor 30 Jahren angegebene Pajaneelin bis jetzt der einzige Hinweis auf das Vorkommen von Fructose in Zuckerestern der Hydroxyzimtsäuren geblieben.

Von den bisher nicht in Pflanzen aufgefundenen Hydroxycinnamylglucosen stellten Birkofer und Mitarb. (*24*) die isomeren Mono-*p*-cumaryl- und Mono-ferulyl-D-glucopyranosen mit dem Acylrest in 2-, 3-, 4- oder 6-Stellung des Zuckers dar, wobei unterschiedliche Gemische der α- und β-D-Glucosen entstanden. Die Verbindungen wurden papierchromatographisch und papierelektrophoretisch untersucht:

(72) 2-O-*p*-Cumaryl-D-glucose, Schmp. 195—197°, α_D = +29° → 85° (80%ig. wäßr. Dioxan)

(73) 3-O-*p*-Cumaryl-D-glucose, Schmp. 180—181°, α_D = +42° → 53° (80%ig. wäßr. Dioxan)

(74) 4-O-*p*-Cumaryl-D-glucose, Schmp. 208—211°, α_D = +8 → +2° (80%ig. wäßr. Dioxan)

(75) 6-O-*p*-Cumaryl-D-glucose, Schmp. 186—188°, α_D = 32° → +39° (80%ig. wäßr. Dioxan); α_D = +25° (Methanol), λ_{max} 313 nm, ε 25 000 (Äthanol) (*293*)

(76) 2-O-Ferulyl-D-glucose, Schmp. 150—153°

(77) 3-O-Ferulyl-D-glucose, Schmp. 150—153°, α_D = +32° → 40° (80%ig. wäßr. Dioxan)

(78) 4-O-Ferulyl-D-glucose, Schmp. 158—160°, α_D = +15° → 0° (80%ig. wäßr. Dioxan)

(79) 6-O-Ferulyl-D-glucose, Schmp. 198—201°, α_D = +33° → +38° (80%ig. wäßr. Dioxan); Schmp. 218°, α_D = +30° (Methanol), λ_{max} 328°, ε 23 000 (Äthanol) (*293*)

Zur Synthese wurden Zuckerderivate mit leicht abspaltbaren Schutzgruppen, die in der gewünschten Position noch freie OH-Gruppen besaßen, z. B. Acetale und Acetate, mit *p*-O-Acetyl-cinnamyl-chlorid bzw. O-Acetyl-ferulylchlorid verestert. Aus den so gewonnenen Zwischenprodukten wurden die Schutzgruppen unter Erhaltung der neu geknüpften Esterbindung durch saure Hydrolyse (*24*) oder Acetylschutzgruppen durch katalytische Umesterung mit Natriummethylat in wasserfreiem Methanol entfernt (*293*).

Weiterhin sind u. a. durch Synthese bekannt:

1-O-*p*-Cumaryl-D-xylopyranose, Schmp. 227—228° (*17*),

2-O-*p*-Cumaryl-D-xylopyranose, Schmp. 222—224° (*17*),

3-O-*p*-Cumaryl-D-xylopyranose, Schmp. 108—110° (*17*),

5-O-*p*-Cumaryl-D-xylofuranose, Schmp. 147—149° (*21*),

5-O-*p*-Cumaryl-L-rhamnofuranose, Schmp. 87—90°, $\alpha_D = +54° \rightarrow +44°$ (90%ig. wäßr. Dioxan) (*24*),

4-O-(*p*-Cumaryl)-L-rhamnopyranosyl(1→6)-D-glucose, Schmp. 138° (Acetat Schmp. 95°) (*293*),

4-O-(*p*-Cumaryl)-methyl-L-rhamnopyranosid, Schmp. 88—90°, $\alpha_D = -24°$ (Dioxan) (*24*).

2. Weitere Ester der Hydroxyzimtsäuren

Aus Samen des Reis *(Oryza sativa)* wurde eine Verbindung, Schmp. 184—188°, isoliert, die durch *Aspergillus oryzae* zu 2 Molekülen Ferulasäure, 3 Molekülen Glucose, 1 Molekül Glutaminsäure und 2 Atomen Calcium abgebaut wurde. Daneben konnten chromatographisch 6 weitere Verbindungen erhalten werden (*310*), die aus folgenden Komponenten bestehen:

Verbindung	enthaltene Phenolsäuren	enthaltene Zucker und Aminosäuren
III-1	Ferulasäure, Sinapinsäure	Fructose oder Mannose, Alanin, Glutaminsäure, Asparaginsäure
III-2	Ferulasäure	Asparaginsäure
III-4	Ferulasäure	Glucose, Glutaminsäure, Asparaginsäure
III-5	Ferulasäure, Sinapinsäure	Fructose oder Mannose, Glutaminsäure, Asparaginsäure
III-6	Ferulasäure	Glucose, Asparaginsäure
III-7	Ferulasäure, Sinapinsäure	Glucose, Glutaminsäure, Asparaginsäure.

In den Wurzeln von *Polygala senega* wurden Ester aufgefunden, die Derivate der *p*-Cumar-, *p*-Methoxy-zimt-, Ferula-, Sinapin- oder 3,4,5-Trimethoxy-zimtsäure sind. In diesen Estern sind eine oder zwei Typen dieser Säuren mit mehreren Glucose-Molekülen verestert (*51*). In Farnblättern einer Reihe von *Adiantum*-Species und in *Pteridium aquilinum* wurden 6′-O-Sulfatester der 1-O-*p*-Cumaryl-glucose und 1-O-Caffeylglucose aufgefunden (*50*).

Literaturverzeichnis: SS. 118—132

(89) Linocinnamarin = 4-O-β-D-Glucopyranosyl-cumarsäuremethylester, Schmp. 167°, $\alpha_D = -73°$ (Methanol) (*152*).

(90) Linocaffein = 4-O-β-D-Glucopyranosyl-kaffeesäuremethylester, Schmp. 206—208°, $\alpha_D = -88°$ (Methanol) (*151*).

Linocinnamarin und Linocaffein kommen in den Samen des Leins (*Linum usitatissimum*) vor und wurden auch synthetisch dargestellt (*151, 152*).

Kaffeesäure-4-O-glucosid (**88**) wurde nach papierchromatographischer Auftrennung der phenolischen Inhaltsstoffe der Kulturpreiselbeeren (*Vaccinium macrocarpon*) angegeben (*45*).

Synthetisch wurde u. a. Ferulasäure-4-O-β-D-glucosid (**91**), Schmp. 198—199°, $\alpha_D = -37°$ (Äthanol) dargestellt (*105*).

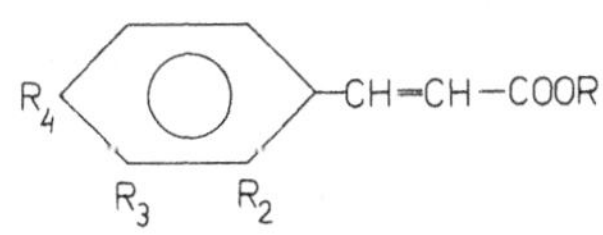

		R_2	R_3	R_4	R
(91)	Ferulasäure-4-O-β-D-glucosid	H	OCH_3	O-glucosid	H
(87)	Kaffeesäure-3-O-β-D-glucosid	H	O-glucosid	OH	H
(88)	Kaffeesäure-4-O-β-D-glucosid	H	OH	O-glucosid	H
(89)	Linocinnamarin	H	H	O-glucosid	CH_3
(90)	Linocaffein	H	OH	O-glucosid	CH_3
(86)	*o*-Cumarsäure-glucosid	O-glucosid	H	H	H

b) Glykoside der Hydroxybenzoesäuren:

Bei den bisher aufgeklärten Verbindungen der Hydroxybenzoesäuren handelt es sich, wenn man von der Gallussäure absieht, hauptsächlich um Glykoside, wobei solche des Salicylsäuremethylesters, Gentisinsäurebenzylesters und der Gentisinsäure bisher zahlenmäßig überwiegen. Als Bestandteil ätherischer Öle ist das Primverosid (β-D-Xylopyranosyl-(1 → 6)-D-glucosid) des Salicylsäuremethylesters (Monotropitin, Gaultherin) (**92**) z. B. in *Gaultheria*-Species und in Ericaceen aufgefunden worden (*280*). Weiterhin ist aus *Viola*-Arten Violutosid, das Vicianosid (= L-Arabinosyl-(1 → 6)-D-glucosid) des Salicylsäuremethylesters (**93**) isoliert worden (*202, 211*). Wegen der Salicylsäure enthaltenden Verbindungen der *Salix*- und *Populus*-Species siehe S. 103.

Von der Gentisinsäure sind bisher folgende Glykoside bekannt:

(94) Gentisinsäure-5-O-β-D-glucopyranosid, Schmp. 124—126°, $\alpha_D = -63°$ (Methanol), λ_{max} 235 nm, ε 4 200 (Methanol). Vorkommen: Stamm von *Prunus yedoensis* (*82*); Blätter der Sonnenblume (*Helianthus annuus*), die unter Bormangel aufgewachsen sind (*283, 312*); auch synthetisch dargestellt (*82*).

(95) Gentisinsäure-5-O-L-rhamnopyranosyl-(1 → 2)-β-D-glucopyranosid, Schmp. 154—156°, $\alpha_D = -10°$ (Methanol). Vorkommen: Stamm von *Prunus yedoensis* (*83*).

(**96**) Trichocarpin = 2-O-β-D-Glucopyranosid des Gentisinsäurebenzylesters, Schmp. 134—136°, $\alpha_D = -46°$ (Methanol). Vorkommen: Blätter und Rinden von *Populus trichocarpa* und *Populus candicans* zu etwa 1—2% bzw. 2—4% (*166*) und Rinde von *Populus balsamifera* zu 0,6% (*200*).

(**97**) Trichosid = 2-Methoxy-5-O-β-D-glucopyranosyl-benzoesäurebenzylester, Schmp. 163—165°. Vorkommen: Rinde von *Populus trichocarpa* (*196*).

Weiterhin kommt Primulaverin (**98**), das Primverosid des 2-Hydroxy-5-methoxy-benzoesäuremethylester, Schmp. 194—196°, $\alpha_D = -64°$ (Wasser), in den unterirdischen Organen von *Primula veris, Primula elatior* und *Primula vulgaris* vor (*281*). In *Primula veris* und *Primula kewensis* tritt Primverin (**99**), das Primverosid des 2-Hydroxy-4-methoxy-benzoesäuremethylester auf (vgl. *281*).

Von anderen verbreitet auftretenden Hydroxybenzoesäuren sind bisher nur Protocatechusäure-3-O-glucosid (**100**) in Blättern (*41, 42*) und Rinde (*41*) von *Pyrus calleryana* und Blättern von *Pyrus koehnei* (*41*) und Syringasäure-4-O-β-D-glucopyranosid (**101**), Schmp. 206—207°, das auch synthetisch gewonnen wurde, im Stamm von *Anodendron affine* (*240*) bekannt geworden, außerdem 2-Hydroxy-3-O-β-D-glucosylbenzoesäure (**102**) in Blättern von *Vinca minor* (*150*) und vermutlich auch in Blättern der Sonnenblume, die unter Bormangel aufgewachsen sind (*283*). Gentisinsäure-2-O-β-D-glucopyranosid, Schmp. 128—129°, $\alpha_D = -68°$ (Methanol), ist nur synthetisch bekannt (*83*). Weiterhin wurde u. a. über die Darstellung der 2- und 5-β-D-Glucopyranoside verschiedener Gentisinsäureester berichtet (*289*).

c) Weitere Angaben über Glykoside der Hydroxybenzoesäuren und Hydroxyzimtsäuren:

In den Nadeln von *Abies sibirica* und *Abies nephrolepsis* wurden die β-Glucoside der *p*-Hydroxybenzoesäure, Vanillinsäure, Protocatechusäure sowie der *p*-Cumarsäure (*172*) und in den Nadeln von *Larix sibirica* die gleichen β-Glucoside mit Ausnahme der Protocatechusäure (*173*) nach präparativer Chromatographie an Polyamid, saurer Hydrolyse (HCl) der einzelnen Fraktionen und Nachweis der Spaltprodukte mittels gas- und dünnschichtchromatographischer Prüfung angegeben. Die Glucoside waren gegen Alkali beständig und wurden durch entsprechende Enzyme hydrolysiert. Für die Nadeln von *Picea ajanensis* wurden die β-Glucoside der *p*-Hydroxybenzoe-, Vanillin-, Protocatechu-, *p*-Cumar- und Ferulasäure und für die Nadeln von *Picea obovata* die gleichen außer dem der Protocatechusäure mitgeteilt (*137*). In den Nadeln von *Larix laricina* wurden die β-Glucoside der Vanillinsäure und *p*-Cumarsäure und das α-Glucosid der *p*-Hydroxybenzoesäure (mit Emulsin nicht spaltbar) auf Grund des chromatographischen

Verhaltens und der Spektren nach Standardmethoden identifiziert (*184*). Auch im Phloem von *Picea*-Species wurden β-D-Glucoside der *p*-Hydroxybenzoesäure, Vanillinsäure, Protocatechusäure, *p*-Cumarsäure und Ferulasäure nachgewiesen (*97*).

V. Ester der Hydroxyzimtsäuren und Hydroxybenzoesäuren mit n-Alkoholen, n-ω-Alkandiolen, Glycerin und ω-Hydroxyfettsäuren

1. Ester mit Methanol und Äthanol

Verbindungen der Hydroxyzimtsäuren wie der meisten Hydroxybenzoesäuren mit kurzkettigen *n*-Alkoholen, wie Methanol und Äthanol, sind bisher in der Natur relativ selten aufgefunden worden. Dies gilt vor allem für Hydroxyzimtsäure-Verbindungen. Wenn man vom Linocinnamarin und Linocaffein absieht (vgl. S. 93), ist u. W. nur das Vorkommen des Ferulasäuremethylesters im Reiskleieöl (*273*), des Kaffeesäureäthylesters in Wurzeln von *Verbesina angustifolia* (*30*) und des *p*-Cumarsäuremethylesters in Blättern von *Grevillea robusta* (*39a*) bekannt.

Im Gegensatz dazu scheinen Methylester der Salicylsäure zum Teil als Glykoside (vgl. S. 93) häufiger in der Natur aufzutreten. So wurde Salicylsäuremethylester meist in gaschromatographischen Arbeiten als Bestandteil von Aromastoffen und ätherischen Ölen (*90*), z. B. von Pfirsichen, Schwarzen Johannisbeeren, Preiselbeeren, Tomaten, Passionsfrüchten (*305*), Kartoffeln, Schwarzem Tee aufgefunden. Neben dem verbreitet auftretenden Salicylsäuremethylester (als Glykosid) ist über das Vorkommen von Salicylsäureäthylester, -benzylester und -geranylester in ätherischen Ölen von Pflanzen berichtet worden (*90*). Wegen der Gentisinsäureester siehe S. 93.

Weiterhin wurde Gallussäuremethylester, Schmp. 201° C, ebenfalls wiederholt nachgewiesen, z. B. in Erdbeeren und Samenhäutchen der Walnüsse. *p*-Hydroxybenzoesäure-methylester soll in Passionsfrüchten vorkommen (*305*). Protocatechusäure-methylester, Schmp. 135—136°, und Phloroglucinsäure-methylester, Schmp. 179—180°, wurden aus den Schalen der Zwiebel *(Allium cepa)* (*122*) und 3,4-Dihydroxy-5-methoxybenzoesäure-methylester, Schmp. 117°, aus dem Kraut von *Myricaria alopecuroides* isoliert (*46*).

Eigene gaschromatographische Untersuchungen an Obst und Gemüse lassen vermuten, daß Methyl- und Äthylester der Hydroxybenzoesäuren häufiger vorkommen dürften, z. B. der *p*-Hydroxybenzoesäure, Salicylsäure und Gallussäure sowie der bekannteren Hydroxyzimtsäuren.

2. Ester mit langkettigen n-Alkoholen und n-ω-Alkandiolen

Hydroxyzimtsäuren wurden häufiger mit langkettigen *n*-Alkoholen verestert aufgefunden, vor allem Ester der Ferulasäure in der Rinde zahlreicher Pinaceen, z. B. mit C_{18}-, C_{20}-, C_{22}-, C_{24}- und C_{26}-*n*-Alkoholen in *Pinus banksiana* (*213*), mit C_{20}-, C_{22}-, C_{23}- und C_{24}-*n*-Alkoholen in *Picea abies* (*186*), mit C_{13} bis C_{22}-*n*-Alkoholen in *Pinus silvestris* (*187*) und mit C_{16}- bis C_{26}-*n*-Alkoholen in Rinde und Kernholz verschiedener russischer Pinaceen (*160*). In höheren Konzentrationen kommen vor allem Ester der Ferulasäure mit C_{22}- und C_{24}-*n*-Alkoholen vor (*160, 186, 213*).

Isoliert wurden 1-O-Ferulyl-eicosanol (C_{20}), Schmp. 61—62°, aus dem Kernholz von *Larix laricina* (*183*) und Schmp. 53—54° aus den Samenölen des Raps (*Brassica napus*) und Leins (*Linum usitatissimum*) (*270*) sowie 1-O-Ferulyl-octacosanol (C_{28}), Schmp. 80—82°, aus den Wurzeln von *Lannea grandis* und dem Kernholz von *Gmelina arborea* (*95*). 1-O-Ferulyl-hexacosanol (C_{26}), Schmp. 70—71°, wurde synthetisch dargestellt und kommt vermutlich im Hafer vor (*68*).

1-O-Caffeyl-hexacosanol (Synthese: Schmp. 118—119°) und 1-O-Caffeyl-octacosanol wurden im Hafer (*Avena sativa*) aufgefunden (*68*). Weiterhin wurde 1-O-Caffeyl-dodecanol, Schmp. 107—108°, synthetisiert (*68*).

Ester der *p*-Cumarsäure mit C_{18}- bis C_{28}-*n*-Alkoholen, hauptsächlich C_{20}- und C_{22}-, wurden im Kraut von *Artemisia campestris* (*287*) und mit C_{18}-, C_{20}-, C_{22}- und C_{24}-*n*-Alkoholen, hauptsächlich C_{22}- und C_{24}-, in der Rinde von *Picea abies* (*186*) nachgewiesen.

Hafer (*Avena sativa*) enthält folgende Diole: 1,26-Di-O-caffeyl-hexacosandiol, Schmp. 140—141° (*66*), 1,26-Di-O-ferulyl-hexacosandiol, Schmp. 98—99° (*67, 68*) und 1-O-Caffeyl-26-O-ferulyl-hexacosandiol, Schmp. 85—86° (*66*), zusammen mit den entsprechenden Derivaten des 1,28-Octacosandiols (*67*). Diese Verbindungen zeigen antioxydative Wirkungen (*65*). Auch kommt vermutlich der Monoferulasäureester des 1,26-Hexacosandiols, Schmp. 84—85°, im Hafer vor (*68*). (1,26-Hexacosandiol = $HOCH_2-(CH_2)_{24}-CH_2OH$).

3. Ester mit Glycerin

Schließlich wurde aus Hafer ein Glycerinester isoliert, der bei der Hydrolyse Kaffeesäure, Ferulasäure, Glycerin und ω-Hydroxyfettsäuren im molaren Verhältnis 2 : 1 : 1 : 1 ergab. Die Hydroxyfettsäuren bestehen aus C_{26}- und C_{28}-Säuren im ungefähren Verhältnis 2 : 1 neben etwa 5% C_{22}-Säure (*69*).

Literaturverzeichnis: SS. 118—132

1,3-Di-O-*p*-cumaryl-2-O-acetyl-glycerin, Schmp. 158—159°, λ_{max} 318 nm, log ε 4,55, wurde im lipophilen Exkret der Winterknospen von *Populus lasiocarpa* aufgefunden und durch chemische (Hydrolyse, Acetylierung, Hydrierung) und spektroskopische Methoden (UV, IR, NMR, MS) identifiziert (*8*).

SCHMIDT und BLANK (*230*) stellten 1-O-Gallyl-glycerin, Schmp. 190°, 2-O-Gallyl-glycerin, Schmp. 179°, 1,2-Di-O-gallyl-glycerin, Schmp. 125°, 1,3-Di-O-gallyl-glycerin, Schmp. 256°, und Tri-O-gallyl-glycerin, Schmp. 155°, her. THEWLIS (*275*) synthetisierte 1-O-Caffeyl-glycerin, Schmp. 163—164°, 1-O-Palmityl-2,3-di-O-caffeyl-glycerin, Schmp. 119—120°, und Tri-O-caffeyl-glycerin, Schmp. 138—140°. Mono- und Tri-O-caffeyl-glycerin besitzen eine ähnliche antioxydative Wirkung, gemessen in Haferöl, wie Kaffeesäuremethylester und -propylester und etwa die halbe Aktivität wie der für Lebensmittel häufig verwendete Gallussäurepropylester (*275*).

4. Verbindungen der Hydroxyzimtsäuren mit Hydroxyfettsäuren

Hafer (*Avena sativa*) enthält ca. 24 antioxydativ wirkende Phenole, wovon bisher Ester der Ferulasäure und der Kaffeesäure mit 1,26- und 1,28-*n*-Alkandiolen und mit Glycerin aufgeklärt wurden; vgl. S. 96. Weiterhin kommen 26-O-Caffeyl-hexacosansäure und in geringerer Menge 28-O-Caffeyl-octacosansäure vor, während 26-O-Ferulyl-hexacosansäure, Schmp. 110—111°, synthetisch dargestellt wurde (*68*). Sie kommt vermutlich auch im Hafer vor. Aus dem Kork der Korkeiche (*Quercus suber*) wurde Phellochrysein, der Ferulasäureester des 22-Hydroxy-docosansäuremethylesters, Schmp. 69°, gewonnen (*100*).

Weiter wurde berichtet (*209*), daß *p*-Cumarsäure und Ferulasäure im Cutin, der Strukturkomponente der Pflanzenkutikula, und im Suberin unterirdischer Organe, kovalent gebunden auftritt. Cutin selbst ist ein Polymeres von Fettsäuren und Hydroxyfettsäuren. Nach Behandlung mit $LiAlH_4$ oder $LiAlD_4$ wurden die Reduktionsprodukte beider Hydroxyzimtsäuren durch gekoppelte Gaschromatographie/Massenspektrometrie identifiziert. So wurden beide Hydroxyzimtsäuren im Cutin der Birne und des Pfirsichs, *p*-Cumarsäure allein im Cutin der Äpfel, der Apfelblätter und der Tomaten nachgewiesen. Die Suberine von Kartoffeln, Süßkartoffeln, Möhren, Speiserüben, Kohlrüben und Roten Rüben enthielten andererseits nur Ferulasäure.

VI. Ester der Hydroxyzimtsäuren und Hydroxybenzoesäuren mit Naturstoffen außer Flavonoiden und Glucosiden

Häufig wurden Phenolsäuren mit Alkoholen, wie Cholin und Sterinen, oder Phenolglucosiden in der Natur aufgefunden. Verständlicherweise ist hier in Anbetracht der Tatsache, daß eine Reihe von Hydroxyzimtsäuren und Hydroxybenzoesäuren in der Natur sehr verbreitet vorkommen, mit einer Vielfalt von Verbindungen zu rechnen. Daher kann trotz unseres Bemühens die eine oder andere Verbindung in der Literatur übersehen worden sein, besonders, wenn der Titel der Arbeit oder das Referat keinen Hinweis auf Phenolsäuren enthielt, z. B. bei Alkaloiden und anderen komplizierter gebauten Naturstoffen.

1. Ester mit Cholin

Sinapin (Sinapyl-cholin), das in freier Form sehr unbeständig, dagegen als Salz beständig ist, ist schon seit langem, vor allem in den fetthaltigen Samen der Cruciferen, bekannt (*236*).

Isoferulyl-cholin, Schmp. 152—153°, wurde im Samen von *Sibara virginica* (Cruciferae) aufgefunden (*92*) und später synthetisch dargestellt (*175*).

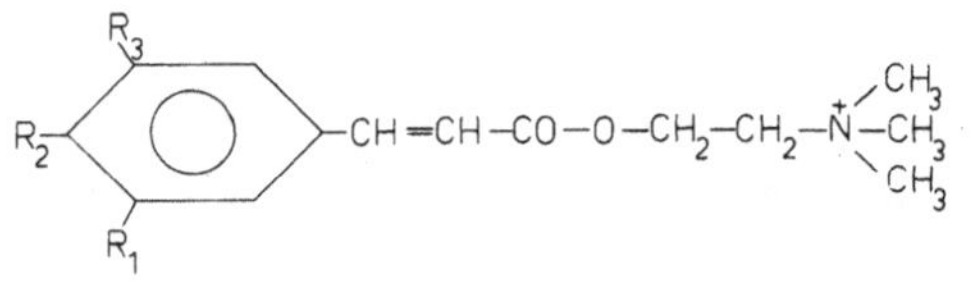

	R_1	R_2	R_3	
(103)	OCH_3	OH	OCH_3	Sinapin (Sinapylcholin)
(104)	OH	OCH_3	H	Isoferulylcholin

2. Phenolsäuren mit Lignin

Im Lignin einiger Pflanzen, besonders von Gräsern (Gramineen), sind Phenolsäuren aufgefunden worden, die mit dem Lignin durch Esterbindung verknüpft sind. Bei den Gramineen (*128, 241*), z. B. Bambus (*Bambusa*-Species) (*241*) und Mais (*Zea mays*) (*157*) handelt es sich hauptsächlich um *p*-Cumarsäure, in geringerem Maße um Ferulasäure. Das Lignin von *Populus tremula* enthält *p*-Hydroxybenzoat-Gruppen (*248*).

3. Ester der Ferulasäure mit Sterinen und Triterpenen

Durch japanische Arbeiten wurden im Reiskleieöl *(Oryza sativa)* neben Ferulasäuremethylester (S. 95) u. a. Ferulyl-β-sitosterin, $C_{39}H_{58}O_4$, Schmp. 131°, $\alpha_D = +11°$ (Chloroform) (*272*), Ferulyl-cycloartanol, $C_{40}H_{58}O_4$, Ferulyl-24-methylencycloartanol, $C_{41}H_{60}O_4$, Schmp. 162—164/193—194°, $\alpha_D = +36°$ (Chloroform) (*189*) und Ferulyl-stigmasterin aufgefunden. Ferulyl-dihydro-β-sitosterin, $C_{39}H_{60}O_4$, Schmp. 156—157°, $\alpha_D = +131°$, Vorkommen: Maiskeimöl (*269*), Ferulyl-cholesterin, $C_{37}H_{54}O_4$, Schmp. 165—166°, und Ferulyl-stigmasterin, $C_{39}H_{56}O_4$, Schmp. 147—148°, wurden u. a. synthetisch gewonnen (*270*). Weizenkeimöl enthält Ferulyl-dihydro-γ-sistosterin, Schmp. 155—156° (*268*). Unlängst wurde über die Massenspektren von Ferulyl-cholesterin, -campesterin, -β-sitosterin und -stigmasterin berichtet (*271*).

In *Melianthus major* (Sapindaceae) wurde 30-Caffeyl-queretarolsäure (3β-Hydroxy-30-caffeyloxy-olean-28-ol-säure), $C_{39}H_{54}O_7$, Schmp. 230—232°, $\alpha_D = +152°$, λ_{max} 333 nm, log ε 4,346, nachgewiesen (*3*).

4. Ester mit Terpenalkoholen

In Wurzeln von *Ferula*-Species sind bisher aufgefunden und auf Grund von Hydrolysen und spektroskopischen Untersuchungen (UV, IR, NMR) identifiziert worden:

D-Chimgin = *p*-Hydroxybenzoyl-D-borneol, $C_{17}H_{22}O_3$, Schmp. 155°, $\alpha_D = +5°$ (Äthanol) (*139, 215*),

L-Chimgin = *p*-Hydroxybenzoyl-L-borneol, $C_{17}H_{22}O_3$, Schmp. 158—159°, $\alpha_D = -60°$ (Äthanol) (*142*),

D-Chimganin = Vanillyl-D-borneol, $C_{18}H_{24}O_4$, Schmp. 85°, $\alpha_D = +5°$ (Äthanol) (*139*),

L-Chimganin = Vanillyl-L-borneol, $C_{18}H_{24}O_4$, Schmp. 85—86°, $\alpha_D = -42$ (Äthanol) (*148*),

Ferungin = *p*-Hydroxybenzoyl-ferutinol, Schmp. 80—81°, $\alpha_D = +83°$ (Äthanol) (*142*),

Ferutinin = 6-O-(*p*-Hydroxybenzoyl)-ferutinol, $C_{22}H_{30}O_4$, Schmp. 120—121°, $\alpha_D = +66°$ (Äthanol) (*217, 218*),

Ferutidin = 6-O-(*p*-Methoxybenzoyl)-ferutinol, $C_{23}H_{32}O_4$, Schmp. 102—103°, $\alpha_D = +104°$ (*217*),

Teferin = 6-O-Vanillyl-ferutinol, $C_{23}H_{32}O_5$, Schmp. 78—80°, $\alpha_D = +87°$ (Chloroform) (*149*),

Ferutin = 6-O-Isovanillyl-ferutinol, $C_{23}H_{32}O_5$, Schmp. 130—131°, $\alpha_D = +100°$ (Äthanol), (*216, 218*),

Ferulin = *p*-Hydroxybenzoyl-angrendiol, $C_{22}H_{30}O_4$, Schmp. 191—192°, $\alpha_D = -90°$ (Chloroform) (*143, 148, 215*),

Federin = *p*-Hydroxybenzoyl-acetyl-angrendiol, $C_{24}H_{32}O_5$, Schmp. 179—180°, $\alpha_D = -91°$ (Äthanol) (*143*),

Chimganidin = Vanillyl-angrendiol, $C_{23}H_{32}O_5$, Schmp. 140—141°, $\alpha_D = -98°$ (Chloroform) (*138, 215*).

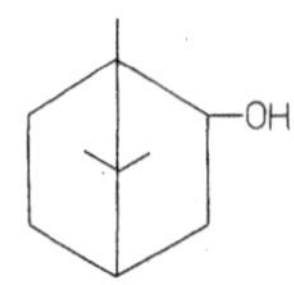

(**105**) Borneol

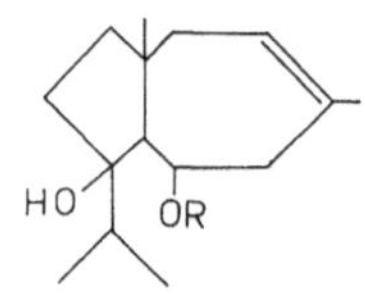

(**106**) Ferutinol = 4,6-Dihydroxydauc-8-en (R = H)

Auch in anderen Pflanzen treten entsprechende Terpenester auf. So enthalten Wurzeln von *Seseli mucronatum* und *Seseli asperulum trans-p*-Cumaryl-D-borneol, $C_{19}H_{24}O_3$, Schmp. 151—152°, $\alpha_D = -28°$ (Äthanol) (*74*). Aus den Blättern von *Verbesina rupestris* wurden *p*-Cumaryl-(−)-borneol, Schmp. 153—154°, und Ferulyl-(−)-borneol als gummiartige Masse isoliert (*33*); sie kommen auch in den Wurzeln von *Verbesina greenmannii* vor (*30*). Die oberirdischen Teile dieser Species enthalten außer *p*-Cumarylborneol zusätzlich Caffeyl-(−)-borneol, das als farbloses Öl, $\alpha_{578} = -20°$ (Chloroform) gewonnen wurde.

Außerdem wurden *p*-Cumaryl-α-verbesinol und *p*-Cumaryl-β-verbesinol, $C_{24}H_{35}O_3$, Schmp. 116—122°, $\alpha_D = +49°$ in der Wurzel von *Verbesina virginica* (*85*) aufgefunden.

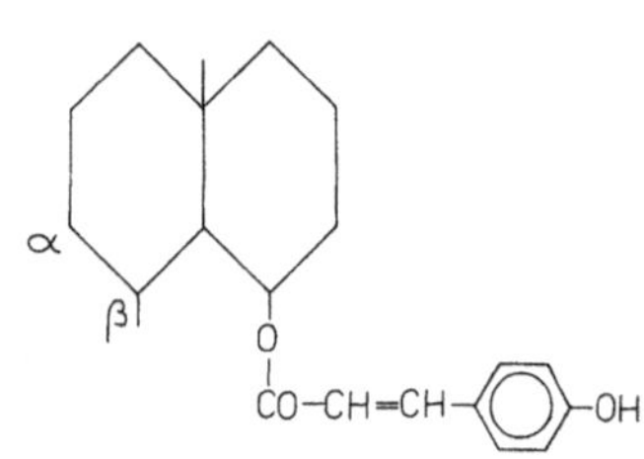

(**108**) *p*-Cumarylverbesinol

5. Ester mit Catalpol

Catalpol ist das Glucosid eines C_9-Iridoids, das bisher nicht frei, sondern stets als Ester der Zimtsäure oder von Phenolsäuren in der Natur aufgefunden worden ist. So wurden 6-(*p*-Hydroxybenzoyl)-catalpol (Catalposid), Schmp. 215—216,5°, $\alpha_D = -184°$ (Methanol), λ_{max} 260 nm, log ε 4,27, aus der Frucht von *Catalpa bignonioides* und anderer *Catalpa*-Species (*28, 29*), 6-Vanillylcatalpol (Picrosid-II), amorph, $\alpha_D =$

−174° (Methanol) aus *Picrorhiza kurrooa* (*295*) und 10-(*p*-Cumaryl)-catalpol (Scutellariosid-II), amorph, $\alpha_D = -80°$ (Äthanol) aus dem frischen Kraut von *Scutellaria altissima* (*296*) isoliert. Zur Konstitutionsbestimmung ist die NMR- und Massenspektroskopie der kristallisierten Peracetate geeignet (*296*). Ester aromatischer Carbonsäuren mit Catalpol dürften in der Natur verbreitet sein.

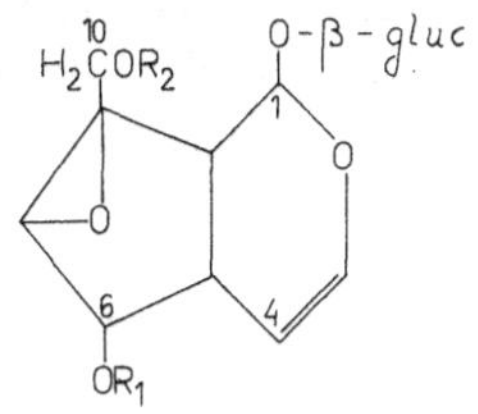

	R_1	R_2	
(**109**)	*p*-Hydroxybenzoyl	H	6-(*p*-Hydroxybenzoyl)-catalpol
(**110**)	Vanillyl	H	6-Vanillylcatalpol
(**111**)	H	*p*-Cumaryl	10-(*p*-Cumaryl)-catalpol

Aus zwei anderen Iridoidglucosiden, Grandiflorosid und Methylgrandiflorosid aus *Anthocleista grandiflora*, wurde bei der Verseifung Kaffeesäure bzw. Ferulasäure abgespalten (*44*).

6. Ester mit weiteren Naturstoffen und einige acylierte Glykoside

An Alkaloiden, die mit Hydroxyzimtsäuren verestert sind, sind uns bekannt:

(**112**) (−)-(*trans-p*-Cumaryl)-lupinin, hochviskoses Öl, das auch synthetisch dargestellt wurde; Vorkommen: Sämlinge von *Lupinus luteus* (*178*).

(**113**) ω-Ferulyloxylupinan, Schmp. 130°, $\alpha_D = -12°$ (Benzol); Vorkommen: Blätter von *Lupinus luteus* (*203*).

An Hydroxycumarinen wurde Secorin (=*p*-Cumaryl-nodakenetin) (**114**), $C_{23}H_{20}O_6$, Schmp. 212—213°, $\alpha_D = -220°$ (Äthylacetat), in *Seseli coronatum* und in den Wurzeln von *Libanotis transcaucasica* aufgefunden (*284*).

Aus Früchten und Blättern von *Vitex agnus castus* wurde Agnusid (**115**), Schmp. 145—146°, $\alpha_D = -92°$ (Äthanol), λ_{max} 258 nm, ε 14.200 (Wasser) erhalten und als O-*p*-Hydroxybenzoyl-aucubin erkannt, wobei die Stellung des Acylrestes nicht festgelegt werden konnte (*304*). Wegen der Formel des Aucubins vgl. (*16*).

Schließlich wurde erstmals Dopaldehyd (3,4-Dihydroxyphenylacetaldehyd) in Form seines (Z)-Enol-(E)-Kaffeesäureesters als sehr labiles gelbes Öl aus den gelben Drüsenfarbstoffen der Labiate *Plectranthus caninus* isoliert und durch Fragmentierung und NMR-Spektroskopie identifiziert (*5*).

Aus *Aconium lindleyi* (ganze Pflanzen) (Fam. Crassulaceae) wurde Lindleyin = 4-(4′-Hydroxyphenyl)-2-butanon-4′-O-β-D-(6″-O-gallyl)-glucosid (**116**), Schmp. 210—211°, $\alpha_D = +14°$ (Pyridin), durch Säulen- und Trockensäulenchromatographie an Kieselgel isoliert. Die Aufklärung erfolgte durch UV-, IR- und NMR-Spektren, Methanolyse und Permethylierung der Trimethylgallyl-Verbindungen sowie saurer und enzymatischer Hydrolyse nach Abspaltung des Gallylrestes (*94*).

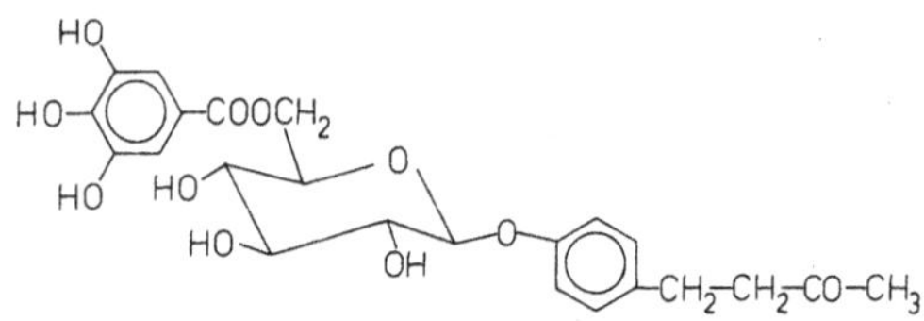

(**116**) Lindleyin

Aus *Portulaca oleracea* wurden mit Ferulasäure acyliertes Betanidin-5-O-cellobiosid und Isobetanidin-5-O-cellobiosid erhalten (*135*).

VII. Ester der Hydroxyzimtsäuren und Hydroxybenzoesäuren mit Phenolglucosiden, Cyclohexanolglucosiden und ähnlichen Naturstoffen

1. Ester mit Arbutin (4-Hydroxyphenyl-β-D-glucopyranosid)

1-O-Gallyl-arbutin, Schmp. 138—140°, $\alpha_D = -20°$, 2-O-Gallyl-arbutin, Schmp. 164—166°, $\alpha_D = -8°$, und 6-O-Gallyl-arbutin, Schmp. 227—229°, $\alpha_D = -31°$ (jeweils wasserfrei in Aceton–Wasser 43 : 1), wurden nach chromatographischer Reinigung und Auftrennung durch Gegenstromverteilung zwischen Essigsäureäthylester und Wasser aus den Blättern von *Arctostaphylos uva-ursi* isoliert; sie kommen ebenfalls in Blättern von *Bergenia crassifolia* und *Bergenia cordifolia* vor (*36*). Die drei Species zählen zu den Ericaceen. 6-O-Gallyl-arbutin wurde auch in den Wurzeln von *Rhodiola coccinea* aufgefunden (*155*). 1- und 6-O-Gallyl-arbutin wurden synthetisch gewonnen und damit wurde deren Konstitution bestätigt (*36*).

Aus Blättern der Preiselbeeren konnte 2-O-Caffeyl-arbutin, Schmp. 165°, isoliert werden (*120*). In Blättern von *Vaccinium arctostaphylos* wurden Caffeyl- und *p*-Cumaryl-arbutin angegeben, wobei die Position der Acylgruppe nicht geklärt ist (*180*).

Literaturverzeichnis: SS. 118—132

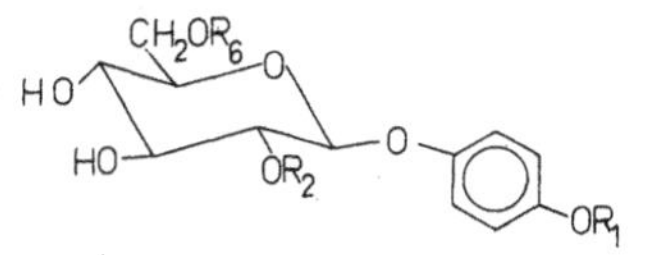

	R_1	R_2	R_6	
(117)	Gallyl	H	H	1-O-Gallylarbutin
(118)	H	Gallyl	H	2-O-Gallylarbutin
(119)	H	H	Gallyl	6-O-Gallylarbutin
(120)	H	Caffeyl	H	2-O-Caffeylarbutin

2. Ester mit Calleryanin (3-Hydroxy-4-O-β-D-glucopyranosyl-benzylalkohol)

p-Hydroxybenzoyl-, Protocatechuyl-, Vanillyl- und Caffeylcalleryanin wurden in Blättern (*42*) und Rinde (*41*) von *Pyrus calleryana* und Blättern von *Prunus lusitanica* (*42*) durch Gegenstromverteilung und anschließend papierchromatographisch aufgetrennt und durch ihre UV-Spektren nach Standardmethoden und Hydrolyse nachgewiesen. Caffeylcalleryanin (**124**) ist weiterhin in der Rinde von *Pyrus communis var. jaspoideae* und *Pyrus ussuriensis,* die anderen 3 Verbindungen in Blättern von *Pyrus koehnei* enthalten (*41*).

(121) *p*-Hydroxybenzoyl-calleryanin, Schmp. 115—118°, $\alpha_D = +37°$ (Methanol), λ_{max} 259 und 275 nm, log ε 4,256 bzw. 4,109 (Methanol); Hexaacetat, Schmp. 124—126°, $\alpha_D = -21°$ (Aceton), wurde in der Wurzelrinde von *Protea cynaroides* zu 5—8% der Trockensubstanz aufgefunden. Es zeigt eine starke antifungale Wirkung gegen *Phytophthora cinnamomi* (*308*).

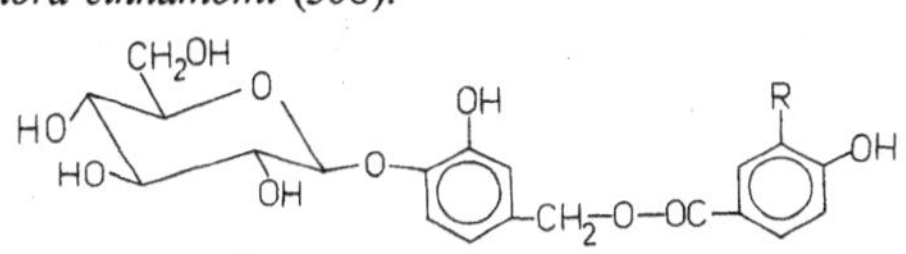

	R	
(121)	H	*p*-Hydroxybenzoyl-calleryanin
(122)	OH	Protocatechuyl-calleryanin
(123)	OCH_3	Vanillyl-calleryanin

3. Ester mit Salicin (2-O-β-D-Glucopyranosyl-benzylalkohol)

(125) ω-Salicylsalicin, Schmp. 166—169°, $\alpha_D = -43°$ (80%ig. Aceton) (*194*), wurde u. a. in Rinden von *Salix purpurea* (*198*) und Blättern von *Populus balsamifera* und *Populus trichocarpa* (*199*) aufgefunden.

(127) ω-Salicylsalicin-2-benzoat, Schmp. 191—193°, $\alpha_D = -31°$ (Aceton), wurde aus Blättern von *Populus grandidentata* und anderen *Populus*-Species (*194*) und aus Rinden von *Salix purpurea* (*198*) isoliert. Es wird leicht in

(128) ω-Salicylsalicin-6-benzoat, Schmp. 180—181°, $\alpha_D = -16°$ (Aceton), umgeestert (*194*).

Weiterhin wurde Populosid (ω-Caffeylsalicin) (**126**), Schmp. 186—188°, $\alpha_D = -38°$, in Rinden von *Populus grandidentata* nachgewiesen und die Konstitution durch Hydrolyse und Massenspektrometrie aufgeklärt (*75*).

Aus Rinden von *Populus trichocarpa* wurde Trichocarposid (**129**) (6-O-*p*-Cumarylsalicin), Schmp. 180—182°, $\alpha_D = -11°$ (80%ig. Aceton), isoliert und auf Grund des IR-Spektrums, der alkalischen und enzymatischen Hydrolyse und der Perjodat-Oxydation identifiziert (*76*).

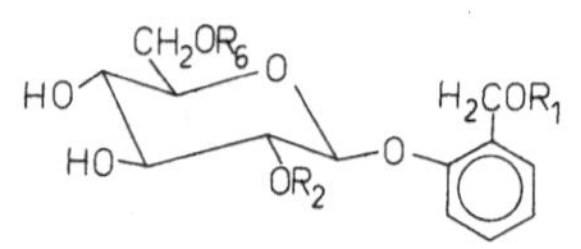

	R_1	R_2	R_6	
(**125**)	Salicyl	H	H	ω-Salicylsalicin
(**126**)	Caffeyl	H	H	Populosid (ω-Caffeylsalicin)
(**127**)	Salicyl	Benzoyl	H	ω-Salicylsalicin-2-benzoat
(**128**)	Salicyl	H	Benzoyl	ω-Salicylsalicin-6-benzoat
(**129**)	H	H	*p*-Cumaryl	6-O-*p*-Cumarylsalicin

4. Ester mit Grandidentin (*cis*-2-Hydroxycyclohexyl-β-D-glucopyranosid)

Aus Rinden von *Populus grandidentata* wurden Grandidentatin (**130**) = 2′-O-*p*-Cumarylgrandidentin, Schmp. 201—202°, $\alpha_D = -66°$ (Methanol (*193*), und Grandidentosid (**131**) = 2′-O-Caffeylgrandidentin, Schmp. 196—197° (*75*), isoliert.

Grandidentatin wurde auch in anderen *Populus*- und *Salix*-Species aufgefunden (*193*).

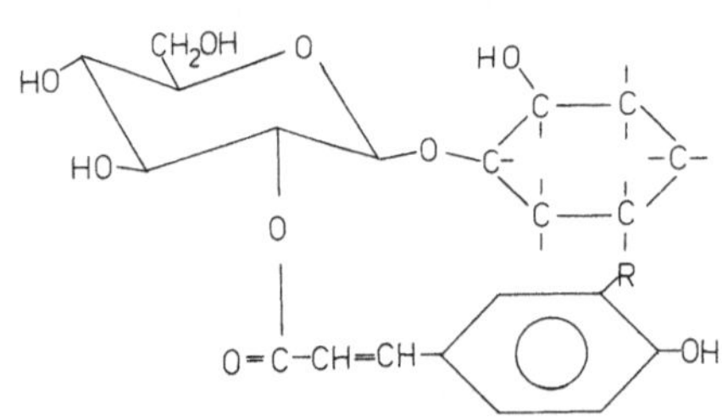

(**130**)	R = H	Grandidentatin
(**131**)	R = OH	Grandidentosid

Aus frischer Juni-Rinde von *Salix purpurea* wurde Purpurein (**132**), Schmp. 112—113°, $\alpha_D = -39°$ (80%ig. wäßr. Aceton), isoliert und durch alkalische und enzymatische Hydrolyse und Massenspektrometrie als Diastereomeres des oben angegebenen Grandidentatins erkannt, wobei die Substitution am 2-OH der Glucose auf Grund von NMR-Spektren festgestellt wurde (*197*).

5. Acylierte 3,4-Dihydroxy-β-phenyläthyl-β-D-glucopyranoside

Aus den Blüten des weißen Flieders *(Syringa vulgaris)* gelang Birkofer u. Mitarb. (*26*), sechs Caffeylglykosen zu isolieren, von denen zwei in kristalliner Form erhalten wurden. Ihre Strukturaufklärung erfolgte durch Permethylierung, saure und alkalische Hydrolyse sowie auf Grund der NMR-, UV- und IR-Spektren:

(**133**) Acteosid = (3,4-Dihydroxy-β-phenyläthyl)-O-α-L-rhamnopyranosyl(1 → 3)-β-D-(4″-O-caffeyl)-glucopyranosid, Schmp. 147—150°, $\alpha_D = -86°$ (Methanol).

(**134**) Neoacteosid = (3,4-Dihydroxy-β-phenyläthyl)-O-α-L-rhamnopyranosyl(1 → 3)-O-(β-D-glucopyranosyl(1 → 6)-β-D-(2″-O-caffeyl)-glucopyranosid, Schmp. 160°.

Bei den übrigen vier Verbindungen konnte papierchromatographisch ermittelt werden, daß es sich um 3,4-Dihydroxy-β-phenyl-äthylglucoside handelt, bei denen die Kaffeesäure ebenfalls esterartig an der Glucose sitzt. Sie unterscheiden sich durch Art und Anzahl der anderen Zuckerkomponenten (*26*).

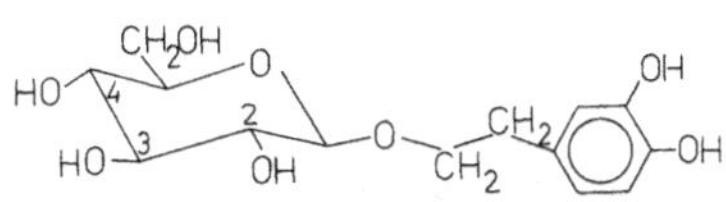

(3,4-Dihydroxy-β-phenyläthyl)-glucosid

Schon früher wurde aus den Wurzeln von *Echinacea angustifolia* Echinacosid, Schmp. um 220°, $\alpha_D = -74°$ (Methanol), isoliert, das aus mit Kaffeesäure acyliertem 3,4-Dihydroxy-β-phenyläthyl-glucosid und je einem Molekül D-Glucose und L-Rhamnose aufgebaut ist (*262*). Schließlich wurde von Harborne (*108*) berichtet, daß „Orobanchin", das aus 3,4-Dihydroxyphenyläthanol, Kaffeesäure und den Zuckern Glucose und Rhamnose besteht, durch chromatographische Prüfung in einer größeren Zahl von Pflanzen verbreitet aufgefunden wurde. Daraus könnte geschlossen werden, daß solche Verbindungen, wie sie Birkofer u. Mitarb. (*26*) in weißem Flieder nachwiesen, im Pflanzenreich häufiger vorkommen.

VIII. Ester der Hydroxyzimtsäuren und Hydroxybenzoesäuren mit Flavonoiden

1. Ester mit Catechinen

Die Blätter des Teestrauchs *(Camellia sinensis)*, die reich an Catechinen sind (bis 30% der Trockensubstanz), enthalten in frischem Zustand hauptsächlich (−)-Epicatechin-3-gallat, (**135**), Schmp. 253°, $\alpha_D = -190°$ (Äthanol), λ_{max} 279 nm, ε 14.000 (Äthanol), und (−)-Epigallocatechin-3-gallat (**136**), Schmp. 218°, $\alpha_D = -185°$ (Äthanol), λ_{max} 275 nm, ε 11.500 (Äthanol) (*288*, vgl. *294*), daneben auch (−)-Epicatechin-3,5-digallat (**137**), $\alpha_D = -9°$, und (−)-Epigallocatechin-3,5-digallat (**138**), $\alpha_D = -13°$ (*59*). Weiterhin wurde (−)-Gallocatechin-gallat (**139**), Schmp. 210—212°, $\alpha_D = -37°$ (Aceton), aus grünem Tee isoliert, das vermutlich als Artefakt bei der Gewinnung der Catechine durch Erhitzen von (−)-Epigallocatechin-gallat entsteht. Es tritt schließlich bei der Epimerisierung von (−)-Epigallocatechingallat auf (*301*). Bei der Teefermentation, d. h. der Herstellung des schwarzen Tees, werden Theaflavine und dabei auch Theaflavin-3-gallat, Theaflavin-3′-gallat und Theaflavin-3,3′-digallat gebildet (*58*).

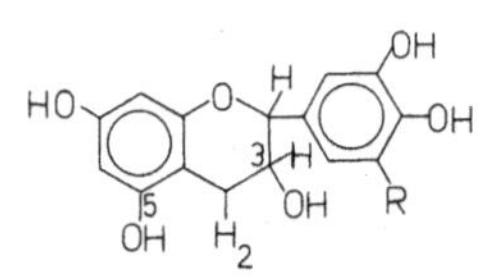

R = H Catechin, Epicatechin
R = OH Gallocatechin, Epigallocatechin

(−)-Epicatechin-3-gallat scheint häufiger in der Natur vorzukommen, z. B. in Rhizomen von *Rheum palmatum* (*80*) und anderen *Rheum*-Species, in Rinde und Blättern von *Hamamelis virginiana* (*81*), in Blättern verschiedener *Rosa*-Species (*192*) und in Blättern einiger *Acacia*-Species (*277*). Letztere enthalten wie *Hamamelis* auch (−)-Epigallocatechin-3-gallat. (+)-Catechin-3-gallat (**140**), $\alpha_D = +56°$ (Äthanol) (vgl. *288*), wurde in den Wurzeln von *Bergenia crassifolia* und *Bergenia cordifolia* nachgewiesen (*116*).

2. Ester mit Flavonen

Während die Phenolsäuren in allen anderen bisher bekannten Fällen als Acylrest mit dem Zucker der Flavon(ol)e verbunden sind (vgl. S. 107), wurde aus dem Arbeitskreis LITVINENKO über Labiaten berichtet, bei denen Hydroxyzimtsäuren mit dem 4′-Hydroxyl von Flavonen verestert sein sollen. So wurden Menthosid = 4′-O-Caffeyl-apigenin-7-rutinosid, Schmp. 271—273°, $\alpha_D = -89°$ (Dimethylformamid), aus Blüten und Blättern (*89*) sowie Piperitosid = 4′-O-Caffeyl-apigenin-7-β-D-glucopyranosid, Schmp. 251—253°, $\alpha_D = -63°$, aus Blüten (*88*) von *Mentha piperita* gewonnen. Das entsprechende 4′-O-*p*-Cumaryl-apigenin-7-β-D-glucopyranosid (Quinquelosid), Schmp. 265—267°, $\alpha_D = -100°$ (Pyridin), wurde im Kraut von *Leonuris quinquelobatus* aufgefunden (*201*). Weiter wurde der papierchromatographische Nachweis von 4′-O-Caffeyl-scutellarein in einer Reihe von *Thymus*-Species angegeben (*165*). (Scutellarein = 5,6,7,4′-Tetrahydroxyflavon; Flavonformeln S. 109.)

IX. Mit Hydroxyzimtsäuren oder Hydroxybenzoesäuren acylierte Flavonoidglykoside

Zahllose Naturstoffe enthalten glykosidisch gebundene Zucker. An die verbliebenen OH-Gruppen der Zucker können Hydroxyzimtsäuren oder Hydroxybenzoesäuren esterartig gebunden werden. Gerade in letzter Zeit sind zunehmend solche Naturstoffe in Pflanzen aufgefunden worden. Es ist zu vermuten, daß diese Bauprinzipe häufiger vorkommen, als es bis vor kurzem den Anschein hatte. So werden immer neue Verbindungen isoliert, deren glykosidisch gebundene Zucker mit Phenolsäuren esterartig verknüpft sind. Daneben kommen andere Säuren als Acylester vor, wie Essigsäure, Malonsäure u. a. Dagegen wurde nur in wenigen Labiaten über eine direkte Veresterung der Phenolsäure mit dem Flavon-Aglykon berichtet; vgl. „Ester mit Flavonen“, S. 106.

Bisher sind neben wenigen acylierten Flavon-C-glykosiden eine beträchtliche Zahl O-acylierter Anthocyanidin- und Flavon(ol)glykoside und vereinzelt andere acylierte Phenol-glykoside in der Natur aufgefunden worden. Letztere sind auf S. 102—105 behandelt. Als Acylreste sind bisher die Hydroxyzimtsäuren *p*-Cumarsäure, Ferulasäure, Isoferulasäure, Kaffeesäure und Sinapinsäure und die Hydroxybenzoesäuren *p*-Hydroxybenzoesäure und Gallussäure bekannt geworden.

Der Acylrest ist bei den Anthocyanidin- und Flavonolglykosiden an den mit der C_3-Position des Flavonoids verknüpften Zucker gebunden (*106*). Durch Säurehydrolyse und H_2O_2-Oxidation nach der Methode von CHANDLER und HARPER (*43*) werden die intakten Zuckerester erhalten. Auch wurde keiner der Zuckerester durch β-Glucosidase gespalten (*106*). Daher kann das reduzierende Ende des Zuckermoleküls nicht an den Acylsubstituenten gebunden sein, wie es bei den natürlich vorkommenden Hydroxyzimtsäurezuckerestern der Fall ist.

Von BIRKOFER u. Mitarb. wurden die verschiedenen *p*-Cumaryl- und Ferulylglucosen synthetisch hergestellt (vgl. S. 89). Durch Vergleich mit diesen Zuckerestern war es möglich, bei einer Reihe von acylierten Flavonoidglykosiden die Verknüpfung des Acylrestes zu beweisen (*18—21*). Eine andere Möglichkeit bietet die NMR-Spektroskopie. Bei einer größeren Zahl der nachgewiesenen Verbindungen ist die Verknüpfung des Acylrestes noch unbekannt.

1. Acylierte Anthocyanidinglykoside

Seit WILLSTÄTTER und MIEG 1915 (*302*) das erste acylierte Anthocyanidinglykosid auffanden, ist eine größere Zahl dieser Verbindungen im Pflanzenreich nachgewiesen worden, vor allem in Vitaceen, Solanaceen und Cruciferen, wobei *p*-Cumarsäure, Ferulasäure und Kaffeesäure, gelegentlich auch Sinapinsäure, als Acylrest auftreten.

So kommen in Weintrauben *(Vitis vinifera)* 3-Glucoside des Malvidins, Delphinidins, Päonidins und Petunidins vor, die mit 1 Mol *p*-Cumarsäure oder Kaffeesäure acyliert sind (*249*). Gefärbte Kartoffelknollen *(Solanum tuberosum)* enthalten 3-Rutinosyl-5-glucosyl-malvidin, -petunidin, -delphinidin, -pelargonidin und -päonidin, acyliert mit *p*-Cumarsäure (*106*); (Rutinose = α-L-Rhamnopyranosyl-(1→6)-D-glucose). In Blüten von *Petunia hybrida* sind 3-Rutinosyl-5-glucosyl-päonidin, -cyanidin, -malvidin, -petunidin und -delphinidin, acyliert mit 1 Mol. *p*-Cumarsäure, teilweise mit 1 Mol. Ferulasäure oder Kaffeesäure, enthalten (*22*). Durch Vergleich mit den entsprechenden synthetisch hergestellten Acylzuckern und durch Abbaureaktionen konnte gezeigt werden, daß zumindest in den Petunidin-, (Schmp. 194—196°), Malvidin- und Päonidinglykosiden (Schmp. 195—198°) die *p*-Cumarsäure mit dem C_4 der Rhamnose verknüpft ist (*20*). Das gleiche ist beim Nasurin der Auberginen *(Solanum melongena)* = 3-O-α-L-(4‴-O-*p*-cumaryl)-rhamnopyranosyl-(1→6)-β-D-glucopyranosyl-5-O-β-D-glucopyranosyl-delphinidin der Fall (*293*).

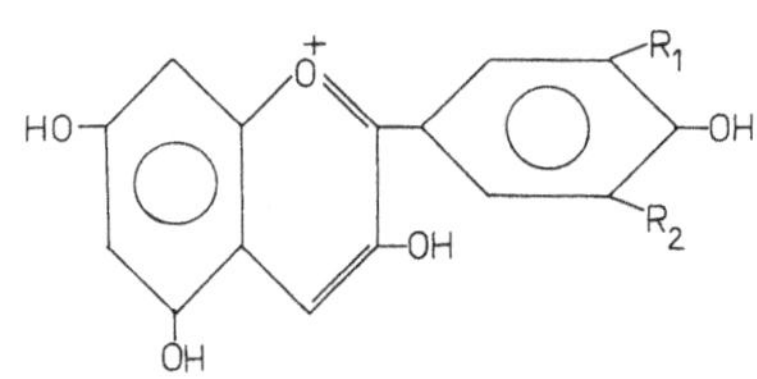

Anthocyanidine

	R_1	R_2	
(141)	H	H	Pelargonidin
(142)	OH	H	Cyanidin
(143)	OH	OH	Delphinidin
(144)	OCH_3	H	Päonidin
(145)	OCH_3	OH	Petunidin
(146)	OCH_3	OCH_3	Malvidin

Von den Farbstoffen der Radieschen *(Raphanus sativus var. sativus)* ist Raphanusin A = 3-O-D-glucopyranosyl-(1 → 2)-β-D-(6″-O-*p*-cumaryl)-glucopyranosyl-5-O-β-D-glucopyranosyl-pelargonidin; Raphanusin B besitzt an Stelle der *p*-Cumarsäure die Ferulasäure (*20*). Blüten von

Salvia splendens enthalten Monardaein = 3-O-β-D-(6''-O-*p*-cumaryl)-glucopyranosyl-5-O-β-D-glucopyranosyl-pelargonidin und Salvianin, bei dem es sich aus Analogiegründen um 3-O-β-D-(6''-O-caffeyl)-glucopyranosyl-5-O-β-D-glucopyranosyl-pelargonidin handeln dürfte (*20*). In *Perilla oscimoides var. crispa* kommt Shisonin = 3-O-β-D-(6''-O-*p*-cumaryl)-glucopyranosyl-5-O-β-D-glucopyranosyl-cyanidin vor (*293*).

Wegen weiterer acylierter Anthocyanidinglykoside sei auf HARBORNE (*106, 111*) verwiesen; wegen der in Rotkohl, Radieschen und Weintrauben auftretenden acylierten Anthocyanidinglykoside siehe bei HERRMANN (*125*).

2. Acylierte Flavon(ol)glykoside

Auf diese Verbindungen ging unlängst WAGNER (*290*) in seinem Beitrag über Flavonoid-Glykoside ein. Als erstes wurde das Tilirosid als ein Kämpferol-3-O-β-D-(*p*-cumaryl)-glucopyranosid, Schmp. 247—256°, $\alpha_D = -63°$ (Methanol) (*106, 130*) aus Blüten von *Tilia argentea* isoliert (*130*). Es wurde auch in *Platanus acerifolia, Fagus silvatica* und einem Farn *Pteridium aquilinum* (*292*) aufgefunden. Kürzlich wurde mitgeteilt, daß im Tilirosid aus den Blüten von *Anaphalis contorta* die *p*-Cumarsäure mit dem Hydroxyl am C_6 der Glucose verestert sein soll. Außerdem wurde Kämpferol-3-O-β-D-(6''-O-*p*-cumaryl)-monoacetylglucosid, Schmp. 178—182°, aufgefunden (*163*).

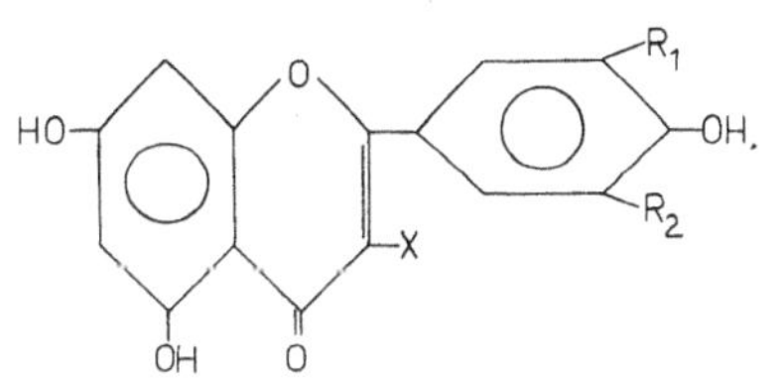

		Flavonole	X = OH	
(147)	Kämpferol	$R_1 = H$	$R_2 = H$
(148)	Quercetin	$R_1 = OH$	$R_2 = H$
(149)	Myricetin	$R_1 = OH$	$R_2 = OH$
(150)	Isorhamnetin	$R_1 = OCH_3$	$R_2 = H$
	Flavone	X = H	
(151)	Apigenin	$R_1 = H$	$R_2 = H$
(152)	Luteolin	$R_1 = OH$	$R_2 = H$
(153)	Chrysoeriol	$R_1 = OCH_3$	$R_2 = H$

Als bisher völlig aufgeklärte Verbindungen sind uns bekannt:

Tribulosid = Kämpferol-3-O-β-D-(6″-O-*p*-cumaryl)-glucopyranosid, Schmp. 224—226°, α_D = −81° (Methanol). Vorkommen: Blätter und Früchte von *Tribulus terrestris* (*15*).

Petunosid = Kämpferol-3-O-β-D-(2‴-O-ferulyl)-glucopyranosyl-(1 → 2)-β-D-glucopyranosid, Schmp. 201—203°, α_D = −64° (90%ig. Äthanol). Vorkommen: *Petunia hybrida* (*18, 23*).

Helichrysosid = Quercetin-3-O-β-D-(6″-O-*p*-cumaryl)-glucopyranosid, Schmp. 181—184°, α_D = −44° (Methanol). Vorkommen: Blütenköpfe von *Helichrysum kraussii* (*39*).

Tellimosid = Quercetin-3-O-β-D-(6″O-gallyl)-glucopyranosid, Schmp. 199—200°. Vorkommen: *Tellima grandiflora* (*48*). — Außerdem sind die entsprechenden Derivate des Kämpferols und Myricetins sowie die drei 3-O-(6″-O-Gallyl)-galactoside des Kämpferols, Quercetins und Myricetins in *Tellima grandiflora* enthalten (*48*).

Quercetin-3-O-β-D-(2″-O-gallyl)-galactopyranosid, Schmp. 180—181°. Vorkommen: *Euphorbia verrucosa* (*182*).

Quercetin-3-O-β-D-(6″-O-gallyl)-galactopyranosid, Schmp. 215—218°. Vorkommen: *Arctostaphylos uva ursi* (*86*) und *Euphorbia platiphyllos* (*182*), Schmp. 202—204°.

Neopetunosid = Quercetin-3-O-β-D-(2‴-O-ferulyl)-glucopyranosyl-(1 → 2)-β-D-glucopyranosid, Schmp. 208°. Vorkommen: Blüten von *Petunia hybrida* (*21*).

Apigenin-7-O-β-D-(4″-O-*p*-cumaryl)-glucopyranosid, Schmp. 239—242°. Vorkommen: Blätter von *Salix alba* (*144*).

Terniflorin = Apigenin-7-O-β-D-(6″-O-*p*-cumaryl)-glucopyranosid, Schmp. 266—267°. Vorkommen: Blüten von *Clematis terniflora var. robusta* (*6, 144*).

Bignonosid = Luteolin-7-O-β-D-(6″-O-*p*-cumaryl)-glucopyranosid, Schmp. 182°. Vorkommen: Blätter von *Catalpa bignonioides* (*19*).

Neobignonosid = Luteolin-7-O-β-D-(6″-O-*p*-hydroxybenzoyl)-glucopyranosid. Vorkommen: Blätter von *Catalpa bignonioides* (*21*).

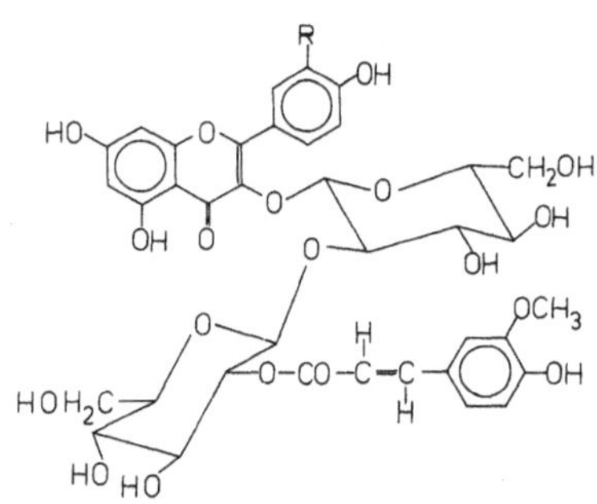

R = H Petunosid
R = OH Neopetunosid
als Beispiele acylierter Diglykoside

Darüber hinaus wurde vor allem in jüngster Zeit eine Reihe acylierter Flavonoide aufgefunden, bei denen der Ort der Veresterung ungeklärt ist. So wurden an acylierten *Flavonolen* z. B. 3-O-(Ferulyl)-glucoside des Kämpferols, Quercetins, Isorhamnetins und Syringetins (3,5,7,4′-Tetrahydroxy-3′,5′-dimethoxyflavon) in Nadeln von *Larix decidua* (*185*) und ein Kämpferol-3-O-(*p*-cumaryl)-arabinosid in den Nadeln von

Larix decidua und *Larix laricina* aufgefunden (*185*), vermutlich auch in *Bryophyllum daigremontianum* (*146*). Blüten von *Narcissus poeticus* enthalten ein 3-O-β-D-(*p*-Hydroxybenzoyl)-glucopyranosyl-kämpferol und ein 3-O-β-D-(*p*-Hydroxybenzoyl)-glucopyranosyl-7-O-β-D-glucopyranosyl-kämpferol (*235*), Blüten von *Tamarix aphylla* ein Quercetin-3-O-β-D-(isoferulyl)-glucuronid, Schmp. 225° (*4*) und *Sedum selskianum* ein Myricetin-3-O-α-L-(gallyl)-rhamnopyranosid (Gallomyricitrin), Schmp. 214—216° (*242*).

Weiterhin wurde aus Knospen und Blättern von *Fagus sylvatica* eine Verbindung, Schmp. 241—246°, isoliert, die ein Kämpferol-3-O-(*p*-cumaryl)-diglucosid darstellen könnte (*72*). In Erbsenpflanzen (*Pisum sativum*) wurden ein Kämpferol-3-O-(*p*-cumaryl)-triglucosid und ein Quercetin-3-O-(*p*-cumaryl)-triglucosid (*84*) und in Blütenblättern von *Helleborus foetidus* ein 3-O-Caffeyl-sophorosyl-7-O-β-D-glucopyranosyl-quercetin (*107*) nachgewiesen. Blätter von *Evonymus europaea* enthalten Bereskletin = 3-O-α-L-rhamnopyranosyl-7-O-β-D-glucopyranosyl-4′-O-β-D-caffeylgalactosyl-quercetin (*237*), bei dem der acylierte Zucker im Gegensatz zu allen bisher bekannten Flavonglykosiden nicht am C_3 des Flavonols sitzen würde.

An *Flavonen* enthalten Blüten von *Tanacetum corymbosum* Chrysoeriol-7-(*p*-cumaryl)-glucosyl-glucuronid (*110*) und bei den Labiaten Blüten von *Scutellaria polyodon* Scutellarein-β-D-glucuronid, acyliert mit Sinapinsäure (*9*), und *Stachys annua* 4′-Methoxyscutellarein-7-O-β-D-mannopyranosyl-(1 → 2)-β-D-glucopyranosid, acyliert mit *p*-Cumarsäure, Schmp. 260—263°, $\alpha_D = -90°$ (*239*). (Scutellarein = 5,6,7,4′-Tetrahydroxyflavon.)

Schließlich sind auch acylierte *C-Glykosylflavone* bekannt geworden. So wurde aus dem Holz von *Vitex lucens* 2″-O-*p*-Hydroxybenzoylvitexin = 8-C-β-D-(2″-O-hydroxybenzoyl)-glucopyranosyl-apigenin, Schmp. 204°, isoliert und durch NMR-Spektren der Hepta-O-methyl- und Hepta-O-acetyl-Derivate aufgeklärt (*133*). 2″-O-*p*-Cumarylvitexin = 8-C-β-D-(2″-O-*p*-Cumaryl)-glucopyranosyl-apigenin, Schmp. 195—198°, wurde aus Samen von *Trigonella foenum graecum* gewonnen und die Konstitution im wesentlichen auf spektroskopischem Wege gesichert (*254*). In Blättern der Melone (*Cucumis melo*) wurden die Kaffeesäureester des 6-C-Diglucosyl-apigenins und 6-C-Diglucosyl-luteolins aufgefunden (*176*).

3. Acylierte Flavanonglykoside

Bisher wurde über Prunin-6″-O-gallat (**154**) = (−)-5,7,4′-Trihydroxyflavanon-7-O-β-D-(6″-O-gallyl)-glucopyranosid, Schmp. 203—206°, $\alpha_D = -97°$ (Pyridin), berichtet, das in den Früchten von *Acacia farnesiana* vorkommt (*246*).

X. Amide der Hydroxyzimtsäuren und Hydroxybenzoesäuren

Phenolsäuren können mit der Aminogruppe anderer Verbindungen unter Bildung von Säureamiden reagieren. In erster Linie ist dabei an Aminosäuren und damit Proteine zu denken. Solche Verbindungen der Hydroxyzimtsäuren sind in letzter Zeit wiederholt nachgewiesen worden. Auch mit anderen Naturstoffen, die Aminogruppen enthalten, wie Putrescin und Glucosamin, sind Amide der Hydroxyzimtsäuren bekannt. Dagegen sind entsprechende Derivate der Hydroxybenzoesäuren bisher noch nicht in der Natur aufgefunden worden, wenn man vom menschlichen und tierischen Harn absieht. Dies schließt nicht aus, daß sie ebenfalls im Pflanzenreich vorkommen können.

1. Verbindungen mit Aminosäuren bzw. Proteinen

N-Ferulyl-glycyl-L-phenylalanin wurde aus den Globulinen der Gerste *(Hordeum vulgare)* von van Sumere u. Mitarb. (*265*) durch partielle Hydrolyse mit 4 *n*-HCl isoliert. Die Verbindung stellt eine Sequenz des Gersteneiweißes dar. In Zusammenhang mit den erwarteten N-Ferulylproteinen der Gerste wurden weiterhin folgende N-Acylaminosäuren nach üblichen Verfahren synthetisch hergestellt: N-Ferulylglycin, Schmp. 215—216°; N-Ferulyl-DL-alanin, Schmp. 157°; N-Ferulyl-DL-valin, amorph; N-Ferulyl-L-leucin, amorph, $\alpha_D = -15°$ (Äthanol) (*204*); N-Ferulylglycyl-DL-phenylalanin, Schmp. 205°; N-Ferulylglycyl-L-phenylalanin, amorph, $\alpha_D = +31°$ (95%ig. Äthanol) (*205*); sowie N-(Ferulyl-2-^{14}C-)-glycin-2-^{3}H (*207*).

Im Rotklee *(Trifolium pratense)* wurden Clovamide (N-Caffeyl-3-(3,4-Dihydroxyphenyl)-L-alanin) aufgefunden, die neben anderen Pflanzenphenolen, wie Phaseolussäure (S. 86), an der Widerstandsfähigkeit gegen Pflanzenkrankheiten beteiligt sind. Neben *trans*-Clovamid, Öl, $\alpha_D = -31°$ (Wasser), λ_{max} 292 bzw. 326 nm, ε 15.500 bzw. 14.700 (Äthanol), wurde *cis*-Clovamid, Öl, $\alpha_D = +29°$ (Wasser), λ_{max} 288 bzw. 320 nm, ε 6.700 bzw. 5.500 (Äthanol) isoliert (*309*).

Schließlich wurden Verbindungen der Hydroxyzimtsäuren mit Proteinen nachgewiesen, z. B. ein Chlorogensäure-haltiges Protein in Sonnenblumensamen *(Helianthus annuus)* (*214*) und ein Kaffeesäure-haltiges in den Umbelliferenfrüchten Anis *(Pimpinella anisum)* und Kümmel *(Carum carvi)* (*35*). Solche Verbindungen scheinen in der Natur verbreitet zu sein (*265*). Auch im Blatt dürfte Chlorogensäure zum Teil an Proteine gebunden sein, was ihre Bestimmung erschwert (*158*).

Literaturverzeichnis: SS. 118—132

Im menschlichen Harn wurden auf Grund papierchromatographischer Untersuchungen u. a. *p*-Cumarylglycin, Ferulylglycin und Vanillylglycin angegeben (*7*).

2. Amide mit Putrescin (1,4-Butandiamin)

(155) *p*-Cumarylputrescin = 1-(*p*-Cumarylamino)-4-aminobutan, Schmp. 182—183°,

(156) Ferulylputrescin = 1-Ferulylamino-4-aminobutan, Schmp. 171—172° (*297*), Schmp. des Hydrochlorids 151—152°,

(157) Caffeylputrescin = 1-Caffeylamino-4-aminobutan, Schmp. des Hydrochlorids 245—250°

wurden aus Callusgewebekulturen von *Nicotiana tabacum* nach säulenchromatographischer Auftrennung und Reinigung kristallisiert gewonnen (*174*). Später wurde der Hauptbestandteil Caffeylputrescin auch in Blüten von *Nicotiana tabacum* aufgefunden (*38*). Es kommt weiterhin in den Samen von *Pentaclethra macrophylla* vor, die als „Ugba"-Samen ein Nahrungsmittel in Nigeria darstellen (*171*). Ferulylputrescin ist in Blättern und Früchten der Grapefruits und Orangen, nicht aber in Tangerinen und Zitronen (Limonen) enthalten, z. B. im Saft von 4 Grapefruitsorten 22—34 mg/l und in Blättern von 4 Grapefruitsorten 100—1 100 mg/kg (*297*). Es zeigt eine hypotensive Wirkung (vgl. *297*).

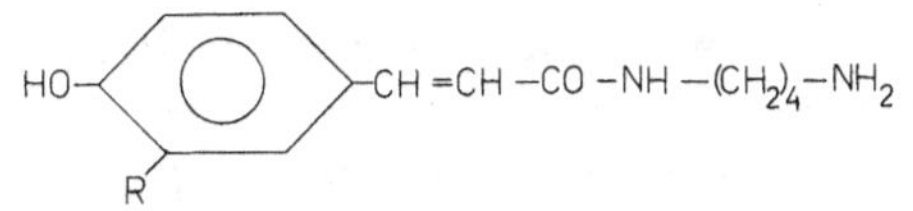

	R	
(155)	H	*p*-Cumarylputrescin
(156)	OCH_3	Ferulylputrescin
(157)	OH	Caffeylputrescin

N-(*p*-Cumaryl)-2-hydroxyputrescin und N-Ferulyl-2-hydroxyputrescin wurden aus Weizenblättern isoliert, die mit *Puccinia graminis f. sp. tritici,* aber auch mit *Pyrenophora triticirepentis* oder virulenten oder avirulenten Species von *Pseudomonas* infiziert worden waren (*219*).

In Blütenknospen von *Nicotiana tabacum* wurden ferner Caffeyl- und Dicaffeyl-spermidin aufgefunden; sie werden bei Nährstoffmangel in den Blättern angereichert. [Spermidin = $H_2N(CH_2)_3$–$NH(CH_2)_4NH_2$] (*70*.)

3. Weitere Hydroxyzimtsäureamide

(158) *p*-Cumaryl-agmatin = 1-(*trans*-*p*-cumarylamino)-4-guanidinobutan, Schmp. des Pikrates 215—218°, kommt in jungen Schößlingen der Gerste *(Hordeum vulgare)* vor. Die Verbindung weist eine fungicide Wirkung gegenüber *Monilinia fructicola* auf (*260*).

(**159**) N-Caffeyl-glucosamin = N-Caffeyl-2-amino-2-desoxy-D-glucose wurde in Tumoren von *Nicotiana*-Species nachgewiesen (*37*). Schon zuvor war über das Vorkommen von Glucosaminestern der *p*-Cumarsäure, Ferulasäure und Kaffeesäure in Gewebekulturen von *Nicotiana tabacum* berichtet worden (*14*).

(**160**) 1-(*p*-Cumarylamino)-2-(4-hydroxyphenyl)-äthan, Schmp. 252—255° (Diacetat Schmp. 175—177°), wurde aus der Rinde von *Evodia belahe* (Rutaceae) isoliert (*212*).

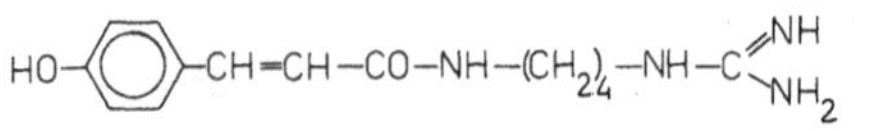

(**158**) *p*-Cumaryl-agmatin

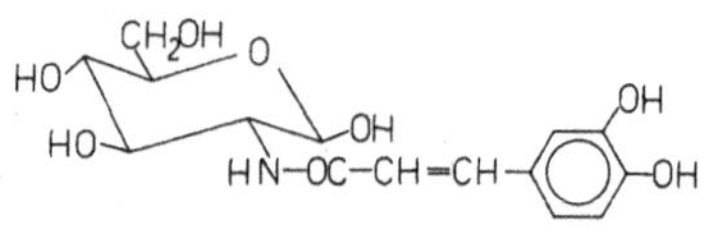

(**159**) N-Caffeylglucosamin

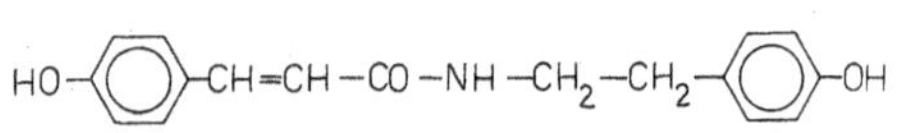

(**160**) 1-*p*-Cumarylamino-2-(4-hydroxyphenyl)-äthan

XI. Isolierung, qualitativer Nachweis und quantitative Bestimmung

Zur Isolierung bzw. Anreicherung oder Auftrennung der Verbindungen der Hydroxyzimtsäuren und Hydroxybenzoesäuren aus pflanzlichem Material können die Säulenchromatographie (Cellulose, Kieselgel, Polyamid oder andere Materialien), die Gegenstromverteilung (z. B. *55, 57, 119*) und die präparative Papierchromatographie mit Erfolg verwendet werden. Zur Vorreinigung eignen sich häufig Polyamidsäulen. Auch wird gelegentlich die Bleifällung benutzt (*224, 309*). Geeignete Vorschriften sollten gegebenenfalls den Arbeiten entnommen werden, die sich mit der Isolierung der interessierenden Verbindungen befassen (siehe Kapitel III bis X).

Zum qualitativen Nachweis ist die Dünnschichtchromatographie, z. B. auf Celluloseplatten mit unterschiedlichen Fließmitteln, günstig. Als Farbreagentien stehen u. a. alkalische Diazobenzolsulfonsäurelösung (100 mg in 50 ml 0,5 nKOH) und andere Diazo-Verbindungen zur Unterscheidung einwertiger Phenole; 1%ige methanolische $FeCl_3$-Lösung; Lösung von 1 g Natriumnitrit und 2,5 g Essigsäure in 100 ml Methanol + 0,5 n methanol. KOH zum Nachweis von 1,2-Dihydroxybenzol-Verbindungen (meist rot); Folin-Reagens (Phosphorwolframsäure) + 0,5 n methanol. KOH zur Verfügung.

Die Hydroxyzimtsäuren und ihre Ester fluoreszieren unter der UV-Lampe meist deutlich blau bis hellblau. Beim Bedampfen mit Ammoniak

Literaturverzeichnis: SS. 118—132

färben sich z. B. die Ester der Kaffeesäure grün, während die freie Säure blau bleibt.

Bei Verwendung wäßriger Fließmittel tritt bei den Hydroxyzimtsäuren und ihren Verbindungen *cis-trans*-Isomerie auf. Beide Isomere haben zum Teil recht unterschiedliche Rf-Werte (vgl. S. 77).

Zur Analyse wird ebenfalls die Gaschromatographie, gegebenenfalls gekoppelt mit der Massenspektrometrie, herangezogen (z. B. *177*).

Über die quantitative Bestimmung von Phenolsäuren ist in der Vergangenheit wiederholt berichtet worden. Insbesondere wurden Methoden zur quantitativen Bestimmung der Chlorogensäure im Kaffee, in dem Chlorogensäuren mit etwa 5—8% einen Hauptbestandteil darstellen, seit etwa 50 Jahren immer wieder beschrieben. Als großer Fortschritt erwies sich die Ausarbeitung spektralphotometrischer Verfahren in 1948 und später durch Häusermann und Brandenberger (*102, 247*). Auch sei auf ein Verfahren verwiesen, das mittels Metaperjodat- und Molybdat-Reagentien gestattet, Chlorogensäuren und Ferulyl-chinasäuren im Kaffee zu bestimmen (*47*). Diese Methoden, die für Gehalte im Prozentbereich ausgearbeitet wurden, eignen sich verständlicherweise nicht ohne weiteres zur exakten Bestimmung wesentlich geringerer Konzentrationen in anderen Objekten. In Obst, Gemüse und Kartoffeln wurde ab den 60er Jahren der Gehalt an Phenolsäuren und Chlorogensäuren meist nach chromatographischer Vorreinigung eines Methanol- oder Aceton-Extraktes durch spektralphotometrische Messung bestimmt.

Je geringer die Konzentrationen sind, desto aufwendiger werden die Reinigungsverfahren und desto größer auch die Fehlermöglichkeiten. Soweit relativ niedrige Konzentrationen in unterschiedlichen Pflanzen und Pflanzenteilen bestimmt werden sollen, sind fast alle bekannten Methoden entweder relativ aufwendig oder mehr oder weniger unzuverlässig. Die in letzter Zeit für Hydroxyzimtsäuren und Chlorogensäuren empfohlene Absorptionsmessung (ca. 325 nm) erfordert eine sehr gute Reinigung der Extrakte und führt damit leicht zu erheblichen Verlusten. Ist dagegen die Reinigung unbefriedigend, wie bei den meisten bisher mitgeteilten und verwendeten Verfahren, können beträchtlich zu hohe Werte die Folge sein. Schließlich ist im Falle der Chlorogensäure-Bestimmungen zu berücksichtigen, daß nicht nur Chlorogensäuren, sondern in der Tat recht unterschiedliche Kaffeesäure-Verbindungen in den einzelnen Pflanzen vorkommen, wie dieser Beitrag zeigt.

Häufig stößt es auf beträchtliche Schwierigkeiten, insbesondere bei geringen Konzentrationen, die natürlich vorliegenden Verbindungen quantitativ zu bestimmen. Hier erscheint es uns günstig, aus den Naturstoffen die zugrunde liegenden Phenolsäuren freizusetzen und diese direkt zu bestimmen. Dabei können Ester z. B. mit Bariumhydroxid (1%) und Glykoside mit Schwefelsäure (1%) oder beide mit geeigneten Enzymen aufgespalten werden. Eine Methode zur Bestimmung von in geringen Konzentrationen vorliegenden Hydroxyzimtsäuren wie Hydroxybenzoesäuren teilten SCHMIDTLEIN und HERRMANN (*233*) mit, bei der die Vorreinigung der hydrolysierten Extrakte über eine Polyamidsäule und die anschließende Chromatographie auf Kieselgel-G-Platten erfolgte. Schließlich wurden die Substanzen abgelöst und im Maximum der Absorption gemessen. Zum Schutze der Kaffeesäure wurde $NaBH_4$ zugesetzt. Ebenso können Phenolsäuren in ausreichend vorgereinigten Extrakten gaschromatographisch bestimmt werden (z. B. *40, 168*). Schließlich wurde auch die Hochdruck-Flüssigkeits-Chromatographie empfohlen (*307*).

Zur quantitativen Bestimmung der Chlorogensäure und anderer Kaffeesäure-Verbindungen eignet sich die Dünnschichtchromatographie auf Cellulose nach geeigneter Vorreinigung, z. B. über eine Polyamidsäule, bzw. die Gaschromatographie (z. B. *78*). Eine Methode der Direktauswertung von Cellulose-Dünnschichtchromatogrammen teilten HANEFELD und HERRMANN (*103, 104*) mit.

Im Arbeitskreis von WENDER wurde die Chlorogensäure gaschromatographisch als Trimethylderivat nach Vorreinigung über eine Polyvinylpyrrolidonsäule quantitativ bestimmt (*303*). Zur Erfassung aller Chlorogensäuren diente eine spektralphotometrische Methode nach papierchromatographischer Auftrennung gereinigter Extrakte (*153*).

XII. Biologische Eigenschaften und Bedeutung

Über die pharmakologischen Wirkungen der Hydroxyzimtsäuren ist bekannt, daß die Chlorogensäure bzw. die Chlorogensäuren neben dem Coffein, dem früher allein die anregende Wirkung zugeschrieben wurde, an der nach Kaffeegenuß auftretenden zentralen Erregbarkeit mitbeteiligt sein dürften. So bewirkte Chlorogensäure nach peroraler Zufuhr in Dosen von 3,75—15 mg/kg eine Steigerung der zentralen Erregbarkeit der Ratte, die verglichen mit Coffein etwa $^1/_6$ so stark war (*62*). Am Magen-Darm-Kanal zeigte Chlorogensäure ebenfalls eine anregende Wirkung (Zunahme der Darmperistaltik). Beim Menschen konnte nach oralen Gaben von 250 mg Chlorogensäure oder Kaffeesäure eine deutliche Zunahme der Magensekretion mit Steigerung der Salzsäurebildung nachgewiesen werden, wobei die Pulsfrequenz vermindert wurde (*63*).

Die Galleausscheidung wird durch eine Reihe von Phenolsäuren angeregt, wobei Ferulasäure die stärkste Wirkung zeigte. Die Wirkung nahm in der Reihenfolge Kaffeesäure, Chlorogensäure, Cynarin, Neochlorogensäure ab (*64*). Wegen der physiologischen und pharmakologischen Wirkungen der Chlorogensäure siehe auch CZOK (*61*).

Phenole mit OH-Gruppen in *ortho*-Stellung (Kaffeesäure, Chlorogensäure, Protocatechusäure und Gallussäure) zeigten eine deutliche Antithiaminwirkung; Phenole mit einer OH-Gruppe in *para*-Stellung eine mittlere Wirkung, z. B. Syringasäure eine schwächere Wirkung (*250*). Dagegen wurden Vermutungen, daß Chlorogensäure z. B. in Kaffeefabriken durch Allergie Erkrankungen der Haut hervorrufen könnte, nicht bestätigt (*159*).

Gut bekannt ist die antioxydative Wirkung der Hydroxyzimtsäuren, besonders Kaffeesäure, Chlorogensäure und Rosmarinsäure, und der Hydroxybenzoesäuren sowie deren Ester (*127*). Hierüber gibt es eine Reihe von Patenten. So wirken z. B. Rosmarinsäure in Gewürzkräutern der Labiaten, Chlorogensäure in Kartoffeln (Schalenschicht) und Hydroxyzimtsäure-Verbindungen im Hafer (vgl. S. 96 und 97) als natürliche Antioxydantien (*127*). Von erheblicher praktischer Bedeutung als Antioxydantien für die Lebensmittelindustrie sind einige Gallussäureester, z. B. Propyl-, Octyl- und Dodecyl-gallat. Aber auch von anderen Hydroxybenzoesäuren ist über eine antioxydative Wirkung berichtet worden.

Von Hydroxybenzoesäuren ist eine zum Teil beträchtliche antimikrobielle Wirkung bekannt. So werden *p*-Hydroxybenzoesäure-äthylester und -propylester als Konservierungsstoffe für Lebensmittel verwendet („PHB"-Ester); sie zählen in Deutschland zu den wenigen gesetzlich zugelassenen Konservierungsstoffen.

Hydroxyzimt- und Hydroxybenzoesäuren können als Wachstumsregulatoren wirken. So wurde gefunden, daß z. B. *p*-Cumarsäure, Ferulasäure, Kaffeesäure und Chlorogensäure, Salicylsäure, *p*-Hydroxybenzoesäure und Gallussäure Wachstumsprozesse inhibieren können (*98, 147*). In Gerste verzögert Ferulasäure die Keimung (*264*).

Schließlich ist eine Fülle von Arbeiten über eine Bedeutung von Phenolsäuren, besonders der Chlorogensäure, für die Widerstandsfähigkeit der Kulturpflanzen gegen pathogene Mikroben berichtet worden; z. B. über die Phytophthora der Kartoffel in Verbindung mit der enthaltenen Chlorogensäure. Ein gut untersuchtes Beispiel ist der beträchtliche Gehalt an Protocatechusäure in den trockenen äußeren Schalen gefärbter Zwiebeln. Über die Bedeutung der Phenolsäuren in der Phytopathologie ist in einer Reihe von Übersichtsarbeiten (z. B. *77, 154, 156, 261*) referiert worden, auf die verwiesen sei.

Literaturverzeichnis

1. ABE, H., and S. MARUMO: Identification of auxin-active substances as ethyl chlorogenate and indolyl-3-acetic acid in immature seeds of *Helianthus annuus*. Agric. Biol. Chem. (Tokyo) **36,** 42—46 (1972).
2. ACHMEDOV, S. G., und V. I. LITVINENKO: (Harashengin, eine neue phenolische Verbindung aus *Scrophularia grosshelmii.*) Ber. Akad. Wiss. Azerb. SSR **24,** Nr. 9, 89—93 (1968).
3. AGARWAL, J. S., and R. P. RASTOGI: Queretaroic (30) caffeate and other constituents of *Melianthus major*. Phytochemistry **15,** 430—431 (1976).
4. EL ANSARI, M. A., M. A. M. NAWWAR, A. EL DEIN, A. EL SHERBEINY, and H. I. EL SISSI: A sulphated kaempferol 7,4′-dimethylether and a quercetin isoferulylglucuronide from the flowers of *Tamarix aphylla*. Phytochemistry **15,** 231—232 (1976).
5. ARIHARA, S., P. RÜEDI und C. H. EUGSTER: Dopaldehyd: erstmalige Isolierung aus einer Pflanze in Form seines (Z)-Enol-(E)-Kaffeesäureesters. Helv. Chim. Acta **58,** 447—453 (1975).
6. ARITOMI, M.: Terniflorin, a new flavonoid compound in flowers of *Clematis terniflora var. robusta*. Chem. Pharm. Bull. (Tokyo) **11,** 1225—1228 (1963). Chem. Abstr. **60,** 4457h (1964).
7. ARMSTRONG, M. D., K. N. F. SHAW, and P. E. WALL: The phenolic acids of human urine. Paper chromatography of phenolic acids. J. biol. Chem. **218,** 293—303 (1956).
8. ASAKAWA, Y., and E. WOLLENWEBER: A novel phenolic acid derivative from buds of *Populus lasiocarpa*. Phytochemistry **15,** 811—812 (1976).
9. BANDYUKOVA, V. A., and A. BOIKOVA: An investigation of the flavonoids of *Scutellaria polyodon* L. Khim. Prir. Soedin. - Chem. nat. compounds **5,** 596—597 (russ.) 513 (engl.) (1969).
10. BATE-SMITH, E. C.: The commoner phenolic constituents of plants and their systematic distribution. Sci. Proc. Royal Dublin Soc. **27,** 165—176 (1956).
11. — Chromatography and systematic distribution of ellagic acid. Chem. and Ind. (London), B.I.F. Review **1956,** R32—R33.
12. — The phenolic constituents of plants and their taxonomic significance. J. Linnean Soc. (Bot.) **58,** 95—173 (1962).
13. BERGER, W.-G., und K. HERRMANN: Catechine und deren Abbauprodukte in Weintrauben, Trestern, Essigweinen und daraus hergestellten Essigen. III. Nachweis von Phloroglucin, Phloroglucincarbonsäure und Methylphloroglucin in Essigweinen. Modellversuche zur Entstehung dieser Stoffe in Trauben bzw. Trestern. Nachweis der Verfälschung von Essigweinen durch Trester- und Nachweine. Z. Lebensm.-Unters.-Forsch. **147,** 1—8 (1971).
14. BERGMANN, L., W. THIES und K. ERDELSKY: Das Vorkommen von Glucosaminestern der Hydroxyzimtsäuren in Gewebekulturen von *Nicotiana tabacum*. Z. Naturforsch. **20b,** 1297—1301 (1965).
15. BHUTANI, S. P., S. S. CHIBBER, and T. R. SESHADRI: Flavonoids of the fruits and leaves of *Tribulus terrestris:* Constitution of tribuloside. Phytochemistry **8,** 299—303 (1969).
16. BIRCH, A. J., J. GRIMSHAW, and H. R. JUNEJA: Aucubin. J. Chem. Soc. (London) **1961,** 5194—5198.
17. BIRKOFER, L., und G. FALSONE: D-Xyloseester der *p*-Cumarsäure. Liebigs Ann. Chem. **752,** 175—181 (1971).
18. BIRKOFER, L., und C. KAISER: Petunosid, ein acyliertes Flavon aus Petunien. Z. Naturforsch. **17b,** 279—280 (1962).
19. BIRKOFER, L., C. KAISER und F. BECKER: Bignonosid, ein acyliertes Flavon aus *Catalpa bignonioides*. Z. Naturforsch. **20b,** 923 (1965).

20. Birkofer, L., C. Kaiser, M. Donike und W. Koch: Acylierte Anthocyane. III. Konstitution von Acyl-anthocyanen. Z. Naturforsch. **20b**, 424—428 (1965).

21. Birkofer, L., C. Kaiser, B. Hillges und F. Becker: NMR-spektroskopische Untersuchungen von Acyl-glykosen. Liebigs Ann. Chem. **725**, 196—202 (1969).

22. Birkofer, L., C. Kaiser, W. Koch, M. Donike und D. Wolf: Acylierte Anthocyane. II. *cis-trans*-Isomerie bei Acyl-anthocyanen. Z. Naturforsch. **18b**, 631—634 (1963).

23. Birkofer, L., C. Kaiser und H. Kosmol: Konstitution des Petunosids. Z. Naturforsch. **20b**, 605—606 (1965).

24. Birkofer, L., C. Kaiser, H. Kosmol, G. Romussi, M. Donike und G. Michaelis: D-Glucose- und L-Rhamnoseester der *p*-Cumar- und Ferulasäure. Liebigs Ann. Chem. **699**, 223—231 (1966).

25. Birkofer, L., C. Kaiser, W. Nouvertné und U. Thomas: Natürlich vorkommende Zuckerester von Phenolcarbonsäuren. Z. Naturforsch. **16b**, 249—251 (1961).

26. Birkofer, L., C. Kaiser und U. Thomas: Acteosid und Neoacteosid; Zuckerester aus *Syringa vulgaris* (L.). Z. Naturforsch. **23b**, 1051—1058 (1968).

27. Blundstone, H. A. W.: Hydroxycinnamic acid derivatives of *Rheum rhaponticum*. Phytochemistry **9**, 1677—1679 (1970).

28. Bobbitt, J. M., H. Schmid, and T. B. Africa: Catalpa glycosides. I. The characterization of catalposide. J. org. Chem. **26**, 3090—3094 (1961).

29. Bobbitt, J. M., D. W. Spiggle, S. Mahboob, W. von Philipsborn, and H. Schmid: Catalpa glycosides. II. The structure of catalposide. Tetrahedron Letters **1962**, 321—329.

30. Bohlmann, F., und C. Zdero: Neue Terpen-Inhaltsstoffe aus Verbesina-Arten. Phytochemistry **15**, 1310—1311 (1976).

31. Bohm, B. A.: Phenolic compounds in ferns. III. An examination of some ferns for caffeic acid derivatives. Phytochemistry **7**, 1825—1830 (1968).

32. Bohm, B. A., and R. M. Tryon: Phenolic compounds in ferns. I. A survey of some ferns for cinnamic and benzoic acid derivatives. Canad. J. Botany **45**, 585—593 (1967).

33. Box, V. G. S., and W. R. Chan: Terpenoids from *Verbesina rupestris*. Phytochemistry **14**, 583 (1975).

34. van Bragt, J., L. M. Rohrbaugh, and S. H. Wender: Identification of neochlorogenic and 3-O-feruloylquinic acids in tomato plants. Phytochemistry **4**, 977—979 (1965).

35. Brieskorn, C. H., und A. Mosandl: Ein Kaffeesäure enthaltendes Proteid aus Umbelliferenfrüchten. Tetrahedron Letters **1970**, 109—111.

36. Britton, G., and E. Haslam: Gallotannins. XII. Phenolic constituents of *Arctostaphylos uva-ursi* L. *Spreng*. J. Chem. Soc. (London) **1965**, 7312—7319.

37. Buta, J. G.: A glucosamine derivative of caffeic acid in Nicotiana nodal tumors. Phytochemistry **9**, 1143—1144 (1970).

38. Buta, J. G., and R. R. Izac: Caffeoylputrescine in *Nicotiana tabacum*. Phytochemistry **11**, 1188—1189 (1972).

39. Candy, H. A., M. Laing, and C. M. Weeks: The crystal and molecular structure of helichrysoside, a new acylated flavonoid glycoside from *Helichrysum kraussii*. Tetrahedron Letters **1975**, 1211—1214.

39a. Cannon, J. R., P. W. Chow, M. W. Fuller, B. H. Hamilton, B. W. Metcalf, and A. J. Power: Phenolic constituents of *Grevillea robusta* (Proteaceae). The structure of robustol, a novel macrocyclic phenol. Austral. J. Chem. **26**, 2257—2275 (1973).

40. Casteele, K. van de, H. de Pooter, and C. F. van Sumere: Gas chromatographic separation and analysis of trimethylsilyl derivatives of some naturally occurring nonvolatile phenolic compounds and related substances. J. Chromatogr. **121**, 49—63 (1976).

41. CHALLICE, J. S.: Phenolic compounds of the genus Pyrus. VI. Distribution of phenols amongst the various tissues of the Pyrus stem. J. Sci. Food Agric. **24**, 285—293 (1973).

42. CHALLICE, J. S., and A. H. WILLIAMS: Phenolic compounds of the genus Pyrus. I. The occurrence of flavones and phenolic acid derivatives of 3,4-dihydroxybenzyl alcohol 4-glucoside in *Pyrus calleryana*. Phytochemistry **7**, 119—130 (1968).

43. CHANDLER, B. V., and K. A. HARPER: Identification of saccharides in anthocyanins and other flavonoids. Austral. J. Chem. **14**, 586—595 (1961).

44. CHAPELLE, J.-P.: Grandifloroside et methylgrandifloroside, glucosides iridoides nouveaux d'*Anthocleista grandiflora*. Phytochemistry **15**, 1305—1307 (1976).

45. CHU, N. T., F. M. CLYDESDALE, and F. J. FRANCIS: Isolation and identification of some fluorescent phenolic compounds in cranberries. J. Food Sci. **38**, 1038—1042 (1973).

46. CHUMBALOV, T. K., T. N. BIKBULATOVA, and M. I. IL'YASOVA: Polyphenols of *Myricaria alopecuroides*. I. Khim. Prir. Soedin. — Chem. nat. compounds **10**, 421 (russ.), 435 (engl.) (1974).

47. CLIFFORD, M. N., and J. WIGHT: The measurement of feruloylquinic acids and caffeoylquinic acids in coffee beans. Development of the technique and its preliminary application to green coffee beans. J. Sci. Food Agric. **27**, 73—84 (1976).

48. COLLINS, F. W., B. A. BOHM, and C. K. WILKINS: Flavonol glycoside gallates from *Tellima grandiflora*. Phytochemistry **14**, 1099—1102 (1975).

49. COOPER-DRIVER, G., J. J. CORNER-ZAMODITS, and T. SWAIN: The metabolic fate of hydroxybenzoic acids in plants. Z. Naturforsch. **27b**, 943—946 (1972).

50. COOPER-DRIVER, G., and T. SWAIN: Sulphate esters of caffeyl- and *p*-coumarylglucose in ferns. Phytochemistry **14**, 2506—2507 (1975).

51. CORNER, J. J., J. B. HARBORNE, S. G. HUMPHRIES, and W. D. OLLIS: Plant polyphenols. VII. The hydroxycinnamoyl esters of *Polygala senega* root. Phytochemistry **1**, 73—77 (1962).

52. CORSE, J.: A new isomer of chlorogenic acid from peaches. Nature (London) **172**, 771—772 (1953).

53. CORSE, J., L. L. LAYTON, and D. C. PATTERSON: Isolation of chlorogenic acids from roasted coffee. J. Sci. Food Agric. **21**, 164—168 (1970).

54. CORSE, J., R. E. LUNDIN, E. SONDHEIMER, and A. C. WAISS, JR.: Conformation analyses of D-(−)-quinic acid and some of its derivatives by nuclear magnetic resonance. Phytochemistry **5**, 767—776 (1966).

55. CORSE, J., R. E. LUNDIN, and A. C. WAISS, JR.: Identification of several components of isochlorogenic acid. Phytochemistry **4**, 527—529 (1965).

56. CORSE, J., and D. C. PATTERSON: 3-O-Sinapoylquinic acid. Phytochemistry **8**, 203—205 (1969).

57. CORSE, J., E. SONDHEIMER, and R. LUNDIN: 3-Feruloylquinic acid. A 3'-methyl ether of chlorogenic acid. Tetrahedron **18**, 1207—1210 (1962).

58. COXON, D. T., A. HOLMES, W. D. OLLIS, and V. C. VORA: The constitution and configuration of the theaflavin pigments of black tea. Tetrahedron Letters **1972**, 5237—5240.

59. COXON, D. T., A. HOLMES, W. D. OLLIS, V. C. VORA, M. S. GRANT, and J. L. TEE: Flavanol digallates in green tea leaf. Tetrahedron **28**, 2819—2826 (1972).

60. CULBERSON, CH. F.: Chemical and Botanical Guide to Lichen Products. Chapel Hill: The Univ. of North Carolina Press 1969.

61. CZOK, G.: Untersuchungen über die Wirkung von Kaffee. Suppl. 5 zur Z. Ernährungswiss. (1966).

62. CZOK, G., und K. LANG: Zur erregenden Wirkung von Chlorogensäure. Arzneimittel-Forschung **11**, 448—450 (1961).

63. — — Chlorogensäure-Wirkungen am Magen-Darmkanal. Arzneimittel-Forschung **11,** 545—549 (1961).

64. Czok, G., W. Midani und R. I. Finke: Beeinflussung der Cholerese und Darmmotorik durch Phenole und Phenolkarbonsäuren. Suppl. 14 zur Z. Ernährungswiss. 68—77 (1972).

65. Daniels, D. G. H., H. G. C. King, and H. F. Martin: Antioxidants in oats: Esters of phenolic acids. J. Sci. Food Agric. **14,** 385—390 (1963).

66. Daniels, D. G. H., and H. F. Martin: Structures of two antioxidants isolated from oats. Chem. and Ind. (London) **1964,** 2058.

67. — — Antioxidants in oats: Diferulates of long-chain diols. Chem. and Ind. (London) **1965,** 1763.

68. — — Antioxidants in oats: Mono-esters of caffeic and ferulic acids. J. Sci. Food Agric. **18,** 589—595 (1967).

69. — — Antioxidants in oats: Glyceryl esters of caffeic and ferulic acids. J. Sci. Food Agric. **19,** 710—712 (1968).

70. Delétang, J.: Presence de caffeoyl putrescine, de caffeoyl spermidine et de dicaffeoyl spermidine chez *Nicotiana tabacum.* Annal Tabac, Sect. 2, **11,** 123—130 (1974); Chem. Abstr. **84,** 147656 (1976).

71. Dickinson, D., and J. H. Gawler: The chemical constituents of Victoria plums: Chrysanthemin, acid and pectin contents. J. Sci. Food Agric. **7,** 699—705 (1956).

72. Dietrichs, H. H., und E. Schaich: Flavonolglucoside in den Knospen und Blättern der Rotbuche (*Fagus sylvatica* Linn.). Naturwiss. **50,** 478 (1963).

73. Dranik, L. I., and T. A. Shubert: *cis-trans*-Isomerization of esters of hydroxycinnamic acids. Khim. Prir. Soedin. — Chem. nat. compounds **6,** 264—265 (russ.), 257 (engl.) (1970).

74. Dukhovlinova, L. I., Yu. E. Sklyar, and L. I. Sdobnina: Borneol *trans-p*-hydroxycinnamate from *Seseli mucronatum* and *S. asperulum.* Khim. Prir. Soedin. — Chem. nat. compounds **11,** 99—100 (russ.), 109—110 (engl.) (1975).

75. Erickson, R. L., I. A. Pearl, and S. F. Darling: Populoside and grandidentoside from the bark of *Populus grandidentata.* Phytochemistry **9,** 857—863 (1970).

76. Estes, T. K., and I. A. Pearl: Hot-water extractives of the green bark of *Populus trichocarpa.* Techn. Assoc. Pulp Paper Ind. **50,** 318—323 (1967).

77. Farkas, G. L., and Z. Kiraly: Role of phenolic compounds in the physiology of plant diseases and disease resistance. Phytopathol. Z. **44,** 105—150 (1962).

78. Felice, L. J., W. P. King, and P. T. Kissinger: A new liquid chromatography approach to plant phenolics. Application to the determination of chlorogenic acid in sunflower meal. J. Agric. Food Chem. **24,** 380—382 (1976).

79. Feucht, G., K. Herrmann und W. Heimann: Über das Vorkommen der Hydroxyzimtsäuren im Gemüse. III. Über die im Kopfsalat hauptsächlich vorkommende Kaffeesäure-Verbindung. Z. Lebensm.-Unters.-Forsch. **145,** 206—212 (1971).

80. Friedrich, H., und J. Höhle: Untersuchung der Polyphenole in der Wurzel von *Rheum palmatum* L. Archiv Pharmaz. (Weinheim) **299,** 857—866 (1966).

81. Friedrich, H., und N. Krüger: Neue Untersuchungen über den Hamamelis-Gerbstoff. I. Der Gerbstoff der Rinde von *H. virginiana.* II. Der Gerbstoff der Blätter von *H. virginiana.* Planta Medica **25,** 138—148 (1974); **26,** 327—332 (1974).

82. Fujii, S., H. Aoki, M. Komoto, and K. Munakata: Fluorescent glycosides accumulated in *Taphrina wiesneri*-infected cherry stems. I. Contents of glycosides, and the isolation and identification of gentisic acid-5-β-D-glucoside. Agric. Biol. Chem. (Tokyo) **35,** 1526—1534 (1971).

83. — — — — Fluorescent glycosides accumulated in *Taphrina wiesneri*-infected cherry stems. II. Isolation and characterization of FG-2, a new gentisic acid-glycoside. Agric. Biol. Chem. (Tokyo) **35,** 1535—1541 (1971).

84. FURUYA, M., and A. W. GALSTON: Flavonoid complexes in *Pisum sativum* L. I. Nature and distribution of the major components. Phytochemistry **4**, 285—296 (1965).

85. GARDNER, P. D., G. J. PARK, and C. C. ALBERS: α- and β-Verbesinol. Sesquiterpene alcohols of the *cis*-decalin series. J. Amer. chem. Soc. **83**, 1511—1512 (1961).

86. GEIGER, H., U. SCHÜCKER, H. WALDRUM, G. VAN DER VELDE und T. J. MABRY: Quercetin-3β-(6-O-galloyl-galactosid), ein Inhaltsstoff von *Arctostaphylos uva ursi* (L.) Spreng. (Ericaceae). Z. Naturforsch. **30c**, 296 (1975).

87. GEISSMANN, T., and H. NEUKOM: A note on ferulic acid as a constituent of the water-insoluble pentosans of wheat flour. Cereal Chemistry **50**, 414—416 (1973).

88. GELLA, E. V., G. V. MAKAROVA und YU. G. BORISYUK: (Flavonoide von *Mentha piperita*. III.) Farm. Zh. (Kiev) **22**, Nr. 4, 80—85 (1967). Chem. Abstr. **68**, 49985 (1968).

89. GELLA, E. V., G. V. MAKAROVA, YU. G. BORISYUK und V. I. LITVINENKO: (Flavonoide von *Mentha piperita*. II.) Farm. Zh. (Kiev) **21**, Nr. 3, 58—66 (1966). Chem. Abstr. **65**, 13810 (1966).

90. GILDEMEISTER, E., und F. HOFFMANN: Die ätherischen Öle, 4. Aufl., Band IIId, S. 561 bis 565, 574, 576. Berlin: Akademie-Verlag. 1966.

91. GLASS, A. D. M., and B. A. BOHM: A further survey of ferns for cinnamic and benzoic acids. Phytochemistry **8**, 629—632 (1969).

92. GMELIN, R., and A. KJAER: Isoferulic acid choline ester: a new plant constituent. Phytochemistry **9**, 667—669 (1970).

93. GOLDSCHMID, O., and H. L. HERGERT: Examination of western hemlock for lignin precursors. Techn. Assoc. Pulp Paper Ind. **44**, 858—870 (1961).

94. GONZÁLEZ, A. G., C. G. FRANCISCO, R. FREIRE, R. HERNÁNDEZ, J. A. SALAZAR, and E. SUÁREZ: Lindleyin, a new phenolic gallylglucoside from *Aconium lindleyi*. Phytochemistry **15**, 344—346 (1976).

95. GOVINDACHARI, T. R., P. C. PARTHASARATHY, H. K. DESAI, and P. A. MOHAMED: Cluytyl ferulate, a new long chain ester from *Gmelina arborea Linn.* and *Lannea grandis (Denust.)*. Indian J. Chem. **9**, 1027 (1971).

96. GRISEBACH, H., und W. BARZ: Biochemie der Flavonoide. Naturwiss. **56**, 538—544 (1969).

97. GROMOVA, A. S., V. I. LUTSKII, and N. A. TYUKAVKINA: Phenolic acids and their glycosides from the phloem of *Picea jezoensis* and *P. koraiensis*. Khim. Prir. Soedin. — Chem. nat. compounds **10**, 798—799 (russ.), 825 (engl.) (1974).

98. GROSS, D.: Growth regulating substances of plant origin. Phytochemistry **14**, 2105 to 2112 (1975).

99. GSTIRNER, F., und G. KORF: Über Inhaltsstoffe des Rhizoms von *Polygonum bistorta* L. Arch. Pharmaz. (Weinheim) **299**, 640—646 (1966).

100. GUILLEMONAT, A., et J.-C. TRAYNARD: Sur la constitution chimique du liège. IV. Structure de la phellochryséine. Bull. Soc. chim. France **1963**, 142—144.

101. GUMENYUK, L. A.: Phenolic carboxylic acids from *Sempervivum ruthenicum*. Khim. Prir. Soedin. — Chem. nat. compounds **7**, 525 (russ.), 503 (engl.) (1971).

102. HÄUSERMANN, M., und H. BRANDENBERGER: Über phenolische Inhaltsstoffe. IV. Spektrophotometrische Bestimmungsmethode für Chlorogensäuren und ihre Anwendung auf pflanzliche Extrakte. Z. Lebensm.-Unters.-Forsch. **115**, 516—527 (1961).

103. HANEFELD, M., and K. HERRMANN: Quantitative determination of caffeic acid esters and catechins by direct measurement on thin-layer chromatography. J. Chromatogr. **123**, 391—395 (1976).

104. HANEFELD, M., und K. HERRMANN: Die quantitative Direktauswertung von Dünnschichtchromatogrammen am Beispiel der Chlorogensäure und anderer Kaffeesäureester in Obst und Gemüse. Dtsch. Lebensm.-Rdsch. **72**, 383—388 (1976).

105. HANN, R. M.: The synthesis of glucosidoferulic acid. J. Amer. Chem. Soc. **52,** 5049—5051 (1930).

106. HARBORNE, J. B.: Plant polyphenols. XI. The structure of acylated anthocyanins. Phytochemistry **3,** 151—160 (1964).

107. — Plant polyphenols. XIV. Characterization of flavonoid glycosides by acidic and enzymic hydrolyses. Phytochemistry **4,** 107—120 (1965).

108. — Caffeic acid ester distribution in higher plants. Z. Naturforsch. **21b,** 604—605 (1966).

109. HARBORNE, J. B., and J. J. CORNER: Plant polyphenols. IV. Hydroxycinnamic acid-sugar derivatives. Biochem. J. **81,** 242—250 (1961).

110. HARBORNE, J. B., V. H. HEYWOOD, and N. A. M. SALEH: Chemosystematics of the Compositae: Flavonoid patterns in the Chrysanthemum complex of the tribe Anthemideae. Phytochemistry **9,** 2011—2017 (1970).

111. HARBORNE, J. B., T. J. MABRY, and H. MABRY: The Flavonoids. London: Chapman & Hall. 1975.

112. HARBORNE, J. B., C. A. WILLIAMS, J. GREENHAM, and P. MOYNA: Distribution of charged flavones and caffeylshikimic acid in Palmae. Phytochemistry **13,** 1557—1559 (1974).

113. HARTLEY, R. D.: Carbohydrate esters of ferulic acid as components of cell-walls of *Lolium multiflorum.* Phytochemistry **12,** 661—665 (1973).

114. HARTLEY, R. D., E. C. JONES, and T. M. WOOD: Carbohydrates and carbohydrate esters of ferulic acid released from cell walls of *Lolium multiflorum* by treatment with cellulolytic enzymes. Phytochemistry **15,** 305—307 (1976).

115. HASLAM, E.: Chemistry of Vegetable Tannins. London-New York: Academic Press. 1966.

116. — (+)-Catechin-3-gallate and a polymeric proanthocyanidin from *Bergenia* species. J. Chem. Soc. (London) (C) **1969,** 1824—1828.

117. HASLAM, E., R. D. HAWORTH, and D. A. LAWTON: Gallotannins. VIII. The preparation and properties of some galloyl esters of quinic acid. J. Chem. Soc. (London) **1963,** 2173—2181.

118. HASLAM, E., R. D. HAWORTH, and G. K. MAKINSON: Synthesis of 3-O-*p*-coumaroyl-quinic acid. J. Chem. Soc. (London) **1961,** 5153—5156.

119. HASLAM, E., G. K. MAKINSON, M. O. NAUMANN, and J. CUNNINGHAM: Synthesis and properties of some hydroxycinnamoyl esters of quinic acid. J. Chem. Soc. (London) **1964,** 2137—2146.

120. HASLAM, E., M. O. NAUMANN, and G. BRITTON: Phenolic constituents of *Vaccinium vitis idaea* L. J. Chem. Soc. Suppl. I (London) **1964,** 5649—5654.

121. HERRMANN, K.: Über die Bildung der Purpurogallincarbonsäure durch fermentative Oxydation der Gallussäure. Arch. Pharmaz. (Weinheim) **287,** 497—503 (1954).

122. — Über die Flavonole und Phenole der Zwiebel (*Allium cepa* L.). Arch. Pharmaz. (Weinheim) **291,** 238—247 (1958).

123. — Über Oxyzimtsäuren mit Ausnahme der Kaffeesäure und der Chlorogensäuren. Pharmazie **13,** 266—276 (1958).

124. — Über den „Gerbstoff" der Labiatenblätter. Arch. Pharmaz. (Weinheim) **293,** 1043—1048 (1960).

125. — Über das Vorkommen und die Bedeutung der Anthocyanine in Lebensmitteln. Z. Lebensm.-Unters.-Forsch. **148,** 290—302 (1972).

126. — Über das Vorkommen der Hydroxybenzoesäure-Verbindungen im Pflanzenreich und in pflanzlichen Lebensmitteln. Z. Lebensm.-Unters.-Forsch. **153,** 170—176 (1973).

127. — Phenolische Pflanzeninhaltsstoffe als natürliche Antioxidantien. Fette — Seifen — Anstrichmittel **75,** 499—504 (1973).

128. HIGUCHI, T., Y. ITO, M. SHIMADA, and I. KAWAMURA: Chemical properties of milled wood lignin of grasses. Phytochemistry **6,** 1551—1556 (1967).

129. HILLER, K., und N. KOTHE: Chlorogen- und Rosmarinsäure — Vorkommen und quantitative Verteilung in Pflanzen der Saniculoideae. Pharmazie **22,** 220—221 (1967).

130. HÖRHAMMER, L., L. STICH und H. WAGNER: Über die Flavonolglykoside der Lindenblüten I. Arch. Pharmaz. (Weinheim) **294,** 685—692 (1961).

131. HÖRHAMMER, L., H. WAGNER, J. IZQUIERDO und H. ENDRES: Isolierung eines Gallussäuremethylesters aus *Cotinus Coggygria Scop.* (*Rhus cotinus* L.) durch Chromatographie an Polyamid. Arch. Pharmaz. (Weinheim) **291,** 269—273 (1958).

132. HORMAN, I., and R. VIANI: The nature and conformation of the caffeine-chlorogenate complex of coffee. J. Food Sci. **37,** 925—927 (1972).

133. HOROWITZ, R. M., and B. GENTILI: Long range proton shielding in C-glycosyl compounds: Structure of some new C-glycosylflavones. Chem. and Ind. (London) **1966,** 625—627.

134. HUNECK, S.: Chemie und Biosynthese der Flechtenstoffe. Fortschr. Chem. Organ. Naturstoffe **29,** 209—306 (1971).

135. IMPERATO, F.: Acylated betacyanins of *Portulaca oleracea.* Phytochemistry **14,** 2091—2092 (1975).

135a. — 1-(3-Glucosyloxy-4-hydroxycinnamyl)glucose from *Lantana hybrida.* Phytochemistry **15,** 1786 (1976).

136. IMPERATO, F., C. DI LEO, and P. TROVATO: 1-Caffeylrhamnose from *Lantana hybrida.* Phytochemistry **14,** 2725 (1975).

137. IVANOVA, S. Z., S. A. MEDVEDEVA, N. A. TYUKAVKINA, and N. D. ZELENIKINA: Phenolic acids and their glycosides from the oleoresin of *Picea obovata* and *Picea ajanensis.* Khim. Prir. Soedin. — Chem. nat. compounds **11,** 415—416 (russ.), 424 (engl.) (1975).

138. KADYROV, A. SH., T. KH. KHASANOV, A. I. SAIDKHODZHAEV, and G. K. NIKONOV: New phenolic compounds of the roots of *Ferula tschimganica.* Khim. Prir. Soedin. — Chem. nat. compounds **8,** 808—809 (russ.), 796—797 (engl.) (1972).

139. KADYROV, A. SH., and G. K. NIKONOV: The structures of chimganin and chimgin. Khim. Prir. Soedin. — Chem. nat. compounds **8,** 59—63 (russ.), 53–56 (engl.) (1972).

140. KAHNT, G.: Isolierung des *trans-* und *cis-o-*Oxyzimtsäureglucosids aus Steinkleeblättern *(Melilotus albus)* und Umwandlung der *trans-*Form in die isomere Verbindung *in vivo* und *in vitro* durch Sonnenlicht. Naturwiss. **49,** 207—208 (1962).

141. KAMESWARAMMA, A., and T. R. SESHADRI: Pajaneelin, a bitter component of *Pajaneelia rheedii.* Proc. Indian Acad. Sci., Sect. A **25,** 43—50 (1947).

142. KAMILOV, KH. M., and G. K. NIKONOV: Phenols of *Ferula prangifolia.* Khim. Prir. Soedin. — Chem. nat. compounds **8,** 718—721 (russ), 702—704 (engl.) (1972).

143. KAMILOV, KH. M., G. K. NIKONOV, and N. N. SHARAKHIMOV: Esters of the roots of *Ferula fedtschenkoana.* Khim. Prir. Soedin. — Chem. nat. compounds **10,** 527 (russ.), 540 (engl.) (1974).

144. KARL, C., G. MÜLLER und P. A. PEDERSEN: Flavonoide aus *Salix alba.* Die Struktur des Terniflorins und eines weiteren Acylflavonoides. Phytochemistry **15,** 1084—1085 (1976).

145. KARRER, W.: Konstitution und Vorkommen der organischen Pflanzenstoffe, 2. Aufl., S. 197—198. Basel-Stuttgart: Birkhäuser. 1976.

146. KARSTEN, U.: Bryophyllosid, ein neues Kämpferolglykosid aus *Bryophyllum daigremontianum* (R. HAMET et PERR.) Berg. Naturwiss. **52,** 84—85 (1965).

147. KEFELI, V. I., and C. S. KADYROV: Natural growth inhibitors, their chemical and physiological properties. Ann. Rev. Plant Physiol. **22,** 185—196 (1971).

148. KHASANOV, T. KH., A. I. SAIDKHODZHAEV, and G. K. NIKONOV: Components of the roots of *Ferula pallida.* Khim. Prir. Soedin. — Chem. nat. compounds **8,** 807—808 (russ.), 794—795 (engl.) (1972).

149. — — — Structure of teferin — a new ester from the roots of *Ferula tenuisecta.* Khim. Prir. Soedin. — Chem. nat. compounds **10,** 528 (russ.), 542 (engl.) (1974).

150. KING, F. E., J. H. GILKS, and M. W. PARTRIDGE: A glycosidic constituent of *Vinca minor* and its identification as 3-β-D-glucosyloxy-2-hydroxybenzoic acid. J. Chem. Soc. (London) **1955,** 4206—4215.

151. KLOSTERMAN, H. J., and R. Z. MUGGLI: The glucosides of flax-seed. II. Linocaffein. J. Amer. Chem. Soc. **81,** 2188—2191 (1959).

152. KLOSTERMAN, H. J., F. SMITH, and C. O. CLAGETT: The constitution of linocinnamarin. J. Amer. Chem. Soc. **77,** 420—421 (1955).

153. KOEPPE, D. E., L. M. ROHRBAUGH, and S. H. WENDER: The effect of varying UV intensities on the concentration of scopolin and caffeoylquinic acids in tobacco and sunflower. Phytochemistry **8,** 889—896 (1969).

154. KOSUGE, T.: The role of phenolics in host response to infection. Ann. Rev. Phytopathol. **7,** 195—222 (1969).

155. KRASNOV, E. A., T. G. KHORUZHAYA, L. I. DRANIK, V. G. GORDIENKO, and I. P. KOVALEV: 6-O-Galloylarbutin from *Rhodiola coccinea*. Khim. Prir. Soedin. — Chem. nat. compounds **11,** 474—478 (russ.), 492—494 (engl.) (1975).

156. KUC, J.: Phytoalexins. Ann. Rev. Phytopathol. **10,** 207—232 (1972).

157. KUC, J., and O. E. NELSON: The abnormal lignins produced by the brown-midrib mutants of maize. I. The brown-midrib-1 mutant. Arch. Biochem. Biophysics **105,** 103—113 (1964).

158. LAHIRY, N. L., L. D. SATTERLEE, H. W. HSU, and G. W. WALLACE: Characterization of the chlorogenic acid binding fraction in leaf protein concentration. J. Food Sci. **42,** 83—85 (1977).

159. LAYTON, L. L., F. C. GREENE, J. W. CORSE, and R. PANZANI: Pure chlorogenic acid not allergenic in atopy to green coffee: a specific protein probably is involved. Nature (London) **203,** 188—189 (1964).

160. LEONT'EVA, V. G., A. S. GROMOVA, V. I. LUTSKII, L. D. MODONOVA, and N. A. TYUKAVKINA: Alkyl ferulates from the bark and wood of plants of the Pinaceae family. Khim. Prir. Soedin. — Chem. nat. compounds **10,** 240—241 (russ.) 244—245 (engl.) (1974).

161. LEVAND, O., and R. M. HEINICKE: Ferulic acid as a component of a complex carbohydrate polymer of bromelain. Phytochemistry **7,** 1659—1662 (1968).

162. LEWIN, G., N. KUNESCH, A. CAVE, T. SEVENET et J. POISSON: Alcaloides d'*Alstonia lanceolifera*. Phytochemistry **14,** 2067—2071 (1975).

163. LIN, J.-H., Y.-M. LIN, and F. C. CHEN: A new kaempferol-3-β-D-(6''-O-*p*-coumaroyl)-glucoside from flowers of *Anaphalis contorta* Hooker. J. Chin. Chem. Soc. (Tapei) **23,** 57—59 (1976); Chem. Abstr. **85,** 10346 (1976).

164. LITVINENKO, V. I., T. P. POPOVA, A. V. SIMONJAN, I. G. ZOZ und V. S. SOKOLOV: „Gerbstoffe" und Oxyzimtsäureabkömmlinge in Labiaten. Planta Medica **27,** 372—380 (1975).

165. LITVINENKO, V. I., und I. G. ZOZ: (Chemotaxonomische Untersuchung der Arten von Thymus L. aus der Ukraine). Rastit. resursy **5,** 481—495 (1969).

166. LOESCHCKE, V., and H. FRANCKSEN: Trichocarpin, ein neues als Resistenzfaktor bedeutsames Phenolglykosid aus Pappelrinde. Naturwiss. **51,** 140 (1964).

167. MACHEIX, J.-J.: Présence dans les pommes de plusieurs dérivés de l'acide *para*-coumarique. C. R. Acad. Sci. Ser. **D 272,** 1097—1100 (1972).

168. MAGA, J. A., and K. LORENZ: Phenolic acid composition and distribution in wheat flours and various triticale milling fractions. Lebensm.-Wiss. u. -Technol. **7,** 273—278 (1974).

169. MAIER, V. P., D. M. METZLER, and A. F. HUBER: 3-O-Caffeoylshikimic acid (dactyliferic acid) and its isomers, a new class of enzymic browning substrates. Biochem. Biophysic. Res. Commun. **14,** 124—128 (1964).

170. MARKWALDER, H. U., and H. NEUKOM: Diferulic acid as a possible cross-link in hemicelluloses from wheat germ. Phytochemistry **15,** 836—837 (1976).

171. MBADIWE, E. I.: Caffeoylputrescine from *Pentaclethra macrophylla.* Phytochemistry **12,** 2546 (1973).

172. MEDVEDEVA, S. A., S. Z. IVANOVA, V. I. LUTSKII, V. V. KEIKO, and N. A. TYUKAVKINA: Phenolic acids and their glycosides from the needles of *Abies sibirica* and *A. nephrolepsis.* Khim. Prir. Soedin. — Chem. nat. compounds **10,** 404 (russ.), 407 (engl.) (1974).

173. MEDVEDEVA, S. A., N. A. TYUKAVKINA, and S. Z. IVANOVA: Glycosides of phenolic acids of the needles of *Larix sibirica.* Khim. Prir. Soedin. — Chem. nat. compounds **7,** 844 (russ.), 827 (engl.) (1971).

174. MIZUSAKI, S., Y. TANABE, M. NOGUCHI, and E. TAMAKI: *p*-Coumaroylputrescine, caffeoylputrescine and feruloylputrescine from callus tissue culture of *Nicotiana tabacum.* Phytochemistry **10,** 1347—1350 (1971).

175. MÖHRLE, H., und G. FANGERAU: Die Synthese des Isoferulasäure-cholinesters. Arch. Pharmaz. (Weinheim) **306,** 489—492 (1973).

176. MONTIES, B., M.-L. BOUILLANT et J. CHOPIN: C-Diholosylflavones dans les feuilles du melon *(Cucumis melo).* Phytochemistry **15,** 1053—1056 (1976).

177. MORITA, H.: Identification of phenolic acids by gas chromatography-mass spectrometry. J. Chromatogr. **71,** 149—153 (1972).

178. MURAKOSHI, I., K. SUGIMOTO, J. HAGINIWA, S. OHMIYA, and H. OTOMASU: (−)-*trans*-4′-Hydroxycinnamoyl-lupinine, a new alkaloid in Lupinus seedlings. Phytochemistry **14,** 2714—2715 (1975).

179. MUSSO, H.: Phenol coupling. In W. I. TAYLOR and A. R. BATTERSBY: Oxidative Coupling of Phenols, S. 1—94. New York: Marcel Dekker. 1967.

180. MZHAVANADZE, V. V., I. L. TARGAMADZE, and L. I. DRANIK: Phenolic compounds of the leaves of *Vaccinium arctostaphylos.* Khim. Prir. Soedin. — Chem. nat. compounds **8,** 124 (russ.), 125 (engl.) (1972).

181. NAGELS, L., and F. PARMENTIER: On the metabolism of cinnamic acid in *Cestrum poepigii.* Identification of *o*-coumaric acid and an *o*-coumaroylquinic acid. Arch. Int. Physiol. Biochem. **81,** 733—736 (1973).

182. NAHRSTEDT, A., K. DUMKOW, B. JANISTYN, and R. POHL: Quercetin-galaktosid-gallate in Euphorbiaceen. Tetrahedron Letters **1974,** 559—562.

183. NAIR, G. V., and E. VON RUDLOFF: The chemical composition of the heartwood extractives of tamarack [*Larix laricina* (Du Roi) K. Koch]. Canad. J. Chem. **37,** 1608—1613 (1959).

184. NIEMANN, G. J.: Phenolic glucosides from needles of *Larix laricina.* Phytochemistry **8,** 2101—2103 (1969).

185. — Acylated flavonol glycosides from Larix needles. Phytochemistry **14,** 1437—1438 (1975).

186. NORIN, T., and B. WINELL: Extractives from the bark of common spruce, *Picea abies* L. Karst. Acta chem. Scand. **26,** 2289—2296 (1972).

187. — — Extractives from the bark of scots pine, *Pinus silvestris* L. Acta chem. Scand. **26,** 2297—2304 (1972).

188. OETTMEIER, W., and A. HEUPEL: *p*-Coumaroyl-*meso*-tartaric acid from spinach chloroplast preparations. Z. Naturforsch. **27b,** 586—587 (1972).

189. OHTA, G., und M. SHIMIZU: (Ein neuer Triterpenoidalkohol, 24-Methylencycloartanol, als Ferulasäureester aus Reiskleieöl). Chem. pharmaz. Bull. (Tokyo) **6,** 325—326 (1958). Chem. Zbl. **1961,** 11726.

190. PANIZZI, L., e M. L. SCARPATI: Isolamento e constituzione del principio attivo del carciofo. Gazz. chim. ital. **84,** 792—805 (1954).

191. — — Sugli acidi 1,4- ed 1,5-dicaffeil-chinici. Gazz. chim. ital. **95,** 71—82 (1965).

192. PANKOV, YU. A., und T. P. SAFRONOVA: (Catechine von Rosa-Species der fernöstlichen Gebiete der UdSSR). Rastit. Resur. **11,** 520—523 (1975). Chem. Abstr. **84,** 86721 (1976).

193. PEARL, I. A., and S. F. DARLING: Grandidentatin, a new glucoside from the bark of *Populus grandidentata*. J. org. Chem. **27**, 1806—1809 (1962).

194. — — Migration of acyl groups during isolation of glycosides from *Populus grandidentata leaves*. Arch. Biochem. Biophysics **102**, 33—38 (1963).

195. — — Further studies on the bark of triploid *Populus tremuloides*. Techn. Assoc. Pulp Paper. Ind. **50**, 324—329 (1967).

196. — — Trichoside, a new glucoside from the bark of *Populus trichocarpa*. Phytochemistry **7**, 825—829 (1968).

197. — — Purpurein, a new glucoside from the bark of *Salix purpurea*. Phytochemistry **9**, 853—856 (1970).

198. — — Phenolic extractives of *Salix purpurea* bark. Phytochemistry **9**, 1277—1281 (1970).

199. — — Phenolic extractives of the leaves of *Populus balsamifera* and of *P. trichocarpa*. Phytochemistry **10**, 2844—2847 (1971).

200. PEARL, I. A., and C. R. POTTENGER: Studies on the hot water extractives of the green bark of *Populus balsamifera*. Techn. Assoc. Pulp Paper Ind. **49**, 152—155 (1966).

201. PETRENKO, V. V.: Quinqueloside — a new flavonoid glycoside from *Leonuris quinquelobatus* Gilib. Khim. Prir. Soedin. — Chem. nat. compounds **1**, 414—419 (russ.), 326—329 (engl.) (1965).

202. PICARD, P.: Le violutoside, nouveau glucoside à salicylate de méthyle, retiré du *Viola cornuta* L. C. R. Acad. Sci. **182**, 1167—1169 (1926).

203. PODKOWINSKA, H., and M. WIEWIOROWSKI: ω-Feruloyloxylupinan, a new lupin alkaloid. Bull. Acad. polon. Sci., Ser. Sci. Biol. **13**, 623—627 (1965).

204. POOTER, H. DE, H. ALI, and C. F. VAN SUMERE: N-Acylamino acids and peptides. I. The synthesis of N-feruloylamino acids. Bull. Soc. Chim. Belg. **82**, 243—257 (1973).

205. — — — N-Acylamino acids and peptides. II. The synthesis of N-feruloylglycyl-(L and DL)-phenylalanine. Bull. Soc. Chim. Belg. **82**, 259—269 (1973).

206. POOTER, H. DE, J. DE BRUCKER, and C. F. VAN SUMERE: Revised synthesis of 3-O-cinnamyl-, 3-O-*p*-coumaryl- and 3-O-isoferulyl-D-(−)-quinic acid. Bull. Soc. Chim. Belg. **84**, 835—843 (1975).

207. POOTER, H. DE, I. PÉ, and C. F. VAN SUMERE: N-Acylamino acids and peptides. III. The synthesis of N-(feruloyl-2-^{14}C)-glycine-2-^{3}H. J. Lab. Compounds **10**, 135—141 (1974).

208. RIBÉREAU-GAYON, P.: Identification d'esters des acides cinnamiques et de l'acide tartique dans les limbes et les baies de *V. vinifera*. C. R. Acad. Sci. **260**, 341—343 (1965).

209. RILEY, R. G., and P. E. KOLATTUKUDY: Evidence for covalently attached *p*-coumaric and ferulic acid in cutins and suberins. Plant Physiol. **56**, 650—654 (1975).

210. ROBERTS, E. A. H., and M. MYERS: Theogallin, a polyphenol occurring in tea. II. Identification as a galloylquinic acid. J. Sci. Food Agric. **9**, 701—705 (1958).

211. ROBERTSON, A., and R. B. WATERS: Syntheses of glucosides. IX. Methylsalicylate vicianoside (? violutoside). J. Chem. Soc. (London) **1932**, 2770 2771.

212. RONDEST, J., M. B. C. DAS et J. POLONSKY: Sur un nouvel amide naturel, le N-(*p*-hydroxy-phényl)-β-éthyl *p*-hydroxycinnamide. Bull. Soc. Chim. France **1968**, 2411—2414.

213. ROWE, J. W., C. L. BOWER, and E. R. WAGNER: Extractives of jack pine bark: Occurrence of *cis*- and *trans*-pinosylvin dimethyl ether and ferulic acid esters. Phytochemistry **8**, 235—241 (1969).

214. SABIR, M. A., F. W. SOSULSKI, and A. J. FINLAYSON: Chlorogenic acid-protein interactions in sunflower. J. Agric. Food Chem. **22**, 575—578 (1974).

215. SAGITDINOVA, G. V., A. I. SAIDKHODZHAEV, G. K. NIKONOV, and U. RAKHMANKULOV:

Esters from the roots of *Ferula lapidosa*. Khim. Prir. Soedin. — Chem. nat. compounds **11**, 115 (russ.), 137 (engl.) (1975).

216. SAIDKHODZHAEV, A. I., and G. K. NIKONOV: The structure of ferutin. Khim. Prir. Soedin. — Chem. nat. compounds **8**, 559—563 (russ.), 544—547 (engl.) (1972).

217. — — Esters of the roots of *Ferula kuhistanica*. Khim. Prir. Soedin. — Chem. nat. compounds **10**, 525 (russ.), 538 (engl.) (1974).

218. — — Components of the roots of *Ferula ovina*. Khim. Prir. Soedin. — Chem. nat. compounds **10**, 526 (russ.), 539 (engl.) (1974).

219. SAMBORSKI, D. J., and R. ROHRINGER: Abnormal metabolites of wheat: Occurrence, isolation and biogenesis of 2-hydroxyputrescine amides. Phytochemistry **9**, 1939—1945 (1970).

220. SCARPATI, M. L., e A. D'AMICO: Isolamento dell'acido monocaffeiltartarico dalla cicoria *(Chicorium intybus)*. Ric. scient. **30**, 1746—1748 (1960).

221. SCARPATI, M. L., e P. ESPOSITO: Struttura dell'acido neoclorogenico e dalla "banda 510". Ann. Chim. (Rom) **54**, 35—50 (1964).

222. — — Isomerizzazione per transesterificazione degli acidi 3-, 4- e 5-mono-caffeilchinici e del loro metileteri. Ann. Chim. (Rom) **54**, 51—55 (1964).

223. SCARPATI, M. L., and M. GUISO: Structure of the three dicaffeoyl-quinic acids of coffee (isochlorogenic acid). Tetrahedron Letters **1964**, 2851—2853.

224. SCARPATI, M. L., and G. ORIENTE: Chicoric acid (dicaffeyltartic acid): its isolation from chicory *(Chicorium intybus)* and synthesis. Tetrahedron **4**, 43—48 (1958).

225. — — Isolamento e constituzione dell'acido rosmarinico (dal *Rosmarinus off.*). Ric. scient. **28**, 2329—2333 (1958).

226. — — Isolamento dal fagiolo *(Phaseolus vulgaris)* dell'acido faselico, sua constituzione e sintesi. Gazz. chim. ital. **90**, 212—219 (1960).

227. — — Constituzione stereochimica dell'acido β-(3,4-diossifenil)-α-lattico dal *Rosmarinus off.* Ric. scient. **30**, 255—259 (1960).

228. SCARPATI, M. L., C. TROGOLO e L. PANIZZI: Sintesi di alcuni acidi caffeil-chinici. Ann. Chim. (Rom) **54**, 56—65 (1964).

229. SCHMIDT, O. TH.: Gallotannine und Ellagengerbstoffe. Fortschr. Chem. Organ. Naturstoffe **13**, 70—136 (1956).

230. SCHMIDT, O. TH., und W. BLANK: Die Gallussäureester des Glycerins. Chem. Ber. **89**, 283—290 (1956).

231. SCHMIDT, O. TH., und H. REUSS: 2-(*p*-Hydroxy-benzoyl)-glucose und 2-Galloylglucose. Liebigs Ann. Chem. **649**, 137—148 (1961).

232. SCHMIDT, O. TH., und H. SCHMADEL: Synthesen des α-Glucogallins und neue Synthesen des β-Glucogallins. Liebigs Ann. Chem. **649**, 149—157 (1961).

233. SCHMIDTLEIN, H., and K. HERRMANN: Quantitative analysis for phenolic acids by thin-layer chromatography. J. Chromatogr. **115**, 123—128 (1975).

234. — — Über die Phenolsäuren des Gemüses I. II. IV. Z. Lebensm.-Unters.-Forsch. **159**, 139—148, 213—218, 255—263 (1975).

235. SCHÖNSIEGEL, I., K. EGGER und M. KEIL: Acylierte Kämpferolglykoside in *Narcissus poeticus* L. Z. Naturforsch. **24b**, 1213—1214 (1969).

236. SCHULTZ, O. E., und R. GMELIN: Papierchromatographie der Senfölglucosid-Drogen. Z. Naturforsch. **7b**, 500—506 (1952).

237. SERGEEVA, N. V., and A. L. SHINKARENKO: Glycosides of the leaves of *Evonymus europaea*. Khim. Prir. Soedin. — Chem. nat. compounds **4**, 324—325 (russ.), 275—276 (1968).

238. SHATOKHINA, R. K., L. I. DRANIK, and K. F. BLINOVA: Phenolic compounds of *Antitoxicum sibiricum*. Khim. Prir. Soedin. — Chem. nat. compounds **10**, 518—519 (russ.), 525 (engl.) (1974).

239. SHEREMET, I. P., and N. F. KOMISSARENKO: Stachannoaciside from *Stachys annua*.

Khim. Prir. Soedin. — Chem. nat. compounds **7**, 845—846 (russ.), 829 (engl.) (1971).

240. SHIMA, K., S. HISADA, and I. INAGAKI: Isolation of glucosyringic acid from *Anodendron affine*. Phytochemistry **10**, 894—895 (1971).

241. SHIMADA, M., T. FUKUZUKA, and T. HIGUCHI: Ester linkages of *p*-coumaric acid in bamboo and grass lignins. Techn. Assoc. Pulp Paper Ind. **54**, 72—78 (1971).

242. SHNYAKINA, G. P., and G. G. ZAPESOCHNAYA: Flavonoids of some species of *Sedum*. Khim. Prir. Soedin. — Chem. nat. compounds **11**, 92—93 (russ.), 97—98 (engl.) (1975).

243. SIMMONDS, N. W., and R. STEVENS: Occurrence of the methylene-dioxy bridge in the phenolic components of plants. Nature (London) **178**, 752—753 (1956).

244. SIMONYAN, A. V., and V. I. LITVINENKO: Rosmarinic acid in representatives of the genus *Thymus*. Khim. Prir. Soedin. — Chem. nat. compounds **8**, 797 (russ.), 776 (engl.) (1972).

245. SIMONYAN, A. V., V. I. LITVINENKO, and A. L. SHINKARENKO: Phenolcarboxylic acids of the genus *Thymus*. Khim. Prir. Soedin. — Chem. nat. compounds **8**, 383 (russ.), 372 (engl.) (1972).

246. SISSI, H. I. EL, N. A. M. SALEH, S. I. EL NEGOUMY, H. WAGNER, M. A. IYENGAR und O. SELIGMANN: Prunin-O-6″-gallat, aus *Acacia farnesiana*. Phytochemistry **13**, 2843—2844 (1974).

247. SLOMAN, K. G., and M. PANIO: Determination of chlorogenic acid in coffee. Advanc. automat. Anal. **2**, 83—85 (1969).

248. SMITH, D. C. C.: *p*-Hydroxybenzoate groups in the lignin of aspen *(Populus tremula)*. J. Chem. Soc. (London) **1955**, 2347—2351.

249. SOMERS, T. C.: Grape phenolics: The anthocyanins of *Vitis vinifera*, variety Shiraz. J. Sci. Food Agric. **17**, 215—219 (1966).

250. SOMOGYI, J. C., and R. BÖNICKE: Connection between chemical structure and antithiamine activity of various phenol derivatives. Intern. Z. Vitaminforsch. **39**, 65—73 (1969).

251. SONDHEIMER, E.: On the distribution of caffeic acid and the chlorogenic acid isomers in plants. Arch. Biochem. Biophysics **74**, 131—138 (1958).

252. SONDHEIMER, E., F. COVITZ, and M. J. MARQUISEE: Association of naturally occurring compounds, the chlorogenic acid-coffeine complex. Arch. Biochem. Biophysics **93**, 63—71 (1961).

253. SONDHEIMER, E., C. D. SZYMANSKI, and J. W. CORSE: Isolation of chlorogenic acid and its isomers from coffee. J. Agric. Food Chem. **9**, 146—149 (1961).

254. SOOD, A. R., B. BOUTARD, M. CHADENSON, J. CHOPIN, and P. LEBRETON: A new flavone C-glycoside from *Trigonella foenum graecum*. Phytochemistry **15**, 351—352 (1976).

255. STAGG, G. V., and D. SWAINE: The identification of theogallin as 3-galloylquinic acid. Phytochemistry **10**, 1671—1673 (1971).

256. STÖHR, H., und K. HERRMANN: Die phenolischen Inhaltsstoffe des Obstes. V. Die phenolischen Inhaltsstoffe der Erdbeeren und deren Veränderungen während Wachstum und Reife der Früchte. Z. Lebensm.-Unters.-Forsch. **158**, 341—348 (1975).

257. — — Die phenolischen Inhaltsstoffe des Obstes. VI. Die phenolischen Inhaltsstoffe der Johannisbeeren, Stachelbeeren und Kulturheidelbeeren. Veränderungen der Phenolsäuren und Catechine während Wachstum und Reife von Schwarzen Johannisbeeren. Z. Lebensm.-Unters.-Forsch. **159**, 31—37 (1975).

258. — — Über die Phenolsäuren des Gemüses. III. Hydroxyzimtsäuren und Hydroxybenzoesäuren des Wurzelgemüses. Z. Lebensm.-Unters.-Forsch. **159**, 219—224 (1975).

259. STÖHR, H., H.-D. MOSEL, and K. HERRMANN: The phenolics of fruits VII. The phenolics of cherries and plums and the changes in catechins and hydroxycinnamic acid derivatives during the development of fruits. Z. Lebensm. Unters.-Forsch. **159,** 85—91 (1975).

260. STOESSL, A.: The antifungal factors in barley. III. Isolation of *p*-coumaroylagmatine. Phytochemistry **4,** 973—976 (1965).

261. — Antifungal compounds produced by higher plants. Recent Advanc. Phytochem. **3,** 143—180 (1970).

262. STOLL, A., J. RENZ und A. BRACK: Isolierung und Konstitution des Echinacosids, eines Glykosids aus den Wurzeln von *Echinacea angustifolia* D. C. Helv. Chim. Acta **33,** 1877—1893 (1950).

263. STOLL, A., und E. SEEBECK: Die Isolierung von Hesperitinsäure, Behensäure und von zwei unbekannten Säuren aus Baldrian. Liebigs Ann. Chem. **603,** 158—168 (1957).

264. SUMERE, C. F. VAN, J. COTTENIE, J. DE GREEF, and J. KINT: Biochemical studies in relation to the possible germination regulatory role of naturally occurring coumarins and phenolics. Recent Advanc. Phytochem. **4,** 165—221 (1972).

265. SUMERE, C. F. VAN, H. DE POOTER, H. ALI, and M. DEGRAUW-VAN BUSSEL: N-Feruloylglycyl-L-phenylalanine. A sequence in barley proteins. Phytochemistry **12,** 407—411 (1973).

266. SUZUKI, Y., M. SHIMADA, K. TADERA, F. KAWAI, and H. MITSUDA: Studies on a new *p*-coumaryl derivative isolated from spinach leaves. I. Its isolation and some physicochemical properties. Agric. Biol. Chem. (Tokyo) **34,** 511—516 (1970).

267. TADERA, K., Y. SUZUKI, F. KAWAI, and H. MITSUDA: Studies on a new *p*-coumaryl derivative isolated from spinach leaves. II. Determination of chemical structure. Agric. Biol. Chem. (Tokyo) **34,** 517—522 (1970).

268. TAMURA, T., T. HIBINO, K. YOKOYAMA und T. MATSUMOTO: (Das Vorkommen von Dihydro-γ-sitosterylferulat im Weizenöl). Nippon Kagaku Zasshi **80,** 215—217 (1959). Chem. Abstr. **55,** 4889 (1961).

269. TAMURA, T., N. SAKAEDANI und T. MATSUMOTO: (Isolierung von Dihydro-β-sitosterylferulat aus Maiskeimöl). Nippon Kagaku Zasshi **79,** 1011—1014 (1958). Chem. Abstr. **54,** 24857 (1960).

270. TAMURA, T., M. TAKEDA, K. MURAKAMI, A. YOSHIDA und T. MATSUMOTO: (Ferulates). Yukagaku **11,** 454—458 (1962). Chem. Abstr. **59,** 5230 (1963).

271. TANAKA, A., und A. KATO: (Massenspektren verschiedener Ferulate). Yukagaku **24,** 570—574 (1975). Chem. Abstr. **84,** 44526 (1976).

272. TANAKA, A., A. KATO und T. TSUCHIYA: (Die Isolierung von β-Sitosterylferulat aus Reiskleiöl). Yukagaku **13,** 260 (1964). zit. n. Fette — Seifen — Anstrichmittel **68,** 155 (1966).

273. — — — Isolation of methyl ferulate from rice bran oil. J. Amer. Oil Chem. Soc. **48,** 95—97 (1971).

274. TANGUY, J., et C. MARTIN: Sur quelques constituants phénoliques des feuilles cotylédonnaires de plantules de *Phaseolus vulgaris* variété Pinto. C. R. Acad. Sci. Ser. D **274,** 3402—3404 (1972).

275. THEWLIS, B. H.: New synthetic antioxidants based on caffeic acid. J. Food Technol. **2,** 83—87 (1967).

276. THIEME, H., und R. BENECKE: Isolierung und Charakterisierung von Kaffeesäure-3-β-D-glucosid. Pharmazie **24,** 292 (1969).

277. THIEME, H., und A. KHOGALI: Über das Vorkommen von Flavonoiden und Gerbstoffen in den Blättern einiger afrikanischer Acacia-Arten. Pharmazie **30,** 736—743 (1975).

278. THIEME, H., und R. RICHTER: Isolierung eines neuen Phenolglykosids aus *Populus tremula* L. Pharmazie **21,** 251 (1966).

279. — — Über das Vorkommen von 1-*p*-Cumaroyl-β-D-glucose in den Blättern von *Populus candicans* Ait. Pharmazie **21,** 252—253 (1966).

280. THIEME, H., und H.-J. WINKLER: Die Phenolglykoside der Ericaceen. I. Untersuchungen über Vorkommen und Verbreitung. Pharmazie **26**, 235—243 (1971).

281. — — Über Vorkommen und Akkumulation von Phenolglykosiden in der Familie der Primulaceen. Pharmazie **26**, 434—436 (1971).

282. TOWERS, G. H. N., and W. S. G. MAAS: Phenolic acids and lignins in the Lycopodiales. Phytochemistry **4**, 57—66 (1965).

283. TROITSKAYA, E. A., L. I. DRANIK, and M. YA. SHKOL'NIK: Phenol composition of sunflower leaves during boron deficiency. Fiziol. Rast. — Soviet Plant Physiol. **18**, 393—396 (russ.), 328—330 (engl.) (1971).

284. TURABELIDZE, D. G., G. K. NIKONOV, and E. P. KEMERTELIDZE: Coumarins of the roots of *Libanotis transcaucasica*. Khim. Prir. Soedin. — Chem. nat. compounds **10**, 402 (russ.), 405 (engl.) (1974).

285. TUTSCHEK, R.: Isolierung und Charakterisierung der *p*-Hydroxy-β-(carboxy-methyl)-zimtsäure (Spaghnumsäure) aus der Zellwand von *Sphagnum magellanicum* Brid. Z. Pflanzenphysiol. **76**, 353—365 (1975).

286. URITANI, I., and M. MIYANO: Derivatives of caffeic acid in sweet potato attacked by black rot. Nature (London) **175**, 812 (1955).

287. VAJS, V., D. JEREMIC, M. STEFANOVIC, and S. MILOSAVLJEVIC: *p*-Coumaric esters and fatty alcohols from *Artemisia campestris*. Phytochemistry **14**, 1659—1660 (1975).

288. VUATAZ, L., H. BRANDENBERGER, and R. H. EGLI: Plant phenols. I. Separation of the leaf polyphenols by cellulose column chromatography. J. Chromatogr. **2**, 173—187 (1959).

289. WAGNER, G.: Über 2- und 5-β-Glucopyranoside verschiedener Gentisinsäureester. Arch. Pharmaz. (Weinheim) **291**, 278—287 (1958).

290. WAGNER, H.: Flavonoid-Glykoside. Fortschr. Chem. Organ. Naturstoffe **31**, 153—216 (1974).

291. WAISS, A. C., JR., R. E. LUNDIN, and J. CORSE: The NMR spectra of chlorogenic acids. Chem. and Ind. (London) **1964**, 1984—1985.

292. WANG, C.-Y., A. M. PAMUKCU, and G. T. BRYAN: Isolation of fumaric acid, succinic acid, astragalin, isoquercitrin and tiliroside from *Pteridium aquilinum*. Phytochemistry **12**, 2298—2299 (1973).

293. WATANABE, S., S. SAKAMURA, and Y. OBATA: The structures of acylated anthocyanins in eggplant and *Perilla* and the position of acylation. Agric. Biol. Chem. (Tokyo) **30**, 420—422 (1966).

294. WEINGES, K., W. BÄHR, W. EBERT, K. GÖRITZ und H.-D. MARX: Konstitution, Entstehung und Bedeutung der Flavonoid-Gerbstoffe. Fortschr. Chem. Organ. Naturstoffe **27**, 158—260 (1969).

295. WEINGES, K., P. KLOSS und W.-D. HENKELS: Picrosid-II, ein neues 6-Vanilloyl-catalpol aus *Picrorhiza kurrooa* Royle und Benth. Liebigs Ann. Chem. **759**, 173—182 (1972).

296. WEINGES, K., K. KÜNSTLER, G. SCHILLING und H. JAGGY: Naturstoffe aus Arzneipflanzen. XX. Scutellariosid-I und -II, ein 10-Cinnamoyl- und 10-(4-Hydroxycinnamoyl)-catalpol aus *Scutellaria altissima*. L. Liebigs Ann. Chem. **1975**, 2190—2194.

297. WHEATON, T. A., and I. STEWART: Feruloylputrescine: Isolation and identification from citrus leaves and fruit. Nature (London) **206**, 620—621 (1965).

298. WHEELER, A. W., and H. G. C. KING: Conversion of tryptophan to auxin by phenolic esters from leaves of dwarf french bean *(Phaseolus vulgaris* L.*)*. Phytochemistry **7**, 1057—1063 (1968).

299. WHITING, G. C., and R. A. COGGINS: 4-*p*-Coumarylquinic acid in apple fruits. Phytochemistry **14**, 593 (1975).

300. WHITMORE, F. W.: Phenolic acids in wheat coleoptile cell walls. Plant Physiol. **53**, 728—731 (1974).

301. WILKINS, C. K., J. DE BRUIJN, O. KORVER, D. J. FROST, and K. WEINGES: Isolation and identification of (−)-gallocatechin gallate and circular dichroism of green tea flavanols. J. Sci. Food Agric. **22,** 480—484 (1971).

302. WILLSTÄTTER, R., und W. MIEG: Über ein Anthocyan des Rittersporns. Liebigs Ann. Chem. **408,** 61—82 (1915).

303. WILSON, J. L., W. J. DUNLAP, and S. H. WENDER: Quantitative determination of chlorogenic acid in plant tissue by combined polyvinylpyrrolidone column and gas chromatography. J. Chromatogr. **35,** 329—335 (1968).

304. WINDE, E., und R. HÄNSEL: Die Pseudoindikane von *Vitex agnus castus* L. Arch. Pharmaz. (Weinheim) **293,** 556—567 (1960).

305. WINTER, M., und R. KLÖTI: Über das Aroma der gelben Passionsfrucht *(Passiflora edulis f. flavicarpa)*. Helv. Chim. Acta **55,** 1916—1921 (1972).

306. WÖLDECKE, M., und K. HERRMANN: D-(+)-Dikaffeoyl-Weinsäure aus Endivien *(Cichorium endivia* L.*)*. Z. Naturforsch. **29c,** 360—361 (1974).

307. WULF, L. W., and C. W. NAGEL: Analysis of phenolic acids and flavonoids by high-pressure liquid chromatography. J. Chromatogr. **116,** 271—279 (1976).

308. WYK, P. S. VAN, and B. H. KOEPPEN: *p*-Hydroxybenzoylcalleryanin: An antifungal substance in the root bark of *Protea cynaroides*. South African J. Sci. **70,** 121 (1974).

309. YOSHIHARA, T., H. YOSHIKAWA, S. SAKAMURA, and T. SAKUMA: Clovamides: L-Dopa conjugated with *trans*- and *cis*-caffeic acids in red clover *(Trifolium pratense)*. Agric. Biol. Chem. (Tokyo) **38,** 1107—1109 (1974).

310. YOSHIZAWA, K., S. KOMATSU, I. TAKAHASHI, and K. OTSUKA: Phenolic compounds in the fermented products. I. Origin of ferulic acid in sake. Agric. Biol. Chem. (Tokyo) **34,** 170—180 (1970).

311. ZANE, A., W. STECK, and S. H. WENDER: Identification of 4-O-caffeoylquinic acid and related depsides in tobacco from cigarettes. Tobacco Sci. **9,** 85—87 (1965).

312. ZANE, A., and S. H. WENDER: Identification of gentisic acid-5-β-D-glucoside in boron-deficient sunflower plants. Chem. and Ind. **1964,** 1835—1836.

313. — — A synthesis of 1-O-, 4-O- and 5-O-feruloylquinic acids. Chem. and Ind. (London) **1965,** 1034—1035.

314. — — Depsides in sunflower leaves. Nature (London) **209,** 80—81 (1966).

(Received September 28, 1976)

Natural 4-Ylidenebutenolides and 4-Ylidenetetronic Acids

By G. PATTENDEN, Department of Chemistry, The University, Nottingham, U. K.

Contents

I. Introduction

4-Ylidenebutenolides and 4-ylidenetetronic acids comprise a group of natural products many members of which have been discovered only in the past decade. The term "butenolide" describes the unsaturated γ-lactone system (**1**) and "tetronic acid" is the trivial name for 3-hydroxy-but-2-enolide (**2**). The corresponding 4-ylidene derivatives contain an additional exocyclic carbon-to-carbon double bond at C-4, and hence formulation (**3**) describes the 4-ylidenebutenolide, and (**4**) the 4-ylidenetetronic acid, ring systems*. An excellent and comprehensive account of natural 4-ylidenebutenolides and 4-ylidenetetronic acids known up to 1962 has been given by DEAN in his book (*1*). At this time the list numbered just thirteen metabolites, and included protoanemonin and matricaria lactone found in higher plants, patulin and dehydrocarolic acid isolated from fungi, and eight lichen pulvinic acid pigments. Today approximately seventy natural metabolites characterized structurally by the presence of one of the ring systems (**3**) or (**4**) are known. The early chemistry of butenolides has been well reviewed by RAO (*2*) and a useful account of tetronic acids is also available (*3*).

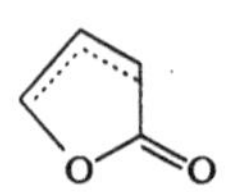

"Butenolide" (**1**)

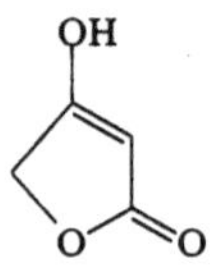

Tetronic acid (**2**)

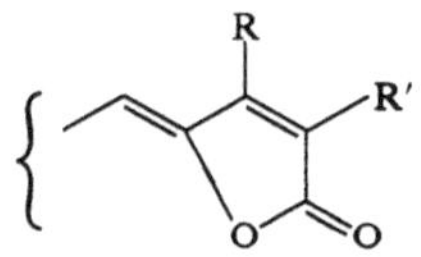

4-Ylidenebutenolide (**3**)

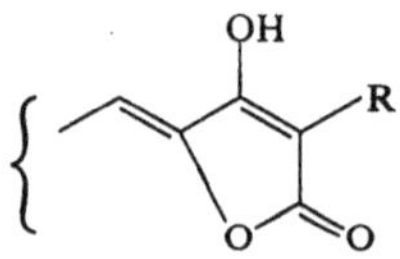

4-Ylidenetetronic acid (**4**)

The rapid expansion in the number of structurally novel 4-ylidenebutenolides and 4-ylidenetetronic acids since 1962, many displaying interesting biological properties, has prompted considerable activity into the chemical synthesis and biosynthesis of these metabolites. This review aims to reflect the structure and distribution of natural 4-ylidenebutenolides and 4-ylidenetetronic acids and to discuss developments in their chemical synthesis and biosynthesis.

II. Structure and Distribution

1. 4-Ylidenebutenolides

Natural 4-ylidenebutenolides vary enormously in complexity of structure, ranging from the simplest possible, protoanemonin (**5**), isolated from members of the Ranunculaceae, to the complex quinone methide

* 4-Ylidenebutenolides can also be regarded as either the enol lactones of 4-oxo-2-enoic acids, and named as such, or as furanone derivatives, *viz* 5-ylidene-2-5*H*-furanones.

xylerythrin (**7a**) found in wood attacked by the fungus *Peniophora sanguinea* Bres. Protoanemonin (**5**) is present in Nature as the corresponding glycoside ranunculin (**6**), and was one of the first 4-ylidenebutenolides to be isolated and characterized (*4*). It has been isolated from the steam distillate of buttercups, and from many species of *Ranunculus, Anemone* and *Clematis.* Several members of the Ranunculaceae have powerful vesicant properties, and this property is reported to be due to protoanemonin which is liberated enzymatically from ranunculin in the plants on crushing the tissues (*5*). Excellent accounts of the isolation, synthesis and chemistry of protoanemonin have been published previously (*1, 3*).

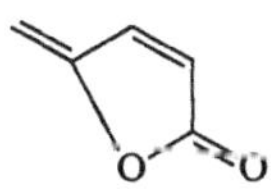

Protoanemonin (**5**)

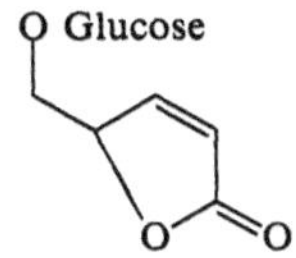

Ranunculin (**6**)

Xylerythrin (**7a**), described as a black crystalline material with a green lustre, is the principal pigment isolated from decaying wood attacked by the fungus *Peniophora sanguinea* (*6*). The infected wood becomes coloured dark red by the mycelium of the fungus. Structure (**7a**) followed for the pigment from X-ray measurements on the corresponding *bis*-bromoacetate derivative. The closely related pigments peniophorin (**7b**), peniophorinin (**8**) and peniosanguin (**9**) co-occur with xylerythrin in *P. sanguinea* (*7, 8, 9*).

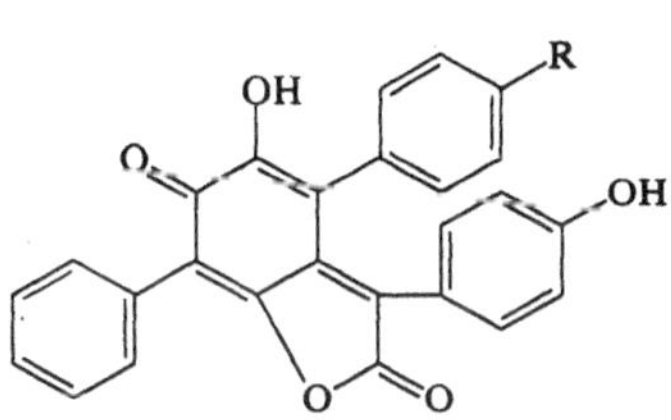

Xylerythrin (**7a**) (R=H)
Peniophorin (**7b**) (R=OH)

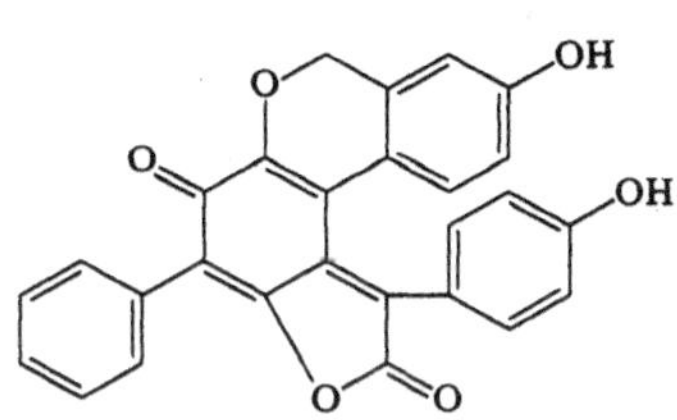

Peniophorinin (**8**)

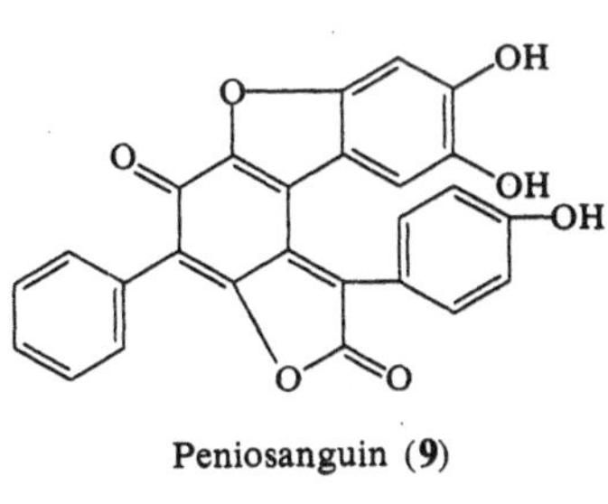

Peniosanguin (**9**)

The ylidenebutenolide patulin (**10**) occupies a somewhat special place in organic chemistry. This structurally simple molecule, which is produced by the mold *Penicillium urticae* (synonym, *P. patulum, P. flexuosum*) was isolated independently by several groups of workers during the early 1940's. It aroused immediate and widespread interest because of its potent antibiotic properties, but this interest evaporated almost as

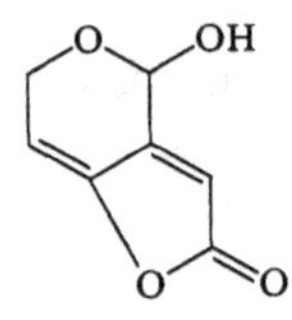

Patulin (**10**)

quickly on account of its high toxicity to animals. Recently the molecule has featured prominently as a useful fungal toxin in food (*10*) and as a general plant toxin (*11*). The very interesting chemistry of patulin (also called "clavicin", "clavatin", "claviformin" and "expansin"), which ultimately led to the structure (**10**) for the metabolite, has been summarized by Dean (*1*). Above all else, the detailed biosynthesis of patulin has sustained the interests of many natural product chemists in this molecule. This interest has its origins in the application of patulin as a convenient and relatively simple model compound in which to examine the detailed enzymology of polyketide biogenesis (see later discussion).

Several members of the Umbelliferae contain alkylidenephthalides and reduced forms of alkylidenephthalides as important odoriferous principles. The essential oil of seed from the celery plant *Apium graveolens*, for example, is the main source of celery odour and flavour in the perfumery

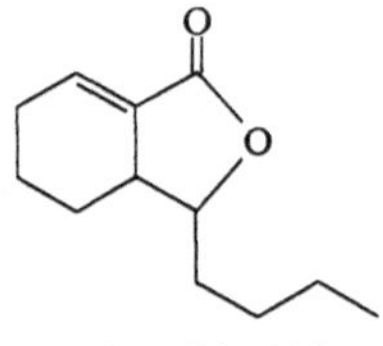

Sedanolide (**11**)

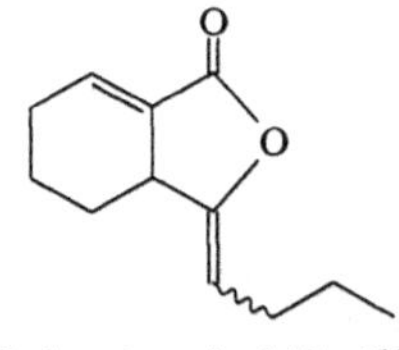

Sedanoic anhydride (**12**)

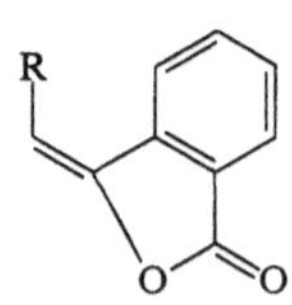

(**13**)

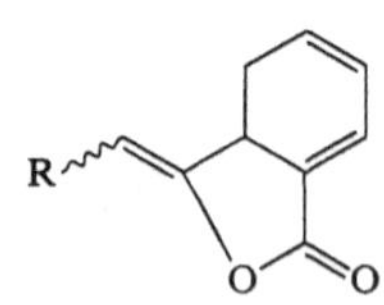

(**14**)

a) $R = i-C_3H_7$
b) $R = i-C_4H_9$
c) $R = n-C_3H_7$

and flavour industries (*12*). The compounds responsible for the characteristic odour of celery oil are sedanolide (**11**) and "sedanoic anhydride" (**12**) (*13, 14*), whereas the alkylidenephthalides (**13a**) and (**13b**), and the related dihydro-derivatives (**14a**) and (**14b**) are the main odoriferous principles in the stem and leaf portions of the plant (*15, 16*). The closely related butylidenephthalide (**13c**) and dihydro-analogue (**14c**) occur together in livèche, *Levisticum officinale Koch* (*17*), *Ligusticum acutilobun* Sieb. *et* Zucc. (*18*) and in *Cnidium officinale* (*19*), also members of the Umbelliferae. Although the geometrical configurations about the exocyclic double bonds in (**12**) and (**14**) are not known, the natural alkylidenephthalides (**13a**) and (**13b**) are thought to have the *E*-stereostructures shown. These configurational assignments were suggested from synthesis of the *Z*- and *E*-forms of the molecules and comparison of spectral and physical data with the natural products (*20*). The highly desirable odoriferous properties of alkylidenephthalides and related molecules have prompted several studies of the odour characteristics of a number of analogues (*21*). Paradoxically, the ylidenebutenolide "bovolide" (**15**) is a characteristic component of butter flavour. It is reported to have a strong odour reminiscent of celery, and occurs to the extent of about 0.5 mg per kg in butter (*22*).

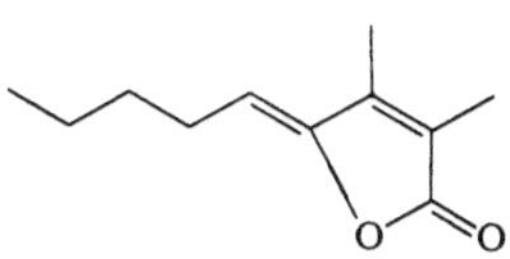

Bovolide (**15**)

Over the past decade or so, Bohlmann and his co-workers have systematically examined hundreds of species of the Compositae for polyacetylenes and related metabolites. These studies have uncovered a whole range of highly unsaturated (acetylenic, allenic and olefinic) and novel 4-ylidendebutenolides in Compositae (Table 1). The first 4-ylidenebutenolide of this class to be found in Nature was the acetylenic compound "matricaria lactone" (**18**). This butenolide was first reported in leaves of *Matricaria inodora* L., and was briefly investigated by Sörensen and Stavholt in 1950 (*23*). A few years later a compound showing similar chemical and spectral properties was isolated along with "matricaria ester" (**16**) from the roots of *Chrysanthemum viscidiflorus* (Hooker). In their contemporaneous studies of the synthesis of acetylenes, Jones and Whiting and their co-workers prepared the *Z*-2, *Z*-8-isomer (**17**) of matricaria acid and showed that in sodium hydrogen carbonate it was smoothly isomerized into the lactone (**18**), identical with the natural metabolite (*24*). The more recent studies of Bohlmann and co-workers

Table 1. *Naturally Occurring Ylidenebutenolides Co-Occurring With Polyacetylenes*

Compound	Occurrence
Matricaria lactone (**18**) 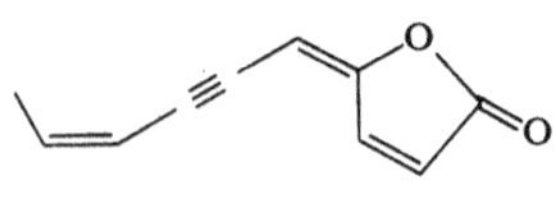 (**19**)	Widespread amongst the Astereae *(Solidago, Happlopappus, Amellus, Aster, Diplopappus, Conyza, Baccharis* species) and the Anthemideae *(Anthemis, Matricaria* species) (*25*). *Matricaria caucasica* Poir (*26*); *Solidago virgaurea* L. (*27*).
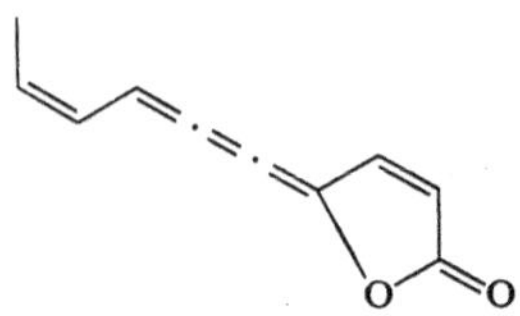 (**20**)	Some Astereae *(Conyza, Boltonia, Erigeron* species) (*25, 28*)
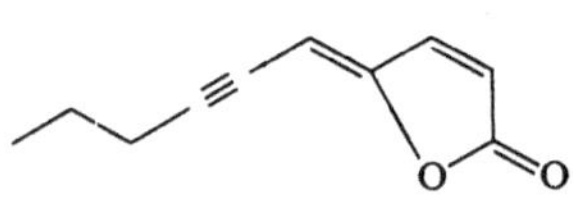 (**21**)	Some Astereae *(Aster, Boltonia* and *Conyza* species) (*25, 29*).
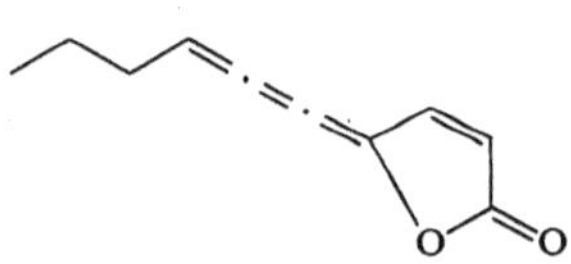 (**22**)	*Erigeron canadensis* L. (*30*) and other Astereae (*25*).
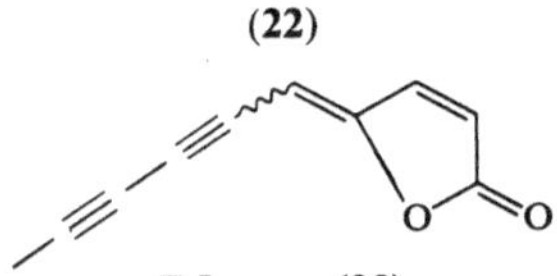 *Z*-Isomer (**23**) *E*-Isomer (**24**)	Widespread amongst the Anthemideae *(Anthemis, Artemisia, Matricaria* and *Tanacetum* species) (*31, 32*).
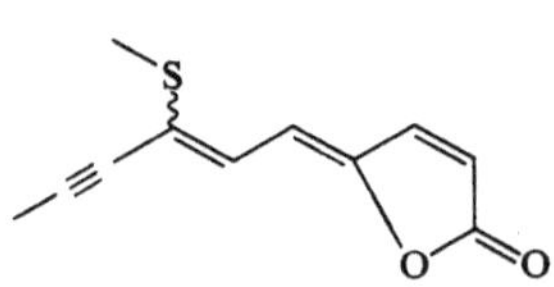 *Z*-Isomer (**25**) *E*-Isomer (**26**)	*Anthemis cupaniara* (*31*); *A. punctata; A. maritima; A. rigescens; A. kelevays* (*25*).
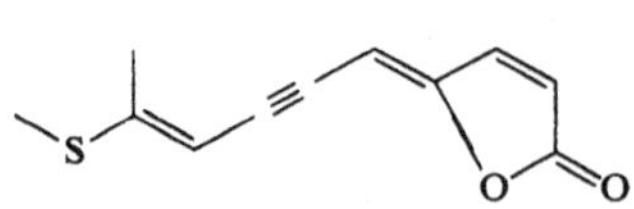 (**27**)	*A. rigescens* (*31*); *A. melanoloma* (*25*).

Table 1 (continued)

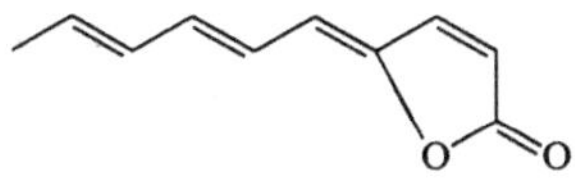

(28)

Matricaria caucasica (Poir) (*26*).

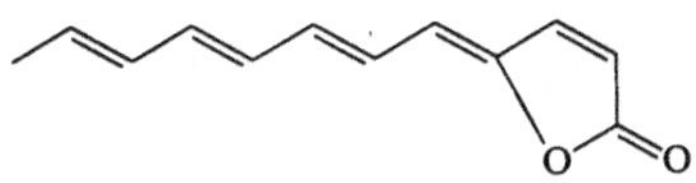

(29)

Some Astereae (*Aster, Boltonia, Conyza* and *Townsendia* species) (*25*).

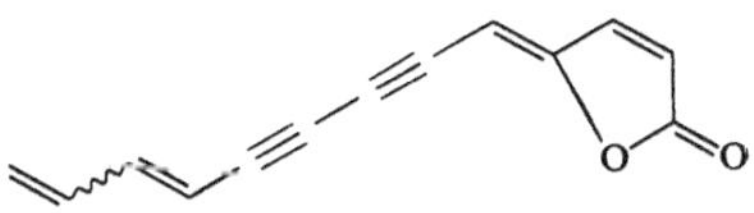

(30)

Carlina vulgaris (*33*).

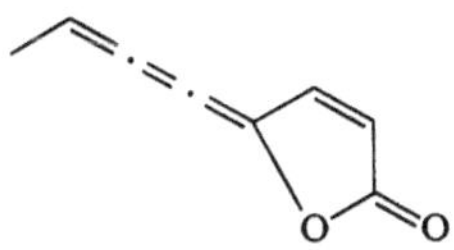

(31)

Some members of *Erigeron*; *Happlopappus brandegei*, *H. clementis* (*25*).

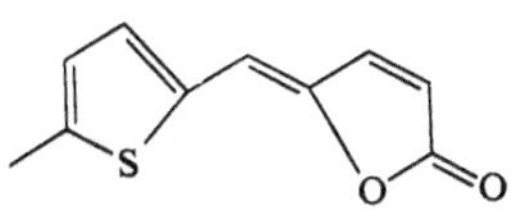

(32)

Anthemis nobilis (*34*); *Artemisia verlotorum* (*25*).

(*25*) have demonstrated that matricaria lactone is particularly widespread amongst the Astereac and Anthemideae. Although the *E*-4, *Z*-8-isomer **(19)** has been reported to co-occur with **(18)** in the aerial parts of *M. caucasica* Poir (*26*) and also in *Solidago virgaurea* (*27*) the isomer may be an artifact produced by photochemical *E*-*Z*-isomerization.

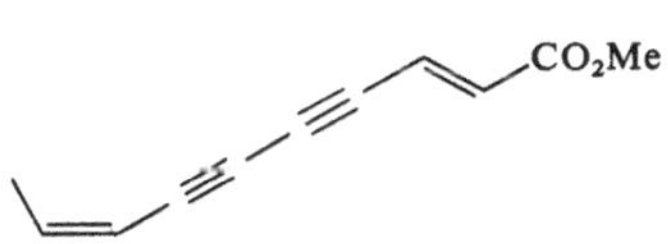

Matricaria ester **(16)**

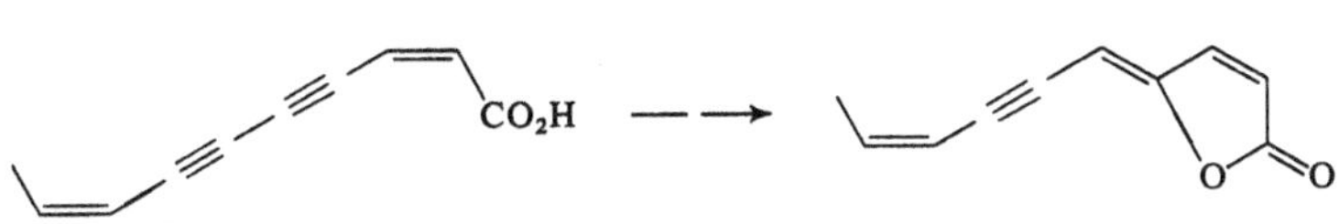

(17) Matricaria lactone **(18)**

The cumulenic butenolide (**20**), also isomeric with matricaria lactone, has been isolated from several species of *Conyza, Boltonia* and *Erigeron* (*25, 28*); this butenolide was the first cumulene to be found in Nature. A reduced form of matricaria lactone (*viz.* **21**) occurs in several species of *Aster, Boltonia* and *Conyza* (*25, 29*) and the cumulene (**22**) related to (**21**) has been isolated from the aerial parts of *Erigeron canadensis* L. (*30*).

Particularly widespread amongst members of the *Anthemideae* are the *Z,E*-acetylenic butenolides (**23**) and (**24**) (*31, 32*). The structurally novel thio-ethers (**25**), (**26**) and (**27**) co-occur with (**23**) in the roots of *Anthemis rigescens* (*31*). *Z*-Butenolide (**23**) has also been isolated from *M. caucasica* and from several species of *Artemisia* and *Tanacetum* (*25*).

The conjugated polyolefinic butenolide (**28**) is found with matricaria lactone (**18**) in *M. caucasica* (Poir) (*26*), whereas pentaene (**29**) has been isolated from several members of the Astereae (*25*). An unusual C_{13}-diene-diyne-butenolide has been extracted from *Carlina vulgari*[3], and the molecule has been assigned constitution (**30**) (*33*). The C_8-cumulene (**31**) is commonly found in *Erigeron* species (*25*), but it is also present in *Happlopappus brandegei* and *H. clementis.* Finally, in this group the novel thiophene-butenolide (**32**) has been isolated along with polyacetylenes from *Anthemis nobilis* (*34*) and *Artemisia verlotorum* (*25*).

An acetylenic 4-ylidenebutenolide of a quite different constitution is freelingyne (**33**). This highly unsaturated furano-sesquiterpene is found in the heartwood of the Australian shrub *Eremophila freelingii,* and it was the first acetylenic terpenoid to be isolated (*35*). Dihydrofreelingyne (**34**) has also been isolated from *E. freelingii* (*36*) and the unusual diterpenoid eremolactone (**35**) occurs in the leaves of the same species (*37*). The complete stereochemistry of freelingyne has been established by direct X-ray methods (*38*).

One of the more complex molecules which incorporate a 4-ylidenebutenolide system is the C_{37}-carotenoid pigment peridinin (**36**). This orange-red coloured pigment is the principal carotenoid of dinoflagellates

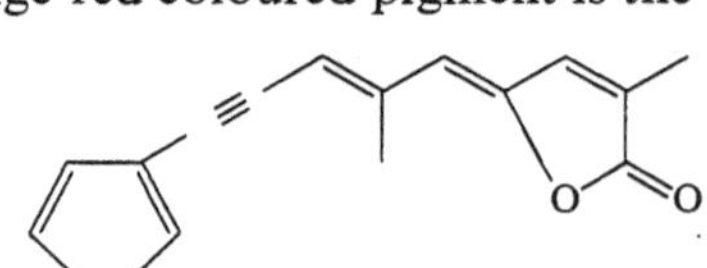

Freelingyne (**33**)

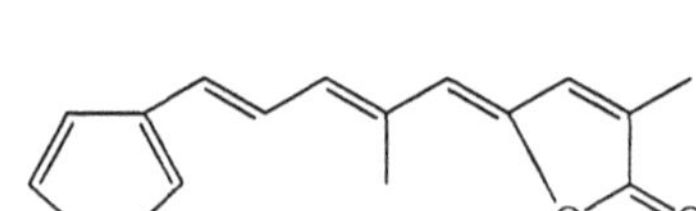

Dihydrofreelingyne (**34**)

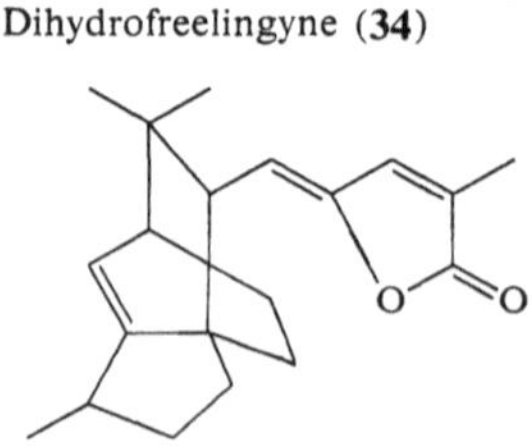

Eremolactone (**35**)

and was first isolated in 1890. It has since been found in several marine and fresh-water dinoflagellates, including endozooic symbionts of marine corals, clams and sea anemones. The unique C_{37}-tricyclic formulation (**36**) for peridinin was published in 1971 (*39*). In an investigation of the photosynthetic dinoflagellates *Amphidinium carterae, Glenodinium splendens, G. nelsoni* and *G. dorsum,* LIAAEN-JENSEN and co-workers have also isolated peridinenol (**37**), pyrrhoxanthin (**38**) and pyrroxanthinol (**39**) which represent structural variations of peridinin (*40*). The C_{37}-metabolites (**36—39**) occur together with carotenoids possessing the ordinary C_{40}-carotenoid skeleton, which makes the problem of the biogenesis of the unusual C_{37}-skeleton a particularly intriguing one (see later discussion).

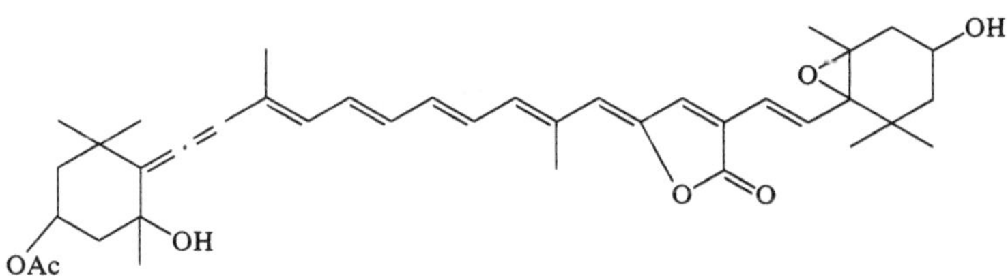

Peridinin (**36**)

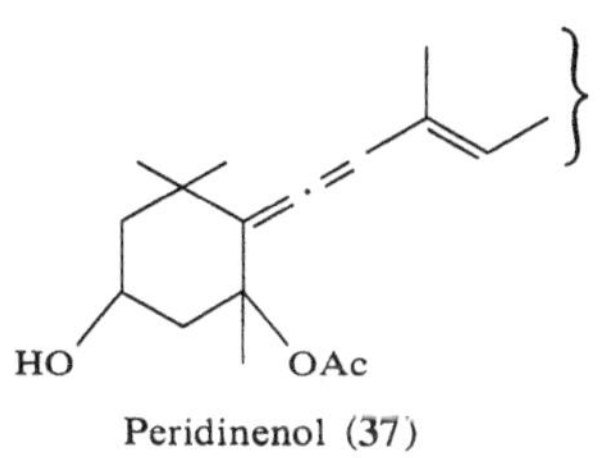

Peridinenol (**37**)

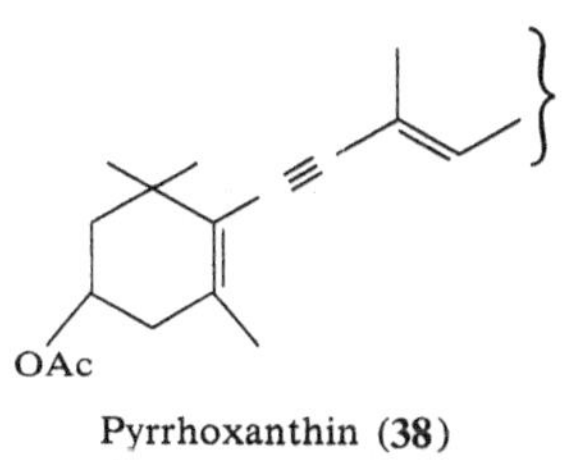

Pyrrhoxanthin (**38**)

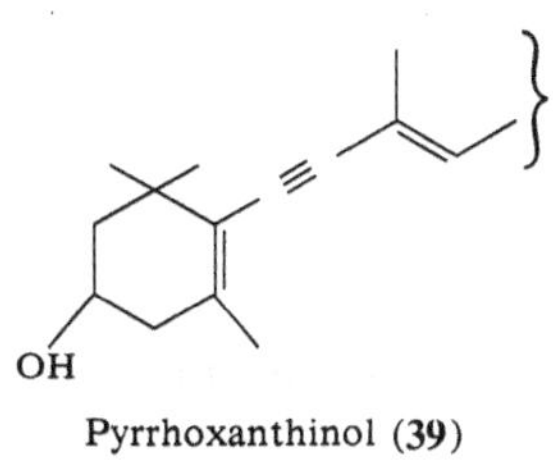

Pyrrhoxanthinol (**39**)

Tetrenolin, which has been formulated as (**40**), is another example of a natural 4-ylidenebutenolide displaying extended conjugation at C-2 (*41*). This yellow crystalline butenolide was isolated in 1969 from cultures of *Micropolyspora venezuelensis* Thiemann 1969 and shows antibacterial properties (*42*). The novel sesterterpenoid ylidenebutenolides (**41 a**) and (**41 b**) have been isolated from the Australian sponge *Thorecta marginalis* (*43*). These metabolites are closely related structurally to several sesterterpenoid ylidenetetronic acids found earlier in other members of the sponge family (Dictyoceratida) (see later discussion).

4-Ylidenebutenolides of general constitution (**42**) have been reported amongst the products of metabolism of the hormone herbicides 4-chloro-, 4-chloro-2-methyl-, and 2,4-dichloro-phenoxyacetic acids (*44, 45, 46*).

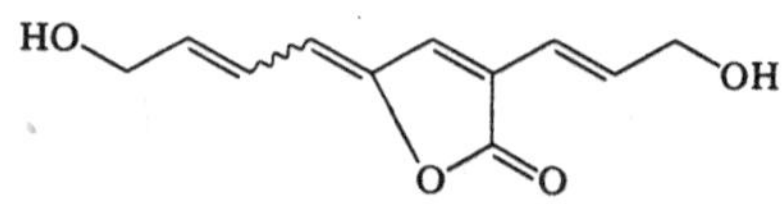

Tetrenolin (**40**)

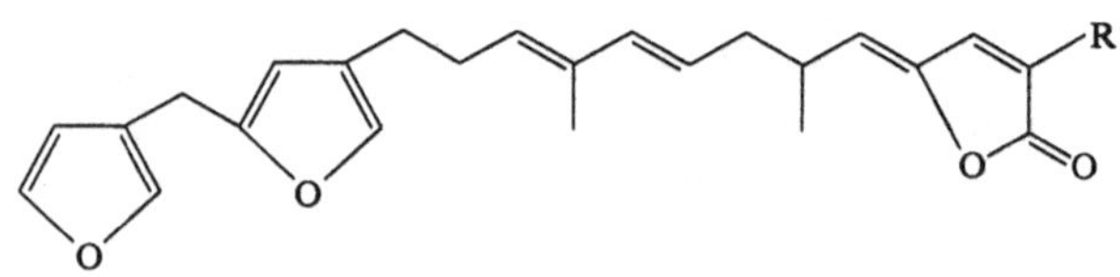

(**41**) *a)* R = Me
b) R = CH_2OH

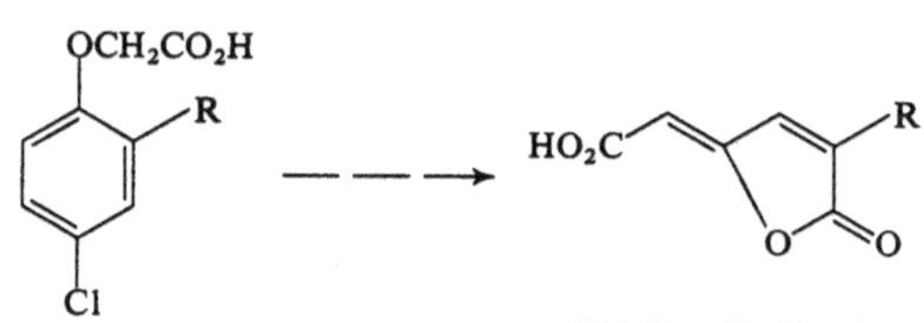

(**42**) R = H, Me, Cl

2. 4-Ylidenetetronic Acids

Natural products which incorporate a 4-ylidenetetronic acid ring system have been known for almost a century. The ring system is a characteristic structural feature in the "pulvinic acid" group of yellow and red pigments found in lichens and known since 1883. In more recent years pulvinic acids have also been isolated from certain higher fungi belonging to the Boletaceae and Gomphidiaceae. Their production in lichens is almost certainly associated entirely with the fungal partner.

Formulation (**43**) shows the basic pulvinic acid carbon skeleton, and all known natural pulvinic acids show variations which include either hydroxylation of methoxylation of the benzene rings or lactonisations involving the carboxy function (Table 2). In addition, a few methyl ester and amide derivatives of the acids occur naturally. Early studies of the constitutions of pulvinic acids found in lichens have been summarized by DEAN (*1*), and a useful account of pigments found in mushrooms, which includes pulvinic acid pigments, has been presented recently by STEGLICH (*47*).

Table 2. *Pulvinic Acids of Lichen and Fungi*

Compound	Source
Pulvinic Acid (*43*)	*Sticta coronata* (*48*).
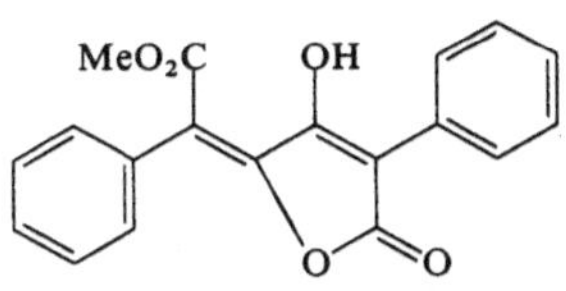 Vulpinic acid (**44**)	*Lepraria, Evernia, Centraria, Cyphelium, Calycium* and *Parmelia* species of Lichen (*1*).
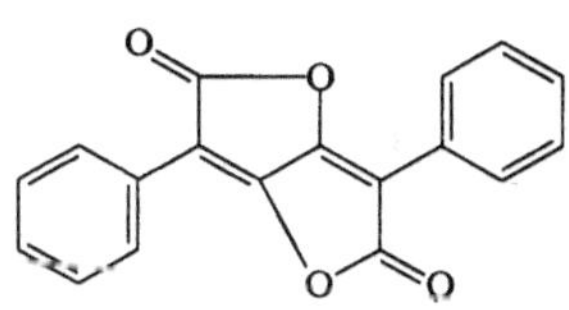 Pulvinic anhydride	*Sticta, Candelaria,* and *Gyalolechia* species of Lichen (*1*).
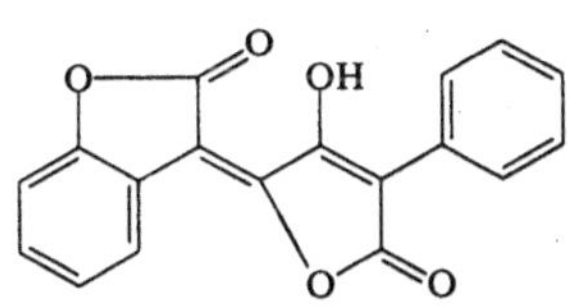 Calycin (**46**)	*Sticta, Candelaria Gyalolechia, Lepraria, Gasparrima, Physcia, Chrysothrix* and *Callopsima* species of Lichen (*1, 50*).
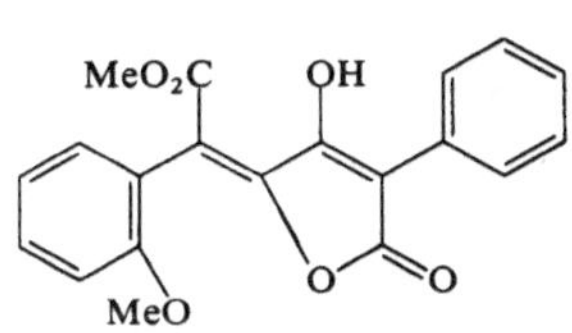 Leprapinic acid	*Lepraria citrina* Schaer (*1*).
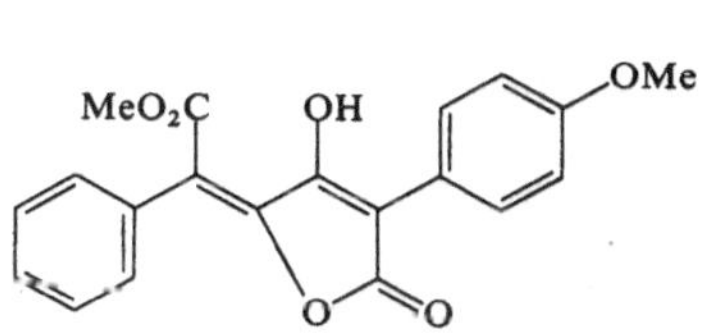 Pinastric acid (**47**)	*Lepraria, Cetraria* and *Calycium* species of Lichen (*1, 52*).
Variegatic acid (**50**)	Several *Boletus* species (*55*); *Xerocomus chrysenteron* (*56*); *Gomphidius glutinosus* (*57*); *Rhizopogen roseolus* (*58*).

Table 2 (continued)

Variegatorubin (**52**)

Suillus piperatus (60); Rhizopogon roseolus (58); Boletus rubellus (62).

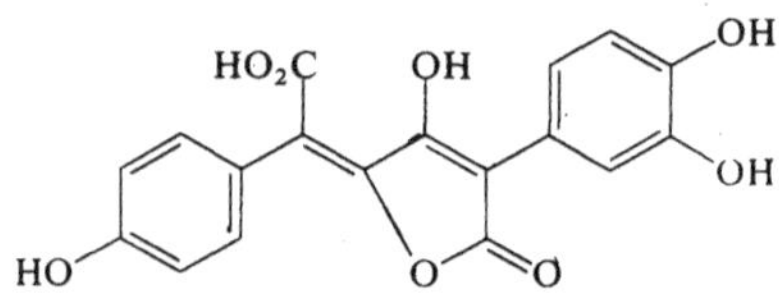

Xerocomic acid (**53**)

Xerocomus chrysenteron (56); Gomphidius glutinosus (57); G. rutilus (62); Paxillus atrotomentosus (61).

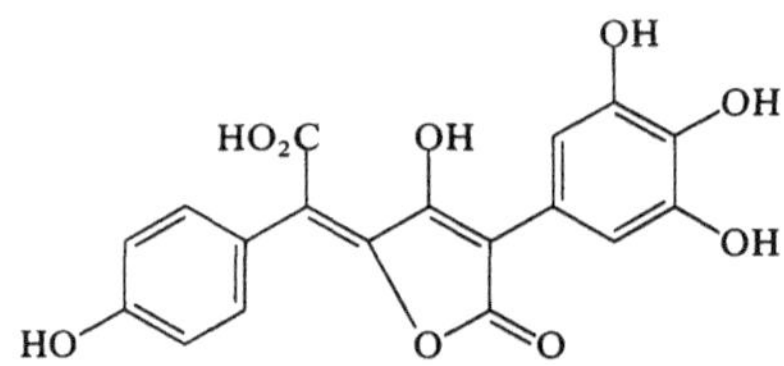

Gomphidic acid (**56**)

Gomphidius glutinosus (57).

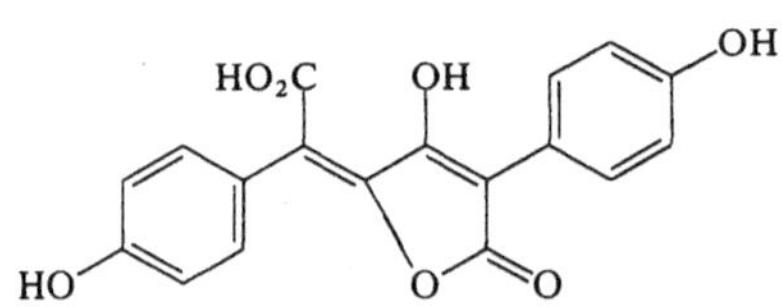

Atromentic acid (**57**)

Xerocomus chrysenteron (56); Paxillus atrotomentosus (61); Clitocyne illudens (64); Leccinum aurantiacum (65).

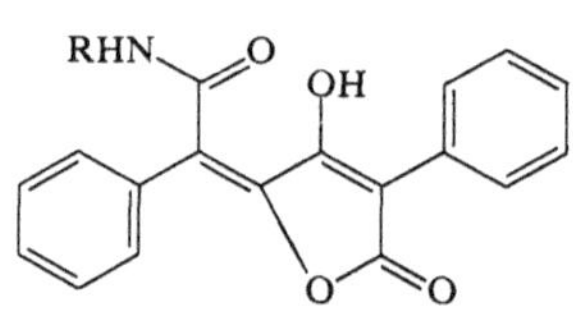

Pulvinamide (**58**)
R = H

Several species of *Pseudocyphellaria* (*66*).

Epanorin (**58**)
$R = CH(CO_2Me)CH_2CHMe_2$

Epanora lecanora (1, 67).

Rhizocarpic acid (**58**)
$R = CH(CO_2Me)CH_2Ph$

Rhizocarpon geographicum (1).

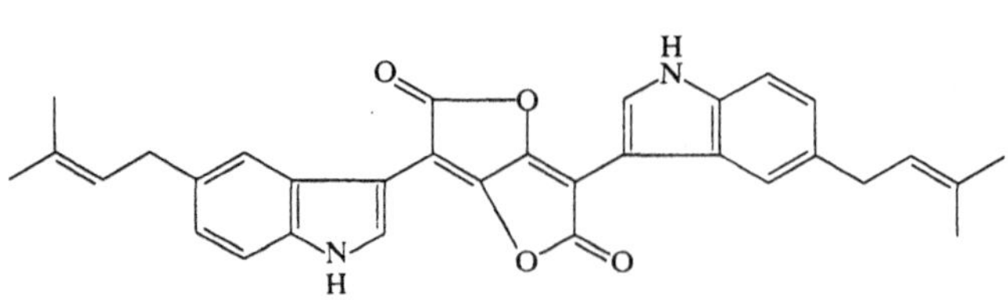

Cochliodinone (**59**)

Chaetomium globosum; C. cochliodes (68).

References, pp. 189—198

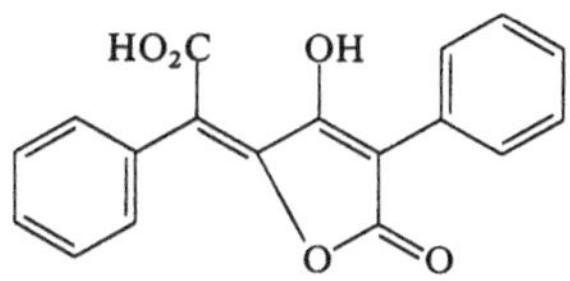

Pulvinic acid (**43**)

Pulvinic acid itself (**43**) was not found in nature until 1952 (*48*), although the corresponding methyl ester, "vulpinic acid" (**44**) has been known since 1831. The deep-red coloured pigment "calycin" was first isolated from the lichen *Lepraria candelaris* Shaer in 1880. Degradative and synthetic work reported by Asano and Kameda led to the *bis*-lactone formulation (**45**) for the pigment (*49*), but the more recent investigations by Åkermark conclude that the pigment actually possesses the mono-lactone structure (**46**) (*50*). Indeed, structural elucidation amongst the pulvinic acids has frequently presented problems particularly in those instances where the benzene rings are unsymmetrically substituted (*51*). During a period covering almost a decade Seshadri and co-workers, for example, examined the structure of pinastric acid (**47**), a yellow pigment found in a number of *Lepraria, Cetraria* and *Calycium* species of lichen. Extensive degradative, synthetic and chemical studies led to either of the formulations (**47**) or (**48**) for the metabolite. The former structure was ultimaltely favoured (*52*) which endorsed conclusions arrived at some thirty years earlier by Asano and Kameda (*49*). The structure (**47**) for pinastric acid has recently been confirmed by total synthesis (*53*).

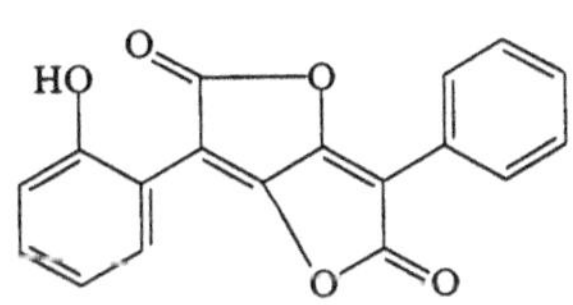

(**45**)

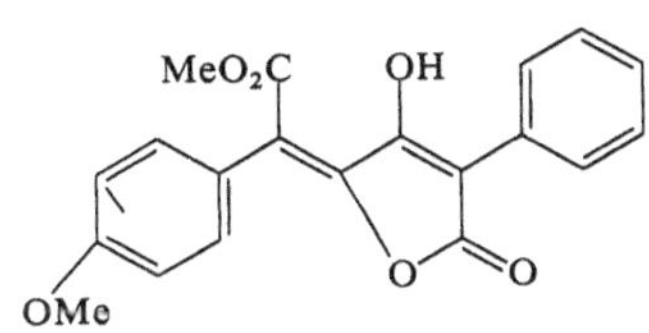

(**48**)

The difficult problem of assigning structures to unsymmetrically substituted pulvinic acids has received recent attention. Edwards and Gill (*51*) have made the interesting observation that when fully methylated derivatives of pulvinic acids are treated with alkali, they are converted into 2-aryl-3-methoxylated maleic anhydrides as indicated in Scheme 1. Anhydrides of the type (**49**) can be compared with authentic synthetic specimens, and based on the outcome of these comparisons, structures can be assigned to the original pulvinic acids. In contemporaneous investigations, Steglich and co-workers (*54*) have demonstrated that comparative shift data in the p.m.r. spectra of isomeric pulvinic acids can provide strong clues to the structure of unsymmetrical pulvinic acids.

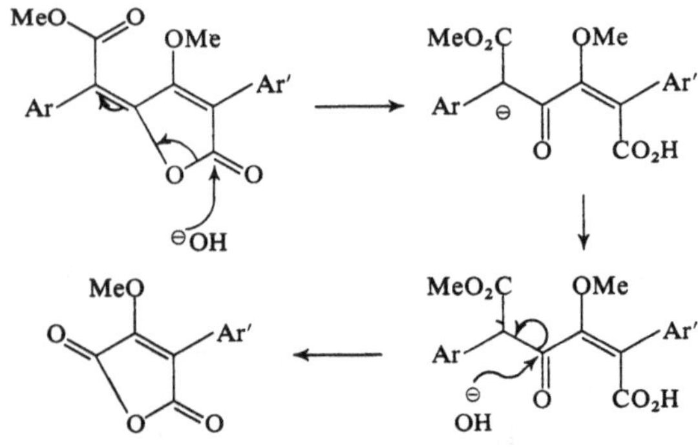

(49)

Scheme 1. Alkaline degradation of permethylated pulvinic acids

The tetra-hydroxy substituted pulvinic acid, variegatic acid (**50**), was the first example of a 4-ylidenetetronic acid to be isolated from a Basidiomycete. First isolated by EDWARDS and co-workers in 1967 (*55*) from the fungus *Suillus (Boletus) variegatus* (Swartz *ex* Fr) it has now been reported in several members of Boletus, and also in *Xerocomus* (*56*), *Gomphidius* (*57*) and *Rhizopogon* (*58*). The "blueing" of the flesh of boleti caused by bruising has been known for a long time and the phenomenon has attracted considerable interest and attention. In this respect it is interesting that dilute solutions of variegatic acid (λ_{max} ~ 415 nm) rapidly become blue (λ_{max} ~ 595 nm) in the presence of dilute alkali or

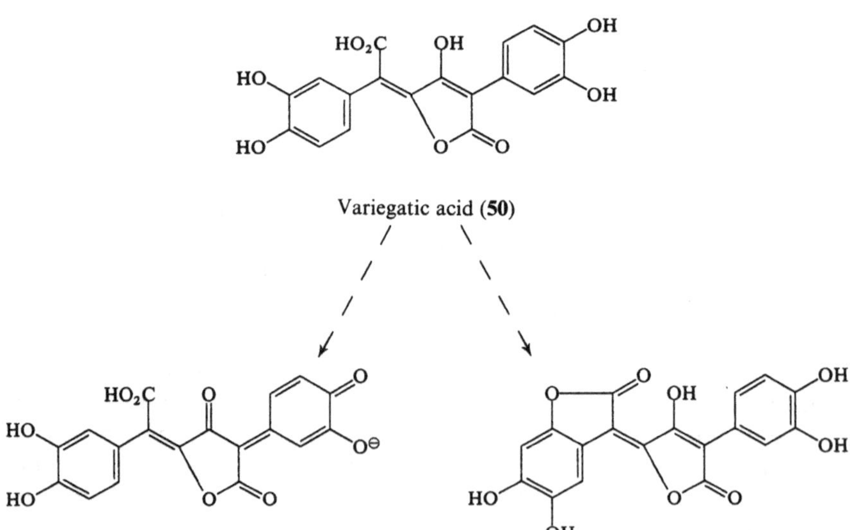

(**51**) (λ_{max} 595 nm)

Variegatorubin (**52**) (λ_{max} 510 nm)

when potato extract is added, the colour closely resembling that which occurs when the fungus is bruised. The particular reaction has been investigated by STEGLICH and co-workers (*59*) who have shown that the blueing is due to the production of the quinone-methide chromophore shown in (**51**). The closely related red-coloured variegatorubin (**52**) ($\lambda_{max} \sim 510$ nm) has been found in *Suillus piperatus* (*60*) and other Boletaceae, and can be produced *in vitro* when variegatic acid is treated with copper acetate in acetic acid.

The structure of xerocomic acid (**53**), found in a variety of fungi (*56, 57, 61, 62, 63*), has been the subject of considerable investigation. Constitution (**53**) followed for the metabolite, largely as a result of studies of the oxidative degradation of the two synthetic permethylated isomers (**54**) and (**55**), and comparison between these data and those obtained for the natural metabolite (*63*). Evidence for the orientation of the hydroxy substituents in the related gomphidic acid (**56**), found in *Gomphidius glutinosus* (*57*), until very recently, rested solely on comparative p.m.r. shift data with model compounds (*54*). A total synthesis of permethylated gomphidic acid has now confirmed the structure (**56**) (*53*).

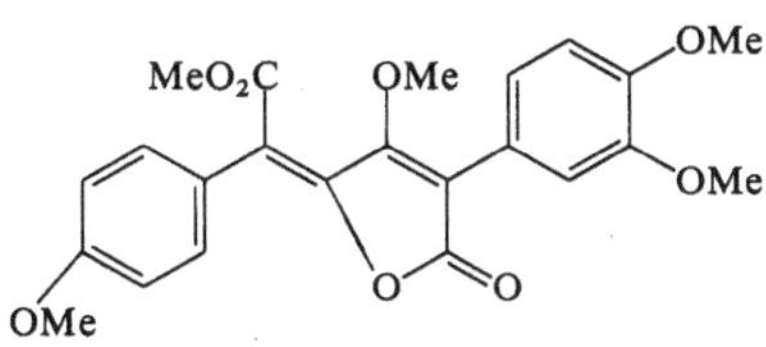

(**54**)

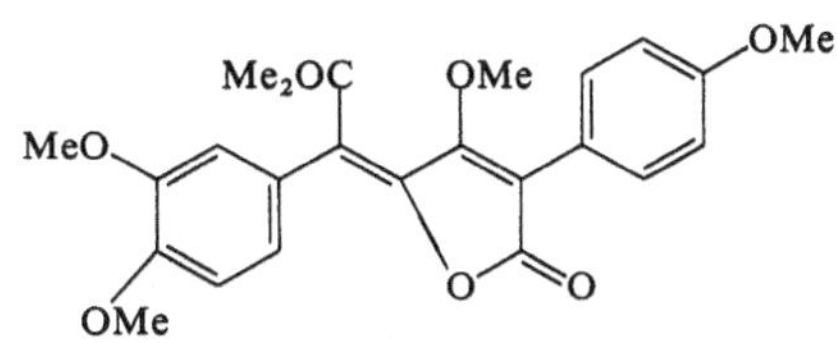

(**55**)

Variegatic and xerocomic acids co-occur with atromentic acid (**57**) in *Xerocomus chrysenteron* (*56*), and atromentic acid has also been isolated from cultures of the mushrooms *Paxillus atrotomentosus* (*61*), *Clitocyre illudens* (*64*) and *Leccinum aurantiacum* (*65*).

A number of nitrogen-containing pulvinic acid lichen pigments are known. Three of these, designated pulvinamide, epanorin and rhizocarpic acid are amides (*viz* **58**) formally derived from combination of pulvinic acids with ammonia, L-leucine and L-phenylalanine respectively. Pulvinamide (**58**, R = H) was found only recently in nature (*66*) whereas epanorin and rhizocarpic acid have been known for almost a century (*1*). The indole analogue (**59**) of pulvinic anhydride is unique in that it represents the only example of a natural "pulvinic acid" containing a heterocyclic ring. This orange-coloured fluorescent metabolite, called cochliodinone, is produced by cultures of *Chaetomium globosum* and *C. cochliodes* (*68*).

The group of 4-ylidenetetronic acid pigments known as "pulvinones" are closely related to the pulvinic acids and possess the general structure (**60**). It is interesting that pulvinones have only very recently been reported in nature, whereas the basic pulvinone carbon-skeleton has been known since about 1895. During early investigations of the constitutions of lichen pulvinic acids, a convenient degradation process was to effect decarboxylation of the metabolites by heating with, for example, copper chromite (*69*, *70*). This procedure not only led to the anticipated "pulvinone" product (*viz* **60**) but also to a diarylcyclopentanetrione (**62**) and to an isomeric pulvinone (**61**) produced by thermal equilibration (Scheme 2). The interesting findings were later exploited by CLAISEN (*71*) in the development of a useful synthesis of the pulvinone skeleton based on condensation between diethyl oxalate and *bis*-benzyl ketones followed by thermal equilibration of the resulting cyclopentanetriones (see Synthesis section).

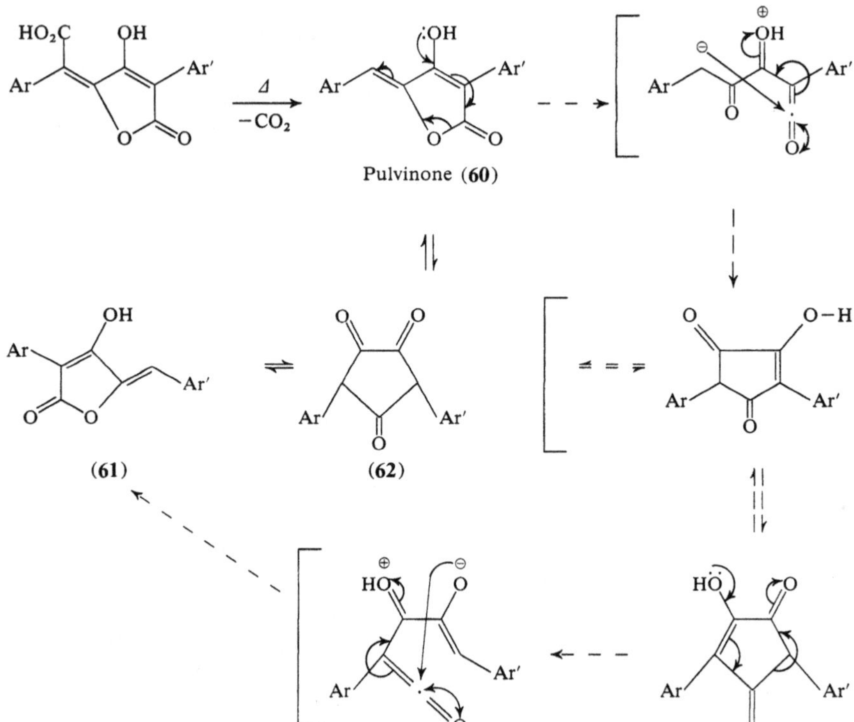

Scheme 2. Equilibration of pulvinones *via* diarylcyclopentan-2,3,5-triones

During 1973, EDWARDS and GILL (*70*) reported the isolation of the trihydroxypulvinone (**63**) from the stems of the common larch mushroom (*Suillus grevillei* (Klotsch) Sing.), and SETO and his co-workers found the dihydroxy compound (**64**) in cultures of *Aspergillus terreus* (*72*). Two years later SETO *et al.* (*73*) reported the presence of six additional pulvinones in *A. terreus*, some containing prenylated aromatic rings*). The structures first assigned to five of these metabolites were subsequently shown to be incorrect. Synthetic studies by KNIGHT and PATTENDEN (*74, 75*), and application of X-ray methods (*76*) have now established constitutions (**65**), (**66**), (**67**), (**68**) and (**69**) for five of the metabolites; the other pulvinone, an oxidized derivative of (**66**), has one of the constitutions (**70**), (**71**) or (**72**) and most probably the latter. The presence of pulvinone (**64**) in *A. terreus* has been confirmed by other workers (*77*), and recent enzymatic studies of the biosynthesis of the prenylated pulvinones (**68**) and (**69**) in *A. terreus* have demonstrated the presence of the two additional and biosynthetically significant metabolites (**73**) and (**74**) in young cultures (*75, 78, 79*).

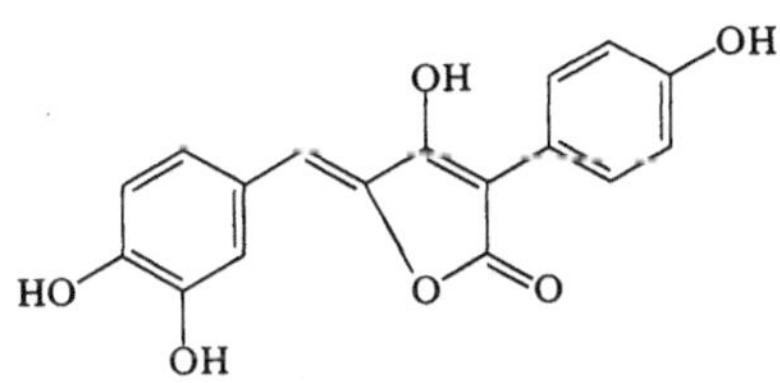

(**63**)

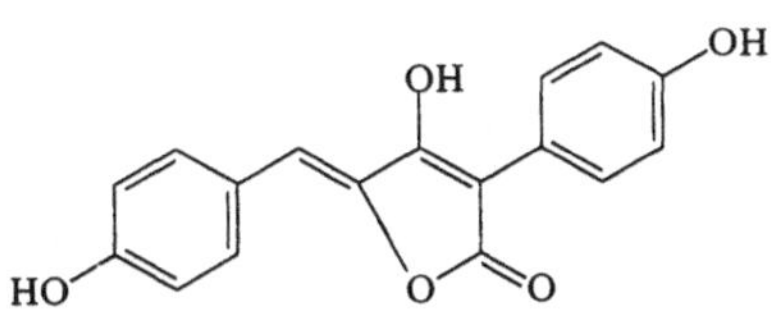

(**64**)

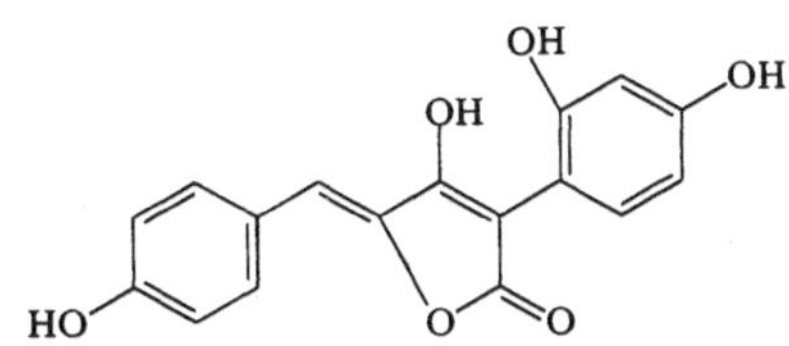

(**65**)

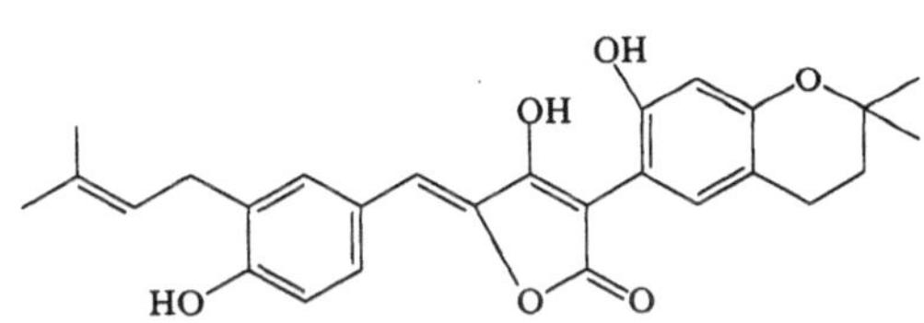

(**66**)

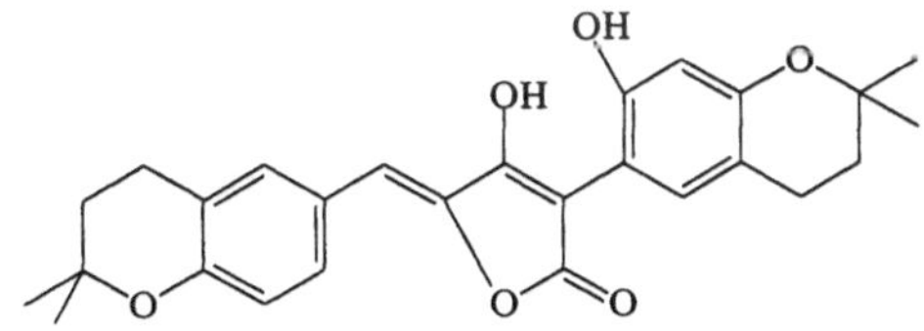

(**67**)

* The name "aspulvinone" has been suggested to designate the family of pulvinone metabolites isolated from *A. terreus* (*79*).

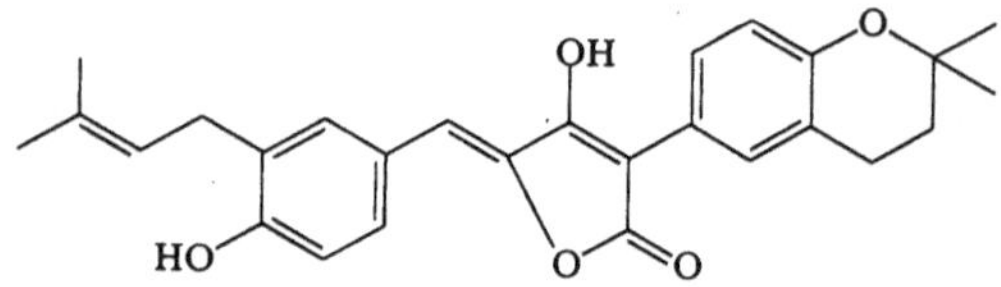

(68)

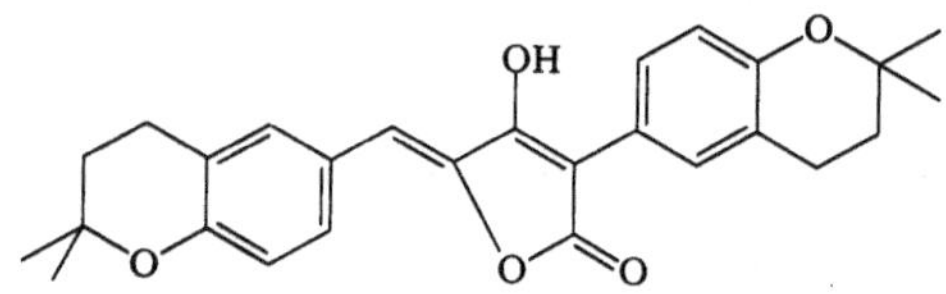

(69)

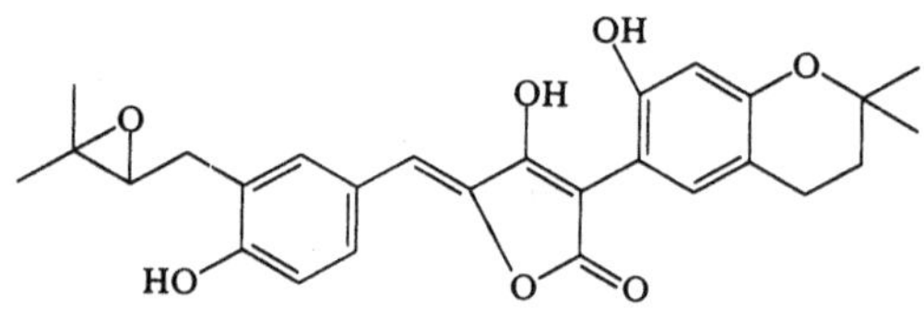

(70)

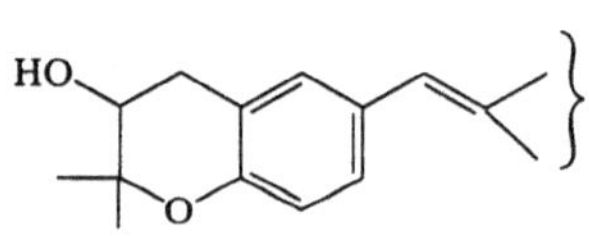

(71)

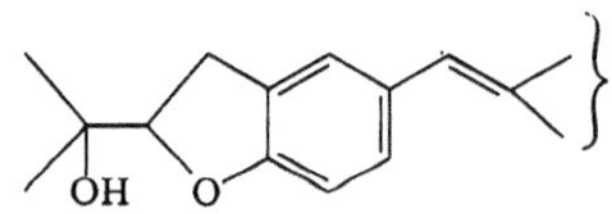

(72)

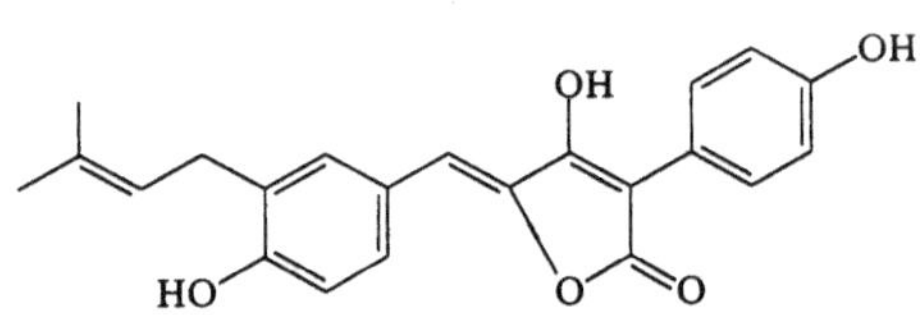

(73)

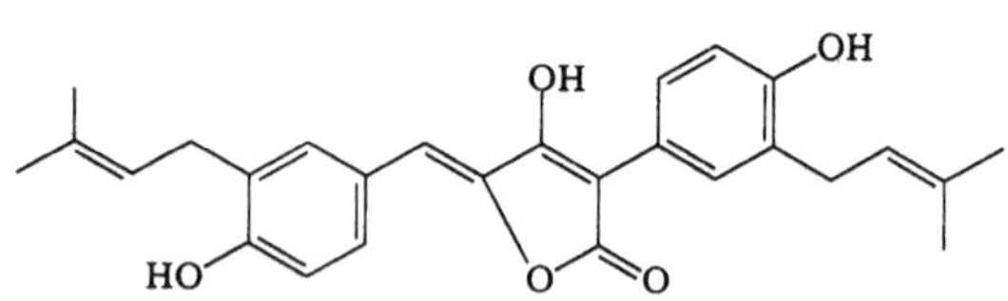

(74)

Aspulvinone pigments from *Aspergillus terreus*

A small group of novel 4-ylidenetetronic acid derivatives has been isolated from *Penicillium* species. The first of these to be discovered was dehydrocarolic acid from *P. cinerascens* Biourge in 1947 (*80*). The structure (**75**) of this metabolite resembles closely those of carolic acid (**76a**), carolinic acid (**76b**), carlic acid (**77a**), and carlosic acid (**77b**) from *P. charlesii* G. Smith, and of terrestric acid (**76c**) from *P. terrestre* Jensen (*1*). It is most probable that dehydrocarolic acid exists as the hydrated form (**75**) only in solution, and that the crystalline compound has the ether structure (**78**) (*81*). Multicolic (**79**) and multicolosic acid (**80**) have been found only recently in *P. multicolor* (IMI 104602) (*82*).

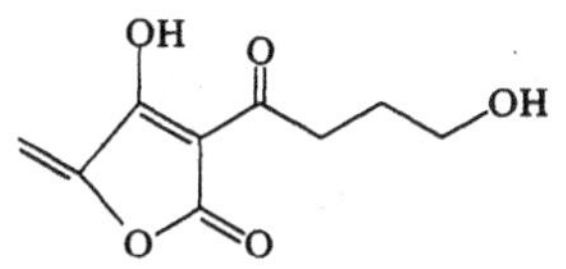

Dehydrocarolic acid (**75**)

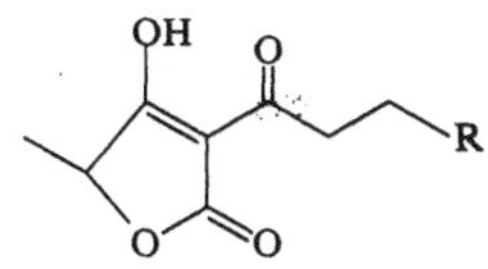

Carolic acid (**76a**) (R = CH_2OH)
Carolinic acid (**76b**) (R = CO_2H)
Terrestric acid (**76c**) [R = CH(OH)Et]

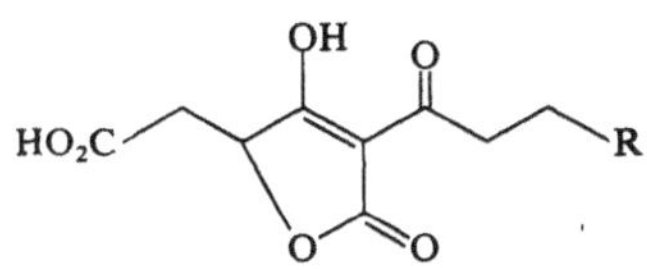

Carlic acid (**77a**) (R = CH_2OH)
Carlosic acid (**77b**) (R = CH_3)

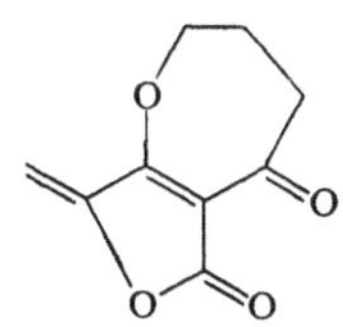

(**78**)

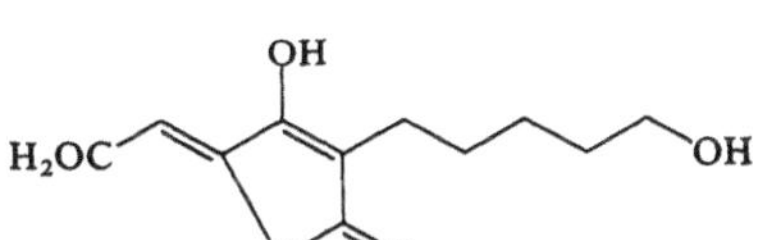

Multicolic acid (**79**)

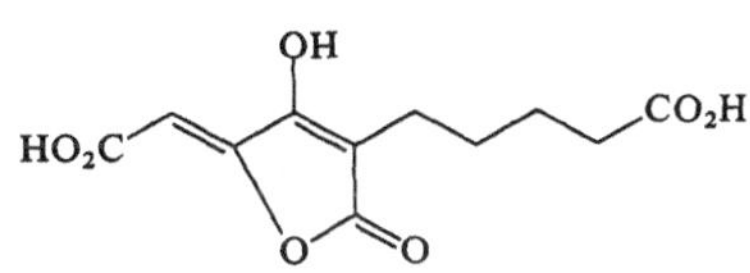

Multicolosic acid (**80**)

The extensive investigations of natural products from marine sources during the past decade have uncovered the presence of a number of interesting ylidenetetronic acids from marine sponges (Table 3). Constitutionally these are all sesterterpenes, and display furan ring systems in addition to the tetronic acid ring systems. The first of these metabolites to be characterized was fasciculatin (**81**) from the sponge *Ircinia fasciculata* in 1972 (*83*). Ircinins-1 and -2, (**82**) and (**83**) respectively, were reported soon after in *I. oros* (*84*) and FAULKNER described the isolation of variabilin (**84**) from *I. variabilis* in 1973 (*85*). Strobilinin (**85**), a positional

Table 3. *Ylidenetetronic Acid Sesterterpenoids from Marine Sponges*

Compound	Source
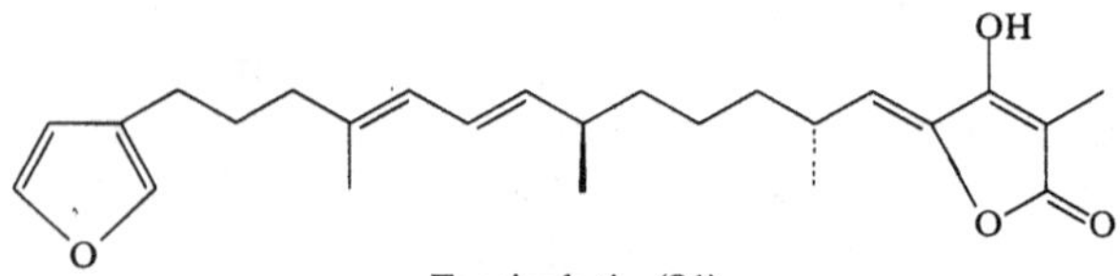 Fasciculatin (**81**)	*Ircinia fasciculata* (*83*).
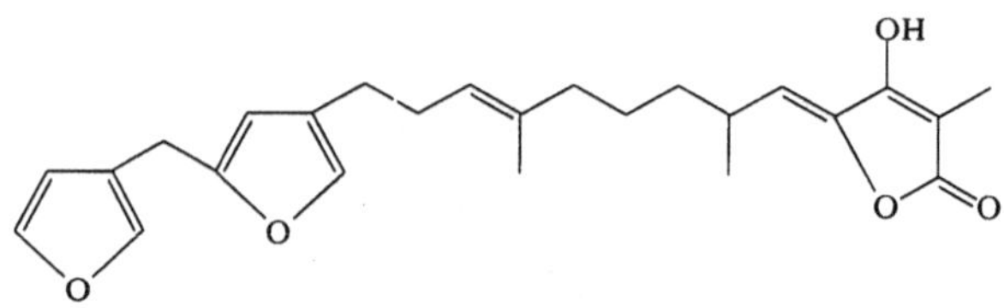 Ircinin-1 (**82**)	*Ircinia oros* (*84*).
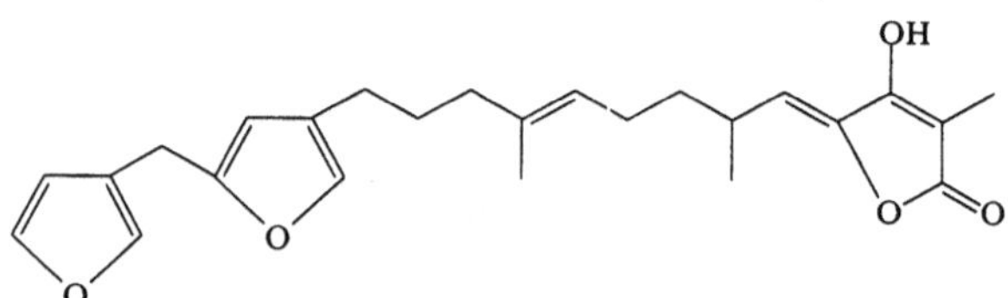 Ircinin-2 (**83**)	*I. oros* (*84*).
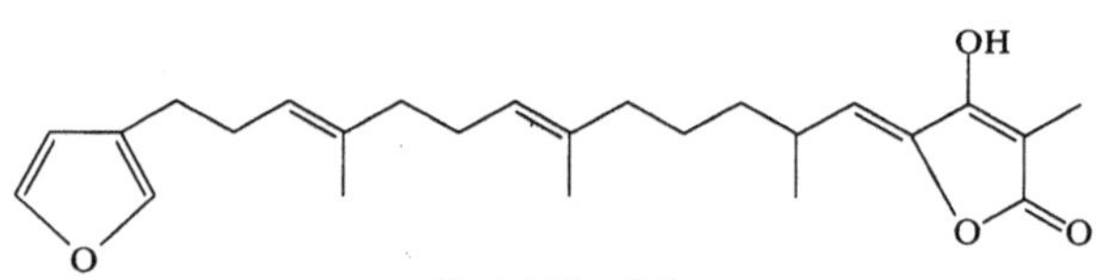 Variabilin (**84**)	*I. variabilis* (*85*); *I. strobilina* (*86*); *Fasciospongia fovea* (*43*).
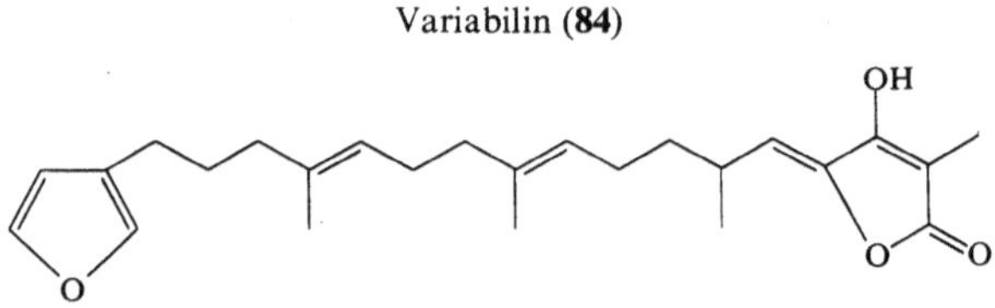 Strobilinin (**85**)	*I. strobilina* (*86*).

isomer of variabilin, has been isolated with variabilin from *I. strobilina* (*86*). The ircinins (**82**) and (**83**) show a close structural similarity to ircinolide (**41 a**) and 24-hydroxyircinolide (**41 b**) found later in the Australian sponge *Thorecta marginalis* (*43*).

Piperolide (**86**) which has been found in the roots and stems of *Piper sanctum* (Mig.) Schlecht was the first example of a 4-ylidenetetronic acid derivative to be characterized from a higher plant (*87*). The corresponding methylenedioxy- and epoxy-piperolides (**87**) and (**88**) respectively, have also been isolated from the same plant (*88*).

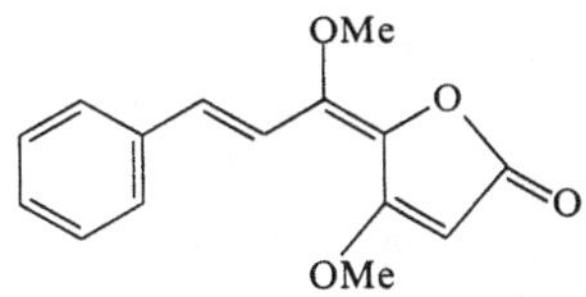

Piperolide (**86**)

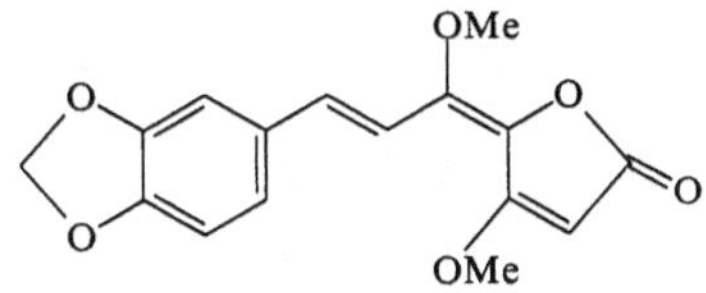

(**87**)

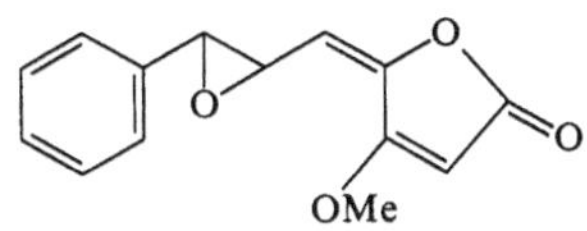

(**88**)

III. Synthesis

In general terms, two broadly based approaches towards the synthesis of 4-ylidenebutenolides and 4-ylidenetetronic acids can be discerned. The first of these approaches involves, as a key stage, the intramolecular cyclisation of an acyclic precursor molecule, frequently in the form of a 4-oxo-2-enoic acid (*viz* **89**) or a *Z*-2-en-4-ynoic acid (*viz* **90**) (Scheme 3).

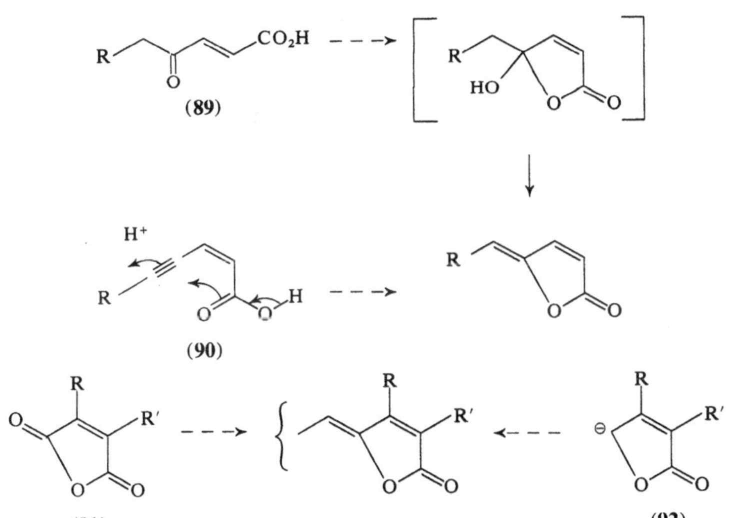

Scheme 3. Synthetic approaches to 4-ylidenebutenolides and 4-ylidenetetronic acids

In the second approach, a substrate incorporating an intact but-2-enolide system is employed as a pivotal intermediate. Thus, either of the procedures involving addition of carbon nucleophiles to substituted maleic anhydrides (*viz* **91**) or addition of carbanions produced from simple butenolides (*viz* **92**) to carbonyl compounds, followed by appropriate 1,2-eliminations, provide expeditious routes to the corresponding 4-ylidene systems (Scheme 3). The present discussion will be made with these broadly-based approaches in mind. Synthetic approaches which are not accommodated comfortably within this framework are collected in a miscellany.

1. From 4-Oxo-2-enoic Acids

Although a number of synthetic approaches have been made towards protoanemonin (**5**), one of the most convenient starts from readily available levulinic acid which is first converted into acetylacrylic acid (**93**) by successive bromination and dehydrobromination. When the acrylic acid is warmed with a mixture of acetic anhydride and acetic acid containing a small amount of sulphuric acid, proteoanemonin is secured in approximately 30% yield (*89*); studies show that the last reaction proceeds by way of the 4-acetoxy derivative (**94**). Two useful variations to this general approach to protoanemonin and analogues are worthy of mention. In the first, GRUNDMANN and KOBER (*90*) have shown that when levulinic acid is first converted into the corresponding unsaturated lactone (**95**) ("α-angelica lactone") subsequent bromination and dehydrobromination leads to considerably enhanced yields of protoanemonin. A later publication by WALTON describes the application of the Diels-Alder adduct (**96**), obtained from maleate, in the synthesis of

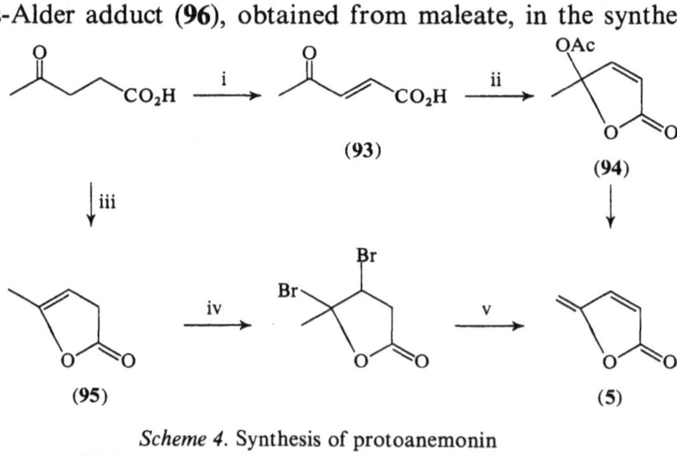

Scheme 4. Synthesis of protoanemonin

Reagents: i, $Br_2(-HBr)$; ii, $Ac_2O-HOAc-H_2SO_4$; iii, $\Delta(-H_2O)$; iv, Br_2; v, quinoline.

protoanemonin and several analogues (*91*). Controlled saponification of (**96**), followed by conversion to the keto acids (**98**) *via* the corresponding acid chlorides (**97**) and lactonisation produced the lactones (**99**), which by a retro Diels-Alder reaction led to the ylidenebutenolides (Scheme 5); this sequence, although protracted, is particularly useful where the corresponding 4-oxo-2-enoic acids are difficultly accessible.

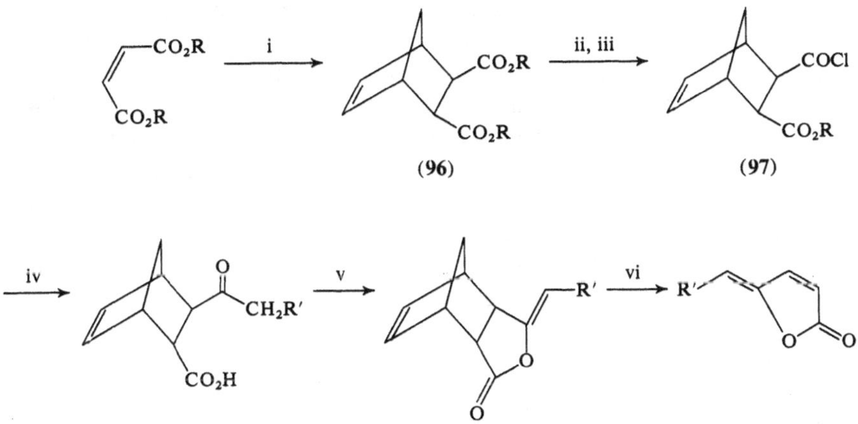

(**96**) (**97**) (**98**) (**99**)

Scheme 5. Synthesis of protoanemonin analogues
Reagents: i, C_5H_6; ii, NaOH; iii, $SOCl_2$; iv, $R'CH_2ZnCl$; v, Ac_2O-HCl; vi, Δ, 200°.

In what amounts to a biogenetically patterned approach, a synthesis of xylerythrin (**7a**) has been reported involving condensation between polyporic acid (**100**) and *p*-hydroxyphenylacetic acid (*6*); this Perkin-type reaction proceeds *via* the 4-oxo-2-enoic acid intermediate (**101**).

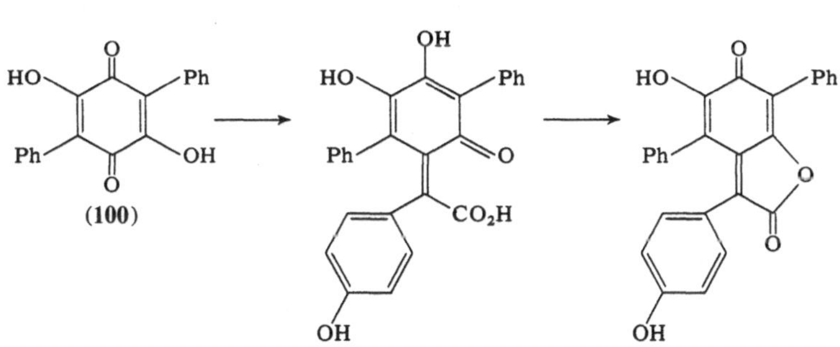

(**100**) (**101**) (**7a**)

The first synthesis of the carbon skeleton found in the lichen pulvinic acids (*viz* **43**) was described by VOLHARD (*92*) as early as 1894. The synthesis begins with a substituted phenylacetonitrile which is first converted into the *bis*-ketonitrile (**102**) by condensation with diethyl oxalate. Acid hydrolysis of (**102**) then leads to the *bis*-lactone (**104**) by way of the double keto acid (**103**) as transient intermediate. Mild saponification of the *bis*-lactone (**104**) with barium hydroxide then produces the corresponding pulvinic acid (**105**). A number of symmetrical and unsymmetrically substituted pulvinic acids have been prepared by this general procedure (*49, 50, 55, 56, 60, 63, 93—95*)*. In the case of the synthesis of unsymmetrically substituted pulvinic acids by this general approach, an excess of diethyl oxalate is employed in the first stage of the sequence, which is interrupted at the pyruvate stage (**106**). Further condensation of the latter with a second, different phenylacetonitrile and hydrolysis and saponification as before then produces a mixture of isomeric pulvinic acids [*viz* (**107**) and (**108**)]. The isomers are then separated by fractional crystallisation and/or chromatography. The general method, which suffers from the obvious disadvantage of producing a mixture of isomers, is illustrated in Scheme 6 for the synthesis

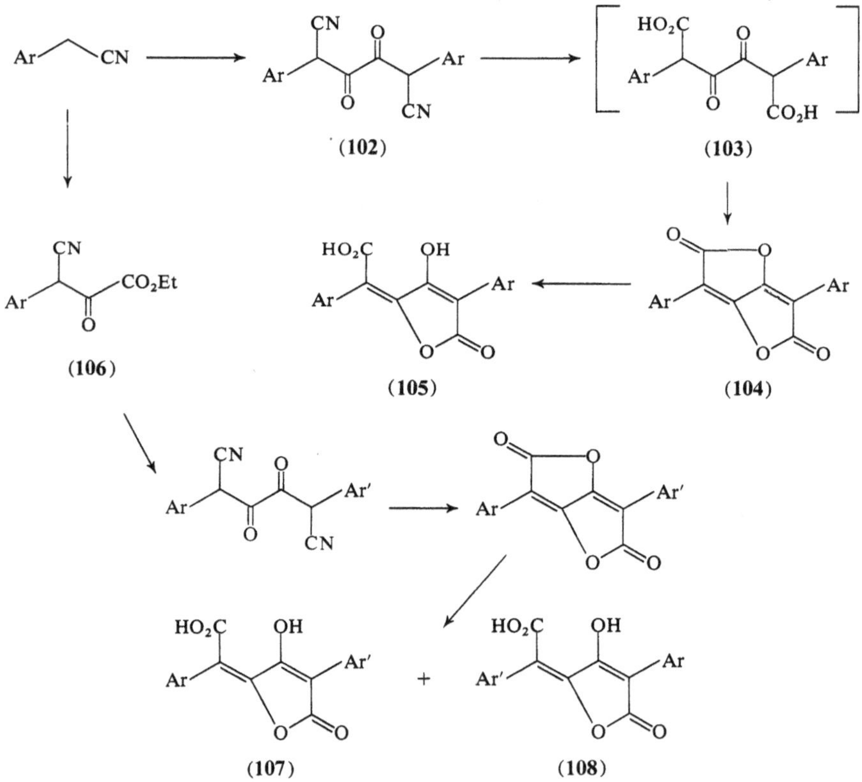

* In addition, several analogues of the pulvinic acids, required for biological evaluation have been prepared by way of VOLHARD's method (*96—106*).

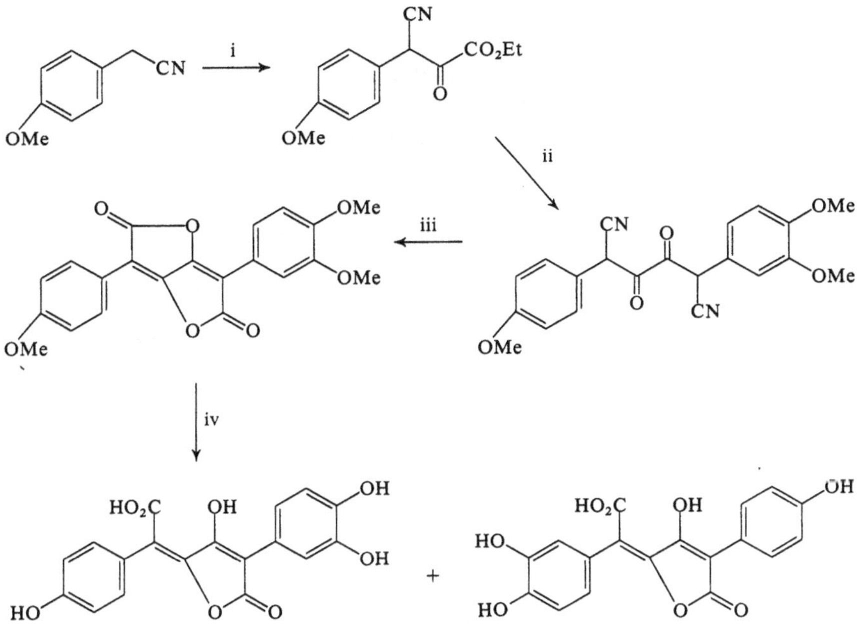

(53)

Scheme 6. Synthesis of xerocomic acid
Reagents: i, NaH, $(CO_2Et)_2$; ii, NaH-3,4-di-OMe$C_6H_3CH_2CN$; iii, $Ac_2O-H_2SO_4$; iv, Ac_2O-HI.

of xerocomic acid (**53**) found in a variety of higher fungi (see Table 2) (*56, 63*). It is only very recently that a new and unambiguous total synthesis of unsymmetrically substituted pulvinic acids has been described (see later discussion).

In other investigations of the synthesis of ylidenebutenolides from 4-oxo-2-enoic acid derivatives, Elvidge and co-workers (*107*) have shown that (**110**) and (**112**) are obtained, albeit in low yields, when the corresponding keto acids (**109**) and (**111**) respectively, are treated with acetic anhydride-hydrogen bromide.

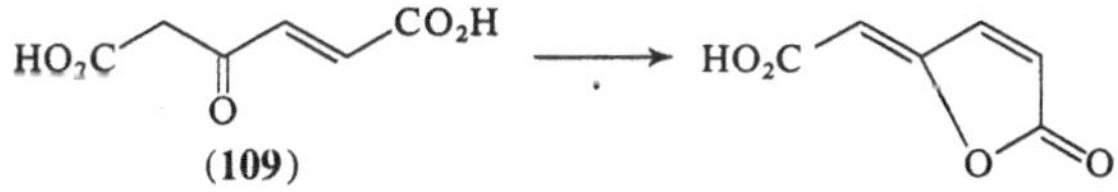

(109) (110)

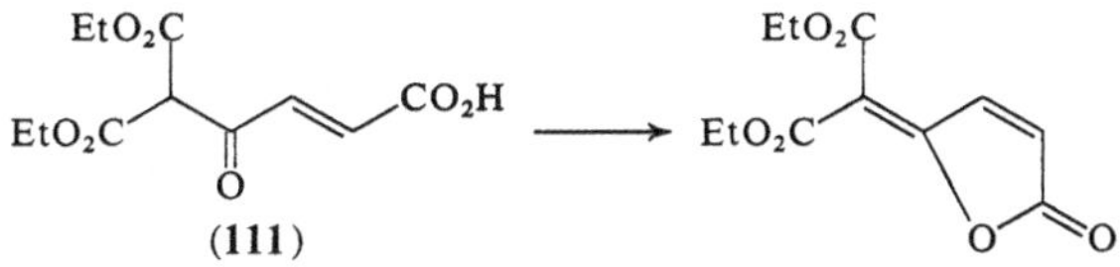

(111) (112)

2. From Z-2-En-4-ynoic Acids

The formation of an ylidenebutenolide from cyclisation of a Z-2-en-4-ynoic acid was first observed by JONES and WHITING and their co-workers (*24, 108*) during investigations of the synthesis of the polyacetylene matricaria acid (**17**). These workers found that when the acid (**17**) was treated with dilute sodium hydrogen carbonate solution at pH 7, it was converted into the corresponding butenolide (**18**), "matricaria lactone",

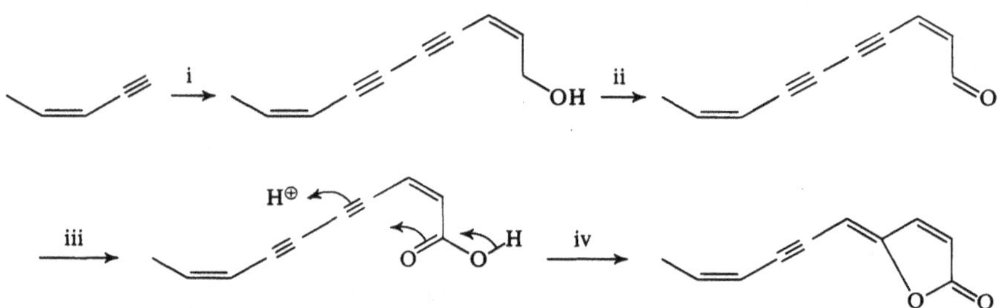

(**17**) (**18**)

Scheme 7. Synthesis of matricaria lactone
Reagents: i, HC≡C·CH=CH·CH_2OH; ii, MnO_2; iii, CrO_3; iv, pH 7.

identical with the natural product (named previously "Composit-Cumulene I") found in *Matricaria inodora* L. (Scheme 7). Acetylenic 4-ylidene-butenolides like matricaria lactone (**18**) co-occur with the corresponding acyclic Z-enynoic acid isomers (*viz* **17**) in nature and it is clear that the latter are the immediate biogenetic precursors of the butenolides. The general route from Z-2-en-4-ynoic acids frequently produces a mixture of Z- and E-isomers about C-4 of the 4-ylidenebutenolides which can be

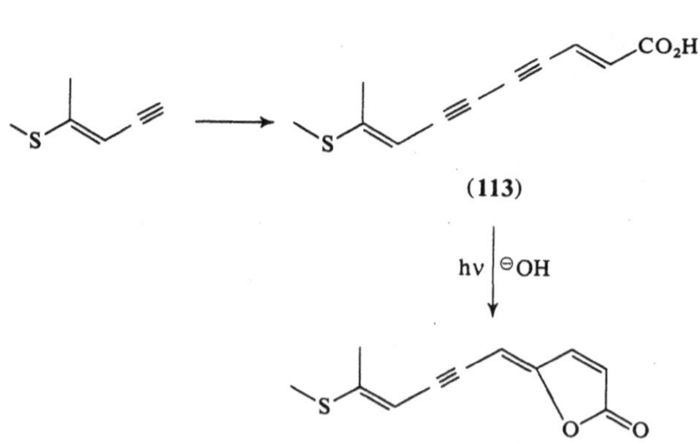

(**113**)

(**27**)

distinguished by comparative p.m.r. shift data (*109*). It would appear that a limitation of this method of synthesis is the general inaccessibility of an enynoic acid with an appropriate *Z*-geometry. Bohlmann and Reinecke have shown that when this is the case however, the corresponding *E*-isomer (*viz* **113**) will suffice. Irradiation of an alkaline solution of the *E*-acid (**113**) for example, promotes *E*-*Z*-isomerisation and produces the thioenoletherbutenolide (**27**) identical with the natural metabolite from *Anthemis* species (*109*).

A particular application of the synthesis of 4-ylidenebutenolides *via* acetylenic precursors is shown in Scheme 8 for the case of derivatives of the fungal metabolite patulin (**10**). Reaction between the acetylenic Grignard reagent (**114**) and keto-acetal (**115**) led first to (**116**) which by Wittig condensation with methoxycarbonylmethylenetriphenylphosphorane was converted into a mixture of *Z*- and *E*-isomers of ester (**117**). Saponification of the mixture of isomers then led to an isomeric mixture of the corresponding carboxylic acids, one of which, the *Z*-isomer (**118**) was cyclized to the butenolide (**119**). Acidic hydrolysis of (**119**) in the presence of hydroxylamine hydrochloride then gave the oxime derivative (**120**) of natural patulin (*110*).

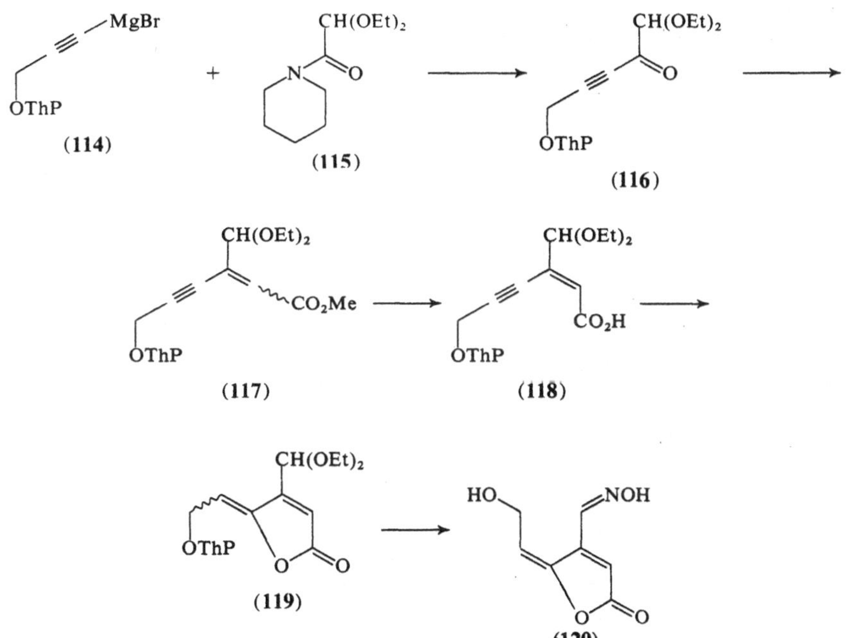

Scheme 8. Synthesis of patulin oxime

Phenylpropargylidenemalonic acid (**121**) can be isomerized to the corresponding ylidenebutenolide (**122**) under purely thermal conditions (*111*), whereas alkali treatment of enol *p*-bromobenzenesulphonates derived from the related phenylpropioloylmalonates (**123**) provides an expeditious route to the 4-ylidenetetronic acid system (**124**) (*112*). Considerably improved yields of the ylidenetetronic acid (**124**) are realized when the enol sulphonate (**126**) from cinnamoylmalonate is treated sequentially with bromine and alkali (*112*). In the case of crotonoylmalonate (**127**), even the enol sulphonate is unnecessary, and simple addition of bromine followed by treatment with aqueous alkali leads to the corresponding ylidenetetronic acid (**128**) in 46% yield (*113*).

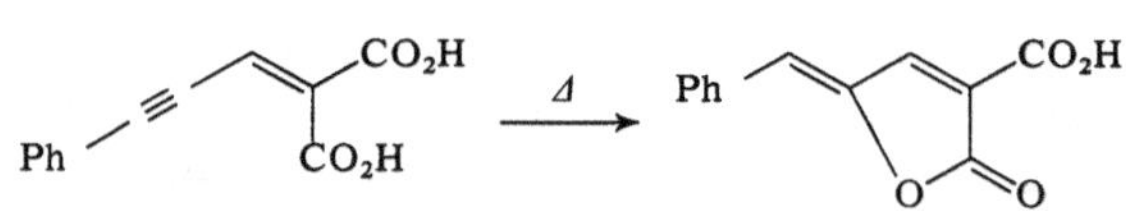

(**121**) (**122**)

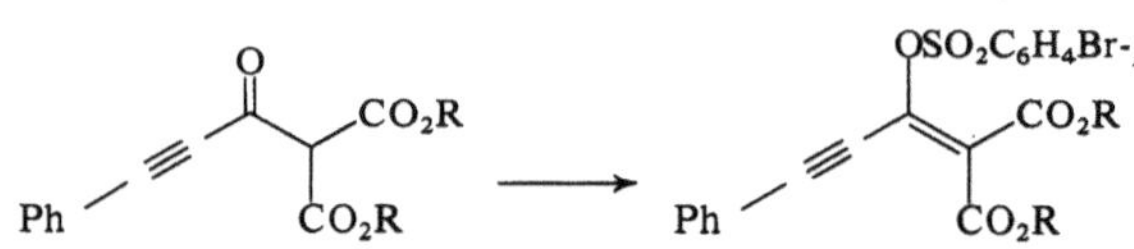

(**123**)

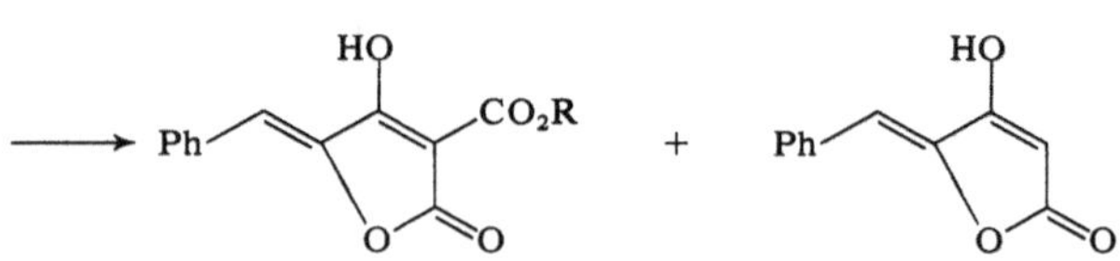

(**124**) (**125**)

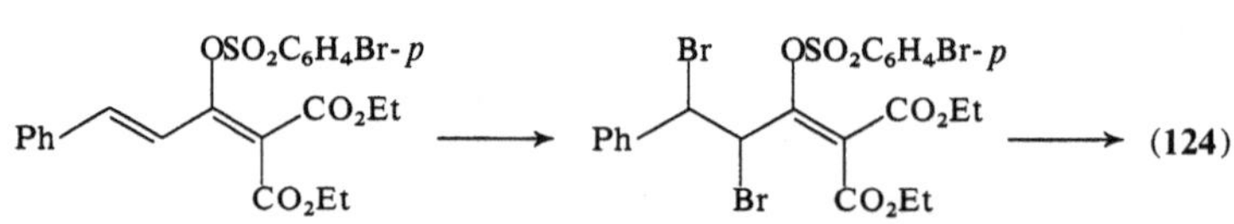

(**126**)

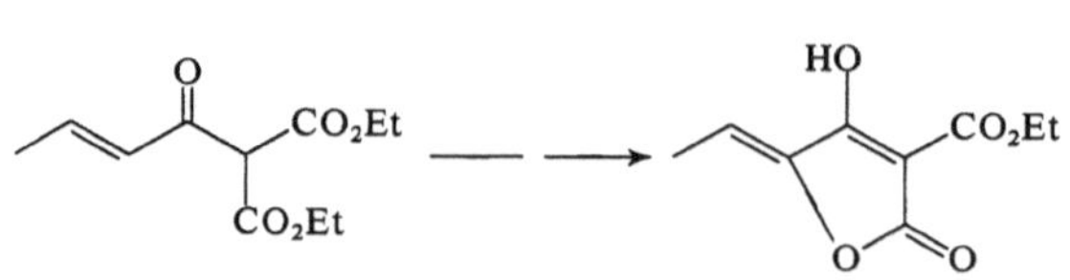

(**127**) (**128**)

References, pp. 189—198

3. From But-2-enolides

Aldol-type condensations between but-2-enolides and carbonyl compounds constitute probably the most flexible routes to 4-ylidenebutenolides and 4-ylidenetetronic acids. One of the first applications of this general approach was in the preparation of (**130**) from but-2-enolide (**129a**) employing aqueous sodium hydrogen carbonate as base (*114*). BOHLMANN and ZDERO (*34*) later described a synthesis of the naturally occurring thiophene (**131**) involving condensation between butenolide (**129b**) and 5-methylthiophene-2-aldehyde in the presence of piperidine acetate in glacial acetic acid at 80°.

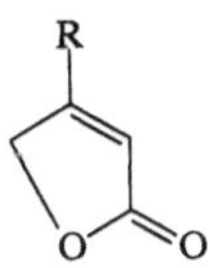

(**129**) *a)* R = Me
b) R = H.

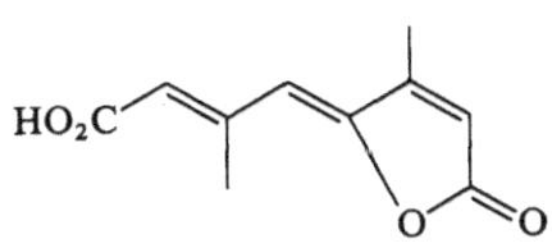

(**130**)

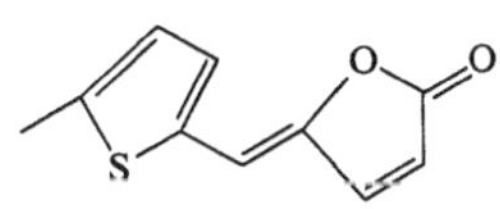

(**131**)

In more recent years the same general approach has been employed in the first unambiguous syntheses of both the unsymmetrically substituted "pulvinone" (*viz* **60**) and "pulvinic acid" (*viz* **43**) carbon skeletons (*53, 74, 75*). The total synthesis of the permethylated derivative of the novel prenylated-pulvinone (**68**) found in *Aspergillus terreus* is illustrated in Scheme 9 (*75*). 4-Hydroxybenzaldehyde was first converted into the 3-prenylated derivative (**133**) following conversion into the allyl ether (**132**) and Claisen rearrangement. Cyclisation in acid led next to chroman (**134**) which was then converted into the nitrile (**135**) *via* the corresponding benzyl alcohol and benzyl chloride. Condensation between the anion produced from (**135**) and diethyl oxalate then gave the pyruvate (**136**), which by methylation followed by acid hydrolysis led to the maleic anhydride (**137**). Reduction of (**137**) with lithium aluminium hydride at low temperatures was regioselective and produced the key butenolide intermediate (**138**). Metallation of this butenolide at −78° in tetrahydrofuran using lithium N-cyclohexyl-N-isopropylamide, followed by treatment with the methyl ether derivative of (**133**) and dehydration of the intermediate carbinol, gave the pulvinone (**139**), identical with the permethylated metabolite found in *A. terreus*. In a similar manner KNIGHT and PATTENDEN have effected total syntheses of permethyl pulvinone (**63**), found in the larch mushroom *Suillus grevillei* (*74*) and also of permethylated derivatives of pulvinones (**64**), (**68**) and (**69**) found in *A. terreus* (*74, 75*).

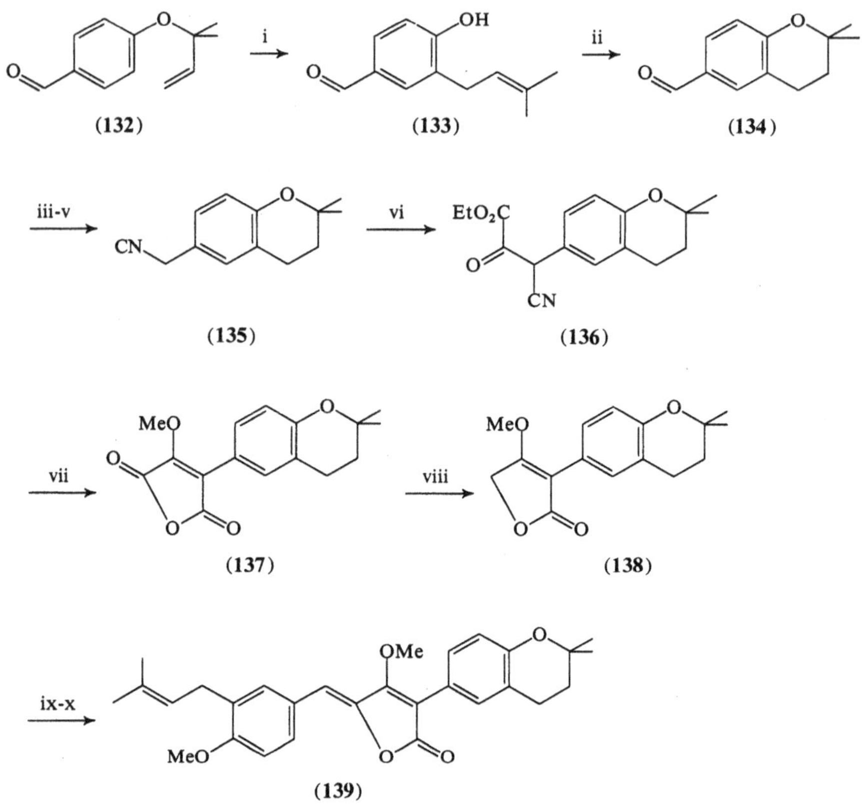

Scheme 9. Synthesis of prenylated aspulvinones
Reagents: i, Δ, $PhNEt_2$; ii, H^+; iii, $LiAlH_4$; iv, $SOCl_2$; v, NaCN; vi, NaH, $(EtO_2C)_2$; vii, $HOAc-H_2SO_4$, methylation; viii, $LiAlH_4$; ix, $C_6H_{11}\bar{N}Pr^iLi^+$, O-Me-(**133**); x, H^+

In developments of the approach leading to the pulvinone carbon skeleton, KNIGHT and PATTENDEN have also described the first unambiguous synthesis of unsymmetrically substituted pulvinic acids. Thus, condensation between methyl 4-methoxybenzoylformate and the anion produced from butenolide (**140**) was found to be largely regioselective producing *tert*-carbinol (**141**). Dehydration of (**141**) in a hot suspension of phosphorus pentoxide and dry benzene then led to permethylated gomphidic acid (**142**) whose spectral features established structure (**56**) for the natural metabolite isolated from *Gomphidius glutinosus* (*53*). In a similar sequence the butenolide (**143**) and methyl benzoylformate gave O-methyl pinastric acid (**144**) identical with the O-methyl derivative of pinastric acid (**47**) found in *Lepraria flava* and other lichens.

References, pp. 189—198

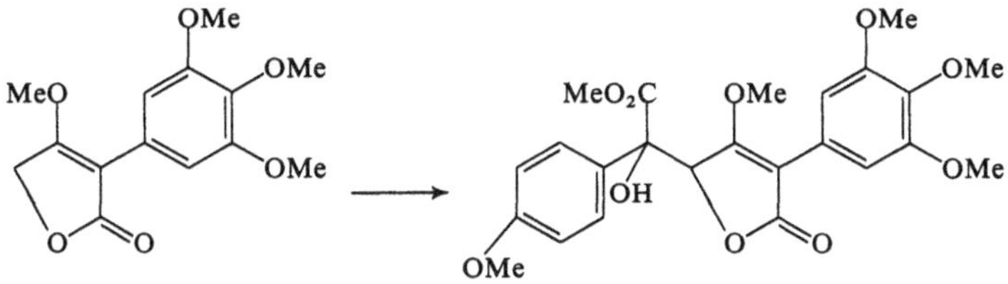

(140) (141)

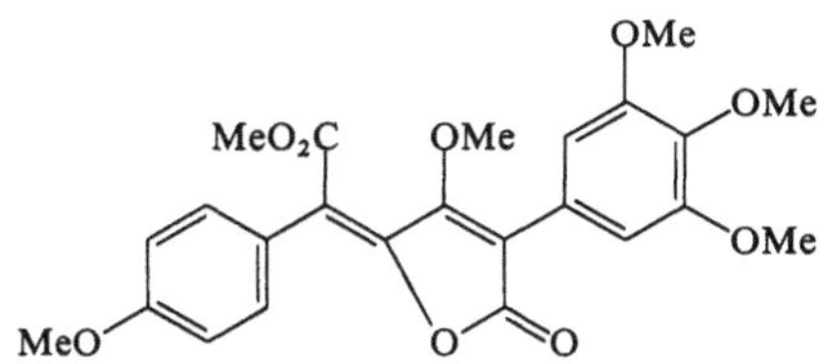

(142)

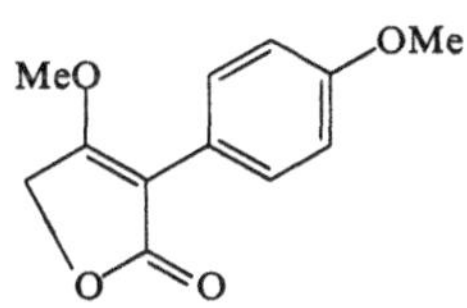

(143)

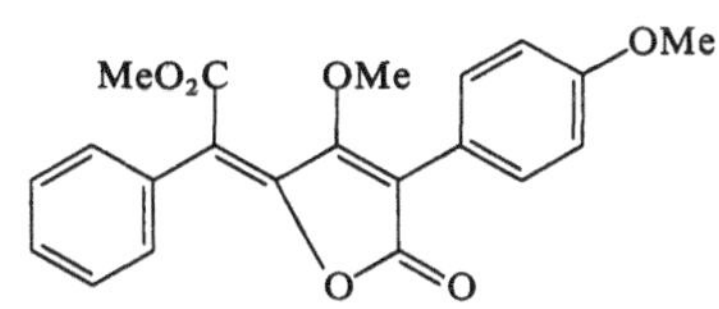

(144)

Phosphoranylides have enjoyed widespread use as intermediates in the synthesis of olefinic bonds over the past two decades. It is perhaps not surprising then that ylides of general structure (**145**) have been investigated in the synthesis of 4-ylidenebutenolides and 4-ylidenetetronic acids (*20, 36, 115, 116*). The first application in the synthesis of a naturally occurring 4-ylidenebutenolide was described by KNIGHT and PATTENDEN (*117, 118*) for the case of the acetylenic sesquiterpene freelingyne (**33**) found in *Eremophila freelingii* (Scheme 10). The phosphonium salt precursor (**147**) of the ylide (**149**) was prepared from the readily available lactol (**146**) either directly using triphenylphosphine hydrobromide or by way of the corresponding bromide (**148**). A Wittig condensation between the *E*-enynal (**150**), prepared as outlined in the Scheme, and the ylide (**149**) then produced an easily resolved mixture of *Z*-*E*-(C-4)-geometrical isomers (**33**) and (**151**) of freelingyne, one of which, the *Z*-2,*E*-4 isomer (**33**) was shown to be identical with the natural metabolite. Later workers have reported the application of the same butenolide ylide in a synthesis of dihydrofreelingyne (**34**) (*36*).

(145)

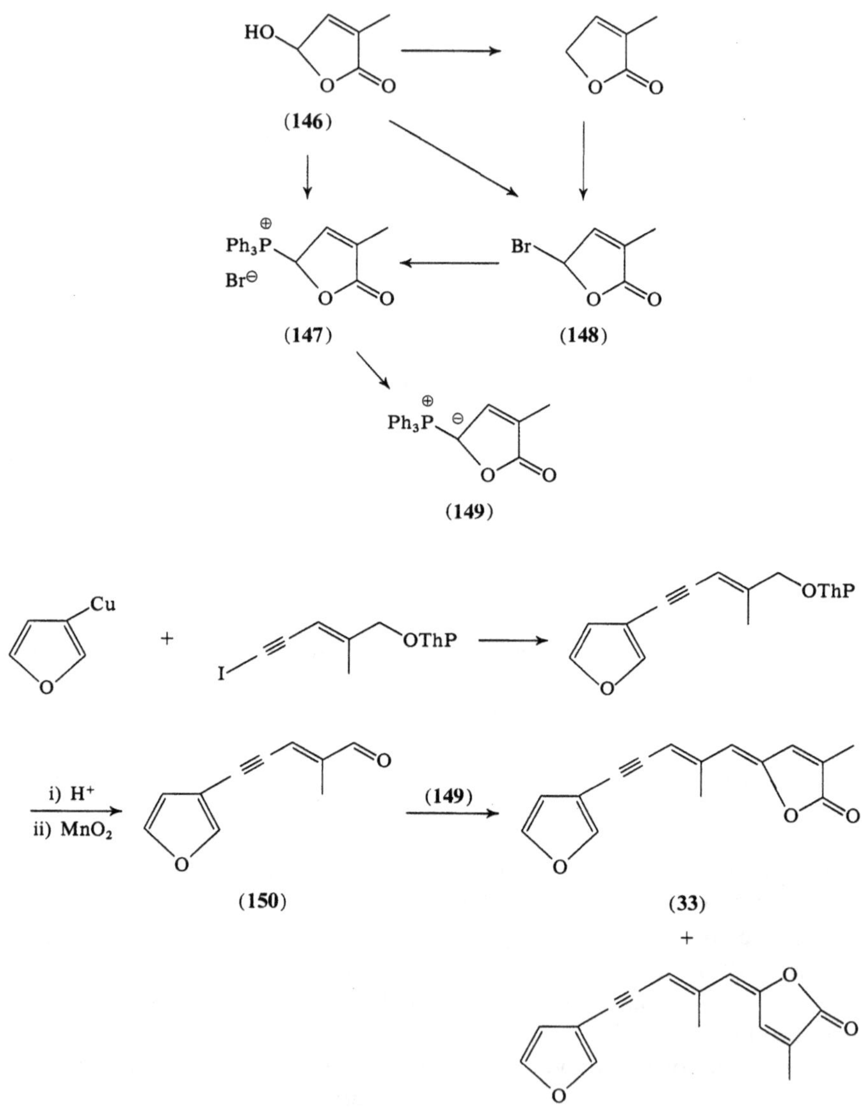

(146) (147) (148) (149) (150) (33) (151)

Scheme 10. Synthesis of freelingyne

One advantage that the application of phosphoranylides (**145**) has in the synthesis of 4-ylidenebutenolides is that a mixture of *Z*- and *E*-isomers about the newly formed double bond is invariably produced. This feature is particularly important in cases where the geometrical configuration of a naturally occurring ylidenebutenolide is in doubt, since comparative p.m.r. data with synthetic *Z*-*E*-isomers usually provide an unambiguous distinction. An illustration of this is provided by the alkylidenephthalide odoriferous principles (**13a**) and (**13b**) found in the celery plant *Apium graveolens*. Early syntheses of alkylidenephthalides have involved Perkin- and Grignard-type reactions with phthalic anhydride leading to the thermodynamically more stable *Z*-isomers (*viz* **153**) (*20*). Wittig condensations involving the corresponding phthalide-ylide (**152**) and aldehydes, however, produce predominantly the *E*-isomers (*13*). This feature of the reaction is particularly important since the *E*-isomers (**13a**) and (**13b**) are those present in the celery plant (*15, 16*).

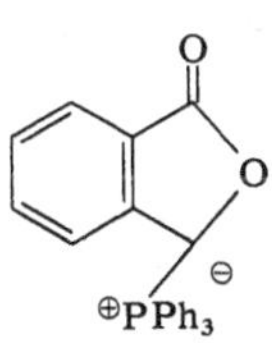

(152)

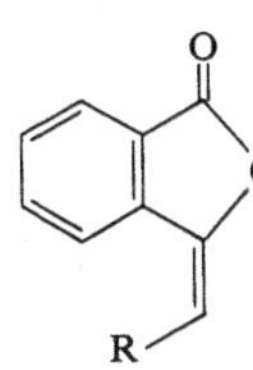

(13)

a) R = *i*-Pr.
b) R = *i*-Bu.

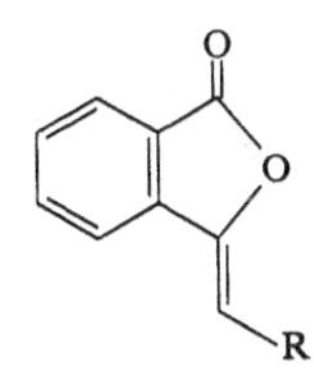

(153)

The phthalide ylide (**152**) also condenses with phenylglyoxal and with methyl benzoylformate to produce ylidenebutenolides (**154**) and (**155**) respectively, and the phosphonate carbanion corresponding to (**152**) has been employed in synthesis (*119*).

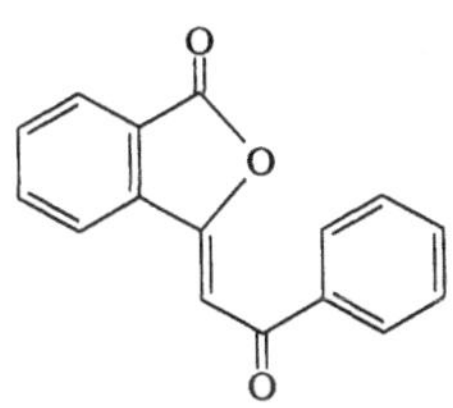

(154)

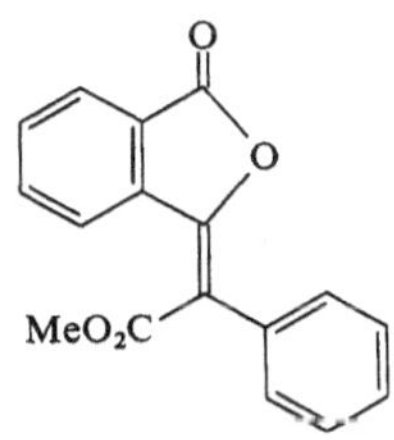

(155)

4. From Substituted Maleic Anhydrides

Condensation reactions between phthalic anhydride and aryl or alkyl anhydrides in a Gabriel modification of the Perkin reaction (Scheme 11) have provided an exceptionally flexible synthetic route to a number of alkylidene and arylidenephthalides (*15, 120—127*). These reactions generally produce only the *Z*-ylidenephthalide (*viz* **153**) although BERTI (*128*) has been able to separate the corresponding *E*-isomers from certain condensations. The addition of Grignard reagents or organocadmium reagents to phthalic anhydrides, followed by dehydration of the intermediate *tert*-alcohols, provides a useful alternative route to ylidenephthalides (*21*).

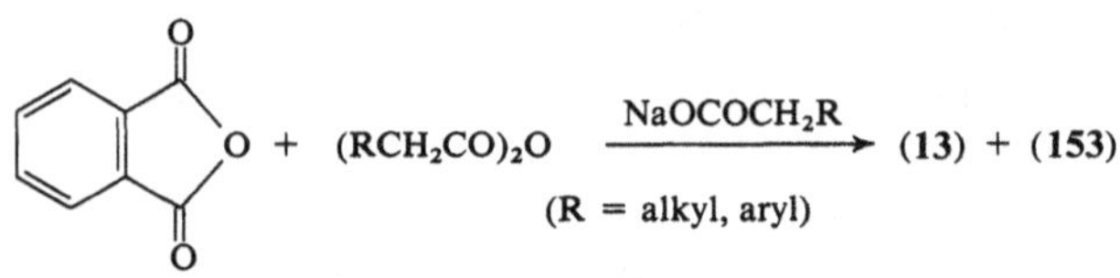

Scheme 11. Perkin-style synthesis of alkylidenephthalides

In a series of communications, a Unilever group has outlined the synthesis of a whole range of ylidenebutenolides of general structure (**156**) from addition of alkyl Grignard reagents to various 2,3-dialkylmaleic anhydrides at −70°, followed by dehydration (*129—131*). These compounds are useful as flavouring agents and as agents in perfumery. The general method has been used in the synthesis of bovolide (**15**), an odoriferous component found in butter (*22*). A limitation of the method

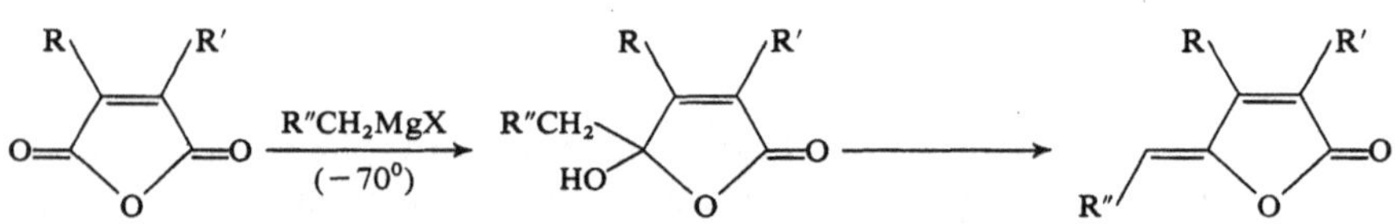

(R = R′ = R″ = alkyl) (**156**)

Scheme 12. Synthesis of bovolide and analogues

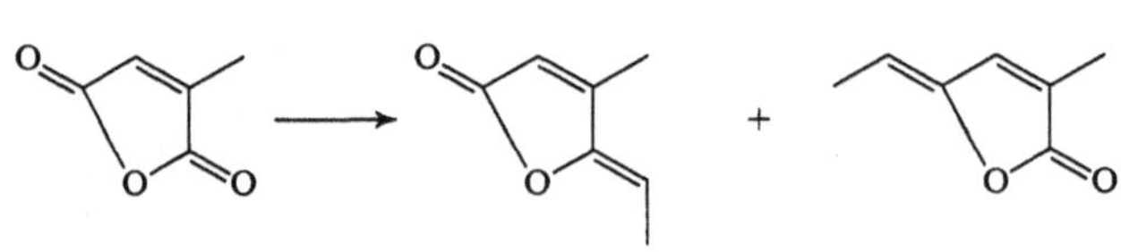

(**157**) (**158**)

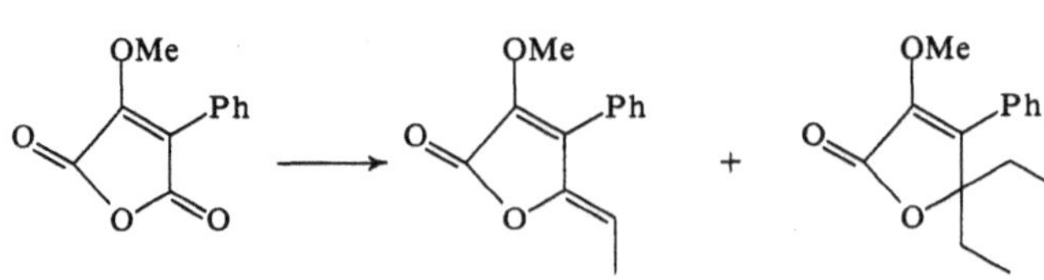

(**159**) (**160**)

is that, with unsymmetrically substituted maleic anhydrides, a mixture of isomeric ylidenebutenolides is invariable produced. Thus reaction between 2-methylmaleic anhydride and ethylmagnesium bromide has been shown to lead to a mixture of the ylidenebutenolides (**157**) and (**158**) present in the approximate proportions 6 : 4 (*141, 142*). A similar reaction with 2-methoxy-3-phenylmaleic anhydride produces only one ylidenebutenolide (**159**), but (**159**) is also accompanied by the butenolide (**160**) produced by addition of a second molecule of the Grignard reagent (*132*).

A complementary procedure for the preparation of ylidenebutenolides from maleic anhydrides, which has several advantages over the Grignard- and Perkin-type approaches, employs phosphoranylides. This approach was first applied in the synthesis of alkylidenephthalides (*e.g.* **162**) by reaction between phthalic anhydride and the appropriate triphenylphosphorane (*133*). Interestingly, the condensations involving phosphoranes (**161**) and (**164**, R = Me or Ph) gave predominantly the corresponding *E*-ylidenephthalides (*viz* **162** and **163**) (*134*). In more extensive investigations Massy-Westropp and co-workers (*134*) have demonstrated that

$Ph_3P{:}CHCO_2Et$

(**161**)

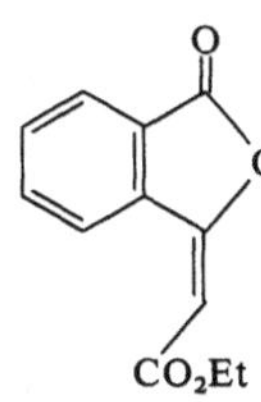

(**162**)

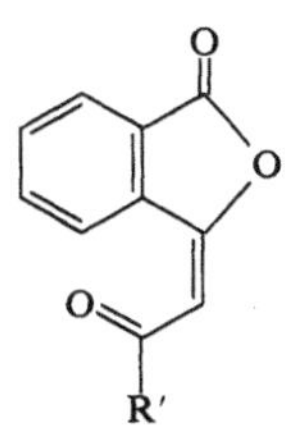

(**163**)

$Ph_3P{:}CHCOR$

(**164**)

the Wittig approach is generally useful for reactions involving weakly nucleophilic phosphoranylides of the type (**161**) and (**164**). Thus, using 2-methyl- and 2,3-dimethylmaleic anhydrides and phosphoranes (**161**) and (**164**), useful routes to *Z*-*E*-isomers of members of the ylidenebutenolides (**165**—**167**) have been accomplished. Ylidenebutenolides were not isolated from similar condensations employing maleic anhydride

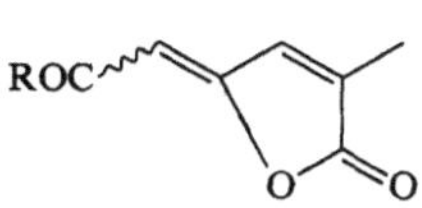

(**165**)

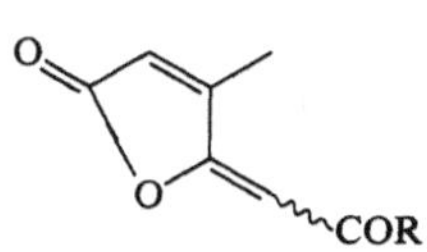

(**166**)
(R = OEt, Me)

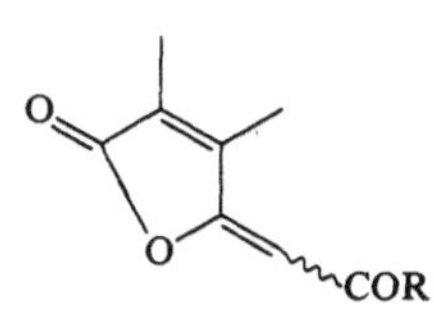

(**167**)

itself; instead largely polymeric material was produced. The *E*-ylidenebutenolide (**169**) could be prepared, however, by reaction between phosphorane (**161**) and the corresponding Diels-Alder adduct (**168**) obtained from maleic anhydride, followed by a retro Diels-Alder reaction.

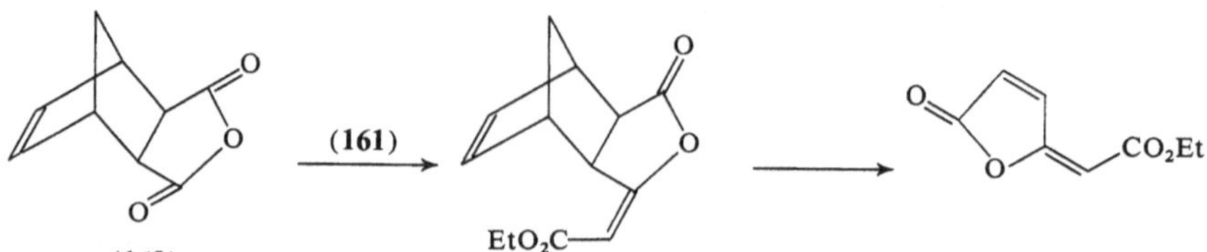

(**168**) (**169**)

MASSY-WESTROPP and co-workers (*135*) have used the *E*-ylidenebutenolide (**165**, R=Me) in an alternative synthesis of freelingyne (**33**). Thus, Reformatsky reaction between the butenolide and the furylpropargyl bromide (**170**) first gave the *tert*-alcohol (**171 a**). Flash vacuum pyrolysis of the derived acetate (**171 b**) then led to a mixture of geometrical isomers of freelingyne, one of which was identical with the natural metabolite.

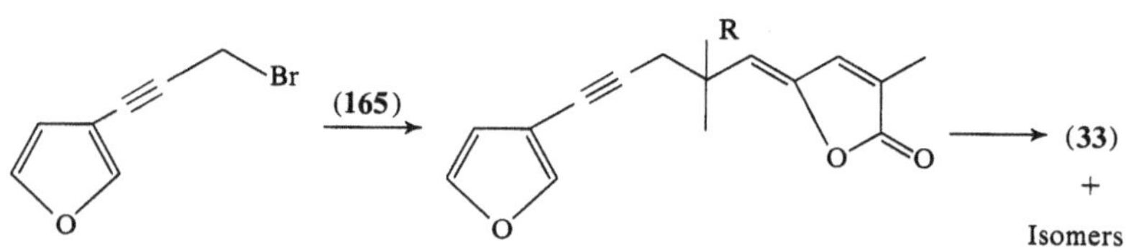

(**170**) (**171**) *a)* R=H
b) R=Ac.

In studies of the reactions between 2-methoxysubstituted maleic anhydrides (*viz* **172**) and the stabilized phosphoranes (**161**) and (**164**), it has been shown that the condensations are regioselective leading to those products (*viz* **173**) resulting from nucleophilic addition to the carbonyl functions nearest to the methoxy-substituent (*136*).

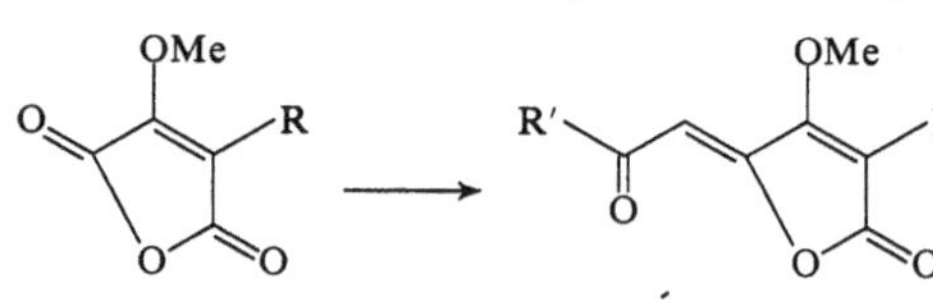

(172)
R = Me or Ar

(173)
R¹ = Me, OMe

5. Miscellaneous Syntheses

The biosynthesis of pulvinic acids in lichens has recently been shown to proceed *via* enzymatic oxidative rearrangement of hydroxyterphenylquinones (see biosynthesis section). As early as 1928, KÖGL had provided a laboratory model for this fascinating rearrangement, when he isolated the pulvinic acid dilactone **(175)** from treatment of naturally occurring atromentin **(174)** with hydrogen peroxide (*137*). More recent investigations

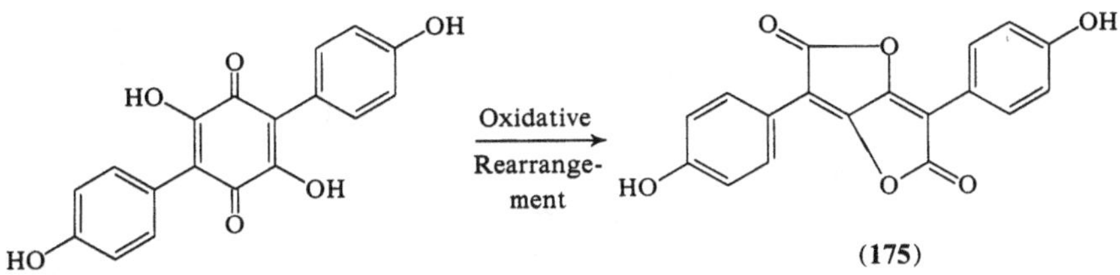

(174) **(175)**

by WIKHOLM and MOORE (*138*) have demonstrated that oxidative rearrangements of this nature are quite general and can be effected in upwards of 95% yield by simply treating the terphenylquinones with a solution of dimethyl sulphoxide — acetic anhydride at 60° for approximately 15 min. Under identical reaction conditions, the corresponding monohydroxyterphenylquinones **(176)** give butenolides of constitution **(177)***. In closely related studies MOORE and his co-workers have shown that the decomposition of azido-1,4-benzoquinones (*viz* **178**) with acid is a highly stereospecific process and produces cyano-butenolides of the general type **(179)** (*144*); an excellent review of these and related transformations has been published (*145*).

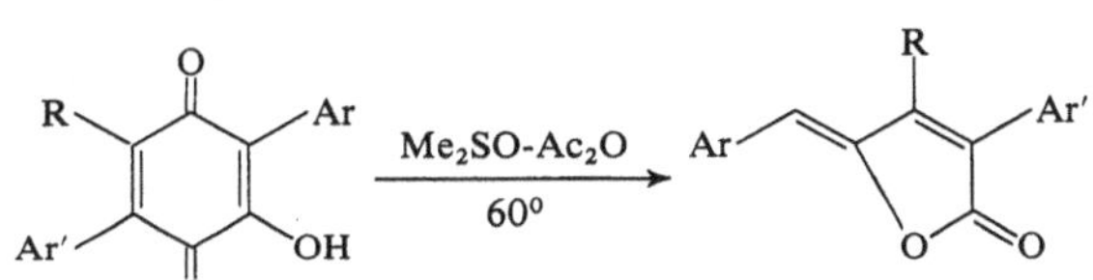

(176) R = Cl, OMe, H **(177)**

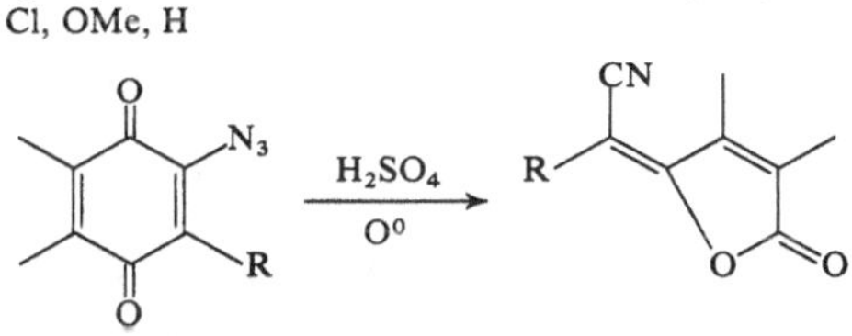

(178) **(179)**

In an earlier section of this review (p. 148), and in another context, the novel thermal equilibration of the pulvinone carbon skeleton **(60)** with the 1,3-*bis*-arylcyclopentan-2,4,5-trione system **(62)** was discussed briefly. The incidence of this equilibrium has inspired a useful synthesis of

* A number of alternative methods of oxidative rearrangement of quinones to ylidenebutenolides have been published (*e. g. 139—143*).

symmetrically substituted pulvinones starting from 1,3-*bis*-arylpropan-2-ones (**180**) which are first converted into triones (**181**) by condensation with diethyl oxalate. On heating the triones to approximately 250°, an equilibrium mixture of the corresponding pulvinone and trione is then produced from which the pulvinone can be separated by extraction and chromatography (*71*); a number of other examples of the equilibration of ylidenetetronic acids with cyclopentan-2,4,5-triones are known (*69, 113, 146*).

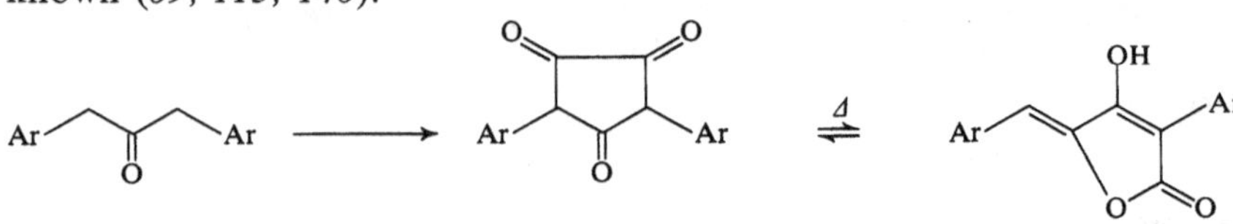

(**180**) (**181**)

During studies directed towards his total synthesis of penicillic acid (**185**), RAPHAEL (*147, 148*) effected a synthesis of the key ylidenetetronic acid intermediate (**184**) by addition of methanol to the acetylenic ester (**182**), leading to (**183**), followed by base-catalyzed isomerization (Scheme 13).

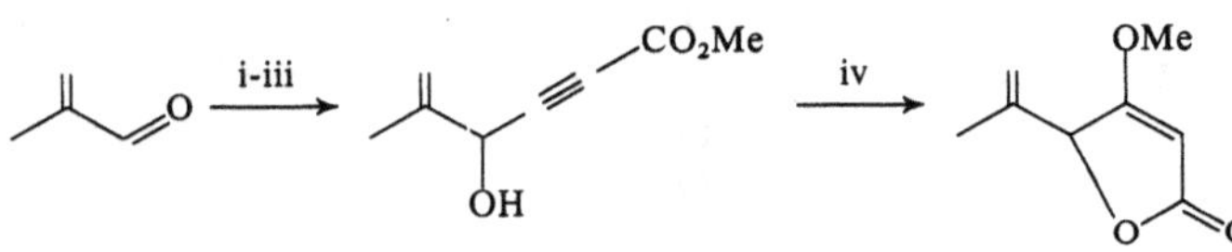

(**182**) (**183**)

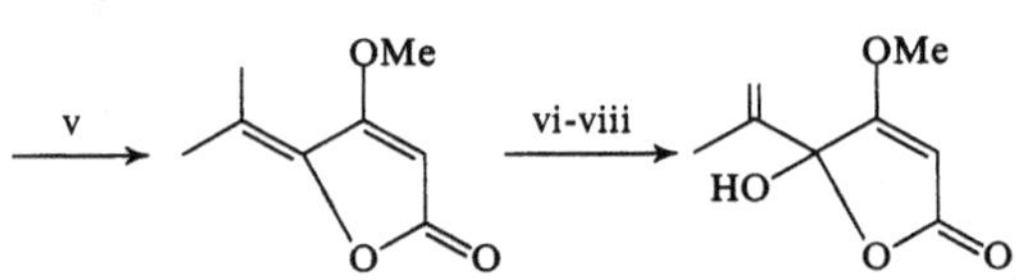

(**184**) (**185**)

Scheme 13. Synthesis of penicillic acid
Reagents: i, $NaC{\equiv}CH$; ii, Na, CO_2; iii, $MeOH-H^+$; iv, $MeOH-BF_3-HgO$; v, NH_3; vi, NBS; vii, Me_3N; viii, Δ, MgO.

An interesting route to ylidenebutenolides described by GILMAN and co-workers (**149**) involves the acid-catalyzed decomposition of 5-halo-furfuryl carbinols (Scheme 14). Arylidenebutenolides can be obtained

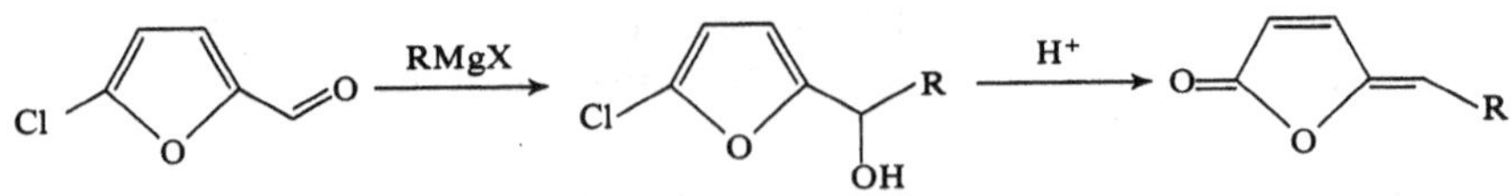

Scheme 14. Synthesis of ylidenebutenolides *via* furans

by the unusual "dehydration" of keto lactones of constitution (**186**) with acetic anhydride in the presence of sodium acetate (*150*). Using this general procedure YAMADA and co-workers (*150*) have effected a synthesis of the naturally occurring thiophenebutenolide (**32**).

(**186**)

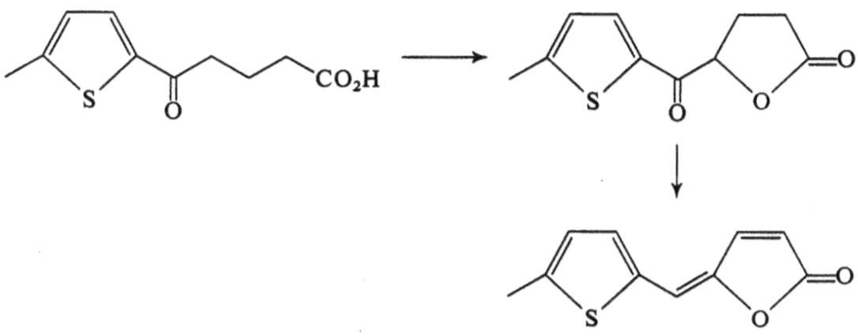

(**32**)

A Friedel-Crafts type reaction between maleic anhydride and isopropenyl acetate in the presence of aluminium chloride leads to the ylidenebutenolide (**187**), albeit in low yields (*151, 152*), and a number of ylidenebutenolides have been produced from reactions between acetylenes and carbon monoxide (*153—157*)*.

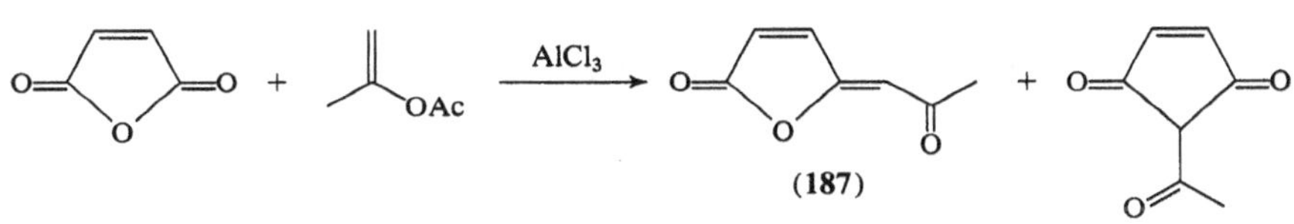

(**187**)

IV. Biosynthesis

The relatively high incorporations of tracer molecules into secondary metabolites in micro-organisms has enabled detailed studies to be made of the biosynthesis of the fungal metabolites patulin (**10**) and multicolic acid (**79**). Carbon-14 work has demonstrated the polyketide origin of patulin, and the biosynthesis of multicolic acid from acetate has been studied by ^{13}C-n.m.r. spectroscopy. The biosynthesis of patulin has been investigated extensively at the enzymology level. In the first

* For other miscellaneous methods leading to 4-ylidenebutenolides see refs. (*158—162*).

application of ^{3}H-n.m.r. spectroscopy to a biosynthetic problem, details of the biosynthesis of the tetronic acid penicillic acid (**185**) from ^{3}H-acetate have been obtained.

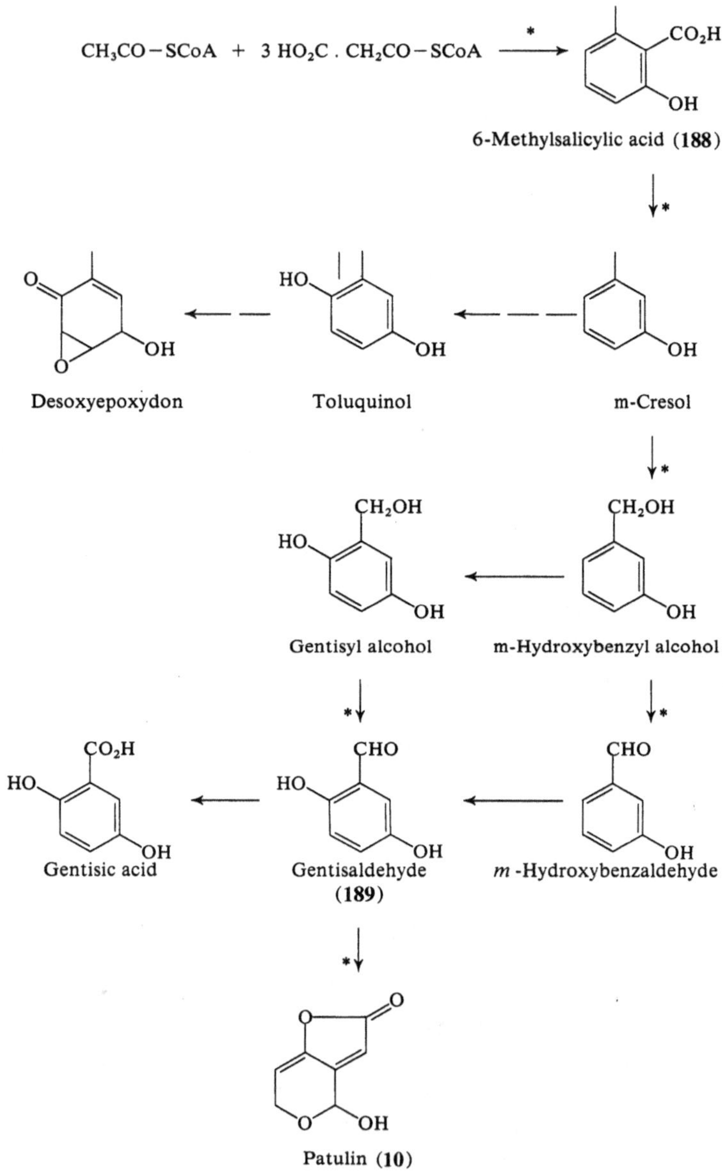

* show major pathways

Scheme 15. Biosynthesis of patulin

Carbon-14 work has shown that the pulvinic acid pigments found in lichens are derived *via* the shikimic acid pathway, and that the unsaturated ylidenebutenolides found with polyacetylenic compounds in the Compositae are derived from oleic acid. No biosynthetic studies have been reported for the terpenoid ylidenebutenolides freelingyne, peridinin and the several sesterterpenes from marine sponges.

1. Fungal Metabolites

The fungal metabolites patulin (**10**) and multicolic acid (**79**) are derived in nature *via* oxidative cleavages of polyketide derived aromatic intermediates; a similar pathway has been established for the tetronic acid penicillic acid (**185**). By contrast, the biosynthesis of carolic acid (**76a**) which is found with dehydrocarolic acid (**75**) in *P. charlesii,* has been shown to occur from Krebs cycle intermediates.

The biosynthesis of patulin (**10**) has been studied extensively, since the molecule represents a relatively simple model system in which to examine the detailed enzymology of polyketide biosynthesis. Patulin is biosynthesized by the fungus *Penicillium patulum via* an oxidative pathway from 6-methylsalicylic acid (**188**) which is synthesized from acetyl-CoA and malonyl-CoA. The major pathway from (**188**) and the biosynthetic relationships of the phenolic secondary metabolites of *P. patulum* are

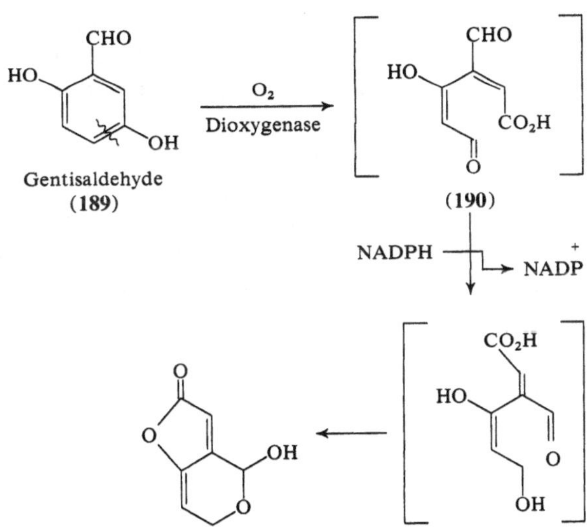

Scheme 16. Biosynthesis of patulin from gentisaldehyde

summerized in Scheme 15 (*163, 164, 165, 166*). Evidence for the sequence has been obtained using various ^{2}H, ^{14}C and $^{14}C/^{3}H$-labelled intermediates in replacement cultures of the organism (*163*), and also by kinetic pulse labelling techniques with radio-labelled acetate and other metabolites (*165*).

In recent years considerable progress has been made in demonstrating, at the enzyme level, some of the individual transformations depicted in Scheme 15. The enzymes 6-methylsalicylic acid synthetase (*167*), 6-methylsalicylic acid decarboxylase (*168*) and *m*-hydroxybenzyl alcohol dehydrogenase (*165, 169*), catalyzing steps one, two and four respectively have all now been purified. Probably the most significant advance at the enzyme level, however, has been the demonstration by SCOTT and BEADLING of the dioxygenase mediated ring cleavage of gentisaldehyde (**189**) (step six, Scheme 15) (*164*). A mechanism involving oxidative cleavage of gentisaldehyde at the 4,5-position by a dioxygenase leading to intermediate (**190**) which then undergoes molecular rearrangement to patulin was proposed (Scheme 16), but neither the order or mechanisms of the final lactonisation and hemiacetal formation stages are clear.

In one of the first demonstrations of the use of doubly ^{13}C-labelled acetate in a biosynthetic problem, HOLKER and co-workers (*82*) have shown that the biosynthesis of multicolic acid (**79**) in *Penicillium multicolor* proceeds *via* oxidative cleavage between C-4 and C-5 of the polyketide derived aromatic acid (**191**). From 1,2-^{13}C-acetate feedings, and the magnitudes of the ^{13}C-^{13}C-couplings in the ^{13}C-n.m.r. spectra of the metabolite, it was established that the acetate residues C-2—C-5, C-6—C-7, C-8—C-9 and C-4—C-10 were incorporated intact into the acid (Scheme 17). These data are only compatible with a biosynthetic origin by oxidative cleavage of the polyketide derived aromatic precursor (**191**).

The biosynthesis of the fungal tetronic acids penicillic acid (**185**) and carolic acid (**76a**), which are closely related structurally to patulin and multicolic acid, have also been studied extensively. Early work with ^{14}C-labelled precursors has clearly demonstrated the polyketide origin of penicillic acid, and also the intermediacy of orsellinic acid

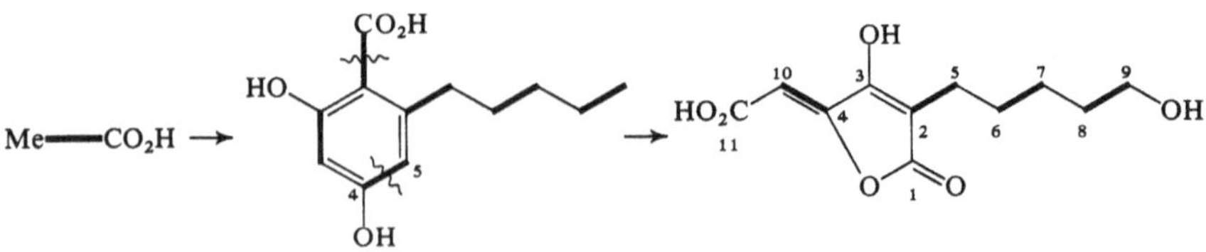

(191) (79)

Scheme 17. Biosynthesis of multicolic acid

(**192**) (*170, 171*). By feeding ^{3}H-acetate of high specific activity to *P. cyclopium*, THOMAS and co-workers (*172*) have confirmed this result and, from the assigned tritium positions in the tritium-labelled metabolite, have tentatively proposed the pathway shown in Scheme 18 for the biosynthesis. In contemporaneous studies, SETO and co-workers (*173*) have

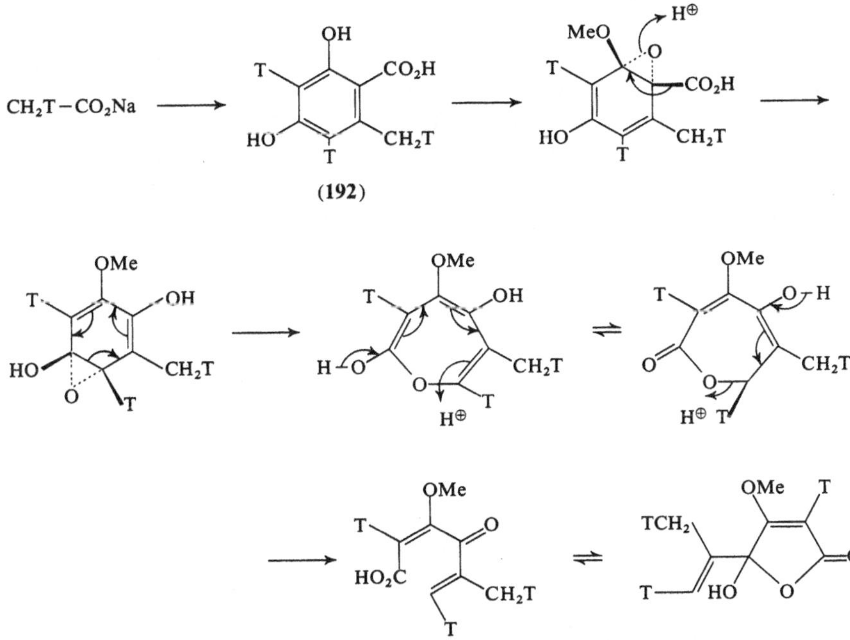

Scheme 18. Biosynthesis of penicillic acid

arrived at a similar biosynthetic scheme based on considerations of ^{13}C-^{13}C-couplings in the ^{13}C-n.m.r. spectra of penicillic acid fed with doubly ^{13}C-labelled acetate. Furthermore, AXBERG and GATENBECK have

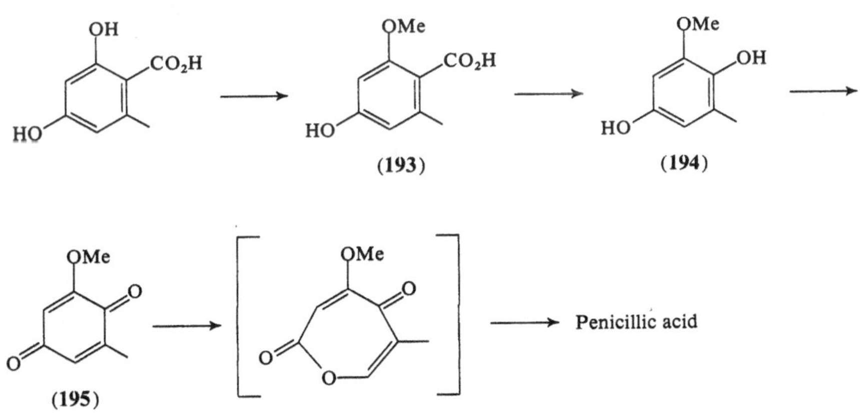

(193) (194) (195)

recently demonstrated the involvement of intermediates (**193**), (**193**) and (**195**) in the biosynthesis of penicillic acid from orsellinic acid (*174*). The interesting Baeyer-Villiger type oxidation in the ring opening reaction in the last step has been confirmed in a cell-free system.

The biosynthesis of the fungal tetronic acid carolic acid (**76a**) in *Penicillium charlesii* is markedly different from that of patulin, multicolic acid and penicillic acid. Carbon-14 work has demonstrated that the carbon sub-unit C-3, C-4, C-9 in the acid is derived from a C-4-dicarboxylic acid such as succinate, whereas the remaining six carbons (C-1, C-2, C-5, C-6, C-7, C-8) are derived from two malonate units (C-5, C-6, C-7, C-8) and just one acetate unit (C-1, C-2) (Scheme 19) (*175*). It is proposed then that the biosynthesis of carolic acid proceeds from a C_4-dicarboxylic acid first to either γ-carboxymethyltetronic acid (**196**) or to carlosic acid (**77b**); hydroxylation of carlosic acid then gives rise to carlic acid (**77a**) which on decarboxylation produces carolic acid (*176*).

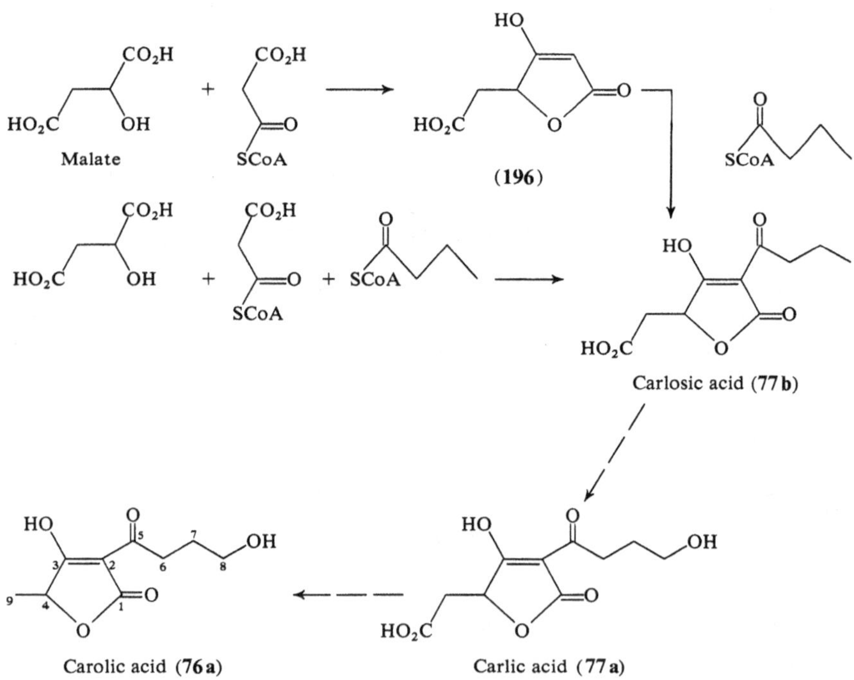

Scheme 19. Biosynthesis of carolic acid

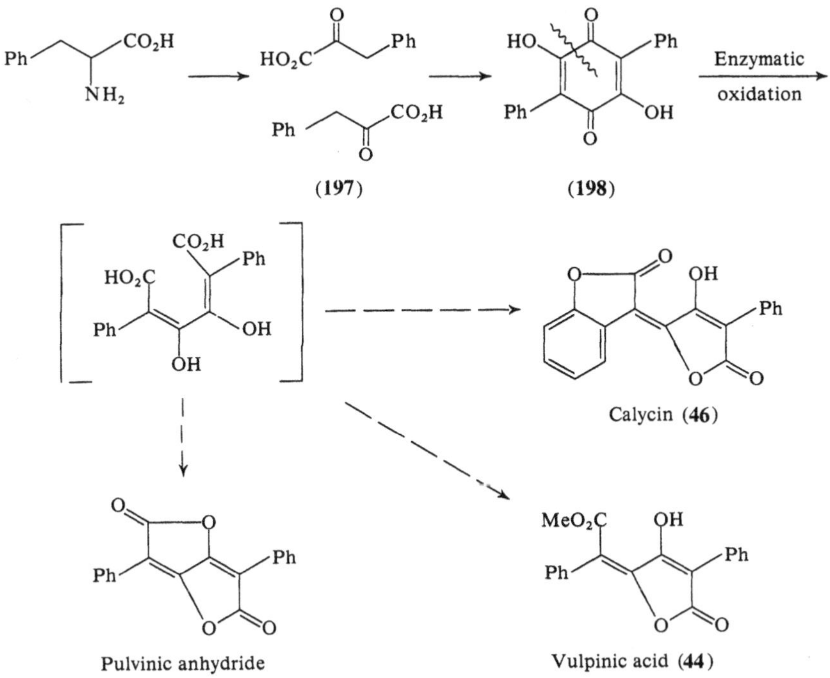

Scheme 20. Biosynthesis of pulvinic acids in lichens

2. Pulvinic Acids and Pulvinones

Experiments with ^{14}C-labelled precursors in the lichens *Pseudocyphellaria crocata* and *Letharia vulpina* show that the basic pulvinic acid carbon skeletons in vulpinic acid (**44**) and calycin (**46**) are derived *via* a "phenylpropanoid" pathway involving polyporic acid (**198**) as the key intermediate (Scheme 20) (*66, 177, 178, 179*). The terphenyl quinone polyporic acid (**198**), which can be produced in artificial culture (*179*), arises from phenylalanine by "dimerisation" of phenylpyruvate (**197**). Enzymatic oxidative rearrangement of the quinone then leads to vulpinic acid (**44**) whereas sequential hydroxylation and rearrangement produces calycin (**46**); efficient non-enzymatic conditions for the oxidative rearrangements of terphenyl quinones have been discussed elsewhere (p. 169). In a similar sequence, variegatorubin (**52**) is derived from tyrosine in *Boletaceae*, probably *via* *p*-hydroxyphenylpyruvate (**199**), the quinone atromentin (**174**) and variegatic acid (**50**) (Scheme 21) (*41, 60*).

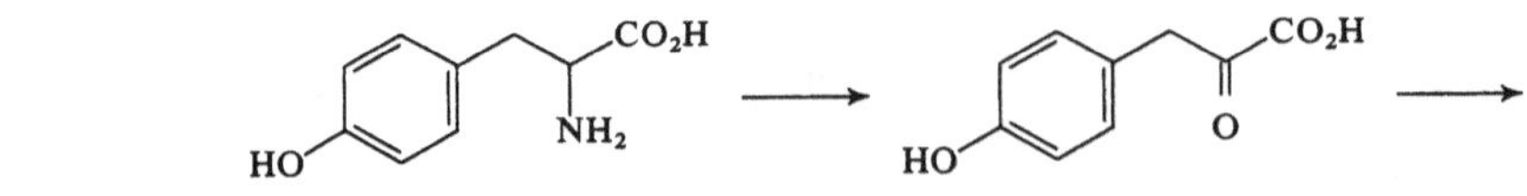

(199)

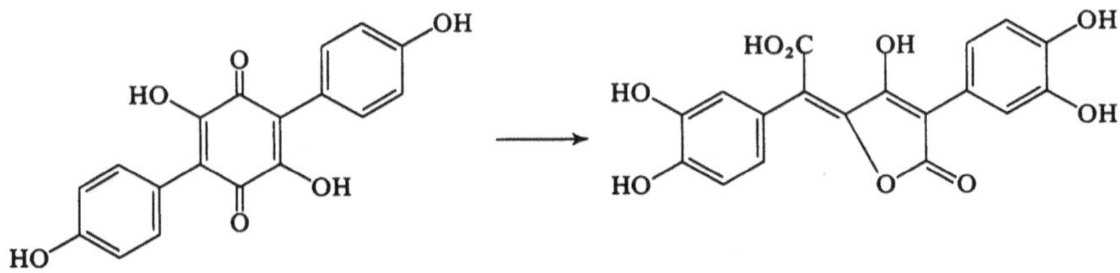

Atromentin (174) Variegatic acid (50)

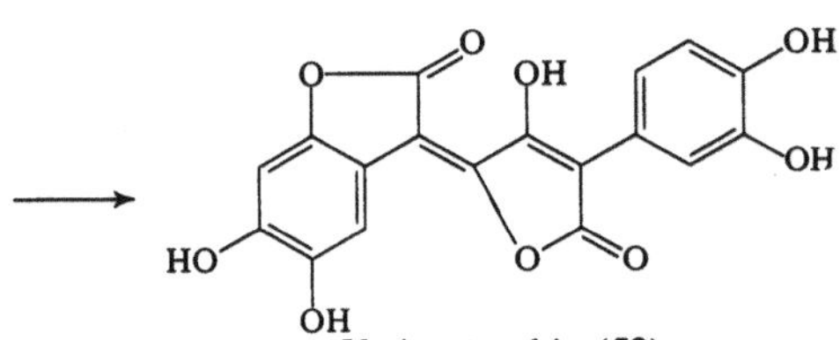

Variegatorubin (52)

Scheme 21. Biosynthesis of variegatorubin

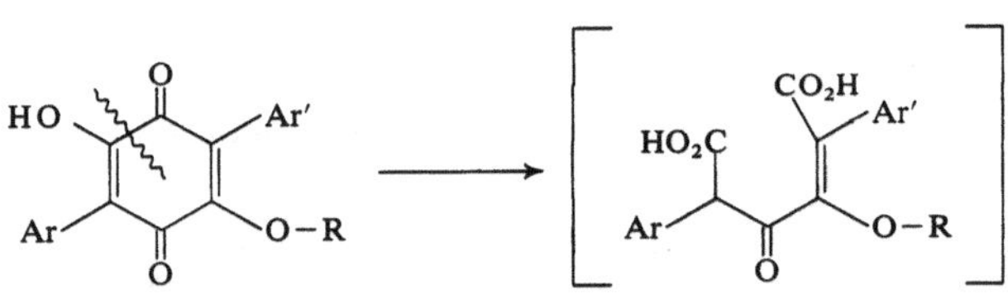

(200)

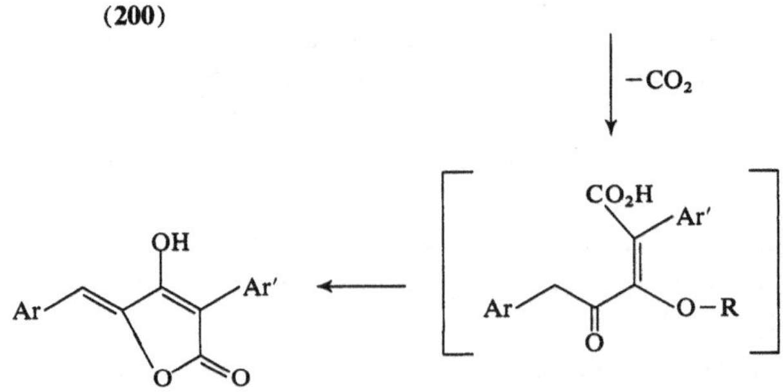

(60)

Scheme 22. Hypothetical biosynthetic scheme to pulvinones

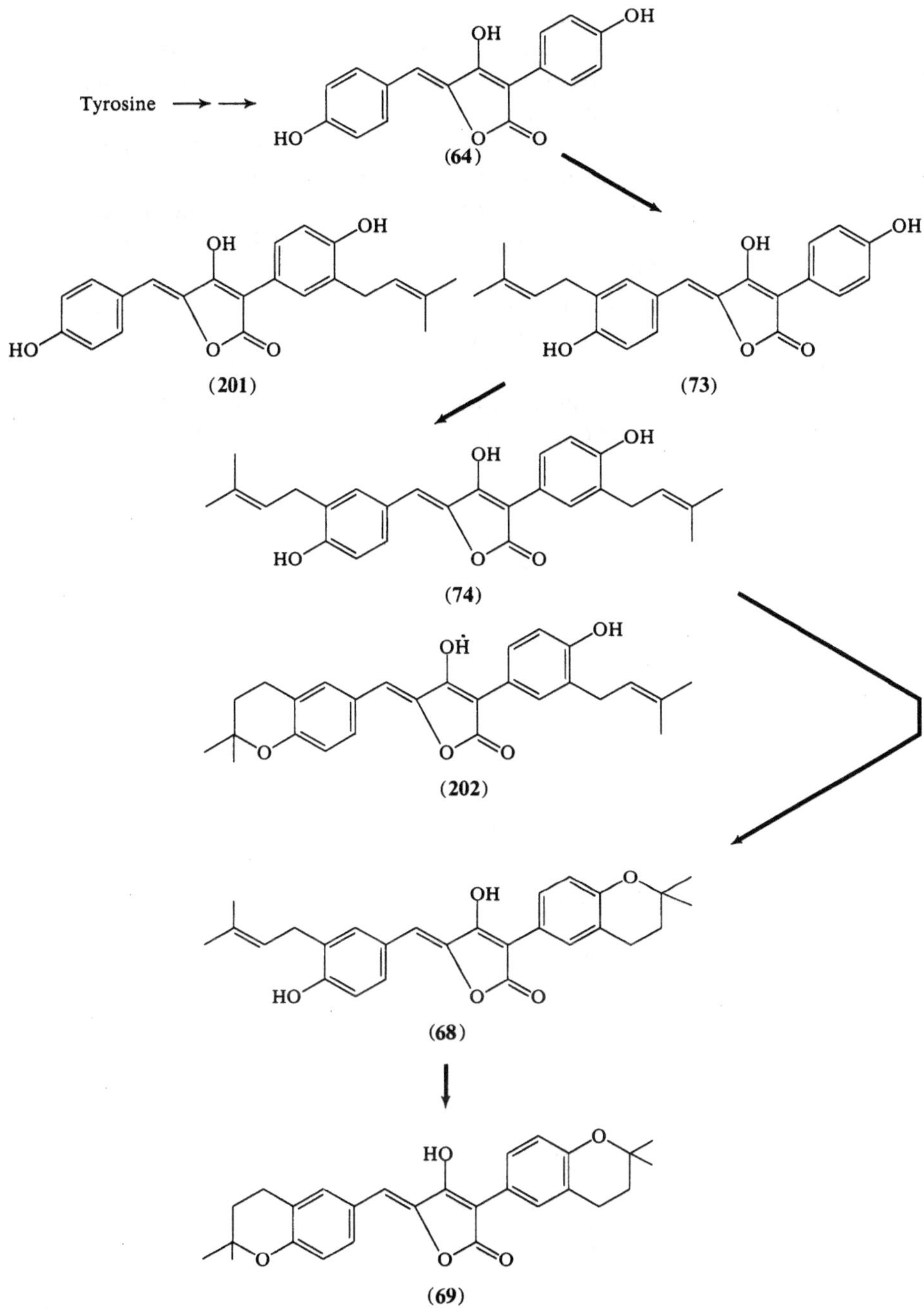

Scheme 23. Specificities of enzymatic prenylation and chromanation in late stages of the biosynthesis of chroman-aspulvinones

Although no studies of the early stages of the biosynthesis of the "pulvinone" pigments (general structure **60**) found in *Aspergillus terreus* and *Suillus grevillei* have been reported, it seems probable that their biosynthesis follows a closely similar pathway to that worked out for the related pulvinic acids. Thus oxidative rearrangement of and decarboxylation from an appropriate phenylpropanoid derived terphenyl quinone could furnish the basic pulvinone carbon skeleton; alternatively, a "half-protected" form of the quinone (*viz* **200**) might be implicated in the sequence (Scheme 22).

In studies of later stages in the biosynthesis of the "chromanated" pulvinone (**69**) from the 4,4′-dihydroxy pulvinone (**64**) in *A. terreus*, SETO and co-workers (*78, 79*) have been able to separate a cell-free enzyme system from the fungus which catalyses the transfer of the prenyl-moiety from 3,3-dimethylallyl pyrophosphate to the aryl rings of

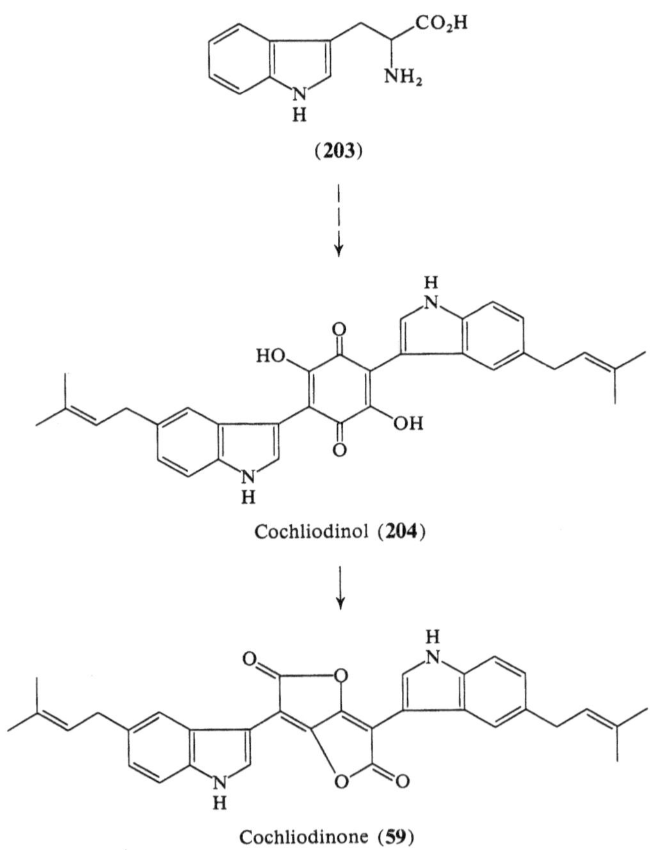

Scheme 24. Biosynthesis of cochliodinone

(**64**). Other studies have shown that the transfer of the prenyl moieties to (**64**) leading to (**74**) is a step-wise process proceeding *via* (**201**) or (**73**). Furthermore the conversion of the *bis*-prenylated metabolite (**74**) to (**69**) ("chromanation") is also a step-wise process proceeding *via* one of the mono-chromanated pulvinones (**202**) or (**68**). In their contemporaneous synthetic studies KNIGHT and PATTENDEN (*75*) were able to distinguish between these alternative possibilities by effecting unambiguous syntheses of permethylated derivatives of the unsymmetrical pulvinones corresponding to (**201**) and (**73**) and to (**202**) and (**68**). These investigations established that the initially formed prenylated metabolite in the biosynthesis of (**69**) is actually represented by (**73**), and that chroman (**68**), not (**202**), is the most likely immediate precursor of (**69**).

The biosynthesis of cochliodinone (**59**) in *Chaetomium globosum* probably takes place from tryptophan (**203**) along closely parallel lines to those of pulvinic acids, *via* the quinone cochliodinol (**204**) which co-occurs with (**59**) in *C. globosum* (Scheme 24) (*68*). It is also plausible that the pigment xylerythrin (**7a**), produced by the fungus *Peniophora sanguinea* (*6*), is formed in nature from condensation (Perkin-type) between polyporic acid (**198**) and *p*-hydroxyphenylacetic acid; polyporic acid has been found in the closely related *Peniophora filamentosa* (B&C) Burt (*180*).

Although the pulvinic acid amide derivatives rhizocarpic acid and epanorin (**58**) could originate from simple coupling between the appropriate amino acid and the pulvinic acid, MAASS (*66*) has proposed a direct route to pulvinamide itself (*viz* **58**, R = H), involving condensation between phenylalanine and phenylpyruvic acid.

3. Acetylenic Ylidenebutenolides in Compositae

Detailed studies of the biosynthesis of polyacetylenic compounds from higher plants have shown that they originate ultimately from oleic acid (**205**) (*25, 181*). The route outlined in Scheme 25 *via* linoleic acid (**206**), crepenynic acid (**207**), dehydrocrepenynic acid (**208**), diyne (**209**) and triyne (**210**) has been suggested for the biosynthesis of the important acetylene precursor (**211**) from oleic, and the chain shortening of (**211**) to the C_{10}-compound dehydromatricaria ester (**212a**) probably occurs as shown. Enzymatic isomerization of the corresponding dehydromatricaria acid (**212b**) then leads to the corresponding butenolide dehydromatricaria lactone (**23**). In a similar manner matricaria acid (**17**) is the immediate precursor of both matricaria lactone (**18**) and of the cumulene (**20**) which co-occur in some Astereae. It has been suggested that the thio-ether butenolides (**25**), (**26**) and (**27**) found in *Anthemis* species are

$Me(CH_2)_7CH{=}CH(CH_2)_7CO_2H$

(205)

↓

$Me(CH_2)_4CH{=}CH.CH_2.CH{=}CH(CH_2)_7CO_2R$

(206)

↓

$Me(CH_2)_4C{\equiv}C.CH_2CH{=}CH(CH_2)_7CO_2R$

(207)

↓

$Me(CH_2)_2CH{=}CH.C{\equiv}C.CH_2CH{=}CH(CH_2)_7CO_2R$

(208)

↓

$Me(CH_2)_2C{\equiv}C{-}C{\equiv}C.CH_2CH{=}CH(CH_2)_7CO_2R$

(209)

↓

$MeC{\equiv}C{-}C{\equiv}C{-}C{\equiv}C.CH_2CH{=}CH(CH_2)_7CO_2R$

(210)

↓

$MeC{\equiv}C{-}C{\equiv}C{-}C{\equiv}C.CH{=}CH.CH(OH)(CH_2)_7CO_2R$

(211)

↓

$MeC{\equiv}C{-}C{\equiv}C{-}C{\equiv}C{-}CH{=}CH.CO(CH_2)_7CO_2R$

↓

$Me(C{\equiv}C)_3CH{=}CH.CO.O(CH_2)_7CO_2R$

↓

$Me(C{\equiv}C)_3CH\overset{cis}{=}CH.CO_2R$ ⟶

(212) *a)* R = Me
b) R = H

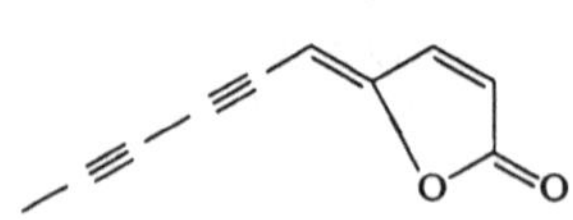

(23)

Scheme 25. Biosynthesis of matricaria lactone

produced by addition of methanethiol to dehydromatricaria lactone (**23**); equally the addition of the thiol group could precede cyclisation, leading first to the thio-ester (**213**), a known natural product. In a similar manner, the biosynthesis of the thiophene-butenolide (**32**), found in *Anthemis nobilis* and *Artemisia verlotorum*, could proceed from dehydromatricaria ester (**212**) either by way of dehydromatricaria lactone (**23**) or the thiophene (**214**).

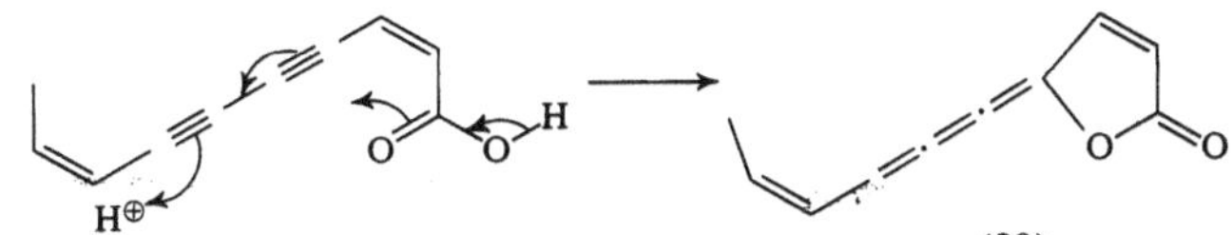

(17) (20) (18)

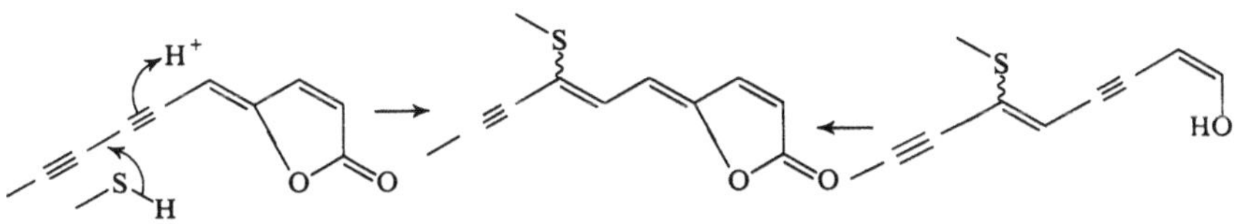

(23) (25/26) (213)

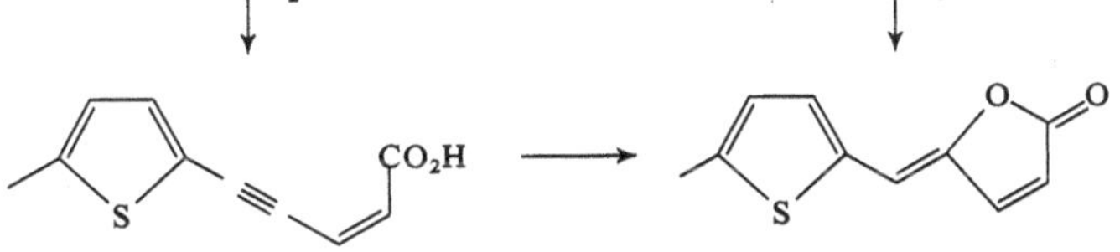

(212) (23) (214) (32)

4. Terpenoid Butenolides

The unique C_{37}-tricyclic carotenoid peridinin (**36**) occurs together with carotenoids possessing the ordinary isoprenoid C_{40}-carbon skeleton in dinoflagellates (*39, 40*). Taking into account that the dinoflagellates are photosynthetic, KJØSEN (*182*) has made the interesting suggestion that the elimination of the C_3-unit from the C_{40}-skeleton may be a photosensitized reaction although enzymatically controlled. Thus a photo-[2+2] reaction involving the C-11,C-13-diene system of the C_{40}-polyene system could lead to a cyclobutene intermediate **(215)** which by retro [2+2] reaction produces **(216)**. Photo-isomerization to **(217)** followed

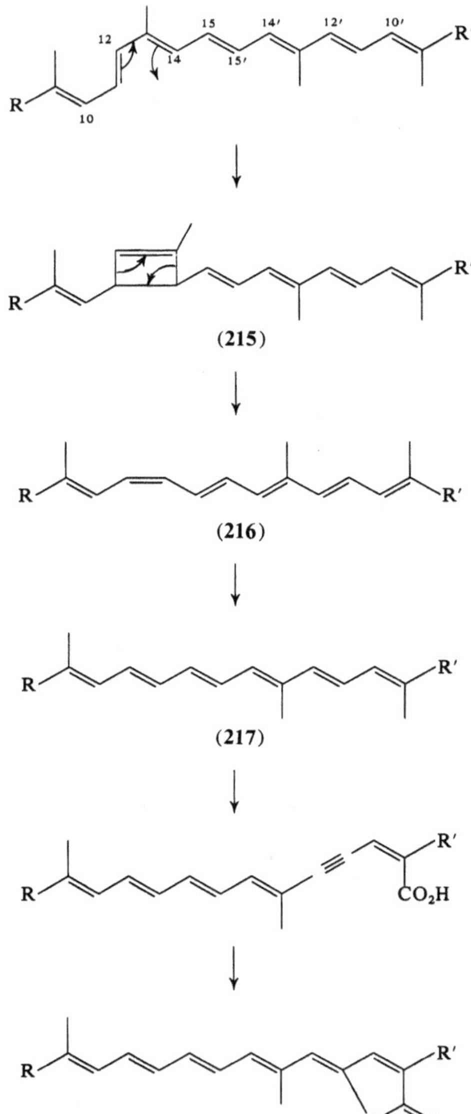

Scheme 26. Hypothetical biosynthetic scheme to peridinin

by oxidation of the 19′-methyl group to a carboxylic acid and lactonisation *via* addition to a carbon-to-carbon triple bond at C-11′,12′, according to Scheme 26, then completes the sequence to peridinin (**36**). The elimination of the C_3-unit has also been rationalized in terms of initial oxidation of the C-20 methyl group followed by rearrangement and elimination (*39*) (Scheme 27).

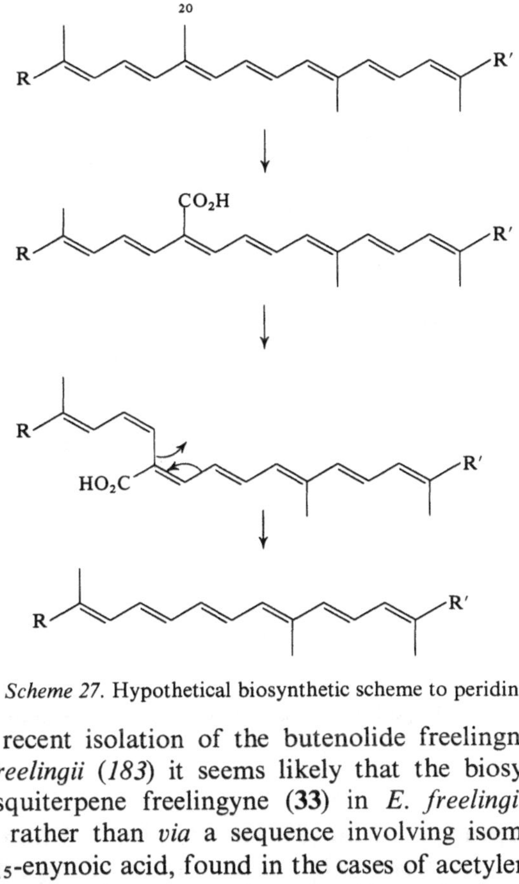

Scheme 27. Hypothetical biosynthetic scheme to peridinin

With the recent isolation of the butenolide freelingnite (**218**) from *Eremophila freelingii* (*183*) it seems likely that the biosynthesis of the acetylenic sesquiterpene freelingyne (**33**) in *E. freelingii* proceeds by way of (**218**) rather than *via* a sequence involving isomerization of a preformed C_{15}-enynoic acid, found in the cases of acetylenic butenolides in Compositae and also proposed in the biosynthesis of peridinin. It has been suggested that the butenolide moiety in freelingnite arises from a hydrocarbon precursor *via* several stages of oxidation as shown in the scheme; alternatively, a terminal *cis*-α-unsaturated carboxylic acid precursor might be implicated. Freelingyne is then formed from freelingnite *via* the dihydro analogue (**34**) by stepwise dehydrogenation. The closely related *bis*-furan (**219**) has been found in *Ursina* species and could also function as an intermediate in the biosynthesis of the butenolides (**218**) and (**33**) (*184, 185*).

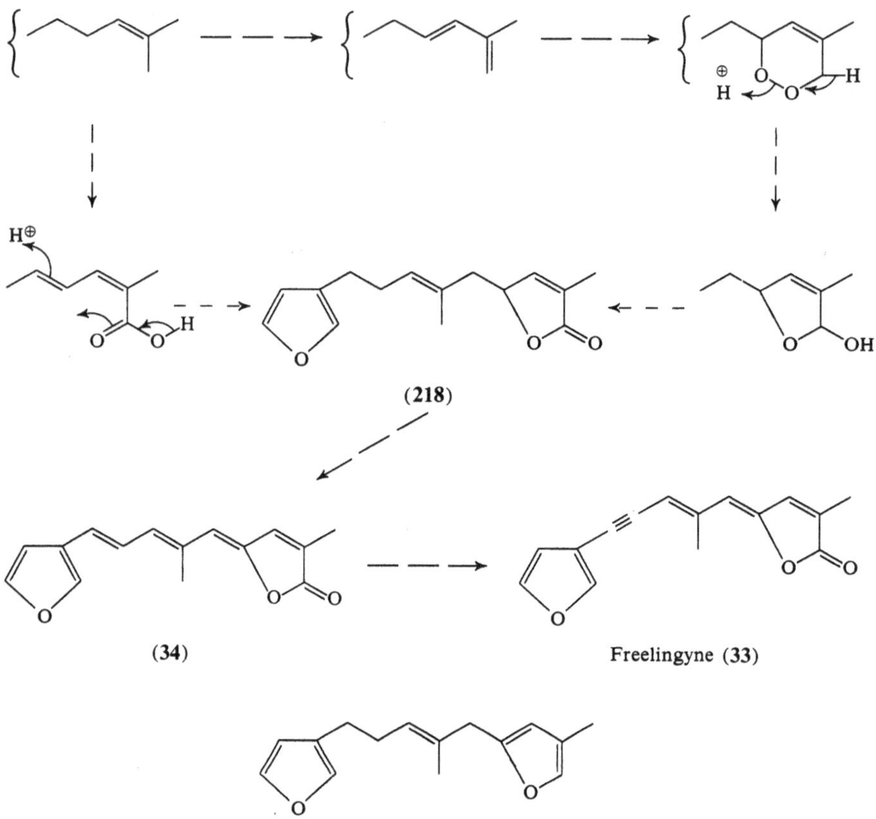

(219)

Scheme 28. Hypothetical biosynthetic scheme to freelingyne

The sesquiterpene variabilin (**84**) has been isolated from the marine sponge *Fasciospongia fovea* (*43*) where it is found together with the furan (**221**) and geranylfarnesol (**220**). This supports the hypothesis that variabilin is derived from geranylfarnesol *via* furan (**221**) in *F. fovea*, and also in sponges of the genus *Ircinia*. The furan (**221**) has also been described from *Ircinia spinosula* (*186*) and geranylfarnesol has previously been found in an insect of the Ophiobalanes (*187*). Variabilin is the most likely precursor of ircinin-1 (**82**), ircinin-2 (**83**) and fasciculatin (**81**) (*188*). The interesting biogenetic interrelationships between 3-substituted furans and the corresponding butenolides co-occurring in the marine sponge *Spongia officinalis* have been investigated in detail (*189*).

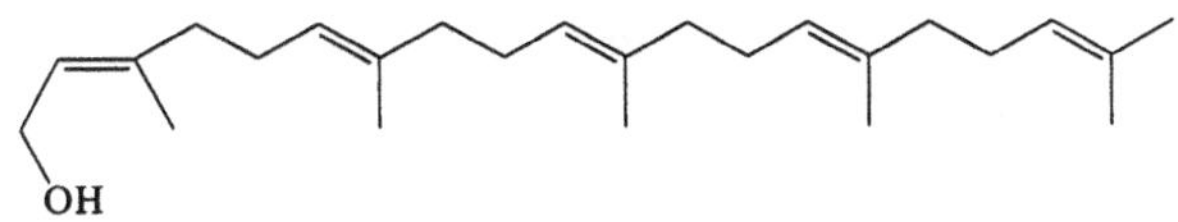

(220)

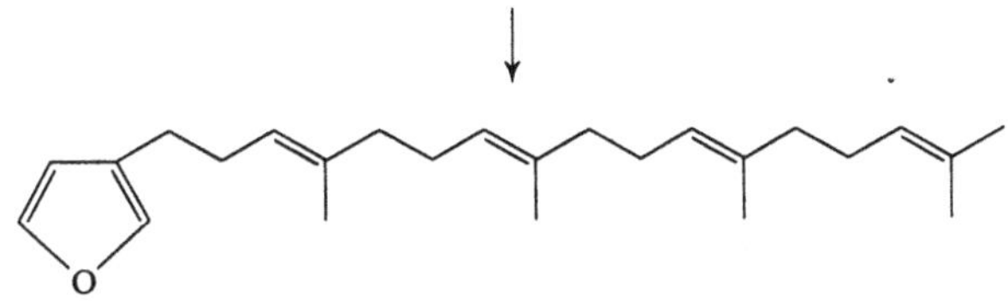

(221)

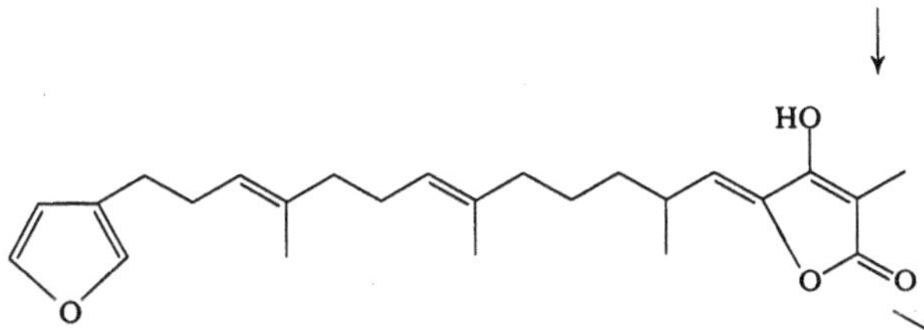

Variabilin (**84**)

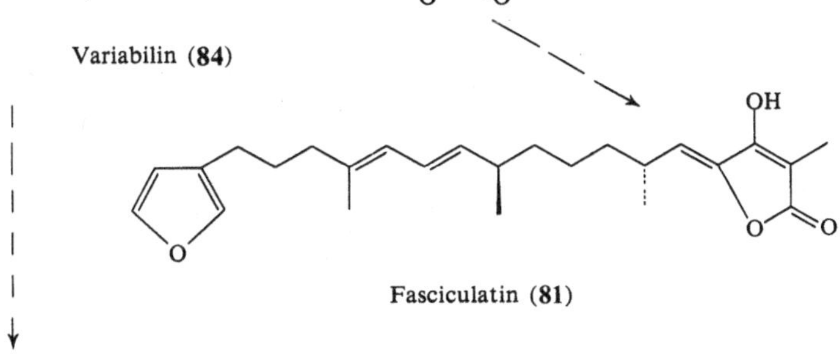

Fasciculatin (**81**)

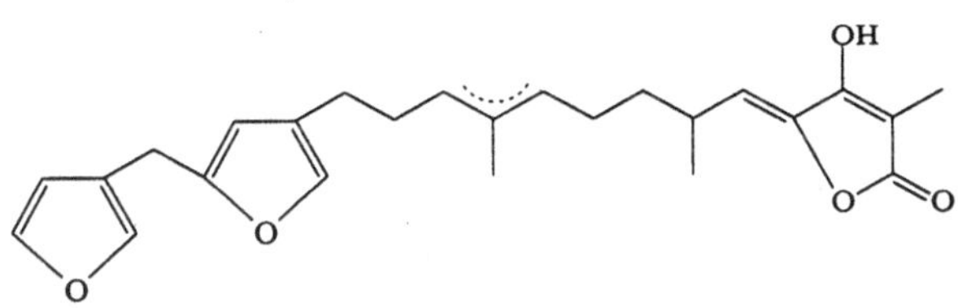

Ircinins 1 and 2

(**82**/**83**)

5. Other Metabolites

Investigations of the biosynthesis of protoanemonin (**5**) in *Ranunculus glaber* have led to the suggestion that the metabolite is derived from malonate and glycerate, according to Scheme 29 (*190*). The naturally occurring alkylidenephthalides, and their dihydro-analogues found in *Apium graveolens* and *Levisticum* species are produced from acetate (*18, 191*).

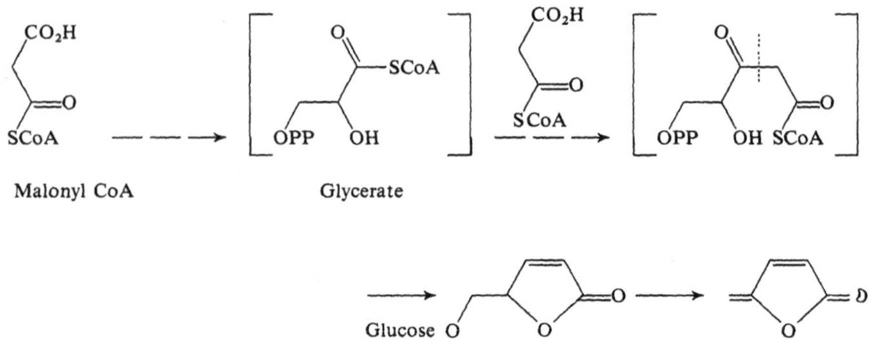

Scheme 29. Biosynthesis of protoanemonin

V. Biological Activity

The but-2-enolide (**1**) and tetronic acid (**2**) ring systems constitute important structural features in a wide range of biologically important natural products, *e.g.* vitamin C (**222**), the cardenolide digitoxigenin (**223**), and strigol (**224**), the germination stimulant for the plant parasite *Striga* (*192, 193*). Two of the first 4-ylidenebutenolides to be found in nature were protoanemonin (**5**) and patulin (**10**). These molecules aroused

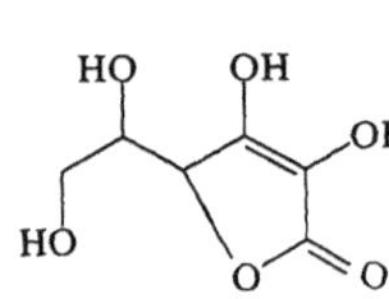

(**222**)

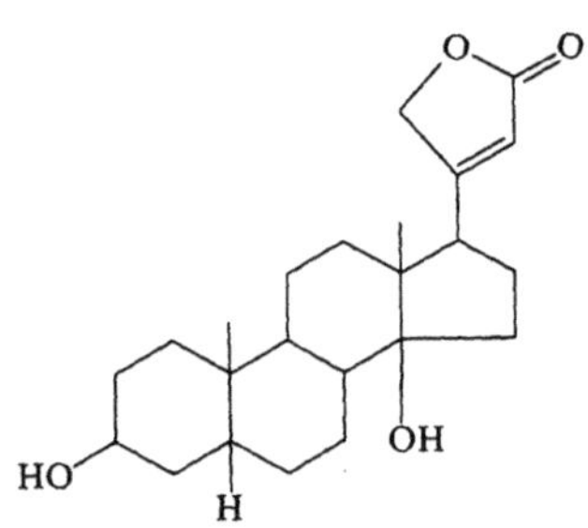

(**223**)

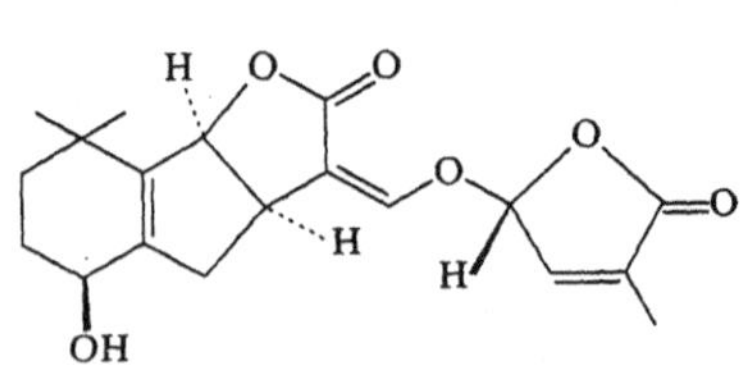

Strigol (**224**)

immediate interest because of their marked antibiotic activity, but their use in therapy was precluded by their acute toxicity to mammals. In spite of this, however, several structure-activity studies amongst protoanemonin analogues have been made (*194*, *195*). Patulin (**10**) has recently been employed as a fungal toxin in food (*10*), and as a general plant toxin (*11*). The 4-ylidenetetronic acids variabilin (**84**) and strobilinin (**85**), found in marine sponges, and tetrenolin (**40**), produced by *Micropolyspora venezuelensis,* also possess strong antibiotic activity.

It is generally known that early Eskimos used lichens to poison wolves, but it was not until 1952, during tests with vulpinic acid (**44**) in guinea-pigs and cats, that the toxicity was associated with the presence of pulvinic acids (**43**) in the lichens (*196*). In more recent years, a number of pulvinic acid analogous have been synthesized, and many have been shown to possess useful anti-inflammatory properties (*96—106, 146, 197*). The polyacetylenic precursors, matricaria ester (**16**), dehydromatricaria ester (**212a**) and lachnophyllum ester (**225**) of the ylidenebutenolides (**18**), (**23**) and (**21**) respectively, found in the Compositae, have recently been shown to have strong growth inhibitory effects on other plants, and the occurrence of these metabolites in nature has been discussed from the view point of allethopathy (*198*). In a similar vein, the synthetic phenacylidenephthalide (**154**) has been shown to be a highly active inhibitor of root geotropism when fed to seeds of cress and ryegrass (*199*).

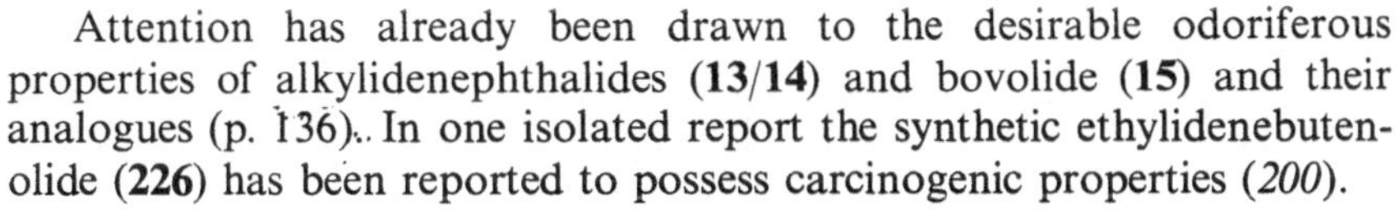

(**16**) R = MeCH=CH–
(**212a**) R = MeC≡C–
(**225**) R = $MeCH_2CH_2$–

(**226**)

Attention has already been drawn to the desirable odoriferous properties of alkylidenephthalides (**13/14**) and bovolide (**15**) and their analogues (p. 136). In one isolated report the synthetic ethylidenebutenolide (**226**) has been reported to possess carcinogenic properties (*200*).

Acknowledgement

I thank David W. Knight for help in collecting information for this review.

References

1. Dean, F. M.: Naturally Occurring Oxygen Ring Compounds. Butterworths. 1963.
2. Rao, Y. S.: Recent Advances in the Chemistry of Unsaturated Lactones. Chem. Rev. **76,** 625 (1976); *idem:* Chemistry of Butenolides, ibid. **64,** 353 (1964).
3. Haynes, L. J., and J. R. Plimmer: Tetronic Acids. Quart. Rev. (Chem. Soc. London) **14**, 292 (1960).

4. TSCHESCHE, R., K. WELMAR, G. WULFF, and G. SNATZKE: Über Folgeprodukte einer unbekannten genuinen Vorstufe des Ranunculins in Ranunculaceen. Chem. Ber. **105,** 290 (1972).
5. HILL, R., and R. VAN HEYNINGEN: Ranunculin: The Precursor of the Vesicant Substance of the Buttercup. Biochem. J. **49,** 332 (1951).
6. GRIPENBERG, J., and J. MARTIKKALA: Fungus Pigments XIX. Xylerythrin and its 5-O-Methyl Derivative. Acta Chem. Scand. **23,** 2583 (1969).
7. — — Fungus Pigments XX. On the Structure of Peniophorin, One of the Pigments Produced by *Peniophora sanguinea* Bres. Acta. Chem. Scand. **24,** 3444 (1970).
8. GRIPENBERG, J.: Fungus Pigments XXI. Peniophorinin, a Further Pigment Produced by *Peniphora sanguinea* Bres. Acta. Chem. Scand **24,** 3449 (1970).
9. — Fungus Pigments XXII. Peniosanguin and its Methyl Ether. Acta. Chem. Scand. **25,** 2999 (1971).
10. SCOTT, P. M., and B. P. C. KENNEDY: Improved Method for the Thin-layer chromatographic determination of Patulin in Apple Juice. J. Ass. Off. Anal. Chem. **56,** 813 (1973).
11. ELLIS, J. R., and T. M. MCCALLA: Effects of patulin and method of application on growth stages of wheat. Appl. Microbiol. **25,** 562 (1973).
12. SMALL, J.: Celery seed. Food **17,** 181 (1948).
13. CIAMICIAN, G., and P. SILBER: Über die hochsiedenden Bestandteile des Sellerieöls. Chem. Ber. **30,** 492 (1897).
14. BARTON, D. H. R., and J. X. DE VRIES: The Constitution of Sedanolide. J. Chem. Soc. (London) 1916 (1963).
15. GOLD, H. J., and C. W. WILSON: III: Alkylidene Phthalides and Dihydrophthalides from Celery. J. Organ. Chem. (U. S. A.) **28,** 985 (1963).
16. — — III: The Volatile Flavour Substances of Celery. J. Food Science **28,** 484 (1963).
17. NAVES, Y. R.: Etudes sur les Matières Végétales Volatiles XXIV. Composition de l'huile Essentielle et du Résinoide de Livèche (*Levisticum officinale* Koch.). Helv. Chim. Acta **26,** 1281 (1943).
18. MITSUHASHI, H., and U. NAGAI: Studies on the Constituents of Umbilliferae Plants VII. Structure of Ligustilide 2. Tetrahedron **19,** 1277 (1963).
19. MITSUHASHI, H., T. MURAMATSU, U. NAGAI, T. NAKANO, and K. UENO: Studies on the Constituents of Umbilliferae Plants VIII. Distribution of Alkylphthalides in Umbelliferae Plants. Chem. Pharm. Bull. (Tokyo) **11,** 1317 (1963).
20. KNIGHT, D. W., and G. PATTENDEN: 4-Phosphoranylidenebutenolide Intermediates in the Synthesis of 4-Ylidenebutenolides and 4-Ylidenetetronic Acids. Synthesis of Alkylidenephthalide Constituents of Celery Odour and Models for Freelingyne. J. Chem. Soc. Perkin I (London) 635 (1975).
21. KARIYONE, T., and S. SHIMIZU: Synthesis of Alkylidenephthalides and their Odor. J. Pharmac. Soc. Japan **73,** 336 (1953).
22. LARDELLI, G., G. DIJKSTRA, P. D. HARKES, and J. BOLDINGH: A New γ-Lactone found in Butter. Rec. trav. chim. Pays-Bas. **85,** 43 (1966).
23. SÖRENSEN, N. A., and K. STAVHOLT: A Hexahydro Matricaria ester — "Composit-Cumulene 1" from Scentless Mayweed (*Matricaria inodora* L.). Acta Chem. Scand. **4,** 1080 (1950).
24. CHRISTENSEN, P. K., N. A. SÖRENSEN, I. BELL, E. R. H. JONES, and M. C. WHITING: Constitution of the so-called "Composit-cumulene 1" from Scentless Mayweed (*Matricaria inodora*). Festschr. Arthur Stoll 545 (1957) [Chem. Abstr. **53,** 346e, (1959)].
25. BOHLMANN, F., T. BURKHARDT, and C. ZDERO: Naturally Occurring Acetylenes. London: Academic Press. 1973.

26. BOHLMANN, F., H. MÖNCH, and P. BLASZKIEWICZ: Die Polyine der Gattung *Matricaria* L. Chem. Ber. **100,** 611 (1967).
27. LAM, J.: Polyacetylenes of *Solidago virgaurea*: Their Seasonal Variation and NMR Longe-range Spin Coupling Constants. Phytochem. **10,** 647 (1971).
28. BOHLMANN, F., H. BORNOWSKI, and C. ARNDT: Über das erste natürlich vorkommende Kumulen. Chem. Ber. **98,** 2236 (1965).
29. CHRISTENSEN, P. K.: Structural and Spectroscopical Studies on Naturally Occurring Acetylenes of Composites. Norges Tek. Vitenkap akad. Ser. **2,** 7 (1959) [Chem. Abstr. **54,** 6531i (1960)].
30. BOHLMANN, F., and C. ZDERO: Über ein neues Kumulen aus *Erigeron canadense* L. (1). Tetrahedron Letters 2465 (1970).
31. BOHLMANN, F., K.-M. KLEINE, C. ARNDT, and S. KÖHN: Neue Inhaltsstoffe der Gattung *Anthemis* L. Chem. Ber. **98,** 1616 (1965).
32. BOHLMANN, F., P. HERBST, C. ARNDT, H. SCHÖNOWSKY, and H. GLEINIG: Über einen Neuen Typ von Polyacetylenverbindungen aus verschiedenen Vertretern des *Tribus Anthemideae* L. Chem. Ber. **94,** 3193 (1961).
33. BOHLMANN, F., and K.-M. RODE: Die Polyine der Gattung *Carlina* L. Chem. Ber. **100,** 1507 (1967).
34. BOHLMANN, F., and C. ZDERO: Über ein Thiophenlacton aus *Chamaemelum nobile* L. Chem. Ber. **99,** 1226 (1966).
35. MASSY-WESTROPP, R. A., G. D. REYNOLDS, and T. M. SPOTSWOOD: Freelingyne, an Acetylenic Sesquiterpenoid. Tetrahedron Letters 1939 (1966).
36. INGHAM, C. F., and R. A. MASSY-WESTROPP: The Stereochemistry of Freelingyne and the Synthesis of Dihydrofreelingyne. Austral. J. Chem. **27,** 1491 (1974).
37. BIRCH, A. J., G. S. R. SUBBA RAO, and J. P. TURNBULL: Eremolactone. Tetrahedron Letters 4749 (1966).
38. BEGLEY, M. J., D. W. KNIGHT, and G. PATTENDEN: Stereochemistry of Freelingyne [(4*Z*,6*E*)-9-(3-Furyl)-2,6-dimethylnona-2,4,6-Trien-8-yn-4-olide] by X-ray Analysis. J. Chem. Soc. Perkin II (London) 1863 (1975).
39. STRAIN, H. H., W. A. SVEC, K. AITZETMÜLLER, M. C. GRANDOLFO, J. J. KATZ, H. KJØSEN, S. NORGÅRD, S. LIAAEN-JENSEN, F. T. HAXO, P. WEGFAHRT, and H. RAPOPORT: The Structure of Peridinin, the Characteristic Dinoflagellate Carotenoid. J. Amer. Chem. Soc. **93,** 1823 (1971).
40. JOHANSEN, J. E., W. A. SVEC, S. LIAAEN-JENSEN, and F. T. HAXO: Carotenoids of the Dinophyceae. Phytochem. **13,** 2261 (1974).
41. GALLO, G. G., C. CORONELLI, A. VIGEVANI, and G. C. LANCINI: The Structure of Tetrenolin, a New Antibiotic Substance. Tetrahedron **25,** 5677 (1969).
42. PAGANI, H., G. LANCINI, G. TAMONI, and C. CORONELLI: Tetrenolin and SS/1018 A, Antibacterial Agents Isolated from a Strain of Actinomycetales. J. Antibiotics **26,** 1 (1973).
43. KAZLAUSKAS, R., P. T. MURPHY, R. J. QUINN, and R. J. WELLS: Two New Sesterterpene Lactones from a Sponge. Tetrahedron Letters 2635 (1976).
44. EVANS, W. C., B. S. W. SMITH, P. MOSS, and H. N. FERNLEY: Bacterial Metabolism of 4-Chlorophenoxyacetate. Biochem. J. **122,** 509 (1971).
45. EVANS, W. C., and J. K. GAUNT: Metabolism of 4-chloro-2-methylphenoxyacetate by a Soil Pseudomonad. Preliminary Evidence for the Metabolic Pathway. Biochem. J. **122,** 519 (1971).
46. EVANS, W. C., B. S. W. SMITH, H. N. FERNLEY, and J. I. DAVIES: Bacterial Metabolism of 2,4-Dichlorophenoxyacetate. Biochem. J. **122,** 543 (1971).
47. STEGLICH, W.: Pilzfarbstoffe. Chem. Unserer Zeit **9,** 117 (1975).
48. MURRAY, J.: Lichens and Fungi. Part I, Polyporic Acid in *Stictae*. J. Chem. Soc. (London) 1345 (1952).

49. ASANO, M., and Y. KAMEDA: Über die Konstitution des Calycins und dessen Synthese (IV. Mitteil. über Flechten-Farbstoffe der Pulvinsäure-Reihe). Chem. Ber. **68,** 1568 (1935).
50. ÅKERMARK, B.: Studies on the Chemistry of Lichens. 14. The Structure of Calycin. Acta Chem. Scand. **15,** 1695 (1961).
51. EDWARDS, R. L., and M. GILL: Constituents of the Higher Fungi. Part XIII. 2-Aryl-3-methoxymaleic Anhydrides from Pulvinic Acid Derivatives. A Convenient Method for Determination of Structure of Fungal and Lichen Pulvinic Acid Derivatives. J. Chem. Soc. (London) 1534 (1973).
52. AGARWAL, S. C., and T. R. SESHADRI: A Reinvestigation of the Structure of Pinastric Acid and Isopinastric Acid. Indian J. Chem. **2,** 17 (1964).
53. KNIGHT, D. W., and G. PATTENDEN: Synthesis of the Pulvinic Acid Pigments of Lichen and Fungi. Chem. Commun. 660 (1976).
54. STEGLICH, W., H. BESL, and K. ZIPFEL: Pigments of Fungi XIX. Structural Determination of Pulvinic Acids by NMR Spectroscopy. Z. Naturforsch. **29b,** 96 (1974).
55. BEAUMONT, P. C., R. L. EDWARDS, and G. C. ELSWORTHY: Constituents of the Higher Fungi. Part VIII. The Blueing of *Boletus* Species. Variegatic Acid, a Hydroxytetronic Acid from *Boletus* Species and a Reassessment of the Structure of Boletol. J. Chem. Soc. (C) (London) 2968 (1968).
56. STEGLICH, W., W. FURTNER, and A. PROX: Neue Pulvinsäure-Derivate aus *Xerocomus chrysenteron* (Bull. ex St. Amans), Quel. und Untersuchungen zur Frage des Vorkommens von Anthrachinonpigmenten bei Boletaceen. Z. Naturforsch. **23b,** 1044 (1968).
57. — — — Xerocomsäure und Gomphidsäure, zwei chemotaxonomisch interessante Pulvinsäure-Derivate aus *Gomphidius glutinosus* (Schff) Fr. Z. Naturforsch. **24b,** 941 (1969).
58. STEGLICH, W., I. PILS, and A. BRESINSKY: Nachweis und chemotaxonomische Bedeutung von Pulvinsäure in *Rhizopogon* (Gasteromycetes). Z. Naturforsch. **26b,** 376 (1971).
59. BESL, H., A. BRESINSKY, W. STEGLICH, and K. ZIPFEL: Über Gyrocyanin, das blauende Prinzip des Kornblumen-Röhrlings *(Gyroporus cyanescens),* und eine oxidative Ringverengung des Atromentins. Chem. Ber. **106,** 3223 (1973).
60. STEGLICH, W., W. FURTNER, and A. PROX: Pigment V. Variegatorubin, ein Oxydationsprodukt der Variegatsäure aus *Suillus piperatus* (Bull. ex Fr.) O. Kuntze und anderen Boletaceen. Z. Naturforsch. **25b,** 557 (1970).
61. GAYLORD, M. C., R. G. BENEDICT, G. M. HATFIELD, and L. R. BRADY: Isolation of Diphenyl-substituted Tetronic Acids from Cultures of *Paxillus atrotomentosus.* J. Pharm. Sci. **59,** 1420 (1970).
62. BEAUMONT, P. C., and R. L. EDWARDS: Constituents of the Higher Fungi, Part XI. Boviquinone-3(2,5-Dihydroxy-3-farnesyl-1,4-benzoquinone), Diboviquinone-3,4, Methylenediboviquinone-3,3, and Xerocomic Acid from *Gomphidius rutilus* Fr. and Diboviquinone-4,4 from Voletus *(Suillus) bovinus* (Linn ex Fr.) Kuntze. J. Chem. Soc. (C) (London) 2582 (1971).
63. EDWARDS, R. L., and M. GILL: Constituents of the Higher Fungi, Part XII. Identification of Involutin as (−)-*cis*-5-(3,4-Dihydroxyphenyl)-3,4-dihydroxy-2-(4-hydroxyphenyl)-cyclopent-2-enone and Synthesis of (±)-*cis*-Involutin Trimethyl Ether from Isoxerocomic Acid Drivatives. J. Chem. Soc. Perkin I (London) 1529 (1973).
64. SINGH, P., and M. ANCHEL: Atromentic Acid from *Clitocybe illudens.* Phytochemistry **10,** 3259 (1971).
65. BRESINSKY, A., H. BESL, and W. STEGLICH: Gyroporin und Atromentinsäure aus *Leccinum-aurantiacum-Kulturen.* Phytochemistry **13,** 271 (1974).
66. MAASS, W. S. G.: Pulvinamide and Possible Biosynthetic Relationships with Pulvinic Acid. Phytochemistry **9,** 2477 (1970).

67. FRANK, R. L., S. M. COHEN, and J. N. COKER: The Structures and Syntheses of Rhizocarpic Acid and Epanorin. J. Amer. Chem. Soc. **72,** 4454 (1950).
68. JERRAM, W. A., A. G. MCINNES, W. S. G. MAASS, D. G. SMITH, A. TAYLOR, and J. A. WALTER: The Chemistry of Cochliodinol, a Metabolite of *Chaetomium spp.* Canad. J. Chem. **53,** 727 (1975).
69. SCHÖNBERG, A., and A. SINA: Molecular Rearrangements. Part V. A Case of a Thermal Reversible α-Diketone-γ-Lactone Rearrangement. Some Experiments with Vulpinic Acid and Related Compounds. J. Chem. Soc. (London) 601 (1946).
70. EDWARDS, R. L., and M. GILL: Constituents of the Higher Fungi. Part XIV. 3′,4′,4-Trihydroxypulvinone, Thelephoric Acid, and Novel Pyrandione and Furanone Pigments from *Suillus grevillei* (Klotsch) Sing [*Boletus elegans* (Schum. *per* Fries)]. J. C. S. Perkin I, 1921 (1973).
71. CLAISEN, L., and T. EWAN: Über die Einwirkung des Oxaläthers auf Dibenzylketon. Ann. **284,** 245 (1895).
72. OJIMA, N., S. TAKENAKA, and S. SETO: New Butenolides from *Aspergillus terreus.* Phytochemistry **12,** 2527 (1973).
73. — — — Structures of Pulvinone Derivatives from *Aspergillus terreus.* Phytochemistry **14,** 573 (1975).
74. KNIGHT, D. W., and G. PATTENDEN: Synthesis of Pulvinones, Metabolites of *Aspergillus terreus* and *Suillus grevillei.* Chem. Commun. 876 (1975).
75. — — Specificities of Enzymatic Prenylation and Chromanation in the Biosynthesis of Aspulvinone Pigments in *Aspergillus terreus.* Chem. Commun. 635 (1976).
76. BEGLEY, M. J., D. W. KNIGHT, and G. PATTENDEN: Structural Revision for Prenylated Pulvinone Metabolites from *Aspergillus terreus.* Tetrahedron Letters 131 (1976).
77. GOLDING, B. T., R. W. RICKARDS, and Z. VANEK: New Metabolites of *Aspergillus terreus*: 3-Hydroxy-2,5-bis(p-hydroxyphenyl)penta-2,4-dien-4-olide and Derivatives. J. C. S. Perkin I, 1961 (1975).
78. OJIMA, N., K. OGURA, and S. SETO: Biosynthesis of Pulvinone Derivatives in *Aspergillus terreus.* Enzymatic Prenylation of Dihydroxypulvinone. Chem. Commun. 717 (1975).
79. OJIMA, N., I. TAKAHASHI, K. OGURA, and S. SETO: New Metabolites from *Aspergillus terreus* related to the Biosynthesis of Aspulvinones. Tetrahedron Letters 1013 (1976).
80. BRACKEN, A., and H. RAISTRICK: Studies in the Biochemistry of Micro-organizms. 75. Dehydrocarolic Acid, a Metabolitic Product of *Penicillium cincerascens* Biourge. Biochem. J. **41,** 569 (1947).
81. PLIMMER, J. R.: The Structure of Carolic Acid. J. Org. Chem. **29,** 511 (1964).
82. GUDGEON, J. A., J. S. E. HOLKER, and T. J. SIMPSON: Use of Singly and Doubly Labelled ^{13}C-Acetate in the Elucidation of the Structures and Biosynthesis of Multicolic and Multicolosic Acids, New Tetronic Acids from *Penicillium multicolor.* Chem. Commun. 636 (1974).
83. CAFIERI, F., E. FATTORUSSO, C. SANTACROCE, and L. MINALE: Fasciculatin, a Novel Sesterterpene from the Sponge *Ircinia fasciculata.* Tetrahedron **28,** 1579 (1972).
84. CIMINO, G., S. DE STEFANO, L. MINALE, and E. FATTORUSSO: Ircinin-1 and -2, Linear Sesterterpenes from the Marine Sponge *Ircinia oros.* Tetrahedron **28,** 333 (1972).
85. FAULKNER, D. J.: Variabilin, an Antibiotic from the Sponge, *Ircinia variabilis.* Tetrahedron Letters 3821 (1973).
86. ROTHBERG, I., and P. SHUBIAK: The Structure of some Antibiotics from the Sponge *Ircinia strobilina.* Tetrahedron Letters 769 (1975).
87. HÄNSEL, R., and A. PELTER: Cinnamylidenbutenolide aus *Piper sanctum.* Phytochemistry **10,** 1627 (1971).
88. — — Epoxipiperolid aus *Piper sanctum.* Z. Naturforsch. **27b,** 1186 (1972).
89. SHAW, E.: A Synthesis of Protoanemonin. The Tautomerism of Acetylacrylic Acid, and of Penicillic Acid. J. Amer. Chem. Soc. **68,** 2510 (1946).

90. GRUNDMANN, C., and E. KOBER: An Improved Synthesis of Protoanemonin. J. Amer. Chem. Soc. **77,** 2332 (1955).
91. WALTON, H. M.: Potential Antimicrobial Agents 11. 4-Hydroxy-2,4-alkadienoic Acid γ-Lactones. J. Org. Chem. **22,** 312 (1957).
92. VOLHARD, J., and F. HENKE: Über die Derivate des Diphenylketipinsäure-dinitrils oder Oxalylbisbenzylcyanürs. Ann. chimie. **282,** 45 (1894).
93. MITTAL, O. P., and T. R. SESHADRI: Synthesis of Leprapinic Acid and Constitution of Pinastric Acid. J. Chem. Soc. (London) 1734 (1956).
94. GROVER, P. K., and T. R. SESHADRI: Preparation and Constitution of Isopinastric Acid. Tetrahedron **4,** 105 (1958).
95. ASANO, M., and S. HUZIWARA: Lichen Pigments of the Pulvinic Acid Series. VI. Synthesis of Atromentinic Acid. J. Pharmac. Soc. Japan **59,** 675 (1939).
96. SUTTON, B. M.: 4-Cyclohexylvulpinic Acid Derivatives. U. S. 3,752.829. Chem. Abstr. **79,** 115428u (1973).
97. — Phenylvulpinic Acid Derivatives. U. S. 3,780,065. Chem. Abstr. **80,** 82624v (1974).
98. — Substituted 2,5-Diphenyl-3,4,6-trihydroxy $\Delta^{2,4}$-Hexadienoic Acid Lactones (1,4). U. S. 3,772,341. Chem. Abstr. **80,** 133047u (1974).
99. — Pyridyl Ketipate Lactones and Derivatives. U. S. 3,816,440. Chem. Abstr. **81,** 91365j (1974).
100. — 4-Cyclohexylvulpinic Acid Derivatives in the Treatment of Arthritis. U.S. 3,821,397. Chem. Abstr. **81,** 105256p (1974).
101. — 3,4,6-Trihydroxy-2,5-Diphenyl-2,4-hexadienoic acid γ-Lactones in the treatment of Arthritis. U. S. 3,821,398. Chem. Abstr. **81,** 120257h (1974).
102. — Pyridyl Ketipate Lactones and Derivatives in the treatment of arthritis. U. S. 3,818,092. Chem. Abstr. **81,** 91367m (1974).
103. WEINSTOCK, J.: Thiopulvinic Acid Derivatives. U. S. 3,780,064. Chem. Abstr. **80,** 95710v (1974).
104. FODEN, F. R., and D. M. O'MANT: 2-[α-(Methoxycarbonyl) benzylidene]-4- phenyl-3-hydroxy-5-oxo-2,5-dihydrofurans. Brit. 1,335,269. Chem. Abstr. **80,** 47828 t (1974).
105. FODEN, F. R., J. MCCORMICK, and D. M. O'MANT: Antiinflammatory, analgesic and antipyretic 2-[α-(methoxycarbonyl) benzylidene]-4-phenyl-3-hydroxy-5-oxo-2,5-dihydrofuran derivatives. Brit. 1,358,382. Chem. Abstr. **81,** 120438t (1974).
106. WEINSTOCK, J., J. E. BLANK, and B. M. SUTTON: Preparation of Some Thiovulpinic Acids. J. Org. Chem. **39,** 2454 (1974).
107. EISNER, U., J. A. ELVIDGE, and R. P. LINSTEAD: Unsaturated Lactones and Related Substances. Part V. Dihydro-β-ketomuconic Acid and Carboxy-lactones of the Protoanemonin Type. J. Chem. Soc. (London) 1501 (1951).
108. BELL, I., E. R. H. JONES, and M. C. WHITING: Researches on Acetylenic Compounds. Part LXI. The Synthesis of Three Polyacetylenic Esters. J. Chem. Soc. (London) 1313 (1958).
109. BOHLMANN, F., and R. REINECKE: Synthese des Thioenolätherbutenolids aus *Anthemis rigescens* Willd. Chem. Ber. **99,** 3437 (1966).
110. SERRATOSA, F.: An Acetylenic Approach to Patulin Derivatives. Tetrahedron **16,** 185 (1961).
111. CASTÃNER, J., and J. PASCUAL: Isomerisation of Phenylpropargylidenemalonic acid to γ-Benzylidene-α-carboxybutenolide. J. Chem. Soc. (London) 3962 (1958).
112. FLEMING, I., and J. HARLEY-MASON: Enol Elimination Reactions. Part II. A New Synthesis of Tetronic Acids. J. Chem. Soc. (London) 4778 (1963).
113. MULHOLLAND, T. P. C., R. FOSTER, and D. B. HAYDOCK: A Synthesis of Tetronic Acid [Furan-2(3*H*),4(5*H*)-dione] and three Analogues. J. Chem. Soc. Perkin I (London) 1225 (1972).

114. Conradie, W. J., C. F. Garbers, and P. S. Steyn: Oxidation of 3-Methylbut-2-enolide to *cis*-β-Formylcrotonic Acid. J. Chem. Soc. (London) 594 (1964).

115. Corrie, J. E. T.: A Facile Synthesis of Ylidenebutenolides. Tetrahedron Letters **4873** (1971).

116. Howe, R. K.: (*E*)-3-Benzylidenephthalides. J. Org. Chem. **38,** 4164 (1973).

117. Knight, D. W., and G. Pattenden: Total Synthesis of the Acetylenic Sesquiterpene Freelingyne. Chem. Commun. 188, (1974).

118. — — Synthesis of Freelingyne, an Acetylenic Sesquiterpene from *Eremophila freelingii.* J. Chem. Soc. Perkin I, 641 (1975).

119. Okazaki, M., and A. Yamaguchi: Modified Wittig Reaction III, Preparation and Some Reactions of Diphenyl 6-Nitrophthalide-3-phosphonate. Nippon Kagaku Kaishi, 110 (1973). Chem. Abstr. **78,** 84494k (1973).

120. Weiss, R.: Benzalphthalide. Org. Synth. Coll. **2,** 61 (1943).

121. Hrnciar, P.: Phthalides and 1,3-Indandiones. XVI. Preparation of 4-Arylmethylene-cinchomeronides and 2-aryl-5-aza-1,3-indandiones. Chem. Zvesti **19,** 360 (1965). Chem. Abstr. **63,** 5616d (1965).

122. Hrnciar, P., and D. Joniak: Phthalides and 1,3-Indandiones. XXI. Perkin Synthesis of 7-Nitro-3-benzalphthalide and 4-Nitro-3-benzalphthalide; their Reduction and Conversion into 4-Acetamino-2-phenylindan-1,3-dione. Chem Zvesti **20,** 336 (1966). Chem. Abstr. **65,** 8832h (1966).

123. Hrnciar, P., and L. Kuruc: Phthalides and 1,3-Indandiones. XXVIII. Preparation of 5- and 6-Substituted 3-Benzalphthalides by the Perkin Reaction. Chem. Zvesti **21,** 267 (1967). Chem. Abstr. **67,** 73304v (1967).

124. Sigal, M. V., P. Marchini, and B. L. Poet: Phthalazine Derivatives. U.S. 3,274,185. Chem. Abstr. **65,** 18600c (1966).

125. Horeau, A., and J. Jacques: Recherches dans la série des isocoumarines. Bull. Soc. Chim. France 53 (1948).

126. Bergmann, E. D.: Fulvenes and Thermochromic Ethylenes. Part 30. Experiments in the 2,3-Diarylindone Series. J. Org. Chem. **21,** 461 (1956).

127. Mowry, D. T., E. L. Ringwald, and M. Renoll: Vinyl Aromatic Compounds. VI. Alkylidenephthalides and Related Compounds. J. Amer. Chem. Soc. **71,** 120 (1949).

128. Berti, G.: Stereoisomers of Benzalphthalide. Gazz. chim. ital. **86,** 655 (1956). Chem. Abstr. **52,** 1958a (1958).

129. Boldingh, J., G. Lardelli, P. Roele, G. Dijkstra, P. Haverkamp-Begemann, P. D. Harkes, and A. S. M. van der Zijden: Dihydrofuranone Derivatives. Ger. 1,072,629. Chem. Abstr. **55,** 19949 (1961).

130. Boldingh, J., G. Dijkstra, P. Roele, P. Haverkamp-Begemann, P. D. Harkes, G. Lardelli, and A. S. M. van der Zijden: Lactones and Their Use as Flavouring Agents. Brit. 893,321. Chem. Abstr. **57,** 9670c (1962).

131. Lardelli, G., P. Haverkamp-Begemann, P. D. Harkes, and A. S. M. van der Zijden: Lactonals. Brit. 893,322. Chem. Abstr. **57,** 9672h (1962).

132. Knight, D. W.: Natural Butenolides, Ph. D. Thesis, University of Nottingham, 1975.

133. Chopard, P. A., R. F. Hudson, and R. J. G. Searle: The Reaction of Stable Phosporanes with Phthalic Anhydride; a Case of cis-trans Isomerism. Tetrahedron Letters 2357 (1965).

134. Ingham, C. F., R. A. Massy-Westropp, G. D. Reynolds, and W. D. Thorpe: A Synthesis of Enol-Lactones. Aust. J. Chem. **28,** 2499 (1975).

135. Ingham, C. F., R. A. Massy-Westropp, and G. D. Reynolds: A Synthesis of Freelingyne. Aust. J. Chem. **27,** 1477 (1974).

136. Begley, M. J., D. W. Knight, and G. Pattenden: Regioselective Nucleophilic Additions to Unsymmetrically Substituted Maleic Anhydrides. Tetrahedron Letters 4279 (1975), and unpublished work.

137. KOGL, F., and H. BECKER: Untersuchungen über Pilzfarbstoffe. VI. Die Konstitution des Atromentins. Ann. Chimie **465,** 211 (1928).

138. WIKHOLM, R. J., and H. W. MOORE: Dimethyl Sulphoxide-Acetic Anhydride Oxidative Rearrangements of Hydroxyterphenylquinones. A Possible Biosynthetic Model. J. Amer. Chem. Soc. **94,** 6152 (1972).

139. CAIN, B. F.: Potential Anti-tumour Agents. Part II. Polyporic Acid Series. J. Chem. Soc. (London) 356 (1963).

140. FRANK, R. L., G. R. CLARK, and J. N. COKER: The Synthesis of Vulpinic Acid from Polyporic Acid. J. Amer. Chem. Soc. **72,** 1824 (1950).

141. KARRER, P., and L. SCHNEIDER: Oxydation von *o*-Dicarbonylverbindungen durch Persäuren. Helv. Chim. Acta **30,** 859 (1947).

142. KARRER, P., and E. TESTA: Oxydativer Abbau des Tetrachlor-o-benzochinons mit Perphtalsäure. Untersuchung der Oxydationsprodukte. Helv. Chim. Acta **32,** 1019 (1949).

143. KARRER, P., and TH. HOHL: Abbau des Tetrabrom-o-benzochinons durch Perphtalsäure zum α,β,δ-Tribrommuconsäure-γ-Lacton und dessen weitere Umsetzungsprodukte. Helv. Chim. Acta **32,** 1028 (1949).

144. MOORE, H. W., H. R. SHELDON, D. W. DETERS, and R. J. WIKHOLM: Rearrangements of Azidoquinones V. Stereoselective Acid-Catalyzed Rearrangements of Azidoquinones to γ-Cyanoalkylidine-(Cyanoarylidine) $\Delta^{\alpha,\beta}$-butenolides. J. Amer. Chem. Soc. **92,** 1675 (1970).

145. MOORE, H. W.: Chemistry of Azido Quinones and Related Compounds. Chem. Soc. Rev. **2,** 415 (1973).

146. O'MANT, D. M.: The Preparation and Chemical Properties of 3-Acyl-5-arylidene-4-hydroxy-2,5-dihydro-2-oxothiophens, a New Class of Immunosuppressive Agent. J. Chem. Soc. C. (London) 1501 (1968).

147. RAPHAEL, R. A.: Compounds Related to Penicillic Acid. Part III. Synthesis of Penicillic Acid. J. Chem. Soc. (London) 1508 (1948).

148. — Compounds Related to Penicillic Acid. Part II. Synthesis of Dihydropenicillic Acid. J. Chem. Soc. (London) 805 (1947).

149. GILMAN, H., R. A. FRANZ, A. P. HEWLETT, and G. F. WRIGHT: Decomposition of 5-Halogeno-2-furylmethyl Ethers to Benzalcrotonolactones. J. Amer. Chem. Soc. **72,** 3 (1950).

150. YAMADA, K., Y. TOGAWA, T. KATO, and Y. HIRATA: A Convenient Method for the Preparation of γ-Arylidene-α,β-Unsaturated γ-Lactones. Application to the Synthesis of the Thiophene Lactone obtained from *Chamaemelum nobile* L. Tetrahedron **27,** 5445 (1971).

151. NILSSON, M.: Preparation of 2-Acetylcyclopent-4-ene-1,3-diones from Maleic Anhydrides and Isopropenyl Acetate. Acta. Chem. Scand. **18,** 441 (1964).

152. FOWLER, J., and S. SELTZER: The Synthesis of Model Compounds for Maleylacetoacetic Acid. Maleylacetone. J. Org. Chem. **35,** 3529 (1970).

153. SAUER, J. C., R. D. CRAMER, V. A. ENGELHARDT, T. A. FORD, H. E. HOLMQUIST, and B. W. HOWK: Bifurandione. I. Preparation and Characterization. J. Amer. Chem. Soc. **81,** 3677 (1959).

154. SAUER, J. C.: Dilactones from Carbon Monoxide and Acetylenes. U.S. 2,840,570. Chem. Abstr. **53,** 4300a (1959).

155. ALBANESI, G., and M. TOVAGLIERI: Reactions between Acetylenic Hydrocarbons and Carbon Monoxide catalyzed by Cobalt Carbonyls. 1. Octatrienediolides. Chim. e. ind. (Milan) **41,** 189 (1959). Chem. Abstr. **53,** 19872d (1959).

156. CHIUSOLI, G. P., and L. CASSAR: Nickel-catalyzed Reactions of Allyl Halides and Related Compounds. Angew. Chem. int. Edit. **6,** 124 (1967).

157. KIJI, J., S. YOSHIKAWA, and J. FURUKAWA: Carbonylation of Phenylacetylene by Rhodium Carbonyl Chloride. Bull. Chem. Soc. Japan **47,** 490 (1974).

158. LINSTEAD, R. P., L. N. OWEN, and R. F. WEBB: Elimination Reactions of Esters. Part III. Derivatives of Dibasic Polyhydroxy-acids. J. Chem. Soc. (London) 1225 (1953).

159. RAMIREZ, F., H. YAMANAKA, and O. H. BASEDOW: New Reactions of Phosphite Esters: The Conversion of Phthalic Anhydride into Biphthalyl by Trialkyl Phosphites and into Phthalide-3-phosphonates by Dialkyl Phosphites. J. Amer. Chem. Soc. **83,** 173 (1961).

160. STAAB, H. A., and J. IPAKTSCHI: Photochemische Reaktionen des Benzocyclobutendions-(1.2). Chem. Ber. **101,** 1457 (1968).

161. EICHER, T., and N. PELZ: Zur Chemie des Phenylcyclopropenons. Tetrahedron Letters 1631 (1974).

162. RAO, Y. S., and R. FILLER: Chemistry of Lactones: A One-step Synthesis of α-Phenyl-γ-Benzylidene-$\Delta^{\alpha,\beta}$-butenolide. Tetrahedron Letters 1457 (1975).

163. SCOTT, A. I., L. ZAMIR, G. T. PHILLIPS, and M. YALPANI: The Biosynthesis of Patulin. Bioorganic Chem. **2,** 124 (1973).

164. SCOTT, A. I., and L. BEADLING: Biosynthesis of Patulin. Dehydrogenase and Dioxygenase Enzymes of *Penicillium patulum.* Bioorganic Chem. **3,** 281 (1974).

165. FORRESTER, P. I., and G. M. GAUCHER: Conversion of 6-Methylsalicylic Acid into Patulin by *Penicillium urticae.* Biochemistry **11,** 1102 (1972).

166. MURPHY, G., G. VOGEL, G. KRIPPAHL, and F. LYNEN: Patulin Biosynthesis. Role of Mixed-Function Oxidases in the Hydroxylation of m-cresol. Eur. J. Phytochem. **49,** 443 (1974).

167. VOGEL, G., and F. LYNEN: 6-Methylsalicylic Acid Synthetase. Methods in Enzymology, Vol. 43, Antibiotics Edit. HASH, J. H., p. 520. Academic Press. 1975.

168. LIGHT, R. J., and G. VOGEL: 6-Methylsalicylic Acid (2,6-Cresotic Acid) Decarboxylase. Methods in Enzymology, Vol. 43, Antibiotics, Edit. HASH, J. H., p. 530. Academic Press. 1975.

169. GAUCHER, G. M.: *m*-Hydroxybenzyl Alcohol Dehydrogenase. Methods in Enzymology, Vol. 43, Antibiotics, Edit. HASH, J. H., p. 540. Academic Press. 1975.

170. MOSBACH, K.: Die Biosynthese der Orsellinsäure und Penicillinsäure (I). Acta. Chem. Scand. **14,** 457 (1960).

171. BENTLEY, R., and J. G. KEIL: Tetronic Acid Biosynthesis in Molds II. Formation of Penicillic Acid in *Penicillium cyclopium.* J. Biol. Chem. **237,** 867 (1962).

172. AL-RAWI, J. M. A., J. A. ELVIDGE, D. K. JAISWAL, J. R. JONES, and R. THOMAS: Use of Tritium Nuclear Magnetic Resonance for the Direct Location of ^{3}H in Biosynthetically-labelled Penicillic Acid. Chem. Commun. 220 (1974).

173. SETO, H., L. W. CARY, and M. TANABE: Utilisation of ^{13}C-^{13}C Coupling in Structural and Biosynthetic Studies. IV. Penicillic Acid. J. Antibiotics **27,** 558 (1974).

174. AXBERG, .K., and S. GATENBECK: Intermediates in the Penicillic Acid Biosynthesis in *Penicillium cyclopium.* Acta. Chem. Scand. **B29,** 749 (1975).

175. BENTLEY, R., D. S. BHATE, and J. G. KEIL: Tetronic Acid Biosynthesis in Molds 1. The Formation of Carlosic and Carolic Acids in *Penicillium charlesii.* J. Biol. Chem. **237,** 859 (1962).

176. BLOOMER, J. L., F. E. KAPPLER, and G. N. PANDEY: Biosynthesis of Carolic Acid in *Penicillium charlesii:* The Intermediate Precursors. Chem. Commun. **243** (1972).

177. MOSBACH, K.: On the Biosynthesis of Lichen Substances. Part 2. The Pulvic Acid Derivative Vulpinic Acid. Biochem. Biophys. Res. Commun. **17,** 363 (1964).

178. MAASS, W. S. G., and A. C. NEISH: Lichen Substances II. Biosynthesis of Calycin and Pulvinic Dilactone by the Lichen, *Pseudocyphellaria crocata.* Canad. J. Botany **45,** 59 (1967).

179. MOSBACH, K., H. GUILFORD, and M. LINDBERG: The Terphenyl Quinone Polyporic Acid: Production Isolation and Characterization. Tetrahedron Letters 1645 (1974)

180. MURRAY, J.: Lichens and Fungi. Part I. Polyporic Acid in *Stictae.* J. Chem. Soc. (London) 1345 (1952).

181. BOHLMANN, F.: Biogenetische Beziehungen der natürlichen Acetylenverbindungen. Fortschritte der Chemie Organischer Naturstoffe, Vol. XXV, p. 1. Wien: Springer. 1967.

182. KJØSEN, H.: Synthetic and Spectroscopic Studies of Carotenoids, The University of Trondheim, Trondheim, 1970.

183. KNIGHT, D. W., and G. PATTENDEN: Freelingnite, a New Furanosesquiterpene from *Eremophila freelingii.* Tetrahedron Letters 1115 (1975).

184. BORNOWSKI, H.: Die Struktur des Lasiospermans. Ein neuer Typ von Furansesquiter: penen. Tetrahedron **27,** 4101 (1971).

185. HIKINO, H., and C. KONNO: Furanosesquiterpenoids. Heterocycles **4,** 817 (1976).

186. CIMINO, G., S. DE STEFANO, and L. MINALE: Polyprenyl Derivatives from the Sponge *Ircimia spinosula,* 2-Polyprenylbenzoquinones, 2-polyprenylbenzoquinols Prenylated Furans and a C-31-Difuranoterpene. Tetrahedron **28,** 1315 (1972).

187. RÍOS, T., and S. PÉREZ: Geranylfarnesol, A New Acyclic C_{25}-Isoprenoid Alcohol isolated from Insect Wax. Chem. Commun. 214 (1969).

188. CORDELL, G. A.: The Occurrence, Structure Elucidation and Biosynthesis of the Sesterterpenes. Phytochemistry **13,** 2343 (1974).

189. CIMINO, G. S., S. DE STEFANO, and L. MINALE: Oxidized Furanoterpenes from the Sponge *Spongia officinalis.* Experientia **30,** 18 (1974).

190. HIRATA, T., and T. SUGA: The Biosynthesis of Protoanemonin in *Ranunculus glaber* Makino. Chem. Letters 637 (1973).

191. MITSUHASHI, H., and M. NOMURA: Studies on the Constituents of Umbelliferae Plants. XII: Biogenesis of 3-Butylphthalide. Chem. Pharm. Bull. (Tokyo) **14,** 777 (1966).

192. COOK, C. E., L. P. WHICHARD, M. E. WALL, G. H. EGLEY, P. COGGAN, P. A. LUCHAN, and A. T. MCPHAIL: Germination stimulants. 2. The Structure of Strigol – a potent seed germination stimulant for Witchweed (*Striga lutea* Lour). J. Amer. Chem. Soc. **94,** 6198 (1972).

193. JOHNSON, A. W., G. ROSEBERY, and C. PARKER: A Novel Approach to *Striga* and *Orobanche* Control using Synthetic Germination Stimulants. Weed Research **16,** 223 (1976).

194. BAER, H., M. HOLDEN, and B. C. SEEGAL: The Nature of the Antibacterial Agent from *Anemone pulsatilla.* J. Biol. Chem. **162,** 65 (1946).

195. ZECHNER, L., and H. WOHLMUTH: Anemonin and Protoanemonin I. New Methods of Isolation of Anemonin from *Ranunculus acer.* Scientia Pharm. **22,** 74 (1954).

196. SODERBERG, U.: Action of Vulpinic Acid. Acta. Physiol. Scand. **27,** 97 (1952).

197. FODEN, F. R., J. MCCORMICK, and D. M. O'MANT: Vulpinic Acids as Potential Anti-inflammatory Agents. 1. Vulpinic Acids with Substituents in the Aromatic Rings. J. Medic. Chem. **18,** 199 (1975).

198. KOBAYASHI, A., Y. SHIBATA, and K. YAMASHITA: A Novel Preparation of 1-Acetoxy-*cis*-pent-2-en-4-yne: A New Synthetic Route to *cis*-Dehydromatricaria Ester. Agr. Biol. Chem. **39,** 911 (1975), and refs. cited therein.

199. BROWN, B. T., O. JOHANSEN, G. F. KATEKAR, and W. H. F. SASSE: The Effect on Root Geotropism of Certain ortho-Carboxyphenylpropanones. Pestic. Sci. **4,** 473 (1973).

200. JONES, J. B., and J. M. YOUNG: Carcinogenicity of Lactones. IV. Alkylation of Analogues of DNA Guanine Groups such as Imidazole, N-Methylimidazole, and Guanosine by α,β-Unsaturated Acids. Canad. J. Chem. **48,** 1567 (1970).

(Received February 8, 1977)

Naturally Occurring Plant Coumarins

By R. D. H. Murray, Chemistry Department,
University of Glasgow, Scotland

Contents

I. Introduction: Scope of the Review

Over the past 150 years, since Vogel in 1820 isolated the parent oxygen heterocycle, coumarin (**1**) from *Coumarouna odorata* = *Dipteryx odorata* (*1*), coumarins have been recognized as widely distributed plant products. The vast majority carry an oxygen substituent at C-7; consequently 7-hydroxycoumarin [umbelliferone, (**2**)], one of the most widely distributed coumarins, is often regarded as the parent, in a structural sense and also biogenetically, of a large number of the structurally more complex coumarins. A number of review articles on natural plant coumarins have appeared (*154, 302, 344, 420, 525, 541, 548*) since the last in this series (*153*), the most comprehensive, albeit dealing solely with coumarins of the Umbelliferae, being that of Nielsen in 1970 (*419*).

Improvements in isolation procedures and structural analysis, especially in the last decade, have led to a marked increase in the number of known plant coumarins many of which are referred to in the literature only by their botanically derived trivial names.

The known monomeric plant coumarins are listed in the accompanying Tables principally according to the number and orientation of oxygen atoms on the benzenoid ring and then by the number of carbon atoms in the generally isoprenoid-derived substituent and therein by the oxidation level of the substituent. For completeness, the much smaller, though rapidly growing, number of similar coumarins in which the pyrone ring is *O*- and/or *C*-substituted have also been included. For every entry, leading references to the isolation, structural and, if relevant, stereochemical elucidation and, where effected, synthesis, of the coumarin are given.

By using the Formula Index and the alphabetically arranged Trivial Name Index, it is hoped that the reader may be able to gain an insight into the fascinating chemistry of the natural coumarins despite the proliferation of trivial names.

This review specifically excludes aryl substituted coumarins, coumarins bearing an additional fused carbocycle, and dimeric coumarins.

II. Structural Types

1. Coumarins With Acyclic Substituents

Oxygenation is known to occur at one or more of the six available positions on the coumarin nucleus, such oxygen atoms being present in phenolic, ethereal and glycosidic groups. A feature common to most coumarins is the presence of isoprenoid chains, either intact or partially degraded, of one, two or three units attached to a carbon atom of the aromatic nucleus, or to an oxygen atom, or to both. Frequently the unit is the simple 3-methylbut-2-enyl(prenyl) group. This can also be encountered not only as the corresponding epoxide or glycol but also in a variety of oxidized and skeletally rearranged forms. Indeed, in the field of natural products, coumarins offer perhaps the best example of a class of compound exhibiting the greatest number of biogenetic modifications of the simple isoprenoid unit.

2. Furano- and Pyrano-Coumarins

By biogenetic oxidative ring closure to neighbouring oxygen, the prenyl group may be encountered as a hydroxyisopropyldihydrofuran unit. Thc latter, probably by loss of acetone, is the known biogenetic precursor of the furan ring. When nuclear prenylation of umbelliferone occurs at C-6, this leads eventually to linear furanocoumarins of the psoralen (**122**) type. However, if an alternative mode of oxidative cyclisation prevails, linear hydroxydihydropyranocoumarins which lead to pyranocoumarins of the xanthyletin (**141**) type are produced. Conversely, alkylation at C-8 leads, through the dihydrofuranocoumarin, to furanocoumarins of the angelicin (**199**) class and *via* the isomeric dihydropyranocoumarin, to angular pyranocoumarins having the seselin (**241**) skeleton.

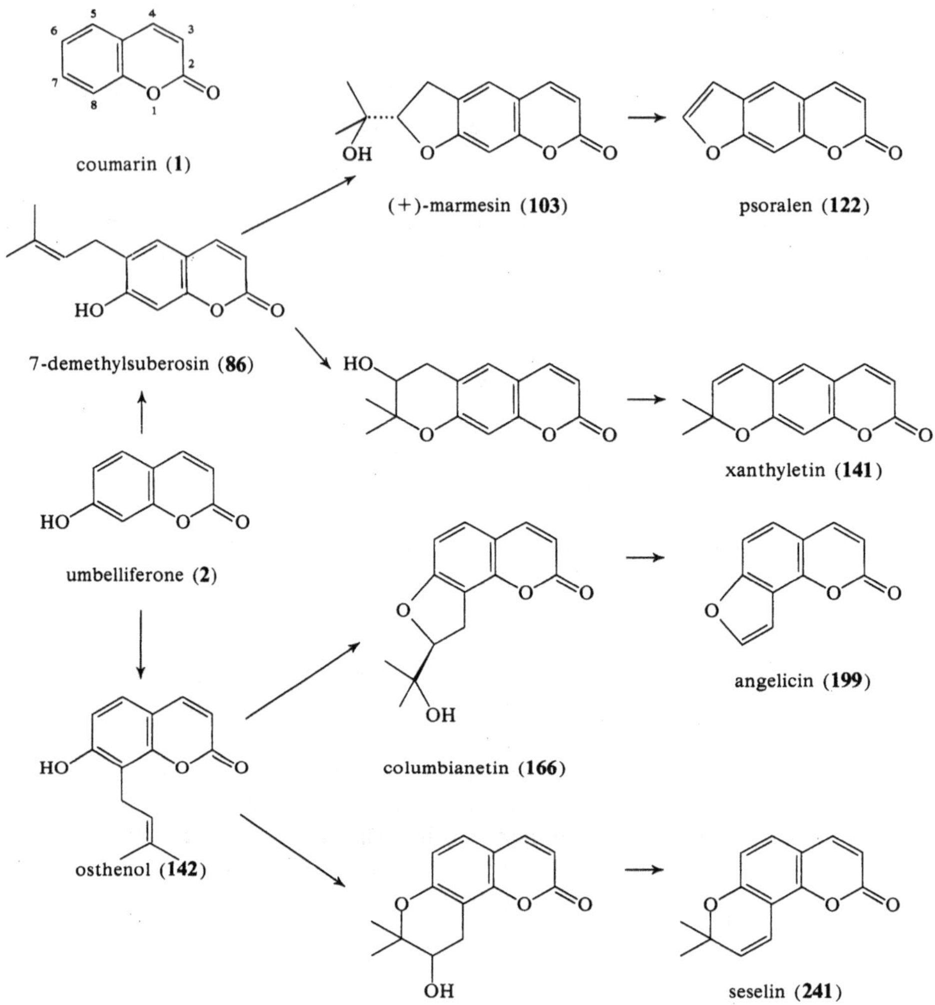

Scheme

Most other reviews have classified coumarins in an arbitrary manner according to whether, for example, particular compounds were derivatives of linear or angular furanocoumarins or pyranocoumarins. In compiling the Tables included in the present review a biogenetically related approach has been adopted which is based firstly upon the number of nuclear oxygen atoms. Within each Table, the entries are presented in the following order — (i) coumarins with acyclic substituents, (ii) dihydrofuranocoumarins, (iii) furanocoumarins, (iv) dihydropyranocoumarins, (v) pyranocoumarins. The coumarins of each sub-class are listed in order of increasing number of carbon atom in the substituent and in increasing oxidation level within that group. Phenols are considered before ethers and glycosides, and alcohols precede glycosides and esters.

III. Nomenclature

Unfortunately, many natural coumarins have been assigned trivial names, mostly botanically derived. Although these can facilitate discussion, to a considerable extent they serve to confuse, especially since many of the endings (*e.g.* -ol, -one) are not at all consistent with structure. Thus umbelliferone (**2**) is a phenol while osthol (**143**) is the methyl ether of osthenol (**142**). Structurally closely related coumarins can have markedly different names; aculeatin (**249**) is the epoxide of toddaculin (**248**), both of which occur in *Toddalia aculeata*, while conversely almost identical trivial names can be given to structurally quite dissimilar coumarins, *e.g.* ferudenol (**339**) and ferulenol (**481**). Not infrequently, the same coumarin isolated by different workers from different plant sources is given two or even three trivial names and even enantiomers are known with distinct trivial names, *e.g.* marmesin (**103**) and nodakenetin (**104**).

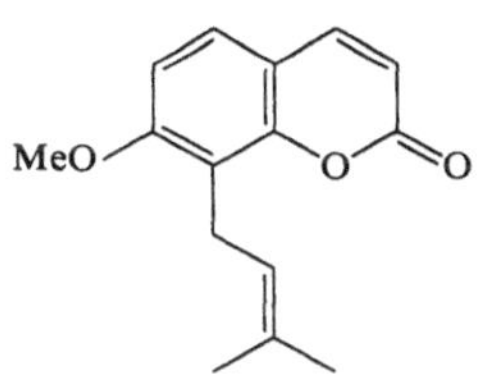

osthol (**143**)

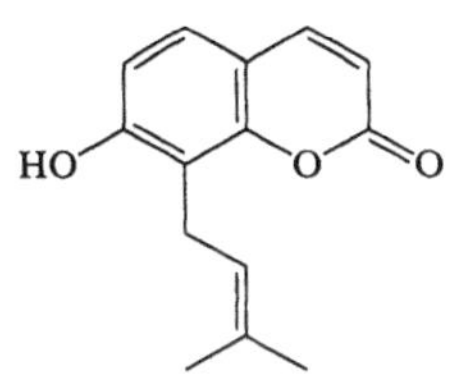

osthenol (**142**)

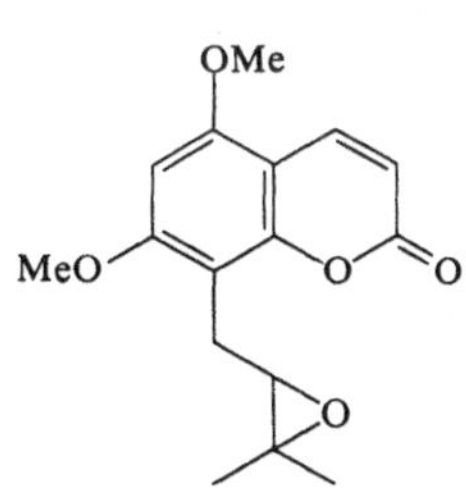

aculeatin (**249**)

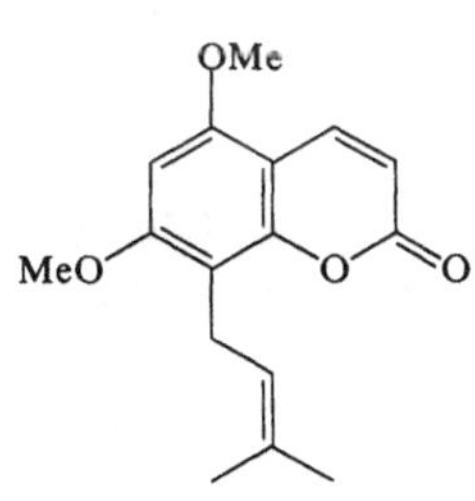

toddaculin (**248**)

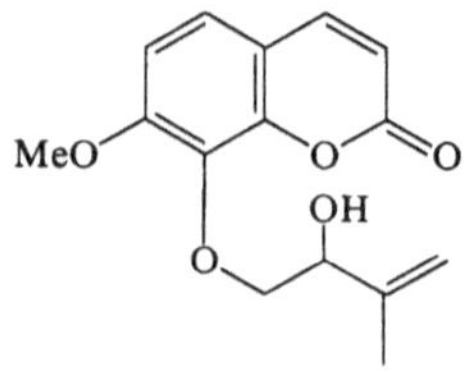

ferudenol (**339**)

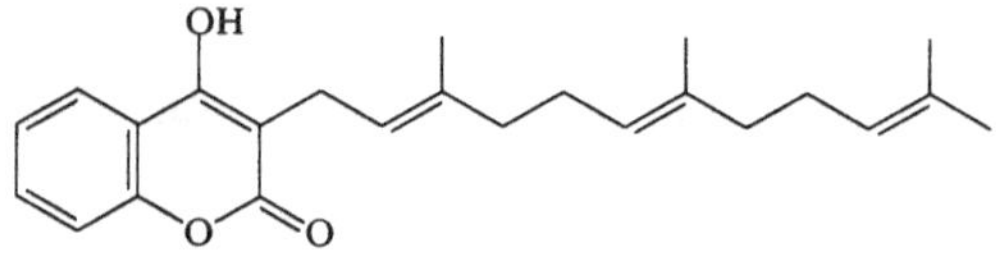

ferulenol (**481**)

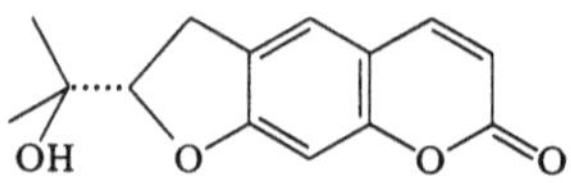

marmesin (**103**)

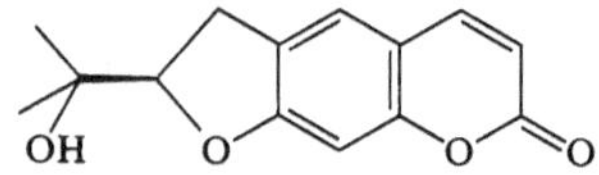

nodakenetin (**104**)

Despite these shortcomings, the main advantage of trivial names is their ease of use when contrasted with rational nomenclature. Thus coumarin itself is systematically named 2-*H*-1-benzopyran-2-one, and psoralen, 7-*H*-furo[3,2-g][1]-benzopyran-7-one. A much simpler system has developed based on a few basic coumarin structural types. It is a straightforward matter for the reader on encountering the name 8-methoxypsoralen in the literature to deduce its structure and by using the relevant Table on 7,8-Dioxygenated Coumarins to discover that this is a synonym for xanthotoxin (**351**). An added benefit is that many such names retain the basic numbering system of the simple coumarin nucleus.

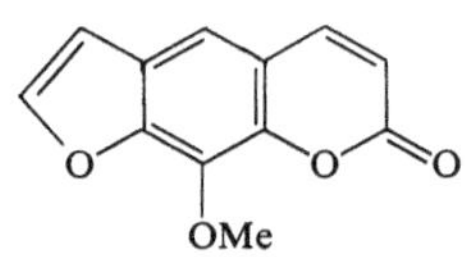

xanthotoxin (**351**)

IV. Isolation and Identification

1. Isolation

Coumarins are widely distributed in the plant kingdom but are particularly associated with the Umbelliferae (*218, 419, 420*) and the Rutaceae (*548*). In plants in which they occur, they are found in all parts from roots to flowers and fruits, often as mixtures.

Until recently, advantage was taken of their lactonic nature to separate them from plant extracts containing other natural products, alkali inducing opening of the lactone ring to give water-soluble alkali coumarinates (*1, 153*). Acidification leads normally to spontaneous regeneration of the coumarin. It is now recognized that many coumarins are acid and/or alkali sensitive which necessarily leads to the isolation of artefacts rather than the original coumarins. Particularly sensitive are benzylic esters of dihydrofurano- and dihydropyrano-coumarins. Consequently efforts should be made to identify coumarins present in extracts before chemical treatment. This is achieved conveniently by paper or thin-layer chromatography taking advantage of the fluorescence of most coumarins under ultraviolet light.

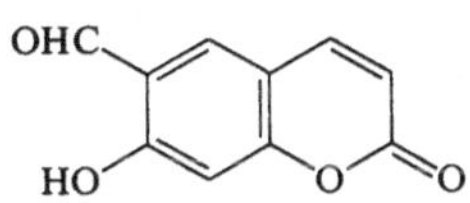

angelical (**80**)
= crenulatin

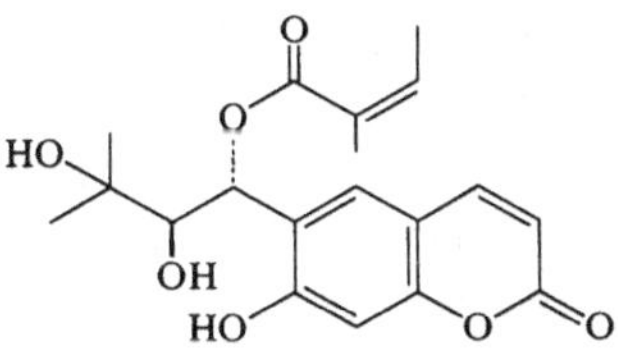

angelol (**98**)

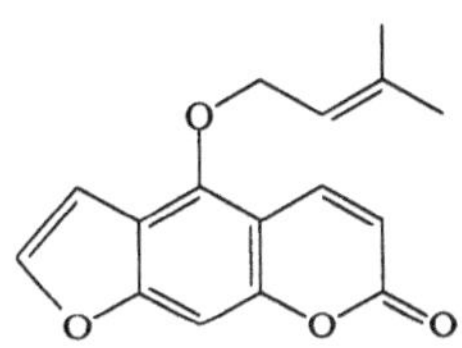

isoimperatorin (**254**)

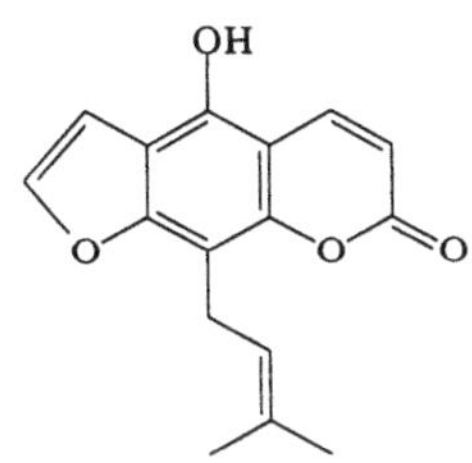

alloisoimperatorin
(**291**)

Many coumarins may be purified by high vacuum sublimation, but this should be avoided in work with potentially new coumarins. Alloisoimperatorin (**291**) is probably an artefact from thermal rearrangement of the prenyl ether, isoimperatorin (**254**) (*153*), while angelical (**80**) is now known to arise from thermal degradation of angelol (**98**) (*253*). It is not yet known whether crenulatin [= angelical, (**80**)], recently isolated from *Hesperathusa crenulata*, is also an artefact.

2. Identification of Known Coumarins

2.1. Paper and Thin-Layer Chromatography

Many known coumarins can be identified in plant extracts from their Rf values on paper or thin-layer chromatograms and the characteristic fluorescence of their spots (*274, 586*).

2.2. Gas-Liquid Chromatography

Gas-liquid chromatography, when combined with ultraviolet analysis (*556*) and additionally with nuclear magnetic resonance spectroscopy (*557*), has been developed into a powerful tool for the characterization of sub-milligram quantities of coumarins from plant extracts. Free hydroxycoumarins are better analyzed as their trimethylsilyl ethers (*203*) though a new technique (*206*) has permitted the gas chromatographic-mass spectrometric analysis of coumarins avoiding the necessity of forming derivatives.

2.3. High Pressure Liquid Chromatography

This newer technique has been successfully applied to furanocoumarins from spring parsley (*564*).

V. Structure Determination

The proof of structure of many of the coumarins in the Tables has come as a result of the ingenious and enlightened use of a variety of organic chemical techniques by many researchers and the reader is referred to the particular references quoted for each compound. Assignment of structure to a new coumarin is a two-fold problem of initially identifying the different substituent groups and then allocating these to their correct position on the coumarin nucleus.

With the development of spectroscopic techniques it is now often a straightforward matter to arrive fairly rapidly at a possible structure. It is one of the main purposes of this review to enable the reader to identify tentatively a coumarin and then, from the Tables, to establish whether or not it is new.

References, pp. 400—429

1. Use of Spectroscopy

1.1. Nuclear Magnetic Resonance Spectroscopy

Proton magnetic resonance spectroscopy is currently the most useful technique for the assignment of structure to a natural coumarin. Spectra-structure correlations have been compiled by STECK and MAZUREK (*561*) to facilitate the rapid identification of the major structural types.

In other important publications, data have been collected and interpreted for a variety of structural types, especially isomers, thereby aiding their identification (*196, 215, 369, 419, 460—463, 473, 484*). Long-range coupling (*283*), solvent- (*103, 241, 413, 414*) and lanthanide-induced shifts (*238, 239*) and, more recently, ^{13}C nuclear magnetic resonance spectroscopy (*151, 362*) have also proved of value in the structural assignment of coumarins.

1.2. Mass Spectrometry

The use of mass spectrometry for the exact determination of molecular weight of compounds in limited supply is a technique routinely applied to new coumarins. The results of a number of investigations have shown that characteristic fragmentation patterns are obtained from differently substituted coumarins. These are particularly useful for distinguishing between isomeric structures (*58, 110, 337, 419, 502, 603, 605—607*).

1.3. Ultraviolet and Infrared Spectroscopy

Coumarins absorb ultraviolet light at ~ 320 nm ($\log \varepsilon \sim 4$), the exact wavelength and complete spectrum depending on the substituents present. In particular, by observing spectral changes in the presence of alkali and complexing agents, it is possible to obtain information on the hydroxylation pattern of the nucleus (*383*).

Coumarins typically exhibit a strong absorption band in the infrared at $\sim 1725\ cm^{-1}$. The most comprehensive compilation of data on ultraviolet and infrared spectroscopy pertaining to coumarins is to be found in NIELSEN's review (*419*).

2. Stereochemistry

The stereochemical features of coumarins vary from simple geometric isomerism to the complexities of the terpenoid portions of the many umbelliferone ethers and encompass coumarins having one or more chiral centres present in acyclic substituents and in dihydrofuran and dihydropyran rings.

A variety of approaches has been employed including chemical correlation with compounds of known absolute configuration, nuclear magnetic resonance spectroscopy for obtaining relative stereochemistry, X-ray diffraction, optical rotatory dispersion and circular dichroism. The attention of the reader is directed to the specific examples in the Tables where stereochemistry is defined and particularly to the review of NIELSEN (*419*),

VI. Synthesis and Biosynthesis

1. Synthesis

The synthesis of natural coumarins has attracted considerable attention (*304, 525*) since in many cases it presents the fascinating challenge of developing methods for the regiospecific introduction of the appropriate substituents into the coumarin nucleus. References to syntheses of specific coumarins where these have been effected are included in the leading references given in the Tables,

2. Biosynthesis

The common structural features of many coumarins argue in favour of a common biogenesis. The biosynthesis of a number of simple coumarins has been studied sufficiently to permit the generalisation that they are derived from the corresponding cinnamic acids, these being formed *via* the shikimic acid pathway (*105, 108, 242, 419, 567*).

There has been much interest in the biosynthesis of coumarins bearing isoprenoid-related substituents; these are now considered to be mevalonate derived (*338, 340, 341*). In part this has stemmed from the important skin-sensitizing activity of some furanocoumarins which has been correlated with their photoreactivity towards pyrimidine bases of DNA (*409*). The proposal by BIRCH (*72*) that the two "extra" carbon atoms of the furan ring are derived by loss of three carbon atoms from an intermediate hydroxyisopropyldihydrofuranocoumarin has received considerable experimental support in recent years (*152, 198, 199*). Indeed, it has now been established that (+)-marmesin (**103**), formed from umbelliferone (**2**) *via* 7-demethylsuberosin (**86**) is a key intermediate in the biosynthesis of linear furanocoumarins (*105, 107,*

178, 560), while an analogous pathway leads from umbelliferone, *via* osthenol (**142**) and columbianetin (**166**) to angular furanocoumarins (*107, 178, 559*) (see Scheme on p. 202).

VII. Introduction to Tables

The natural coumarins are listed numerically in Tables 1 to 11, their order of entry following the guidelines given in sections I and II. In a few cases, *e. g.* aesculetin (**303**) they were isolated as aglycones before being encountered as genuine natural products. In each case, the plant source from which the coumarin was first isolated is given together with the year of isolation, molecular formula and trivial name. An asterisk in the Structure column implies that the absolute stereochemistry is not yet known. The $[\alpha]_{\lambda}^{t}$, solvent, columns refer to the specific rotation at t °C in the given solvent at a given wavelength λ (nm). Where no wavelength is quoted the rotation has been measured at 589 nm.

Many of the coumarins listed in the Tables have been isolated only from the plant source cited. However, coumarin itself and some of the structurally simple coumarins have been found to occur in a wide variety of plants; in this review only the first reported source is normally recorded. Occasionally however, more than one plant source has been cited. In such cases the later publication cited, while referring to the same coumarin as the earlier, assigns a different trivial name and/or melting point and/or records measurement of the specific rotation for the first time.

Table 1. *Coumarin and 7-Oxygenated Coumarins*

Trivial Name(s)	Year iso- lated	Structure
(1) Coumarin	1820	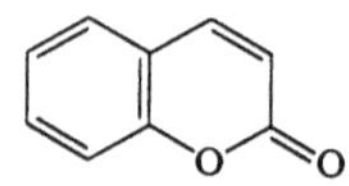
(2) Umbelliferone	1859	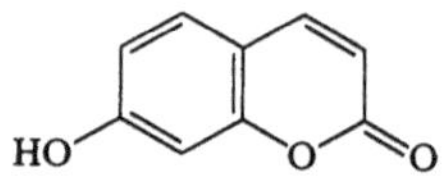
(3) Herniarin	1874	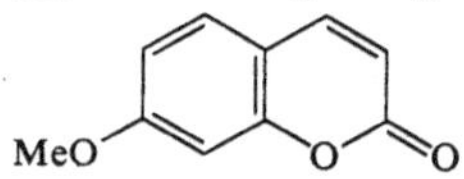
(4) 7-Isopentenyl- oxycoumarin	1966	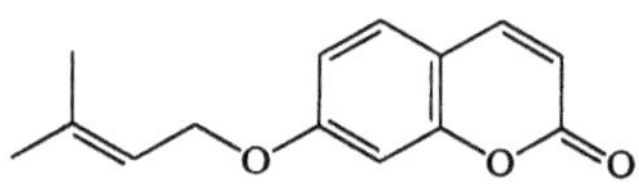
(5)	1972	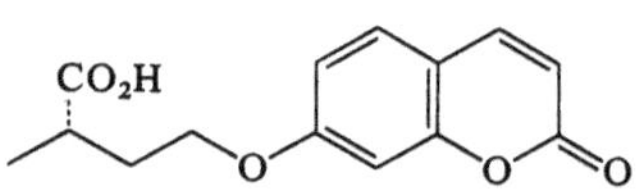
(6)	1972	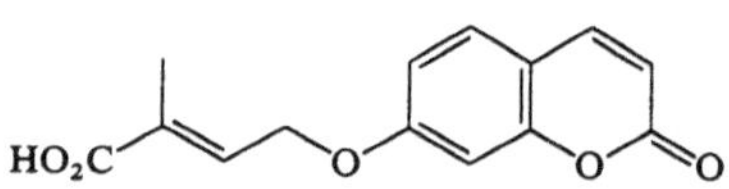
(7) Aurapten	1930	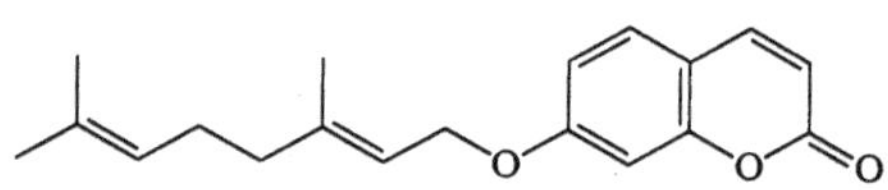
(8) 6′,7′-Epoxy- aurapten	1968	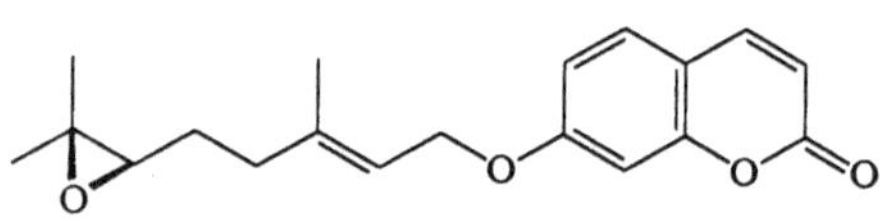
(9) 3′,6′-Epoxy- cyclo-aurapten	1968	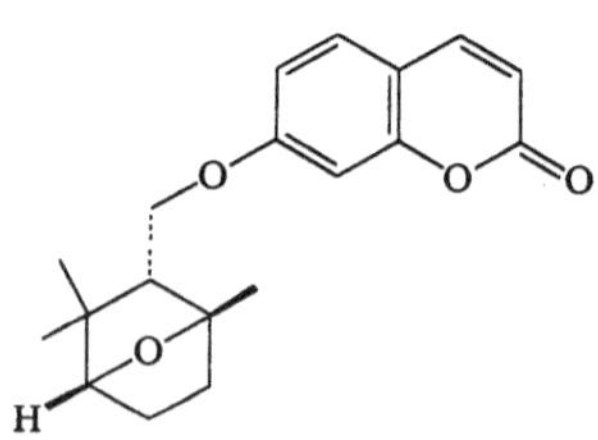

References, pp. 400—429

Formula	M. p.	$[\alpha]_{\lambda}^{t}$	Solvent	Plant source(s)	Leading References
$C_9H_6O_2$	67—68			*Coumarouna odorata* = *Dipteryx odorata*	(*1, 153*)
$C_9H_6O_3$	223—224			*Ferula galbaniflua*	(*153*)
$C_{10}H_8O_3$	117—118			*Herniara glabra*	(*153*)
$C_{14}H_{14}O_3$	75—76			*Libanotis intermedia*	(*73, 365, 476*)
$C_{14}H_{14}O_5$	112—115	—7.8	$CHCl_3$	*Evodia vitiflora*	(365)
$C_{14}H_{12}O_5$	256—262			*Evodia vitiflora*	(*365*)
$C_{19}H_{22}O_3$	68			*Citrus aurantium*	(*301, 332*)
$C_{19}H_{22}O_4$	oil 55.3—57.5	21_{578}^{22}	$CHCl_3$	*Aster yunnanensis*	(*90, 139, 598*)
$C_{19}H_{22}O_4$	187	18.9_{578}^{25}	$CHCl_3$	*Aster yunnanensis*	(*90, 139*)

Table 1 (continued)

	Trivial Name(s)	Year isolated	Structure
(10)	6′-Hydroxy-β-cyclo-aurapten	1968	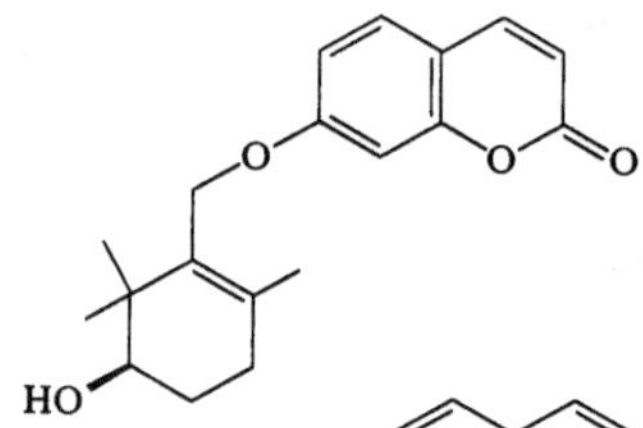
(11)	6′-Acetoxy-β-cyclo-aurapten	1968	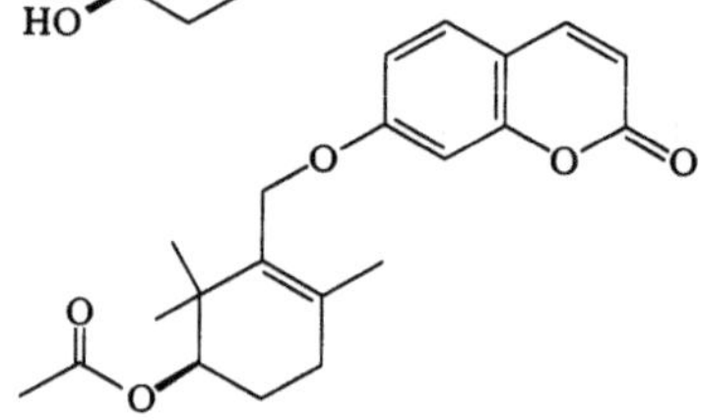
(12)	6′-Hydroxy-β-cyclo-aurapten	1968	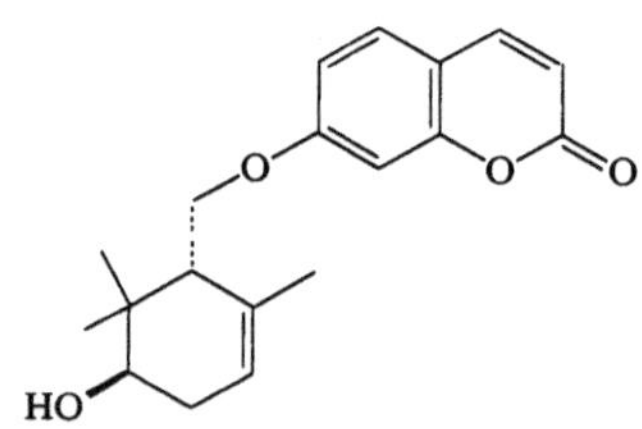
(13)		1968	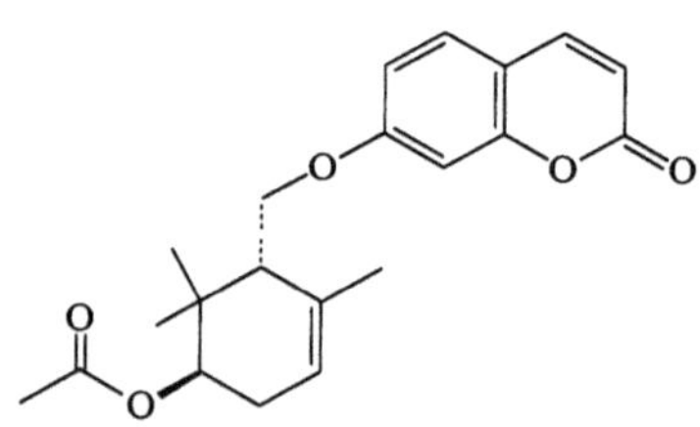
(14)	Diversinin 5′-Oxoaurapten	1975 1976	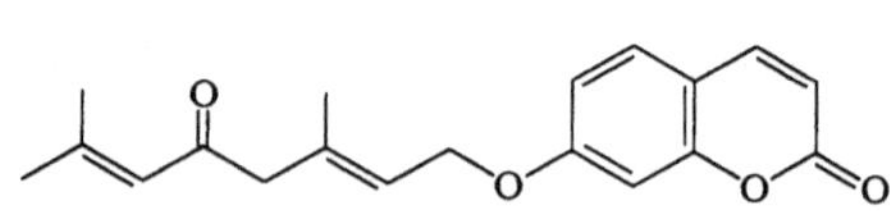
(15)	5′-Oxo-Δ-3′-aurapten	1976	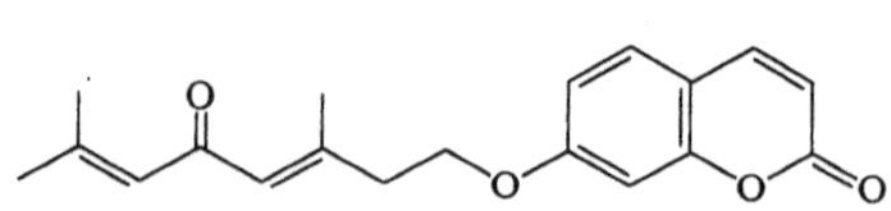

Formula	M. p.	$[\alpha]^t_\lambda$	Solvent	Plant source(s)	Leading References
$C_{19}H_{22}O_4$	74—76	28^{25}_{578}	$CHCl_3$	*Aster yunnanensis*	(*90*)
$C_{21}H_{24}O_5$	oil	13^{22}_{578}	$CHCl_3$	*Aster yunnanensis*	(*90*)
$C_{19}H_{22}O_4$	43—45			*Aster yunnanensis*	(*90, 139*)
$C_{21}H_{24}O_5$				*Aster yunnanensis*	(*90*)
$C_{19}H_{20}O_4$	55—57 63.5			*Ferula diversivittata* *Machaeranthera scabrella*	(*331*) (*81*)
$C_{19}H_{20}O_4$	91			*Machaeranthera scabrella*	(*81*)

Table 1 (continued)

Trivial Name(s)	Year iso-lated	Structure
(16) Diversin	1969	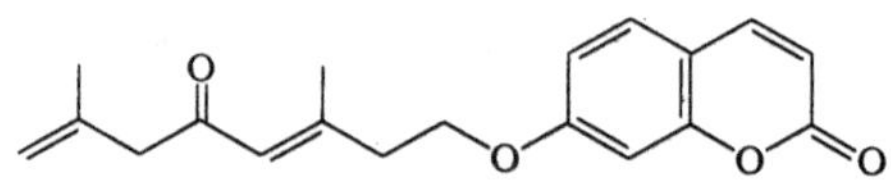
(17)	1970	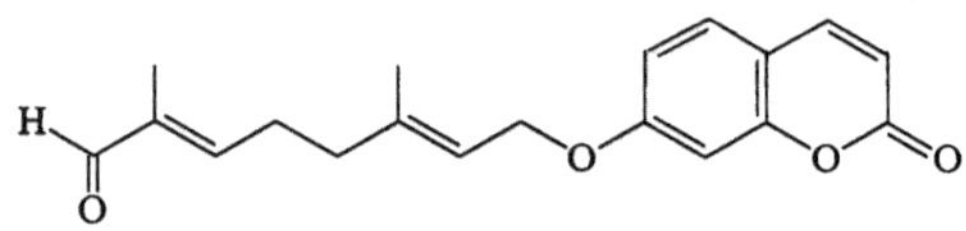
(18) Marmin	1959	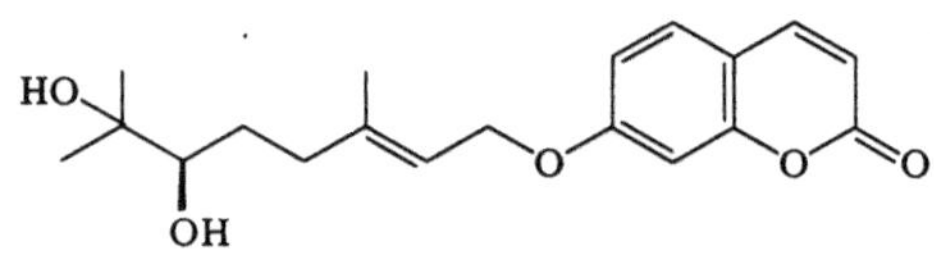
(19) Diversoside	1974	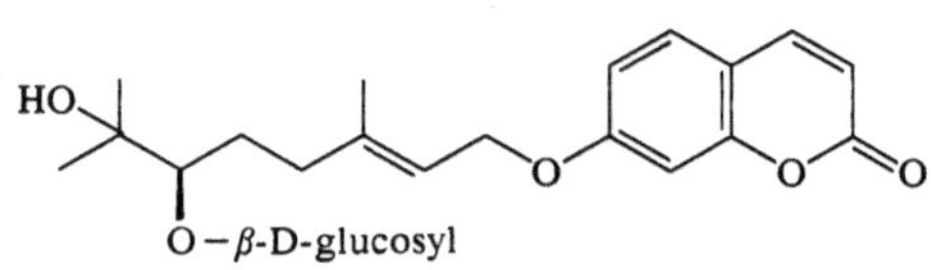
(20)	1972	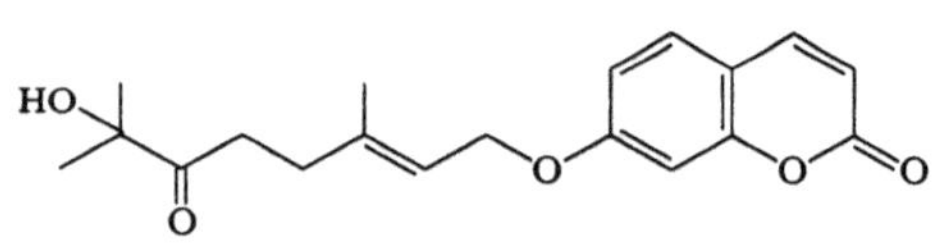
(21) 2′,3′-Dihydro-geiparvarin	1972	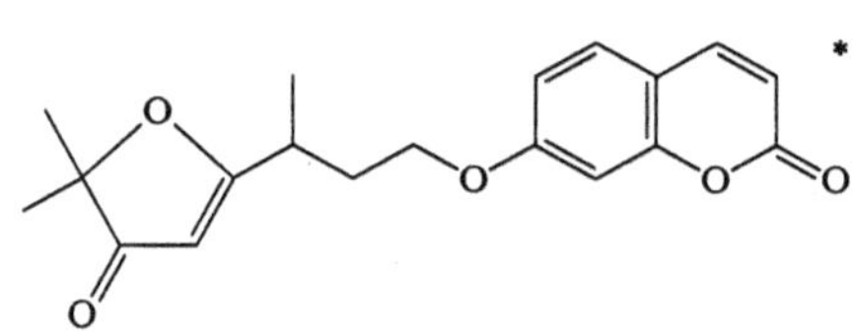
(22) Geiparvarin	1967	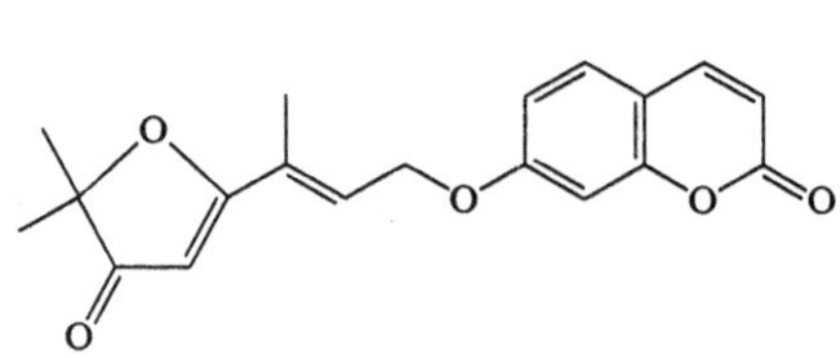

Formula	M. p.	$[\alpha]_\lambda^t$	Solvent	Plant source(s)	Leading References
$C_{19}H_{20}O_4$	97—98.5			*Ferula diversivittata*	(*320, 331*)
$C_{19}H_{20}O_4$	64			*Capnophyllum peregrinum*	(*74, 84*)
$C_{19}H_{24}O_5$	123—124	25^{30}	EtOH	*Aegle marmelos*	(*126, 130, 139,* (*195, 197, 598*)
$C_{25}H_{34}O_{10}$	154—155	10^{20}	EtOH	*Ferula diversivittata*	(*294*)
$C_{19}H_{22}O_5$	76.5—78			*Geijera parviflora*	(*169*)
$C_{19}H_{20}O_5$	123.5—124.5	13	EtOH	*Geijera parviflora*	(*169*)
$C_{19}H_{18}O_5$	160—161			*Geijera parviflora*	(*119, 358*)

Table 1 (continued)

Trivial Name(s)	Year isolated	Structure	
(**23**)	1970	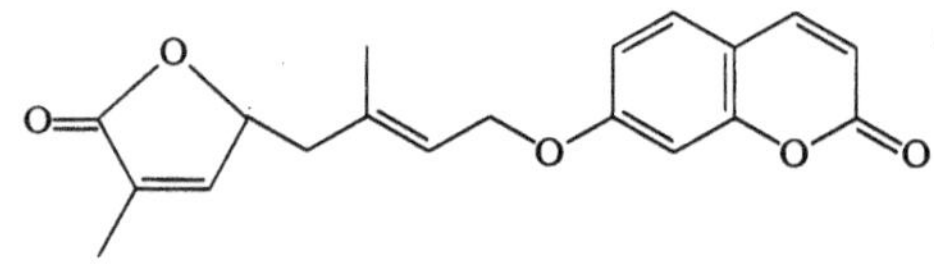	*
(**24**)	1970	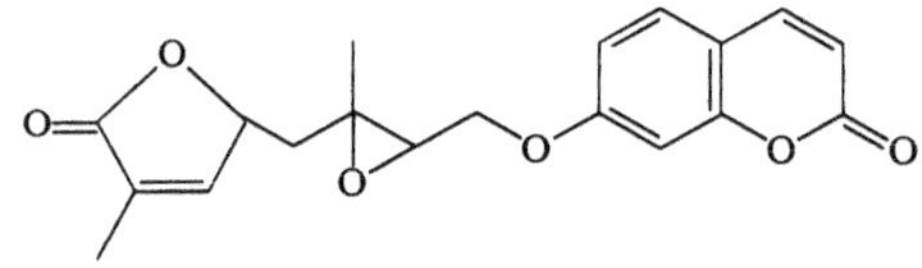	*
(**25**) Tavicone	1969	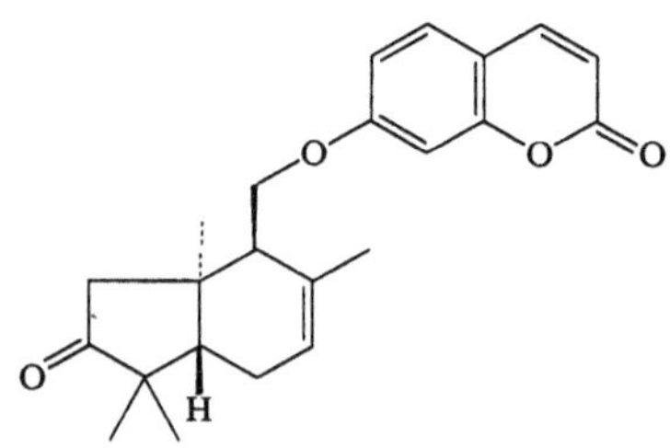	
(**26**) Umbelliprenin	1938	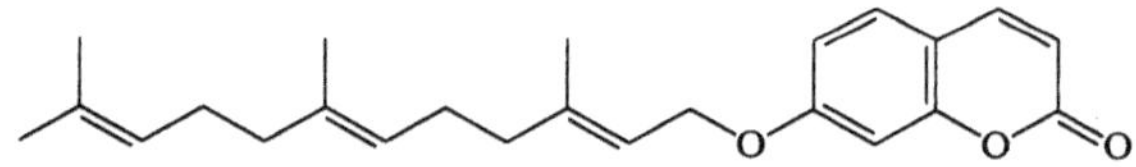	
(**27**) Cocanicin	1961	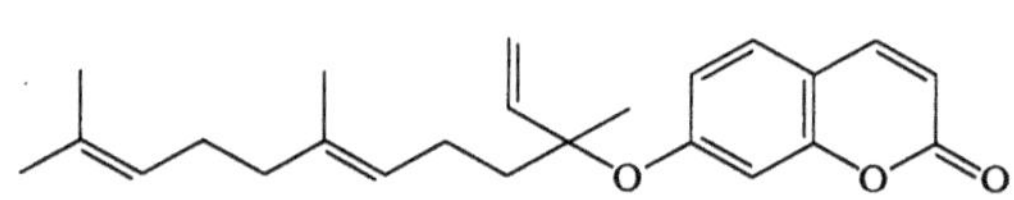	*
(**28**) Karatavicinol	1969	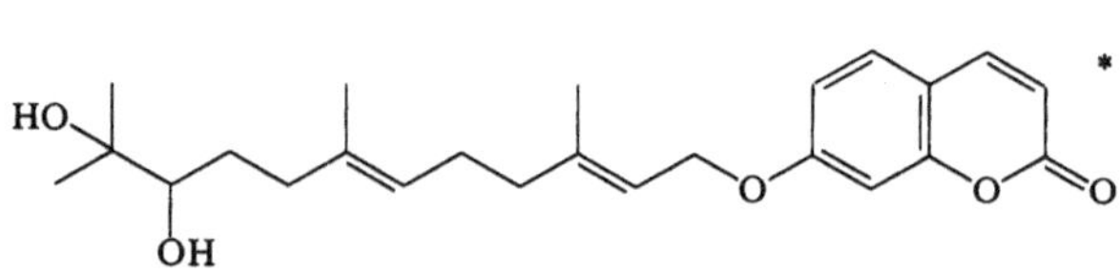	*
(**29**) Reoselin Reoselin A	1963 1975	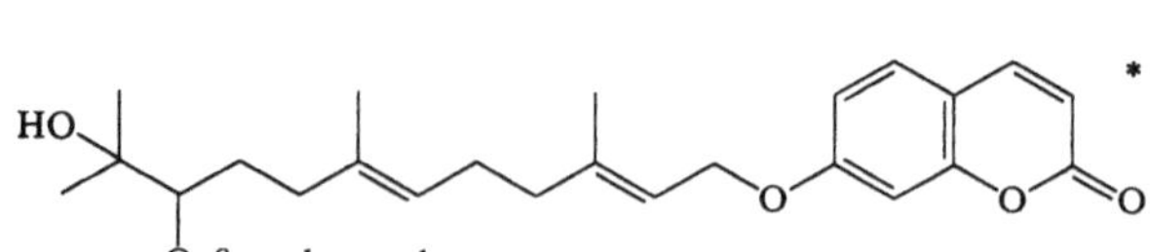	*

Formula	M. p.	$[\alpha]^t_\lambda$	Solvent	Plant Source(s)	Leading References
$C_{19}H_{18}O_5$	126	-26^{25}_{578}		*Capnophyllum peregrinum*	(*74, 84*)
$C_{19}H_{18}O_6$	125			*Capnophyllum peregrinum*	(*74*)
$C_{23}H_{26}O_4$	141—142	-77^{26}	C_6H_6	*Ferula karatavica*	(*44, 48*)
$C_{24}H_{30}O_3$	61—63			*Angelica archangelica*	(*63, 153, 317*)
$C_{24}H_{30}O_3$	34—35	-3	$CHCl_3$	*Ferula cocanica*	(*317*)
$C_{24}H_{32}O_5$	52—53	-12^{20}	EtOH	*Ferula karatavica*	(*323*)
$C_{36}H_{52}O_{15}$	155—156	-24.4	EtOH	*Ferula pseudoreoselinum*	(*296, 318, 327, 328*)
	160—161	-73.5^{20}	MeOH	*Ferula korshinskyi*	(*293*)

Table 1 (continued)

Trivial Name(s)	Year isolated	Structure
(**30**) Karatavicin	1965	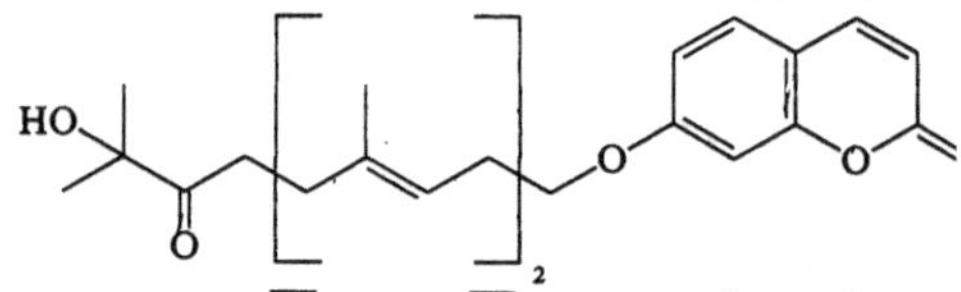
(**31**) Feroside	1975	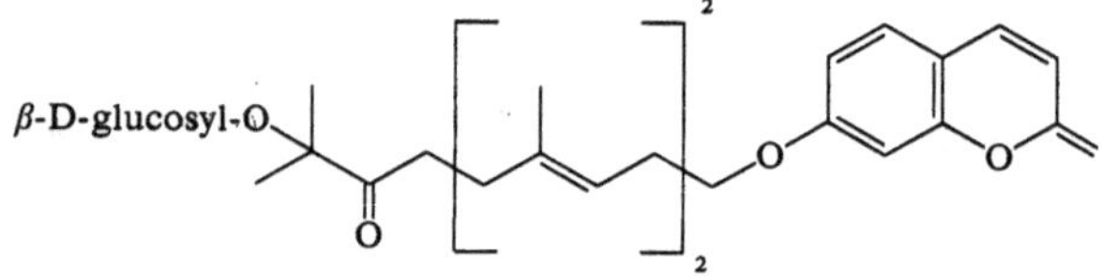
(**32**) Farnesiferol C	1958	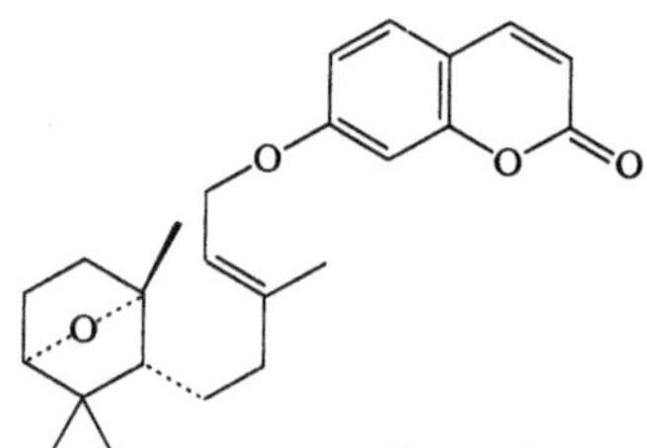
(**33**) Farnesiferol B	1958	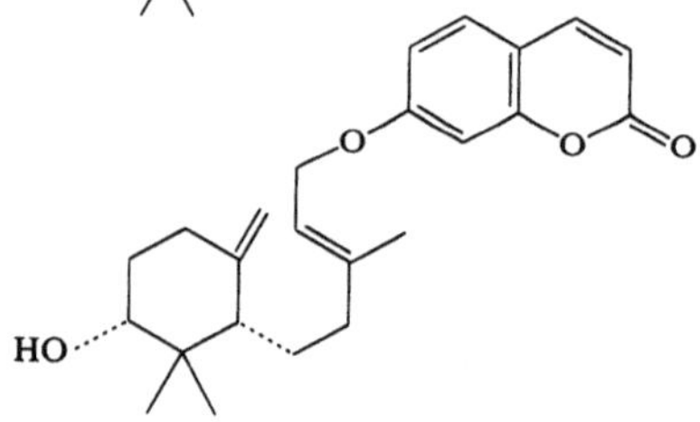
Kopeodin	1974	
(**34**) Kopetdaghin	1974	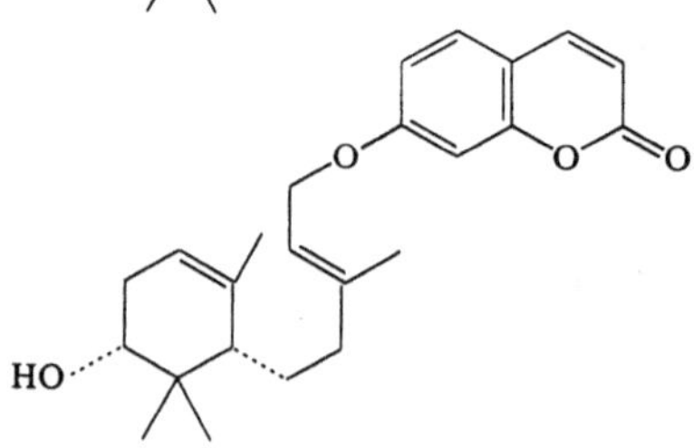
(**35**) Kopeolin	1973	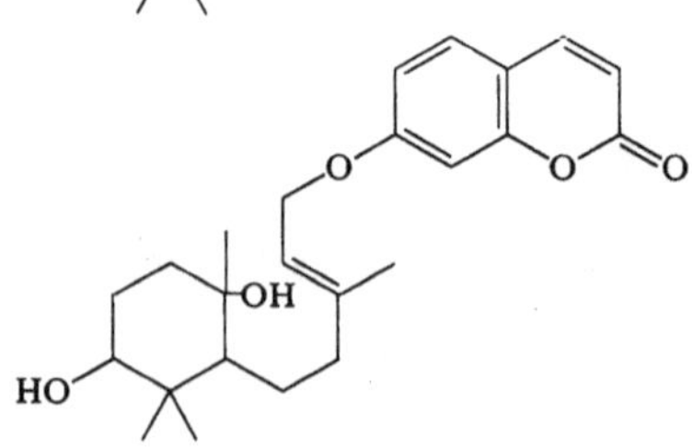

Formula	M. p.	$[\alpha]_{\lambda}^{t}$	Solvent	Plant Source(s)	Leading References
$C_{24}H_{30}O_5$	59—60			*Ferula karatavica*	(*318, 321*)
$C_{30}H_{40}O_{10}$	110—111	18.1^{20}	MeOH	*Ferula korshinskyi*	(*293*)
$C_{24}H_{30}O_4$	84—85	−29	$CHCl_3$	*Ferula assafoetida*	(*111, 112, 587*)
$C_{24}H_{30}O_4$	113.5—114	10	$CHCl_3$	*Ferula assafoetida*	(*111, 112*)
	111—112	14.8^{25}	EtOH	*Ferula kopetdaghensis*	(*295, 298*)
$C_{24}H_{30}O_4$	125—126.5	28^{18}	$CHCl_3$	*Ferula kopetdaghensis*	(*298*)
$C_{24}H_{32}O_5$	146—147	-15.9^{25}	EtOH	*Ferula kopetdaghensis*	(*295*)

Table 1 (continued)

	Trivial Name(s)	Year isolated	Structure	
(**36**)	Kopeoside	1973	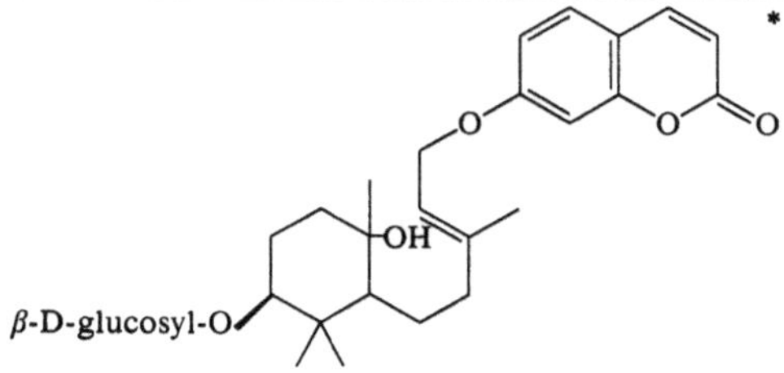	*
(**37**)	Feropolol	1976	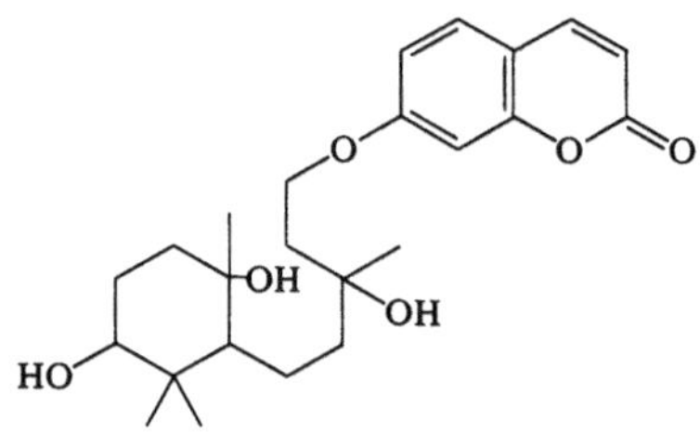	*
(**38**)	Feropolin	1976	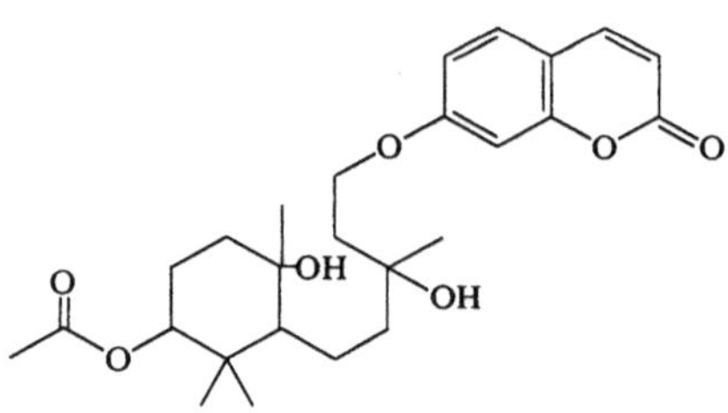	*
(**39**)	Feropolone	1976	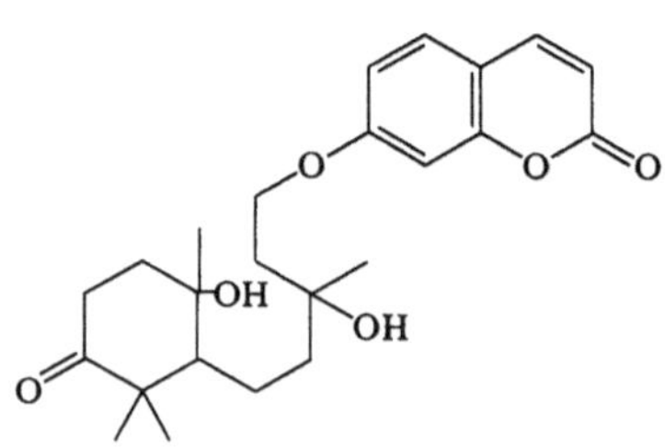	*

Formula	M. p.	$[\alpha]^{t}_{\lambda}$	Solvent	Plant Source(s)	Leading References
$C_{30}H_{42}O_{10}$	177—178	-22.1^{25}	EtOH	*Ferula kopetdaghensis*	(*295*)
$C_{24}H_{34}O_{6}$	96—98	38.2^{20}	$CHCl_3$	*Ferula polyantha*	(*308*)
$C_{26}H_{36}O_{7}$	63—65	85^{20}	$CHCl_3$	*Ferula polyantha*	(*308*)
$C_{24}H_{32}O_{6}$	225—226	-7.5^{20}	$CHCl_3$	*Ferula polyantha*	(*308*)

Table 1 (continued)

	Trivial Name(s)	Year iso-lated	Structure
(**40**)	Karatavic acid	1965	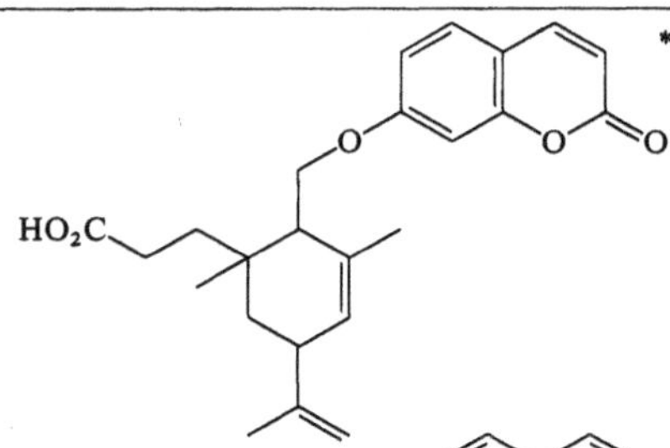
(**41**)	Galbanic acid	1955	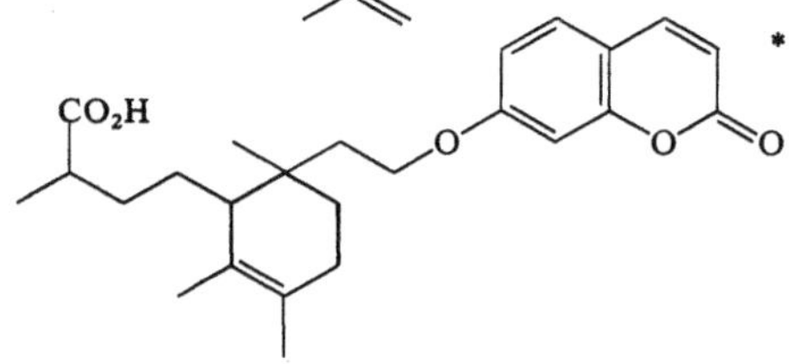
(**42**)	Methyl galbanate	1974	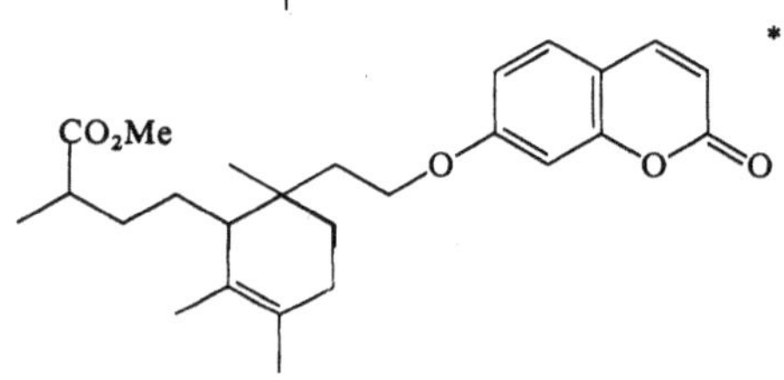
(**43**)	Galbanum acid	1965	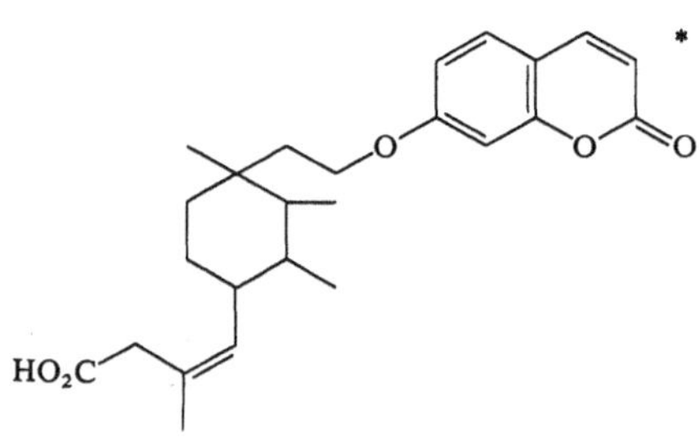
(**44**)	Farnesiferol A	1958	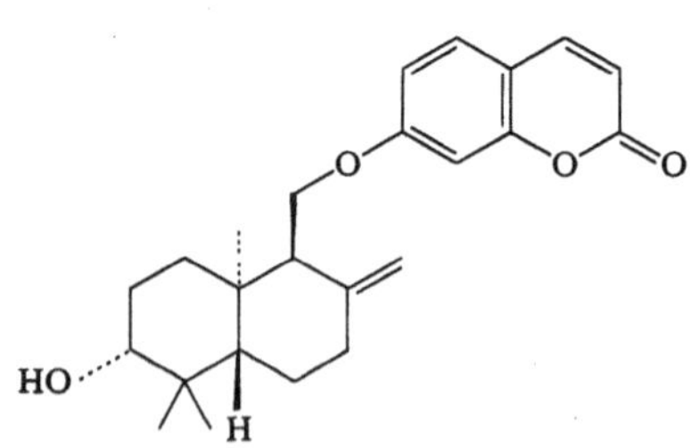
	Mogoltadin	1974	

Formula	M. p.	$[\alpha]_\lambda^t$	Solvent	Plant Source(s)	Leading References
$C_{24}H_{28}O_5$	89—90	−105	EtOH	*Ferula karatavica*	(*47, 318, 322*)
$C_{24}H_{30}O_5$	92—93	-35.2^{20}	MeOH	*Ferula gummosa*	(*96, 468*)
$C_{25}H_{32}O_5$	b. p. 252			*Ferula microloba*	(*94*)
$C_{24}H_{30}O_5$	110—112			*Ferula* species	(*318*)
$C_{24}H_{30}O_4$	155—155.5	−55	$CHCl_3$	*Ferula assafoetida*	(*111, 457, 458, 499, 587*)
	158—159	-55^{20}	EtOH	*Ferula mogoltavica*	(*306, 457*)

Table 1 (continued)

Trivial Name(s)	Year isolated	Structure
(45) Polyanthin	1974	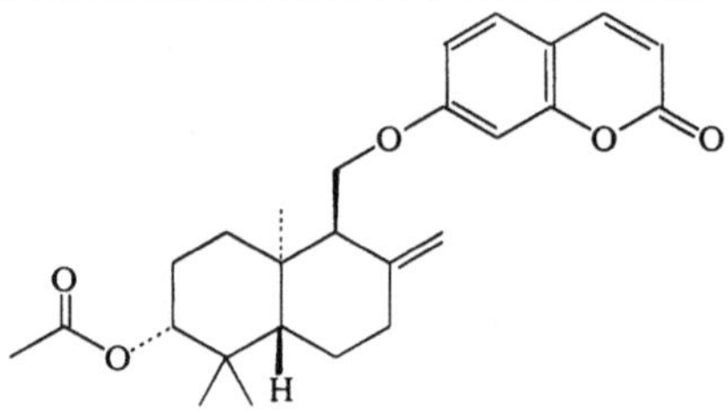
(46) Gummosin	1965	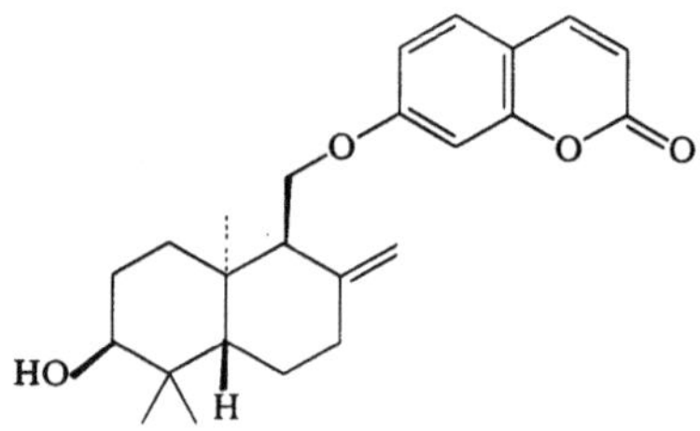
(47) Polyanthinin	1974	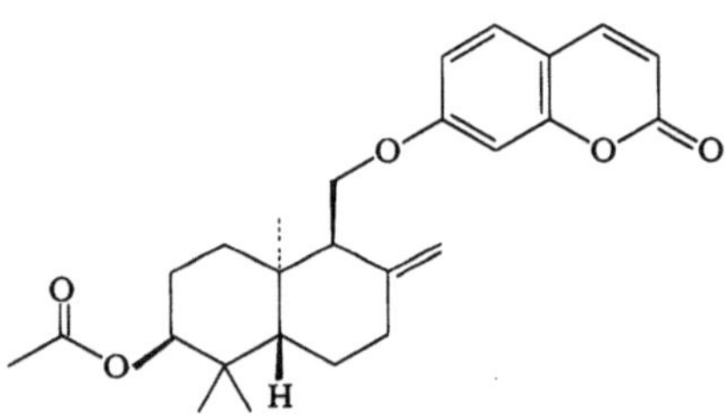
(48) Mogoltadone	1974	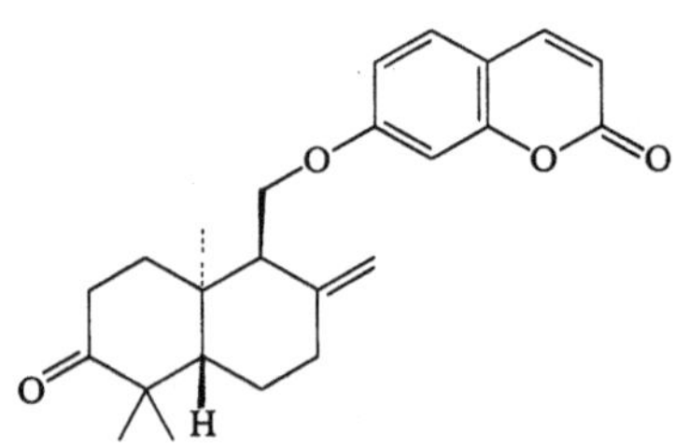
(49) Colladonin	1969	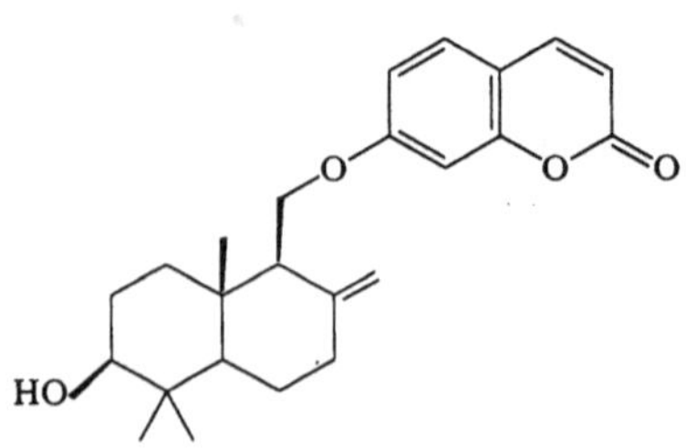

Formula	M. p.	$[\alpha]_{\lambda}^{t}$	Solvent	Plant Source(s)	Leading References
$C_{26}H_{32}O_5$	148—149	-50^{20}	$CHCl_3$	*Ferula polyanthum*	(*307, 455*)
$C_{24}H_{30}O_4$	176—177	-54	$CHCl_3$	*Ferula gummosa*	(*318, 329, 457, 458, 499, 500*)
$C_{26}H_{32}O_5$	127—129	-32^{20}	EtOH	*Ferula polyanthum*	(*307, 455*)
$C_{24}H_{28}O_4$	131—132	-41.7^{21}	$CHCl_3$	*Ferula mogoltavica*	(*306, 457*)
$C_{24}H_{30}O_4$	158.5—160	-50^{22}	$CHCl_3$	*Colladonia triquetra* = *Prangos triquetra*	(*57, 109, 457, 458, 499*)

Table 1 (continued)

	Trivial Name(s)	Year isolated	Structure
(50)	Colladin	1969	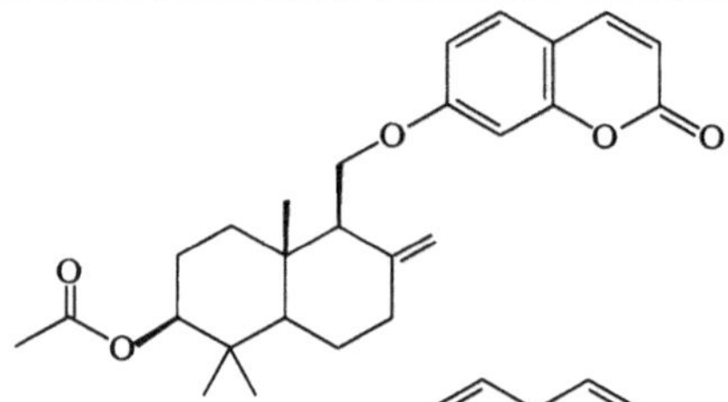
(51)	Badrakemin	1967	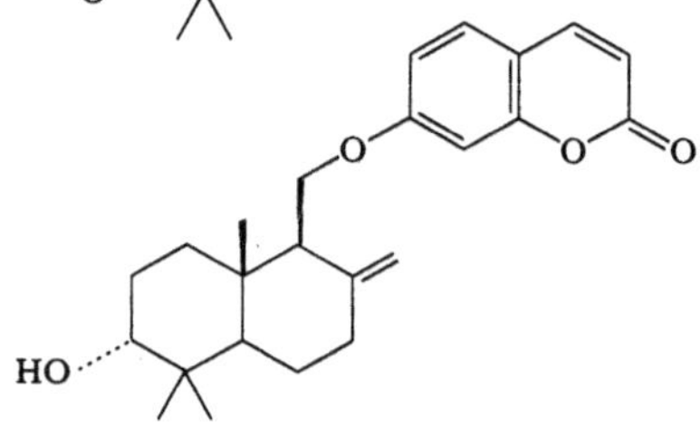
(52)	Badrakemin acetate	1974	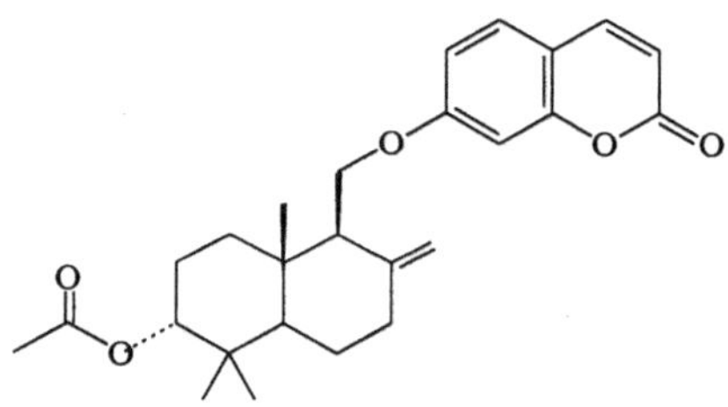
(53)	Badrakemone	1970	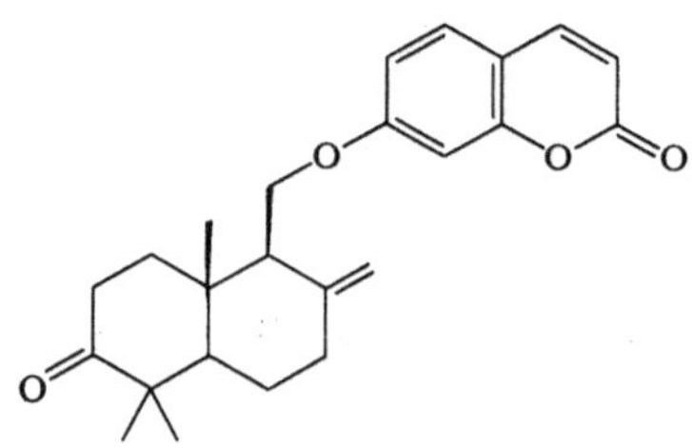
(54)	Feselol Moschatol	1973 1973	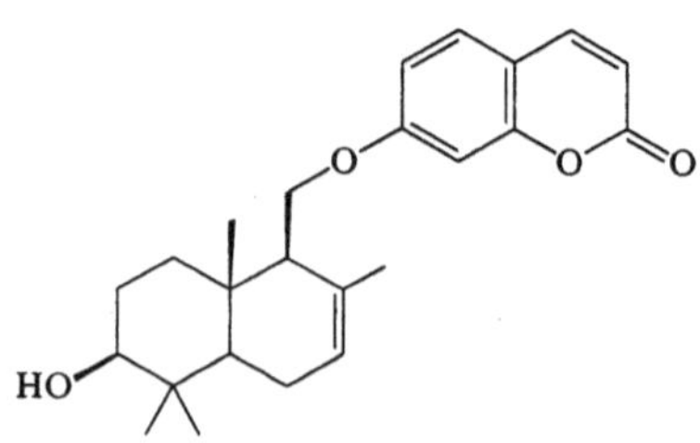

References, pp. 400—429

Formula	M. p.	$[\alpha]_\lambda^t$	Solvent	Plant Source(s)	Leading References
$C_{26}H_{32}O_5$	153—154	-65^{22}	$CHCl_3$	*Colladonia triquetra* = *Prangos triquetra*	(*57, 109*)
$C_{24}H_{30}O_4$	198—199	-64	$CHCl_3$	*Ferula badrakema*	(*46, 319, 457, 458, 499, 500*)
$C_{26}H_{32}O_5$	172.5	-22.28^{20}	$CHCl_3$	*Ferula tuberifera*	(*95*)
$C_{24}H_{28}O_4$	185—186	-42	$CHCl_3$	*Ferula nevskii*	(*42*)
$C_{24}H_{30}O_4$	116.5—117.5	-98.5^{20}	EtOH	*Ferula pseudo-oreoselinum*	(*326*)
	79—80	-77.4^{20}	EtOH	*Ferula moschata*	(*537*)

Table 1 (continued)

Trivial Name(s)	Year isolated	Structure
(55) Conferol	1972	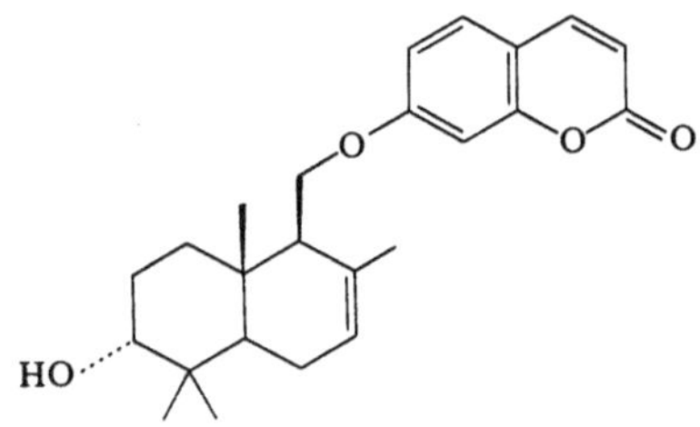
(56) Conferol acetate	1974	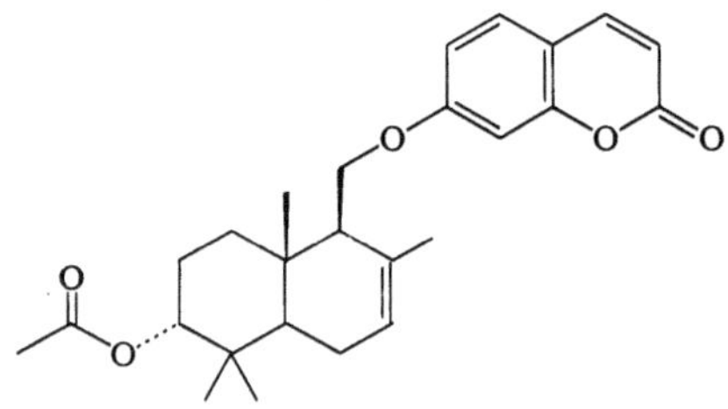
(57) Conferone	1972	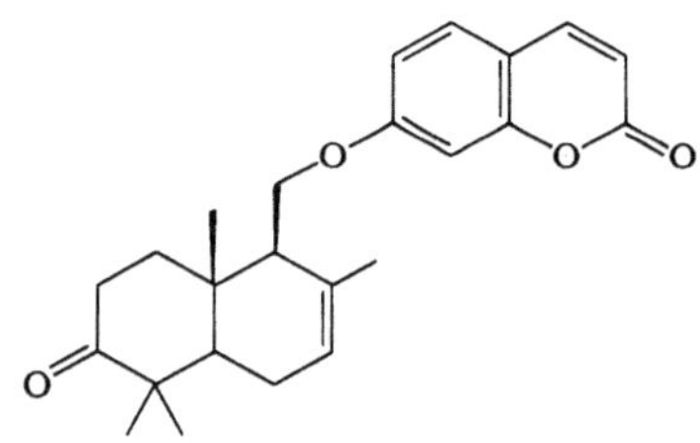
(58) Feropolidin	1976	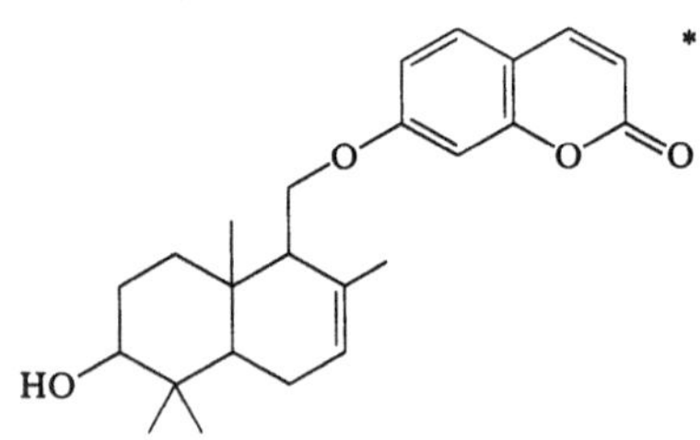
(59) Mogoltacin	1976	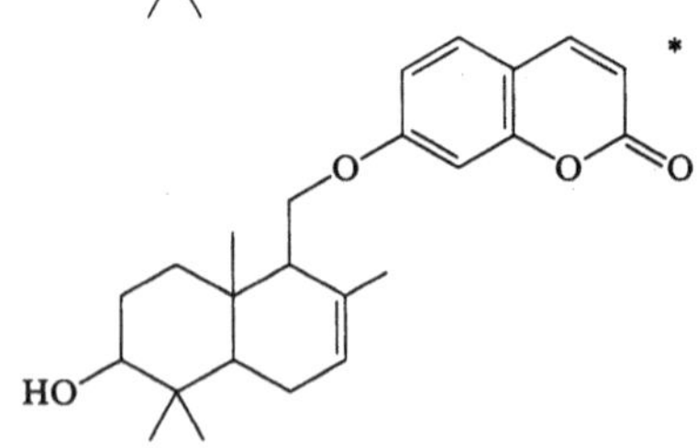

Formula	M. p.	$[\alpha]^t_\lambda$	Solvent	Plant Source(s)	Leading References
$C_{24}H_{30}O_4$	137—138	-84.2^{20}	EtOH	*Ferula conocaula*	(*326, 459, 537, 585*)
$C_{26}H_{32}O_5$	160	-42.39^{20}	$CHCl_3$	*Ferula tuberifera*	(*95*)
$C_{24}H_{28}O_4$	142—142.5	-51^{20}	EtOH	*Ferula cococaula*	(*326, 459, 584*)
$C_{24}H_{30}O_4$	154—156	154^{20}	$CHCl_3$	*Ferula polyantha*	(*308*)
$C_{24}H_{30}O_4$	155—156	-55^{23}	$CHCl_3$	*Ferula mogoltavica*	(*309*)

Table 1 (continued)

Trivial Name(s)	Year isolated	Structure
(**60**) Mogoltin	1964	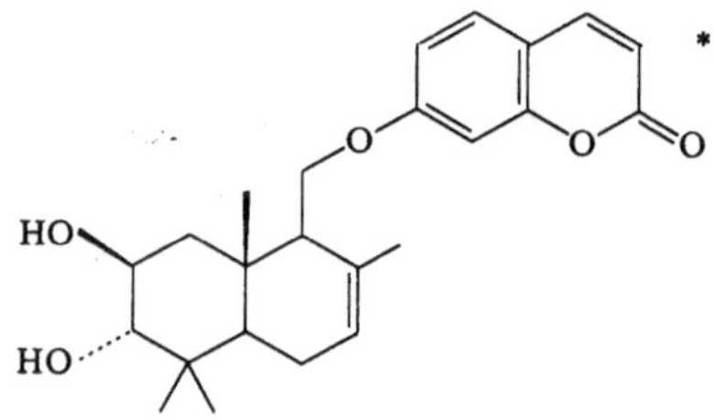 *
(**61**) Mogoltavin	1964	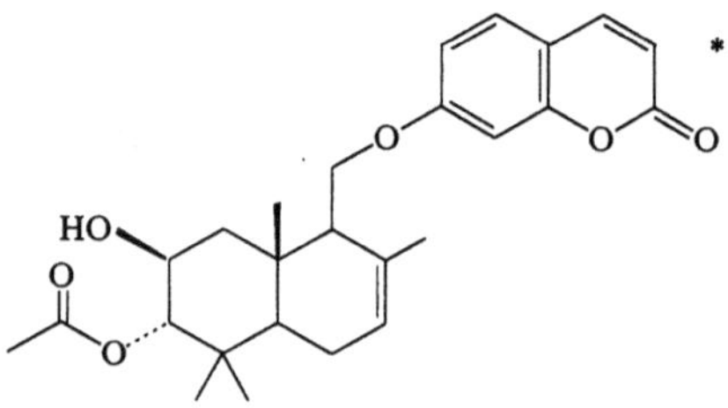 *
(**62**) Mogoltavinin	1964	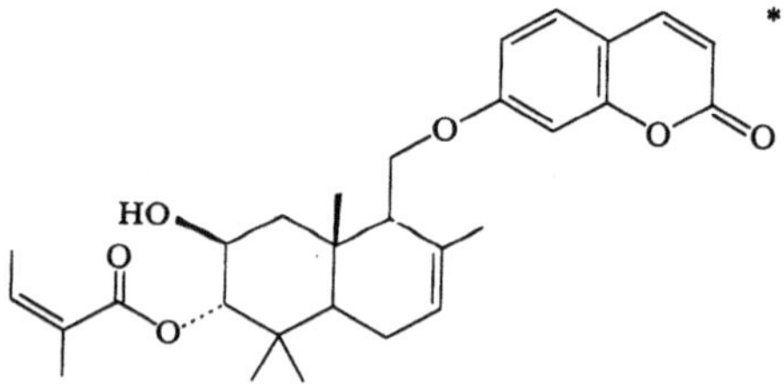 *
(**63**) Conferin	1974	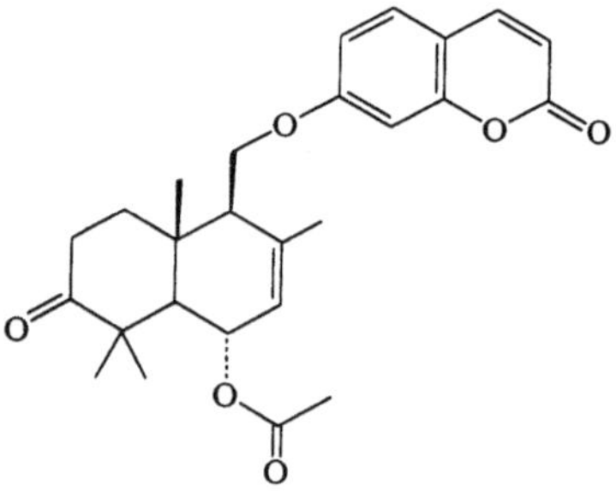
(**64**) Conferdione	1974	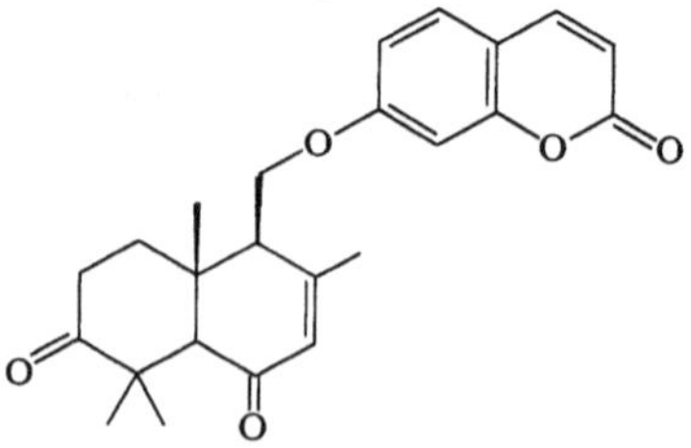

Formula	M. p.	$[\alpha]_{\lambda}^{t}$	Solvent	Plant Source(s)	Leading References
$C_{24}H_{30}O_5$	183—185	-63.3^{22}	$CHCl_3$	*Peucedanum mogoltavicum*	(*437, 444, 608*)
$C_{26}H_{32}O_6$	196—197	-108^{30}	$CHCl_3$	*Peucedanum mogoltavicum*	(*438, 444, 608*)
$C_{29}H_{36}O_6$	180—182	-119.2	$CHCl_3$	*Peucedanum mogoltavicum*	(*439, 444, 608*)
$C_{26}H_{30}O_6$	141—142	-124^{20}	EtOH	*Ferula conocaula*	(*464, 582*)
$C_{24}H_{26}O_5$	150—152	-51.9^{20}	EtOH	*Ferula conocaula*	(*583*)

Table 1 (continued)

	Trivial Name(s)	Year isolated	Structure
(65)	Nevskin Isosamarcandin	1972 1974	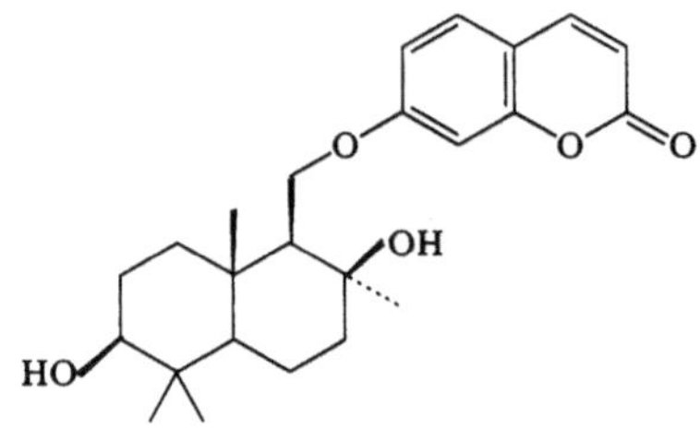
(66)	Isosamarcandin angelate	1972	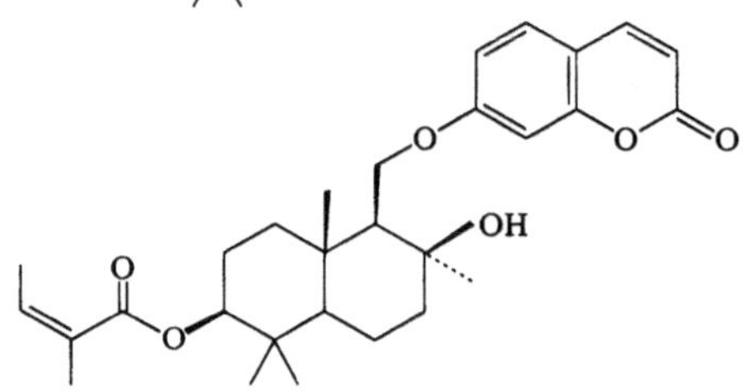
(67)	Samarcandin	1968	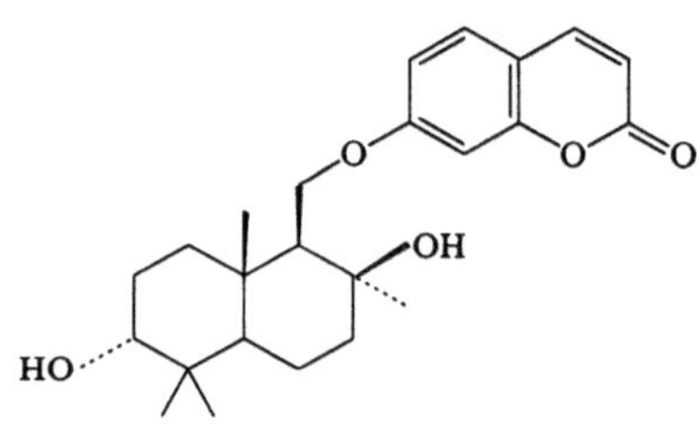
(68)	Samarcandin acetate	1972	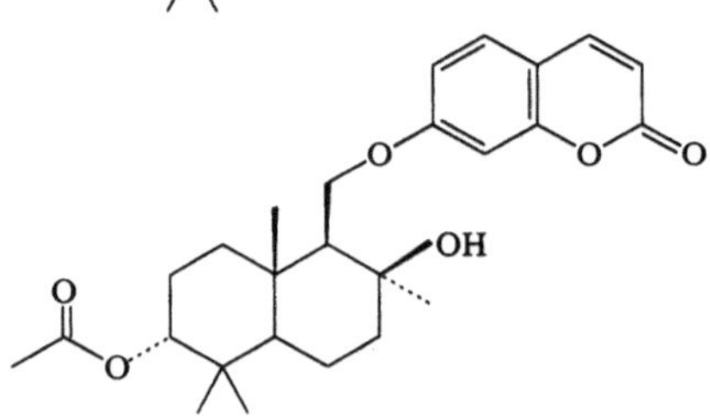
(69)	Samarcandone	1968	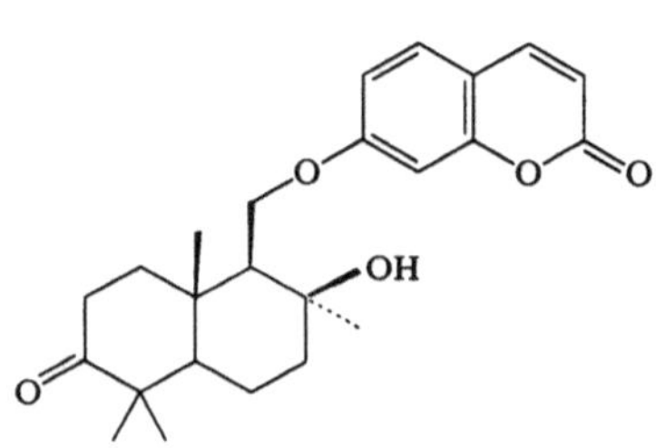

References, pp. 400—429

Formula	M. p.	$[\alpha]_{\lambda}^{t}$	Solvent	Plant Source(s)	Leading References
$C_{24}H_{32}O_5$	193—194 221	-79^{26} 26.75^{20}	pyridine EtOH	*Ferula nevskii* *Ferula microloba*	(*43, 49*) (*93*)
$C_{29}H_{38}O_6$	176—178	-26^{20}	EtOH	*Ferula pseudooreoselinum*	(*324*)
$C_{24}H_{32}O_5$	176—177	30	EtOH	*Ferula samarcandica*	(*43, 45, 57, 330*)
$C_{26}H_{34}O_6$	152—153	29.4^{20}	EtOH	*Ferula pseudooreo-selinum*	(*325*)
$C_{24}H_{30}O_5$	216—217	25	EtOH	*Ferula samarcandica*	(*45, 57, 330*)

Table 1 (continued)

Trivial Name(s)	Year isolated	Structure
(70) Colladocin	1975	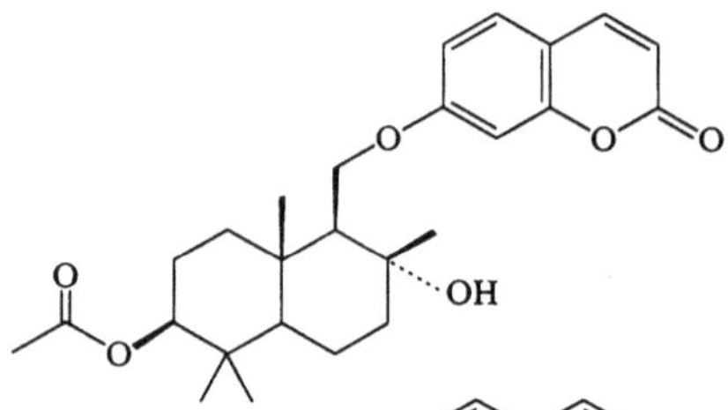
(71) Mogoltavidin	1974	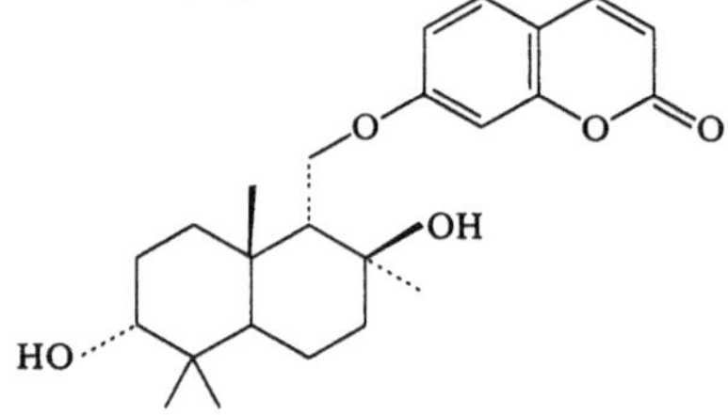
(72) Mogoltavicin	1974	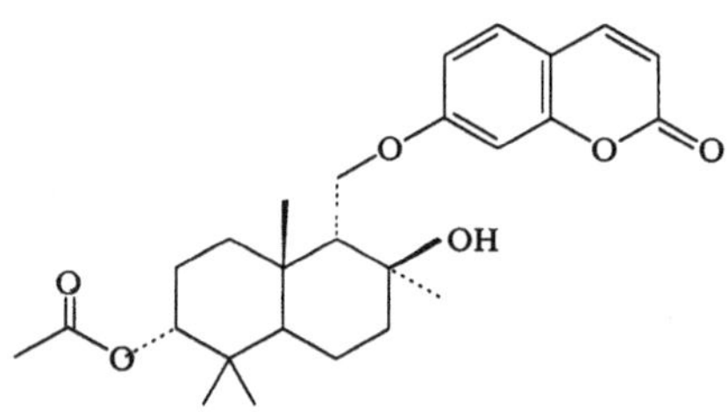
(73) Kellerin	1973	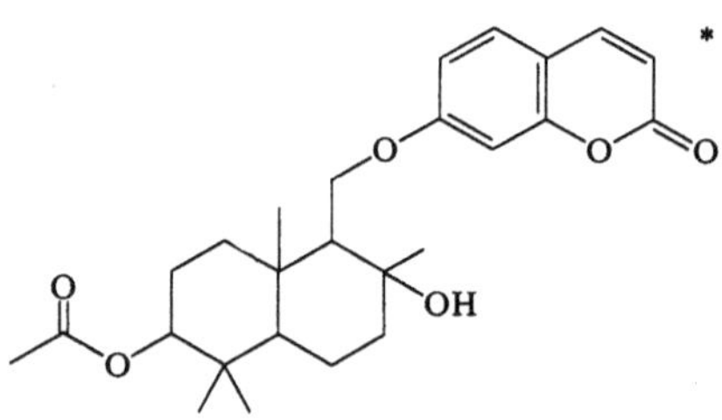
(74) Kamoloi	1966	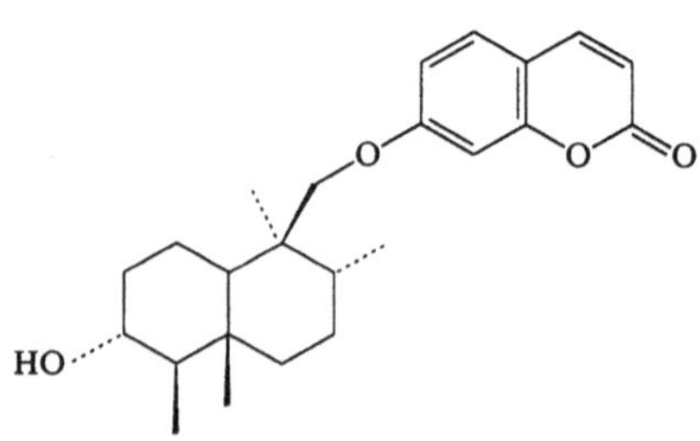

References, pp. 400—429

Formula	M. p.	$[\alpha]_{\lambda}^{t}$	Solvent	Plant Source(s)	Leading References
$C_{26}H_{34}O_6$	219	-4.58^{20}	EtOH	*Colladonia triquetra* = *Prangos triquetra*	(*92*)
$C_{24}H_{32}O_5$	159—161	-16^{23}	EtOH	*Ferula mogoltavica*	(*305*)
$C_{26}H_{34}O_6$	151—152	-12^{23}	EtOH	*Ferula mogoltavica*	(*305*)
$C_{26}H_{34}O_6$	76—78	66.4^{20}	EtOH	*Ferula kelleri*	(*30*)
$C_{24}H_{32}O_4$	141—142			*Ferula penninervis*	(*180, 183, 454, 456, 604*)

Table 1 (continued)

	Trivial Name(s)	Year isolated	Structure
(75)	Kamolone	1966	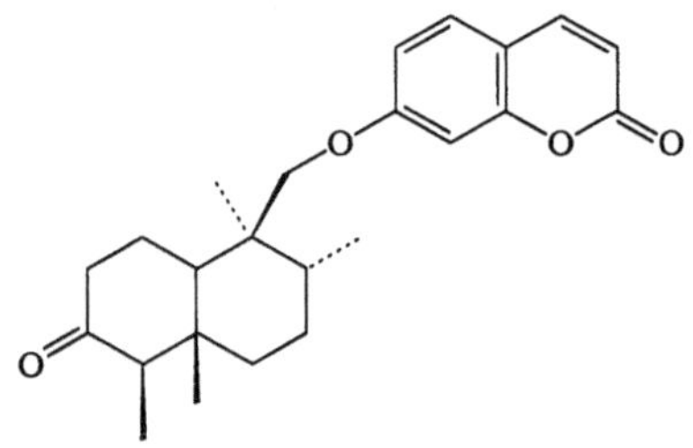
(76)	Gummosinin	1965	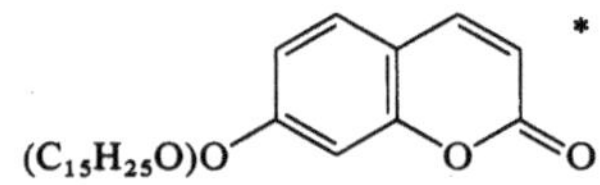
(77)	Kopetin	1973	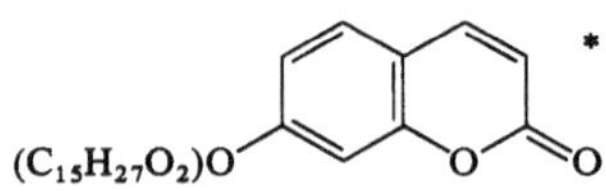
(78)	Skimmin	1884	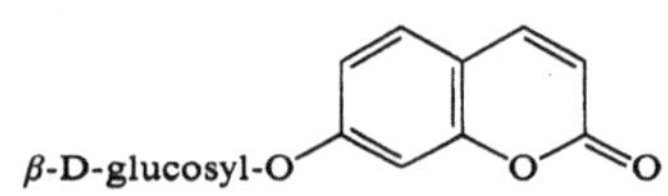
(79)	Neohydrangin	1955	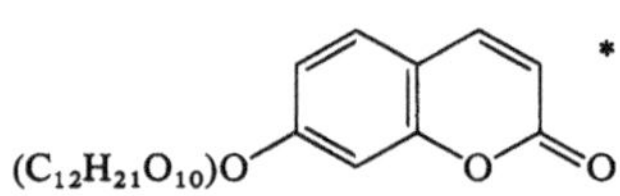

Formula	M. p.	$[\alpha]_\lambda^t$	Solvent	Plant Source(s)	Leading References
$C_{24}H_{30}O_4$	191—192	55^{18}	$CHCl_3$	*Ferula penninervis*	(*180, 183, 297, 454, 456, 604*)
$C_{24}H_{30}O_4$	163—165	−28	$CHCl_3$	*Ferula gummosa*	(*318*)
$C_{24}H_{32}O_5$	145—146	-25.4^{25}	EtOH	*Ferula kopetdaghensis*	(*295*)
$C_{15}H_{16}O_8$	219—221	-79.8^{18}	pyridine	*Skimmia japonica*	(*153*)
$C_{21}H_{26}O_{13}$	204	-130.9^{21}	H_2O	*Hydrangea paniculata*	(*411*)

Table 1.1 *6-Substituted 7-Oxygenated Coumarins*

	Trivial Name(s)	Year iso- lated	Structure
(80)	Angelical (artefact)	1958	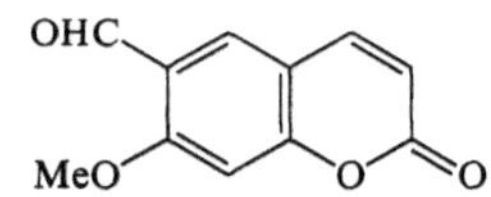
		1967	
		1975	
	Crenulatin	1975	
(81)		1976	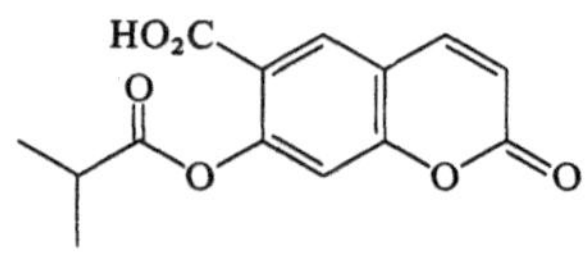
(82)	Peuruthenicin	1973	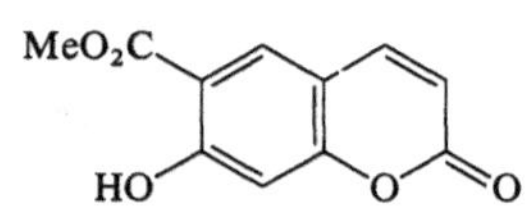
	Officinalin	1976	
(83)	Officinalin acetate	1976	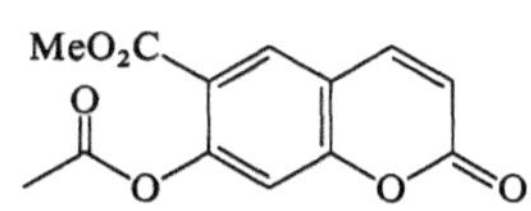
(84)	Officinalin isobutyrate	1976	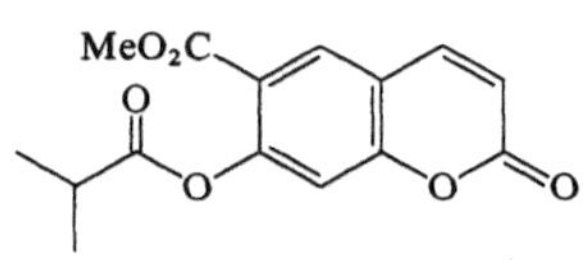
(85)	Suberenon	1972	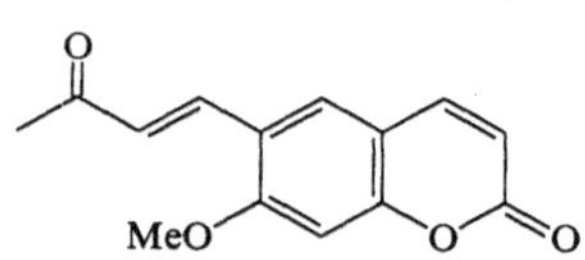
(86)	7-Demethylsuberosin	1954	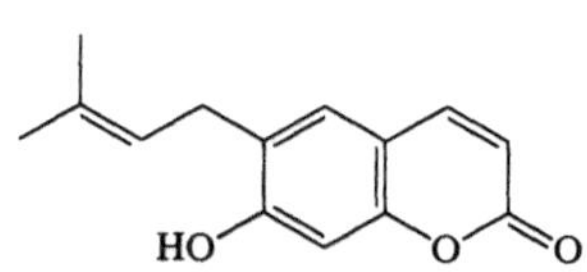
(87)	Suberosin	1950	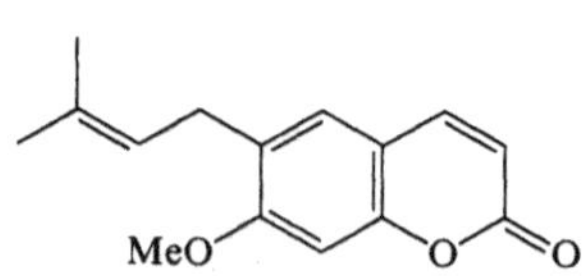

Formula	M. p.	$[\alpha]_\lambda^t$	Solvent	Plant Source(s)	Leading References
$C_{11}H_8O_4$	256—257			*Angelica pubescens*	(*252, 253, 254, 266*)
	255			*Zanthoxylum suberosum*	(*245*)
	250			*Boenninghausenia albiflora*	(*574*)
	252—253			*Hesperathusa crenulata*	(*59*)
$C_{14}H_{12}O_6$	183—185			*Peucedanum stenocarpum*	(*217*)
$C_{11}H_8O_5$	196—197			*Peucedanum ruthenicum*	(*543*)
	162—164			*Peucedanum officinale*	(*219*)
$C_{13}H_{10}O_6$	193.5—195			*Peucedanum officinale*	(*219*)
$C_{15}H_{14}O_6$	118.5—120			*Peucedanum officinale*	(*219*)
$C_{14}H_{12}O_4$	232—234			*Ruta graveolens*	(*489, 574*)
$C_{14}H_{14}O_3$	133.5—134			*Chloroxylon swietenia*	(*38, 315, 400, 401, 553*)
$C_{15}H_{16}O_3$	87—88			*Zanthoxylum suberosum*	(*38, 153, 185*)

Table 1.1 (continued)

Trivial (Name(s)	Year iso- lated	Structure
(88) (±)-Suberosin epoxide	1972	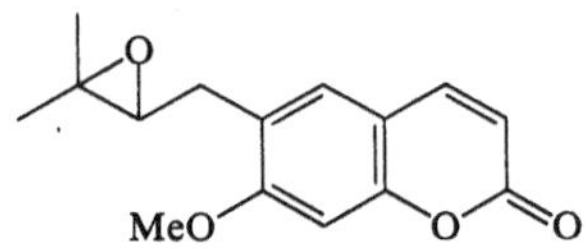 racemic
(89) Suberenol	1967	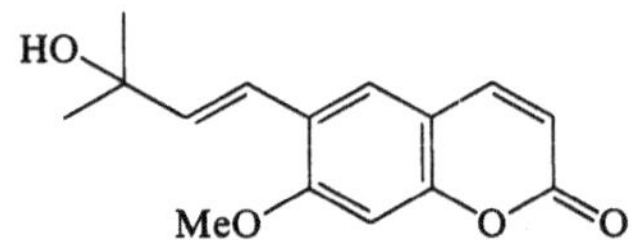
(90) Geijerin	1955	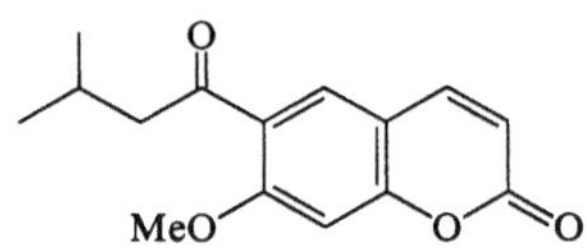
(91) Thamnosmin	1969	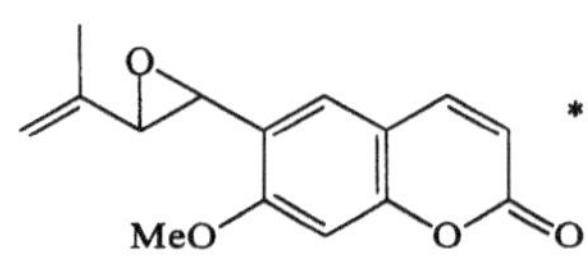 *
(92) Dehydrogeijerin	1967	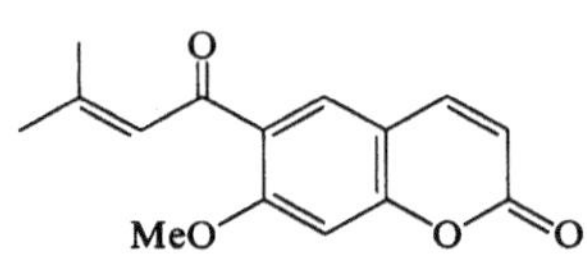
(93) Peucedanol	1968	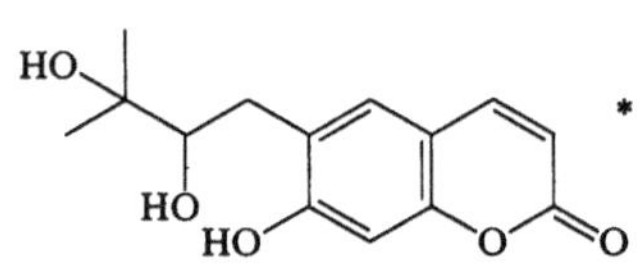 *
(94)	1968	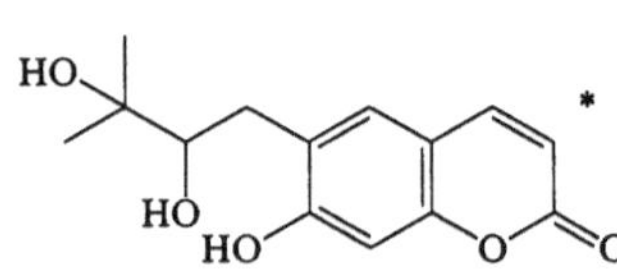 * enantiom[illegible] of **(93)**
(95) Ulopterol 2′,3′-Dihydroxydihydro- suberosin Peucedanol methyl ether	1968 1972 1974 1975	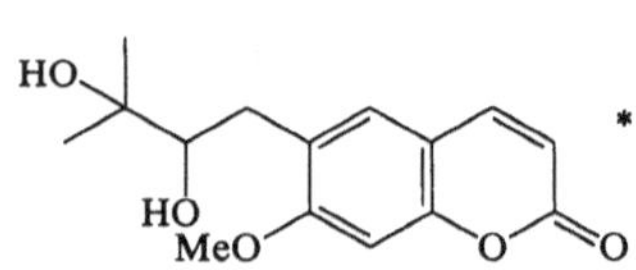 *

Formula	M. p.	$[\alpha]^t_\lambda$	Solvent	Plant Source(s)	Leading References
$C_{15}H_{16}O_4$	121—121.5			*Hesperethusa crenulata*	(*417*)
$C_{15}H_{16}O_4$	172—173			*Zanthoxylum suberosum*	(*8, 245, 417, 479*)
$C_{15}H_{16}O_4$	121			*Geijera salcifolia*	(*360*)
$C_{15}H_{14}O_4$	101—104	-17.3^{23}	$CHCl_3$	*Thamnosma montana*	(*339, 342*)
$C_{15}H_{14}O_4$	130—131			*Geijera parviflora*	(*358*)
$C_{14}H_{16}O_5$	174—175	31.2^{20}	EtOH	*Peucedanum japonicum*	(*262*)
$C_{14}H_{16}O_5$	175	-47	EtOH	*Evodia belahe*	(*496*)
$C_{15}H_{18}O_5$	141.5—142.5			*Prangos uloptera*	(*18, 19*)
	134—135.5			*Coleonema album*	(*171*)
	136—138	81.8^{22}	MeOH	*Xanthoxylum arnottianum*	(*279*)
	139—140	57	$CHCl_3$	*Micromelum pubescens*	(*292*)

Table 1.1 (continued)

	Trivial Name(s)	Year iso- lated	Structure
(**96**)	Lophopterol	1974	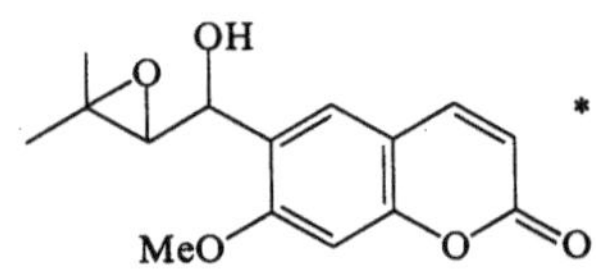
(**97**)	Thamnosmonin	1976	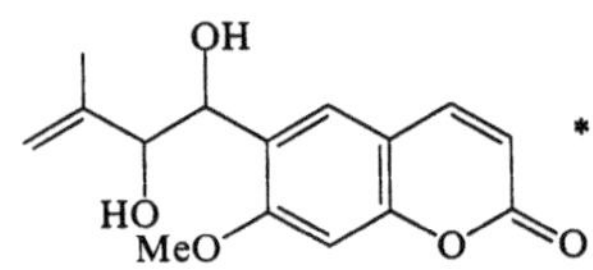
(**98**)	Angelol	1965	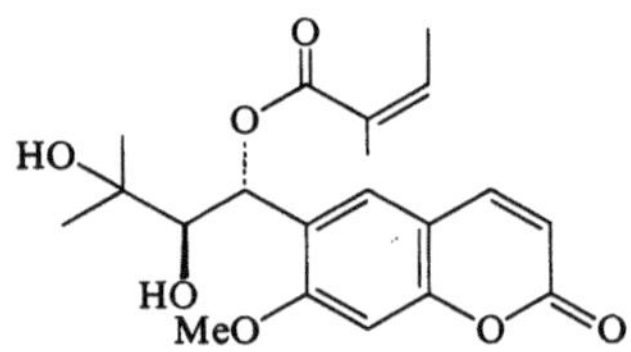
(**99**)	Micromelin Micromelumin	1967 1967	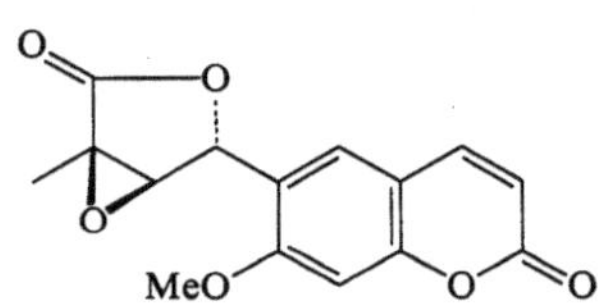
(**100**)	Ostruthin	1874	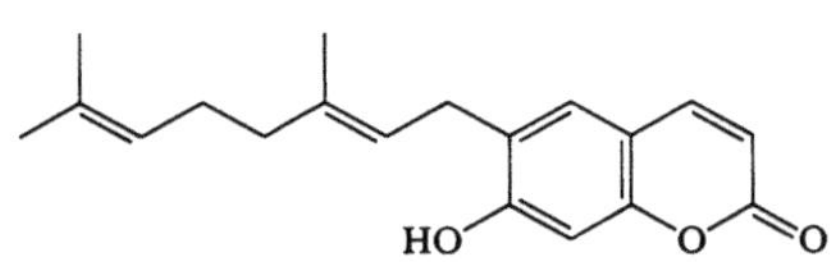
(**101**)	Peucenol Peumorisin	1963 1964	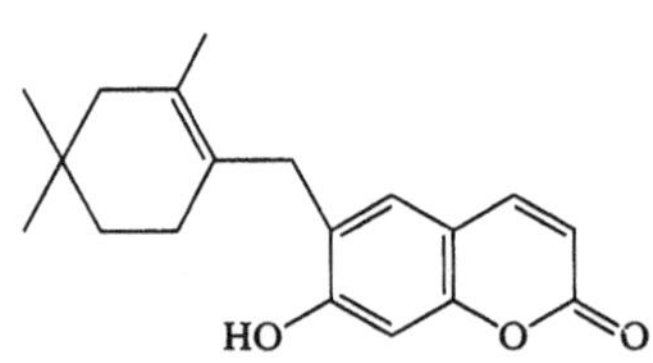
(**102**)	Ammirin (artefact) Isoangenomalin	1974 1975 1975	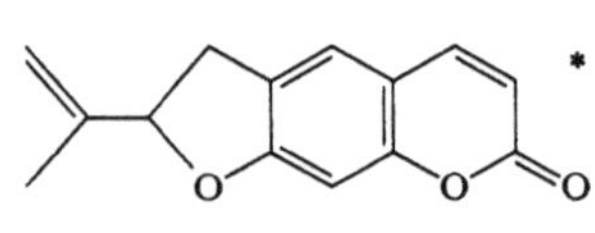

Formula	M. p.	$[\alpha]_\lambda^t$	Solvent	Plant Source(s)	Leading References
$C_{15}H_{16}O_5$	183—184			*Prangos lophoptera*	(*7, 8, 9*)
$C_{15}H_{16}O_5$	162—164	187^{26}	MeOH	*Thamnosma montana*	(*124*)
$C_{20}H_{24}O_7$	104—105	-94.7^{24} -112.9^{23}	$CHCl_3$ MeOH	*Angelica pubescens*	(*252, 253, 254, 255*)
$C_{15}H_{12}O_6$	218—219.5 218—219	-87.4	$CHCl_3$	*Micromelum minutum* *Micromelum pubescens*	(*361*) (*128*)
$C_{19}H_{22}O_3$	117—119			*Imperatoria ostruthium*	(*118, 153*)
$C_{19}H_{22}O_3$	155—157 145.5—146.5			*Peucedanum morisonii* *Peucedanum morisonii*	(*443, 533*) (*610*)
$C_{14}H_{12}O_3$	116—117 109—111 117	 7^{20} 	 $CHCl_3$ 	*Scaevola frutescens* *Ammi majus* *Scaevola lobelia*	(*313*) (*4*) (*83*)

Table 1.1 (continued)

	Trivial Name(s)	Year isolated	Structure
(**103**)	Marmesin	1949	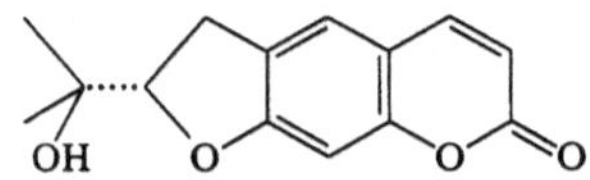
(**104**)	Nodakenetin	1966	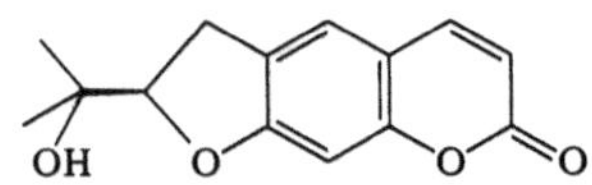
(**105**)	Prangeferol (±)-Marmesin	1972 1972	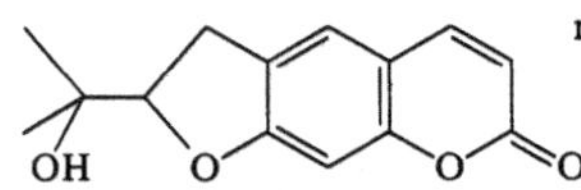 racemic
(**106**)	Ammajin Marmesinin	1958 1961	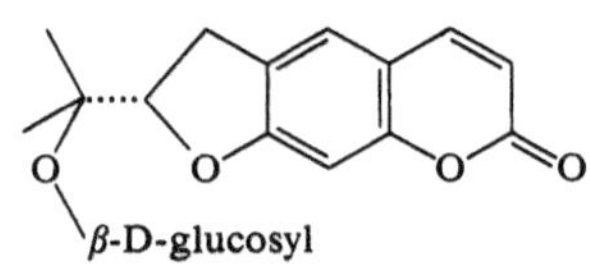
(**107**)	Nodakenin	1927	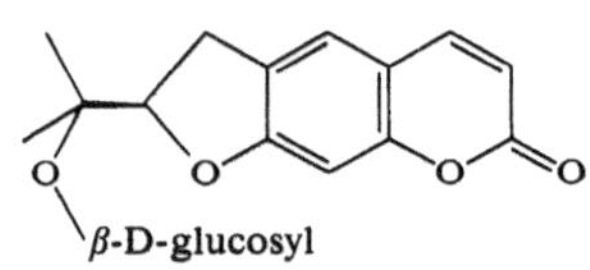
(**108**)	Nodakenetin acetate	1973	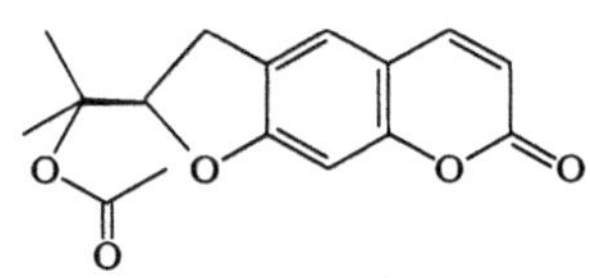
(**109**)	Seseliflorin	1970	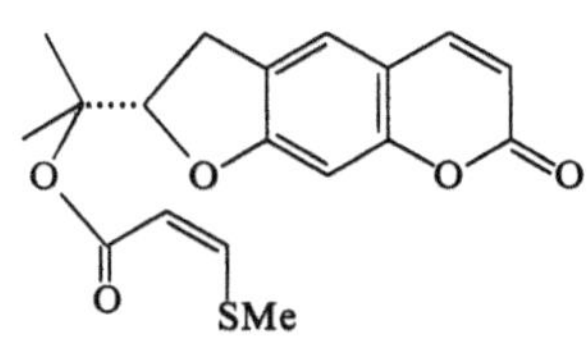
(**110**)	Sprengelianine	1972	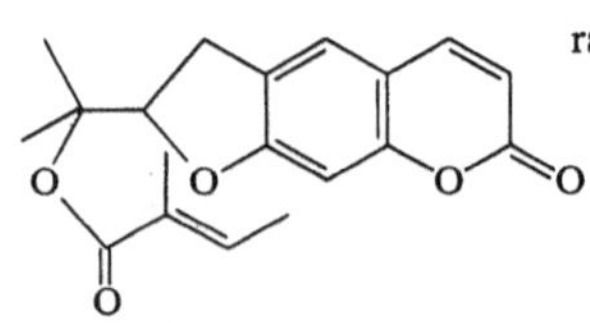 racemic

Formula	M. p.	$[\alpha]_{\lambda}^{t}$	Solvent	Plant Source(s)	Leading References
$C_{14}H_{14}O_4$	189.5—190	26.8[34]	$CHCl_3$	*Aegle marmelos*	(*153, 166, 249, 412, 553*)
$C_{14}H_{14}O_4$	185—192	− 25.4[24]	$CHCl_3$	*Angelica decursiva*	(*153, 249, 264, 553*)
$C_{14}H_{14}O_4$	176.5 166			*Prangos ferulacea* *Heracleum sprengelianum*	(*14*) (*386, 408*)
$C_{20}H_{24}O_9$	216 259—260	− 29 − 60[25]	$CHCl_3$ EtOH: :H_2O (1 : 1)	*Ammi majus* *Ammi majus*	(*4*) (*551*)
$C_{20}H_{24}O_9$	215—219	56.6[30]	H_2O	*Peucedanum decursivum*	(*153*)
$C_{16}H_{16}O_5$	134—135	− 13.7	$CHCl_3$	*Boenninghausenia albiflora*	(*573*)
$C_{18}H_{18}O_5S$	142—144	− 58.4[19]	$CHCl_3$	*Seseli sessiliflorum*	(*505, 507, 510*)
$C_{19}H_{20}O_5$	105			*Heracleum sprengelianum*	(*386*)

Table 1.1 (continued)

Trivial Name(s)	Year isolated	Structure
(111) Deltoin	1964	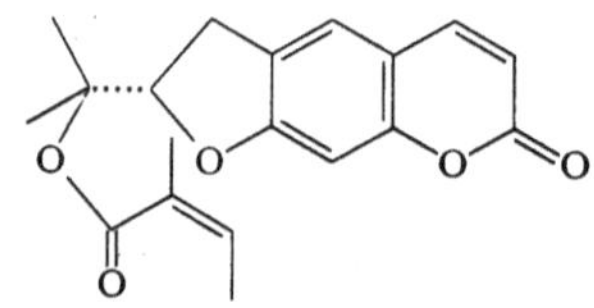
(112) Prantschimgin	1964	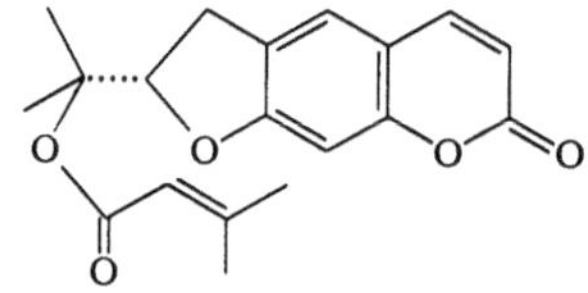
(113) Fedtschin	1964	Mixture of **(111)** and **(112)**
(114) Lindiol	1974	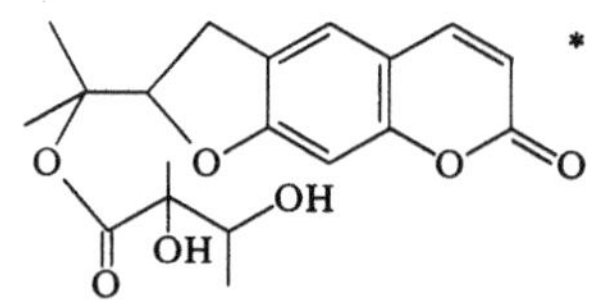
(115) Felamidin	1967	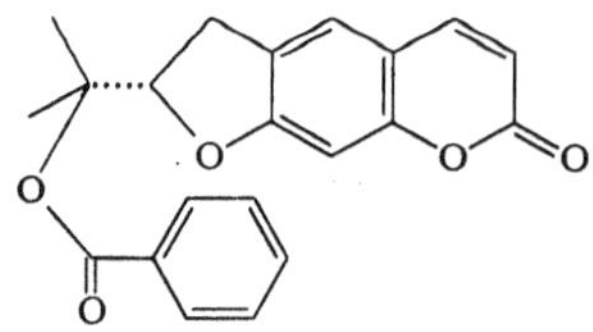
(116) Tschuin	1972	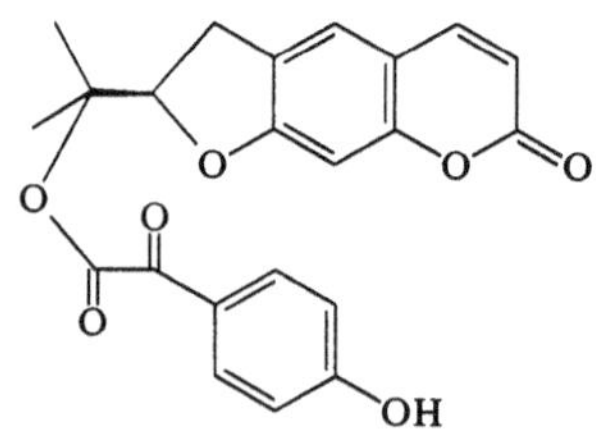
(117) Secorin	1973	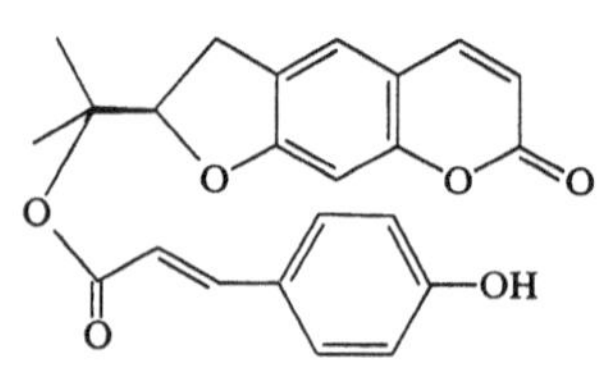

Formula	M. p.	$[\alpha]^t_\lambda$	Solvent	Plant Source(s)	Leading References
$C_{19}H_{20}O_5$	105—106.5	-51.2^{16}	$CHCl_3$	*Peucedanum deltoideum*	(*432*)
$C_{19}H_{20}O_5$	138—140	-23^{20}	$CHCl_3$	*Prangos tschimganica*	(*350, 351, 356, 553*)
	102—103			*Prangos fedtschenkoi*	(*351, 356*)
$C_{19}H_{22}O_7$	112			*Prangos ferulacea*	(*6*)
$C_{21}H_{18}O_5$	132—134	-98.9^{20}	$CHCl_3$	*Ferulago meoides*	(*335, 450, 536*)
$C_{22}H_{18}O_7$	212—213.5	-225^{26}	EtOH	*Seseli tschuense*	(*26, 27*)
$C_{23}H_{20}O_6$	212—213	-220^{20}	EtOAc	*Seseli coronatum*	(*174, 580*)

Table 1.1 (continued)

Trivial Name(s)	Year isolated	Structure
(**118**) Prandiol	1974	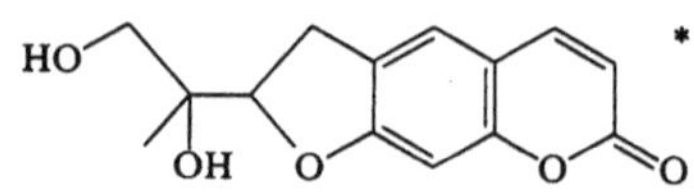
(**119**) Xanthoarnol	1973	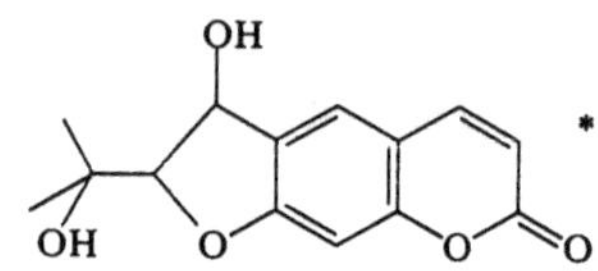
(**120**) Smirniorin	1969	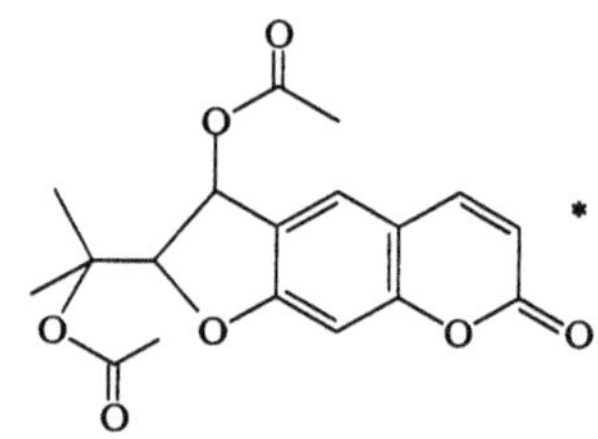
(**121**) Smirnioridin	1970	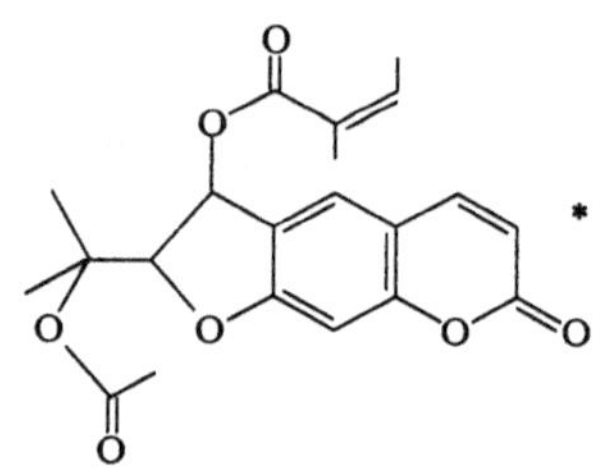
(**122**) Psoralen	1933	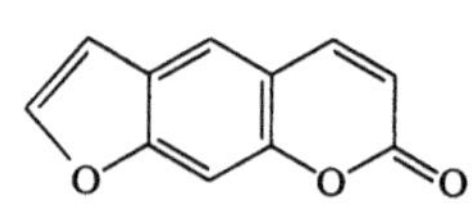
(**123**) Sabandin	1965	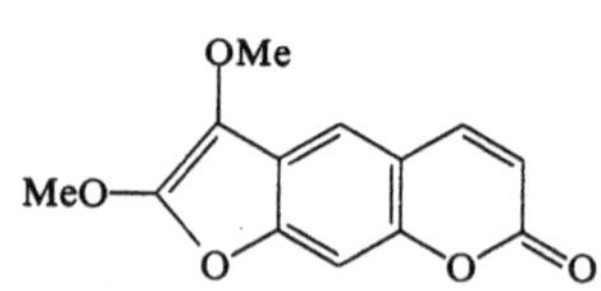
(**124**) Prangosin	1967	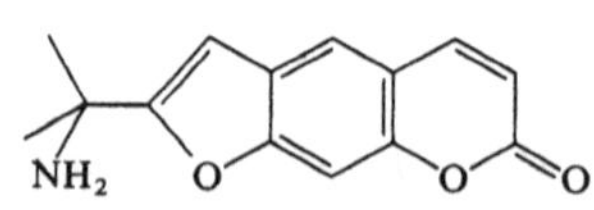

References, pp. 400—429

Formula	M. p.	$[\alpha]^t_\lambda$	Solvent	Plant Source(s)	Leading References
$C_{14}H_{14}O_5$	131—132			*Prangos biebersteinii*	(*10*)
$C_{14}H_{14}O_5$	208—209	$\pm 0^{24}$	MeOH	*Xanthoxylum arnottianum*	(*281*)
$C_{18}H_{18}O_7$	143—145	-138^{20}	EtOH	*Smirniopsis aucheri*	(*506*)
$C_{21}H_{22}O_7$	126—128	-229^{20}	$CHCl_3$	*Smirniopsis aucheri*	(*508*)
$C_{11}H_6O_3$	161—163			*Psoralea corylifolia*	(*153, 518, 597*)
$C_{13}H_{10}O_5$	144—145			*Ruta pinnata*	(*491, 494*)
$C_{14}H_{13}NO_3$	131—132			*Prangos pabularia*	(*23, 394, 395*)

Table 1.1 (continued)

Trivial Name(s)	Year isolated	Structure	
(**125**) Peucedanin	1833	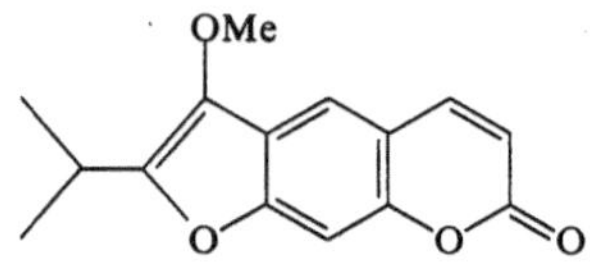	
(**126**) Decursinol	1968	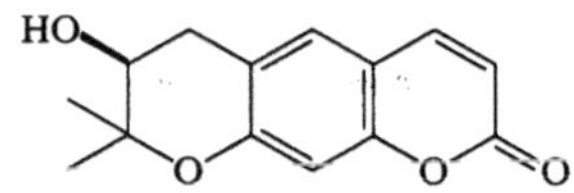	
(**127**) Agasyllin	1966 1968	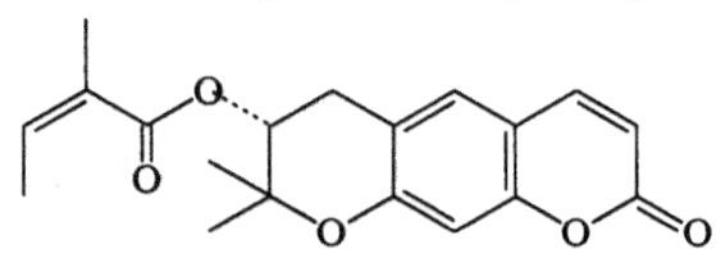	
(**128**) Decursin	1966	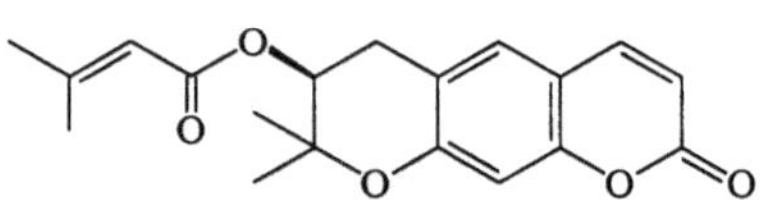	
(**129**) Graveolone	1967	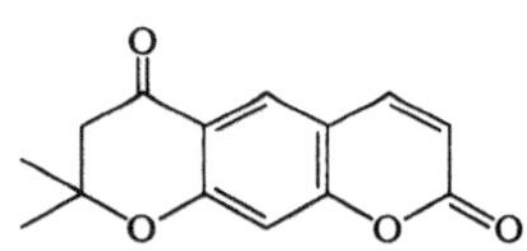	
(**130**) Peuarin	1976	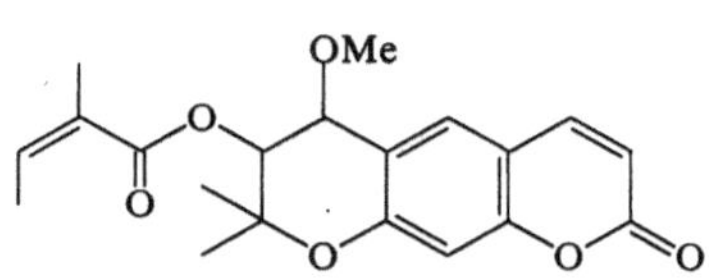	* 3′,4′-*c*
(**131**)	1969	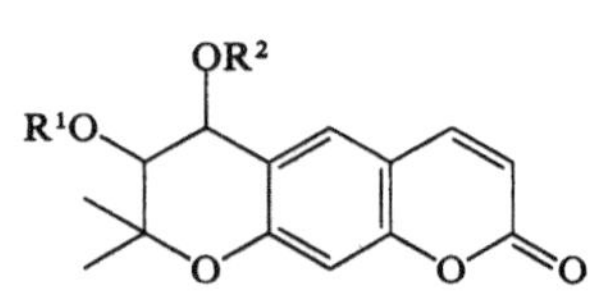 R^1 or R^2 = 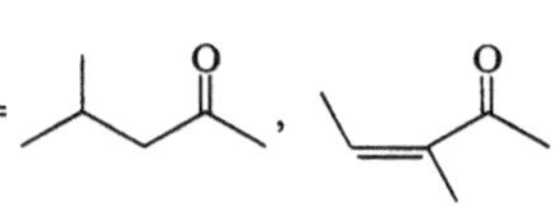	*

Formula	M. p.	$[\alpha]^t_\lambda$	Solvent	Plant Source(s)	Leading References
$C_{15}H_{14}O_4$	109			*Peucedanum officinale*	(*153*)
$C_{14}H_{14}O_4$	178	18^{33} 103.8^{17}	$CHCl_3$ pyridine	*Angelica gigas*	(*264, 334, 408, 553*)
$C_{19}H_{20}O_5$	non-cryst. 78—80	-130^{25} -44^{24}	$CHCl_3$ $CHCl_3$	*Seseli libanotis* *Agasyllis latifolia*	(*371*) (*445, 581*)
$C_{19}H_{20}O_5$	110—111	172.9^{15}	$CHCl_3$	*Angelica decursiva*	(*264, 265, 503, 504, 553*)
$C_{14}H_{12}O_4$	177.5—178			*Anethum graveolens*	(*31, 392, 569*)
$C_{20}H_{22}O_6$	85.5—86.5	-87.3^{25}	$CHCl_3$	*Peucedanum arenarium*	(*612*)
$C_{24}H_{28}O_7$	oil			*Angelica decursiva*	(*265*)

Table 1.1 (continued)

Trivial Name(s)	Year iso-lated	Structure
(**132**) Xanthalin	1968	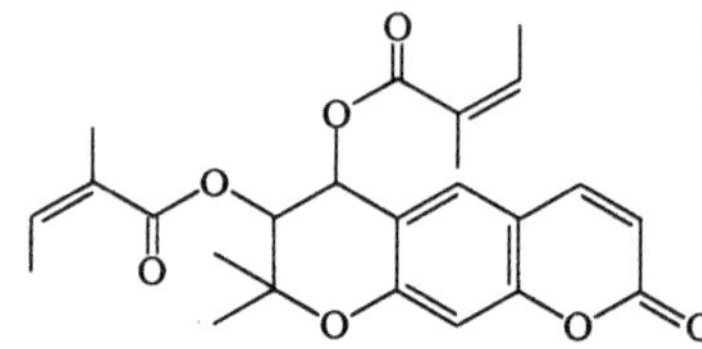 * 3′,4′-*ci*
(**133**) AD-II	1973	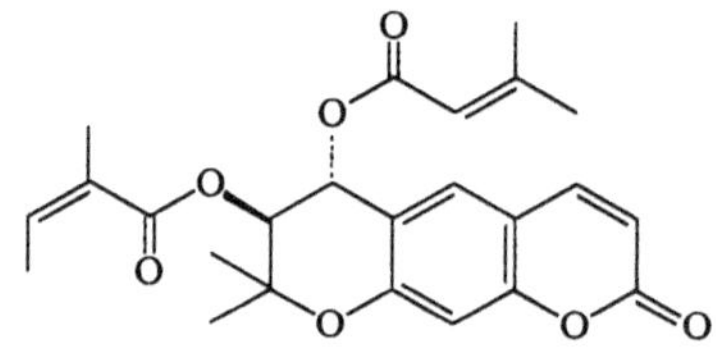
(**134**) Andelin	1970	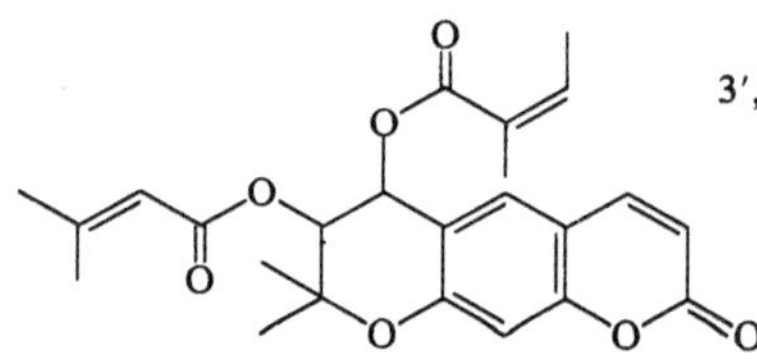 * 3′,4′-*tra*
(**135**) Decursidin	1969	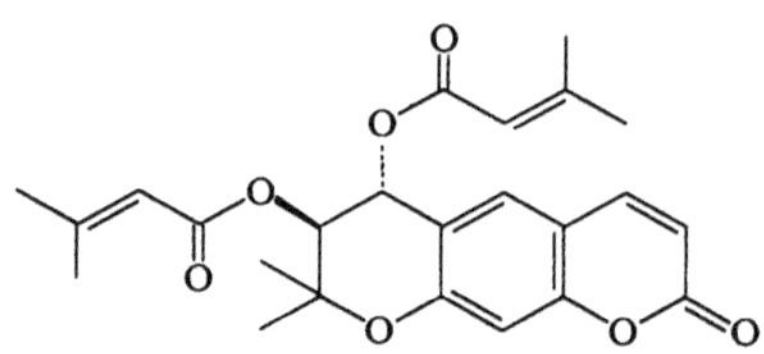
(**136**) Peuarenarine	1971	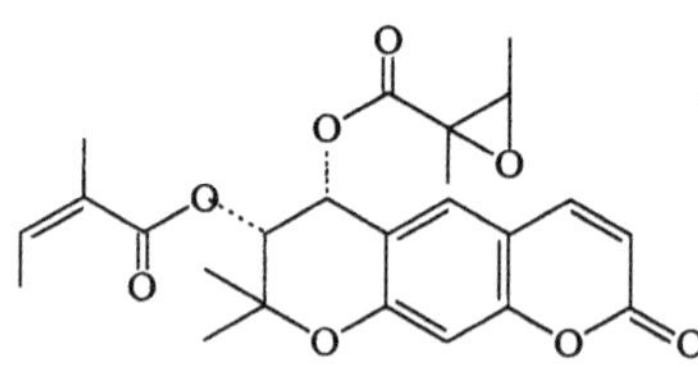 *

Formula	M. p.	$[\alpha]_{\lambda}^{t}$	Solvent	Plant Source(s)	Leading References
$C_{24}H_{26}O_7$	111—113	-164.2^{20}	EtOH	*Xanthogalum purpurascens*	(*176, 544, 545, 614*)
$C_{24}H_{26}O_7$	glass	-24.2	EtOH	*Angelica decursiva*	(*503, 504*)
$C_{24}H_{26}O_7$	glass	-28^{20}	$CHCl_3$	*Angelica decursiva*	(*41*)
$C_{24}H_{26}O_7$	60—62	-44.8^{23}	MeOH	*Angelica decursiva*	(*265, 503, 504*)
$C_{24}H_{26}O_8$	114—116	-75.8^{25}	$CHCl_3$	*Peucedanum arenarium*	(*611, 613, 614*)

Table 1.1 (continued)

Trivial Name(s)	Year iso-lated	Structure
(**137**) Peuarenine	1971	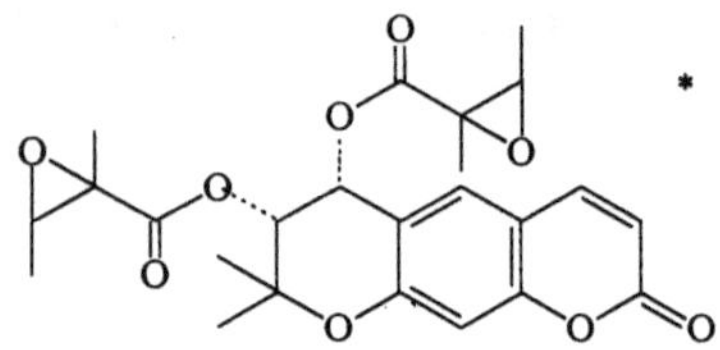 *
(**138**) Peuchlorin	1976	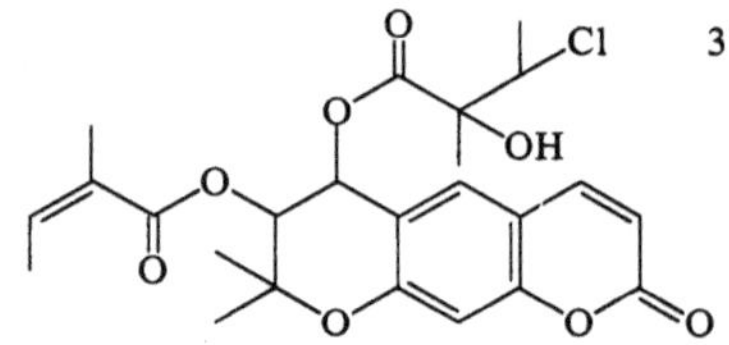 * 3′,4′-c
(**139**) Peuchlorinin	1976	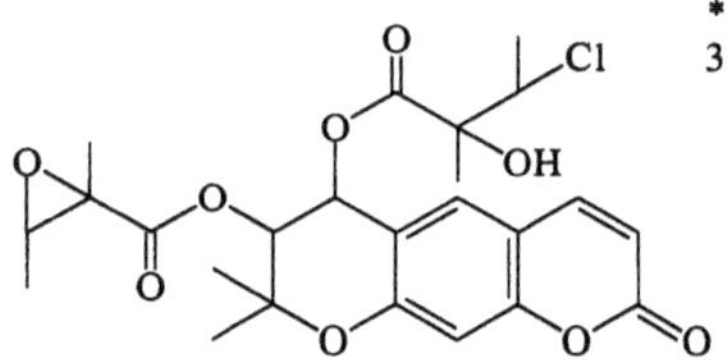 * 3′,4′-c
(**140**) Peuchloridin	1976	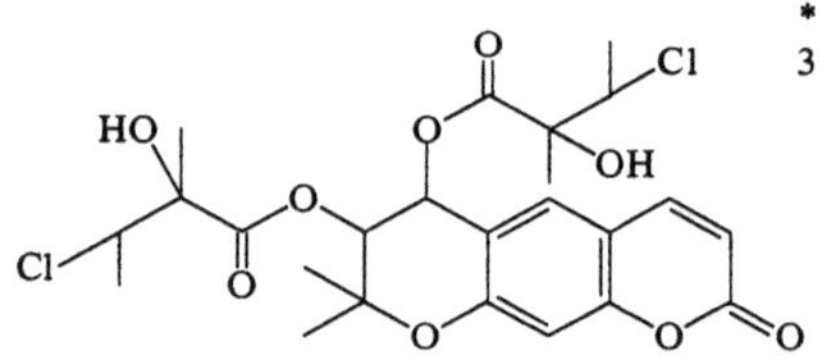 * 3′,4′-c
(**141**) Xanthyletin	1936	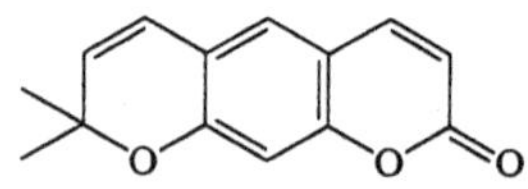

Formula	M. p.	$[\alpha]_{\lambda}^{t}$	Solvent	Plant Source(s)	Leading References
$C_{24}H_{26}O_9$	192—194	51.12^{25}	$CHCl_3$	*Peucedanum arenarium*	(*611, 613, 614*)
$C_{24}H_{27}ClO_8$	88—89			*Peucedanum arenarium*	(*612*)
$C_{24}H_{27}ClO_9$	193—194	6.6^{25}	$CHCl_3$	*Peucedanum arenarium*	(*612*)
$C_{24}H_{28}Cl_2O_9$	174—176	-49.4^{25}	MeOH	*Peucedanum arenarium*	(*612*)
$C_{14}H_{12}O_3$	128.5			*Xanthoxylum americanum*	(*153, 518, 553*)

Table 1.2 *8-Substituted 7-Oxygenated Coumarins*

	Trivial Name(s)	Year isolated	Structure
(142)	Osthenol	1937	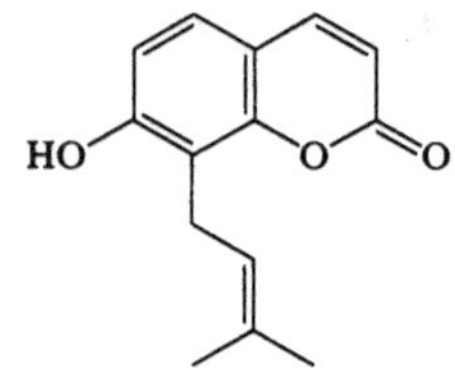
(143)	Osthol	1909	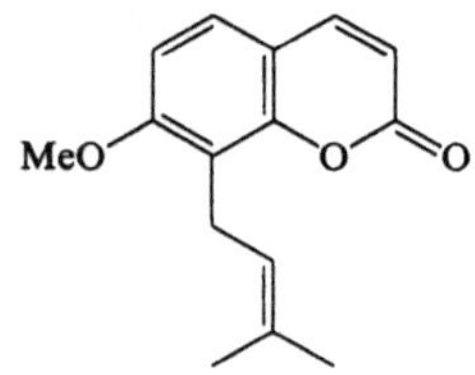
(144)		1972	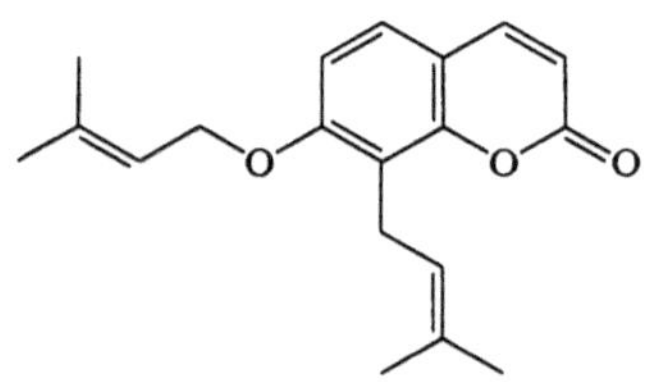
(145)	Vellein	1951	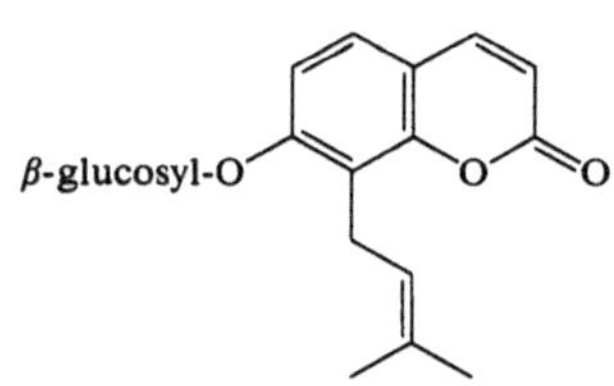
(146)	Meranzin Auraptene	1939 1938	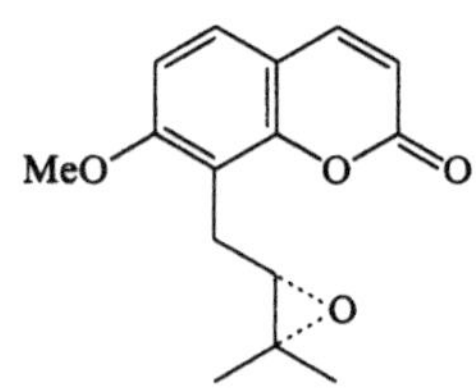
(147)	Poncimarin	1975	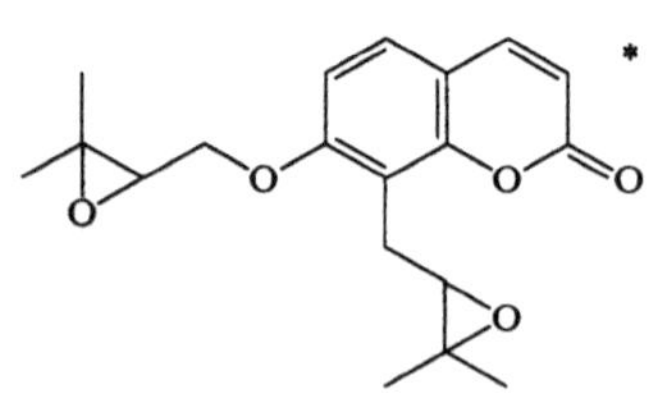

Formula	M. p.	$[\alpha]_{\lambda}^{t}$	Solvent	Plant Source(s)	Leading References
$C_{14}H_{14}O_3$	124—125			*Angelica archangelica*	(*153, 272, 400, 401*)
$C_{15}H_{16}O_3$	83—84			*Imperatoria ostruthium*	(*153*)
$C_{19}H_{22}O_3$	66.5—68			*Choisya arizonica* *Choisya mollis*	(*171*) (*171*)
$C_{20}H_{24}O_8$	187.5—189			*Velleia discophora*	(*153*)
$C_{15}H_{16}O_4$	98 91	 -33.4^{20}	 EtOH	*Citrus aurantium*	(*153, 382*)
$C_{19}H_{22}O_5$	140	-56.2^{25}	$CHCl_3$	*Poncirus trifoliata*	(*243, 495*)

Table 1.2 (continued)

Trivial Name(s)	Year iso-lated	Structure
(**148**) Auraptenol	1965	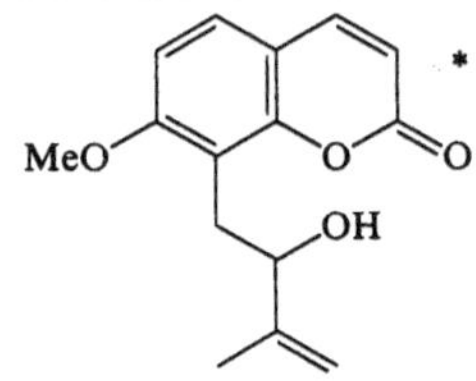
(**149**) Arnottinin	1974	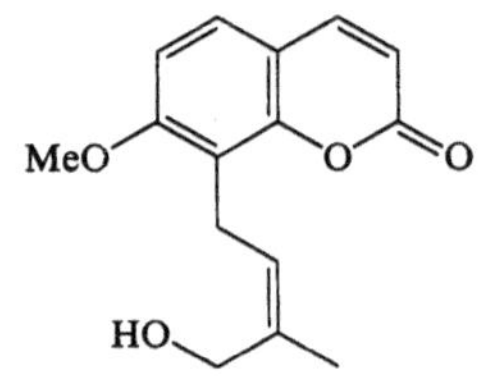
(**150**) Macrocarpin	1973	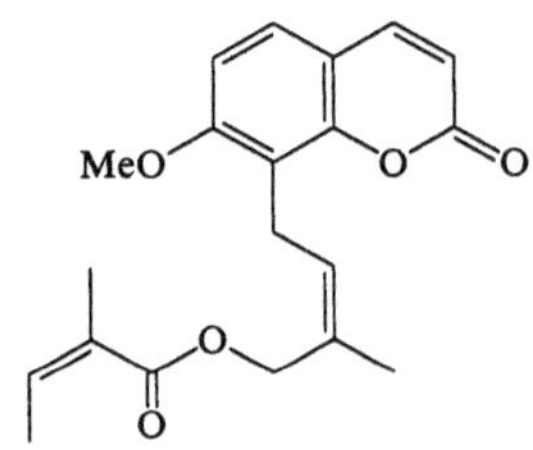
(**151**)	1965	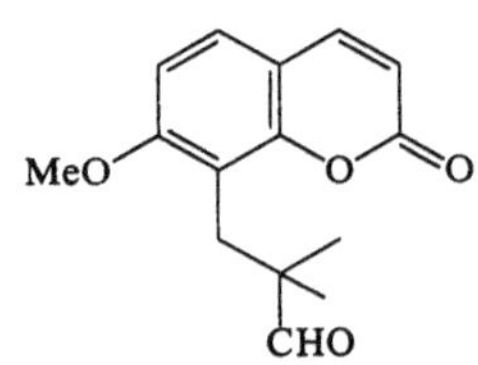
(**152**) Isomeranzin	1974	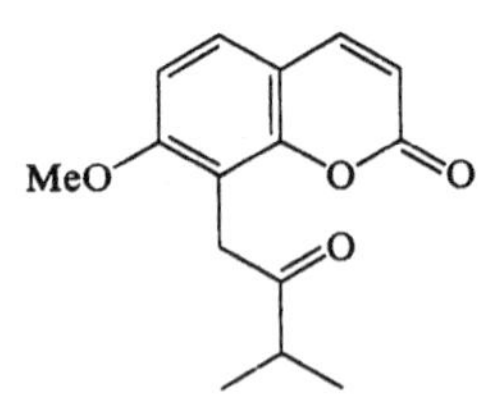
(**153**) Isoponcimarin	1976	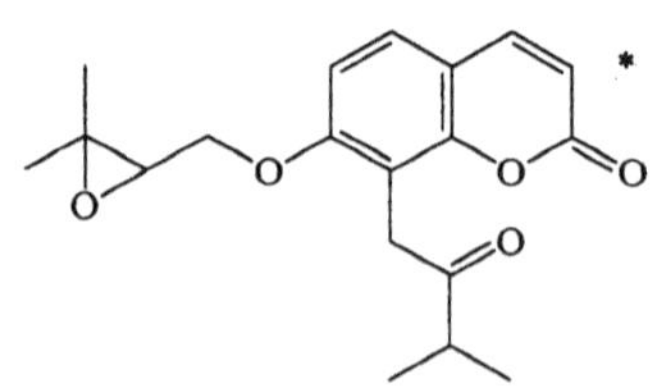

Formula	M. p.	$[\alpha]^t_\lambda$	Solvent	Plant Source(s)	Leading References
$C_{15}H_{16}O_4$	109—110	14^{26}	EtOH	*Citrus aurantium*	(*478, 550*)
$C_{14}H_{14}O_4$	191—193			*Xanthoxylum arnottianum*	(*278, 279, 280*)
$C_{20}H_{22}O_5$	71—73.5			*Lomatium macrocarpum*	(*555*)
$C_{15}H_{16}O_4$	137—138			*Citrus paradisi*	(*132, 195, 196*)
$C_{15}H_{16}O_4$	60—62			*Skimmia japonica*	(*37*)
$C_{19}H_{22}O_5$	85	-6.94^{20}_{465}	$CHCl_3$	*Poncirus trifoliata*	(*244*)

Table 1.2 (continued)

Trivial Name(s)	Year iso- lated	Structure
(154) Meranzin hydrate	1965	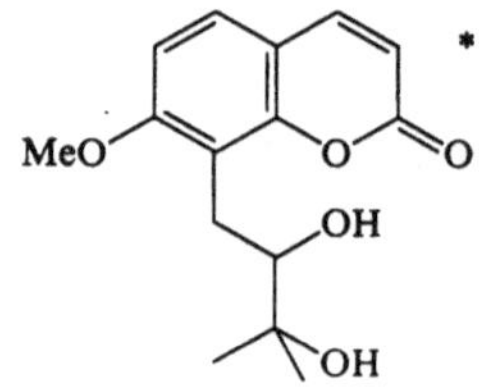 *
(155) Pranferin (artefact?)	1970	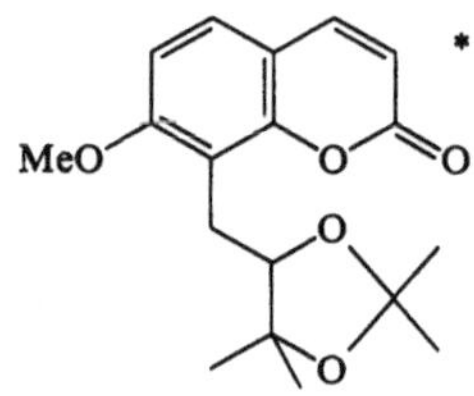 *
(156) Meranzin hydrate acetate	1972 1974	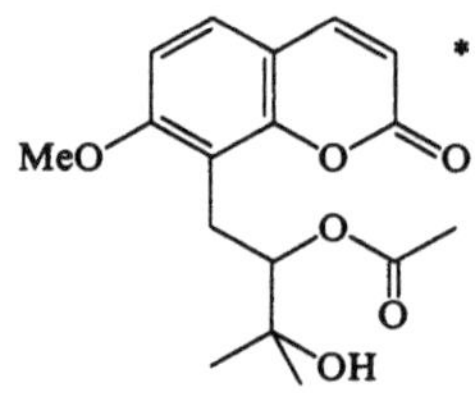 *
(157) Ferudiol	1974	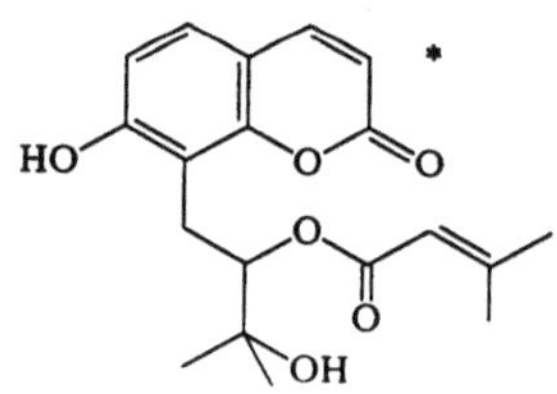 *
(158) *trans*-Dehydroosthol	1972	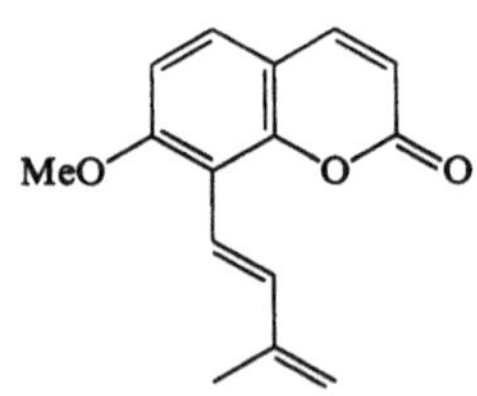
(159) Phebalosin	1966	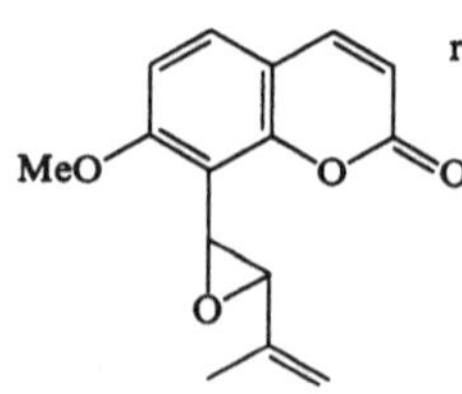 racemic

Formula	M. p.	$[\alpha]_{\lambda}^{t}$	Solvent	Plant Source(s)	Leading References
$C_{15}H_{18}O_5$	128	-53.03^{20}	EtOH	*Prangos ferulacea*	(*345, 346*)
$C_{18}H_{22}O_5$	167.5—169			*Prangos ferulacea*	(*15, 447*)
$C_{17}H_{20}O_6$	135—137 125—129	81.1^{20} 10.5^{23}	$CHCl_3$ $CHCl_3$	*Prangos ferulacea* *Scaevola frutescens*	(*17*) (*313*)
$C_{19}H_{22}O_6$	oil			*Prangos ferulacea*	(*6*)
$C_{15}H_{14}O_3$	84			*Choisya ternata*	(*77*)
$C_{15}H_{14}O_4$	120.5—121.5			*Phebalium tuberculosum*	(*138*)

Table 1.2 (continued)

Trivial Name(s)	Year isolated	Structure
(160) Murralongin	1972	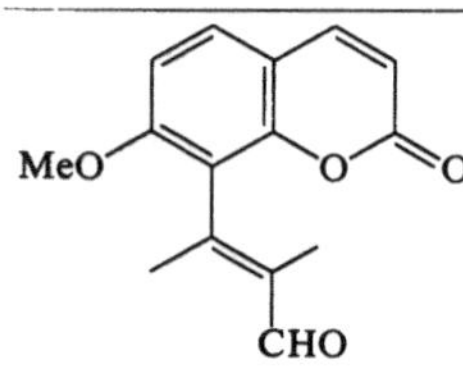
(161)	1973	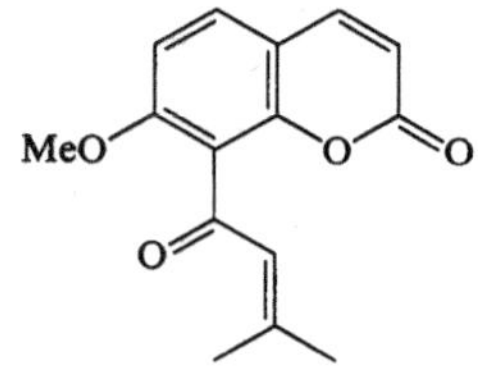
(162) Micropubescin	1967	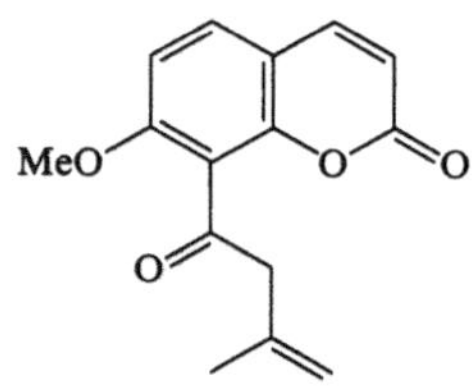
(163) Murrayone	1972	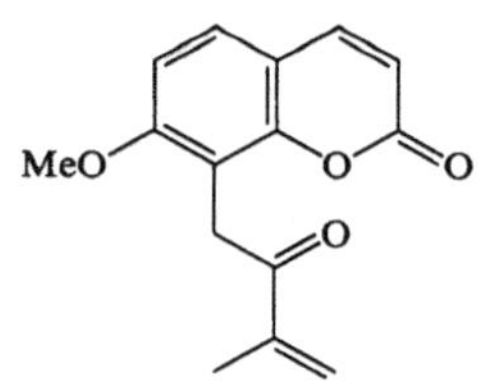
(164) Murrangatin	1972 1973	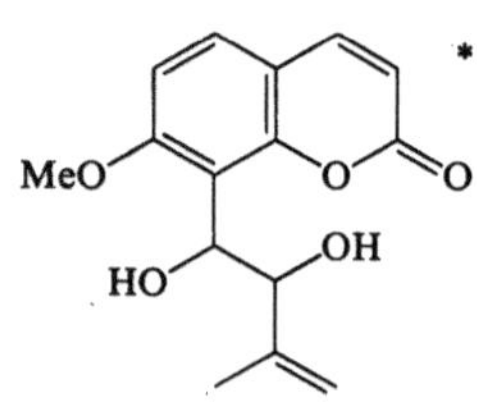
(165) Paniculatin	1972	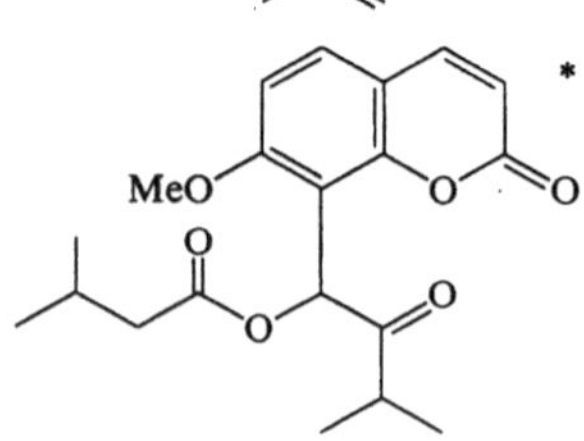

Formula	M. p.	$[\alpha]^t_\lambda$	Solvent	Plant (Source(s)	Leading References
$C_{15}H_{14}O_4$	135			*Murraya elongata*	(*572*)
$C_{15}H_{14}O_4$	131—132			*Ligusticum hultenii*	(*257, 403*)
$C_{15}H_{14}O_4$	130—131			*Micromelum pubescens*	(*128, 129*)
$C_{15}H_{14}O_4$	130			*Murraya exotica*	(*366, 478*)
$C_{15}H_{16}O_5$	131—132			*Murraya paniculata*	(*310, 311*)
	133	−3	$CHCl_3$	*Murraya elongata*	(*571*)
$C_{20}H_{24}O_6$				*Murraya paniculata*	(*554*)

Table 1.2 (continued)

Trivial Name(s)	Year iso-lated	Structure
(166) Columbianetin Zosimol (from zosimin) (+)-Dihydrooroselol	1971 1971	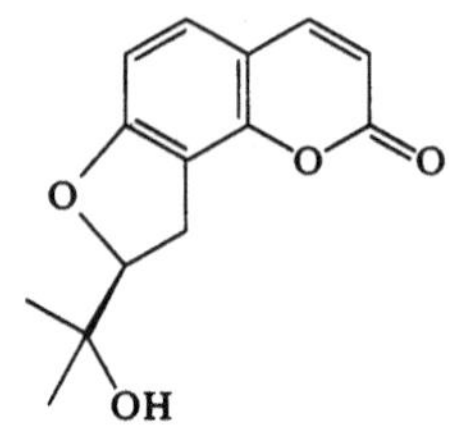
(167) (R)-(−)-Columbianetin	1974	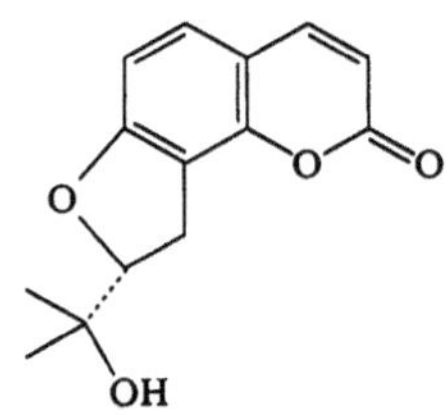
(168) Columbianin	1964	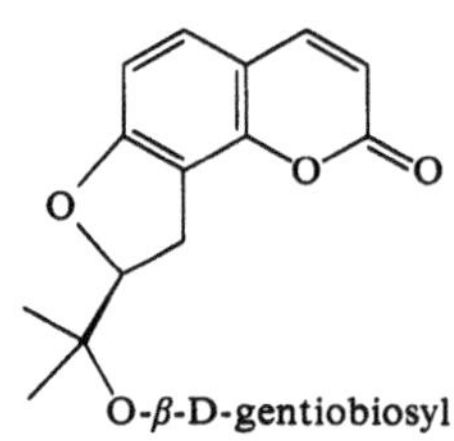
(169)	1971	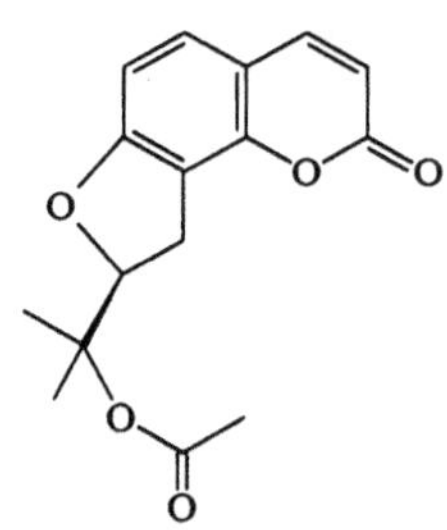
(170) Libanoridin	1969	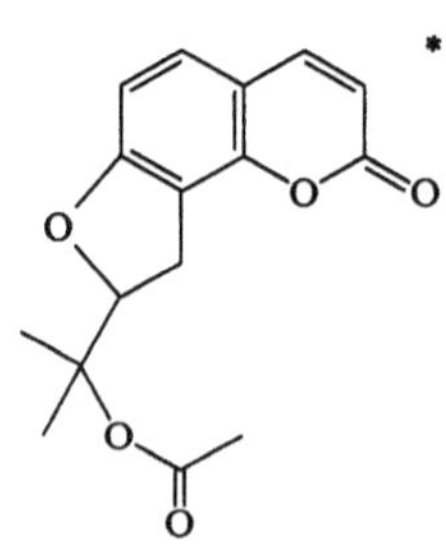

Formula	M. p.	$[\alpha]_{\lambda}^{t}$	Solvent	Plant Source(s)	Leading References
$C_{14}H_{14}O_4$	162—163	166^{29}	$CHCl_3$	*Angelica laxiflora*	(*75, 259, 408,* (*426, 535, 552*)
		250^{24}	MeOH	*Heracleum ceylanicum*	(*386, 387*)
	156—158	209.4	$CHCl_3$		(*440*)
$C_{14}H_{14}O_4$	166—166.5	-188.3^{17}	$CHCl_3$	*Xanthoxylum arnottianum*	(*278*)
		$-262^{16.5}$	MeOH		
$C_{26}H_{34}O_{14}$	275—276	118^{23}	H_2O	*Lomatium columbianum*	(*426, 534, 596*)
$C_{16}H_{16}O_5$	134.5—135.5	263^{20}	MeOH	*Ligusticum seguieri*	(*256, 374*)
$C_{16}H_{16}O_5$	125—126			*Libanotis schrenkiana*	(*56*)

Table 1.2 (continued)

	Trivial Name(s)	Year isolated	Structure
(171)	Cnidiadin	1972	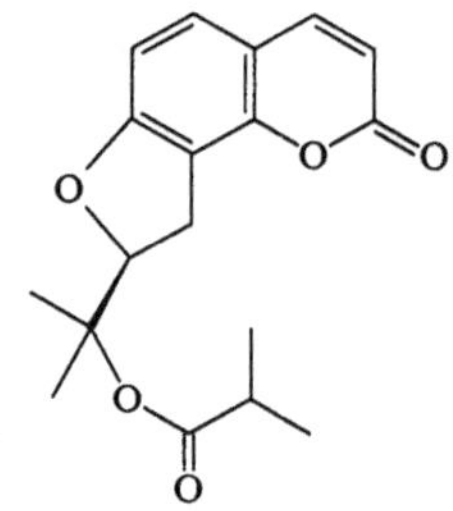
(172)		1970	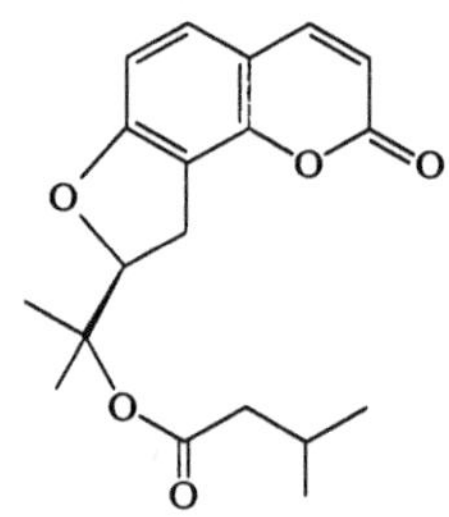
(173)	Columbianadin Zosimin	1964 1964 1964	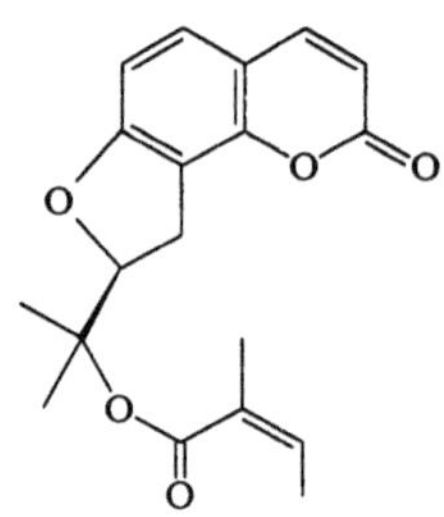
(174)	Libanorin *O*-Senecioyl-dihydrooroselol	1969 1969	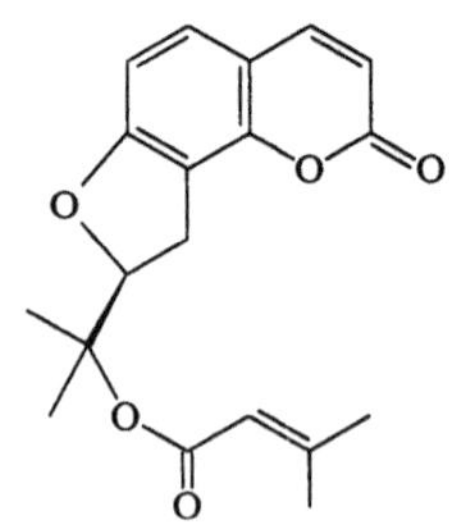

Formula	M. p.	$[\alpha]^t_\lambda$	Solvent	Plant Source(s)	Leading References
$C_{18}H_{20}O_5$	144—145	195^{20}	$CHCl_3$	*Cnidium japonicum*	(*256*)
$C_{19}H_{22}O_5$	76.5—77	305^{20}	MeOH	*Peucedanum oreoselinum*	(*256, 370*)
$C_{19}H_{20}O_5$	121—122	26.5^{27}	dioxan	*Lomatium columbianum*	(*426, 534, 596*)
	119—120	272^{16}	$CHCl_3$	*Zosimia absinthifolia*	(*440, 442*)
	118.5—119	227^{26}	$CHCl_3$	*Peucedanum palustre*	(*425*)
$C_{19}H_{20}O_5$	78—79	213^{16}	MeOH	*Libanotis schrenkiana*	(*181*)
	79	197^{24}_{578}	$CHCl_3$	*Ligusticum pyrenaicum*	(*78*)

Table 1.2 (continued)

	Trivial Name(s)	Year isolated	Structure
(175)	Columbianadinoxide	1965	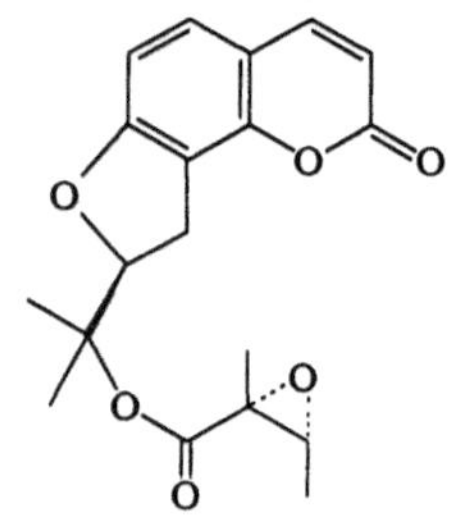
(176)	Peulustrin	1965	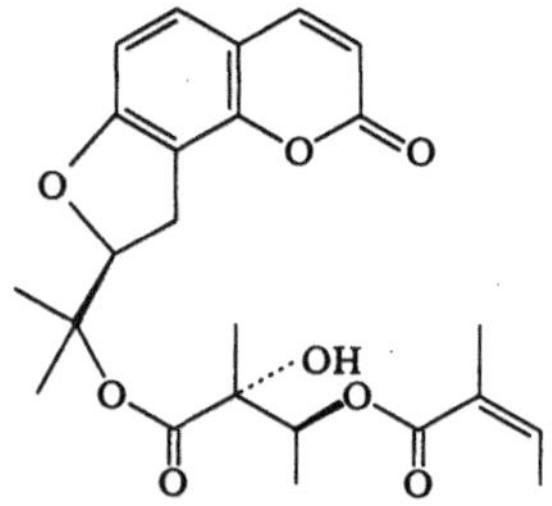
(177)	Isopeulustrin	1965	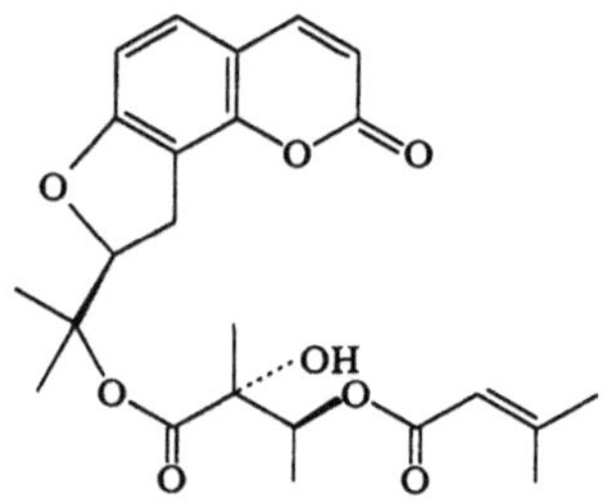
(178)	Angenomalin Masquin Majurin	1967 1967 1971	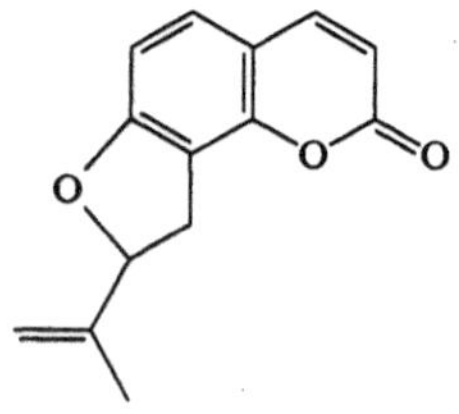

Formula	M. p.	$[\alpha]^t_\lambda$	Solvent	Plant Source(s)	Leading References
$C_{19}H_{20}O_6$	97	305^{28}	MeOH	*Peucedanum palustre*	(*428*)
$C_{24}H_{28}O_8$	129.5	278^{25}	MeOH	*Peucedanum palustre*	(*427*)
$C_{24}H_{28}O_8$	137.5—138	273^{25}	MeOH	*Peucedanum palustre*	(*428*)
$C_{14}H_{12}O_3$	105—106 107—109 94—96	39^{20}	EtOH	*Angelica anomala* *Pimpinella rupicola* *Ammi majus*	(*212, 261*) (*381*) (*2*)

Table 1.2 (continued)

	Trivial Name(s)	Year iso- lated	Structure
(179)	Discophoridin (artefact?)	1963	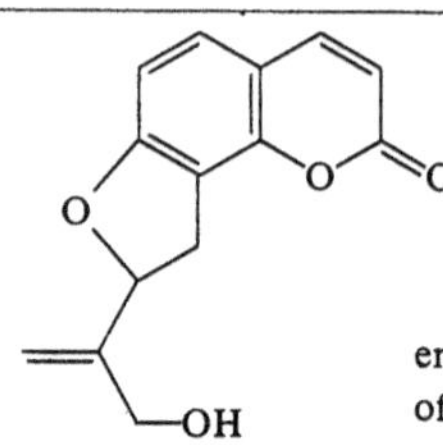 * enantiom[e] of **(180)**
(180)	Sachalinin	1965	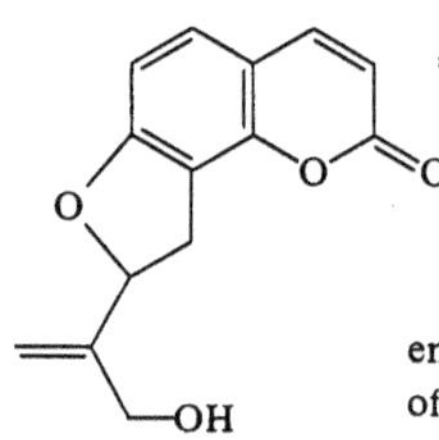 * enantiom[e] of **(179)**
(181)	Vaginidiol Calcicolin?	1971 1962	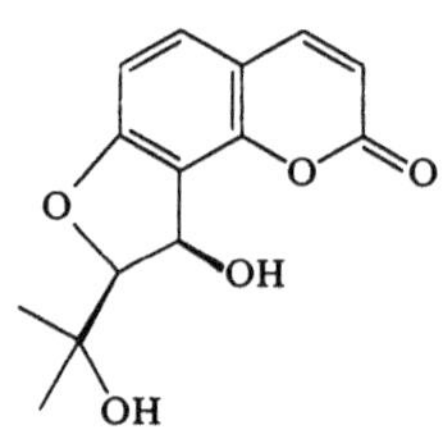
(182)	Vaginol	1970	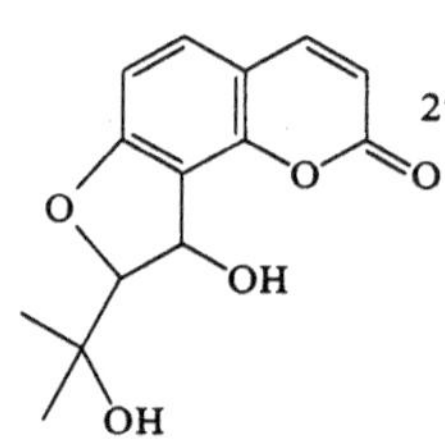 * 2′,3′-*tra[ns]*
(183)	Vaginidin	1967	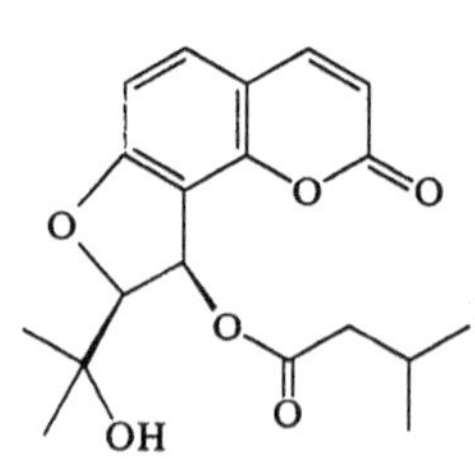

Formula	M. p.	$[\alpha]^t_\lambda$	Solvent	Plant Source(s)	Leading References
$C_{14}H_{12}O_4$	134—135	20.4^{20}	EtOH	*Velleia discophera*	(*97*)
$C_{14}H_{12}O_4$	132—134	-187.1^{25}	EtOH	*Angelica sachalinensis*	(*436, 446*)
$C_{14}H_{14}O_5$	174	234.6^{25}	EtOH	*Selinum vaginatum*	(*524*)
	172	140^{29}	EtOH	*Angelica saxicola*	(*72, 410*)
$C_{14}H_{14}O_5$	164—165	127.6^{24}	EtOH	*Selinum vaginatum*	(*480*)
$C_{19}H_{22}O_6$	133—134			*Selinum vaginatum*	(*370, 520, 522, 524*)
	138.5—139	172^{20}	MeOH		

Table 1.2 (continued)

Trivial Name(s)	Year iso-lated	Structure
(184)	1974	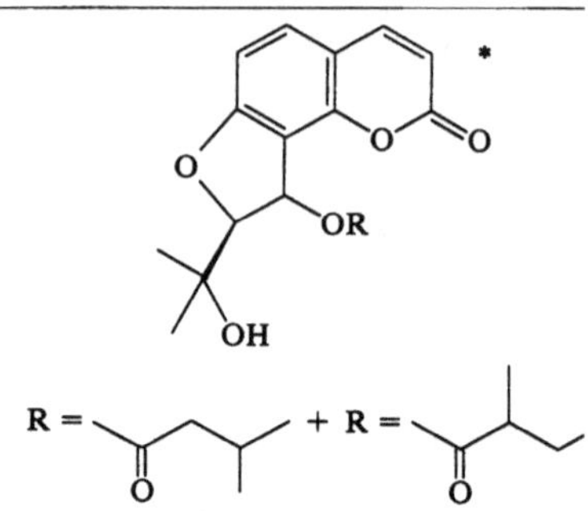
(185) Apterin	1974 1974	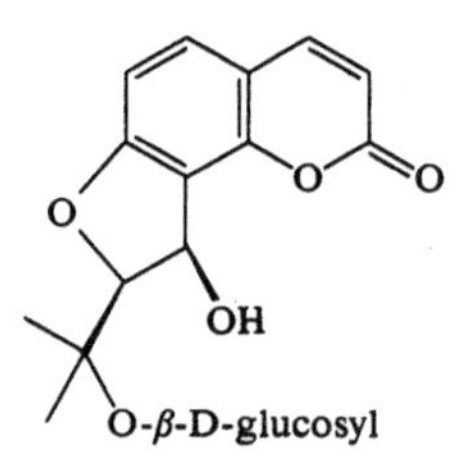
(186)	1969	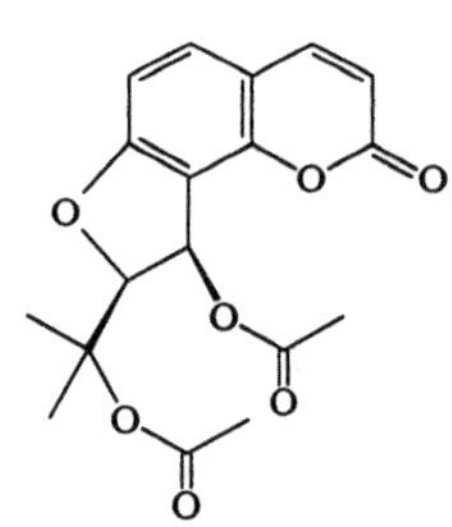
(187)	1972	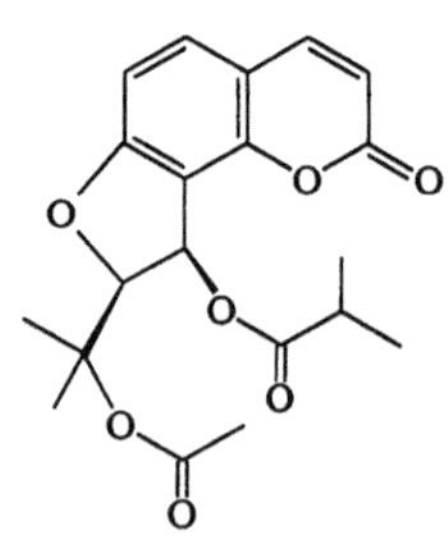

Formula	M. p.	$[\alpha]^t_\lambda$	Solvent	Plant Source(s)	Leading References
$C_{19}H_{22}O_6$				*Zizia aptera*	(*562*)
$C_{20}H_{24}O_{10}$	236—238 250	-229^{25}	H_2O	*Zizia aptera* *Heracleum mantegazzianum*	(*562*) (*191, 192*)
$C_{18}H_{18}O_7$	123			*Ligusticum pyrenaicum*	(*78*)
$C_{20}H_{22}O_7$	153—154	94.4^{20}	$CHCl_3$	*Cnidium japonicum*	(*256*)

Table 1.2 (continued)

	Trivial Name(s)	Year iso- lated	Structure
(188)	Edultin Cnidimin Libanotin	1962 1964 1965	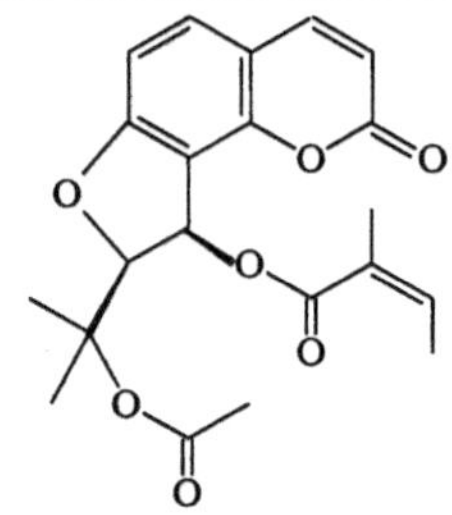
(189)	 Isopeucenidin	1974 1975	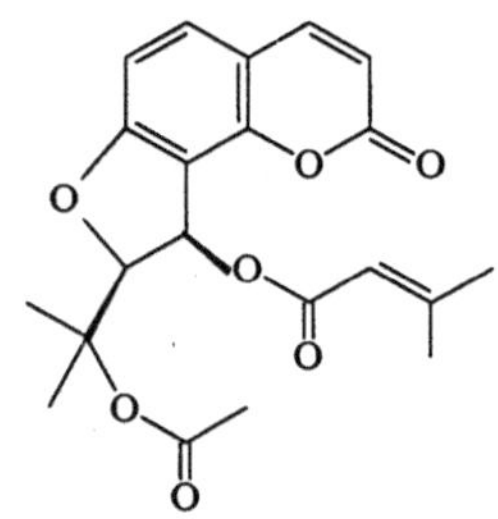
(190)		1973	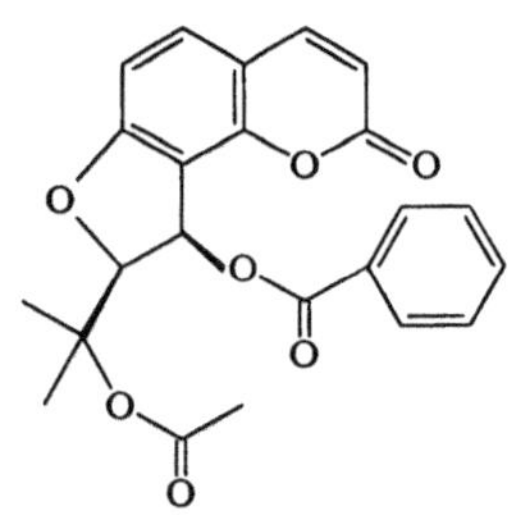
(191)		1976	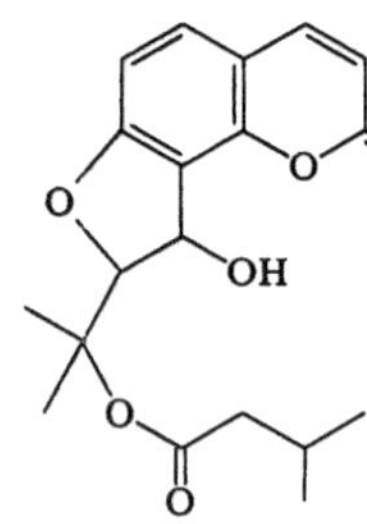 * 2′,3′-*tra*

Formula	M. p.	$[\alpha]_\lambda^t$	Solvent	Plant Source(s)	Leading References
$C_{21}H_{22}O_7$	136—142	$41.5^{10\cdot 7}$	pyridine	*Angelica edulis*	(*383, 384, 416, 511*)
	151—153	113^{18}	EtOH	*Cnidium monnieri*	(*433, 435*)
	160—161	75^{20}	$CHCl_3$	*Libanotis transcaucasica*	(*475*)
$C_{21}H_{22}O_7$	133—133.5	-54.8^{25}	$CHCl_3$	*Zizia aptera*	(*562*)
	132—134	134^{20}	EtOH	*Libanotis montana*	(*29*)
$C_{23}H_{20}O_7$	148.5	-0.5^{24}	$CHCl_3$	*Cnidium silaifolium*	(*79*)
$C_{19}H_{22}O_6$				*Peucedanum bourgaei*	(*216*)

Table 1.2 (continued)

Trivial Name(s)	Year iso- lated	Structure
(192)	1969 1970	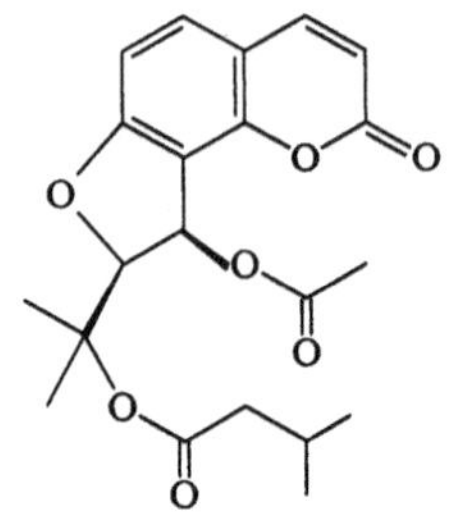
(193) Athamantin	1844	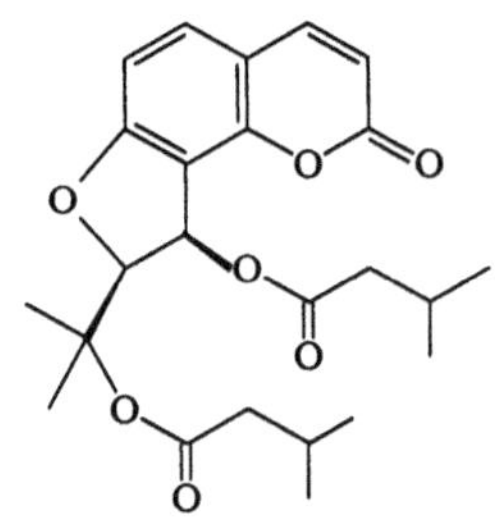
(194) Isoedultin	1973	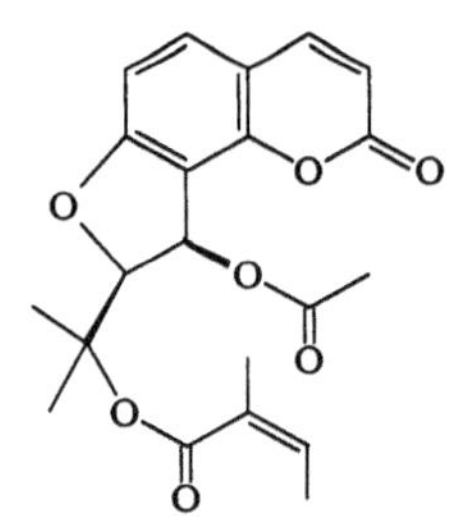
(195) Archangelicin	1964	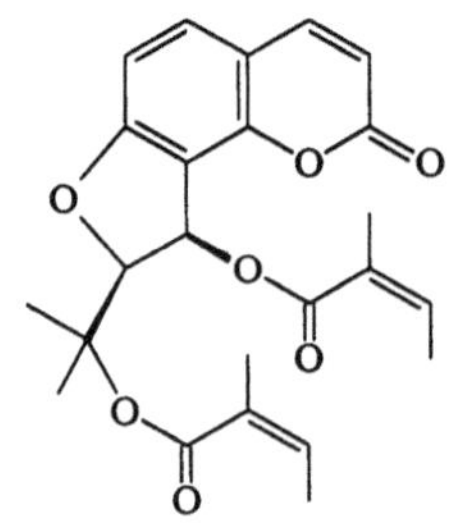

References, pp. 400—429

Formula	M. p.	$[\alpha]_\lambda^t$	Solvent	Plant Source(s)	Leading References
$C_{21}H_{24}O_7$	oil 88—88.5	 49^{20} -0.9^{20}	 MeOH $CHCl_3$	*Ligusticum pyrenaicum* *Peucedanum oreoselinum*	(*78*) (*370*)
$C_{24}H_{30}O_7$	58—60	96^{22}	MeOH	*Peucedanum oreoselinum* (= *Athamanta oreoselinum*)	(*153, 248, 370, 416, 426*)
$C_{21}H_{22}O_7$	89—90	-27.3^{20}	$CHCl_3$	*Angelica longeradiata*	(*258*)
$C_{24}H_{26}O_7$	100.5—102	112.7^{26}	MeOH	*Angelica archangelica*	(*424, 426*)

Table 1.2 (continued)

Trivial Name(s)	Year isolated	Structure
(**196**) Angeladin	1973	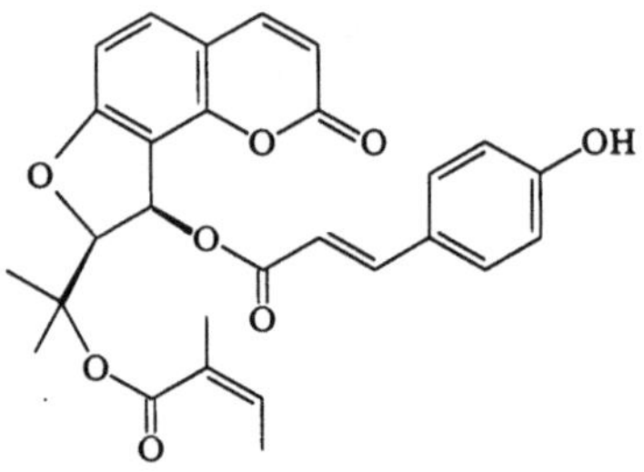
(**197**)	1970	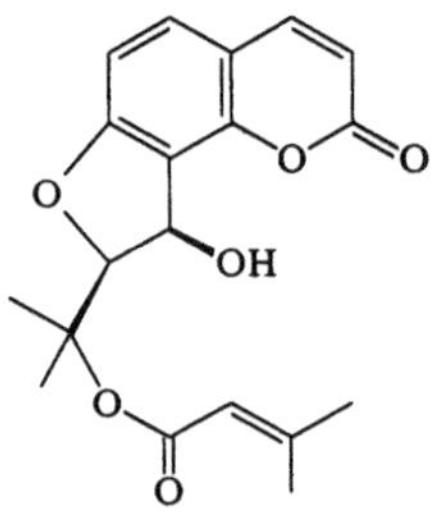
(**198**) Peucenidin	1964	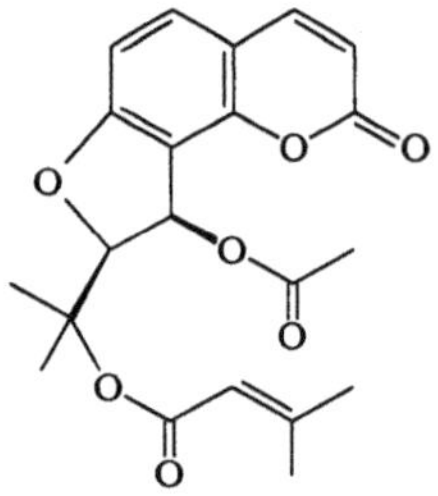
(**199**) Angelicin	1934	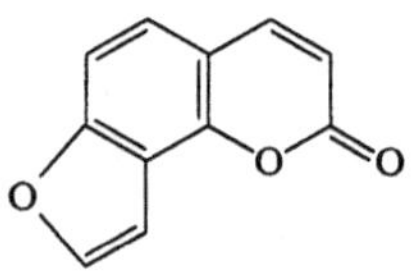
(**200**) Oroselol	1957	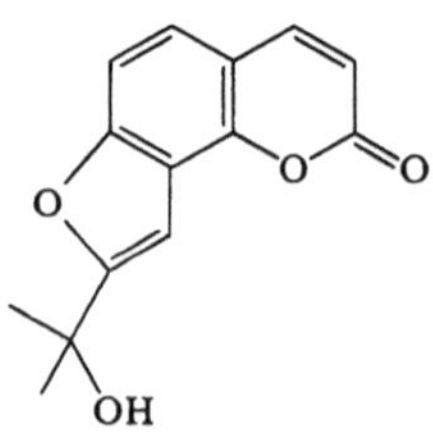

Formula	M. p.	$[\alpha]^t_\lambda$	Solvent	Plant Source(s)	Leading References
$C_{28}H_{26}O_8$	156—158	82.9^{20}	$CHCl_3$	*Angelica longeradiata*	(*258*)
$C_{19}H_{20}O_6$	132—135	113^{20}	MeOH	*Peucedanum oreoselinum*	(*370*)
$C_{21}H_{22}O_7$	124.5—125.5	-46^{20}	$CHCl_3$	*Peucedanum oreoselinum*	(*78, 370, 474, 609*)
$C_{11}H_6O_3$	138—140			*Archangelica officinalis*	(*153*)
$C_{14}H_{12}O_4$	156—157			*Peucedanum oreoselinum*	(*248*)

Table 1.2 (continued)

	Trivial Name(s)	Year iso-lated	Structure
(201)	Archangin (artefact) Oroselol methyl ether	1954	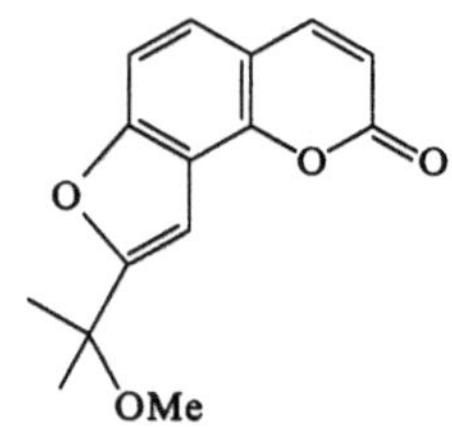
(202)	Oroselone Kvannin	1957 1954	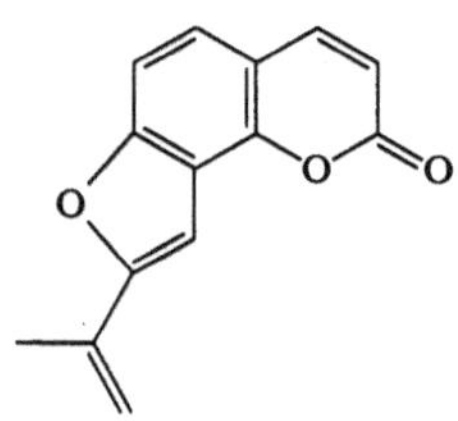
(203)	Lomatin Selinetin, Jatamansinol, Xanthogalol } Products of ester hydrolysis	1964 1964 1964 1964	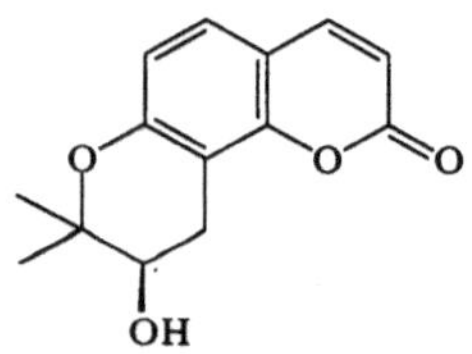
(204)		1968	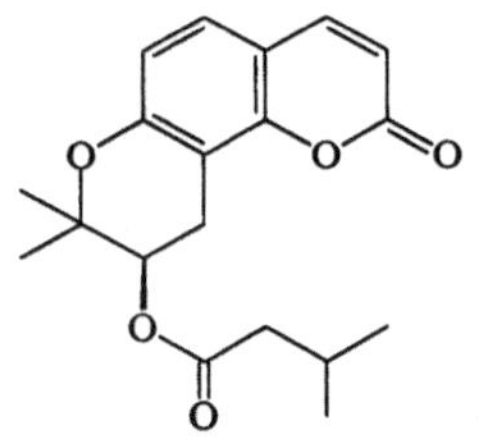
(205)	Selinidin Jatamansin Xanthogalin	1964 1964 1964	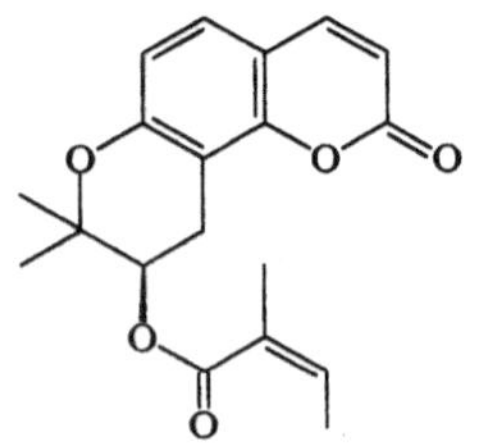

Formula	M. p.	$[\alpha]^t_\lambda$	Solvent	Plant Source(s)	Leading References
$C_{15}H_{14}O_4$	116—118			*Angelica archangelica*	(*424, 566*)
$C_{14}H_{10}O_3$	188—189			*Peucedanum oroselinum*	(*153, 248*)
	170—174			*Angelica archangelica*	(*566*)
$C_{14}H_{14}O_4$	187—188	74.8^{25}	EtOH	*Lomatium nuttalii*	(*75, 372, 373, 408, 542, 552*)
	183—184	17.2^{29}	dioxan		(*519, 521*)
	182—183	7.6^{28}	$CHCl_3$		(*519, 528*)
	183—185	13.7^{16}	$CHCl_3$		(*434*)
$C_{19}H_{22}O_5$	oil	30^{22}_{578}	$CHCl_3$	*Libanotis buchtormensis*	(*85*)
$C_{19}H_{20}O_5$	97—98	20.3^{29}	dioxan	*Selinum vaginatum*	(*519, 520, 521, 522*)
	97—98	-24.06	$CHCl_3$	*Nordostachys jatamansi*	(*519, 527, 528*)
	100—102.5	-41.4^{16}	$CHCl_3$	*Xanthogalum purpurascens*	(*434, 441*)

Table 1.2 (continued)

	Trivial Name(s)	Year iso-lated	Structure
(206)	Nuttalin Buchtormin	1968 1968	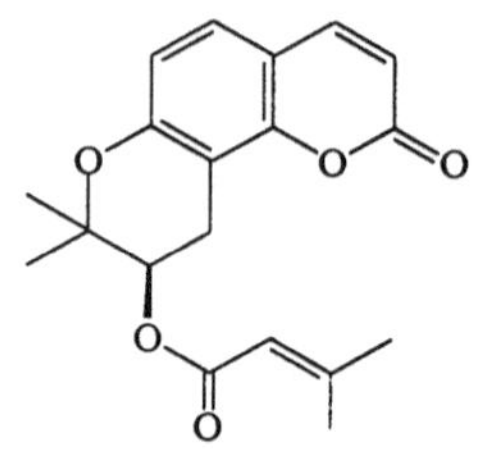
(207)		1970	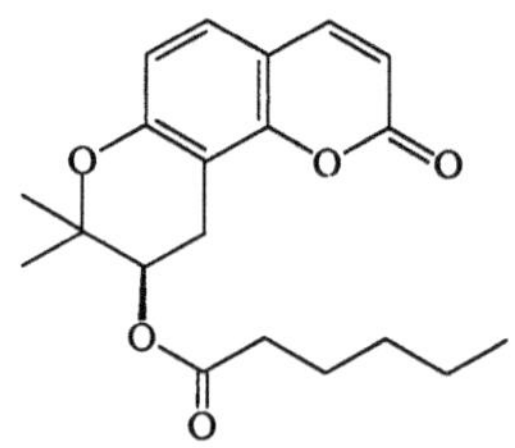
(208)		1970	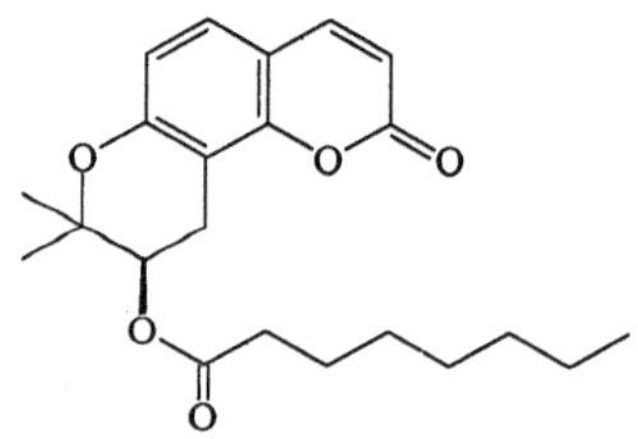
(209)		1970	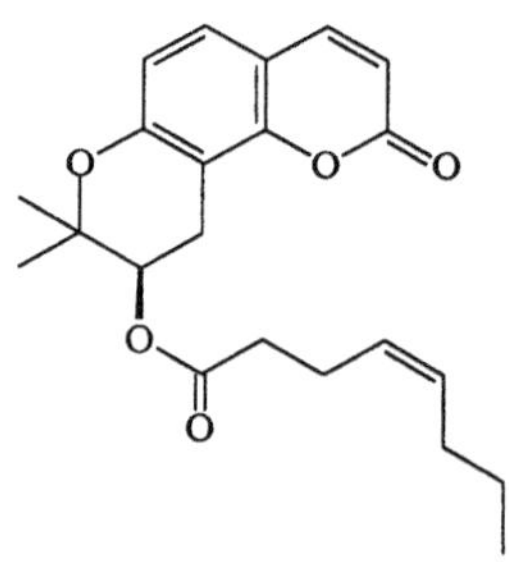

Formula	M. p.	$[\alpha]^t_\lambda$	Solvent	Plant Source(s)	Leading References
$C_{19}H_{20}O_5$	58—62	75^{20}	$CHCl_3$	*Lomatium nuttalii*	(*367*)
	glass	71.8^{20}	$CHCl_3$	*Libanotis buchtormensis*	(*85, 179, 182*)
$C_{20}H_{24}O_5$	oil	30.1^{20}	CCl_4	*Seseli gummiferum*	(*422*)
$C_{22}H_{28}O_5$	oil	32.5^{20}	CCl_4	*Seseli gummiferum*	(*422*)
$C_{22}H_{26}O_5$	oil	30.7^{20}	CCl_4	*Seseli gummiferum*	(*422*)

Table 1.2 (continued)

Trivial Name(s)	Year iso- lated	Structure
(210)	1976	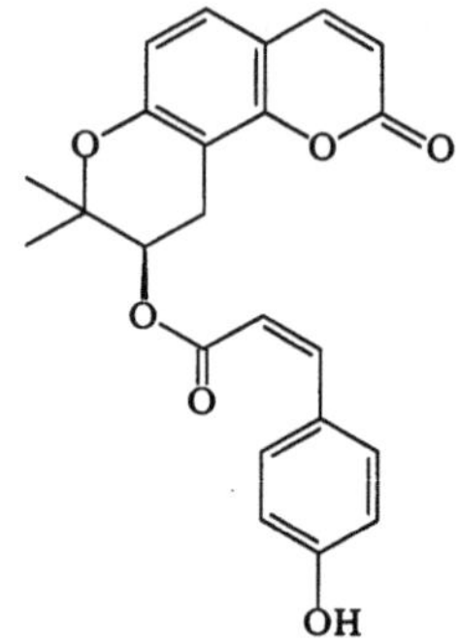
(211)	1976	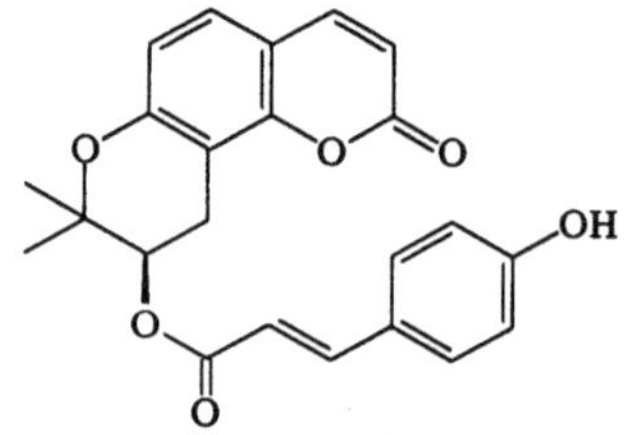
(212) Sechulin	1972	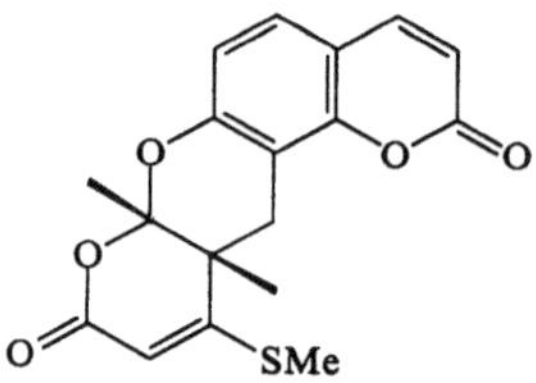
(213) (+)-*cis*-Khellactone	1971	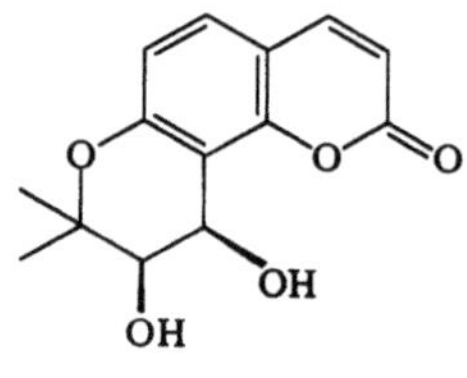
(214) (−)-*trans*-Khellactone	1967	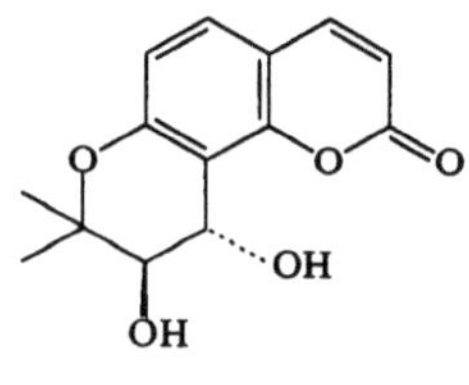

Formula	M. p.	$[\alpha]^t_\lambda$	Solvent	Plant Source(s)	Leading References
$C_{23}H_{20}O_6$	171.5—176	$-12^{25.3}$	$CHCl_3$	*Lomatium columbianum*	(*421*)
$C_{23}H_{20}O_6$	glass	$370^{25.8}$	$CHCl_3$	*Lomatium columbianum*	(*421*)
$C_{18}H_{16}O_5S$	245—246.5	833.3^{20}	$CHCl_3$	*Seseli tschuense*	(*26, 28*)
$C_{14}H_{14}O_5$	174.5—175	82^{25}	$CHCl_3$	*Seseli gummiferum*	(*373, 423, 516*)
$C_{14}H_{14}O_5$	187 185—186	 -18^{23}	 $CHCl_3$	*Angelica purpureafolia*	(*137, 373, 516*)

Table 1.2 (continued)

Trivial Name(s)	Year iso-lated	Structure
(215)	1968 1975	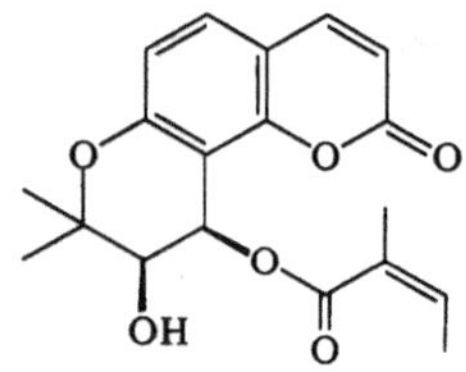
(216) Suksdorfin	1957	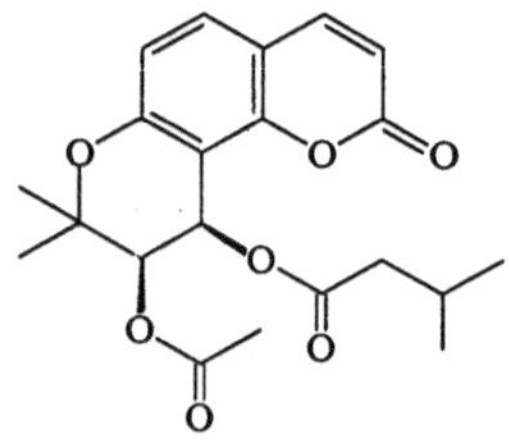
(217) Pteryxin	1958	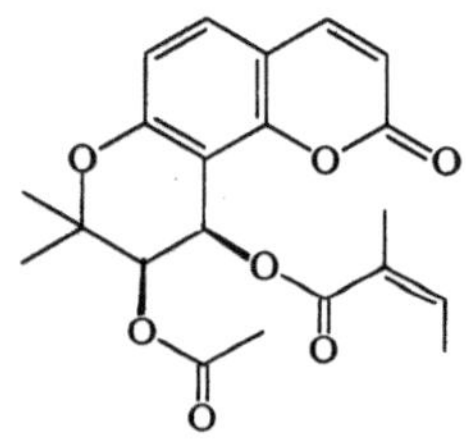
(218) Isosamidin	1966	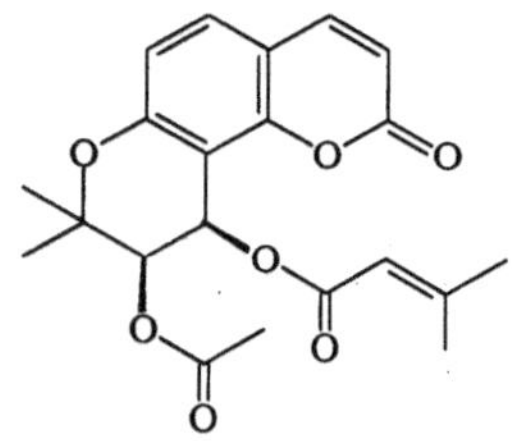
(219) Epoxypteryxin	1970	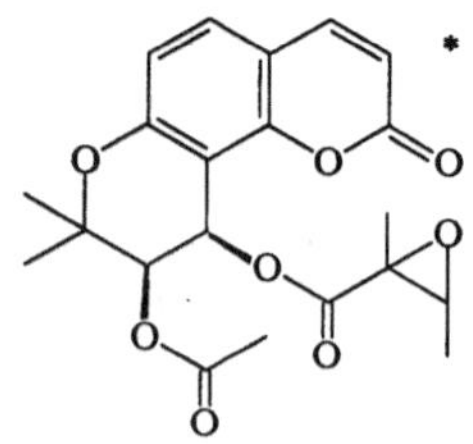

Formula	M. p.	$[\alpha]^t_\lambda$	Solvent	Plant Source(s)	Leading References
$C_{19}H_{20}O_6$				*Angelica ursina*	(*73*)
	powder	-54.5^{23}	$CHCl_3$	*Libanotis lehmanniana*	(*24, 25*)
$C_{21}H_{24}O_7$	140.5—141	4^{24}	EtOH	*Lomatium suksdorfii*	(*113, 595*)
$C_{21}H_{22}O_7$	86—86.5	12.5^{25}	EtOH	*Pteryxia terebinthina*	(*73, 114, 367, 430, 595*)
		-4^{25}	$CHCl_3$		
$C_{21}H_{22}O_7$	120.5—121	-11.9^{25}	EtOH	*Seseli libanotis*	(*371, 373*)
$C_{21}H_{22}O_8$	oil			*Laserpitium archangelica*	(*87*)

Table 1.2 (continued)

Trivial Name(s)	Year iso- lated	Structure
(**220**) Isofloroselin	1974	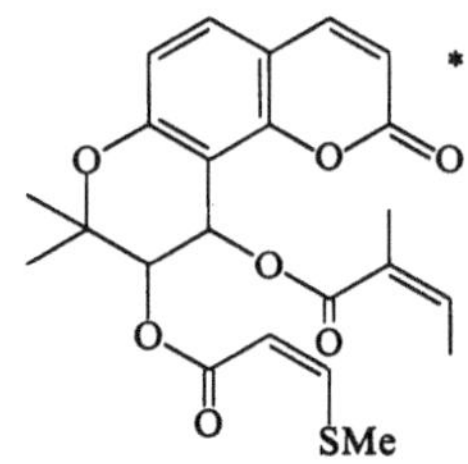 * 3′,4′-c
(**221**)	1968	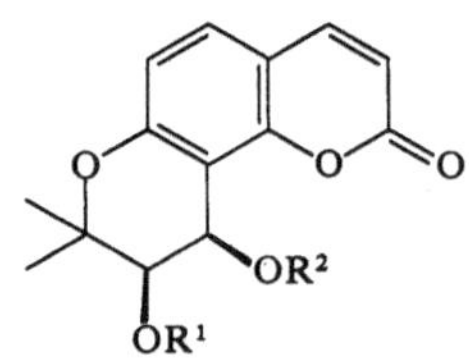 R¹ + R² = 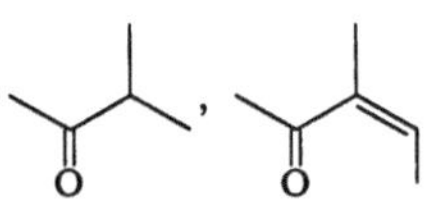
(**222**) Visnadin	1952	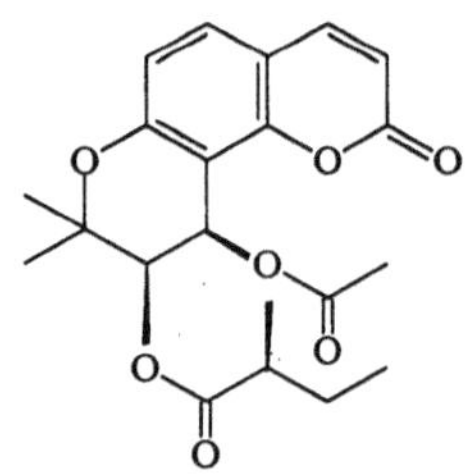
(**223**) Dihydrosamidin	1952	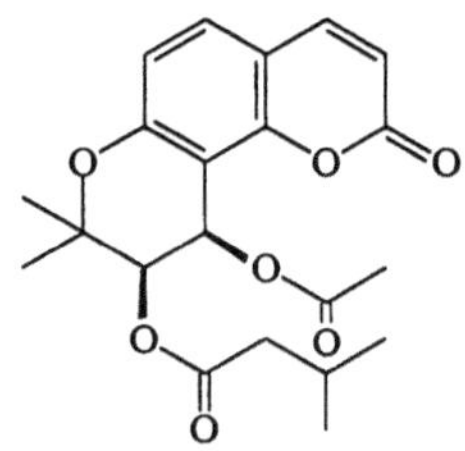

Formula	M. p.	$[\alpha]^t_\lambda$	Solvent	Plant Source(s)	Leading References
$C_{23}H_{24}O_7S$	122—123	-87.3^{20}	EtOH	*Seseli coronatum*	(*175*)
$C_{24}H_{28}O_7$				*Seseli libanotis*	(*73*)
$C_{21}H_{24}O_7$	84—86	38	dioxan	*Ammi visnaga*	(*65, 66, 373, 516, 527, 532, 538, 539, 540*)
$C_{21}H_{24}O_7$	117—119	63	dioxan	*Ammi visnaga*	(*65, 66, 75, 373, 516, 527, 538, 539, 540*)

Table 1.2 (continued)

Trivial Name(s)	Year isolated	Structure
(224)	1974	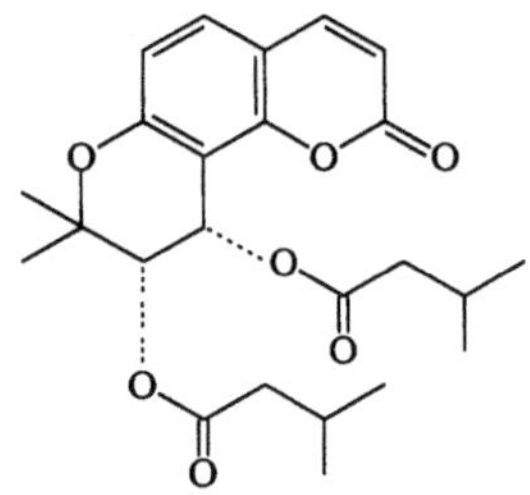
(225)	1971	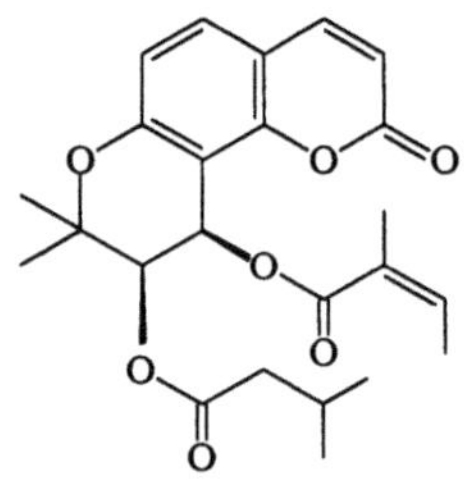
(226)	1968	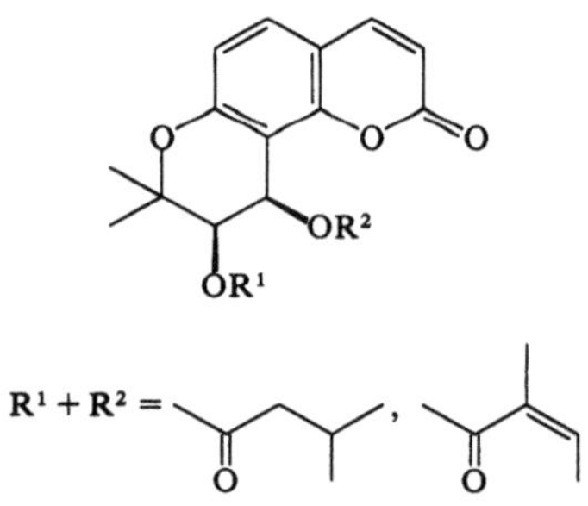
(227)	1968 1975	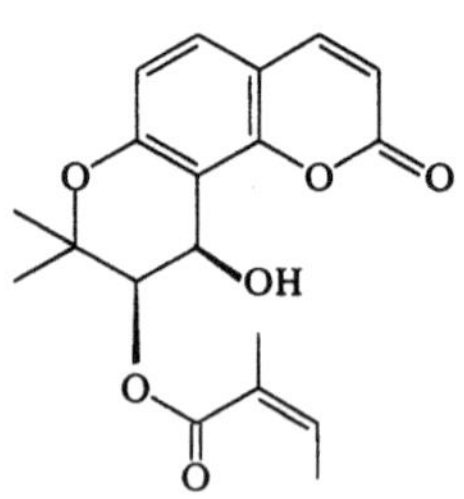

Formula	M. p.	$[\alpha]_{\lambda}^{t}$	Solvent	Plant Source(s)	Leading References
$C_{24}H_{30}O_7$	88—89	−38.8	$CHCl_3$	*Peucedanum japonicum*	(*599*)
$C_{24}H_{28}O_7$	glass	21.4^{25}	$CHCl_3$	*Seseli gummiferum*	(*25, 423*)
$C_{24}H_{28}O_7$				*Seseli libanotis*	(*73*)
$C_{19}H_{20}O_6$				*Angelica ursina*	(*73*)
	148—149	-91.8^{23}	$CHCl_3$	*Libanotis lehmanniana*	(*24, 25*)

Table 1.2 (continued)

Trivial Name(s)	Year isolated	Structure
(**228**) Isopteryxin	1967	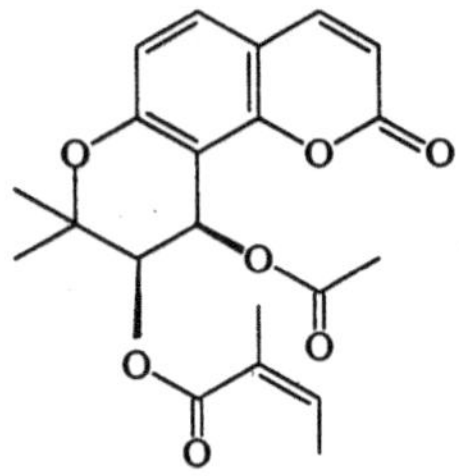
(**229**) Floroselin	1970	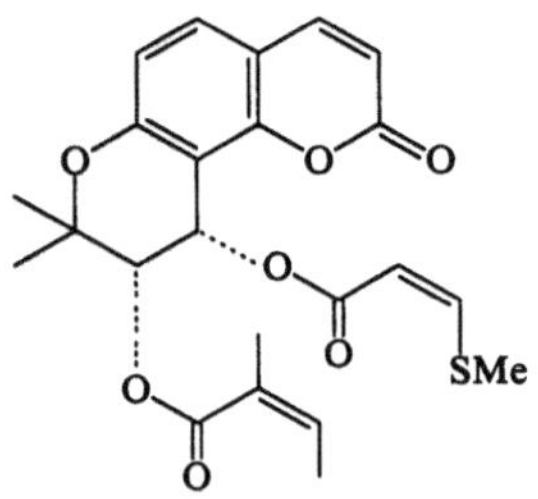
(**230**) Anomalin	1966	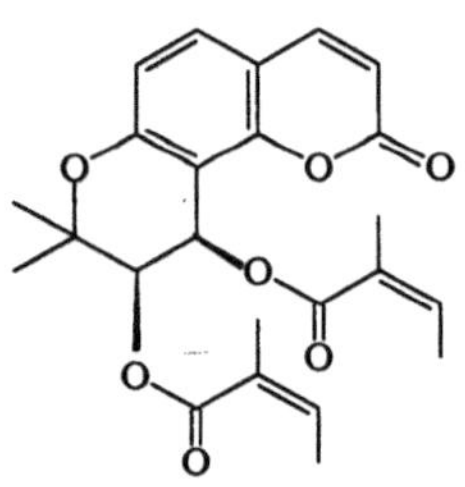
(**231**) (+)-Anomalin	1968	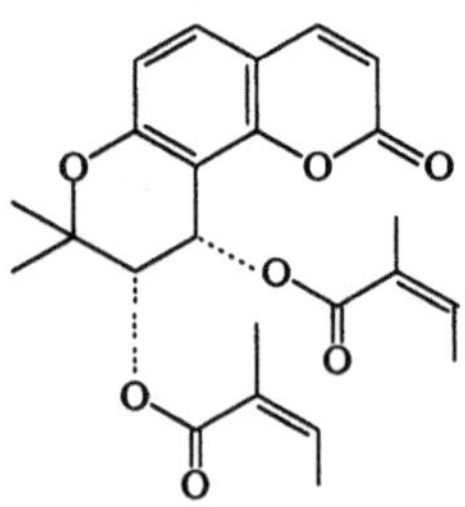

Formula	M. p.	$[\alpha]_\lambda^t$	Solvent	Plant Source(s)	Leading References
$C_{21}H_{22}O_7$	135—135.5	-39^{24} -45^{25}	EtOH $CHCl_3$	*Pteryxia terebinthina*	(*373, 430*)
$C_{23}H_{24}O_7S$	161.5—163	-121.2^{21}	$CHCl_3$	*Seseli sessiliflorum*	(*507, 509, 510*)
$C_{24}H_{26}O_7$	173—174	-37^{27} -78.4^{27}	$CHCl_3$ EtOH	*Angelica anomala*	(*73, 260, 261*)
$C_{24}H_{26}O_7$	171—172	34.6^{24}	$CHCl_3$	*Peucedanum formosanum*	(*262, 373*)

Table 1.2 (continued)

Trivial Name(s)	Year iso- lated	Structure
(232) Calipteryxin	1967	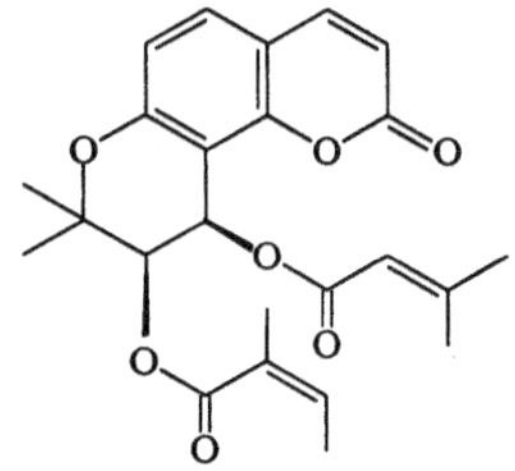
(233) Peuformosin	1966	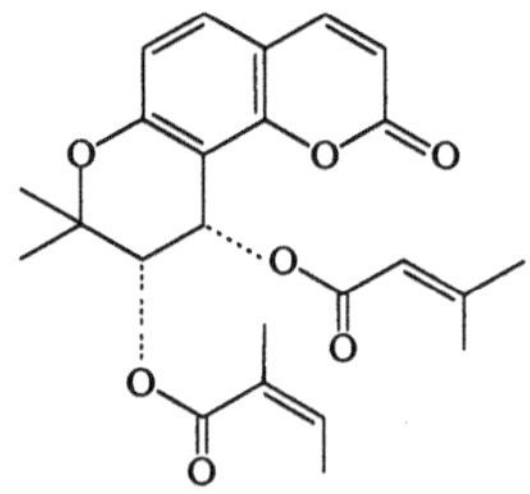
(234)	1970	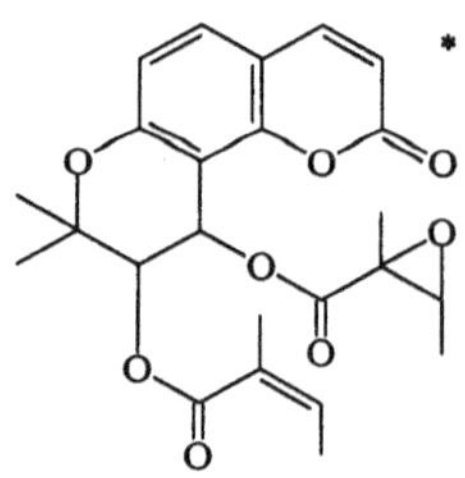
(235) Samidin	1952	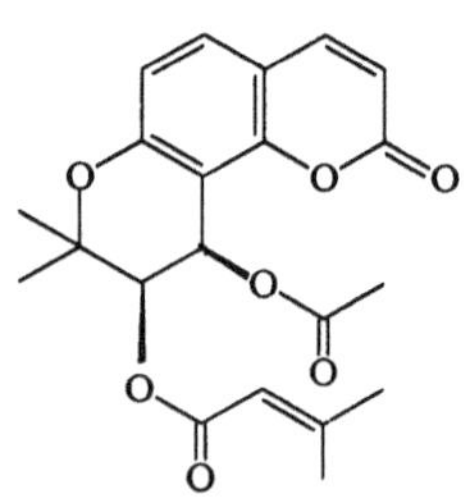

Formula	M. p.	$[\alpha]^t_\lambda$	Solvent	Plant Source(s)	Leading References
$C_{24}H_{26}O_7$	147.5—148	-55^{25}	$CHCl_3$	*Pteryxia terebinthina*	(*373, 430*)
$C_{24}H_{26}O_7$	155—156	67.3^{27}	$CHCl_3$	*Peucedanum formosanum*	(*262, 263, 373*)
$C_{24}H_{26}O_8$				*Laserpitium archangelica*	(*87*)
$C_{21}H_{22}O_7$	135—137	100	dioxan	*Ammi visnaga*	(*65, 66, 373, 516, 538, 539, 540*)

Table 1.2 (continued)

Trivial Name(s)	Year iso- lated	Structure
(236)	1971	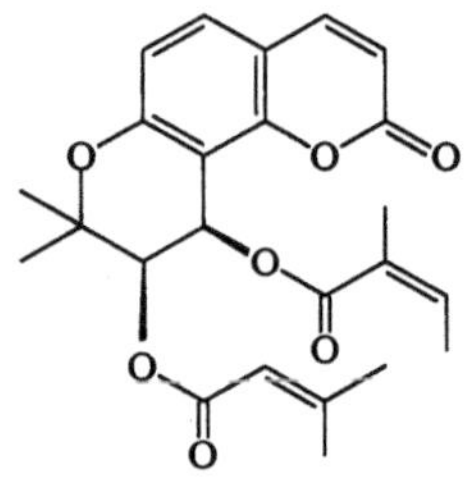
(237)	1966	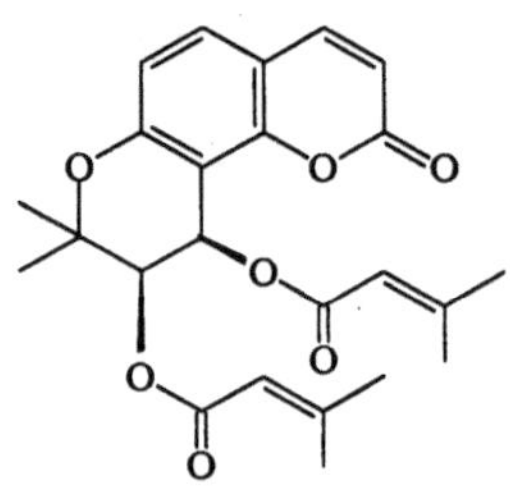
(238)	1974	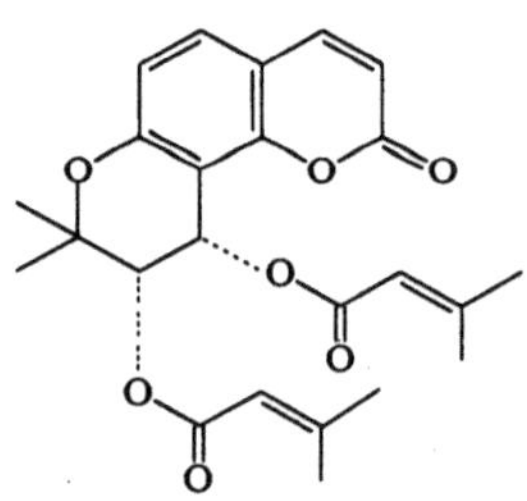
(239)	1970	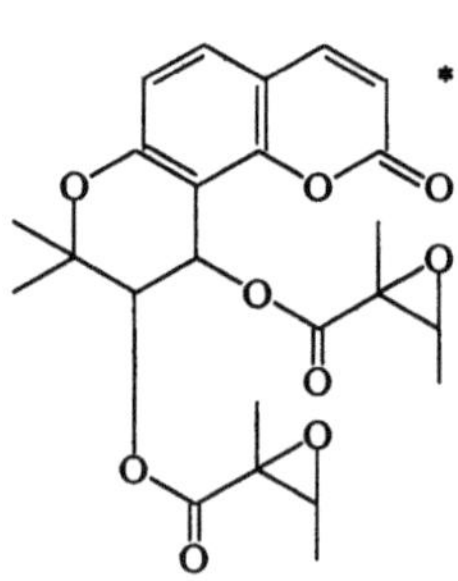

Formula	M. p.	$[\alpha]^t_\lambda$	Solvent	Plant Source(s)	Leading References
$C_{24}H_{26}O_7$	78—79.5	17.9^{22}	$CHCl_3$	*Seseli gummiferum*	(*423*)
$C_{24}H_{26}O_7$	108—108.5	15.8^{25}	$CHCl_3$	*Seseli libanotis*	(*371, 373*)
$C_{24}H_{26}O_7$	112—113	−47.7	$CHCl_3$	*Peucedanum japonicum*	(*599*)
$C_{24}H_{26}O_9$	179			*Laserpitium archangelica*	(*87*)

Table 1.2 (continued)

Trivial Name(s)	Year iso-lated	Structure
(240)	1970	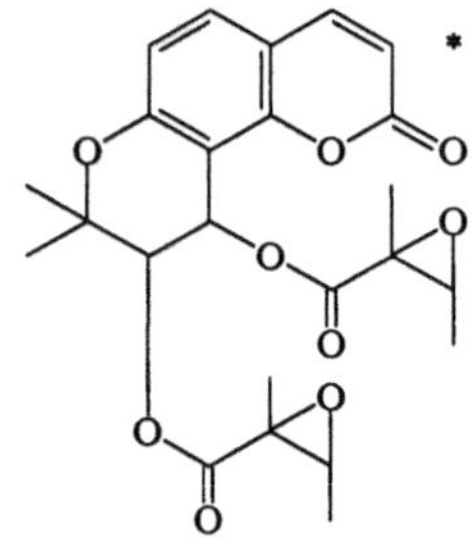
(241) Seselin	1930	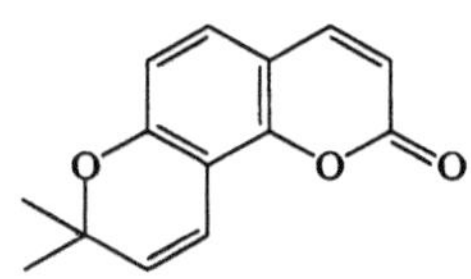

Table 1.3 *6,8-Disubstituted-7-Oxygenated Coumarins*

Trivial Name(s)	Year iso-lated	Structure
(242) Swietenone	1975	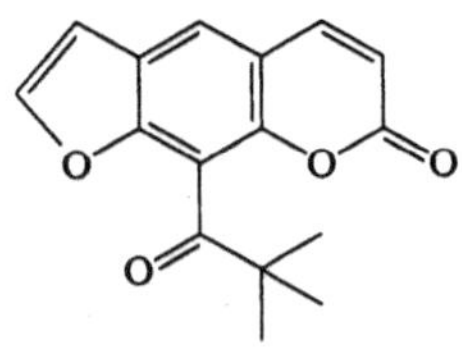

Formula	M. p.	$[\alpha]_{\lambda}^{t}$	Solvent	Plant Source(s)	Leading References
$C_{24}H_{26}O_9$	211			*Laserpitium archangelica*	(*87*)
$C_{14}H_{12}O_3$	119—120			*Skimmia repens* *Seseli indicum*	(*153*)

Formula	M. p.	$[\alpha]_{\lambda}^{t}$	Solvent	Plant Source(s)	Leading References
$C_{16}H_{14}O_4$	160—161			*Chloroxylon swietenia*	(*393*)

Table 2. *5,7-Dioxygenated Coumarins*

	Trivial Name(s)	Year iso-lated	Structure
(243)	Limettin Citropten	1890 1901	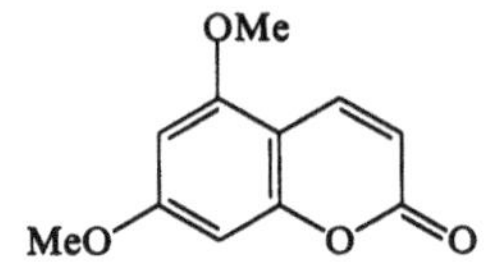
(244)	Lacoumarin	1976	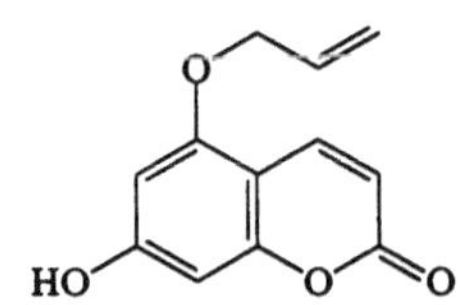
(245)		1957	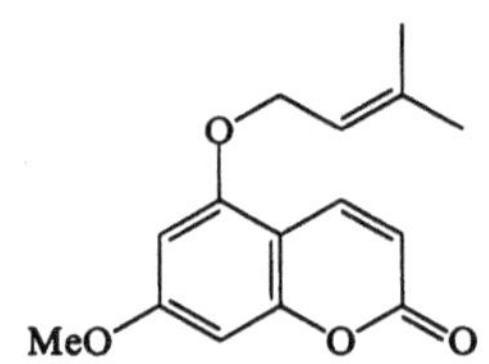
(246)		1945	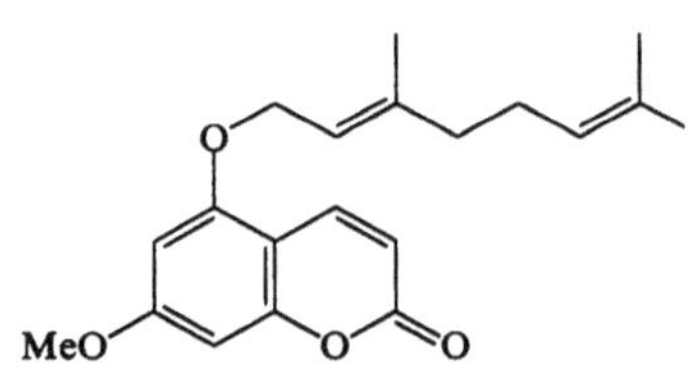
(247)		1971	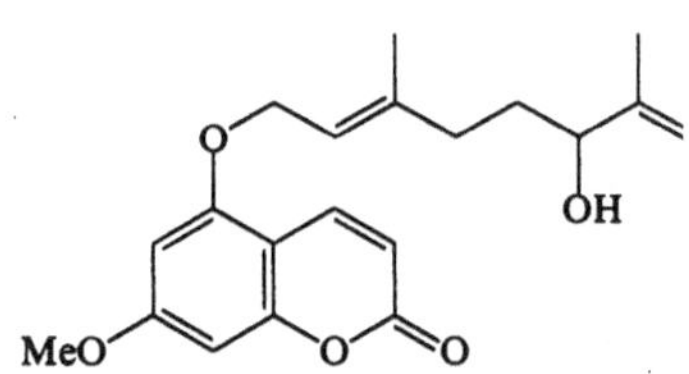
(248)	Toddaculin	1961	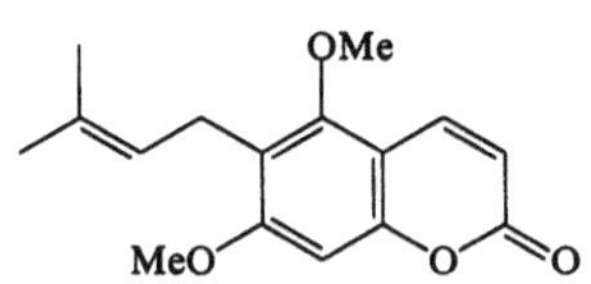

Formula	M. p.	$[\alpha]^t_\lambda$	Solvent	Plant Source(s)	Leading References
$C_{11}H_{10}O_4$	146—147.5			*Citrus limetta*	(*153*)
$C_{12}H_{10}O_4$	162—164			*Lawsonia inermis*	(*71*)
$C_{15}H_{16}O_4$	90—92			*Citrus limonum*	(*399, 549*)
$C_{20}H_{24}O_4$	86—87			*Citrus aurantifolia*	(*153*)
$C_{20}H_{24}O_5$				*Citrus aurantifolia*	(*186*)
$C_{16}H_{18}O_4$	95			*Toddalia aculeata*	(*140, 379, 399, 404*)

Table 2 (continued)

	Trivial Name(s)	Year iso-lated	Structure
(**249**)	Aculeatin	1942	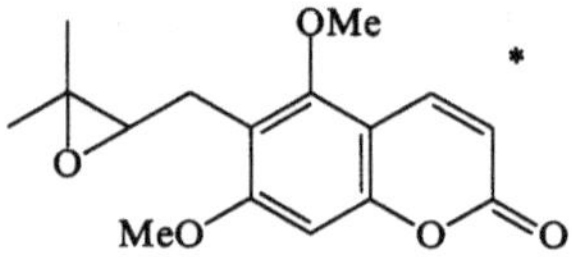
(**250**)	Toddalolactone	1933	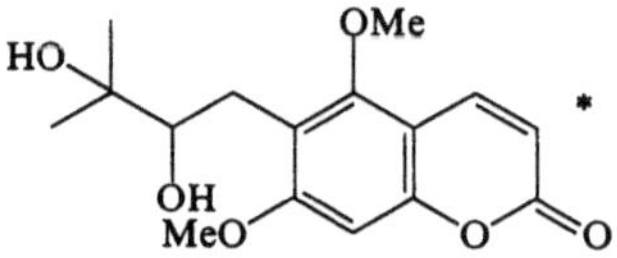
(**251**)	Bergaptol	1934	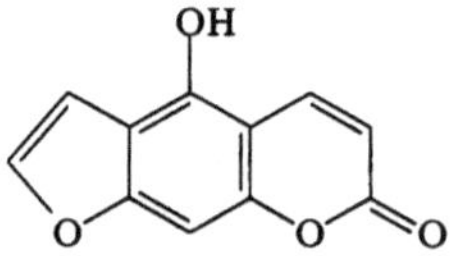
(**252**)	Bergapten Heraclin Majudin	1839 1879 1948	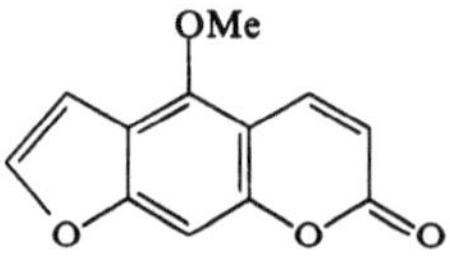
(**253**)	Pranferol	1966	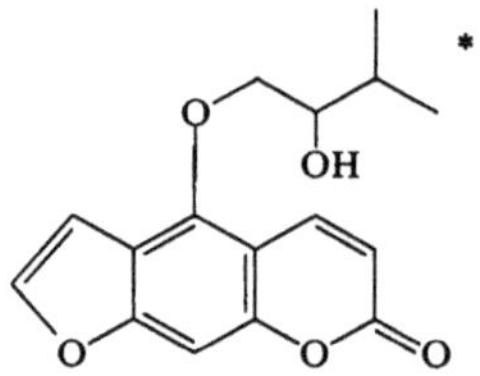
(**254**)	Isoimperatorin Cnidin	1933 1960	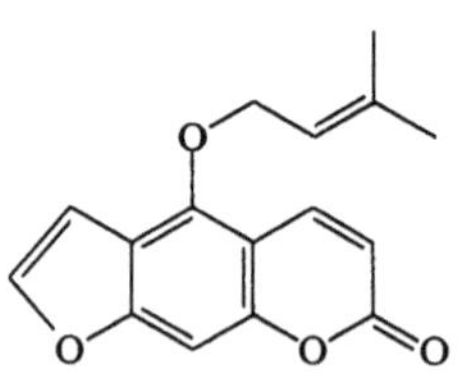

Formula	M. p.	$[\alpha]^t_\lambda$	Solvent	Plant Source(s)	Leading References
$C_{16}H_{18}O_5$	113	-16.8^{24}	EtOAc	*Toddalia aculeata*	(*153*)
$C_{16}H_{20}O_6$	131—132.5	55.9^{30}	$CHCl_3$	*Toddalia aculeata*	(*153*)
$C_{11}H_6O_4$	276—282			*Citrus bergamia*	(*153*)
$C_{12}H_8O_4$	188—191			*Citrus bergamia* *Heracleum giganteum* *Ammi majus*	(*21, 153*) (*302*) (*302*)
$C_{16}H_{16}O_5$	108.5—110	-19.38^{19}	$CHCl_3$	*Prangos ferulacea*	(*13, 347*)
$C_{16}H_{14}O_4$	109 108—110			*Imperatoria ostruthium* *Cnidium dubium*	(*153*) (*135*)

Table 2 (continued)

Trivial Name(s)	Year iso- lated	Structure
(**255**) (±)-Oxypeucedanin	1839	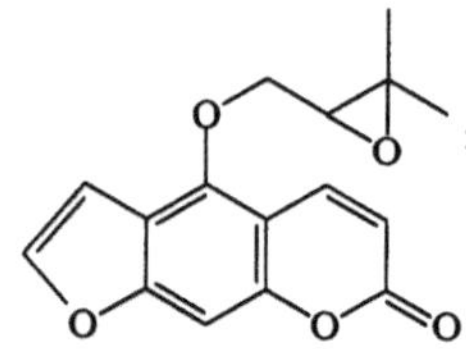 racem
(**256**) Prangolarin (+)-Oxypeucedanin	1963 1964	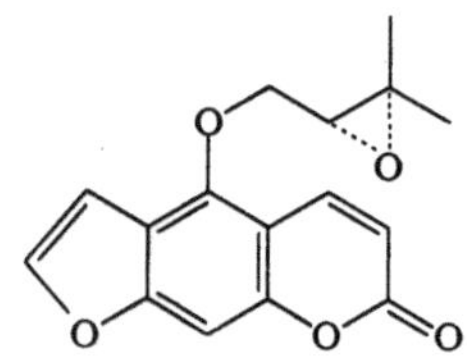
(**257**) (S)-(−)-Oxypeucedanin	1971	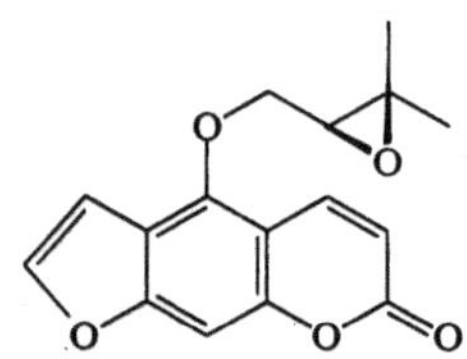
(**258**) Pangeline Pabulenol Gosferol	1971 1971 1972	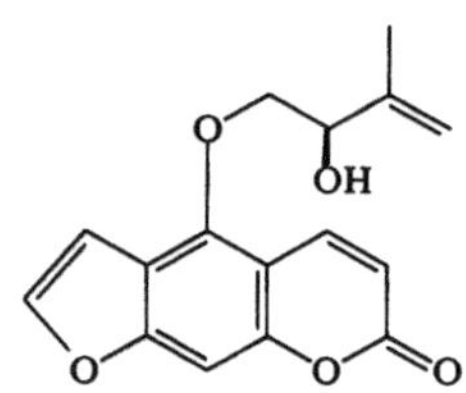
(**259**) Angeloylpangeline	1971	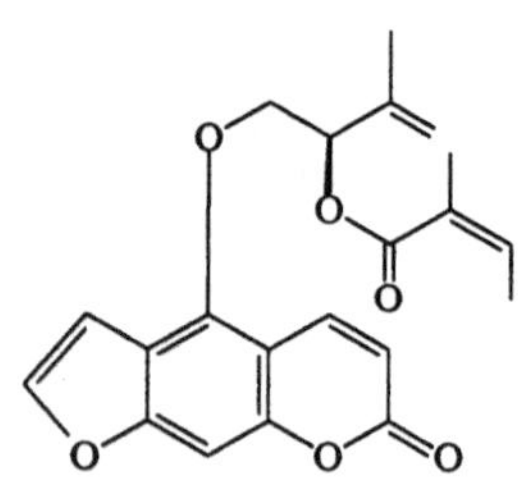

Formula	M.p	$[\alpha]_\lambda^t$	Solvent	Plant Source(s)	Leading References
$C_{16}H_{14}O_5$	142—143			*Peucedanum officinale*	(*153*)
$C_{16}H_{14}O_5$	104—105	20.1^{30}	$CHCl_3$	*Prangos pabularia*	(*210*)
	104	17^{27}	$CHCl_3$	*Peucedanum palustre*	(*167, 425, 429*)
$C_{16}H_{14}O_5$	103—104	-14^{20}	$CHCl_3$	*Ligusticum seguieri*	(*375*)
$C_{16}H_{14}O_5$	122—123	11^{20}		*Angelica pancicii*	(*451, 452*)
	134—135	-3.8^{25}	EtOH	*Prangos pabularia*	(*61, 125*)
	136.5—138.5	$\pm 0^{20}$	EtOH	*Prangos ferulacea*	(*16*)
$C_{21}H_{20}O_6$	87—88	-27^{20}		*Angelica pancicii*	(*451*)

Table 2 (continued)

	Trivial Name(s)	Year isolated	Structure
(260)	Isooxypeucedanin	1963	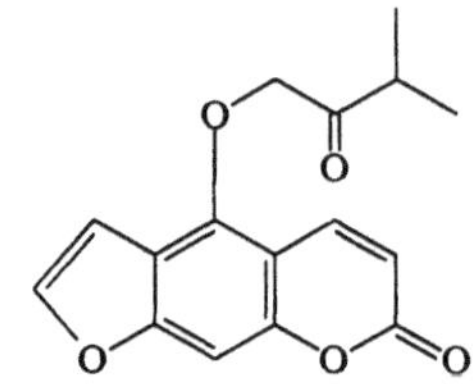
(261)	Pabulenone	1975	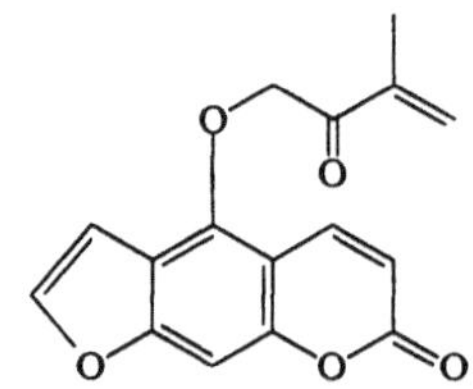
(262)	Prangol Oxypeucedanin hydrate	1964	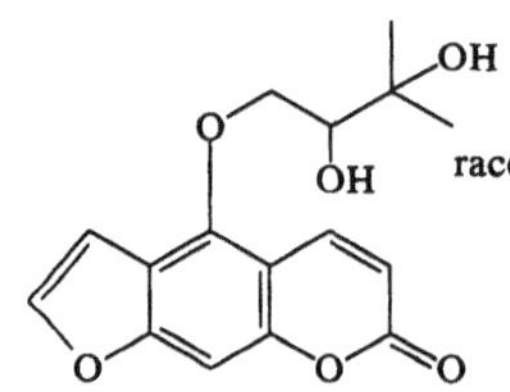 racem
(263)	Aviprin (+)-Oxypeucedanin hydrate	1964 1964 1964	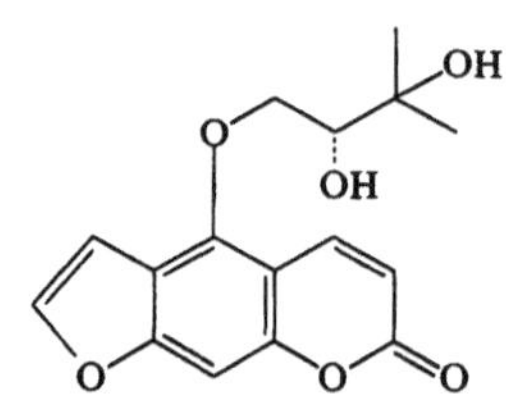
(264)	(+)-Oxypeucedanin methanolate	1972	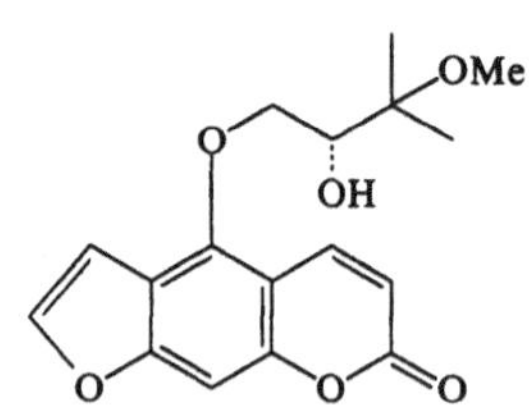

Formula	M.p.	$[\alpha]_\lambda^t$	Solvent	Plant Source(s)	Leading References
$C_{16}H_{14}O_5$	146—148			*Angelica sylvestris*	(*273, 428*)
$C_{16}H_{12}O_5$	119—121			*Peucedanum ostruthium*	(*482*)
$C_{16}H_{16}O_6$	132—134			*Prangos fedtschenkoi*	(*349, 356*)
				Angelica sylvestris	(*116*)
$C_{16}H_{16}O_6$	135—136	-4.36^{24}	pyridine	*Prangos pabularia*	(*247*)
	131—132	15.26^{14}	acetone	*Angelica gmelini*	(*448*)
	131.5—132	18^{24}	acetone	*Peucedanum palustre*	(*425, 429*)
$C_{17}H_{18}O_6$	123—125	12	$CHCl_3$	*Skimmia japonica*	(*37*)

Table 2 (continued)

Trivial Name(s)	Year iso- lated	Structure
(**265**) *Tert-O*-Methyl oxypeucedanin hydrate Alatol	1972 1973	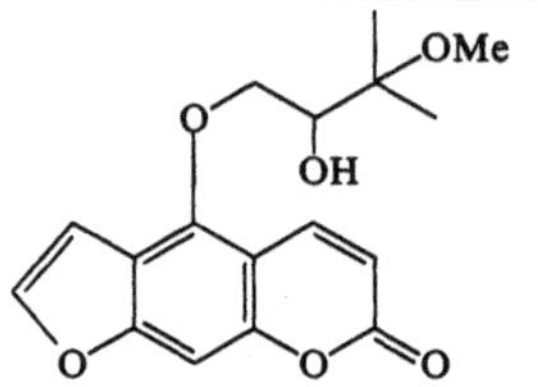
(**266**) Saxalin	1971 1972	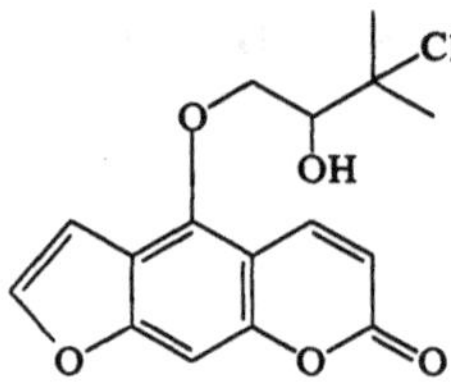
(**267**) Oxypeucedanin D-glucoside	1975	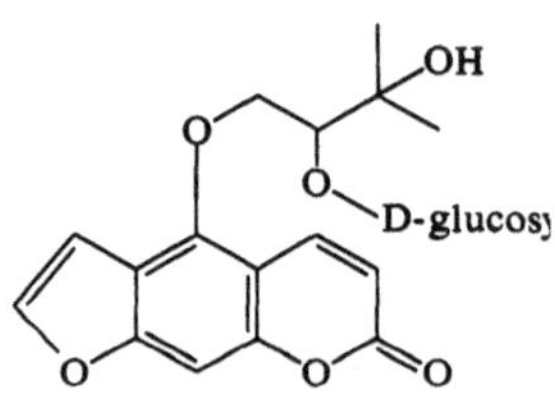
(**268**)	1968	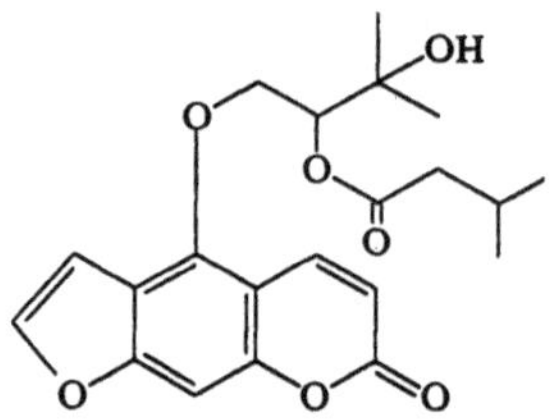
(**269**) Ostruthol	1909	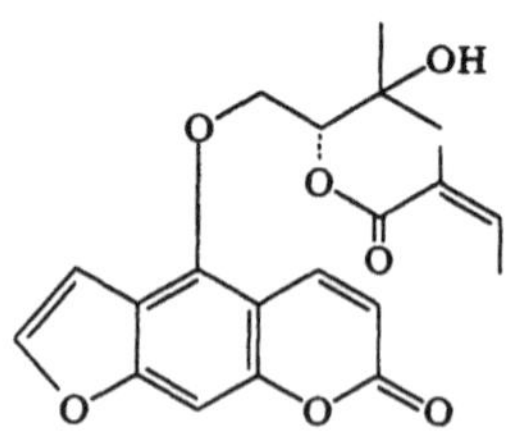

Formula	M. p.	$[\alpha]^t_\lambda$	Solvent	Plant Source(s)	Leading References
$C_{17}H_{18}O_6$	119—120			*Angelica glauca*	(*299*)
	125—127			*Prangos alata*	(*11, 12*)
$C_{16}H_{15}ClO_5$	159—161			*Angelica saxatilis*	(*40*)
		-15.5^{20}	acetone	*Cachrys pubescens*	(*275*)
$C_{22}H_{26}O_{11}$	210—220			*Peucedanum ostruthium*	(*303*)
$C_{21}H_{24}O_7$				*Angelica ursina*	(*73*)
$C_{21}H_{22}O_7$	136—137	8.36^{22}	acetone	*Imperatoria ostruthium*	(*131, 153, 429*)
		-18.3^{15}	pyridine	(= *Peucedanum ostruthium*)	

Table 2 (continued)

	Trivial Name(s)	Year iso-lated	Structure
(270)	Bergamottin Bergaptin	1937 1901	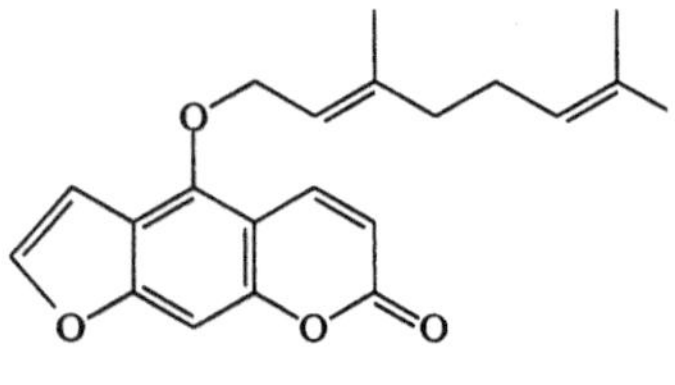
(271)	Archangelin Iselin	1964 1974	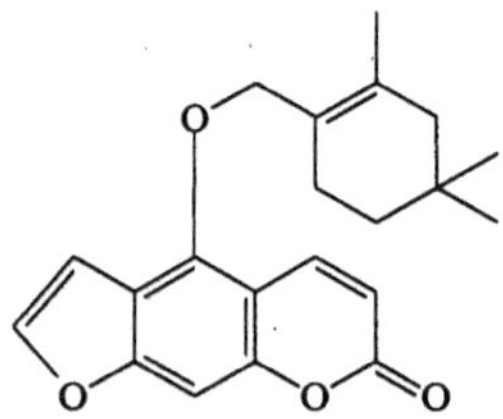
(272)		1965	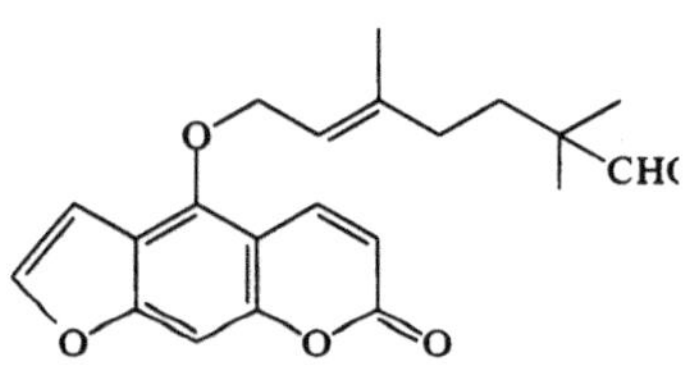
(273)		1973	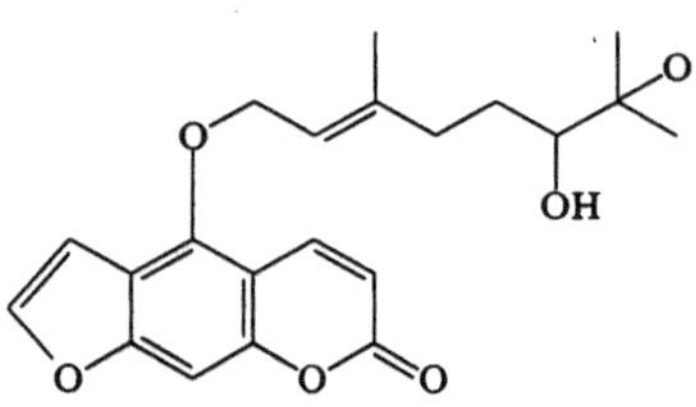
(274)	Clausenin	1966	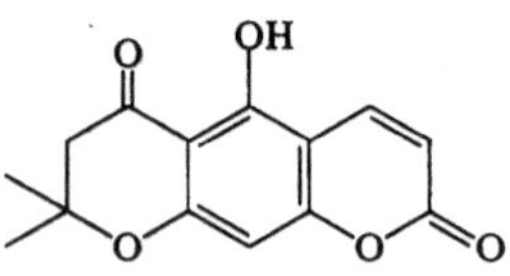
(275)	Eriobrucinol	1973	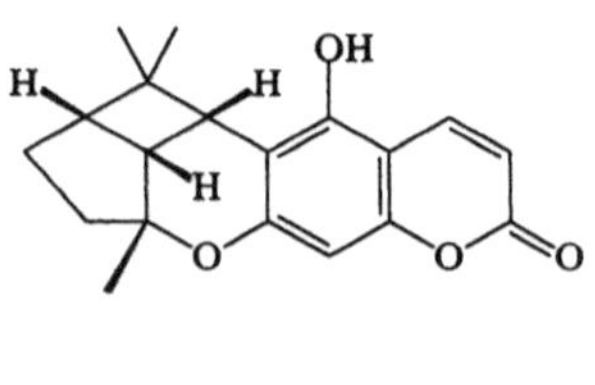

Formula	M.p.	$[\alpha]^t_\lambda$	Solvent	Plant Source(s)	Leading References
$C_{21}H_{22}O_4$	59—61			*Citrus bergamia*	(*63*, *153*)
$C_{21}H_{22}O_4$	132 127—128			*Archangelica officinalis* *Seseli iliense*	(*62*, *131*, *133*) (*173*)
$C_{21}H_{22}O_5$	134—136			*Citrus paradisi*	(*195*, *196*)
$C_{21}H_{24}O_6$	104—106	12.7	$CHCl_3$	*Citrus macroptera*	(*168*)
$C_{14}H_{12}O_5$	156—157			*Clausena heptaphylla*	(*207*, *287*, *291*)
$C_{19}H_{20}O_4$	185	−310	pyridine	*Eriostemon brucei*	(*284*)

Table 2 (continued)

Trivial Name(s)	Year iso-lated	Structure
(276) Hydroxyeriobrucinol	1973	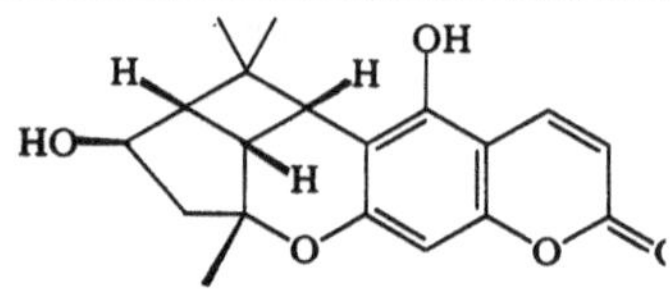
(277) Deoxybruceol	1971	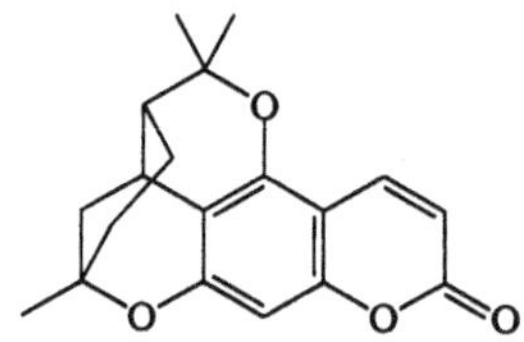
(278) Bruceol	1963	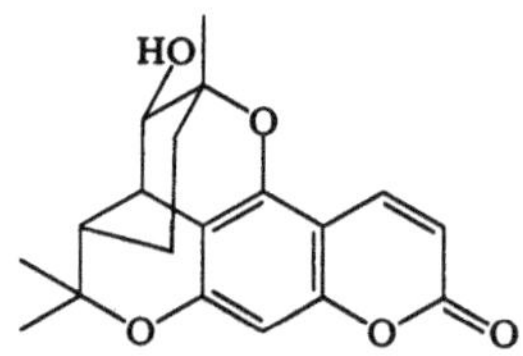
(279) Alloxanthoxyletin	1937	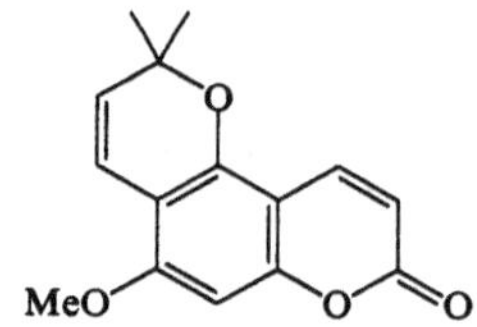
(280) Xanthoxyletin	1829	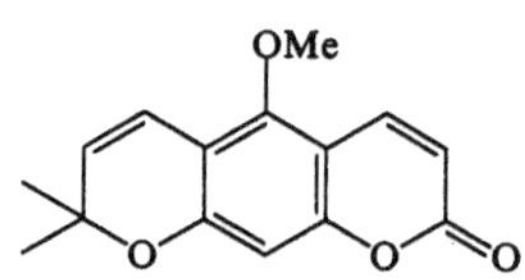
(281) Pinnarin	1970	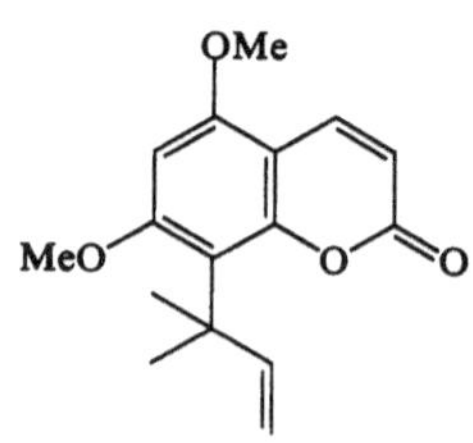

References, pp. 400—429

Formula	M.p.	$[\alpha]_\lambda^t$	Solvent	Plant Source(s)	Leading References
$C_{19}H_{20}O_5$	243			*Eriostemon brucei*	(*284*)
$C_{19}H_{20}O_4$	117—123	-48.19_{349}		*Eriostemon brucei*	(*64, 150*)
$C_{19}H_{20}O_5$	201	-297	$CHCl_3$	*Eriostemon brucei*	(*64, 172*)
$C_{15}H_{14}O_4$	115.5			*Xanthoxylum americanum*	(*153, 207*)
$C_{15}H_{14}O_4$	132			*Xanthoxylum americanum*	(*153, 207, 287*)
$C_{16}H_{18}O_4$	162—163			*Ruta pinnata*	(*398, 492*)

Table 2 (continued)

	Trivial Name(s)	Year isolated	Structure
(**282**)	*O*-Isopentenyllimettin Coumurrayin	1968 1968	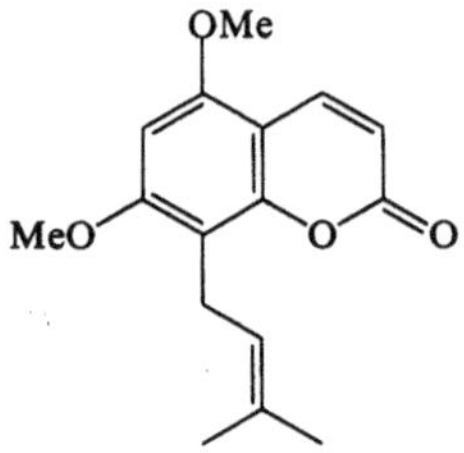
(**283**)	Isoaculeatin Sibiricin	1968 1968	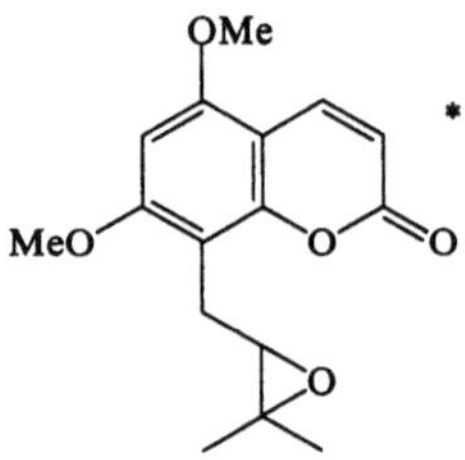
(**284**)		1967	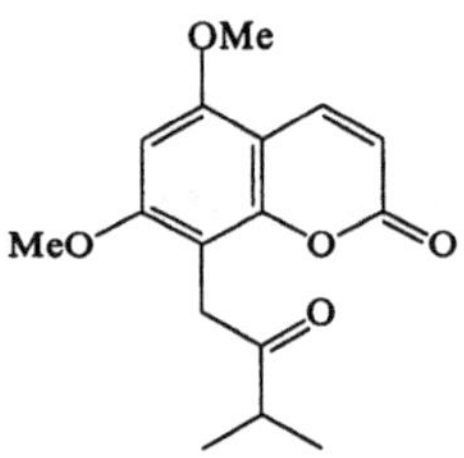
(**285**)	Glabralactone Angelicone	1956 1956	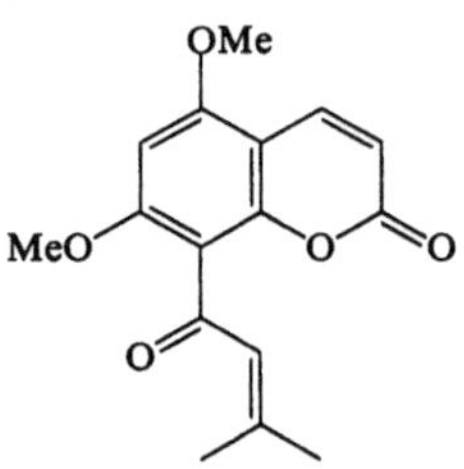
(**286**)	Mexoticin	1967	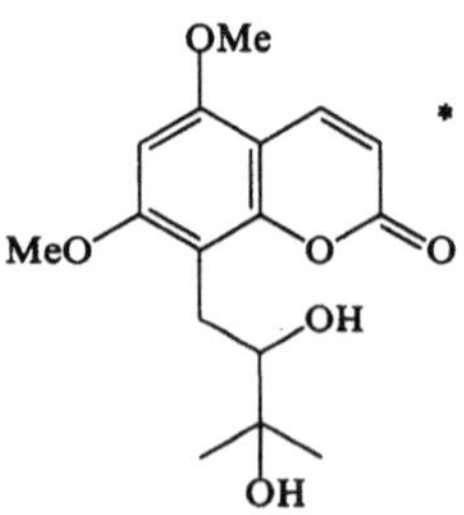

Formula	M.p.	$[\alpha]^t_\lambda$	Solvent	Plant Source(s)	Leading References
$C_{16}H_{18}O_4$	157—158 157			*Murraya paniculata*	(*165, 380, 400, 401, 481, 575*)
$C_{16}H_{18}O_5$	145.5 152—153	60^{22}_{578} 60.4^{30}	$CHCl_3$ dioxan	*Libanotis buchtormensis* *Seseli sibiricum*	(*85*) (*39*)
$C_{16}H_{18}O_5$	128—130			*Severinia buxifolia*	(*164*)
$C_{16}H_{16}O_5$	129—130 130			*Angelica glabra* *Angelica ursina*	(*251, 300, 391, 406*) (*201, 391*)
$C_{16}H_{20}O_6$	185	37.6	$CHCl_3$	*Murraya exotica*	(*123*)

Table 2 (continued)

Trivial Name(s)	Year isolated	Structure
(**287**)	1968	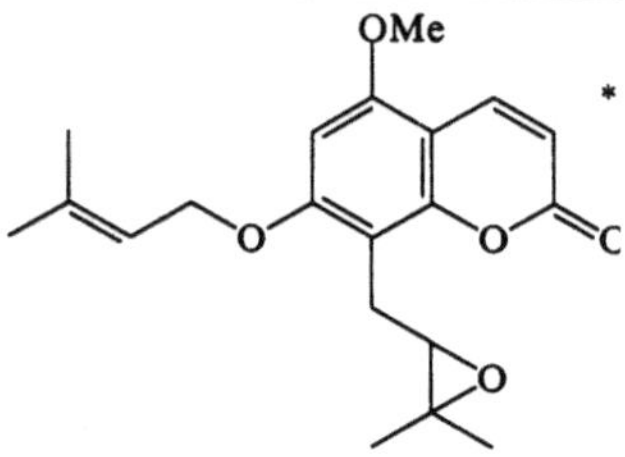
(**288**) Sesibiricin	1970	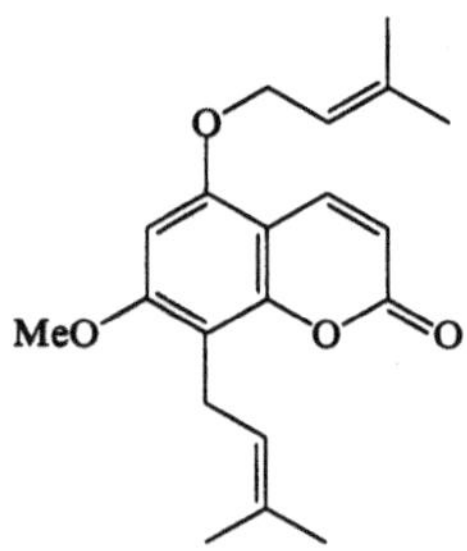
(**289**) Isobergapten	1932	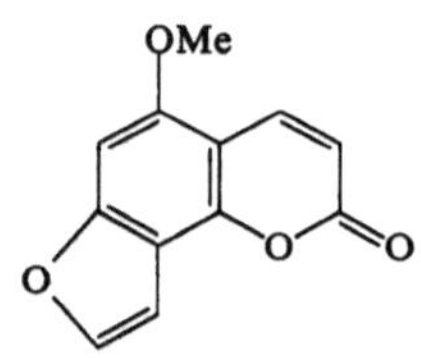
(**290**)	1976	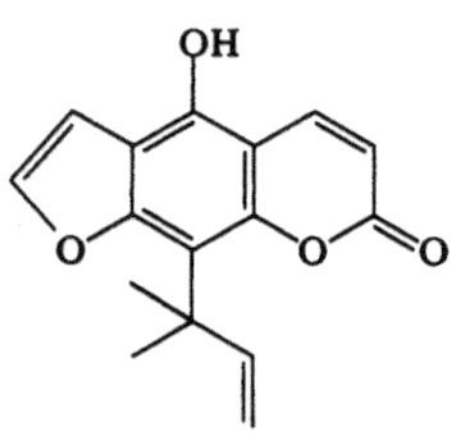
(**291**) Alloisoimperatorin	1971	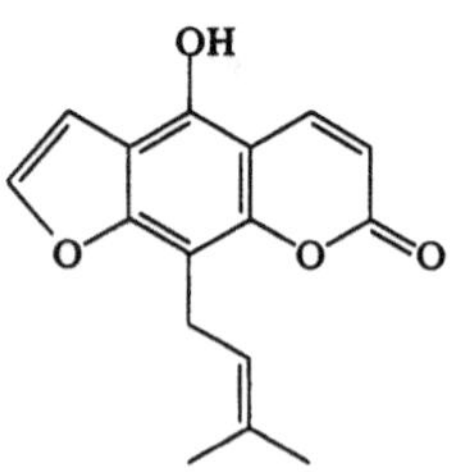

Formula	M.p.	$[\alpha]_{\lambda}^{t}$	Solvent	Plant Source(s)	Leading References
$C_{20}H_{24}O_5$	105.5	52_{578}^{22}	$CHCl_3$	*Libanotis buchtormensis*	(*85*)
$C_{20}H_{24}O_4$	120—122			*Seseli sibiricum*	(*399, 405, 523*)
$C_{12}H_8O_4$	218—222			*Pimpinella saxifraga*	(*153*)
$C_{16}H_{14}O_4$	213—215			*Peucedanum stenocarpum*	(*217*)
$C_{16}H_{14}O_4$	228—230			*Angelica dahurica*	(*501*)

Table 2 (continued)

Trivial Name(s)	Year iso-lated	Structure
(292) Furopinnarin	1970	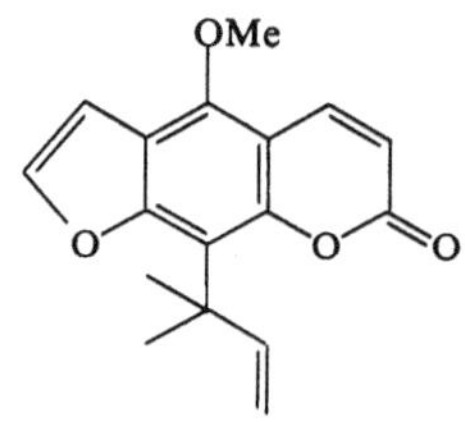
(293) Hortiolone	1976	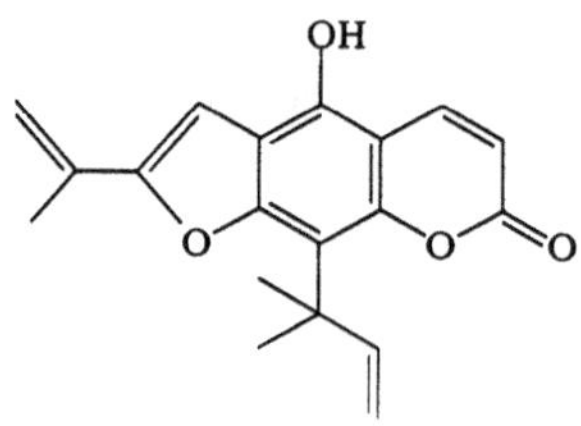
(294) Dipetaline	1976	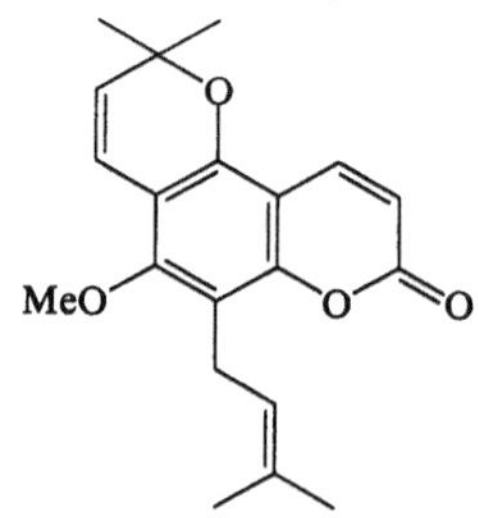
(295) Avicennol	1975	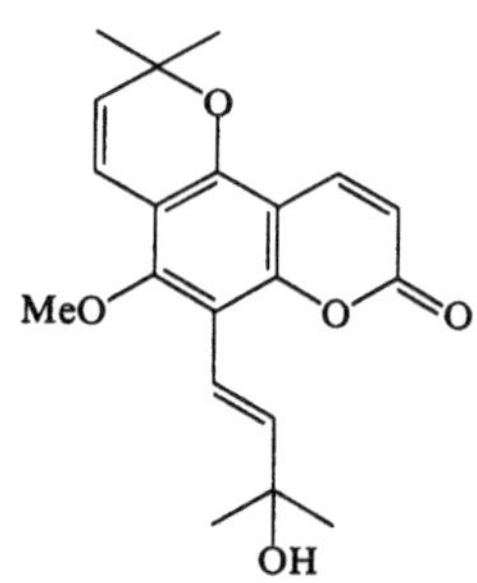
(296) *Cis*-Avicennol	1977	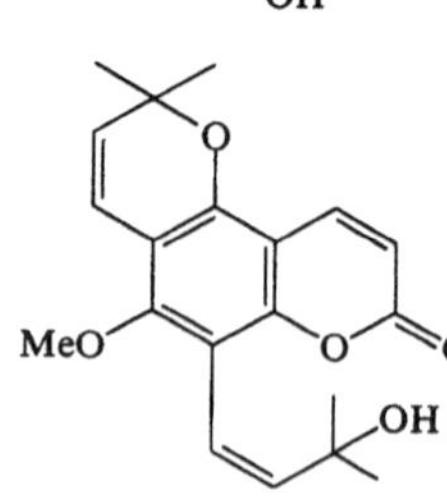

Formula	M. p.	$[\alpha]_{\lambda}^{t}$	Solvent	Plant Source(s)	Leading References
$C_{17}H_{16}O_4$	124—125			*Ruta pinnata*	(*226*, *492*)
$C_{19}H_{18}O_4$	226—227			*Hortia arborea*	(*158*)
$C_{20}H_{22}O_4$				*Zanthoxylum dipetalum*	(*193*, *402*)
$C_{20}H_{22}O_5$	124.5—125			*Zanthoxylum avicennae*	(*239*, *402*)
$C_{20}H_{22}O_5$	oil			*Zanthoxylum elephantiasis*	(*240*)

Table 2 (continued)

	Trivial Name(s)	Year isolated	Structure
(297)	Avicennin	1960	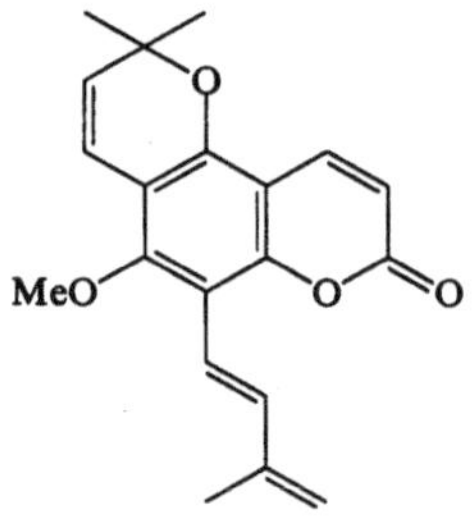
(298)	Dipetalolactone Hortiline	1976 1976	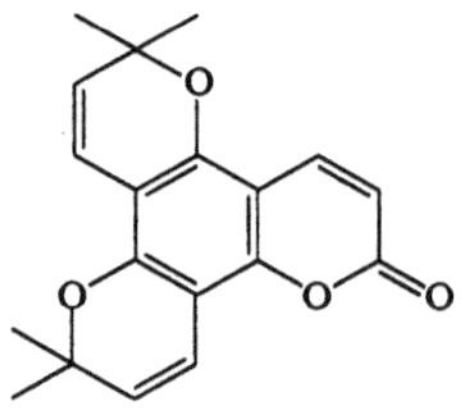
(299)	Clausenidin	1966	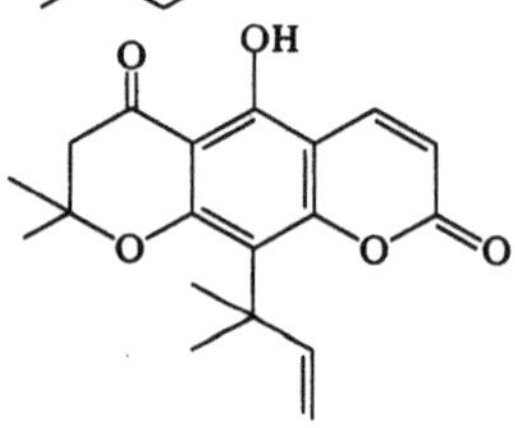
(300)	Nordentatin	1968	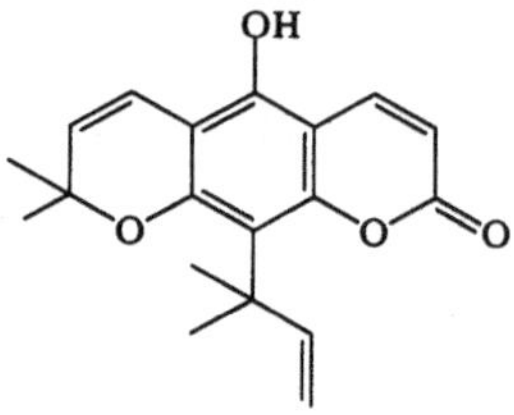
(301)	Trachyphyllin	1969	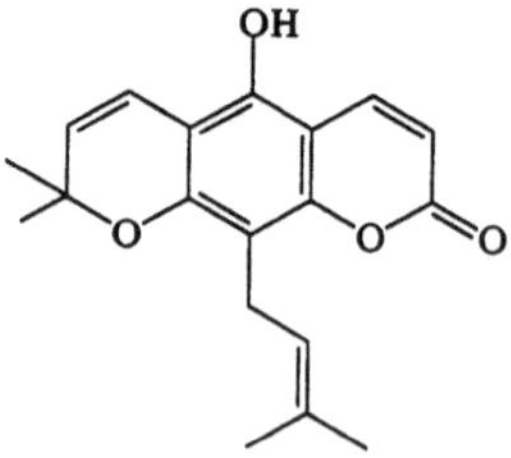
(302)	Poncitrin Dentatin	1969 1968	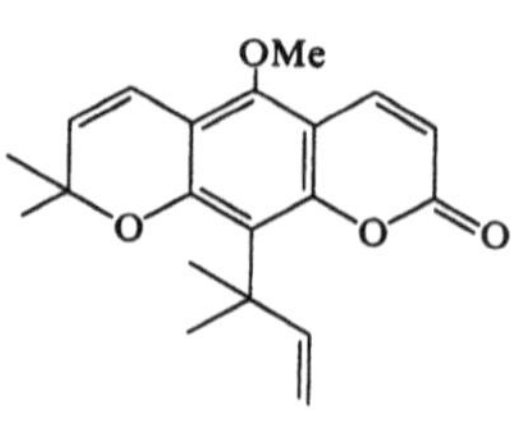

References, pp. 400–429

Formula	M. p.	$[\alpha]^t_\lambda$	Solvent	Plant Source(s)	Leading References
$C_{20}H_{20}O_4$	141—142			*Zanthoxylum avicennae*	(*35, 36, 239*)
$C_{19}H_{18}O_4$	119—120			*Zanthoxylum dipetalum*	(*193, 194, 402*)
	138			*Hortia arborea*	(*158*)
$C_{19}H_{20}O_5$	135—136			*Clausena heptaphylla*	(*200, 287, 291*)
$C_{19}H_{20}O_4$	182			*Clausena dentata*	(*200, 236*)
$C_{19}H_{20}O_4$	213—214			*Eriostemon trachyphyllus*	(*364*)
$C_{20}H_{22}O_4$	93—94			*Poncirus trifoliata*	(*379, 391, 577, 578, 579*)
	95			*Clausena dentata*	(*200, 236, 391*)

Table 3. *6,7-Dioxygenated Coumarins*

Trivial Name(s)	Year isolated	Structure
(303) Aesculetin (aglycone)	1853 1890	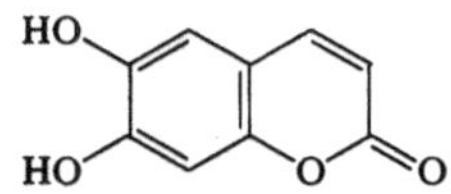
(304) Cichoriin	1876	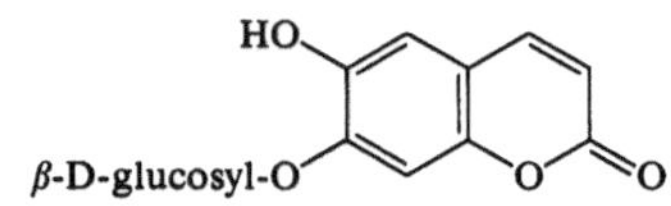
(305) Isoscopoletin	1975 1966	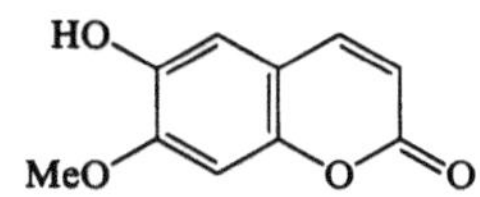
(306) Prenyletin	1967	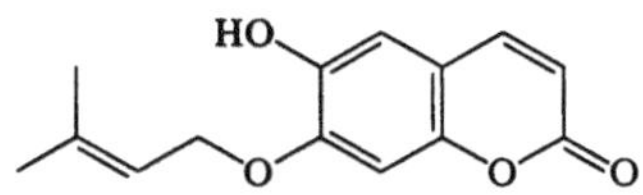
(307) Aesculin	1831	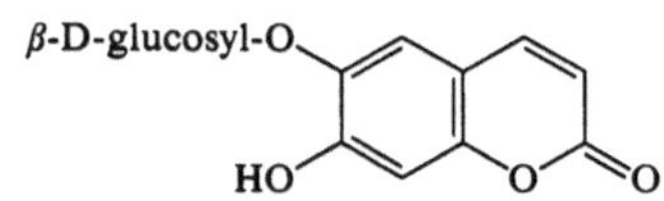
(308) Magnolioside	1968	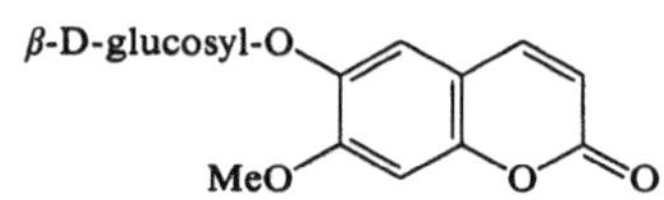
(309) Baisseoside	1970	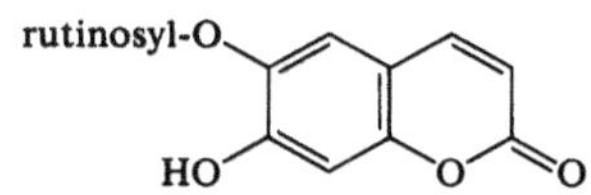
(310) Scopoletin	1884	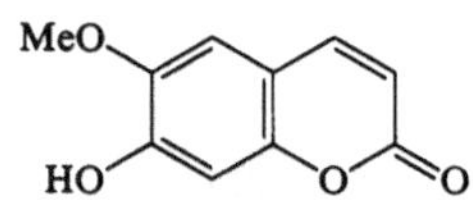
(311) Scopolin	1884	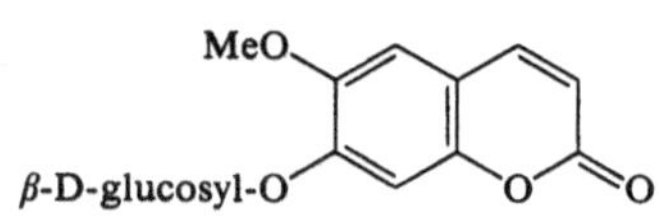
(312) Fabiatrin	1927	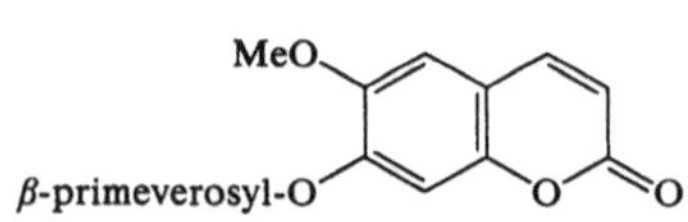

Formula	M. p.	$[\alpha]_{\lambda}^{t}$	Solvent	Plant Source(s)	Leading References
$C_9H_6O_4$	268—272			*Euphorbia lathyris*	(*153*)
$C_{15}H_{16}O_9$	216	-104.5^{18}	dioxan	*Cichorium intybus*	(*153*)
$C_{10}H_8O_4$	185			*Bupleurum frutiscens* *Artemisia messerschmidiana*	(*234, 267*) (*247a*)
$C_{14}H_{14}O_4$	145—146			*Ptaeroxylon obliquum*	(*50, 51, 155, 156*)
$C_{15}H_{16}O_9$	204—205	-146^{15}	MeOH	*Aesculus hippocastanum*	(*153, 267*)
$C_{16}H_{18}O_9$	227	-28	pyridine	*Magnolia macrophylla*	(*469*)
$C_{21}H_{26}O_{13}$	230—232	-166^{20}	MeOH	*Baissea leonensis*	(*472*)
$C_{10}H_8O_4$	204—205			*Scopolia japonica*	(*50, 153*)
$C_{16}H_{18}O_9$	217—219			*Scopolia japonica*	(*153*)
$C_{21}H_{26}O_{13}$	236—238	-140^{20}	H_2O	*Fabiana imbricata*	(*153*)

Table 3 (continued)

	Trivial Name(s)	Year iso- lated	Structure
(313)	Scoparone Scoparin	1954 1928	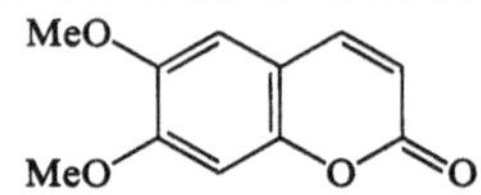
(314)		1967	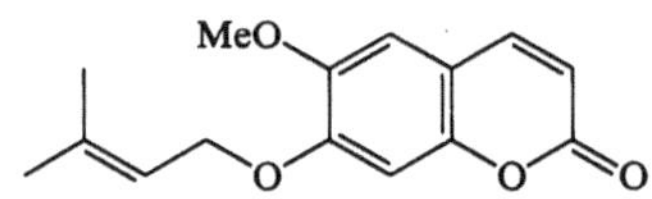
(315)		1970	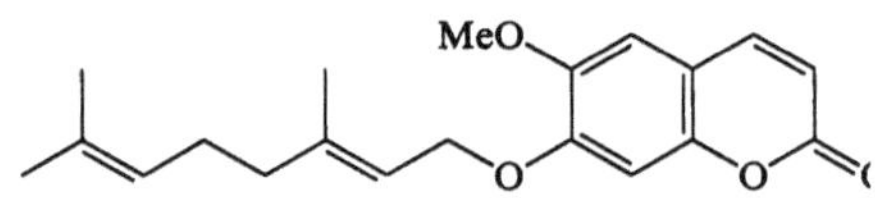
(316)	Pedicellone	1973	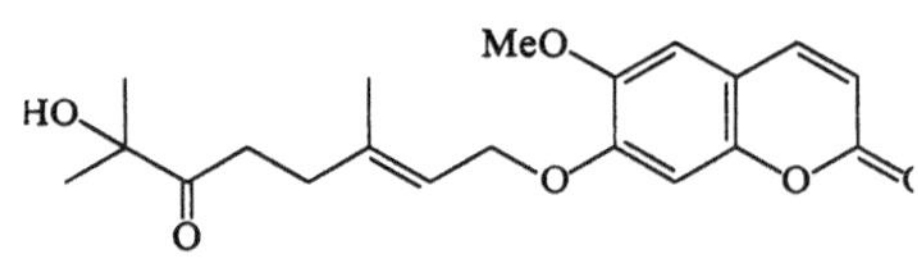
(317)	6-Methoxymarmin	1972	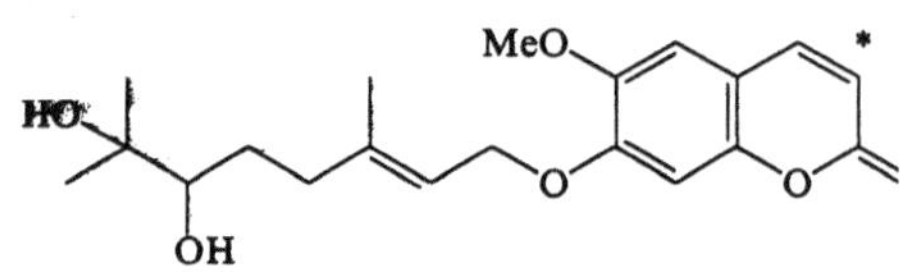
(318)	Ayapin	1937	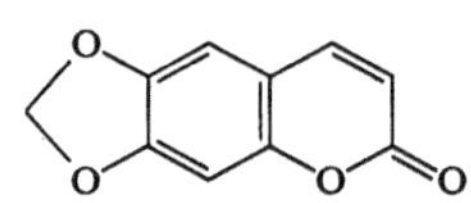
(319)	Obliquin	1966	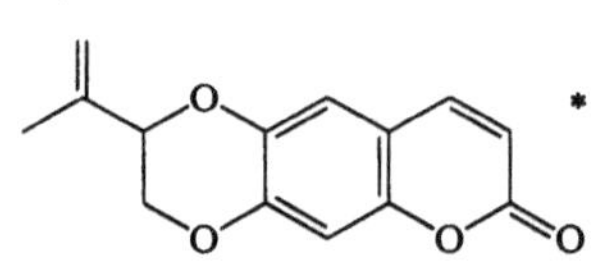
(320)	Obliquol	1967	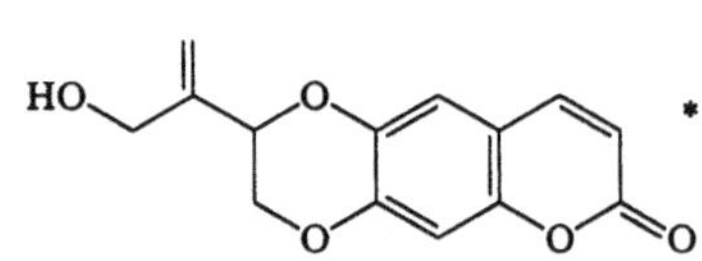
(321)	Bethancorol	1976	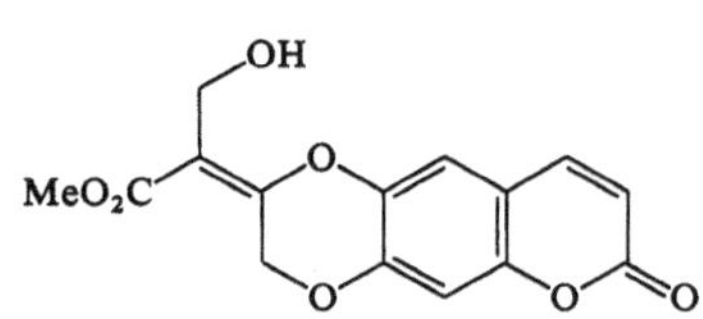

Formula	M. p.	$[\alpha]_\lambda^t$	Solvent	Plant Source(s)	Leading References
$C_{11}H_{10}O_4$	144—146			*Artemisia scoparia* *Zanthoxylum setosum*	(*302*) (*153*)
$C_{15}H_{16}O_4$	81—82			*Ptaeroxylon obliquum*	(*50, 51, 376*)
$C_{20}H_{24}O_4$	84—84.5			*Thapsia garganica*	(*353, 363*)
$C_{20}H_{24}O_6$	90—92			*Haplophyllum pedicellatum*	(*354*)
$C_{20}H_{26}O_6$	119—120	17.6^{19}	EtOH	*Haplophyllum pedicellatum*	(*353*)
$C_{10}H_6O_4$	231—232			*Eupatorium ayapana*	(*153*)
$C_{14}H_{12}O_4$	162	76^{25}	$CHCl_3$	*Ptaeroxylon obliquum*	(*72, 76, 155, 156, 157*)
$C_{14}H_{12}O_5$	191—192	77^{24}	$CHCl_3$	*Ptaeroxylum obliquum*	(*156*)
$C_{15}H_{12}O_7$	190—192			*Cneorum tricoccum*	(*231*)

Table 3 (continued)

	Trivial Name(s)	Year iso- lated	Structure
(**322**)	Obliquetol	1967	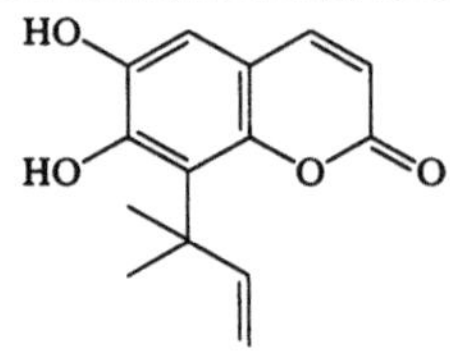
(**323**)	Obliquetin	1967	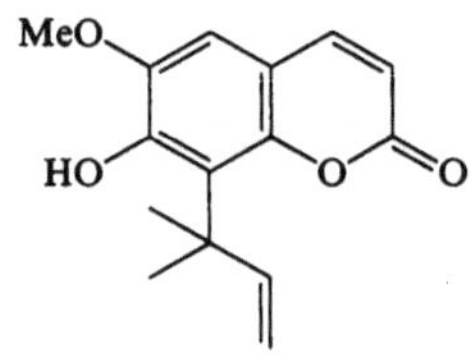
(**324**)	Cedrelopsin Cneorum-coumarin A	1968 1975	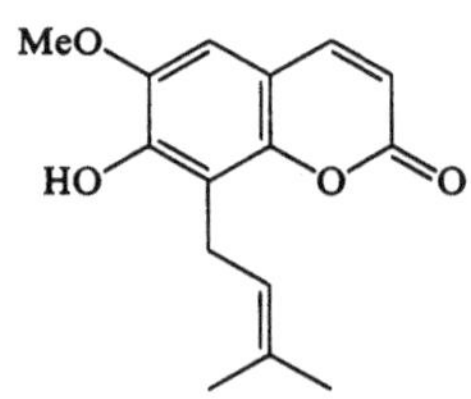
(**325**)	Brayleanin	1949	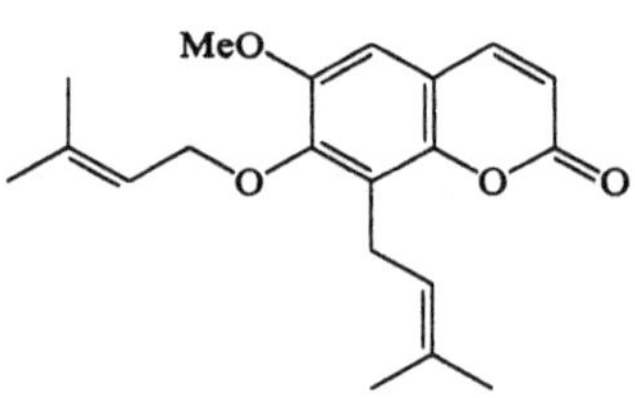
(**326**)	Cneorum-coumarin B	1975	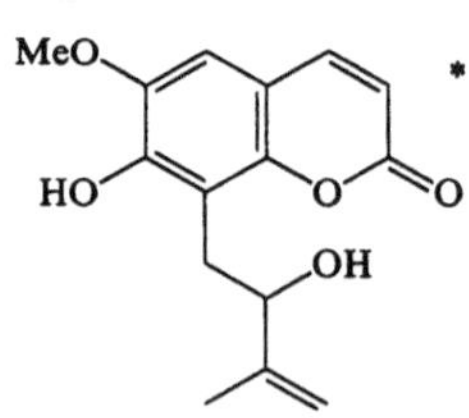 *
(**327**)			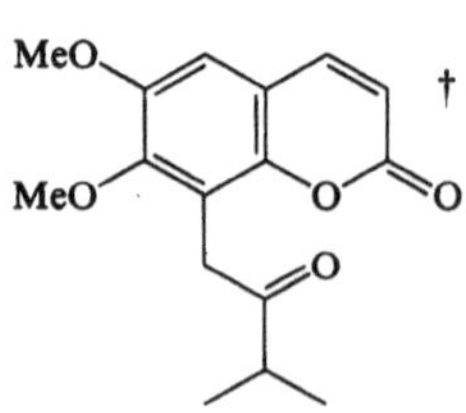 †

Formula	M. p.	$[\alpha]^t_\lambda$	Solvent	Plant Source(s)	Leading References
$C_{14}H_{14}O_4$	217—218			*Ptaeroxylon obliquum*	(*156*)
$C_{15}H_{16}O_4$	138—139			*Ptaeroxylon obliquum*	(*50, 51, 156*)
$C_{15}H_{16}O_4$	170—174 172—174			*Cedrelopsis grevei* *Cneorum pulverulentum*	(*184*) (*389*)
$C_{20}H_{24}O_4$	95			*Flindersia brayleyana*	(*153*)
$C_{15}H_{16}O_5$	175—177			*Cneorum pulverulentum*	(*388*)
$C_{16}H_{18}O_5$					(*218*)

† This compound was inadvertently reported as a natural product in reference (*218*). Professor González has informed me that it appeared through a printer's error.

Table 3 (continued)

	Trivial Name(s)	Year iso-lated	Structure
(**328**)	Nieshoutin Cyclooblliquetin	1967 1967	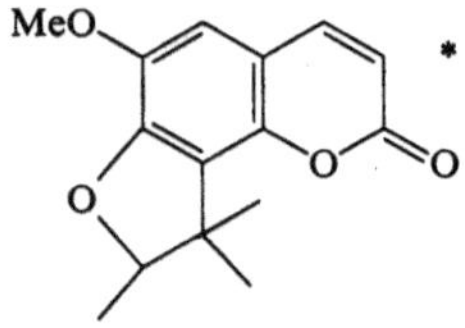 *
(**329**)	Heratomol	1976	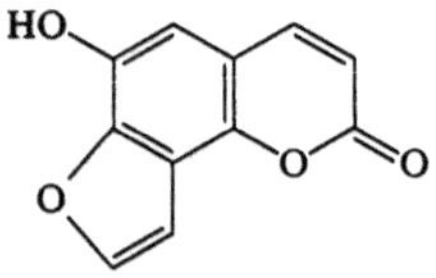
(**330**)	Sphondin	1936	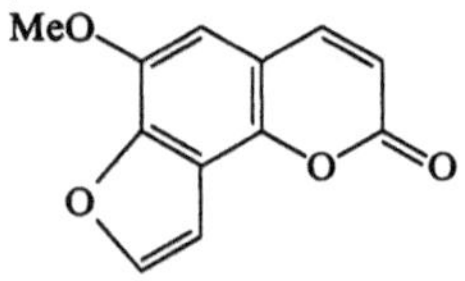
(**331**)	Heratomin	1976	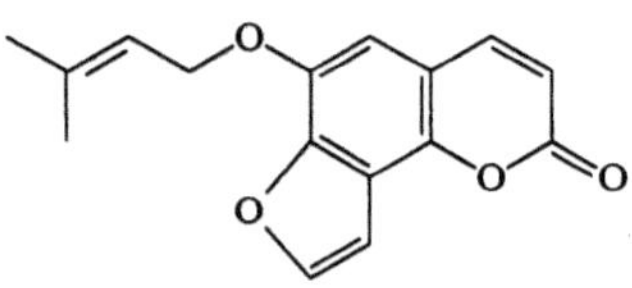
(**332**)	Norbraylin	1976	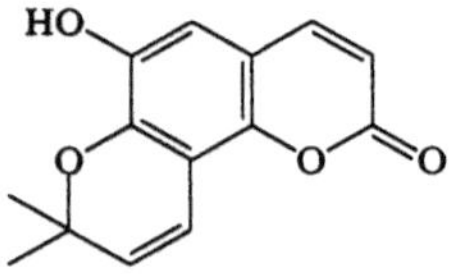
(**333**)	Braylin	1949	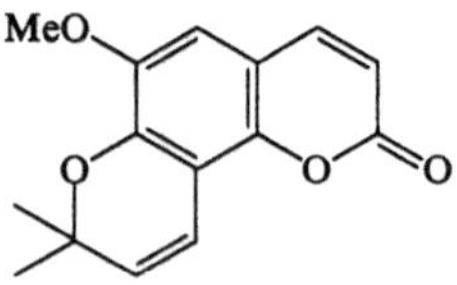

Formula	M. p.	$[\alpha]^t_\lambda$	Solvent	Plant Source(s)	Leading References
$C_{15}H_{16}O_4$	125—127 124			*Ptaeroxylon obliquum*	(*50, 51, 376*) (*156*)
$C_{11}H_6O_4$	253—254			*Heracleum thomsoni*	(*246*)
$C_{12}H_8O_4$	189—191			*Heracleum sphondylium*	(*153, 517*)
$C_{16}H_{14}O_4$	110—111			*Heracleum thomsoni*	(*246*)
$C_{14}H_{12}O_4$	132—134			*Toddalia aculeata*	(*159*)
$C_{15}H_{14}O_4$	150			*Flindersia brayleyana*	(*153*)

Table 4. *7,8-Dioxygenated Coumarins*

Trivial Name(s)	Year isolated	Structure
(**334**) Daphnetin	1822	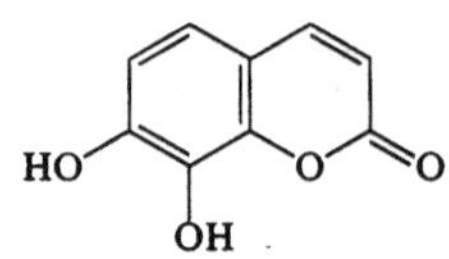
(**335**) Daphnetin glucoside	1933	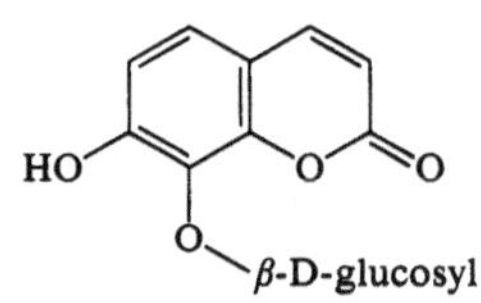
(**336**) Hydrangetin	1961	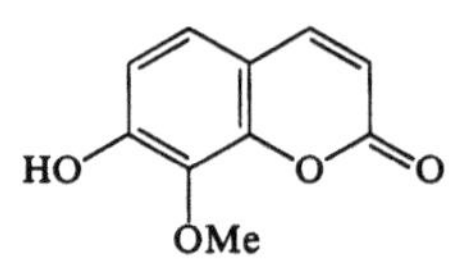
(**337**) Daphnin	1812	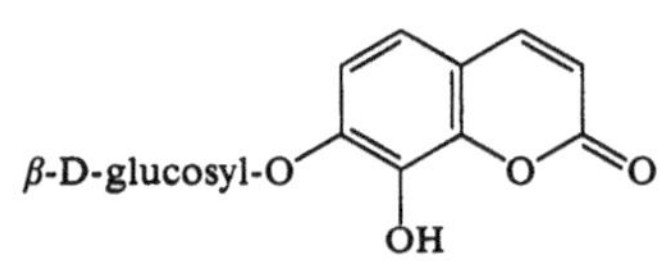
(**338**) Daphnetin 7-methyl ether	1970	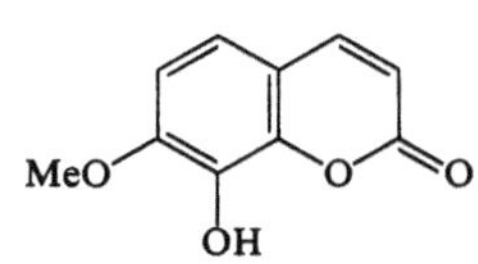
(**339**) Ferudenol	1974	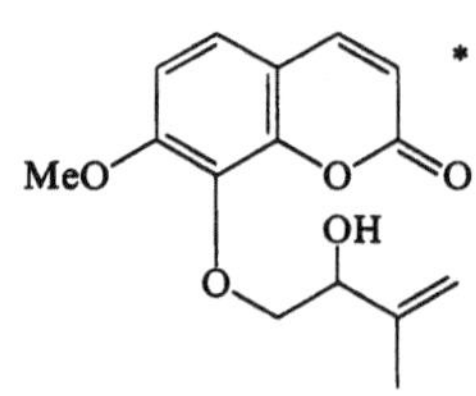
(**340**) Armin	1976	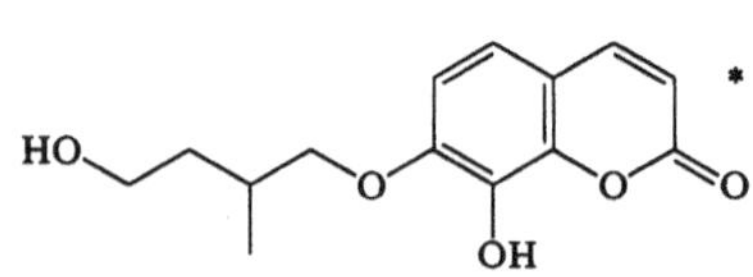

Formula	M. p.	$[\alpha]_{\lambda}^{t}$	Solvent	Plant Source(s)	Leading References
$C_9H_6O_4$	255—256			*Daphne mezereum*	(*153*)
$C_{15}H_{16}O_9$	197—198	4.57^{23} 61.9^{23}	MeOH H_2O	*Gnidia polycephala*	(*153*)
$C_{10}H_8O_4$	186.5			*Hydrangea macrophylla*	(*91, 108*)
$C_{15}H_{16}O_9$	223—224	-114.7^{22}	MeOH	*Daphne odora*	(*153*)
$C_{10}H_8O_4$	169—171			*Artemisia dracunculoides*	(*269*)
$C_{15}H_{16}O_5$	98.5—101			*Prangos ferulacea*	(*6*)
$C_{14}H_{16}O_5$	112—114			*Artemisia armeniaca*	(*497*)

Table 4 (continued)

	Trivial Name(s)	Year isolated	Structure
(**341**)	Collinin	1949	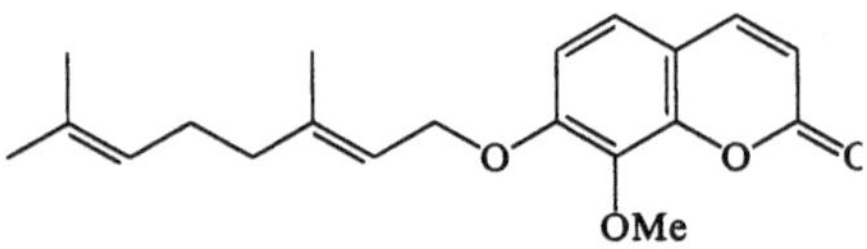
(**342**)	Stenocarpin	1976	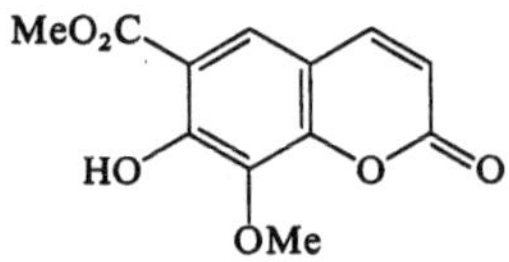
(**343**)	Stenocarpin isobutyrate	1976	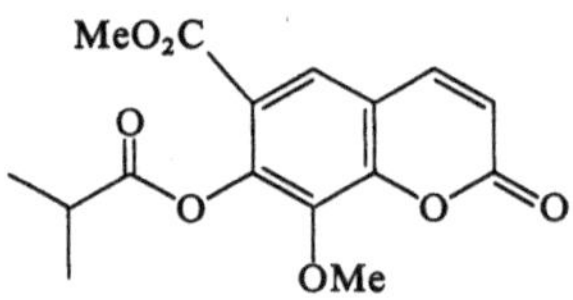
(**344**)	Brosiparin Arnottianol	1970 1974	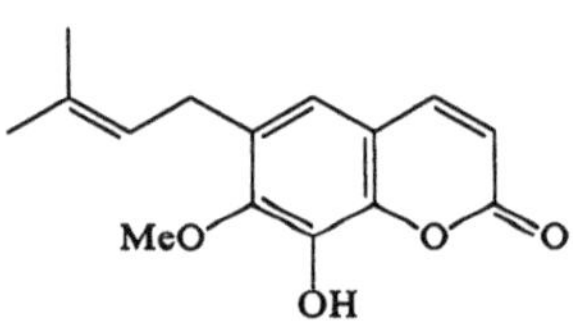
(**345**)	*O*-Prenylbrosiparin		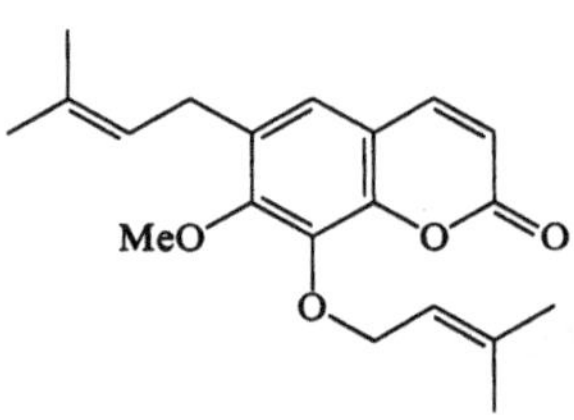
(**346**)	Rutaretin Campesenitin (aglycone)	1967 1973	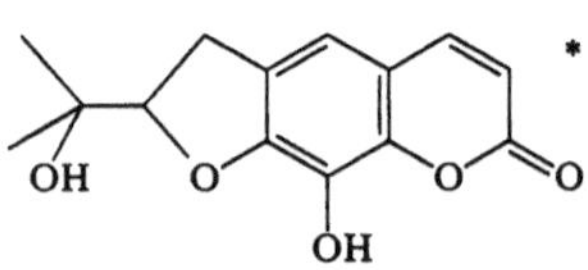
(**347**)	Isorutarin	1974	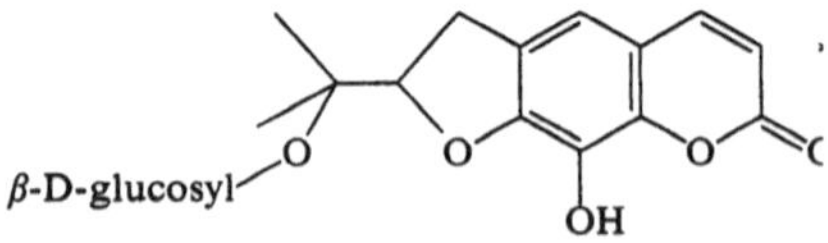

References, pp. 400—429

Formula	M.p.	$[\alpha]^t_\lambda$	Solvent	Plant Source(s)	Leading References
$C_{20}H_{24}O_4$	67—68			*Flindersia collina*	(*153*)
$C_{12}H_{10}O_6$	194.5—197			*Peucedanum stenocarpum*	(*220*)
$C_{16}H_{16}O_7$	116—116.5			*Peucedanum stenocarpum*	(*220*)
$C_{15}H_{16}O_4$	121—122 119—121			*Brosimum paraense* *Xanthoxylum arnottianum*	(*99, 101*) (*278, 279*)
$C_{20}H_{24}O_4$	oil			*Brosimum parinariodes*	(*235*)
$C_{14}H_{14}O_5$	192—193 195.5—196	-33.6^{24} -33.89^{23}	$CHCl_3$ EtOH	*Ruta graveolens*	(*514*) (*526*)
$C_{20}H_{24}O_{10}$	245—248			*Ruta graveolens*	(*588*)

Table 4 (continued)

Trivial Name(s)	Year iso- lated	Structure
(**348**) Rutarin Campesenin	1967 1970 1973	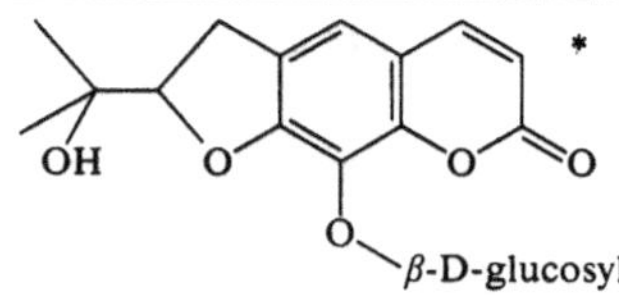 *
(**349**) Rutaretin methyl ether	1974	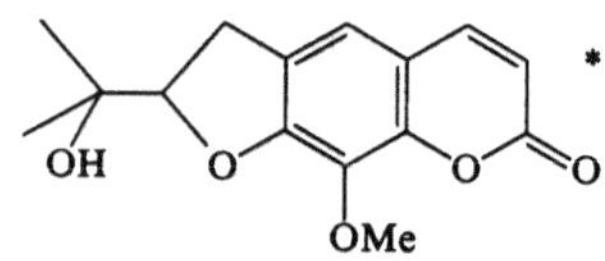 *
(**350**) Xanthotoxol	1933	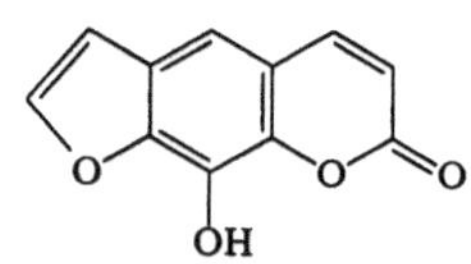
(**351**) Xanthotoxin	1911	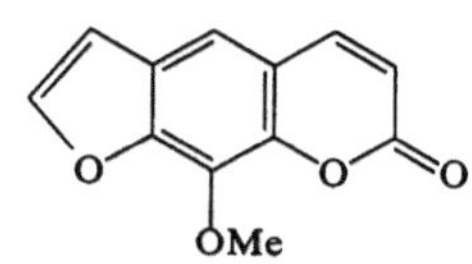
(**352**) Imperatorin	1933	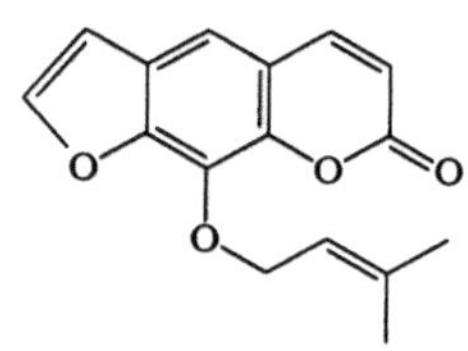
(**353**) Feruliden	1974	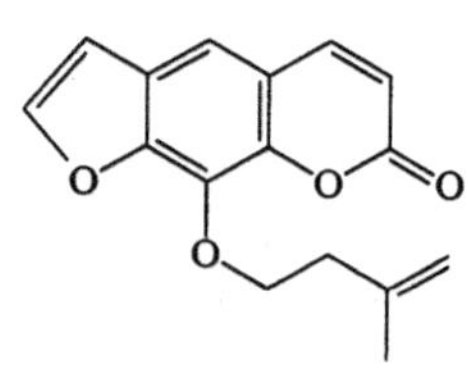
(**354**) Prangenin (±)-Heraclenin	1953 1966	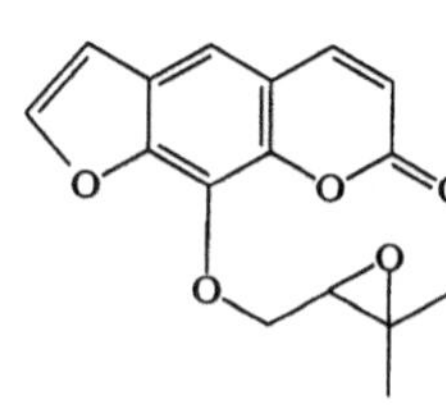 racemic

Formula	M.p.	$[\alpha]^t_\lambda$	Solvent	Plant Source(s)	Leading References
$C_{20}H_{24}O_{10}$	143—144	-95.7^{23}	pyridine	*Ruta graveolens*	(*513*)
	139—140			*Seseli campestre*	(*352*)
	145—148	-47.1^{22}	EtOH		(*526*)
$C_{15}H_{16}O_5$	147—148	10.8^{14}	$CHCl_3$	*Xanthoxylum arnottianum*	(*278, 279*)
		-41^{18}	MeOH		
$C_{11}H_6O_4$	252—253			*Angelica archangelica*	(*153*)
$C_{12}H_8O_4$	145—146			*Xanthoxylum senegalense*	(*153, 563*)
$C_{16}H_{14}O_4$	102—103			*Imperatoria ostruthium*	(*153*)
$C_{16}H_{14}O_4$	119—120			*Prangos ferulacea*	(*6*)
$C_{16}H_{14}O_5$	114—115			*Prangos pabularia*	(*343, 355, 467, 515*)
	113—114.5			*Phebalium drummondii*	(*138*)

Table 4 (continued)

	Trivial Name(s)	Year isolated	Structure
(355)	(+)-Heraclenin (+)-Imperatorin oxide	1964	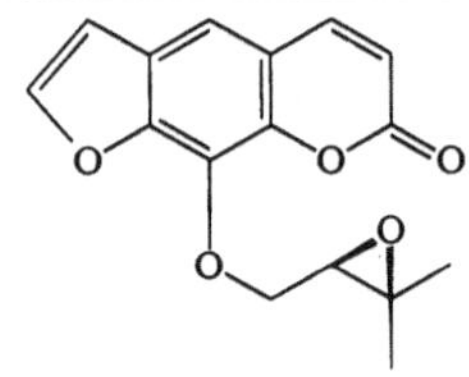
(356)	(−)-Imperatorin oxide (−)-Heraclenin	1966	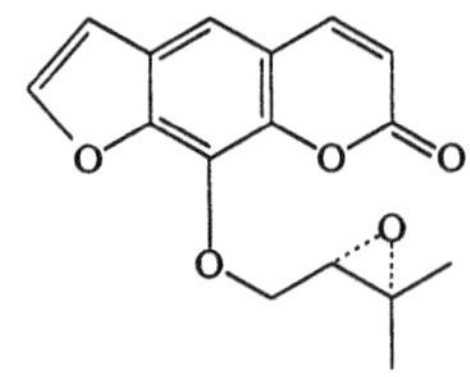
(357)	Isogosferol	1974 1974	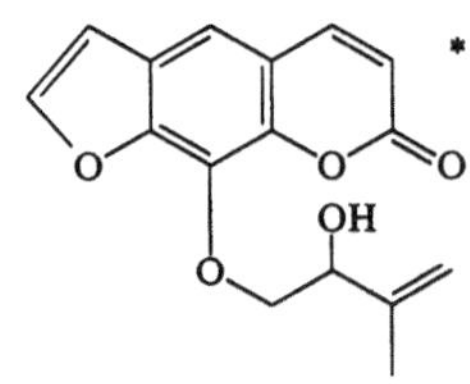
(358)	Prangenone Pabularinone Isoheraclenin Isooxyimperatorin	1966 1970 1970	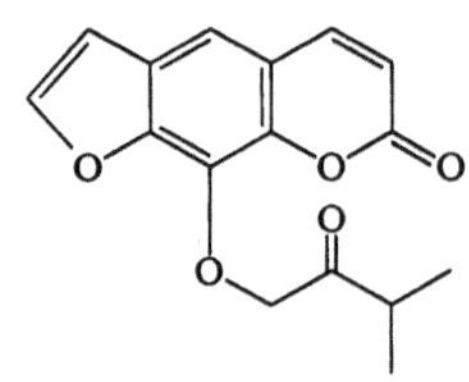
(359)	Komalin	1964	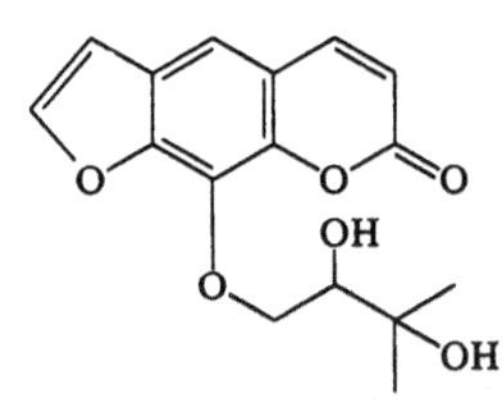 racem
(360)	(+)-Heraclenol	1964	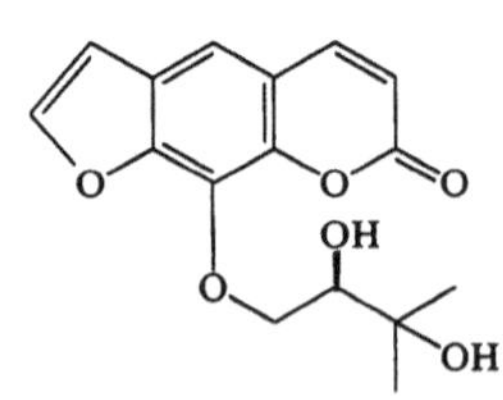

Formula	M. p.	$[\alpha]^t_\lambda$	Solvent	Plant Source(s)	Leading References
$C_{16}H_{14}O_5$	111	22^{32}	pyridine	*Heracleum candicans*	(*166, 429, 530*)
$C_{16}H_{14}O_5$	106.5—108	$-24^{23.5}$	pyridine	*Phebalium drummondii*	(*138*)
$C_{16}H_{14}O_5$	72—73.5 52.5—53.5			*Prangos lophoptera* *Heracleum granatense*	(*5*) (*213*)
$C_{16}H_{14}O_5$	134—135 131—132 135			*Prangos serawschanica* *Prangos pabularia* *Heracleum candicans*	(*357*) (*60, 125*) (*54, 55*)
$C_{16}H_{16}O_6$	127—129	0	$CHCl_3$	*Prangos pabularia*	(*247*)
$C_{16}H_{16}O_6$	117—118	16.5^{32}	pyridine	*Heracleum candicans*	(*429, 529*)

Table 4 (continued)

Trivial Name(s)	Year isolated	Structure
(**361**) *tert-O*-β-Glucosyl-heraclenol	1970	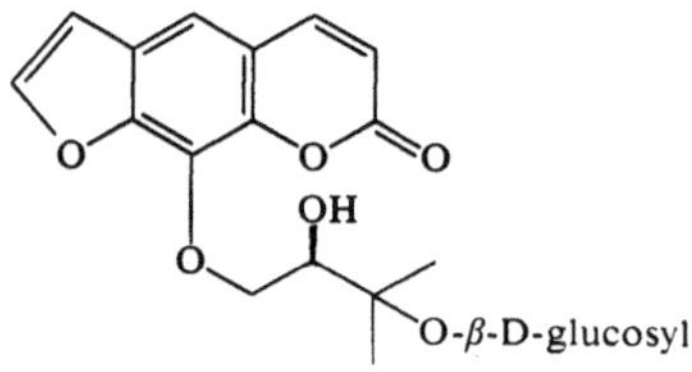
(**362**) *tert-O*-Methyl-heraclenol	1970	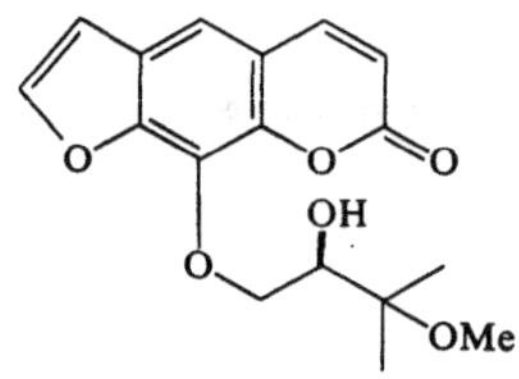
(**363**) (artefact?)	1974	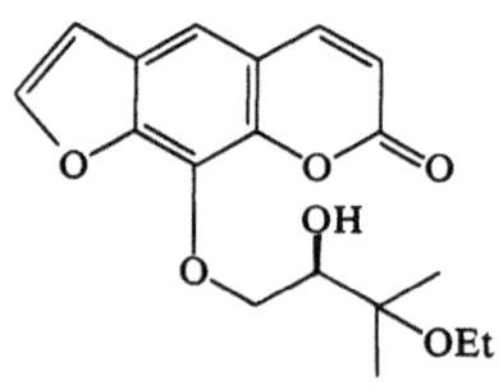
(**364**)	1974	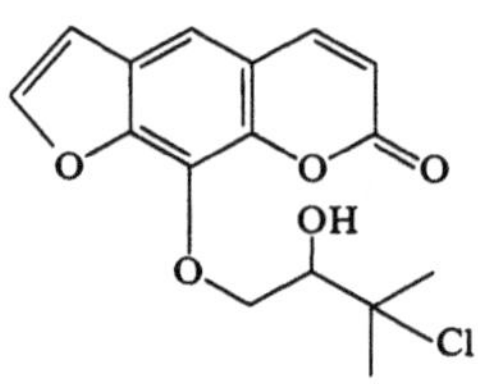 *
(**365**) Tomasin	1969	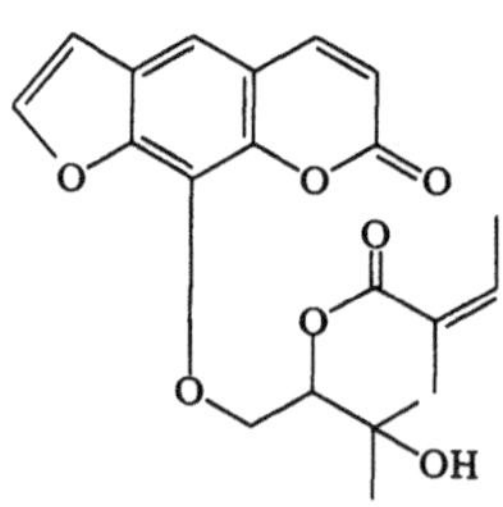 *
(**366**) 8-Geranyloxypsoralen	1957 1966	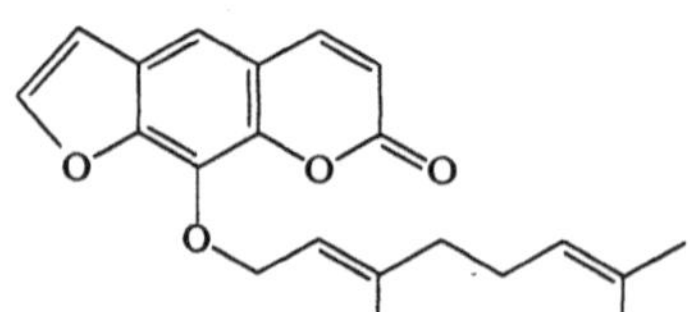

Formula	M.p.	$[\alpha]^t_\lambda$	Solvent	Plant Source(s)	Leading References
$C_{22}H_{26}O_{11}$	198	9^{25}	H_2O	*Heracleum candicans*	(*55*)
$C_{17}H_{18}O_6$	84	42		*Heracleum candicans*	(*52, 54*)
$C_{18}H_{20}O_6$	93—94	24.4	$CHCl_3$	*Heracleum granatense*	(*214*)
$C_{16}H_{15}ClO_5$	103—104	36	$CHCl_3$	*Heracleum granatense*	(*214*)
$C_{21}H_{22}O_7$	104—105	19^{20}	EtOH	*Xanthogalum purpurascens*	(*546*)
$C_{21}H_{22}O_4$	61—62 53—54			*Heracleum candicans*	(*549*) (*531*)

Table 4 (continued)

	Trivial Name(s)	Year isolated	Structure
(367)	Iliensin	1974	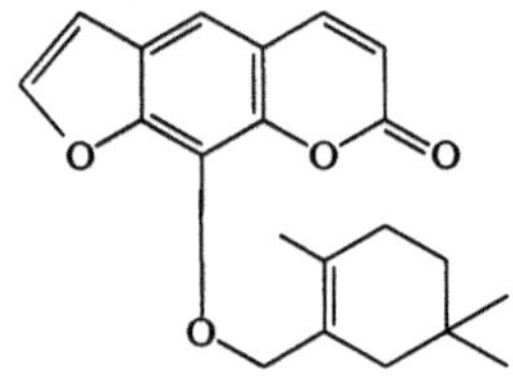
(368)	Arnottianin	1972	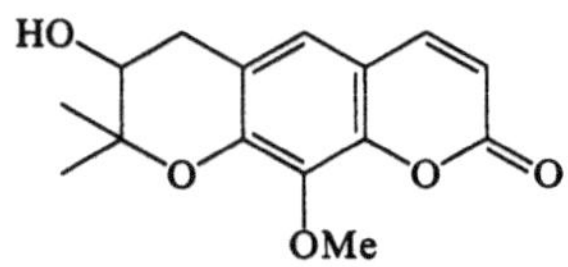 racem
(369)	Luvangetin	1939	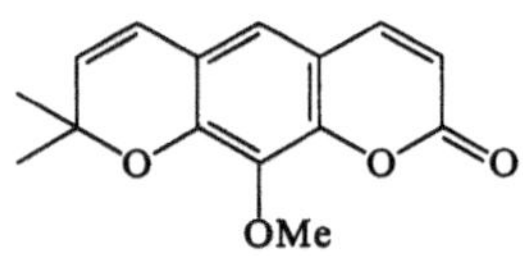
(370)	Brosiprenin	1971	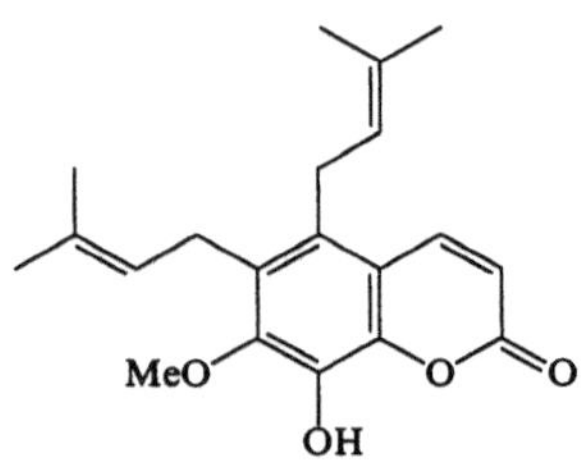
(371)	Alloimperatorin Prangenidin	1957 1965	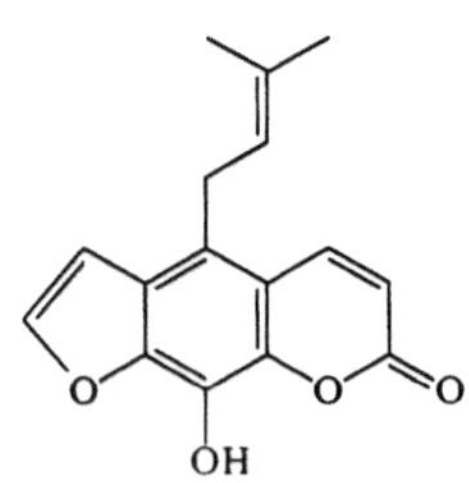
(372)	Benahorin	1971	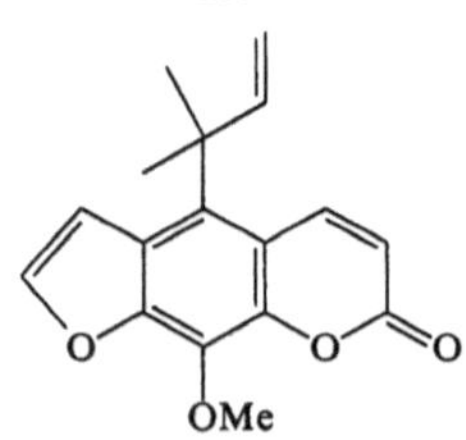

Formula	M. p.	$[\alpha]^t_\lambda$	Solvent	Plant Source(s)	Leading References
$C_{21}H_{22}O_4$	72—73			*Seseli iliense*	(*173*)
$C_{15}H_{16}O_5$	197—198	0^{17}		*Xanthoxylum arnottianum*	(*277, 279*)
$C_{15}H_{14}O_4$	108—109			*Luvanga scandens*	(*153, 270, 271*)
$C_{20}H_{24}O_4$	132—134			*Brosimum rubescens*	(*100, 101*)
$C_{16}H_{14}O_4$	226—227 233			*Aegle marmelos* *Prangos ornata*	(*498*) (*348*)
$C_{17}H_{16}O_4$	88—90			*Ruta pinnata*	(*226, 233*)

Table 4 (continued)

Trivial Name(s)	Year iso-lated	Structure	
(373) Methoxyalloimperatorin	1966	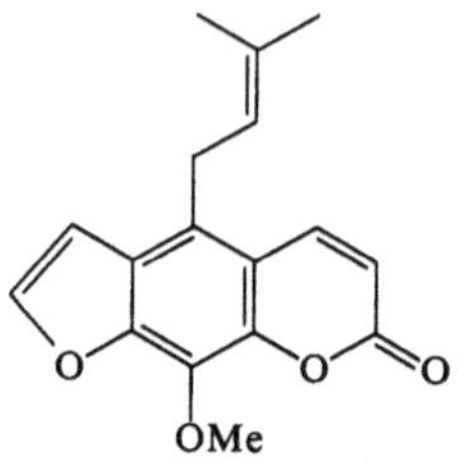	
(374)	1969	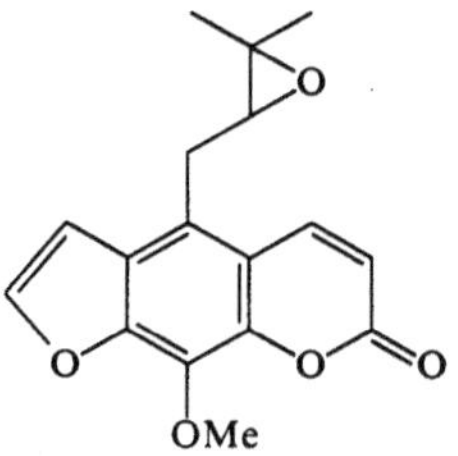	racemi
(375)	1966	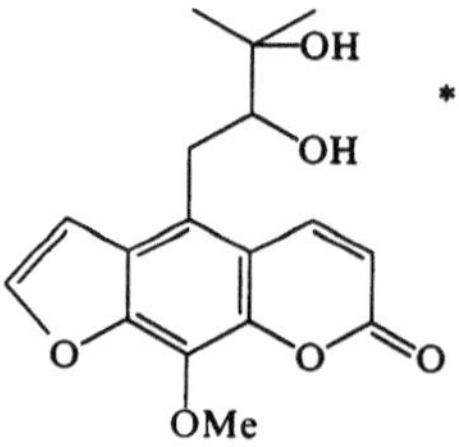	*
(376) Thamontanin	1976	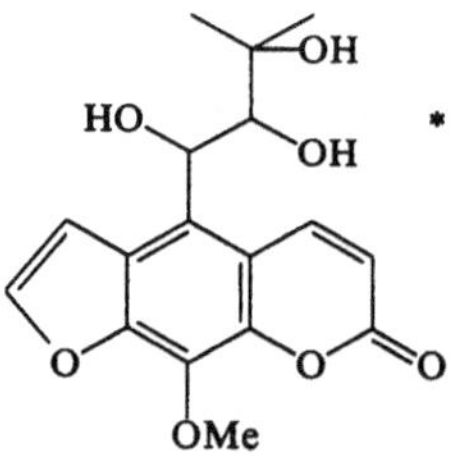	*

Formula	M. p.	$[\alpha]_{\lambda}^{t}$	Solvent	Plant Source(s)	Leading References
$C_{17}H_{16}O_4$	108—110			*Thamnosma montana*	(*163*)
$C_{17}H_{16}O_5$	103—104			*Thamnosma montana*	(*339, 342*)
$C_{17}H_{18}O_6$	174—176	-30.6^{23}	EtOH	*Thamnosma montana*	(*163*)
$C_{17}H_{18}O_7$	202—204	218^{26}	MeOH	*Thamnosma montana*	(*124*)

Table 5. *5,6,7-Trioxygenated Coumarins*

Trivial Name(s)	Year iso- lated	Structure
(377) Isofraxetin	1973	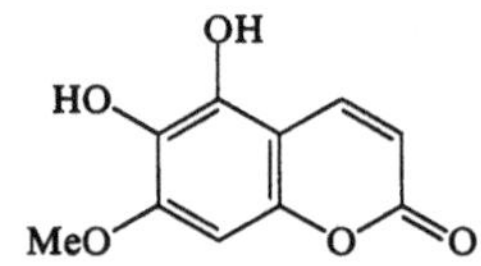
(378) Tomentin (aglycone)	1969	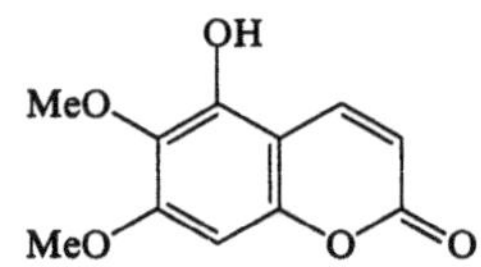
(379) Fraxinol	1937	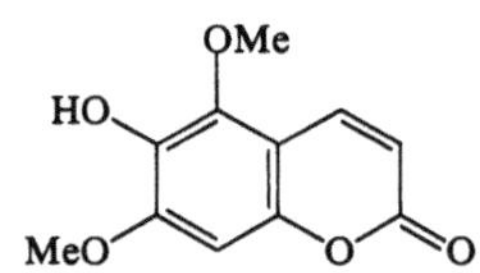
(380) Umckalin	1974	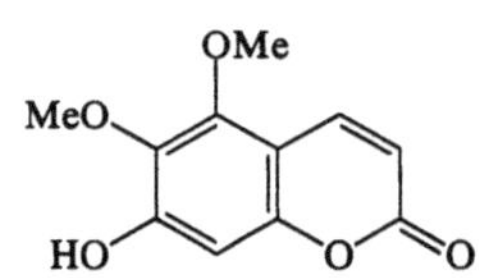
(381)	1975	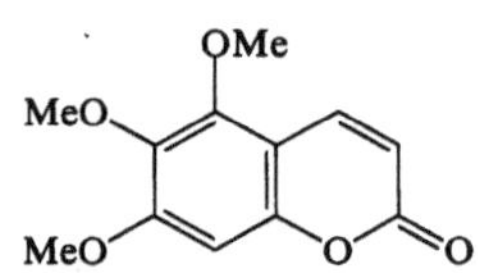
(382)	1975	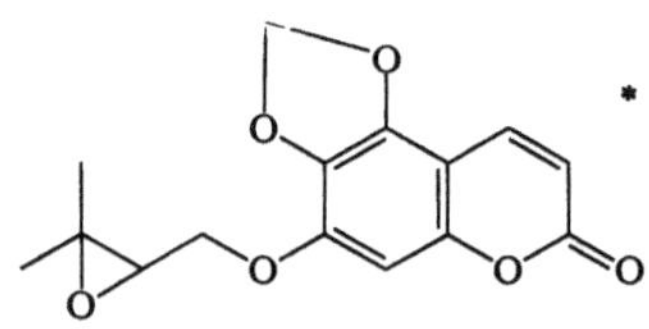
(383) Tomenin	1969	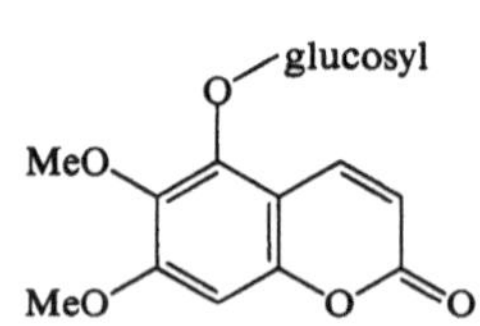

Formula	M.p.	$[\alpha]^t_\lambda$	Solvent	Plant Source(s)	Leading References
$C_{10}H_8O_5$	228—230			*Fraxinus potamophila*	(*32, 34*)
$C_{11}H_{10}O_5$	185				(*250, 407*)
$C_{11}H_{10}O_5$	171—173			*Fraxinus excelsior*	(*153*)
$C_{11}H_{10}O_5$	146—147			*Pelargonium reniforme*	(*592, 593*)
$C_{12}H_{12}O_5$	74—75			*Pelargonium reniforme*	(*592, 593*)
$C_{15}H_{14}O_6$	81	22	$CHCl_3$	*Pteronia glabrata*	(*82*)
$C_{17}H_{20}O_{10}$	207			*Prunus tomentosa*	(*250*)

Table 5 (continued)

Trivial Name(s)	Year iso-lated	Structure
(**384**) Mandshurin	1970	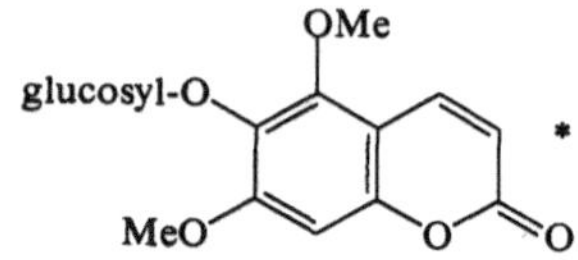
(**385**) Fraxinoside	1973	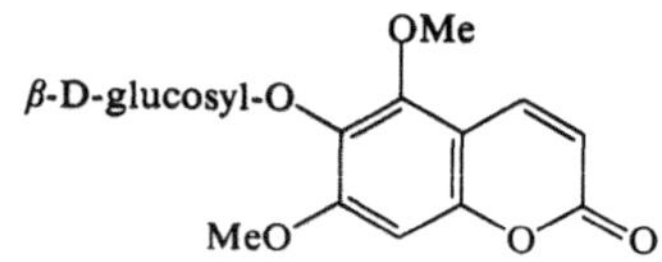
(**386**)	1975	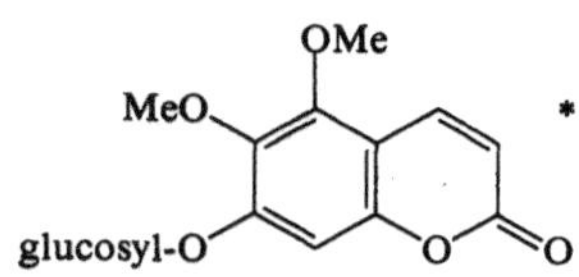
(**387**)	1973	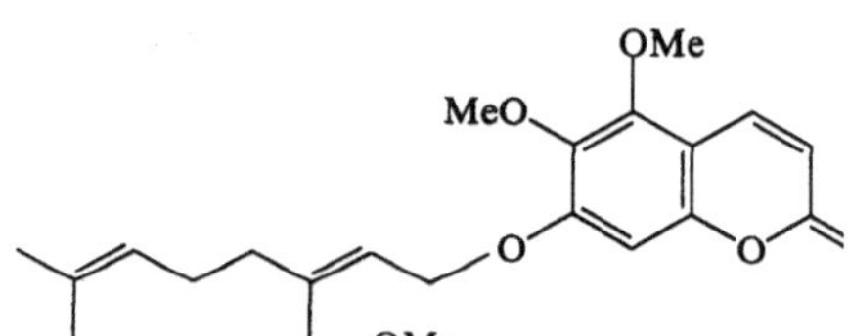
(**388**) Nieshoutol	1967	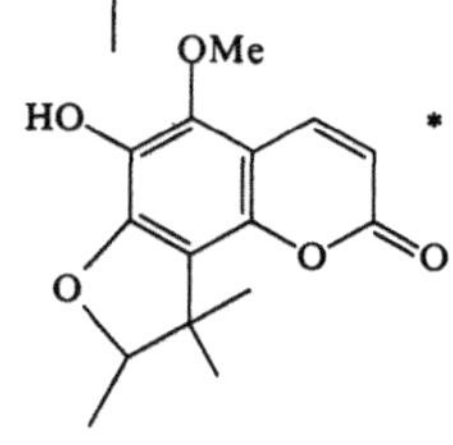
(**389**) Pimpinellin	1872	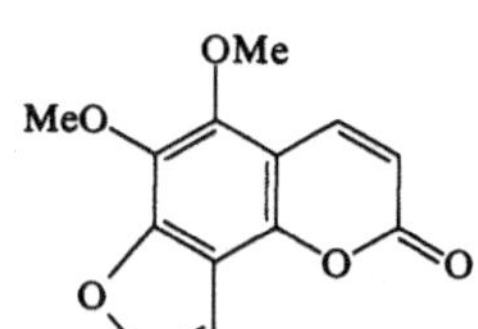
(**390**) 6-Isopentenyloxyisobergapten	1970	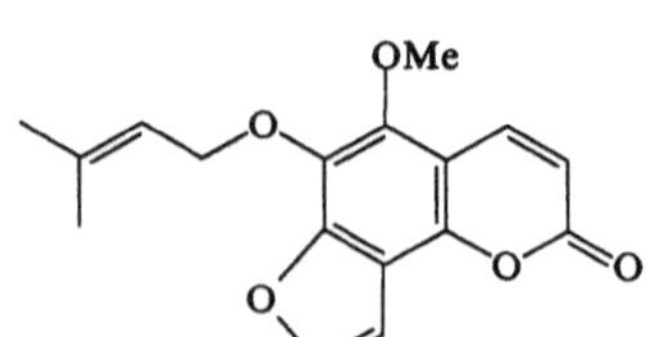

Formula	M. p.	$[\alpha]_{\lambda}^{t}$	Solvent	Plant Source(s)	Leading References
$C_{17}H_{20}O_{10}$	182.5	-26.7^{22}	MeOH	*Fraxinus mandshurica*	(*576*)
$C_{17}H_{20}O_{10}$	134—136			*Fraxinus potamophila*	(*33*)
$C_{17}H_{20}O_{10}$				*Pelargonium reniforme*	(*592, 593*)
$C_{21}H_{26}O_{5}$	oil			*Haplophyllum hispanicum*	(*221*)
$C_{15}H_{16}O_{5}$	143—144	±0	$CHCl_3$	*Ptaeroxylon obliquum*	(*376, 396, 397*)
$C_{13}H_{10}O_{5}$	117—119			*Pimpinella saxifraga*	(*153*)
$C_{17}H_{16}O_{5}$	95—96			*Angelica genuflexa*	(*167*)

Table 6. *5,7,8-Trioxygenated Coumarins*

Trivial Name(s)	Year isolated	Structure
(391)	1976	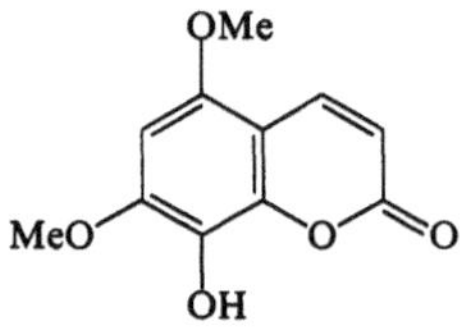
(392)	1976 1976	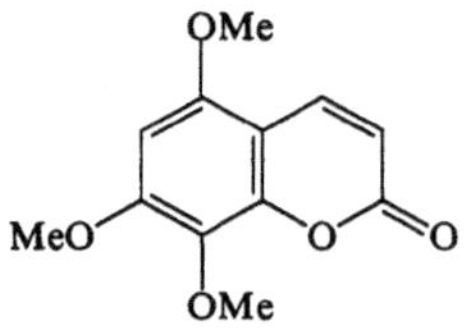
(393) Sabandinin	1971	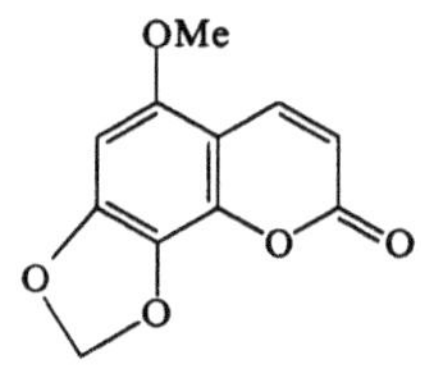
(394) Sabandinone	1970	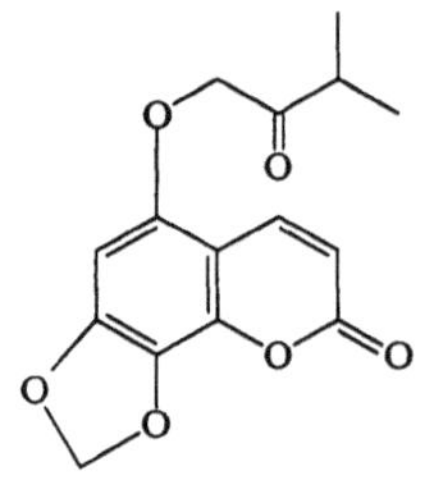
(395) Sabandinol	1973	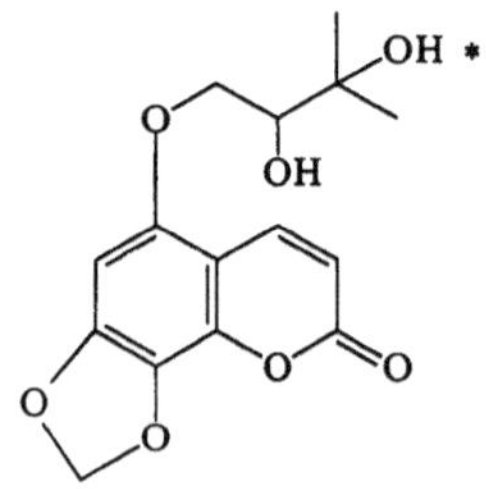
(396)	1956	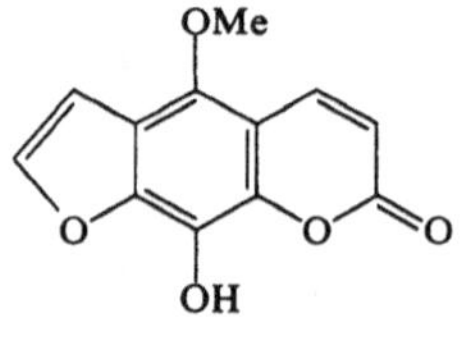

References, pp. 400—429

Formula	M. p.	$[\alpha]^t_\lambda$	Solvent	Plant Source(s)	Leading References
$C_{11}H_{10}O_5$	197—199			*Ruta* species	(*223*)
$C_{12}H_{12}O_5$	179—180 165			*Ruta* species *Toddalia aculeata*	(*223*) (*159*)
$C_{11}H_8O_5$	193—194			*Ruta pinnata*	(*53, 226, 232, 493*)
$C_{15}H_{14}O_6$	154—155			*Ruta pinnata*	(*225, 228, 232*)
$C_{15}H_{16}O_7$	oil			*Ruta pinnata*	(*224*)
$C_{12}H_8O_5$	223—224			*Casimiroa edulis*	(*314*)

Table 6 (continued)

Trivial Name(s)	Year iso- lated	Structure
(**397**) Isopimpinellin	1898	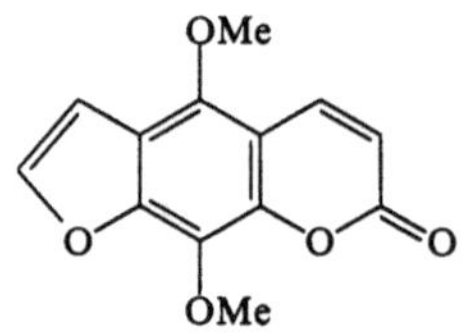
(**398**) Heracol	1970	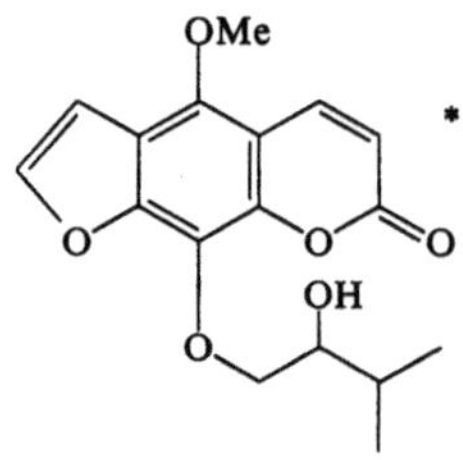
(**399**) 8-(1,1-Dimethylallyloxy)-bergapten	1975	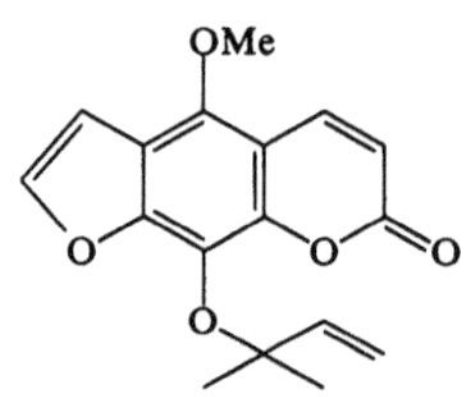
(**400**) Phellopterin	1940	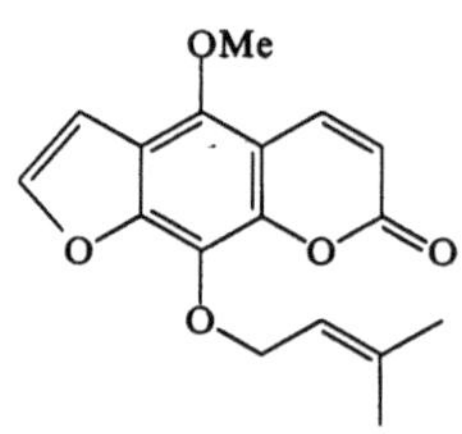
(**401**) Byakangelicol	1938	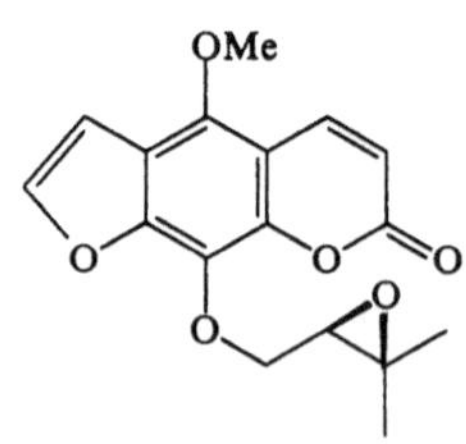

Formula	M.p.	$[\alpha]^t_\lambda$	Solvent	Plant Source(s)	Leading References
$C_{13}H_{10}O_5$	148—151			*Pimpinella saxifraga*	(*22, 153*)
$C_{17}H_{18}O_6$		18^{21}	EtOH	*Heracleum asperum*	(*211*)
$C_{17}H_{16}O_5$	90			*Peucedanum baicalense*	(*82*)
$C_{17}H_{16}O_5$	102			*Phellopterus littoralis*	(*153*)
$C_{17}H_{16}O_6$	106	34.77^{25}	pyridine	*Angelica glabra*	(*153, 429*)

Table 6 (continued)

Trivial Name(s)	Year isolated	Structure
(402) Ferulin	1947	Mixture of **(400)** and **(401)**
(403) Neobyakangelicol	1971	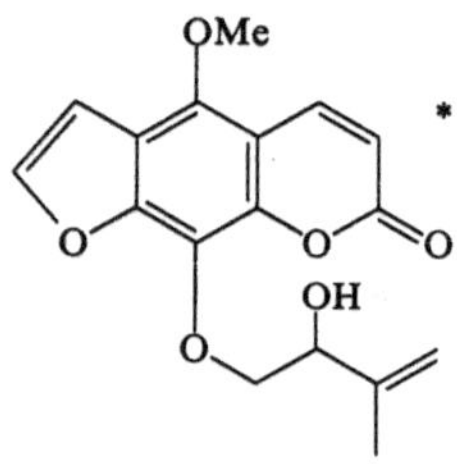
(404) Byakangelicin	1938	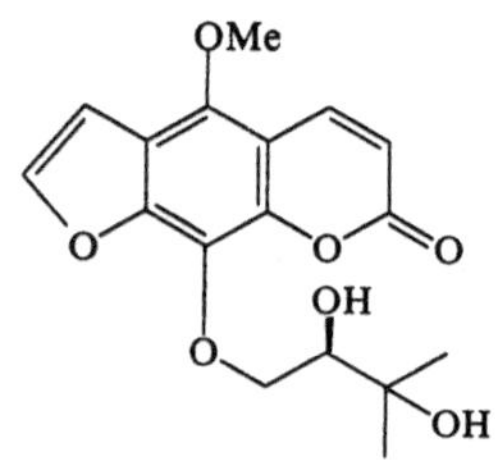
(405)	1968	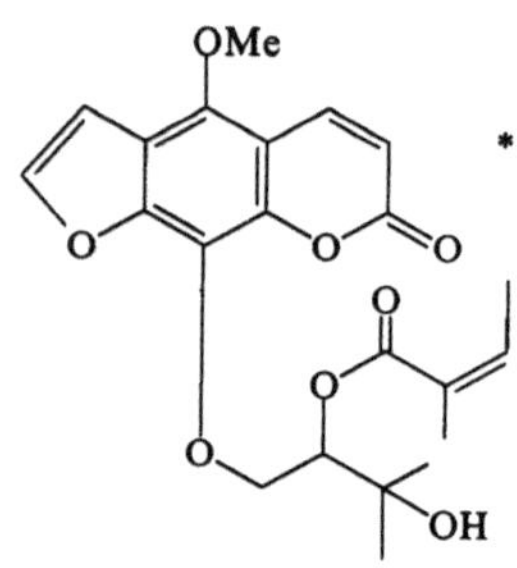
(406)	1976	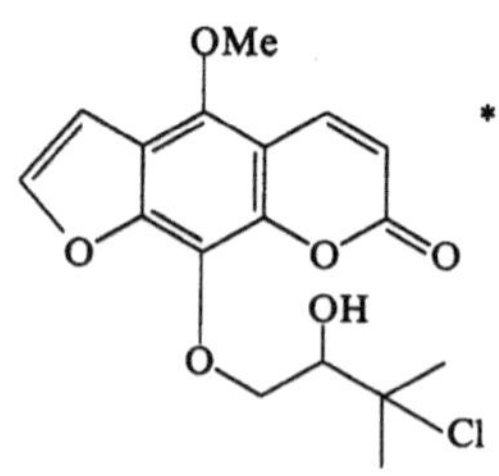

Formula	M. p.	$[\alpha]^{t}_{\lambda}$	Solvent	Plant Source(s)	Leading References
	87	27.31[33]		*Ferula alliacea*	(*127*)
$C_{17}H_{16}O_6$	106—107	−2[20]	EtOH	*Angelica dahurica*	(*501, 568*)
$C_{17}H_{18}O_7$	117—118	24.62[25]	pyridine	*Angelica glabra*	(*153, 429*)
$C_{22}H_{24}O_8$				*Seseli libanotis*	(*73*)
$C_{17}H_{17}ClO_6$	156—158	−28.3	$CHCl_3$	*Heracleum pyrenaicum*	(*212*)

Table 6 (continued)

	Trivial Name(s)	Year isolated	Structure
(**407**)	5-Methoxy-8-geranyloxypsoralen	1976	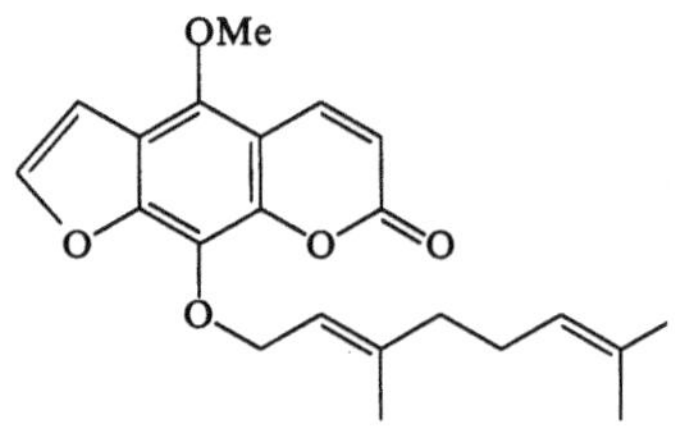
(**408**)	Tederin	1972	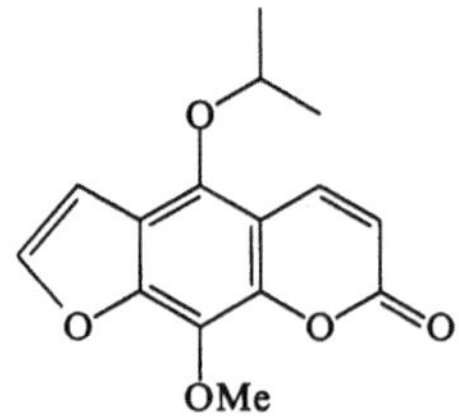
(**409**)	Cnidilin	1960	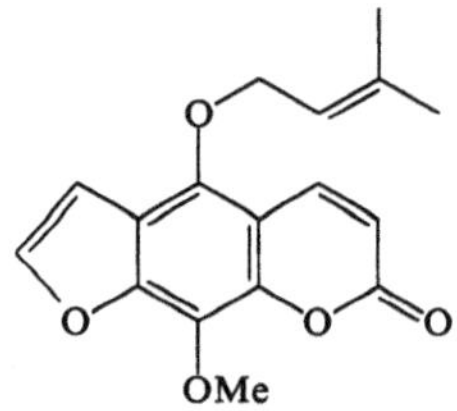
(**410**)	Cnidicin	1960	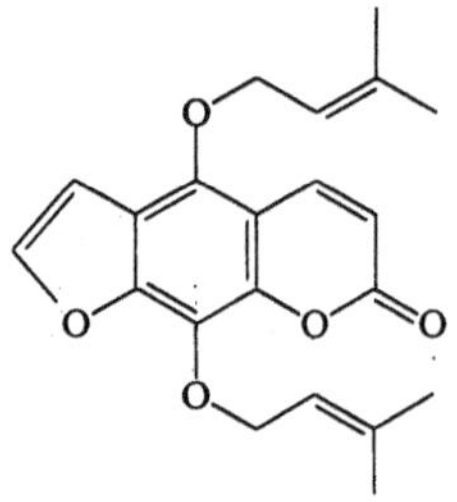

Formula	M.p.	$[\alpha]^t_\lambda$	Solvent	Plant Source(s)	Leading References
$C_{22}H_{24}O_5$				*Komarovia anisospermum*	(*547*)
$C_{15}H_{14}O_5$	126—127			*Ruta pinnata*	(*228*)
$C_{17}H_{16}O_5$	117—118			*Cnidium dubium*	(*135, 333, 368*)
$C_{21}H_{22}O_5$	78—79			*Cnidium dubium*	(*135, 333*)

Table 7. *6,7,8-Trioxygenated Coumarins*

Trivial Name(s)	Year iso-lated	Structure
(411) Arscotin	1973	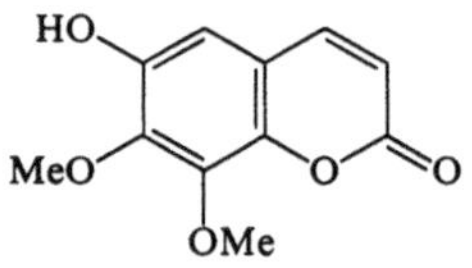
(412) Fraxetin	1857	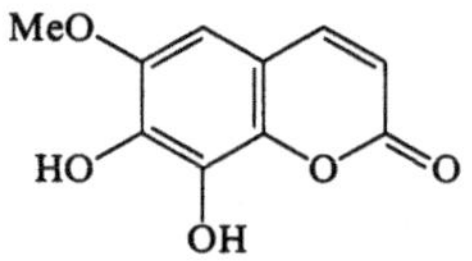
(413) Isofraxidin	1937	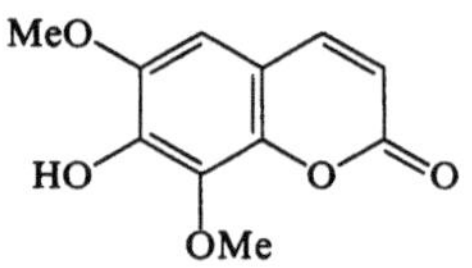
(414) Fraxin	1840	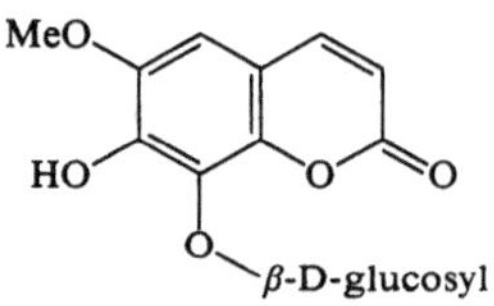
(415) Fraxidin	1937	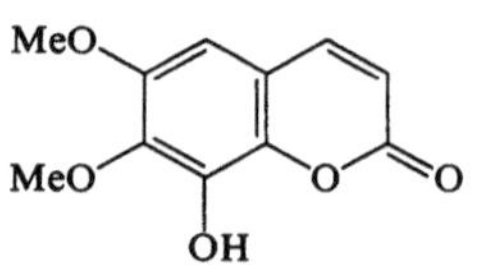
(416)	1970	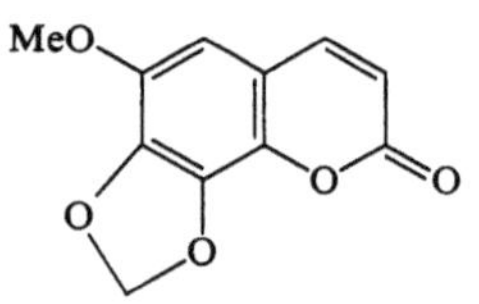
(417)	1954	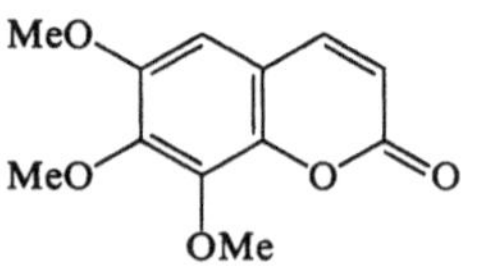

Formula	M.p.	$[\alpha]_{\lambda}^{t}$	Solvent	Plant Source(s)	Leading References
$C_{11}H_{10}O_5$	192—193			*Artemisia scotina*	(*601, 602*)
$C_{10}H_8O_5$	227—228			*Aesculus turbinata*	(*20, 153*)
$C_{11}H_{10}O_5$	148—149			*Fraxinus excelsior*	(*20, 153*)
$C_{16}H_{18}O_{10}$	205			*Fraxinus excelsior*	(*594*)
$C_{11}H_{10}O_5$	196—197			*Fraxinus excelsior*	(*153*)
$C_{11}H_8O_5$	219—221			*Artemisia dracunculoides*	(*269*)
$C_{12}H_{12}O_5$	104			*Fagara macrophylla*	(*315*)

Table 7 (continued)

	Trivial Name(s)	Year iso-lated	Structure
(**418**)	Fraxidin 8-*O*-β-D-glucoside	1976	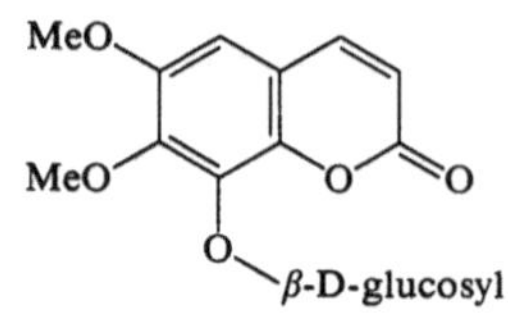
(**419**)	Puberulin	1975 1976	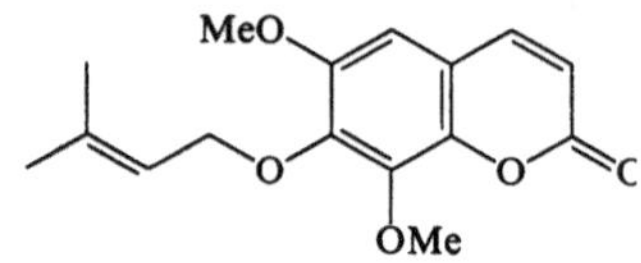
(**420**)	Obtusifol	1974	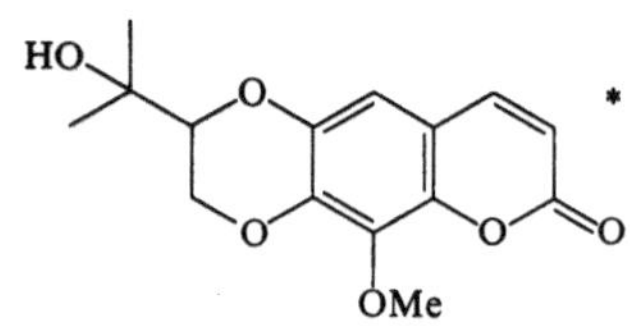
(**421**)	Creticacoumarin	1975	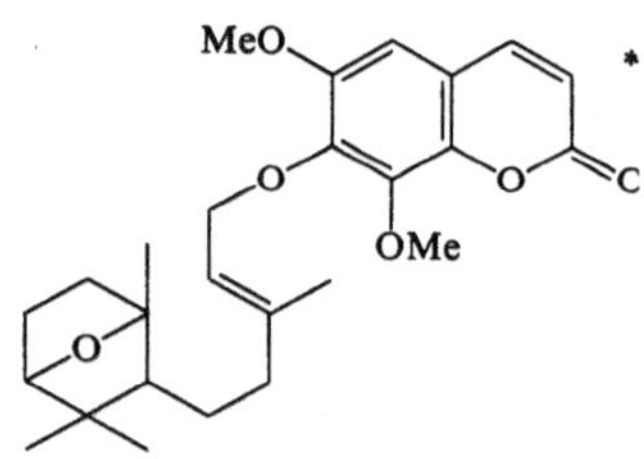
(**422**)	7-*O*-(3-Hydroxydrimen-11-yl)-iso-fraxidin	1974	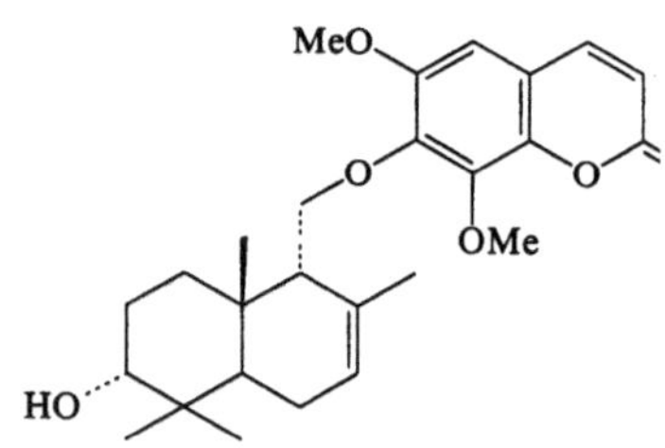

References, pp. 400—429

Formula	M.p.	$[\alpha]_\lambda^t$	Solvent	Plant Source(s)	Leading References
$C_{17}H_{20}O_{10}$	193—195	$-49^{19.5}$	H_2O	*Fraxinus excelsior*	(*285*)
$C_{16}H_{18}O_5$	oil 90—92			*Pteronia ciliata* *Agathosma puberula*	(*82*) (*187*)
$C_{15}H_{16}O_6$	149.5—150	60.39^{19}	EtOH	*Haplophyllum obtusifolium*	(*208*)
$C_{26}H_{34}O_6$	oil			*Anthemis cretica*	(*88*)
$C_{26}H_{34}O_6$	129.5	201^{23}	acetone	*Artemisia pontica*	(*86*)

Table 7 (continued)

Trivial Name(s)	Year iso-lated	Structure
(423)	1975	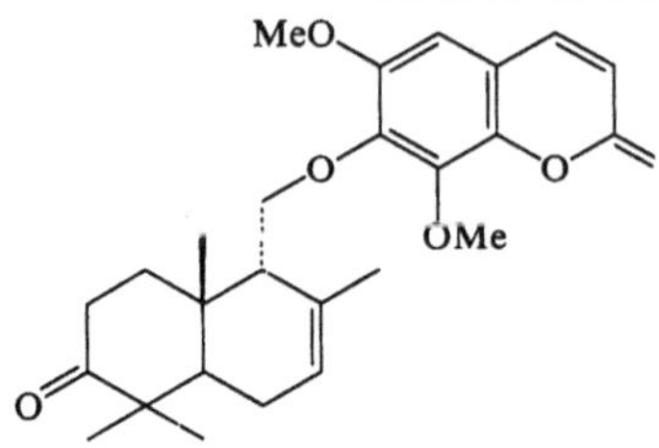
(424) Isofraxoside	1969	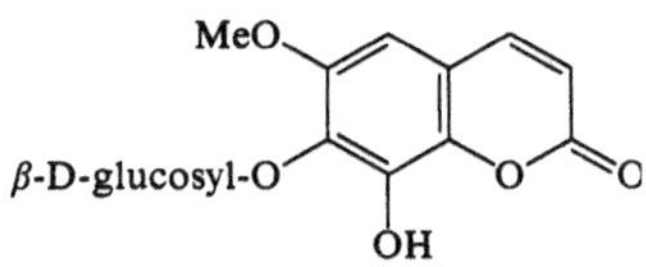
(425) Calycanthoside	1955	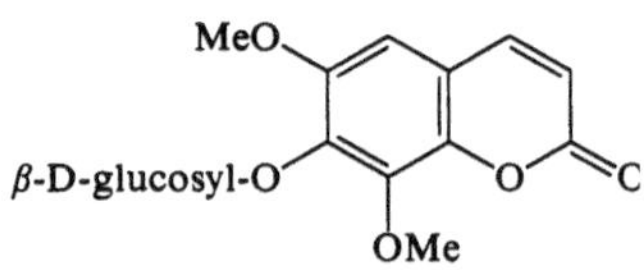
(426) Eleutheroside B_1	1967	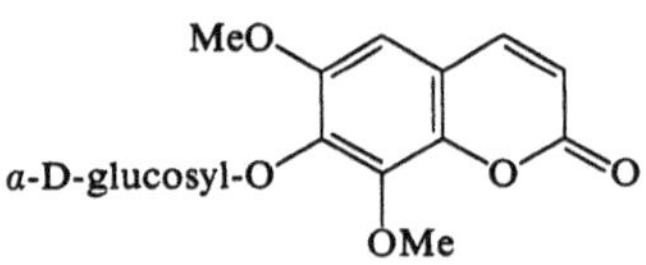

Table 8. *5,6,7,8-Tetraoxygenated Coumarins*

Trivial Name(s)	Year iso-lated	Structure
(427) Artelin	1975	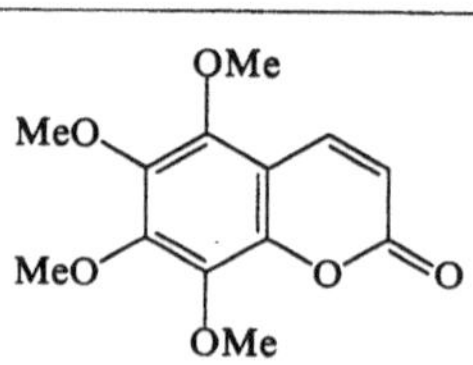

References, pp. 400—429

Formula	M.p.	$[\alpha]_\lambda^t$	Solvent	Plant Source(s)	Leading References
$C_{26}H_{32}O_6$	130	156.3^{22}	$CHCl_3$	*Anthemis aciphylla*	(*86, 88*)
$C_{16}H_{18}O_{10}$	220	-32	pyridine	*Poliothyrsis sinensis*	(*471*)
$C_{17}H_{20}O_{10}$	219–220	-42^{20}	MeOH	*Calycanthus occidentalis*	(*282, 469*)
$C_{17}H_{20}O_{10}$	218	80^{20}	MeOH	*Eleutherococcus senticosus*	(*453*)

Formula	M.p.	$[\alpha]_\lambda^t$	Solvent	Plant Source(s)	Leading References
$C_{13}H_{14}O_6$				*Artemisia tridentata*	(*104*)

Table 9. *3-Substituted Coumarins*

	Trivial Name(s)	Year isolated	Structure
(428)	3-(1′,1′-Dimethylallyl)-herniarin	1968	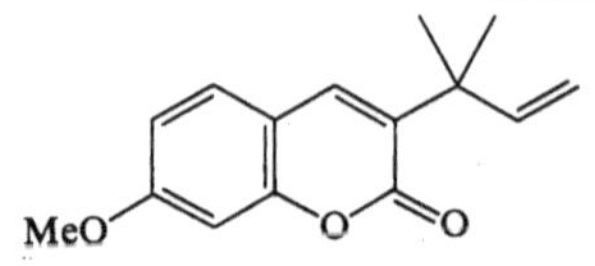
(429)	Pinnaterin	1971	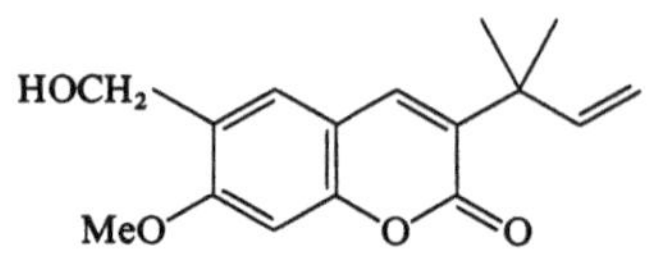
(430)	Gravelliferone	1968	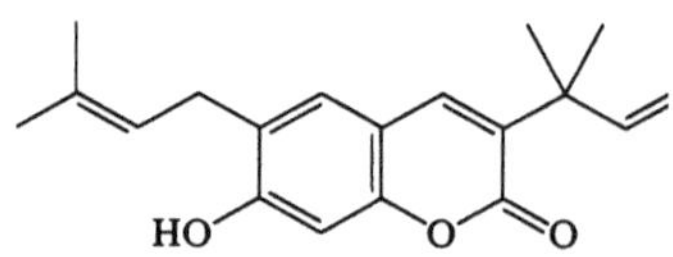
(431)	Gravelliferone methyl ether	1968	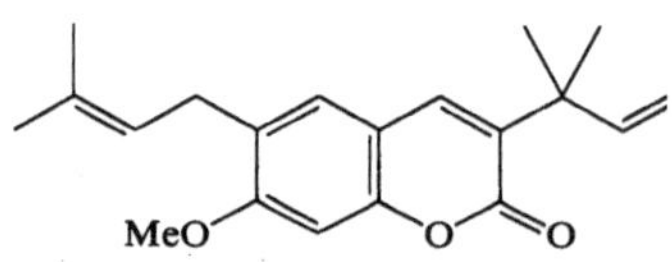
(432)	Heliettin	1967	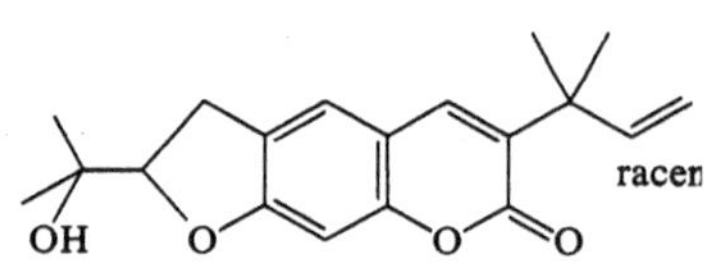
(433)	Chalepin Rutamarin alcohol	1967 1972	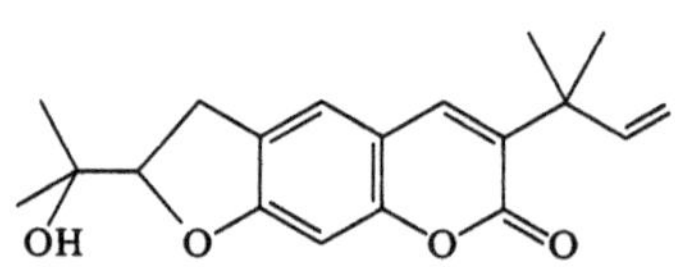
(434)	Chalepin acetate Rutamarin	1967 1967	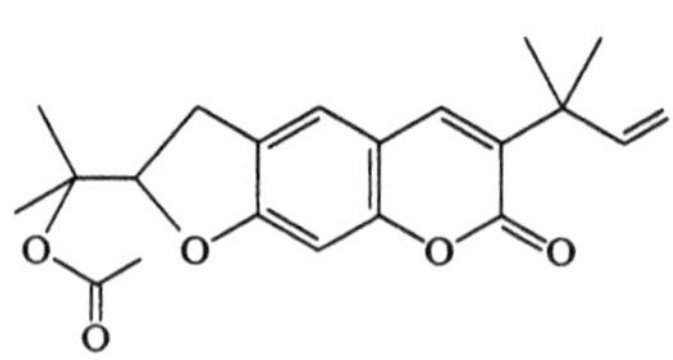

Formula	M. p.	$[\alpha]^{t}_{\lambda}$	Solvent	Plant Source(s)	Leading References
$C_{15}H_{16}O_3$	126—128			*Ruta graveolens*	(*479, 487*)
$C_{16}H_{18}O_4$	130—132			*Ruta pinnata*	(*227*)
$C_{19}H_{22}O_3$	166—168			*Ruta graveolens*	(*486*)
$C_{20}H_{24}O_3$	70—72			*Ruta graveolens*	(*487*)
$C_{19}H_{22}O_4$	165.5—166			*Helietta longifoliata*	(*473*)
	165.5—166	56	$CHCl_3$	*Ruta chalepensis*	(*222*)
$C_{19}H_{22}O_4$	118	31^{24}	$CHCl_3$	*Ruta chalepensis*	(*102*)
	118—119 (aq. EtOH) 165 (C_6H_6)	2.2^{35}	$CHCl_3$	*Ruta graveolens*	(*449, 490*)
$C_{21}H_{24}O_5$	104—105	14.3^{22}	$CHCl_3$	*Ruta chalepensis*	(*102, 222*)
	107—108	10.6^{22}_{578}	$CHCl_3$	*Ruta graveolens*	(*483*)

Table 9 (continued)

	Trivial Name(s)	Year iso-lated	Structure
(435)	Chalepensin Xylotenin	1967 1968 1976	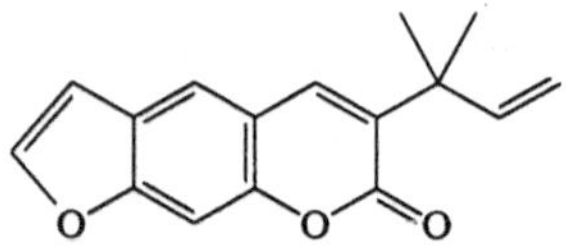
(436)	Rutolide Clausindine	1939 1974	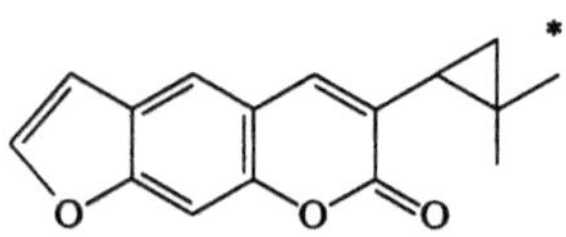
(437)	3-(1,1-Dimethylallyl)-xanthyletin	1974	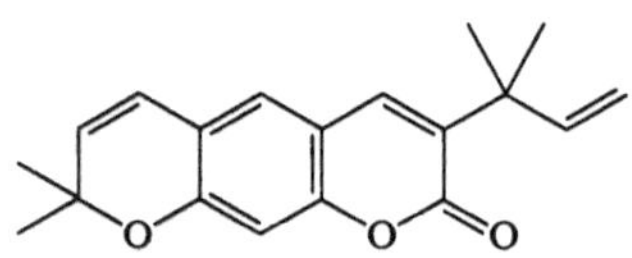
(438)	3-(3,3-Dimethylallyl)-xanthyletin	1974	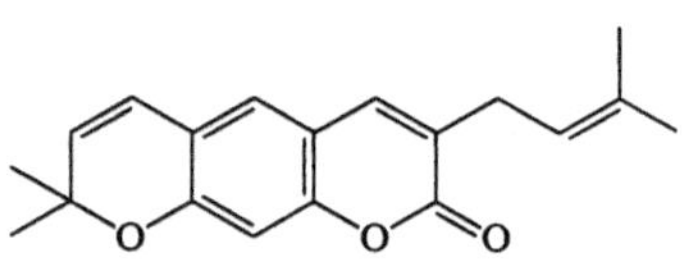
(439)		1970	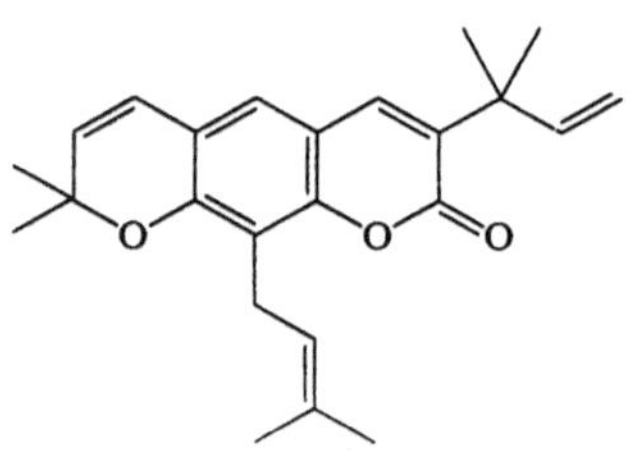
(440)	3-(1,1-Dimethylallyl)-scopoletin	1971	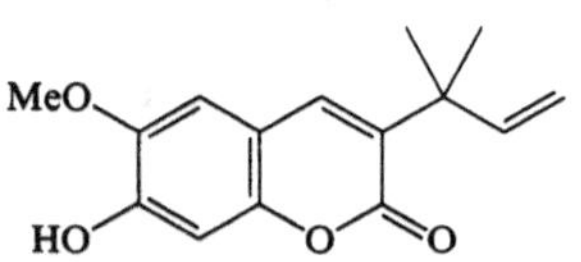
(441)	Rutacultin	1971	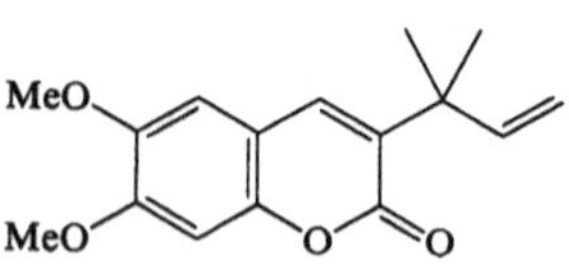

Formula	M.p.	$[\alpha]^t_\lambda$	Solvent	Plant Source(s)	Leading References
$C_{16}H_{14}O_3$	89—90 91—92 86			*Ruta chalepensis* *Chloroxylon swietenia* *Hortia arborea*	(*102, 222*) (*286, 570*) (*158*)
$C_{16}H_{14}O_3$	126—127 128			*Ruta montana* *Clausena indica*	(*336, 466, 485*) (*288, 289*)
$C_{19}H_{20}O_3$	98—99			*Boenninghausenia albiflora*	(*418*)
$C_{19}H_{20}O_3$	103—104			*Amyris simplicifolia*	(*141*)
$C_{24}H_{28}O_3$				*Ruta graveolens*	(*488*)
$C_{15}H_{16}O_4$	132—135			*Ruta graveolens*	(*50, 591*)
$C_{16}H_{18}O_4$	104—105			*Ruta graveolens*	(*50, 449, 490, 558*)

Table 9 (continued)

	Trivial Name(s)	Year iso-lated	Structure
(**442**)		1976	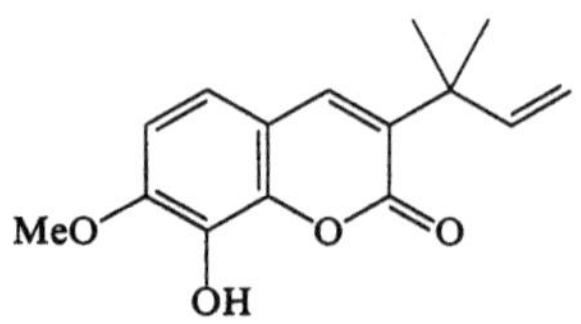
(**443**)	3-(1,1-Dimethylallyl)-daphnetin dimethyl ether	1970	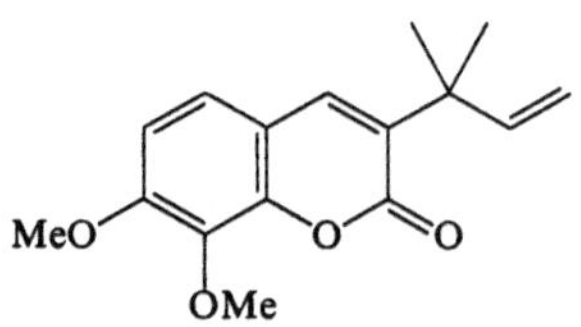
(**444**)	8-Methoxygravelliferone	1970	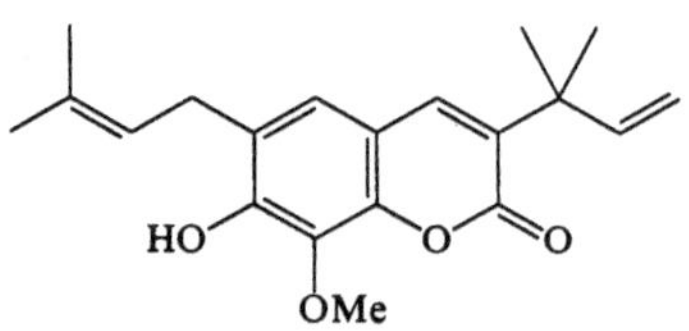

Formula	M.p.	$[\alpha]_\lambda^t$	Solvent	Plant Source(s)	Leading References
$C_{15}H_{16}O_4$	98—100			*Ruta* species	(*223*)
$C_{16}H_{18}O_4$	85—87			*Ruta graveolens*	(*488*)
$C_{20}H_{24}O_4$	131—133			*Ruta graveolens*	(*488*)

Table 10. *4-Substituted Coumarins*

	Trivial Name(s)	Year isolated	Structure
(**445**)		1970	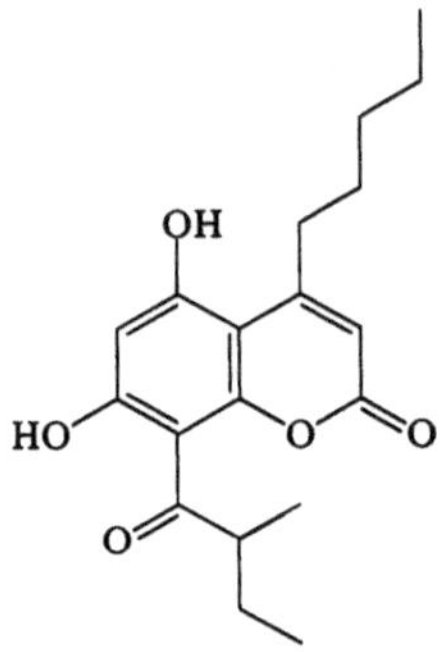 *
(**446**)	Mammea B/BC Normammein	1966 1972	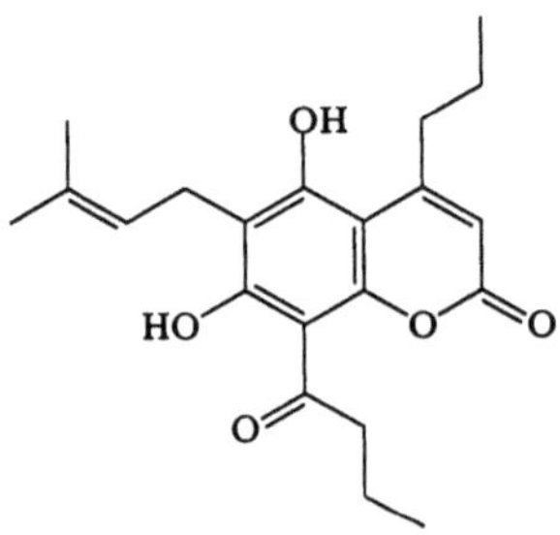
(**447**)		1972	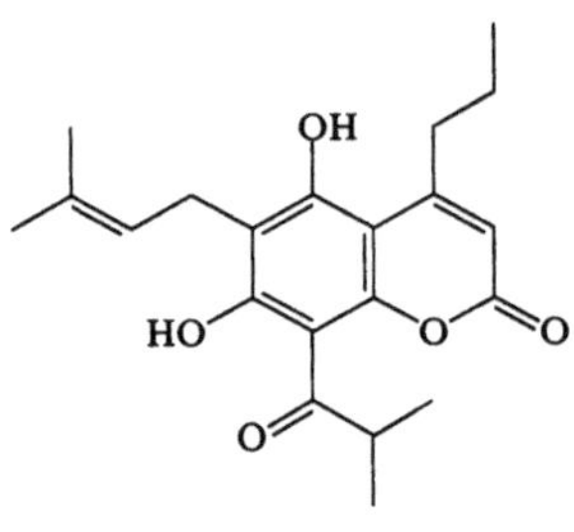
(**448**)	Mammea B/BB Neomammein	1966 1972 1970	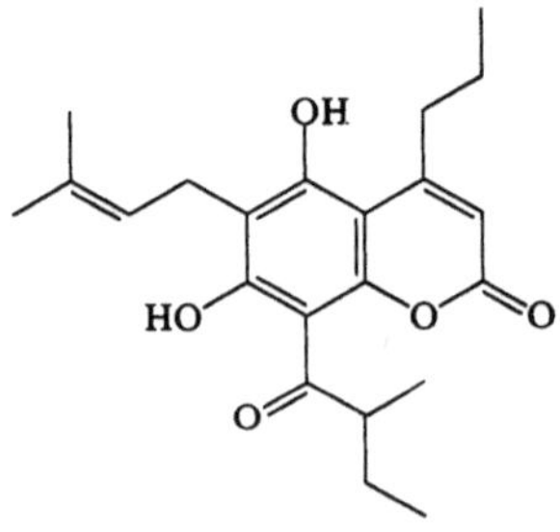

Formula	M.p.	$[\alpha]^t_\lambda$	Solvent	Plant Source(s)	Leading References
$C_{19}H_{24}O_5$	218—220			*Mammea africana*	(*120, 121*)
$C_{21}H_{26}O_5$	132—133			*Mammea americana*	(*148, 149, 204*)
	131—132				(*190*)
$C_{21}H_{26}O_5$				*Mammea americana*	(*204*)
$C_{22}H_{28}O_5$	122			*Mammea americana*	(*148, 149, 204*)
	121—122	$-2.93^{26.5}$	$CHCl_3$		(*190*)
				Mammea africana	(*120, 121*)

Table 10 (continued)

Trivial Name(s)	Year iso-lated	Structure
(**449**) Ferruol A	1967	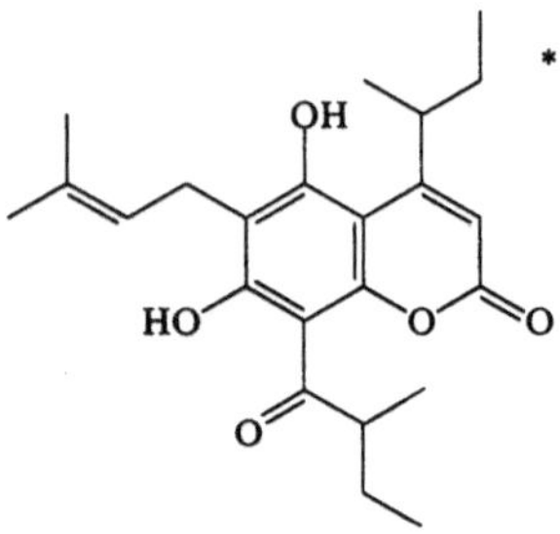
(**450**) Mammea C/BB	1966	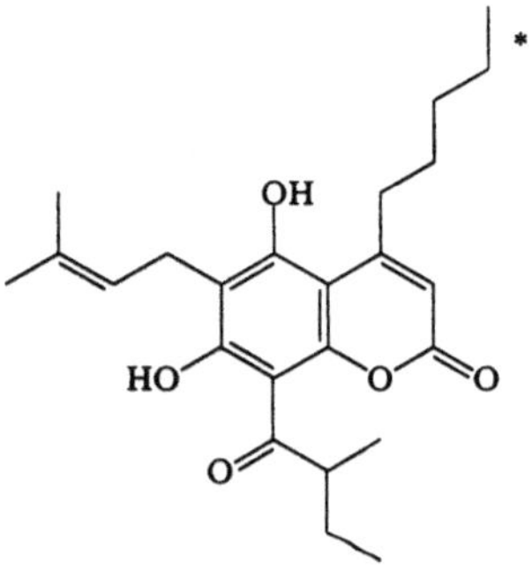
(**451**) Surangin A	1969	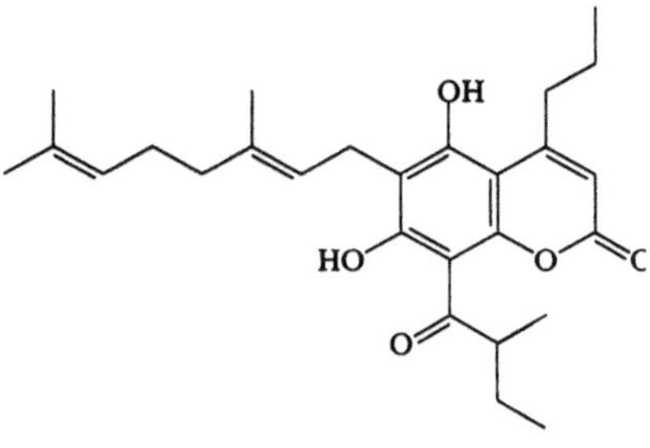
(**452**) Mammein	1953	
Mammea B/BA	1966	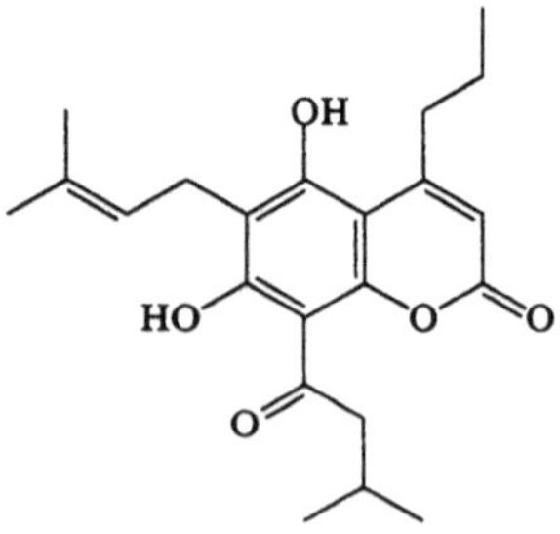

ormula	M. p.	[α[λ	Solvent	Plant Source(s)	Leading References
$_{23}H_{30}O_5$	126.5			*Mesua ferrea*	(*237*)
$_{24}H_{32}O_5$	100—101			*Mammea americana*	(*148, 149, 204*)
$_{27}H_{36}O_5$	83—85	-1.6^{26}	$CHCl_3$	*Mammea longifolia*	(*290*)
$_{22}H_{28}O_5$	128.5—129.5			*Mammea americana*	(*160, 161, 162, 188, 390*)
	127				(*148, 149, 204*)

Table 10 (continued)

Trivial Name(s)	Year iso- lated	Structure
(**453**) Mammea B/AC	1972	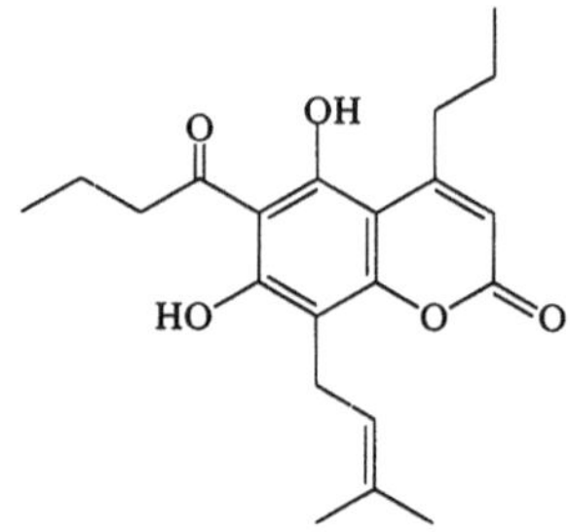
(**454**) Mammea B/AB) MAB-2	1970 1970	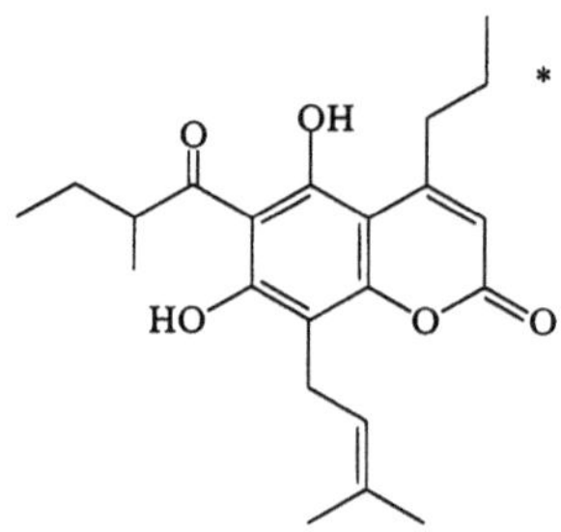
(**455**) Mammea B/AA Isomammein	1970	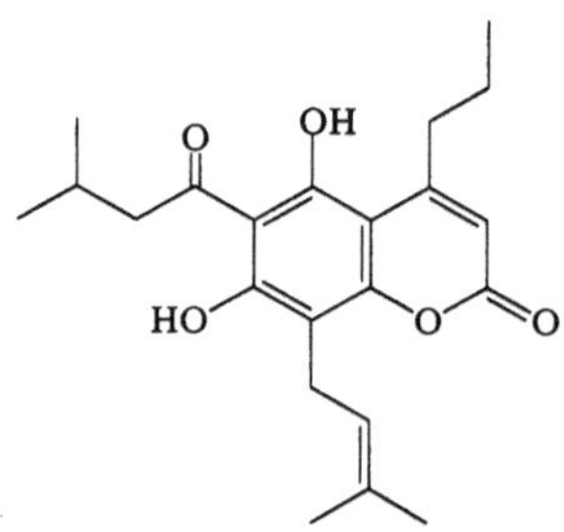
(**456**) Cyclonormammein	1970 1972	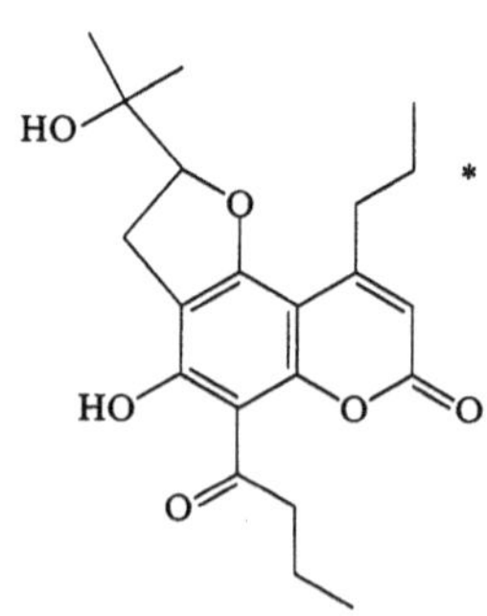

Formula	M. p.	$[\alpha]_{\lambda}^{t}$	Solvent	Plant Source(s)	Leading References
$C_{21}H_{26}O_5$	127.5—128.5			*Mammea americana*	(*144*)
$C_{22}H_{28}O_5$	114			*Mammea americana*	(*144, 147, 205*)
	115—116			*Mammea africana*	(*120, 121*)
$C_{22}H_{28}O_5$	119—121			*Mammea americana*	(*144, 147, 205*)
$C_{21}H_{26}O_6$	129.5—131.5			*Mammea americana*	(*143, 146*)
	126—128				(*190*)

Table 10 (continued)

	Trivial Name(s)	Year iso-lated	Structure
(**457**)	Cycloneomammein	1970 1972	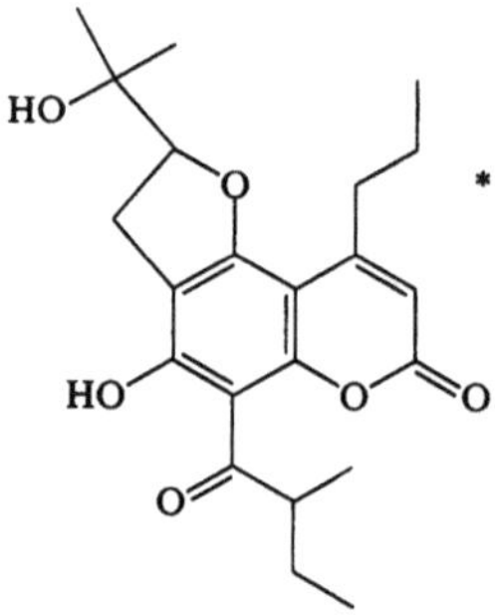
(**458**)	Cyclomammein	1970 1972	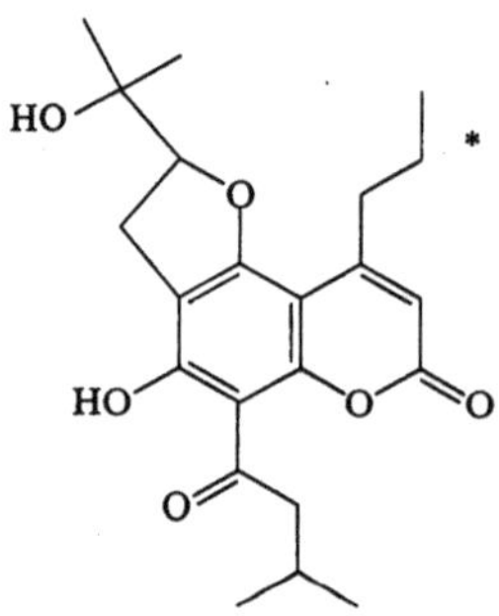
(**459**)		1970	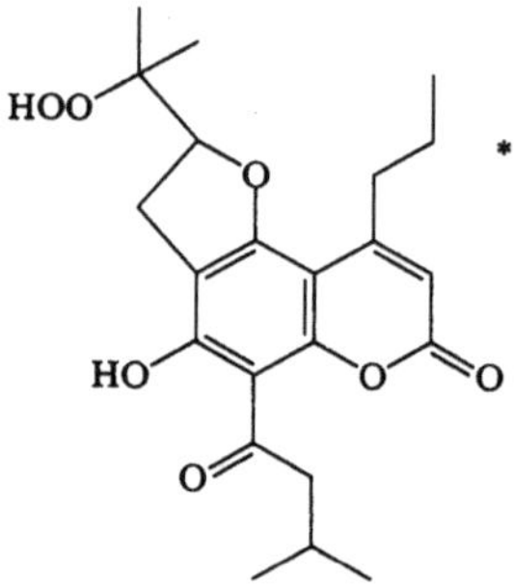
(**460**)		1970	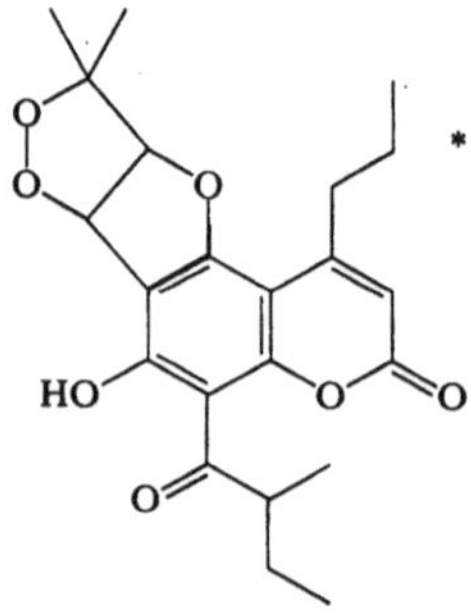

References, pp. 400—429

Formula	M.p.	$[\alpha]_{\lambda}^{t}$	Solvent	Plant Source(s)	Leading References
$C_{22}H_{28}O_6$	130—131 130—131			*Mammea americana*	(*143, 146*) (*190*)
$C_{22}H_{28}O_6$	126—127 128.5—129.5			*Mammea americana*	(*143, 146*) (*190*)
$C_{22}H_{28}O_7$	133	4_{297}		*Mammea americana*	(*143, 146*)
$C_{22}H_{26}O_7$	181—182			*Mammea americana*	(*143, 146*)

Table 10 (continued)

Trivial Name(s)	Year isolated	Structure
(461)	1970	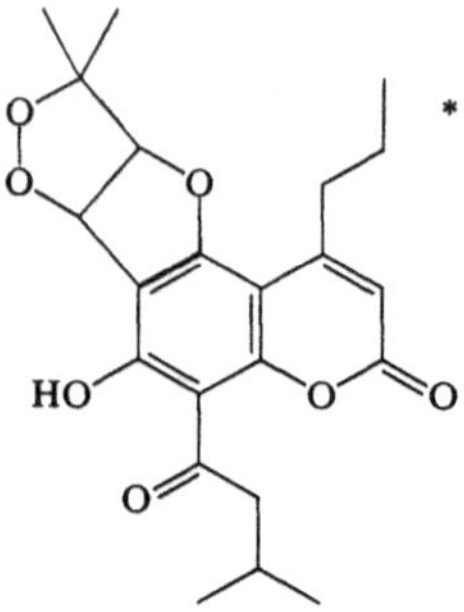 *
(462)	1972	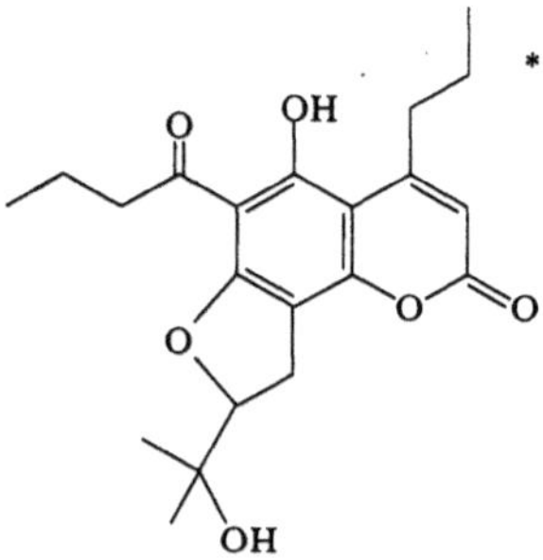 *
(463) MAB-4	1970 1970	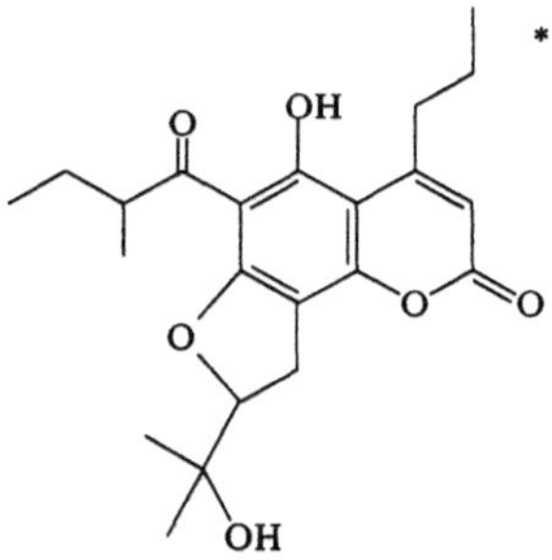 *
(464)	1970	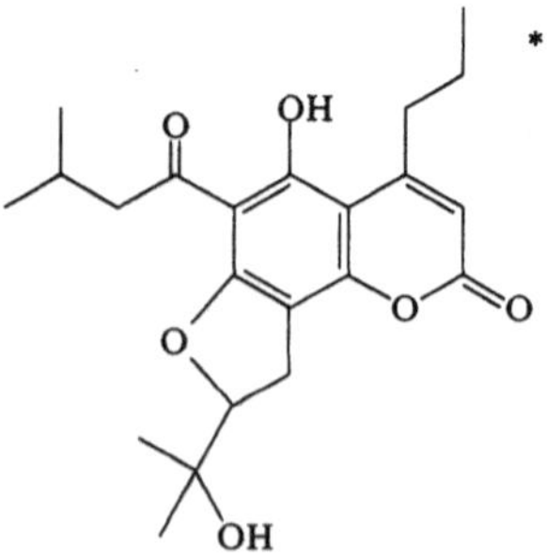 *

References, pp. 400—429

Formula	M. p.	$[\alpha]_\lambda^t$	Solvent	Plant Source(s)	Leading References
$C_{22}H_{26}O_7$	181—182			*Mammea americana*	(*143, 146, 189*)
$C_{21}H_{26}O_6$	75—81			*Mammea americana*	(*144*)
$C_{22}H_{28}O_6$	92—94 83—84			*Mammea americana* *Mammea africana*	(*144, 147*) (*120, 121*)
$C_{22}H_{28}O_6$	72—77			*Mammea americana*	(*144, 147*)

Table 10 (continued)

Trivial Name(s)	Year iso- lated	Structure
(465)	1970	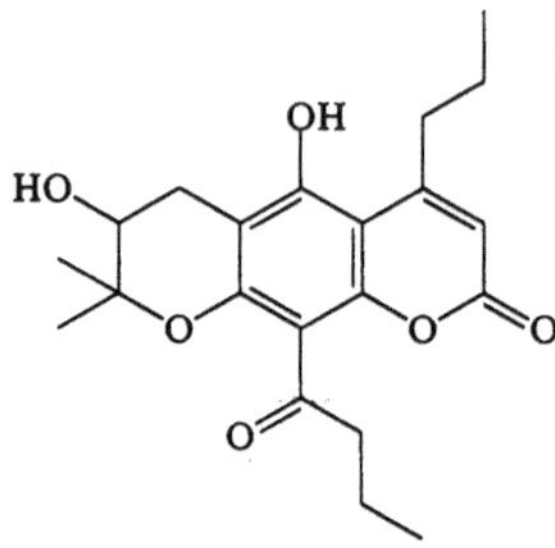 *
(466)	1970	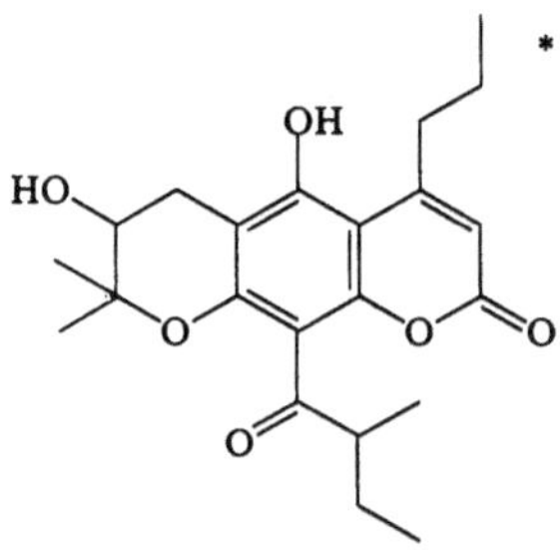 *
(467) M-9	1970 1972	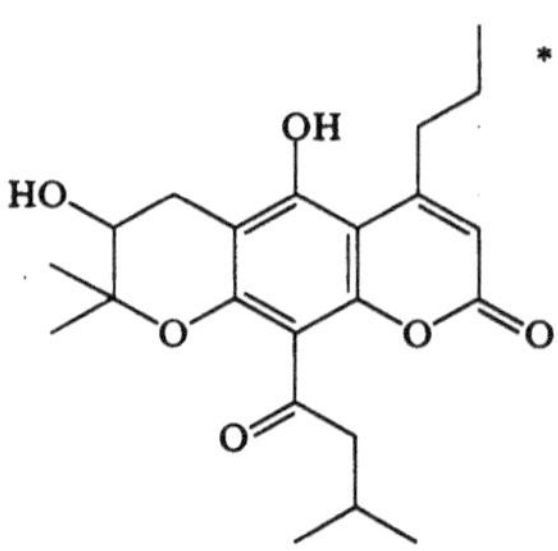 *
(468)	1972	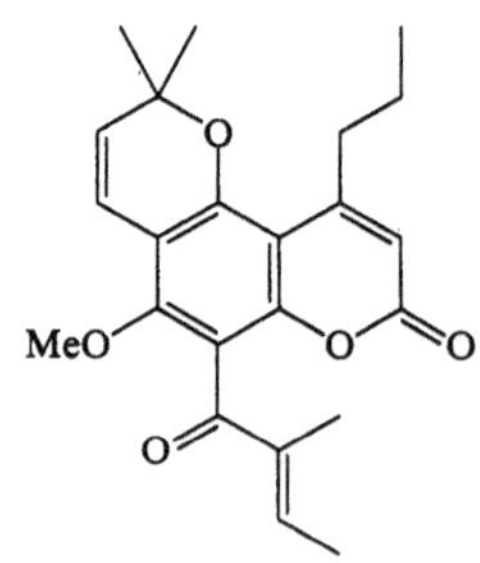

Formula	M.p.	$[\alpha]_{\lambda}^{t}$	Solvent	Plant Source(s)	Leading References
$C_{21}H_{26}O_6$	213—217			*Mammea americana*	(*143, 146*)
$C_{22}H_{28}O_6$	213—217			*Mammea americana*	(*143, 146*)
$C_{22}H_{28}O_6$	209—212 217—219			*Mammea americana*	(*143, 146*) (*189*)
$C_{23}H_{26}O_5$	121—123			*Calophyllum inophyllum*	(*122*)

Table 10 (continued)

Trivial Name(s)	Year iso-lated	Structure
(469)	1972	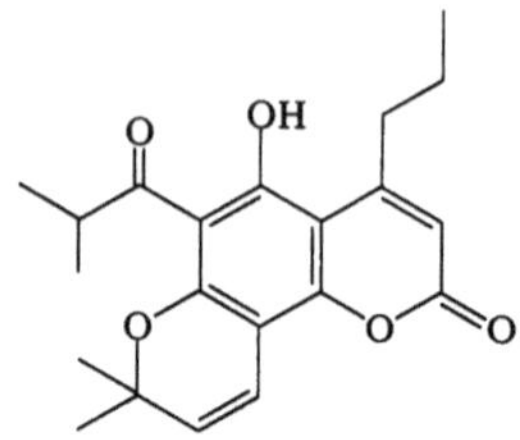
(470)	1972	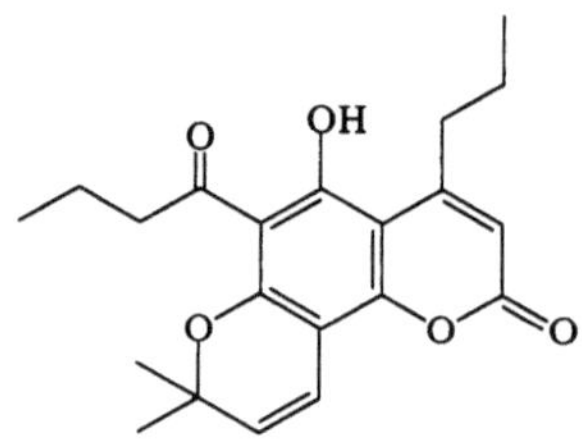
(471) MAB-6	1970 1972	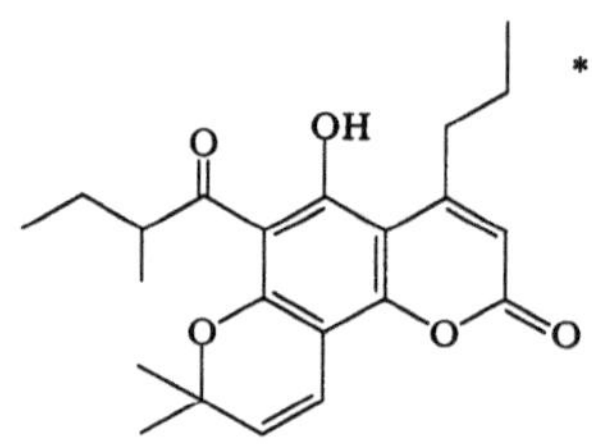
(472)	1972	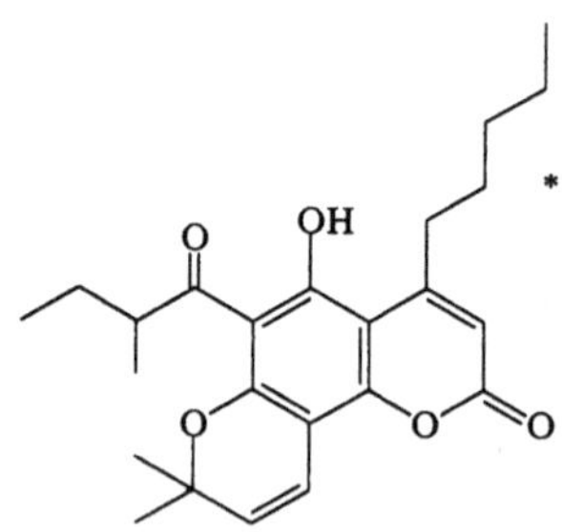
(473)	1972	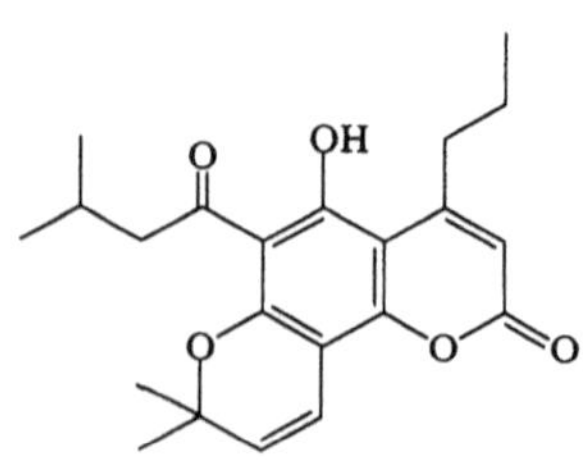

Formula	M. p.	$[\alpha]^t_\lambda$	Solvent	Plant Source(s)	Leading References
$C_{21}H_{24}O_5$				*Mammea americana*	(*204*)
$C_{21}H_{24}O_5$				*Mammea americana*	(*204*)
$C_{22}H_{26}O_5$	gum			*Mammea africana* *Mammea africana*	(*120, 121*) (*204*)
$C_{24}H_{30}O_5$				*Mammea americana*	(*204*)
$C_{22}H_{26}O_5$				*Mammea americana* *Mammea africana*	(*204*)

Table 10 (continued)

Trivial Name(s)	Year iso-lated	Structure
(**474**)	1972	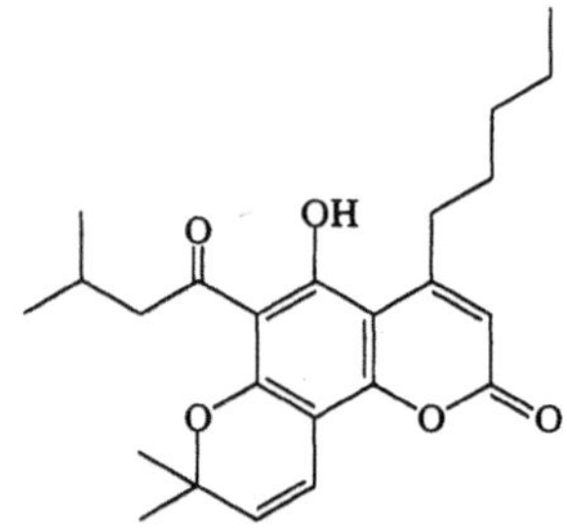
(**475**) Tomentolide B	1967	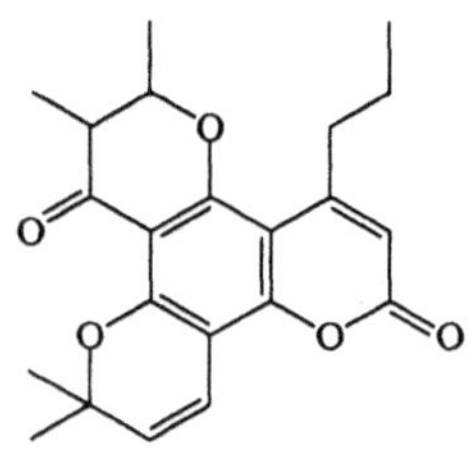
(**476**) Costatolide	1964	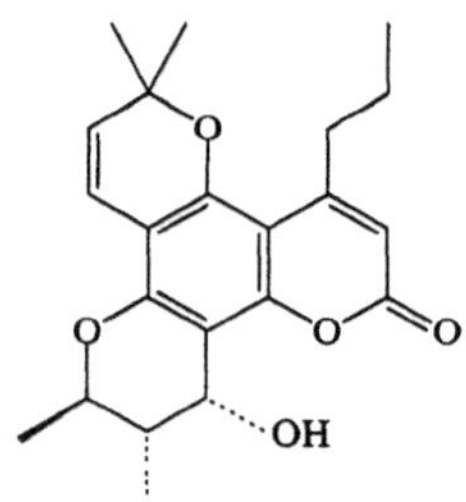
(**477**)	1972	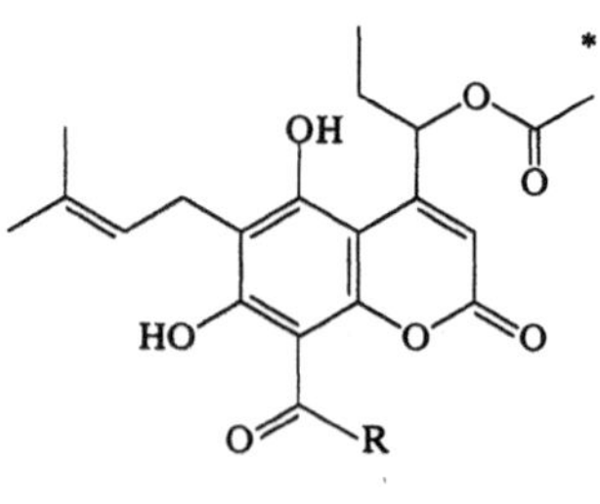 R = *n*-Pr or *i*-Pr

Formula	M. p.	$[\alpha]_\lambda^t$	Solvent	Plant Source(s)	Leading References
$C_{24}H_{30}O_5$				*Mammea americana*	(*204*)
$C_{22}H_{24}O_5$	158—160			*Calophyllum tomentosum*	(*431*)
$C_{22}H_{26}O_5$	181—182			*Calophyllum costatum*	(*565*)
$C_{23}H_{28}O_7$				*Mammea americana*	(*145*)

Table 10 (continued)

Trivial Name(s)	Year isolated	Structure
(**478**)	1970	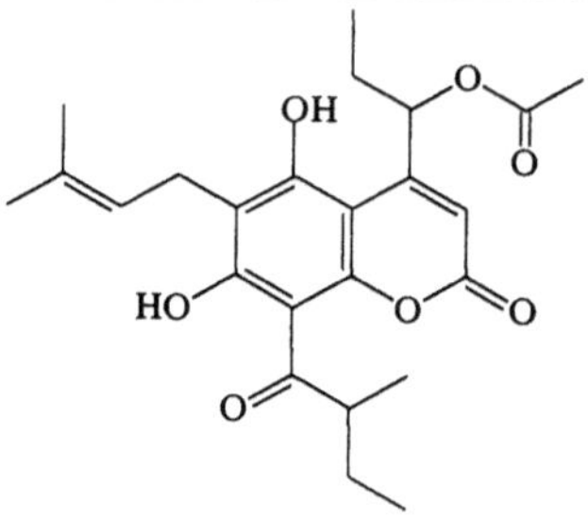 *
(**479**) Surangin B	1969	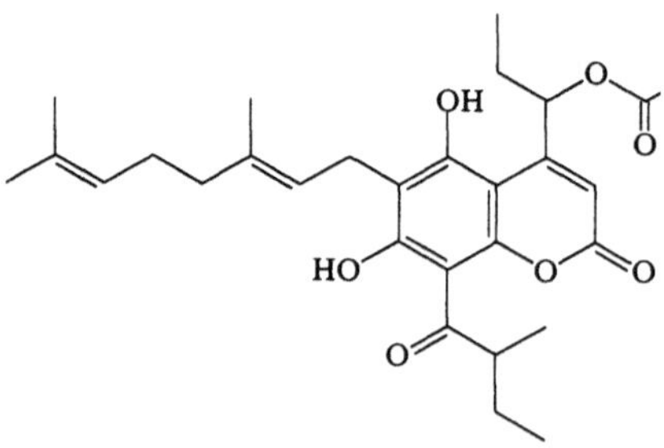
(**480**)	1970	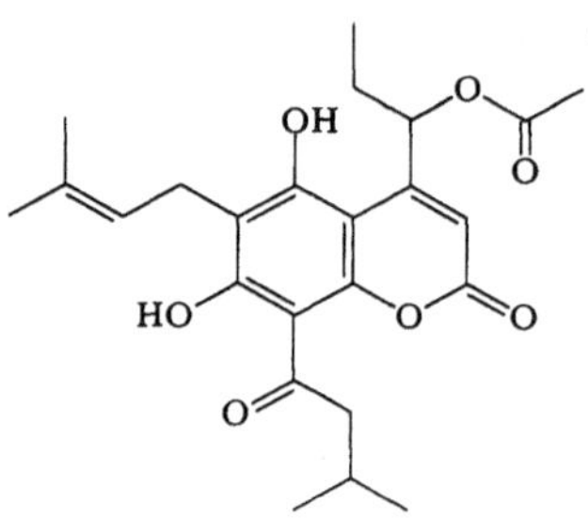 *

Formula	M.p.	$[\alpha]_\lambda^t$	Solvent	Plant Source(s)	Leading References
$C_{24}H_{30}O_7$	50—53	-28.5^{24}	EtOH	*Mammea americana*	(*142, 145*)
$C_{29}H_{38}O_7$	98—100	-30^{24}		*Mammea longifolia*	(*290*)
$C_{24}H_{30}O_7$	50—53	-28.5^{24}	EtOH	*Mammea americana*	(*142, 145*)

Table 11. *Miscellaneous Coumarins*

Trivial Name(s)	Year iso-lated	Structure
(**480a**)	1958	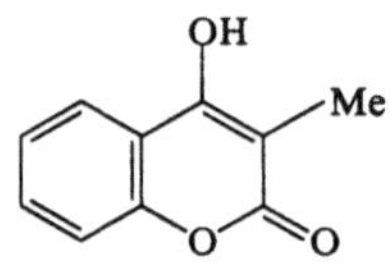
(**481**) Ferulenol	1964	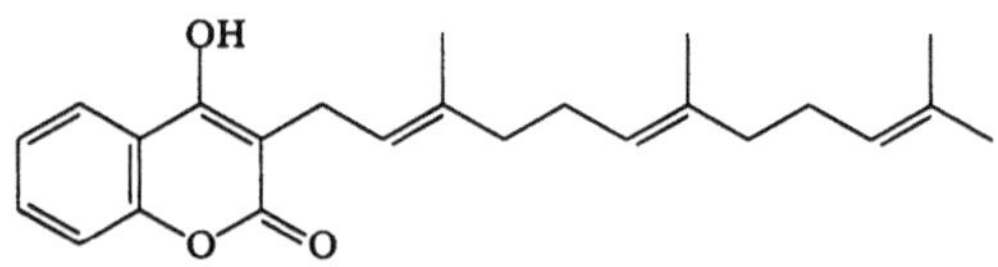
(**482**)	1965	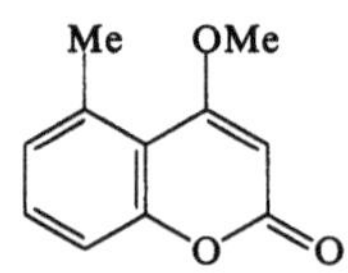
(**483**) Gerberacoumarin	1973	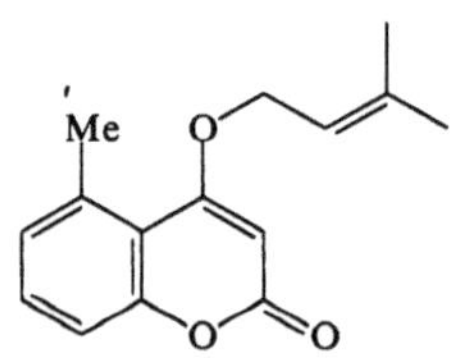
(**484**) Piloselloidal	1975	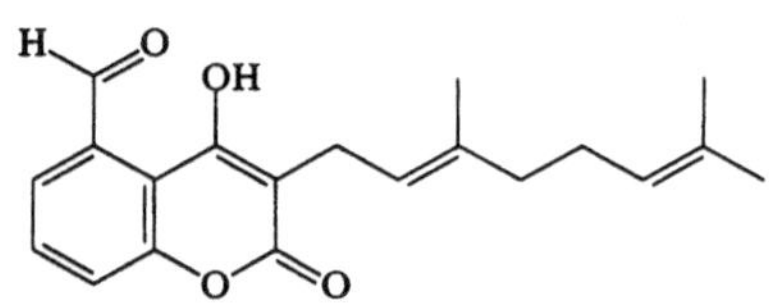
(**485**) Isogerberacoumarin	1973	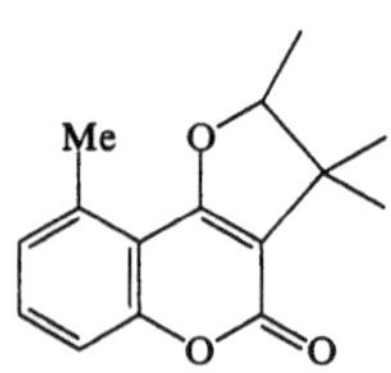 racem:
(**486**)	1964	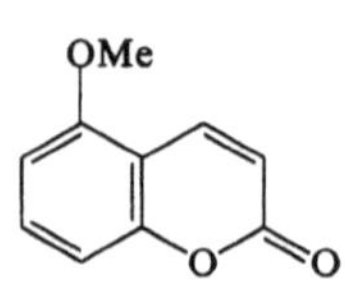

Formula	M.p.	$[\alpha]_{\lambda}^{t}$	Solvent	Plant Source(s)	Leading References
$C_{10}H_8O_3$	220—225			*Ferula communis*	(*115, 209*)
$C_{24}H_{30}O_3$	64—65			*Ferula communis*	(*117*)
$C_{11}H_{10}O_3$	165			*Ekebergia senegalensis*	(*67, 589*)
$C_{15}H_{16}O_3$	99			*Gerbera crocea*	(*89*)
$C_{20}H_{22}O_4$	115.5			*Gerbera piloselloides*	(*80*)
$C_{15}H_{16}O_3$	51			*Gerbera crocea*	(*89*)
$C_{10}H_8O_3$	80—81			*Dalea emoryi*	(*170, 177*)

Table 11 (continued)

	Trivial Name(s)	Year iso-lated	Structure	
(**487**)	Mahaleboside	1960	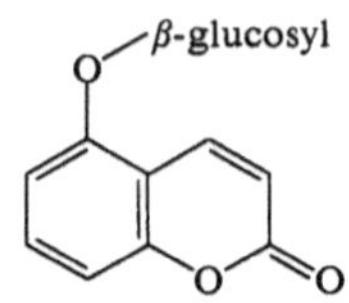	
(**488**)	Liqcoumarin	1976	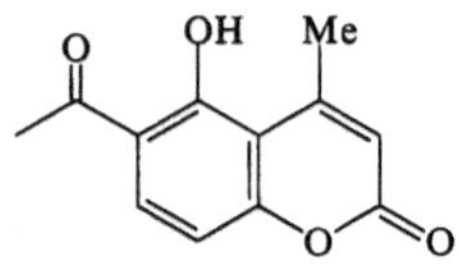	
(**489**)	Armillarisin A	1974	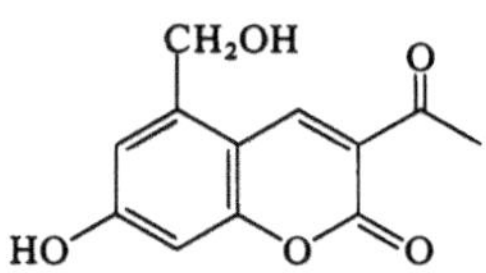	
(**490**)	4,5′,8-Trimethylpsoralen	1963	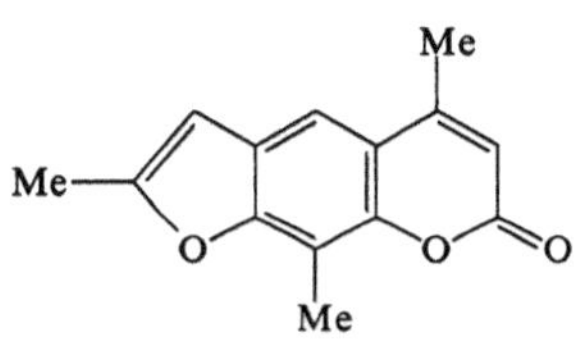	
(**491**)	Glaupalol	1967	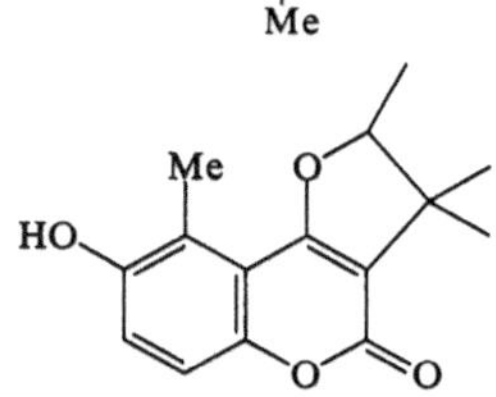	*
(**492**)	Glaupadiol	1973	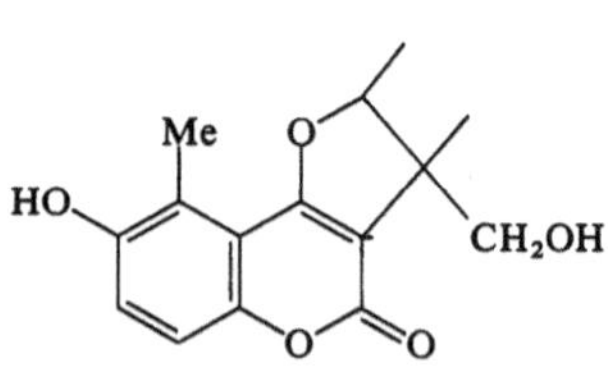	* *cis* + *tran*
(**493**)	Ammoresinol	1925	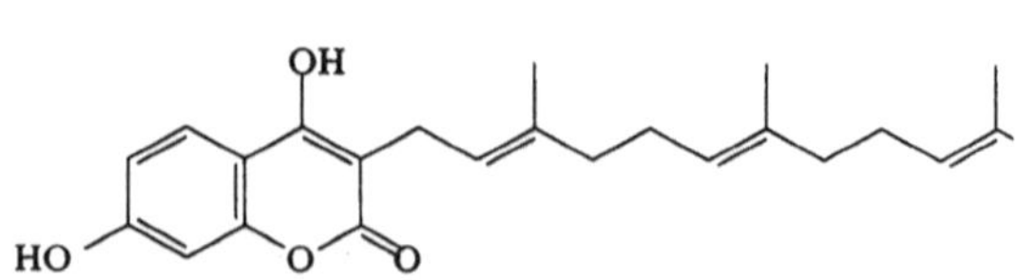	

References, pp. 400—429

Formula	M.p.	$[\alpha]_\lambda^t$	Solvent	Plant Source(s)	Leading References
$C_{15}H_{16}O_8$	268	−36	pyridine	*Prunus mahaleb*	(*470*)
$C_{12}H_{10}O_4$	165—166			*Glycyrrhiza glabra*	(*70*)
$C_{12}H_{10}O_5$	245—246			*Armillariella tabescens*	(*134, 312*)
$C_{14}H_{12}O_3$	226—228			*Apium graveolens* infected with *Sclerotinia sclerotiorum*	(*512*)
$C_{15}H_{16}O_4$	202—204			*Glaucidium palmatum*	(*276, 600*)
$C_{15}H_{16}O_5$	234	−22.5	MeOH	*Glaucidium palmatum*	(*316*)
$C_{24}H_{30}O_4$	107—108			*Dorema ammoniacum*	(*153*)

Table 11 (continued)

Trivial Name(s)	Year iso-lated	Structure
(494)	1972	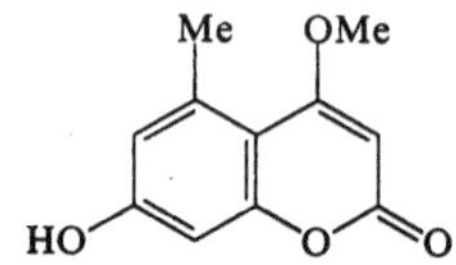
(495) Siderin	1972	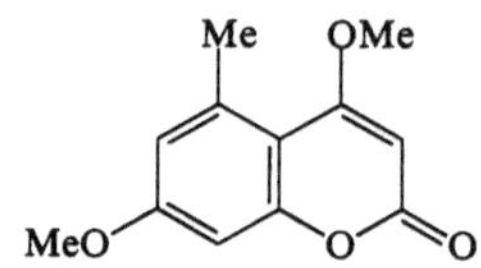
(496)	1972	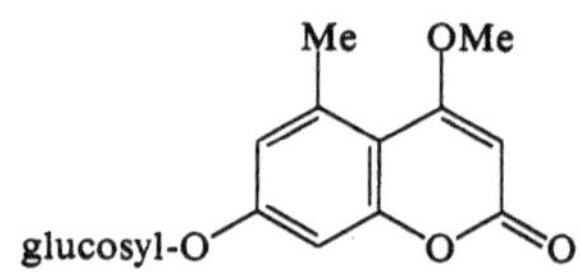
(497) Halkendin	1971	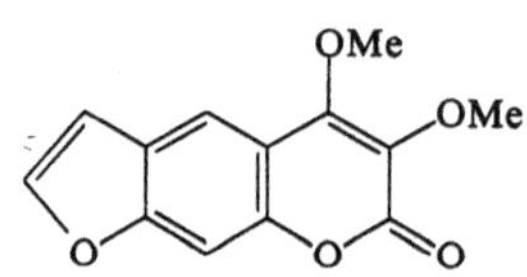
(498)	1976	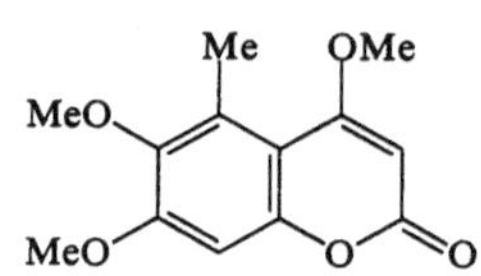
(499)	1972	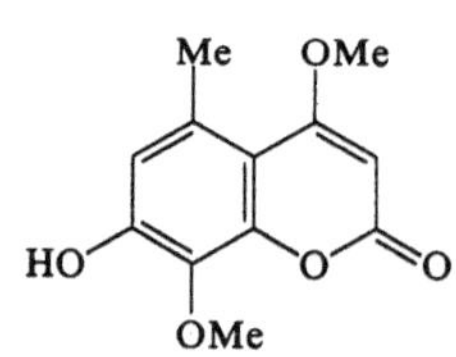

Formula	M. p.	$[\alpha]^t_\lambda$	Solvent	Plant Source(s)	Leading References
$C_{11}H_{10}O_4$	297—300			*Platymiscium praecox*	(*98*)
$C_{12}H_{12}O_4$	196—197			*Sideritis canariensis*	(*136, 229, 230, 362, 590*)
$C_{17}H_{20}O_9$	246—248			*Platymiscium praecox*	(*98*)
$C_{13}H_{10}O_5$	173—173.5			*Halfordia kendack*	(*69, 415*)
$C_{13}H_{14}O_5$	209			*Leonotis nepetaefolia*	(*477*)
$C_{12}H_{12}O_5$	206—208			*Platymiscium praecox*	(*98*)

Table 11 (continued)

Trivial Name(s)	Year iso- lated	Structure
(**500**) Halfordin	1956	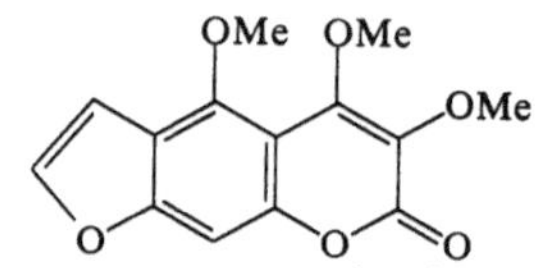
(**501**) Halfordinin	1970	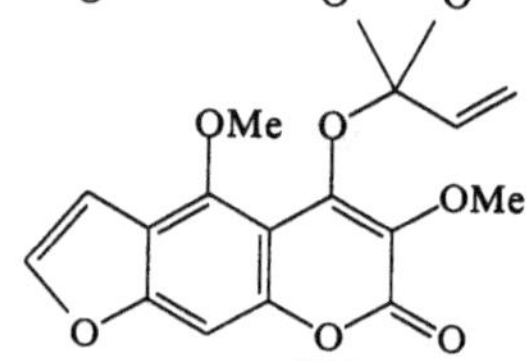
(**502**) Isohalfordin	1956	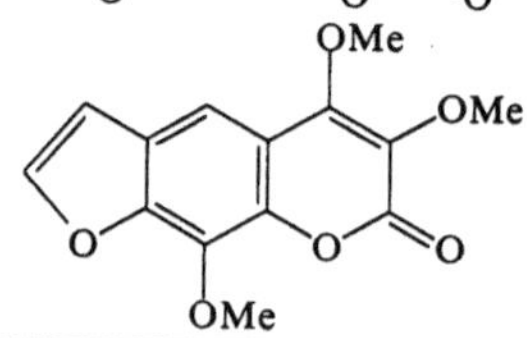

Formula	M. p.	$[\alpha]_\lambda^t$	Solvent	Plant Source(s)	Leading References
$C_{14}H_{12}O_6$	136—137			*Halfordia scleroxyla*	(*68, 268, 359*)
$C_{18}H_{18}O_6$	110			*Halfordia kendack*	(*377, 378*)
$C_{14}H_{12}O_6$	151—152			*Halfordia scleroxyla*	(*202, 268, 359, 414*)

Formula Index

Formula	*Compound Number*
$C_{15}H_{14}O_5$	**(408)**
$C_{15}H_{14}O_6$	**(84), (382), (394)**
$C_{15}H_{16}O_3$	**(87), (143), (428), (483), (485)**
$C_{15}H_{16}O_4$	**(88), (89), (90), (146), (148), (151), (152), (245), (314), (323), (324), (328), (344), (440), (442), (491)**
$C_{15}H_{16}O_5$	**(96), (97), (164), (326), (339), (349), (368), (388), (492)**
$C_{15}H_{16}O_6$	**(420)**
$C_{15}H_{16}O_7$	**(395)**
$C_{15}H_{16}O_8$	**(78), (487)**
$C_{15}H_{16}O_9$	**(304), (307), (335), (337)**
$C_{15}H_{18}O_5$	**(95), (154)**
$C_{16}H_{12}O_5$	**(261)**
$C_{16}H_{14}O_3$	**(435), (436)**
$C_{16}H_{14}O_4$	**(242), (254), (290), (291), (331), (352), (353), (371)**
$C_{16}H_{14}O_5$	**(255), (256), (257), (258), (260), (354), (355), (356), (357), (358)**
$C_{16}H_{15}ClO_5$	**(266), (364)**
$C_{16}H_{16}O_5$	**(108), (169), (170), (253), (285)**
$C_{16}H_{16}O_6$	**(262), (263), (359), (360)**
$C_{16}H_{16}O_7$	**(343)**
$C_{16}H_{18}O_4$	**(248), (281), (282), (429), (441), (443)**
$C_{16}H_{18}O_5$	**(249), (283), (284), (327), (419)**
$C_{16}H_{18}O_9$	**(308), (311)**
$C_{16}H_{18}O_{10}$	**(414), (424)**
$C_{16}H_{20}O_6$	**(250), (286)**
$C_{17}H_{16}O_4$	**(292), (372), (373)**
$C_{17}H_{16}O_5$	**(374), (390), (399), (400), (409)**
$C_{17}H_{16}O_6$	**(401), (403)**
$C_{17}H_{17}ClO_6$	**(406)**
$C_{17}H_{18}O_6$	**(264), (265), (362), (375), (398)**
$C_{17}H_{18}O_7$	**(376), (404)**
$C_{17}H_{20}O_6$	**(156)**
$C_{17}H_{20}O_9$	**(496)**
$C_{17}H_{20}O_{10}$	**(383), (384), (385), (386), (418), (425), (426)**
$C_{18}H_{16}O_5S$	**(212)**
$C_{18}H_{18}O_5S$	**(109)**
$C_{18}H_{18}O_6$	**(501)**
$C_{18}H_{18}O_7$	**(120), (186)**
$C_{18}H_{20}O_5$	**(171)**
$C_{18}H_{20}O_6$	**(363)**
$C_{18}H_{22}O_5$	**(155)**
$C_{19}H_{18}O_4$	(293), (298)
$C_{19}H_{18}O_5$	**(22), (23)**
$C_{19}H_{18}O_6$	**(24)**
$C_{19}H_{20}O_3$	**(437), (438)**
$C_{19}H_{20}O_4$	**(14), (15), (16), (17), (275), (277), (300), (301)**
$C_{19}H_{20}O_5$	**(21), (110), (111), (112), (127), (128), (173), (174), (205), (206), (276), (278), (299)**
$C_{19}H_{20}O_6$	**(175), (197), (215), (227)**
$C_{19}H_{22}O_3$	**(7), (100), (101), (144), (430)**
$C_{19}H_{22}O_4$	**(8), (9), (10), (12), (432), (433)**
$C_{19}H_{22}O_5$	**(20), (147), (153), (172), (204)**
$C_{19}H_{22}O_6$	**(157), (183), (184), (191)**
$C_{19}H_{22}O_7$	**(114)**
$C_{19}H_{24}O_5$	**(18), (445)**
$C_{20}H_{20}O_4$	**(297)**
$C_{20}H_{22}O_4$	**(294), (302), (484)**
$C_{20}H_{22}O_5$	**(150), (295), (296)**
$C_{20}H_{22}O_6$	**(130)**
$C_{20}H_{22}O_7$	**(187)**
$C_{20}H_{24}O_3$	**(431)**
$C_{20}H_{24}O_4$	**(246), (288), (315), (325), (341), (345), (370), (444)**
$C_{20}H_{24}O_5$	**(207), (247), (287)**
$C_{20}H_{24}O_6$	**(165), (316)**
$C_{20}H_{24}O_7$	**(98)**
$C_{20}H_{24}O_8$	**(145)**
$C_{20}H_{24}O_9$	**(106), (107)**
$C_{20}H_{24}O_{10}$	**(185), (347), (348)**
$C_{20}H_{26}O_6$	**(317)**
$C_{21}H_{18}O_5$	**(115)**
$C_{21}H_{20}O_6$	**(259)**
$C_{21}H_{22}O_4$	**(270), (271), (366), (367)**
$C_{21}H_{22}O_5$	**(272), (410)**
$C_{21}H_{22}O_7$	**(121), (188), (189), (194), (198), (217), (218), (228), (235), (269), (365)**
$C_{21}H_{22}O_8$	**(219)**
$C_{21}H_{24}O_5$	**(11), (13), (434), (469), (470)**
$C_{21}H_{24}O_6$	**(273)**
$C_{21}H_{24}O_7$	**(192), (216), (222), (223), (268)**
$C_{21}H_{26}O_5$	**(387), (446), (447), (453)**
$C_{21}H_{26}O_6$	**(456), (462), (465)**
$C_{21}H_{26}O_{13}$	**(79), (309), (312)**
$C_{22}H_{18}O_7$	**(116)**
$C_{22}H_{24}O_5$	**(407), (475)**
$C_{22}H_{24}O_8$	**(405)**
$C_{22}H_{26}O_5$	**(209), (471), (473), (476)**

Trivial Name Index

Name	*Compound Number*
Felamidin	(**115**)
Feropolidin	(**58**)
Feropolin	(**38**)
Feropolol	(**37**)
Feropolone	(**39**)
Feroside	(**31**)
Ferruol A	(**449**)
Ferudenol	(**339**)
Ferudiol	(**157**)
Ferulenol	(**481**)
Feruliden	(**353**)
Ferulin	(**402**)
Feselol	(**54**)
Floroselin	(**229**)
Fraxetin	(**412**)
Fraxidin	(**415**)
Fraxin	(**414**)
Fraxinol	(**379**)
Fraxinoside	(**385**)
Furopinnarin	(**292**)
Galbanic acid	(**41**)
Galbanum acid	(**43**)
Geijerin	(**90**)
Geiparvarin	(**22**)
Gerberacoumarin	(**483**)
Glabralactone	(**285**)
Glaupadiol	(**492**)
Glaupalol	(**491**)
Gosferol	(**258**)
Gravelliferone	(**430**)
Graveolone	(**129**)
Gummosin	(**46**)
Gummosinin	(**76**)
Halfordin	(**500**)
Halfordinin	(**501**)
Halkendin	(**497**)
Heliettin	(**432**)
Heraclenin	(**354**), (**355**), (**356**)
Heraclenol	(**360**)
Heraclin	(**252**)
Heracol	(**398**)
Heratomin	(**331**)
Heratomol	(**329**)
Herniarin	(**3**)
Hortiline	(**298**)
Hortiolone	(**293**)
Hydrangetin	(**336**)
6′-Hydroxy-β-cyclo-aurapten	(**10**), (**12**)
Hydroxyeriobrucinol	(**276**)
Iliensin	(**367**)
Imperatorin	(**352**)
Imperatorin oxide	(**355**), (**356**)
Iselin	(**271**)
Isoaculeatin	(**283**)
Isoangenomalin	(**102**)
Isobergapten	(**289**)
Isoedultin	(**194**)
Isofloroselin	(**220**)
Isofraxetin	(**377**)
Isofraxidin	(**413**)
Isofraxoside	(**424**)
Isogerberacoumarin	(**485**)
Isogosferol	(**357**)
Isohalfordin	(**502**)
Isoheraclenin	(**358**)
Isoimperatorin	(**254**)
Isomammein	(**455**)
Isomeranzin	(**152**)
Isooxyimperatorin	(**358**)
Isooxypeucedanin	(**260**)
Isopeucenidin	(**189**)
Isopeulustrin	(**177**)
Isopimpinellin	(**397**)
Isoponcimarin	(**153**)
Isopteryxin	(**228**)
Isorutarin	(**347**)
Isosamarcandin	(**65**)
Isosamidin	(**218**)
Isoscopoletin	(**305**)
Jatamansin	(**205**)
Jatamansinol	(**203**)
Kamolol	(**74**)
Kamolone	(**75**)
Karatavic acid	(**40**)
Karatavicin	(**30**)
Karatavicinol	(**28**)
Kellerin	(**73**)
cis-Khellactone	(**213**)
trans-Khellactone	(**214**)
Komalin	(**359**)
Kopeodin	(**33**)
Kopeolin	(**35**)
Kopeoside	(**36**)
Kopetdaghin	(**34**)
Kopetin	(**77**)
Kvannin	(**202**)

Name	*Compound Number*
Lacoumarin	**(244)**
Libanoridin	**(170)**
Libanorin	**(174)**
Libanotin	**(188)**
Limettin	**(243)**
Lindiol	**(114)**
Liqcoumarin	**(488)**
Lomatin	**(203)**
Lophopterol	**(96)**
Luvangetin	**(369)**
M-9	**(467)**
MAB-2	**(454)**
MAB-4	**(463)**
MAB-6	**(471)**
Macrocarpin	**(150)**
Magnolioside	**(308)**
Mahaleboside	**(487)**
Majudin	**(252)**
Majurin	**(178)**
Mammea B/AA	**(455)**
Mammea B/AB	**(454)**
Mammea B/AC	**(453)**
Mammea B/BA	**(452)**
Mammea B/BB	**(448)**
Mammea B/BC	**(446)**
Mammea C/BB	**(450)**
Mammein	**(452)**
Mandshurin	**(384)**
Marmesin	**(103)**, **(105)**
Marmesinin	**(106)**
Marmin	**(18)**
Masquin	**(178)**
Meranzin	**(146)**
Meranzin hydrate	**(154)**
Methyl galbanate	**(42)**
Mexoticin	**(286)**
Micromelin	**(99)**
Micromelumin	**(99)**
Micropubescin	**(162)**
Mogoltacin	**(59)**
Mogoltadin	**(44)**
Mogoltadone	**(48)**
Mogoltavicin	**(72)**
Mogoltavidin	**(71)**
Mogoltavin	**(61)**
Molgoltavinin	**(62)**
Mogoltin	**(60)**
Moschatol	**(54)**
Murralongin	**(160)**
Murrangatin	**(164)**
Murrayone	**(163)**
Neobyakangelicol	**(403)**
Neohydrangin	**(79)**
Neomammein	**(448)**
Nevskin	**(65)**
Nieshoutin	**(328)**
Nieshoutol	**(388)**
Nodakenetin	**(104)**
Nodakenin	**(107)**
Norbraylin	**(332)**
Nordentatin	**(300)**
Normammein	**(446)**
Nuttalin	**(206)**
Obliquetin	**(323)**
Obliquetol	**(322)**
Obliquin	**(319)**
Obliquol	**(320)**
Obtusifol	**(420)**
Officinalin	**(82)**
Oroselol	**(200)**
Oroselone	**(202)**
Osthenol	**(142)**
Osthol	**(143)**
Ostruthin	**(100)**
Ostruthol	**(269)**
Oxypeucedanin	**(255)**, **(256)**, **(257)**
Oxypeucedanin D-glucoside	**(267)**
Oxypeucedanin hydrate	**(262)**, **(263)**
Oxypeucedanin methanolate	**(264)**
Pabularinone	**(358)**
Pabulenol	**(258)**
Pabulenone	**(261)**
Pangeline	**(258)**
Paniculatin	**(165)**
Pedicellone	**(316)**
Peuarenarine	**(136)**
Peuarenine	**(137)**
Peuarin	**(130)**
Peucedanin	**(125)**
Peucedanol	**(93)**
Peucenidin	**(198)**
Peucenol	**(101)**
Peuchloridin	**(140)**
Peuchlorin	**(138)**
Peuchlorinin	**(139)**
Peuformosin	**(233)**
Peulustrin	**(176)**
Peumorisin	**(101)**
Peuruthenicin	**(82)**
Phebalosin	**(159)**
Phellopterin	**(400)**

Name	Compound Number
Piloselloidal	(484)
Pimpinellin	(389)
Pinnarin	(281)
Pinnaterin	(429)
Polyanthin	(45)
Polyanthinin	(47)
Poncimarin	(147)
Poncitrin	(302)
Prandiol	(118)
Pranferin	(155)
Pranferol	(253)
Prangeferol	(105)
Prangenidin	(371)
Prangenin	(354)
Prangenone	(358)
Prangol	(262)
Prangolarin	(256)
Prangosin	(124)
Prantschimgin	(112)
Prenyletin	(306)
Psoralen	(122)
Pteryxin	(217)
Puberulin	(419)
Reoselin	(29)
Rutacultin	(441)
Rutamarin	(434)
Rutamarin alcohol	(433)
Rutaretin	(346)
Rutarin	(348)
Rutolide	(436)
Sabandin	(123)
Sabandinin	(393)
Sabandinol	(395)
Sabandinone	(394)
Sachalinin	(180)
Samarcandin	(67)
Samarcandone	(69)
Samidin	(235)
Saxalin	(266)
Scoparin	(313)
Scoparone	(313)
Scopoletin	(310)
Scopolin	(311)
Sechulin	(212)
Secorin	(117)
Selinetin	(203)
Selinidin	(205)
Seseliflorin	(109)
Seselin	(241)
Sesibiricin	(288)
Sibiricin	(283)
Siderin	(495)
Skimmin	(78)
Smirnioridin	(121)
Smirniorin	(120)
Sphondin	(330)
Sprengelianine	(110)
Stenocarpin	(342)
Suberenol	(89)
Suberenon	(85)
Suberosin	(87)
Suksdorfin	(216)
Surangin A	(451)
Surangin B	(479)
Swietenone	(242)
Tavicone	(25)
Tederin	(408)
Thamnosmin	(91)
Thamnosmonin	(97)
Thamontanin	(376)
Toddaculin	(248)
Toddalolactone	(250)
Tomasin	(365)
Tomenin	(383)
Tomentin	(378)
Tomentolide B	(475)
Trachyphyllin	(301)
Tschuin	(116)
Ulopterol	(95)
Umbelliferone	(2)
Umbelliprenin	(26)
Umckalin	(380)
Vaginidin	(183)
Vaginidiol	(181)
Vaginol	(182)
Vellein	(145)
Visnadin	(222)
Xanthalin	(132)
Xanthoarnol	(119)
Xanthogalin	(205)
Xanthogalol	(203)
Xanthotoxin	(351)
Xanthotoxol	(350)
Xanthoxyletin	(280)
Xanthyletin	(141)
Xylotenin	(435)
Zosimin	(173)
Zosimol	(166)

References

1. ABERNETHY, J. L.: The Historical and Current Interest in Coumarin. J. Chem. Ed. **46,** 561 (1969).
2. ABU-MUSTAFA, E. A., F. K. A. EL-BAY, and M. B. E. FAYEZ: Natural Coumarins. XIII. Structure of Majurin, a New Constituent of *Ammi majus* L. Fruits. Tetrahedron Letters **1971,** 1657.
3. — — — Ammirin, a New Coumarin Constituent from *Ammi majus* L. Fruits. Naturwiss. **62,** 39 (1975).
4. ABU-MUSTAFA, E. A., and M. B. E. FAYEZ: Natural Coumarins. I. Marmesin and Marmesinin, Further Products from the Fruits of *Ammi majus* L. J. Org. Chem. **26,** 161 (1961).
5. ABYSHEV, A. Z.: Isogosferol, a New Furocoumarin from the Root of *Prangos lophoptera.* Khim. prirod. Soedinenii **10,** 83 (1974); Chem. Abs. **80,** 120816 (1974).
6. — Structures of Components of the Roots of *Prangos ferulacea.* Khim. prirod. Soedinenii **10,** 568 (1974; Chem. Abs. **82,** 139897 (1975).
7. — Coumarin Composition of the Roots, Stems and Fruit of *Prangos lophoptera.* Khim. prirod. Soedinenii **10,** 708 (1974); Chem. Abs. **82,** 135682 (1975).
8. — Structure of Suberosin Transformation Products and their Possible Biogenesis. Khim. prirod. Soedinenii **11,** 131 (1975); Chem. Abs. **83,** 96944 (1975).
9. — Structure of Lophopterol. Khim. prirod. Soedinenii **12,** 253 (1976); Chem. Abs. **85,** 59601 (1976).
10. ABYSHEV, A. Z., and I. V. BRODSKII: Prandiol, a New Dihydrofurocoumarin from the Roots of *Prangos biebersteinii.* Khim. prirod. Soedinenii **10,** 574 (1974); Chem. Abs. **82,** 121643 (1975).
11. ABYSHEV, A. Z., I. V. BRODSKII, and P. P. DENISENKO: Coumarins of *Prangos alata* Grossh. — *Prangos biebersteinii* Karjag. Khim. prirod. Soedninenii **9,** 269 (1973); Chem. Abs. **79,** 75844 (1973).
12. ABYSHEV, A. Z., I. V. BRODSKII, P. P. DENISENKO, and A. I. ERMAKOV: The Structure of Alatol. Khim. prirod. Soedinenii **9,** 722 (1973); Chem. Abs. **81,** 136018 (1974).
13. ABYSHEV, A. Z., and P. P. DENISENKO: Partial Synthesis of (±)-Pranferol. Khim. prirod. Soedinenii **7,** 522 (1971); Chem. Abs. **75,** 140731 (1971).
14. — — The Structure of Prangeferol: Khim. prirod. Soedinenii **8,** 114 (1972); Chem. Abs. **77,** 88220 (1972).
15. ABYSHEV, A. Z., P. P. DENISENKO, N. P. KOSTYUCHENKO, O. S. ANISIMOVA, A. I. ERMAKOV, and Y. N. SHEINKER: Pranferin, a New Coumarin from the Roots of *Prangos ferulacea.* Khim. prirod. Soedinenii **6,** 675 (1970); Chem. Abs. **74,** 111860 (1971).
16. ABYSHEV, A. Z., P. P. DENISENKO, N. P. KOSTYUCHENKO, A. I. ERMAKOV, and Y. N. SHEINKER: Gosferol, a New Furocoumarin from the Roots of *Prangos ferulacea.* Khim. prirod. Soedinenii **8,** 49 (1972); Chem. Abs. **77,** 58792 (1972).
17. — — — — — Natural Acetate of Meranzin Hydrate, a New Component of the Roots of *Prangos ferulacea.* Khim. prirod. Soedinenii **8,** 608 (1972); Chem. Abs. **78,** 121365 (1973).
18. ABYSHEV, A. Z., and A. M. KUTNEVICH: A Study of the Coumarin Composition of the Roots of *Prangos uloptera.* Khim. prirod. Soedinenii **4,** 378 (1968); Chem. Abs. **70,** 84948 (1969).
19. ABYSHEV, A. Z., A. M. KUTNEVICH, N. P. KOSTYUCHENKO, O. A. ANISIMOVA, A. I. ERMAKOV, and Y. N. SHEINKER: The Structure of Ulopterol. Khim. prirod. Soedinenii **6,** 300 (1970); Chem. Abs. **73,** 76997 (1970).
20. AHLUWALIA, V. K., V. N. GUPTA, C. L. RUSTAGI, and T. R. SESHADRI: Partial Methyl

Ethers of Polyhydroxy Coumarins: Synthesis of Isofraxidin and Fraxetin and Ring Isomeric Change in Coumarins. J. Sci. Ind. Res., India **19B,** 345 (1960).

21. AHLUWALIA, V. K., T. R. SESHADRI, and P. VENKATESWARLU: A Convenient Synthesis of Bergapten. Indian J. Chem. **7,** 831 (1969).

22. — — — A Total Synthesis of Isopimpinellin. Indian J. Chem. **9,** 194 (1971).

23. AKRAMOV, S. T., K. S. MUKHAMEDOVA, M. R. YAGUDAEV, and S. Y. YUNOSOV: NMR Spectrum of Prangosine. Khim. prirod. Soedinenii **3,** 287 (1967); Chem. Abs. **67,** 116998 (1967).

24. AMINOV, A. M., K. B. BIZHANOVA, and G. K. NIKONOV: Coumarins of *Libanotis lehmanniana*. Khim. prirod. Soedinenii **11,** 246 (1975); Chem. Abs. **83,** 144484 (1975).

25. AMINOV, A. M., and G. K. NIKONOV: Coumarins of *Seseli tenuisectum*. Khim. prirod. Soedinenii **8,** 38 (1972); Chem. Abs. **77,** 58793 (1972).

26. — — A Study of the Components of the Roots of *Seseli tschuense*. Khim. prirod. Soedinenii **8,** 799 (1972); Chem. Abs. **78,** 108204 (1973).

27. — — The Structure of Chuin, a New Coumarin from the Roots of *Seseli tschuense*. Khim. prirod. Soedinenii **9,** 487 (1973); Chem. Abs. **80,** 105843 (1974).

28. — — Structure of the Coumarin, Sechulin. Khim. prirod. Soedinenii **10,** 152 (1974); Chem. Abs. **81,** 105334 (1974).

29. ANDRIANOVA, V. B., and Y. E. SKLYAR: Isopeucenidin from *Libanotis montana*. Khim. prirod. Soednineii **11,** 89 (1975); Chem. Abs. **83,** 128636 (1975).

30. ANDRIANOVA, V. B., Y. E. SKLYAR, M. E. PEREL'SON, and M. G. PIMENOV: Kellerin, a New Coumarin from the Roots of *Ferula kelleri*. Khim. prirod. Soedinenii **9,** 795 (1973); Chem. Abs. **81,** 120383 (1974).

31. APLIN, R. T., and C. B. PAGE: The Constituents of Native Umbelliferae, Part 1. Coumarins from Dill (*Anetheum graveolens* L.). J. Chem. Soc. (C) **1967,** 2593.

32. ARTEM'EVA, M. V., M. O. KARRYEV, and G. K. NIKONOV: Isofraxetin, a New Coumarin Extracted from *Fraxinus potamophila*. Izvest. Akad. Nauk Turkm. S.S.R., Ser. Biol. Nauk **1973,** 63; Chem. Abs. **79,** 89463 (1973).

33. — — — Fraxinoside, a New Glucoside from *Fraxinus potamophila*. Izvest. Akad. Nauk Turkm. S.S.R., Ser. Biol. Nauk **1973,** 82; Chem. Abs. **79,** 146800 (1973).

34. ARTEM'EVA, M. V., G. K. NIKONOV, and M. O. KARRYEV: Coumarins of *Fraxinus mandschurica* and *F. potamophila*. Khim. prirod. Soedinenii **9,** 493 (1973); Chem. Abs. **80,** 105844 (1974).

35. ARTHUR, H. C., and C. M. LEE: An Examination of the Rutaceae of Hong Kong. Part V. A New Coumarin, Avicennin, from the Bark of *Zanthoxylum avicennae*. J. Chem. Soc. **1960,** 4654.

36. ARTHUR, H. C., and W. D. OLLIS: A Revised Structure for Avicennin. J. Chem. Soc. **1963,** 3910.

37. ATKINSON, E., D. R. BOYD, and M. F. GRUNDON: Coumarins of *Skimmia japonica*. Phytochemistry **13,** 853 (1974).

38. AUSTIN, P. W., and T. R. SESHADRI: Synthesis of Naturally Occurring Coumarin Derivatives: Part I — Synthesis of Dihydroseselin, Dihydroxanthyletin, Demethylsuberosin and Suberosin. Indian J. Chem. **6,** 412 (1968).

39. AUSTIN, P. W., T. R. SESHADRI, M. S. SOOD, and VISHWAPAUL: Components of *Seseli sibiricum*. Constitution and Synthesis of Sibiricin, a New Coumarin. Tetrahedron **24,** 3247 (1968).

40. AVRAMENKO, L. G., and G. K. NIKONOV: Saxalin, a New Furocoumarin from the Roots of *Angelica saxatilis*. Khim. prirod. Soedinenii **7,** 830 (1971); Chem. Abs. **76,** 138170 (1972).

41. AVRAMENKO, L. G., G. K. NIKONOV, and M. G. PIMENOV: Andelin, a New Dihydropyranocoumarin from the Root of *Angelica decursiva*. Khim. prirod. Soedinenii **6,** 190 (1970); Chem. Abs. **73,** 87814 (1970).

42. BAGIROV, V. Y., and N. P. KIR'YALOV: Coumarins from the Roots of *Ferula nevskii*. Khim. prirod. Soedinenii **6,** 465 (1970); Chem. Abs. **74,** 11044 (1971).

43. — — A Coumarin from the Roots of *Ferula nevskii*. Khim. prirod. Soedinenii **8,** 387 (1972); Chem. Abs. **77,** 139729 (1972).

44. BAGIROV, V. Y., N. P. KIR'YALOV, and V. I. SHEICHENKO: The Coumarin Tavicone from the Roots of *Ferula karatavica*. Khim. prirod. Soedinenii **5,** 591 (1969); Chem. Abs. **73,** 25241 (1970).

45. — — — Structure of Samarcandin. Khim. prirod. Soedinenii **6,** 465 (1970); Chem. Abs. **74,** 8196 (1971).

46. BAGIROV, V. Y., N. P. KIR'YALOV, V. I. SHEICHENKO, and V. V. BOCHKAROV: Structure of Badrakemin. Khim. prirod. Soedinenii **6,** 466 (1970); Chem. Abs. **73,** 124595 (1970).

47. BAGIROV, V. Y., and V. I. SHEICHENKO: The Structure of Karatavic Acid. Khim. prirod. Soedinenii **11,** 700 (1975); Chem. Abs. **84,** 164564 (1976).

48. — — The Structure and Stereochemistry of Tavicone: Khim. prirod. Soedinenii **12,** 452 (1976); Chem. Abs. **85,** 192493 (1976).

49. BAGIROV, V. Y., V. I. SHEICHENKO, and A. I. BAN'KOVSKII: Stereochemistry of Nevskin. Khim. prirod. Soedinenii **12,** 450 (1976).

50. BALLANTYNE, M. M., P. H. MCCABE, and R. D. H. MURRAY: Claisen Rearrangements – II Synthesis of Six Natural Coumarins. Tetrahedron **27,** 871 (1971).

51. BALLANTYNE, M. M., R. D. H. MURRAY, and A. B. PENROSE: An Out-of-Ring Claisen Rearrangement During the Synthesis of Nieshoutin and Obliquetin. Tetrahedron Letters **1968,** 4155.

52. BANDOPADHYAY, M., S. B. MALIK, and T. R. SESHADRI: Synthesis of New Coumarin Components of *Heracleum candicans*. Indian J. Chem. **11,** 530 (1973).

53. BANDOPADHYAY, M., N. P. PARDESHI, and T. R. SESHADRI: Synthesis of Some Naturally Occurring Coumarins. Indian J. Chem. **12,** 23 (1974).

54. BANDOPADHYAY, M., and T. R. SESHADRI: Components of *Heracleum candicans*: Part I. Indian J. Chem. **8,** 855 (1970).

55. — — Components of *Heracleum candicans*: Part II. Indian J. Chem. **8,** 1146 (1970).

56. BAN'KOVSKII, A. I., N. E. ERMATOV, and M. E. PEREL'SON: Libanoridin, a New Coumarin from *Libanotis schrenkiana*. Khim. prirod. Soedinenii **5,** 52 (1969); Chem. Abs. **71,** 7705 (1969).

57. BAN'KOVSKII, A. I., N. E. ERMATOV, M. E. PEREL'SON, L. BUBEVA-IVANOVA, and N. S. PAVLOVA: Structure of the Coumarins Colladin and Colladonin. Khim. prirod. Soedinenii **6,** 173 (1970); Chem. Abs. **73,** 76988 (1970).

58. BARNES, C. S., and J. L. OCCOLOWITZ: The Mass Spectra of Some Naturally Occurring Oxygen Heterocycles and Related Compounds. Austral. J. Chem. **17,** 975 (1964).

59. BASA, S. C.: Crenulatin, a Formyl Coumarin from *Hesperathusa crenulata* (Rutaceae). Austral. J. Chem. **28,** 1159 (1975).

60. BASA, S. C., J. CHATTERJEE, and A. CHATTERJEE: Extractives of Umbelliferae: Pabularirone, a New Furocoumarin from *Prangos pabularia* Lindl. Chem. and Ind. **1970,** 746.

61. — — — Pabulenol — a Biological Transformation Product of Oxypeucedanin. Tetrahedron Letters **1971,** 1977.

62. BATES, R. B., D. J. ECKERT, S. K. PAKNIKAR, and V. P. THALACKER: Structure and Synthesis of Archangelin. Spectral Methods for Distinguishing Bergaptyl from Isobergaptyl Ethers. Tetrahedron Letters **1972,** 3811.

63. BATES, R. B., J. H. SCHAUBLE, and M. SOUČEK: The $C_{10}H_{17}$ Side Chain in Mycelianamide. The Stereochemistry of Bergamottin and Umbelliprenin. Tetrahedron Letters **1963,** 1683.

64. BEGLEY, M. J., L. CROMBIE, D. A. SLACK, and D. A. WHITING: Course of the Condensation between 5,7-Dihydroxycoumarin and Citral: A Reinvestigation of the Constitution of Bruceol and Deoxybruceol from *Eriostemon brucei*. J. C. S. Chem. Comm. **1976**, 140.

65. BENCZE, W., J. EISENBEISS, and H. SCHMID: Die Konstitution des Visamminols. Helv. Chim. Acta **39**, 923 (1956).

66. BERNOTAT-WULF, H., A. NIGGLI, L. ULRICH, and H. SCHMID: Über die absolute Konfiguration der Visnagane. Helv. Chim. Acta **52**, 1165 (1969).

67. BEVAN, C. W. L., and D. E. U. EKONG: Occurrence of 8-Methoxy-4-methyl-coumarin in *Ekebergia senegalensis*. A. Juss. Chem. and Ind. **1965**, 383.

68. BHANU, S., T. SAROJA, T. R. SESHADRI, and S. K. MUKERJEE: A Synthesis of Halfordin. Indian J. Chem. **9**, 380 (1971).

69. — — — — New Syntheses of Halkendin and Isohalkendin. Indian J. Chem. **10**, 16 (1972).

70. BHARDWAJ, D. K., R. MURARI, T. R. SESHADRI, and R. SINGH: Liqcoumarin, a Novel Coumarin from *Glycyrrhiza glabra*. Phytochemistry **15**, 1182 (1976).

71. — — — — Lacoumarin from *Lawsonia inermis*. Phytochemistry **15**, 1789 (1976).

72. BIRCH, A. J., M. MAUNG, and A. PELTER: Studies in Relation to Biosynthesis XL. Some Aspects of the Chemistry of *o*-Isopentenylphenols and Related Compounds. Austral. J. Chem. **22**, 1923 (1969).

73. BOHLMANN, F., V. S. BHASKAR RAO, and M. GRENZ: Über die Cumarine aus *Angelica ursina* und *Seseli libanotis*. Tetrahedron Letters **1968**, 3947.

74. BOHLMANN, F., and E. CLAUSEN: Natürlich vorkommende Cumarin-Derivate, VI. Über die Inhaltsstoffe von *Capnophyllum*-Arten. Chem. Ber. **103**, 3619 (1970).

75. BOHLMANN, F., and H. FRANKE: Natürlich vorkommende Cumarin-Derivate, VII. Synthese von racemischen Lomatin, Columbianetin, Angenomalin und Samidin. Chem. Ber. **104**, 3229 (1971).

76. — — Isolation of obliquin from *Phaenocoma prolifera*. Phytochemistry **12**, 726 (1973).

77. BOHLMANN, F., H. FRANKE, and C. ZDERO: Natürlich vorkommende Cumarin-Derivate, VIII. Struktur und Synthese eines neuen Cumarins aus *Choisya ternata* Kunth. Anales de Quim **68**, 765 (1972).

78. BOHLMANN, F., and M. GRENZ: Natürlich vorkommende Cumarine-Derivate. IV. Über neue Furocoumarine aus *Ligusticum pyrenaicum* Koch. Chem. Ber. **102**, 1673 (1969).

79. — — Natürlich vorkommende Cumarine-Derivate, X. Notiz über die Inhaltsstoffe von *Cnidium silaifolium* (Jacq.). Sim. Chem. Ber. **106**, 1047 (1973).

80. — — Natürlich vorkommende Cumarine-Derivate, XI. Über die Inhaltsstoffe von *Gerbera piloselloides* Cass. Chem. Ber. **108**, 26 (1975).

81. — — Natürlich vorkommende Cumarine-Derivate, XIII. Notiz über zwei neue Cumarine-Derivate aus *Machaeranthera scabrella* (Greene) Shinners. Chem. Ber. **109**, 1584 (1976).

82. BOHLMANN, F., M. GRENZ, and C. ZDERO: Natürlich vorkommende Cumarine-Derivate, XII. Neue Cumarine aus *Peucedanum*- und *Pteronia*-Arten. Chem. Ber. **108**, 2955 (1975).

83. BOHLMANN, F., J. JACOB, and M. GRENZ: Natürlich vorkommende Terpen-Derivate, XLV. Über Inhaltsstoffe von *Scaevola lobelia* (Th.) Murr. Chem. Ber. **108**, 433 (1975).

84. BOHLMANN, F., and D. KÖRNIG: Natürlich vorkommende Terpen-Derivate, XXXVIII. Synthese von Cumarine-Terpenen. Chem. Ber. **107**, 1780 (1974).

85. BOHLMANN, F., and K. M. RODE: Natürlich vorkommende Cumarine-Derivate, II. Über die Cumarine aus *Libanotis buchtormensis*. Chem. Ber. **101**, 2741 (1968).

86. BOHLMANN, F., D. SCHUMANN, and C. ZDERO: Natürlich vorkommende Terpen-Derivate, XXVIII. Über ein neues Sesquiterpen-Derivate aus *Artemisia pontica* L. Chem. Ber. **107**, 644 (1974)

87. BOHLMANN, F., and W. THEFELD: Natürlich vorkommende Cumarine-Derivate, V. Neue Dihydroseselin-Derivate aus *Laserpitium archangelica* Wulf (1). Tetrahedron Letters **1970,** 3577.

88. BOHLMANN, F., and C. ZDERO: Natürlich vorkommende Terpen-Derivate, XLVIII. Über neue Inhaltsstoffe der Gattung *Anthemis.* Chem. Ber. **108,** 1902 (1975).

89. BOHLMANN, F., C. ZDERO, and H. FRANKE: Natürlich vorkommende Cumarine-Derivate, IX. Über die Inhaltsstoffe der Gattung *Gerbera.* Chem. Ber. **106,** 382 (1973).

90. BOHLMANN, F., C. ZDERO, and H. KAPTEYN: Über Cumarine-Derivate aus Pflanzen der Gattung Aster. Annalen **717,** 186 (1968).

91. BOHM, B. A., R. K. IBRAHIM, and G. H. N. TOWERS: The Partial Elucidation of the Structure of a New Coumarin from *Hydrangea macrophylla.* Canad. J. Biochem. Physiol. **39,** 1389 (1961).

92. BORISOV, V. N., A. I. BAN'KOVSKII, N. S. PAVLOVA, L. BUBEVA-IVANOVA. V. I. SHEICHENKO, and V. S. KABANOV: Colladocin, a New Terpenoid Coumarin from *Colladonia triquetra.* Khim. prirod. Soedinenii **11,** 247 (1975); Chem. Abs. **83,** 59096 (1975).

93. BORISOV, V. N., A. I. BAN'KOVSKII, V. I. SHEICHENKO, and V. S. KABANOV: Isosamarcandin from *Ferula microloba.* Khim. prirod. Soedinenii **10,** 786 (1974); Chem. Abs. **82,** 135686 (1975).

94. BORISOV, V. N., A. I. BAN'KOVSKII, V. I. SHEICHENKO, V. S. KABANOV, and P. I. ZAKHAROV: Methyl Galbanate — New Terpenoid Coumarin from *Ferula microloba.* Khim. prirod. Soedinenii **10,** 516 (1974); Chem. Abs. **82,** 14007 (1975).

95. BORISOV, V. N., A. I. BAN'KOVSKII, V. I. SHEICHENKO, and M. G. PIMENOV: Coumarins of *Ferula tuberifera.* Khim. prirod. Soedinenii **10,** 515 (1974); Chem. Abs. **82,** 28588 (1975).

96. BORISOV, V. N., A. I. BAN'KOVSKII, V. I. SHEICHENKO, M. G. PIMENOV, and P. I. ZAKHAROV: The Structure of Galbanic Acid. Khim. prirod. Soedinenii **9,** 429 (1973); Chem. Abs. **79,** 91913 (1973).

97. BOTTOMLEY, W.: Discophoridin, a New Coumarin from *Velleia discophora* F. MUELL. Austral. J. Chem. **16,** 143 (1963).

98. BRAGA DE OLIVEIRA, A., L. G. FONSECA E SILVA, and O. R. GOTTLIEB: Flavonoids and Coumarins from *Platymiscium praecox.* Phytochemistry **11,** 3515 (1972).

99. BRAZ FILHO, R., A. F. MAGALHAES, and O. R. GOTTLIEB: A Quimica de Moráceas Brasileiras I. Novas Cumarinas do *Brosimum paraense.* Anais Acad. brasil Cienc. **42,** 139 (1970); Chem. Abs. **75,** 724586 (1971).

100. — — — Chemistry of Brazilian Moraceae II. Brosiparin and Other Coumarins from *Brosimum rubescens.* Anais Acad. brasil Cienc. **43,** 585 (1971); Chem. Abs. **77,** 161990 (1972).

101. — — — Coumarins from *Brosimum rubescens.* Phytochemistry **11,** 3307 (1972).

102. BROOKER, R. M., J. N. EBLE, and N. A. STARKOWSKY: Chalepensin, Chalepin and Chalepin Acetate, Three Novel Furocoumarins from *Ruta chalepensis.* Lloydia **30,** 73 (1967).

103. BROWN, D., R. O. ASPLUND, and V. A. MCMAHON: An NMR Procedure for Determining Methoxy Positions of Coumarin Trimethyl Ethers using the Benzene Induced Solvent Shift. Phytochemistry **13,** 1923 (1974).

104. — — — Phenolic Constituents of *Artemisia tridentata* spp. *vaseyana.* Phytochemistry **14,** 1083 (1975).

105. BROWN, S. A.: Recent Studies on the Formation of Natural Coumarins. Lloydia **26,** 211 (1963).

106. BROWN, S. A., M. EL-DAKHAKHNY, and W. STECK: Biosynthesis of Linear Furanocoumarins. Canad. J. Biochem. **48,** 863 (1970).

107. BROWN, S. A., and W. STECK: 7-Demethylsuberosin and Osthenol as Intermediates in Furanocoumarin Biosynthesis. Phytochemistry **12,** 1315 (1973).

108. BROWN, S. A., G. H. N. TOWERS, and D. CHEN: Biosynthesis of the Coumarins-V. Pathways of Umbelliferone Formation in *Hydrangea macrophylla.* Phytochemistry **3,** 469 (1964).

109. BUBEVA-IVANOVA, L., and N. S. PAVLOVA: Phytochemical Study of *Colladonia triquetra.* I. Natural Coumarins. Farmatsiya (Bulgaria) **1,** 30 (1969); Chem. Abs. **71,** 57558 (1969).

110. BUDZIKIEWICZ, H., C. DJERASSI, and D. H. WILLIAMS: Structure Elucidation of Natural Products by Mass Spectrometry. **2.** San Francisco: Holden-Day. 1964.

111. CAGLIOTI, L., H. NAEF, D. ARIGONI, and O. JEGER: Zur Kenntnis der Sesquiterpene und Azulene. Über die Inhaltsstoffe der *Asa foetida* I. Farnesiferol A. Helv. Chim. Acta **41,** 2278 (1958).

112. — — — — Zur Kenntnis der Sesquiterpene und Azulene. Über die Inhaltsstoffe der *Asa foetida* II. Farnesiferol B und C. Helv. Chim. Acta **42,** 2557 (1959).

113. CALL, T. G.: Extractive Studies and the Volatile Oil from the Fruits of *Lomatium suksdorfii.* Proc. Montana Acad. Sci. **17,** 37 (1957); Chem. Abs. **52,** 4102 (1958).

114. CALL, T. G., and E. B. FISCHER: Some Pharmacological Effects of the Roots of *Pteryxia terebinthina* var. *terebinthina,* and of Pteryxin, a Substance Isolated from the Roots. Northwest Sci. **32,** 96 (1958); Chem. Abs. **53,** 6429 (1959).

115. CANNAVA, A.: Is 3-Methyl-4-hydroxycoumarin the Active Substance Responsible for the Hypoprothrombinemia Produced by *Ferula communis*? Boll. chim. Farm. **97,** 207 (1958); Chem. Abs. **52,** 18689 (1960).

116. CAPORALE, G., and G. RODIGHIERO: Furocoumarins in the Roots of *Angelica sylvestris.* Ricerca Sci., Rend. Sec. B **1,** 127 (1961); Chem. Abs. **57,** 2596 (1962).

117. CARBONI, S., V. MALAGUZZI, and A. MARSILI: Ferulenol, a New Coumarin Derivative from *Ferula communis.* Tetrahedron Letters **1964,** 2783.

118. CARDILLO, G., R. CRICCHIO, L. MERLINI, and G. NASINI: Poliisoprenifoli naturali. Sintesi della grifolina e della ostrutina. Gazzetta **99,** 308 (1969).

119. CARMAN, R. M., F. N. LAHEY, and J. K. MCLEOD: Geiparvarin: Proton Magnetic Resonance Studies. Austral. J. Chem. **20,** 1957 (1967).

120. CARPENTER, I., E. J. MCGARRY, and F. SCHEINMANN: The Neoflavonoids and 4-Alkylcoumarins from *Mammea africana* G. Don. Tetrahedron Letters, **1970,** 3983.

121. — — — Extractives from Guttiferae. Part XXI. The Isolation and Structure of Nine Coumarins from the Bark of *Mammea africana* G. Don. J. Chem. Soc. (C) **1971,** 3783.

122. CAVE, A., M. DEBRAY, G. HENRY, J. KUNESCH, and J. POLONSKY: Sur la structure d'une nouvelle alkyl-4 coumarine isolée de *Calophyllum inophyllum* Compt. rend. **275,** C, 1105 (1972).

123. CHAKRABORTY, D. P., B. K. CHOWDHURY, and B. C. DAS: Mexoticin, a New Coumarin from *Murraya exotica* L. (1). Tetrahedron Letters **1967,** 3471.

124. CHANG, P. T. O., G. A. CORDELL, G. H. AYNILIAN, H. H. S. FONG, and N. R. FARNSWORTH: Alkaloids and Coumarins of *Thamnosma montana.* Lloydia **39,** 134 (1976).

125. CHATTERJEE, A., J. BANERJI, and S. C. BASA: Lactonic Constituents of *Prangos pabularia* Lindl (Umbelliferae). Tetrahedron **28,** 5175 (1972).

126. CHATTERJEE, A., and A. BHATTACHARYA: The Isolation and Constitution of Marmin, a New Coumarin from *Aegle marmelos* Correa. J. Chem. Soc. **1959,** 1922.

127. CHATTERJEE, A., P. K. BOSE, and S. K. SAHA: Ferulin, the Main Lactone of *Ferula alliacea.* Arch. Pharm. **295,** 248 (1962).

128. CHATTERJEE, A., C. P. DUTTA, and S. BHATTACHARYYA: Micromelumin and Micropubescin — Two New Coumarins from *Micromelum pubescens* L. (Fam. Rutaceae). Sci. Cult. **33,** 371 (1967); Chem. Abs. **69,** 2887 (1968).

129. — — — Structure of Micropubescin, a New Coumarin from *Micromelum pubescens.* Sci. Cult. **34,** 366 (1968); Chem. Abs. **70,** 114946 (1969).

130. CHATTERJEE, A., C. P. DUTTA, S. BHATTACHARYYA, H. E. AUDIER, and B. C. DAS: The Structure of Marmin. Tetrahedron Letters **1967**, 471.

131. CHATTERJEE, A., and S. DUTTA: Structure Investigation of New Lactonic Constituents from the Roots of *Angelica archangelica* Linn. Indian J. Chem. **6**, 415 (1968).

132. CHATTERJEE, A., R. N. REJ, A. BANERJI, and J. BANERJI: Abnormal Rearrangement of Isoprenoid Epoxide: Synthesis of 7-Methoxy-8-(2′-formyl-2′-methylpropyl)-coumarin, the Lactonic Constituent of Grapefruit Peel Oil of *Citrus decumana* Linn. (Rutaceae). Chem. and Ind. **1976**, 410.

133. CHATTERJEE, A., and S. SEN GUPTA: The Constitution of Archangelin, a New Coumarin Isolated from the Root of *Angelica archangelica* Linn. (Umbelliferae). Tetrahedron Letters, **1964**, 1961.

134. Chemical Section, Luminuous Fungus Co-Operative Research Group of Kiangsu: *Armillariella tabescens*. II. Isolation and Determination of the Structure of the Armillarisins. *Wei Sheng Wu Hsueh Po* **14**, 9 (1974); Chem. Abs. **82**, 40463 (1975).

135. CHERNOBAI, V. T., and D. G. KOLESNIKOV: Coumarins of *Cnidium dubium*. Doklady Akad. Nauk S.S.S.R. **133**, 233 (1960); Chem. Abs. **54**, 23188 (1960).

136. CHEXAL, K. K., C. FOUWEATHER, and J. S. E. HOLKER: Biosynthesis of Fungal Metabolites VII. Production and Biosynthesis of 4,7-Dimethoxy-5-methylcoumarin in *Aspergillus variecolor*. J. C. S. Perkin I **1975**, 554.

137. CHI, H.-J.: Components of Umbelliferous Plants in Korea. IV. Coumarin Derivatives of the Root of *Angelica*. Yakhak Hoeji **11**, 36 (1967); Chem. Abs. **71**, 36408 (1969).

138. CHOW, P. W., A. M. DUFFIELD, and P. R. JEFFERIES: The Chemistry of the Western Australian Rutaceae V. The Isolation of Some Coumarins and the Structure of Phebalosin. Austral. J. Chem. **19**, 483 (1966).

139. COATES, R. M., and L. S. MELVIN: Total Synthesis of (±)-Marmin and Related Coumarin Monoterpenes. Tetrahedron **26**, 5699 (1970).

140. COMBES, G., R. PERNET, and R. PIERRE: Sur l'isolement, la structure et les propriétés d'une nouvelle coumarine extraite de *Toddalia aculeata* Pers. Bull. Soc. chim. France **1961**, 1609.

141. CORDOVA B, H. E., and L. N. GARELLI M.: A New Coumarin in *Amyris simplicifolia*. Phytochemistry **13**, 758 (1974).

142. CROMBIE, L., D. E. GAMES, N. J. HASKINS, and G. F. REED: Isolation and Structure of Insecticidal Components from *Mammea americana* L. Tetrahedron Letters **1970**, 251.

143. — — — — Extractives of *Mammea americana* L. Part III. Identification of New Coumarin Relatives of Mammea B/BA, B/BB and B/BC having 5,6-Annulation and Higher Oxidation Levels. J. C. S. Perkin I **1972**, 2241.

144. — — — — Extractives of *Mammea americana* L. Part IV. Identification of New 7,8-Annulated Relatives of the Coumarins Mammea A/AA, A/AB, B/AA, and B/AB, and New Members of the 6-Acyl Family B/AA, B/AB, and B/AC. J. C. S. Perkin I **1972**, 2248.

145. — — — — Extractives of *Mammea americana* L. Part V. The Insecticidal Compounds. J. C. S. Perkin I **1972**, 2255.

146. CROMBIE, L., D. E. GAMES, N. J. HASKINS, G. F. REED, R. A. FINNEGAN, and K. E. MERKEL: Identification of Nine New 5,6-Annulated Relatives of the Mammea coumarins B/BA, B/BB, and B/BC. Tetrahedron Letters **1970**, 3975.

147. — — — — — — Identification of New Mammea Coumarins: Four 7,8-Annulated Relatives of the Mammea Coumarins A/AA-A/AB, B/AA, B/AB, and Two of the 6-Acyl Family B/AA (Isomammein) and B/AB. Tetrahedron Letters **1970**, 3979.

148. CROMBIE, L., D. E. GAMES, and A. MCCORMICK: Isolation and Structure of Mammea B/BA, B/BB, B/BC and C/BB: A Group of 4-*n*-Propyl- and 4-*n*-Amyl-Coumarin Extractives of *Mammea americana* L. Tetrahedron Letters **1966**, 151.

149. — — — Extractives of *Mammea americana* L. Part I. The 4-*n*-Alkylcoumarins. Isolation and Structure of Mammea B/BA, B/BB, B/BC and C/BB. J. Chem. Soc. (C) **1967,** 3545.

150. CROMBIE, L., and R. PONSFORD: Pyridine-Catalyzed Condensation of Citral with Phloroglucinols, a Novel Reaction Leading to Tetracyclic Bis-Ethers and Chromenes. J. Chem. Soc. (C) **1971,** 788.

151. CUSSANS, N. J., and T. N. HUCKERBY: Carbon-13 NMR Spectroscopy of Heterocyclic Compounds — IV. A 20 MHz Study of Chemical Shifts and Carbon-Proton Coupling Constants in a Series of Hydroxy, Methoxy and Glucosyl Coumarins. Tetrahedron **31,** 2719 (1975).

152. DALL'ACQUA, F., A. CAPOZZI, S. MARCIANI, and G. CAPORALE: Biosynthesis of Furocoumarins: Further Studies on *Ruta graveolens*. Z. Naturforsch. **27b,** 813 (1972).

153. DEAN, F. M.: Naturally Occurring Coumarins. Fortschr. Chem. org. Naturstoffe **9,** 225 (1952).

154. — In: Naturally Occurring Oxygen Ring Compounds. London: Butterworths. 1963.

155. DEAN, F. M., and B. PARTON: The Structure and Synthesis of (+)-Obliquin. J. Chem. Soc. (C) **1969,** 526.

156. DEAN, F. M., B. PARTON, N. SOMVICHIEN, and D. A. H. TAYLOR: The Coumarins of *Ptaeroxylon obliquum*. Tetrahedron Letters **1967,** 2147.

157. DEAN, F. M., and D. A. H. TAYLOR: Extractives from East African Timbers. Part II. *Ptaeroxylon obliquum*. J. Chem. Soc. (C) **1966,** 114.

158. DELLE MONACHE, F., F. MARLETTI, and G. MARINI-BETTOLO: Coumarins of *Hortia arborea* (Rutaceae): Hortiline and Hortiolone. Gazzetta **106,** 681 (1976).

159. DESHMUKH, M. N., V. H. DESHPANDE, and A. V. R. RAO: Two New Coumarins from *Toddalia aculeata*. Phytochemistry **15,** 1419 (1976).

160. DJERASSI, C., E. J. EISENBRAUN, R. A. FINNEGAN, and B. GILBERT: Naturally Occurring Oxygen Heterocyclics. V. Mammein. Tetrahedron Letters **1959,** 10.

161. — — — — Naturally Occurring Oxygen Heterocyclics VII. The Structure of Mammein. J. Org. Chem. **25,** 2164 (1960).

162. DJERASSI, C., E. J. EISENBRAUN, B. GILBERT, A. J. LEMIN, S. P. MARFEY, and M. P. MORRIS: Naturally Occurring Oxygen Heterocyclics. II. Characterisation of an Insecticidal Principle from *Mammea americana* L. J. Amer. Chem. Soc. **80,** 3686 (1958).

163. DREYER, D. L.: Constituents of *Thamnosma montana* Torr. and Frem. Tetrahedron **22,** 2923 (1966).

164. — Chemotaxonomy of the Rutaceae — II. Extractives of *Severinia buxifolia* (Poir.) Ten. Tetrahedron **23,** 4613 (1967).

165. — Chemotaxonomy of the Rutaceae. IV. Constituents of *Murraya paniculata* (Linn.) Jack. J. Org. Chem. **33,** 3574 (1968).

166. — Coumarins and Alkaloids of the Genus *Ptelea*. Phytochemistry **8,** 1013 (1969).

167. — Extractives of *Angelica genuflexa* Nutt. J. Org. Chem. **35,** 2294 (1970).

168. DREYER, D. L., and P. F. HUEY: Coumarins of *Citrus macroptera*. Phytochemistry **12,** 3011 (1973).

169. DREYER, D. L., and A. LEE: Extractives of *Geijera parviflora*. Phytochemistry **11,** 763 (1972).

170. DREYER, D. L., K. P. MUDERLOH, and W. E. THIESSEN: Extractives of *Dalea* Species (Leguminosae). Tetrahedron **31,** 287 (1975).

171. DREYER, D. L., M. V. PICKERING, and P. COHAN: Distribution of Limonoids in the Rutaceae. Phytochemistry **11,** 705 (1972).

172. DUFFIELD, A. M., P. R. JEFFERIES, E. N. MASLEN, and A. I. M. RAE: The Structure of Bruceol. Tetrahedron **19,** 593 (1963).

173. DUKHOVLINOVA, L. I., M. E. PEREL'SON, Y. E. SKLYAR, and M. G. PIMENOV: Coumarins of the Roots of *Seseli iliense*. Khim. prirod. Soedinenii **10**, 308 (1974); Chem. Abs. **81**, 105338 (1974).

174. DUKHOVLINOVA, L. I., Y. E. SKLYAR, and M. E. PEREL'SON: Secorin — A New Dihydrofurocoumarin from the Roots of *Seseli coronatum*. Khim. prirod. Soedinenii **9**, 663 (1973).

175. — — — Coumarins of the Roots of *Seseli coronatum*. Khim. prirod. Soedinenii **10**, 782 (1974); Chem. Abs. **82**, 121668 (1975).

176. EL-ANTABLY, S. M., and T. O. SOINE: Coumarins XII: Synthesis of (±)-*cis-trans*-3',4'-Dihydroxy-3',4'-dihydroxanthyletin and Their Diesters. J. Pharm. Sci. **62**, 1643 (1973).

177. EL-BASYOUNI, S., and G. H. N. TOWERS: The Phenolic Acids in Wheat. Natural Occurrence of Orthoferulic Acid. Canad. J. Biochem. **42**, 493 (1964).

178. ELLIS, B. E., and S. A. BROWN: Isolation of Dimethylallylpyrophosphate: Umbelliferone Dimethylallyl Transferase from *Ruta graveolens*. Canad. J. Biochem. **52**, 734 (1974).

179. ERMATOV, N. E., and A. I. BAN'KOVSKII: Chemical Study of *Libanotis buchtormensis*. Trudy Vses. Nauch Issled. Inst. Lek. Aromat. Rast. **15**, 149 (1969); Chem. Abs. **75**, 16083 (1971).

180. ERMATOV, N. E., A. I. BAN'KOVSKII, and M. E. PEREL'SON: Coumarins of *Ferula penninervis*. Khim. prirod. Soedinenii **2**, 158 (1966); Chem. Abs. **65**, 15790 (1966).

181. — — — A New Coumarin from *Libanotis schrenkiana*. Khim. prirod. Soedinenii **5**, 222 (1969); Chem. Abs. **72**, 63616 (1970).

182. ERMATOV, N. E., A. I. BAN'KOVSKII, M. E. PEREL'SON, G. P. SYROVA, and Y. N. SHEINKER: Coumarins of *Libanotis buchtormensis*. Khim. prirod. Soedinenii **4**, 145 (1968); Chem. Abs. **70**, 57702 (1969).

183. — — — — — Structure of Kamolone and Kamolol, New Coumarins from *Ferula penninervis*. Khim. prirod. Soedinenii **5**, 79 (1969); Chem. Abs. **71**, 615712 (1969).

184. ESHIETT, I. T., and D. A. H. TAYLOR: The Isolation and Structure Elucidation of Some Derivatives of Dimethylallyl-Coumarin, Chromone, Quinoline, and Phenol from *Fagara* Species and from *Cedrelopsis grevei*. J. Chem. Soc. (C), **1968**, 481.

185. EWING, J., G. K. HUGHES, and E. RITCHIE: The Chemical Constituents of Australian *Zanthoxylum* Species I. Suberosin, 6-(3,3-Dimethylallyl)-7-methoxycoumarin. Austral. J. Sci. Res. **3A**, 342 (1950).

186. FANGHAENEL, E., R. TAPANES, and J. P. ZAYAS: Analysis of the Coumarins of Centrifuged Essential Oils Produced in Cuba from the Lime (*Citrus aurantifolia*). Cienc. Fis. **3**, 111 (1971); Chem. Abs. **78**, 33779 (1973).

187. FINKELSTEIN, N., and D. E. A. RIVETT: Puberulin, a New Prenyloxy-Coumarin from *Agathosma puberula*. Phytochemistry **15**, 1080 (1976).

188. FINNEGAN, R. A., B. GILBERT, E. J. EISENBRAUN, and C. DJERASSI: Naturally Occurring Oxygen Heterocyclics. VIII. Synthesis of Some Coumarins Related to Mammein. J. Org. Chem. **25**, 2169 (1960).

189. FINNEGAN, R. A., and K. E. MERKEL: Constituents of *Mammea americana* L. IX: Oxidation of Mammein and Mammeisin. J. Pharm. Sci. **61**, 1603 (1972).

190. FINNEGAN, R. A., K. E. MERKEL, and N. BACK: Constituents of *Mammea americana* L. VIII: Novel Structural Variations on the Mammein Theme and Antitumor Activity of Mammein and Related Coumarin and Phloroglucinol Derivatives. J. Pharm. Sci. **61**, 1599 (1972).

191. FISCHER, F. C., P. H. JASPERSE, J. KARLSEN, and A. B. SVENDSEN: Ein neues Furanocoumaringlykosid aus *Heracleum mantegazzianum*. Phytochemistry **13**, 2334 (1974).

192. FISCHER, F. C., and A. B. SVENDSEN: Apterin, a Common Furanocoumarin Glycoside in the Umbelliferae. Phytochemistry **15**, 1079 (1976).

193. FISH, F., A. I. GRAY, R. D. WAIGH, and P. G. WATERMAN: Dipetalolactone: A Novel Pyranocoumarin from the Root Bark of *Zanthoxylum dipetalum*. Phytochemistry **15**, 313 (1976).

194. FISH, F., A. I. GRAY, and P. G. WATERMAN: Alkaloids, Coumarins, Triterpenes, and a Flavanone from the Root of *Zanthoxylum dipetalum*. Phytochemistry **14**, 2073 (1975).

195. FISHER, J. F., and H. E. NORDBY: Isolation and Spectral Characterisation of Coumarins in Florida Grapefruit Peel Oil. J. Food Sci. **30**, 869 (1965).

196. — — Two New Coumarins from Grapefruit Peel Oil. Tetrahedron **22**, 1489 (1966).

197. FISHER, J. F., H. E. NORDBY, A. C. WAISS, and W. L. STANLEY: A New Coumarin from Grapefruit Peel Oil. Tetrahedron **23**, 2523 (1967).

198. FLOSS, H. G.: Biosynthesis of Furanocoumarins. Recent Advan. Phytochem. **4**, 143 (1972).

199. FLOSS, H. G., and U. MOTHES: On the Biosynthesis of Furocoumarins in *Pimpinella magna*. Phytochemistry **5**, 161 (1966).

200. FUHRER, H., T. R. GOVINDACHARI, B. S. JOSHI, and B. R. PAI: Revised Structures of Clausenidin, Dentatin and the Identity of Dentatin with Poncitrin. Indian J. Chem. **8**, 198 (1970).

201. FUJITA, M., and T. FURUYA: Pharmacognostic Study of Crude Drugs Containing Coumarins and Their Derivatives. II. On an Acylcoumarin Newly Isolated from *Angelica* Species in Japan. Yakugaku Zasshi **76**, 538 (1956); Chem. Abs. **50**, 13000 (1956).

202. FUKUI, K., M. NAKAYAMA, S. FUJIMOTO, and O. FUKUDA: The Synthesis of Isohalfordin. Experientia **25**, 354 (1969).

203. FURUYA, T., and H. KOJIMA: Gas-Liquid Chromatography of Coumarins. J. Chromat. **29**, 382 (1967).

204. GAMES, D. E.: Identification of 4-Phenyl- and 4-Alkylcoumarins in *Mammea americana* L., *Mammea africana* G. Don and *Calophyllum inophyllum* by Gas Chromatography-Mass Spectrometry. Tetrahedron Letters **1972**, 3187.

205. GAMES, D. E., and N. J. HASKINS: Synthesis of Some Dimethylpyrano- and 3-Methylbut-2-enyl-, -4-Phenyl- and -4-*n*-Propyl coumarins. J. Chem. Soc. (D) **1971**, 1005.

206. GAMES, D. E., and A. H. JACKSON: Identification of Naturally Occurring Plant Phenolics by Gas Chromatography — Mass Spectrometry. In: A. FRIGERIO (ed.) Proc. Int. Symp. Gas Chromat. Mass Spectrom., p. 261. Milan: Tamburini. 1972; Chem. Abs. **79**, 2178 (1973).

207. GANGULY, A. K., B. S. JOSHI, V. N. KAMAT, and A. H. MANMADE: Synthesis of Clausenin, Xanthoxyletin, Alloxanthoxyletin, Xanthyletin and Nordalbergin. Tetrahedron **23**, 4777 (1967).

208. GASHIMOV, N. F., and G. A. KUZNETSOVA: A New Coumarin, Obtusifol, from *Haplophyllum obtusifolium*. Khim. prirod. Soedinenii **10**, 303 (1974); Chem. Abs. **81**, 105411 (1974).

209. GASTALDI, C.: Active Principles in *Ferula communis*. Ricerca Sci. **28**, 377 (1958); Chem. Abs. **52**, 13011 (1960).

210. GHOSAL, C. R., S. SEN GUPTA, and A. CHATTERJEE: Prangolarine, the Optical Isomer of Oxypeucedanin, Isolated from the Roots of *Prangos pabularia* Lindl. (Umbelliferae). Chem. and Ind. **1963**, 1430.

211. GIORGOBIANI, E. D., N. F. KOMISSARENKO, and E. P. KEMERTELIDZE: Biacangelicin and a New Natural Furocoumarin, Heracol, from the Fruit of *Heracleum asperum*. Soobshch. Akad. Nauk Gruz. S.S.R. **57**, 97 (1970); Chem. Abs. **73**, 38468 (1970).

212. GONZÁLEZ, A. G., J. T. BARROSO, E. D. CHICO, J. R. LUIS, and F. R. LUIS: Componentes de umbeliferas XI. Nueva cumarina del *Heracleum pyrenaicum* Lam. Anales de Quím. **72**, 584 (1976).

213. GONZÁLEZ, A. G., J. T. BARROSO, J. R. LUIS, and F. R. LUIS: Componentes de umbeliferas III. Estructura de una nueva cumarina del *Heracleum granatense* Boiss. et Elench. Anales de Quím. **70**, 369 (1974).

214. — — — — Componentes de umbeliferas VII. Dos nuevas cumarinas del *Heracleum granatense* Boiss. Elench. Anales de Quím. **70**, 856 (1974).

215. GONZÁLEZ, A. G., J. L. BRETON, H. L. DORTA, M. A. M. INIGUEZ, and F. R. LUIS: Aplicacion de la resonancia magnetica nuclear en el estudio de estructuras de cumarinas. Anales de Quim. **69**, 1013 (1973).

216. GONZÁLEZ, A. G., R. J. CARDONA, E. D. CHICO, H. L. DORTA, and F. R. LUIS: Componentes de umbeliferas X. Cumarinas del *Peucedanum bourgaei* Lange in Wilk et Lange. Anales de Quím. **72**, 568 (1976).

217. GONZÁLEZ, A. G., R. J. CARDONA, H. L. DORTA, J. M. MEDINA, and F. R. LUIS: Componentes de umbeliferas XII. Dos nuevas cumarinas del *Peucedanum stenocarpum* Boiss. et Reuter. Anales de Quím. **72**, 588 (1976).

218. GONZÁLEZ, A. G., R. J. CARDONA, H. LOPEZ, J. M. MEDINA, and F. R. LUIS: Distribucion de cumarinas en umbeliferas. Rev. Real Acad. Cienc. Exact., Fis. Nat. **70**, 109 (1976).

219. GONZÁLEZ, A. G., R. J. CARDONA, J. M. MEDINA, and F. R. LUIS: Componentes de umbeliferas VI. Cumarinas del *Peucedanum officinale* L. Anales de Quím. **72**, 60 (1976).

220. — — — — Componentes de umbeliferas VIII. Dos nuevas cumarinas del *Peucedanum stenocarpum* Boiss. et Reuter. Anales de Quím. **72**, 88 (1976).

221. GONZÁLEZ, A. G., R. J. CARDONA, R. M. ORDONEZ, and F. R. LUIS: Nuevas fuentes de cumarinas naturales XXV. Cumarinas del *Haplophyllum hispanicum* Spach. (*H. linifolium* L.). Anales de Quím. **69**, 781 (1973).

222. GONZÁLEZ, A. G., E. D. CHICO, H. L. DORTA, J. R. LUIS, and F. R. LUIS: Nuevas fuentes de cumarinas naturales XXX. Componentes químicos de las *Ruta chalepensis* y del *Dictamnus hispanicus* Webb. Anales de Quím. **73**, 430 (1977).

223. GONZÁLEZ, A. G., E. D. CHICO, H. L. DORTA, J. M. MEDINA, and F. R. LUIS: Nuevas fuentes de cumarinas naturales XXIX. Tres nuevas cumarinas de una especie canaria de *Ruta*. Anales de Quím. **72**, 191 (1976).

224. GONZÁLEZ, A. G., R. ESTEVEZ, J. B. ARENCIBIA, and T. R. PEREZ: Nuevas fuentes de cumarinas naturales XXVI. Cumarins minoritarias de las hojas de la *Ruta pinnata* L. fil; sabandinol, nueva cumarina. Anales de Quim. **69**, 1141 (1973).

225. GONZÁLEZ, A. G., R. ESTEVEZ, and I. JARAIZ: Nuevas fuentes de cumarinas naturales XVIII. Estructura de la sabandinona, nueva cumarina de los frutos de la *Ruta pinnata* L. fil. Anales de Quím. **66**, 1017 (1970).

226. — — — Structure of Sabandinin and Other Coumarins Isolated from the Roots of *Ruta pinnata*. Phytochemistry **10**, 1621 (1971).

227. — — — Pinnaterin, a New Coumarin from the Roots of *Ruta pinnata*. Phytochemistry **10**, 1971 (1971).

228. — — — Nuevas fuentes de cumarinas naturales XXI. Sabandinona y tederina, nuevas cumarinas de los frutos de la *Ruta pinnata* L. fil. Anales de Quím. **68**, 425 (1972).

229. GONZÁLEZ, A. G., B. M. FRAGA, M. G. HERNÁNDEZ, and J. G. LUIS: Siderin, a New Coumarin from *Sideritis canariensis*. Phytochemistry **11**, 2115 (1972).

230. GONZÁLEZ, A. G., B. M. FRAGA, M. G. HERNÁNDEZ, J. G. LUIS, and W. STECK: Siderin: A Revised Structure. Chem. and Ind. **1974**, 166.

231. GONZÁLEZ, A. G., B. M. FRAGA, O. PINO, J. P. DECLERQ, G. GERMAIN, and J. FAYOS: X-Ray Structure of Bethancorol, a New Coumarin from *Cneorum tricoccum*. Tetrahedron Letters **1976**, 1729.

232. GONZÁLEZ, A. G., and F. R. LUIS: Chemistry of the Rutaceae of the Canary Isles. Herba Hung. **10**, 95 (1971); Chem. Abs. **79**, 15854 (1973).

233. GONZÁLEZ, A. G., E. A. MARTINEZ, M. A. M. INIGUEZ, and F. R. LUIS: Nuevas fuentes de cumarinas naturales XVI. Productos de los tallos de la *Ruta pinnata* L. fil. estructura de la benahorina. Anales de Quím. **67,** 441 (1971).

234. GONZÁLEZ, A. G., J. M. TRUJILLO, R. ESTEVEZ, and J. P. PEREZ: Componentes de umbeliferas IV. Lignanos del *Bupleurum fruticescens* L. Anales de Quím. **71,** 109 (1975).

235. GOTTLIEB, O. R., M. LEAO DA SILVA, and J. G. SOARES MAIA: Distribution of Coumarins in Amazonian *Brosimum* Species. Phytochemistry **11,** 3479 (1972).

236. GOVINDACHARI, T. R., B. R. PAI, P. S. SUBRAMANIAM, and N. MUTHUKUMARASWAMY: Coumarins of *Clausena dentata* (Willd.) R. and S. Tetrahedron **24,** 753 (1968).

237. GOVINDACHARI, T. R., B. R. PAI, P. S. SUBRAMANIAM, U. R. RAO, and N. MUTHUKUMARASWAMY: Constituents of *Mesua ferrea* L. — II. Ferruol A, a New 4-Alkylcoumarin. Tetrahedron **23,** 4161 (1967).

238. GRAY, A. I., R. D. WAIGH, and P. G. WATERMAN: Determination of Substitution Pattern in Coumarins Using a Lanthanide Shift Reagent. J. C. S. Chem. Comm. **1974,** 632.

239. — — — Avicennol: A New Pyranocoumarin from the Root Bark of *Zanthoxylum avicennae*, and its Conversion into Avicennin. J. C. S. Perkin I **1975,** 488.

240. — — — *cis*-Avicennol, a New Pyranocoumarin from the Root Bark of *Zanthoxylum elephantiasis*. Phytochemistry **16,** 1017 (1977).

241. GRIGG, R., J. A. KNIGHT, and P. ROFFEY: NMR Solvent Shifts and Structure Elucidation in Coumarins. Tetrahedron **22,** 3301 (1966).

242. GRISEBACH, H., and W. D. OLLIS: Biogenetic Relationship between Coumarins, Flavonoids, Isoflavonoids and Rotenoids. Experientia **17,** 4 (1961).

243. GUIOTTO, A., P. RODIGHIERO, and U. QUINTILY: Poncimarin, a New Coumarin from *Poncirus trifoliata* L. Z. Naturforsch. **30c,** 420 (1975).

244. GUIOTTO, A., P. RODIGHIERO, U. QUINTILY, and G. PASTORINI: Isoponcimarin: New Coumarin from *Poncirus trifoliata*. Phytochemistry **15,** 348 (1976).

245. GUISE, G. B., E. RITCHIE, R. G. SENIOR, and W. C. TAYLOR: The Chemical Constituents of Australian *Zanthoxylum* Species IV. Two New Coumarins from *Z. suberosum* C. T. White (syn. *Z. dominianum* Merr. and Perry; *Z. ovalifolium* Wight). Austral. J. Chem. **20,** 2429 (1967).

246. GUPTA, B. D., S. K. BANERJEE, K. L. HANDA, and C. K. ATAL: Heratomin and Heratomol, New Coumarins from *Heracleum thomsoni*. Phytochemistry **15,** 1319 (1976).

247. GUPTA, B. D., B. K. WALI, VISHWAPAUL, and K. L. HANDA: Coumarins from *Prangos pabularia* Lindl. Indian J. Chem. **2,** 464 (1964).

247a. HAHN, D. R.: Biochemical Studies on the Constituents of *Artemisia messerschmidiana* and their Derivatives. I. Identification of Esculetin Methyl Ethers and their cholagogic Action. Yakhak Hoeji **10,** 20 (1966); Chem. Abs. **68,** 112202 (1968).

248. HALPERN, O., P. WASER, and H. SCHMID: Die Konstitution des Athamantins und des Oroselols. Helv. Chim. Acta **40,** 758 (1957).

249. HARADA, I., Y. HIROSE, and M. NAKAZAKI: The Absolute Configurations of (+)-Marmesin and (−)-Hydroxytremetone. Tetrahedron Letters **1968,** 5463.

250. HASEGAWA, M.: Flavonoids of Various *Prunus* Species X. Wood Constituents of *Prunus tomentosa*. Bot. Mag. Tokyo **82,** 458 (1969); Chem. Abs. **73,** 956 (1970).

251. HATA, K.: Studies on the Chemical Components of Umbelliferous Plants II. Components of *Angelica glabra* Makino (2). Yakugaku Zasshi **76,** 666 (1956); Chem. Abs. **51,** 1152 (1957).

252. HATA, K., and M. KOZAWA: The Constitution of Angelol, a New Coumarin Isolated from the Root of *Angelica pubescens* Maxim. (Umbelliferae). Tetrahedron Letters **1965,** 4557.

253. Hata, K., and M. Kozawa: The Formation of Angelical from Angelol, a New Coumarin Isolated from the Root of *Angelica pubescens* Maxim. (Umbelliferae). Yakugaku Zasshi **87,** 210 (1967); Chem. Abs. **67,** 43633 (1967).

254. — — Studies on Coumarins from the Root of *Angelica pubescens* Maxim I. The Structure of Angelol. Yakugaku Zasshi **88,** 283 (1968); Chem. Abs. **69,** 59041 (1968).

255. — — Studies on Coumarins from the Root of *Angelica pubescens* Maxim II. Absolute Configuration of Angelol. Yakugaku Zasshi **88,** 293 (1968); Chem. Abs. **69,** 59039 (1968).

256. Hata, K., M. Kozawa, and K. Baba: Coumarins from Chinese Crude Drug "She Huangzi", the Fruits of *Cnidium* sp. and from *Cnidium japonicum* Miq. Yakugaku Zasshi **92,** 1289 (1972); Chem. Abs. **78,** 33845 (1973).

257. Hata, K., M. Kozawa, K. Baba, and M. Mitsui: Chemical Components of the Roots of *Ligusticum hultenii* Fernald and *Angelica shikokiana* Makino. Yakugaku Zasshi **93,** 248 (1973); Chem. Abs. **78,** 133334 (1973).

258. — — — — New Ester Coumarins, Angeladin and Isoedultin from *Angelica longeradiata* (Maxim.) Kitagawa. Chem. and Pharm. Bull. (Japan) **21,** 518 (1973).

259. Hata, K., M. Kozawa, K. Baba, and K.-Y. Yen: Coumarins from the Roots of *Angelica laxiflora* Diels. Chem. and Pharm. Bull. (Japan) **19,** 640 (1971).

260. Hata, K., M. Kozawa, and Y. Ikeshiro: The Constitution of Anomalin, a New Coumarin Isolated from the Root of *Angelica anomala* Lall (Umbelliferae). Chem. and Pharm. Bull. (Japan) **14,** 94 (1966).

261. — — — New Coumarins Isolated from the Roots of *Angelica anomala* and *A. cartilaginomarginata.* Yakugaku Zasshi **87,** 1118 (1967); Chem. Abs. **68,** 10231 (1968).

262. Hata, K., M. Kozawa, Y. Ikeshiro, and K.-Y. Yen: New Coumarins Isolated from the Root of *Peucedanum formosanum* Hayata and *Peucedanum japonicum* Thunb. Yakugaku Zasshi **88,** 513 (1968); Chem. Abs. **69,** 96521 (1968).

263. Hata, K., M. Kozawa, and K.-Y. Yen: The Constitution of Peuformosin, a New Coumarin Isolated from the Root of *Peucedanum formosanum* Hayata (Umbelliferae). Chem. and Pharm. Bull. (Japan) **14,** 442 (1966).

264. Hata, K., and K. Sano: The Constitution of Decursin, a New Coumarin Isolated from the Root of *Angelica decursiva* Fr. et Sav. (Umbelliferae). Tetrahedron Letters **1966,** 1461.

265. — — Studies on Coumarins from the Root of *Angelica decursiva* Fr. et Sav. I. The Structure of Decursin and Decursidin. Yakugaku Zasshi **89,** 549 (1969); Chem. Abs. **71,** 38832 (1969).

266. Hata, K., and Y. Tanaka: Chemical Components of Umbelliferous Plants III. Components of the Root of *Angelica pubescens.* I. Yakugaku Zasshi **77,** 937 (1958); Chem. Abs. **52,** 3799 (1960).

267. Head, F. S. H., and A. Robertson: Natural Glucosides. Part II. The Constitution of Aesculin. J. Chem. Soc. **1930,** 2434.

268. Hegarty, M. P., and F. N. Lahey: The Coumarins of *Halfordia scleroxyla* F. Muell. Austral. J. Chem. **9,** 120 (1956).

269. Herz, W., S. V. Bhat, and P. S. Santhanam: Coumarins of *Artemisia dracunculoides* and 3′,6-Dimethoxy-4′,5,7-trihydroxyflavone in *A. arctica.* Phytochemistry **9,** 891 (1970).

270. Hlubucek, J., E. Ritchie, and W. C. Taylor: Synthesis of 2,2-Dimethylchromenes. Tetrahedron Letters **1969,** 1369.

271. — — — Synthesis of 2,2-Dimethylchromenes. Austral. J. Chem. **24,** 2347 (1971).

272. — — — Synthesis of *o*-Isopentenylphenols. Austral. J. Chem. **24,** 2355 (1971).

273. Hörhammer, L., H. Wagner, and W. Eyrich: Über die Inhaltsstoffe der Früchte von *Angelica silvestris* L. Z. Naturforsch. **18b,** 639 (1963).

274. Hörhammer, L., H. Wagner, and D. Kraemer-Heydweiler: Neue Methoden im

Pharmakognostischen Unterricht (12. Mitteilung): Identifizierung von Umbelliferenwurzeln und Nachweis der wichtigsten Verfälschungen mit Hilfe der Dünnschichtchromatographie. Deut. Apotheker-Ztg. **106,** 267 (1966); Chem. Abs. **65,** 2065 (1966).

275. IGNAT'EVA, N. S., V. V. VANDYSHEV, and M. G. PIMENOV: Coumarins from the Roots of *Cachrys pubescens*. Khim. prirod. Soedinenii **8,** 388 (1972); Chem. Abs. **77,** 149704 (1972).

276. IRIE, H., S. UYEO, K. YAMOMOTO, and K. KINOSHITA: The Structure of Glaupalol, a Novel Furocoumarin from *Glaucidium palmatum* Sieb. et Zucc. Chem. Comm. **1967,** 547.

277. ISHII, H., and K. HOSOYA: Arnottianin, a New Dihydropyranocoumarin. Chem. and Pharm. Bull. (Japan) **20,** 860 (1972).

278. ISHII, H., K. HOSOYA, T. ISHIKAWA, and J. HAGINIWA: Studies on the Chemical Constituents of Rutaceous Plants. XX. The Chemical Constituents of *Xanthoxylum arnottianum* Maxim. (1). Isolation of the Chemical Constituents of the Xylem of Roots. Yakugaku Zasshi **94,** 309 (1974); Chem. Abs. **81,** 132752 (1974).

279. ISHII, H., K. HOSOYA, T. ISHIKAWA, E. UEDA, and J. HAGINIWA: Studies on the Chemical Constituents of Rutaceous Plants. XXI. The Chemical Constituents of *Xanthoxylum arnottianum* Maxim. (2). Isolation of the Chemical Constituents of the Xylem of Stems. Yakugaku Zasshi **94,** 322 (1974); Chem. Abs. **81,** 132753 (1974).

280. ISHII, H., and T. ISHIKAWA: Arnottinin: Structure Establishment by Chemical Correlation with Osthenol. Chem. and Pharm. Bull. (Japan) **23,** 934 (1975).

281. ISHII, H., T. ISHIKAWA, H. SEKIGUCHI, and K. HOSOYA: Xanthoarnol: A New Dihydrofuranocoumarin. Chem. and Pharm. Bull. (Japan) **21,** 2346 (1973).

282. JANOT, M. M., J. LE MEN, H. POURRAT, and V. PLOUVIER: Constitution du calycanthoside. Bull. Soc. chim. Biol. **37,** 365 (1955).

283. JARVIS, M. W., and A. G. MORITZ: Long-Range Coupling in Substituted Coumarins. Austral. J. Chem. **21,** 2445 (1968).

284. JEFFERIES, P. R., and G. K. WORTH: The Chemistry of the Western Australian Rutaceae — VI. Two Novel Coumarins from *Eriostemon brucei*. Tetrahedron **29,** 903 (1973).

285. JENSEN, S. R., and B. J. NIELSEN: A New Coumarin, Fraxidin 8-*O*-β-D-glucoside and 10-Hydroxyligstroside from Bark of *Fraxinus exelsior*. Phytochemistry **15,** 221 (1976).

286. JOSHI, B. S., and D. H. GAWAD: Isolation of Some Furanocoumarins from *Clausena indica* and Identity of Chalepensin with Xylotenin. Phytochemistry **10,** 480 (1971).

287. JOSHI, B. S., and V. N. KAMAT: Structures of Clausenin, Clausenidin and a Synthesis of Clausenin and Xanthoxyletin. Tetrahedron Letters **1966,** 5767.

288. JOSHI, B. S., V. N. KAMAT, and D. H. GAWAD: Structure of Clausindine, a New Coumarin from *Clausenia indica* Oliv. Experientia **30,** 223 (1974).

289. — — — Clausidine, a Novel Cyclopropylcoumarin. J. C. S. Perkin I **1974,** 1561.

290. JOSHI, B. S., V. N. KAMAT, T. R. GOVINDACHARI, and A. K. GANGULY: Isolation and Structure of Surangin A and Surangin B, Two New Coumarins from *Mammea longifolia* (Wight) Planch and Triana. Tetrahedron **25,** 1453 (1969).

291. JOSHI, B. S., V. N. KAMAT, and A. K. SAKSENA: Structures of Clausenin and Clausenidin, Two New Pyranocoumarins from the Roots of *Clausena heptaphylla* Wt. and Arn. Tetrahedron **23,** 4785 (1967).

292. JOSHI, B. S., Y. N. SHUKLA, D. S. BHAKUNI, and M. M. DHAR: 6-(2,3-Dihydroxy-3-methylbutyl)-7-methoxycoumarin, a New Coumarin from *Micromelum pubescens* Blume. Indian J. Chem. **13,** 772 (1975).

293. KADYROV, A. S., A. I. SAIDKHODZHAEV, and G. K. NIKONOV: The Structures of Feroside and of Reoselin A, New Glycosides from the Roots of *Ferula korshinskyi*. Khim. prirod. Soedinenii **11,** 574 (1975); Chem. Abs. **84,** 74568 (1976).

294. KAMILOV, K. M., V. V. KISELEVA, and G. K. NIKONOV: Diversoside, a New Coumarin from the Roots of *Ferula diversivitatta*. Khim. prirod. Soedinenii **10,** 781 (1974); Chem. Abs. **82,** 135684 (1975).

295. KAMILOV, K. M., and G. K. NIKONOV: Coumarins of *Ferula kopetdaghensis* and the Structure of Kopeolin and Kopeoside. Khim. prirod. Soedinenii **9,** 308 (1973); Chem. Abs. **79,** 134321 (1973).

296. — — Structure of the Coumarin Reoselin. Khim. prirod. Soedinenii **10,** 84 (1974); Chem. Abs. **80,** 121275 (1974).

297. — — Kamolone from the Roots of *Ferula kopetdaghensis*. Khim. prirod. Soedinenii **10,** 85 (1974); Chem. Abs. **81,** 60806 (1974).

298. — — Coumarins of *Ferula kopetdaghensis* — Kopetdaghin and Kopeodin (Farnesiferol B). Khim. prirod. Soedinenii **10,** 442 (1974); Chem. Abs. **82,** 70266 (1975).

299. KAPOOR, S. L., Y. N. SHARMA, and A. ZAMAN: Extractives of *Angelica glauca*. Phytochemistry **11,** 475 (1972).

300. KARIYONE, T., and K. HATA: Studies on the Chemical Components of Umbelliferous Plants I. Components of the Roots of *Angelica glabra* Makino. Yakugaku Zasshi **76,** 649 (1956); Chem. Abs. **51,** 1152 (1957).

301. KARIYONE, T., and T. MATSUNO: Studies on the Constituents of Orange Oil. I. On the Structure of Auraptene. Pharm. Bull. (Japan) **1,** 119 (1953).

302. KARRER, W.: In: Konstitution und Vorkommen der organischen Pflanzenstoffe. 531. Basel: Birkhäuser. 1958.

303. KHALED, S. A., K. SZENDREI, I. NOVÁK, and J. REISCH: Coumarin Glycosides from *Peucedanum ostruthium*. Phytochemistry **14,** 1461 (1975).

304. KHANNA, P. L., and T. R. SESHADRI: Synthesis of Isoprenoid Polyphenols. J. Sci. Ind. Res., India **33,** 295 (1974).

305. KHASANOV, T. K., A. I. SAIDKHODZHAEV, and G. K. NIKONOV: Structure and Configuration of New Coumarins of the Roots of *Ferula mogoltavica* Khim. prirod. Soedinenii **10,** 10 (1974); Chem. Abs. **80,** 121131 (1974).

306. — — — Structure and Configuration of the Coumarins Mogoltadone and Mogoltadin. Khim. prirod. Soedinenii **10,** 25 (1974); Chem. Abs. **80,** 121138 (1974).

307. — — — Structure and Configuration of Polyanthin and Polyanthinin — New Coumarins from the Roots of *Ferula polyantha*. Khim. prirod. Soedinenii **10,** 517 (1974); Chem. Abs. **82,** 95266 (1975).

308. — — — Structure of Feropolin, Feropolol, Feropolone, and Feropolidin. Khim. prirod. Soedinenii **12,** 91 (1976); Chem. Abs. **85,** 59558 (1976).

309. — — — Structure of Mogoltacin. Khim. prirod. Soedinenii **12,** 95 (1976); Chem. Abs. **85,** 59561 (1976).

310. KHOSA, R. L.: Coumarins from the Leaves of *Murraya paniculata*. Indian J. Pharm. **34,** 47 (1972).

311. — Structure of a New Coumarin from *Murraya paniculata*. Indian J. Pharm. **37,** 43 (1975).

312. Kiangsu Luminescent Bacteria Research Unit: *Armillariella tabescens*. II. Isolation and the Structure of the Armillarisins. Sci. Sinica **17,** 377 (1974); Chem. Abs. **82,** 1919 (1975).

313. KIKUCHI, T., T. YOKOI, K. UMEMOTO, and T. SHINGU: Constituents of *Scaevola frutescens* (Miller) Krause. Yakugaku Zasshi **94,** 1616 (1974); Chem. Abs. **82,** 135677 (1975).

314. KINCL, F. A., J. ROMO, G. ROSENKRANZ, and F. SONDHEIMER: The Constituents of *Casimiroa edulis* Llave et Lex. Part 1. The Seed. J. Chem. Soc. **1956,** 4163.

315. KING, F. E., J. R. HOUSLEY, and T. J. KING: The Chemistry of Extractives from Hardwoods. Part XVI. Coumarin Constituents of *Fagara macrophylla, Zanthoxylum flavum*, and *Chloroxylon swietenia*. J. Chem. Soc. **1954,** 1392.

316. KINOSHITA, K., and S. MURASE: The Structure of *cis*- and *trans*-Glaupadiol. (Studies on the Constituents of the Rhizomes of *Glaucidium palmatum* Sieb. and Zucc.) Yakugaku Zasshi **93,** 202 (1973); Chem. Abs. **78,** 136129 (1973).

317. KIR'YALOV, N. P.: The Structure of Cocanicine and Umbelliprenin, Components of the Neutral Part of Oil from *Ferula cocanica*. Trudy bot. Inst. Akad. Nauk S.S.S.R. **8,** 7 (1961); Chem. Abs. **58,** 498 (1963).

318. — Coumarins from Plants of the Genus *Ferula*. Trudy bot. Inst. Akad. Nauk S.S.S.R. **12,** 82 (1965); Chem. Abs. **64,** 3461 (1966).

319. — Isolation of the Coumarin Badrakemin from the Roots of *Ferula badrakema*. Khim. prirod. Soedinenii **3,** 363 (1967); Chem. Abs. **69,** 35865 (1968).

320. — The Coumarin Diversine from the Roots of *Ferula diversivittata*. Khim. prirod. Soedinenii **5,** 51 (1969); Chem. Abs. **70,** 118026 (1969).

321. KIR'YALOV, N. P., and V. Y. BAGIROV: Structure of Karatavicin. Khim. prirod. Soedinenii **3,** 223 (1967); Chem. Abs. **68,** 68829 (1968).

322. — — The Structure of Karatavic Acid. Khim. prirod. Soedinenii **4,** 283 (1968); Chem. Abs. **70,** 77715 (1969).

323. — — The Structure of Karatavicinol. Khim. prirod. Soedinenii **5,** 225 (1969); Chem. Abs. **72,** 55164 (1970).

324. KIR'YALOV, N. P., and T. V. BUKREEVA: A New Coumarin from the Roots of *Ferula pseudooreoselinum*. Khim. prirod. Soedinenii **8,** 643 (1972); Chem. Abs. **78,** 108210 (1973).

325. — — Samarcandin Acetate from the Roots of *Ferula pseudooreoselinum*. Khim. prirod. Soedinenii **8,** 798 (1972); Chem. Abs. **78,** 108250 (1973).

326. — — Coumarins from the Roots of *Ferula pseudooreoselinum*. Khim. prirod. Soedinenii **9,** 425 (1973); Chem. Abs. **79,** 134324 (1973).

327. KIR'YALOV, N. P., T. V. BUKREEVA, V. A. GINDIN, G. MAMATOV, and I. S. KOZHINA: The Structure of the Coumarin Glucoside Reoselin from the Roots of *Ferula pseudooreoselinum*. Khim. prirod. Soedinenii **11,** 87 (1975); Chem. Abs. **83,** 97826 (1975).

328. KIR'YALOV, N. P., and S. D. MOVCHAN: A New Glycoside, Reoseline from the Rosin of the Roots of *Ferula pseudooreoselinum*. Doklady Akad. Nauk. S.S.S.R. **148,** 1081 (1963); Chem. Abs. **59,** 5255 (1963).

329. — — The Structure of Gummosin. Khim. prirod. Soedinenii **2,** 383 (1966); Chem. Abs. **67,** 11394 (1967).

330. — — The Structure of Samarcandin and Samarcandone, Coumarin Compounds from *Ferula samarcandica*. Khim. prirod. Soedinenii **4,** 73 (1968); Chem. Abs. **69,** 59040 (1968).

331. KISELEVA, V. V., G. K. NIKONOV, and M. O. KARRYEV: Structure of Diversin and Diversinin, Coumarins of *Ferula diversivittata*. Khim. prirod. Soedinenii **11,** 344 (1975); Chem. Abs. **84,** 28001 (1976).

332. KOMATSU, S., S. TANAKA, S. OZAWA, R. KUBO, Y. ONO, and Z. MATSUDA: Biochemical Studies on Grape Fruits, *Citrus aurantium* L. J. Chem. Soc. Japan **51,** 478 (1930).

333. KOMISSARENKO, N. F., and V. T. CHERNOBAI: Furocoumarins of *Cnidium dubium*. Khim. prirod. Soedinenii **2,** 375 (1966); Chem. Abs. **67,** 11395 (1967).

334. KONOSHIMA, M., H.-J. CHI, and K. HATA: Coumarins from the Root of *Angelica gigas* Nakai. Chem. Pharm. Bull. (Japan) **16,** 1139 (1968).

335. KRIVUT, B. A., L. G. AVRAMENKO, M. E. PEREL'SON, and G. K. NIKONOV: Preparation and Quantitative Determination of Benzoylmarmesin. Khim. prirod. Soedinenii **6,** 526 (1970); Chem. Abs. **74,** 53420 (1971).

336. KUFFNER, F., A. NIKIFOROV, and G. SCHULZ: Über das Rutolid. Monatsh. **104,** 911 (1973).

337. KUTNEY, J. P., G. EIGENDORF, T. INABA, and D. L. DREYER: Mass Spectral Fragmentation Studies in Monomeric and Dimeric Coumarins. Org. Mass Spectrom. **5,** 249 (197().

338. KUTNEY, J. P., P. J. SALISBURY, and A. K. VERMA: Biosynthetic Studies in the Coumarin Series — III. Studies in Tissue Cultures of *Thamnosma montana* Torr. and Frem. The Role of Mevalonate. Tetrahedron **29,** 2673 (1973).

339. KUTNEY, J. P., A. K. VERMA, and R. N. YOUNG: Studies on Constituents of *Thamnosma montana* Torr. and Frem. The Structure of Thamnosmin, a Novel Coumarin Epoxide. Tetrahedron **28,** 5091 (1972).

340. — — — Biosynthetic Studies in the Coumarin Series — I. Studies in Plants of *Thamnosma montana* Torr. and Frem. The Role of Mevalonate. Tetrahedron **29,** 2645 (1973).

341. — — — Biosynthetic Studies in the Coumarin Series — II. Studies in Plants of *Thamnosma montana* Torr. and Frem. The Role of Acetate and Glycine. Tetrahedron **29,** 2661 (1973).

342. KUTNEY, J. P., R. N. YOUNG, and A. K. VERMA: Novel Epoxides from *Thamnosma montana* Torr. and Frem. Tetrahedron Letters **1969,** 1845.

343. KUZNETSOVA, G. A.: Discovery of Imperatorin Oxide in Plants. Zhur. obshchei Khim. **35,** 403 (1965); Chem. Abs. **62,** 15071 (1965).

344. — Natural Coumarins and Furocoumarins: Leningrad: Nauka. 1967; Chem. Abs. **67,** 108551 (1967).

345. KUZNETSOVA, G. A., and A. Z. ABYSHEV: Natural (−)-7-Methoxy-8-(β,γ-dihydroxyisopentyl)-coumarin. Khim. prirod. Soedinenii **1,** 283 (1965); Chem. Abs. **64,** 683 (1966).

346. — — Merancin Hydrate — A Component of the Roots of *Prangos ferulacea* (L.) Lindl. Zhur. priklad. Khim. **38,** 2370 (1965); Chem. Abs. **64,** 2326 (1966).

347. KUZNETSOVA, G. A., A. Z. ABYSHEV, M. E. PEREL'SON, Y. N. SHEINKER, and G. Y. PEK: A New Coumarin, Pranferol, from the Roots of *Prangos ferulacea.* Khim. prirod. Soedinenii **2,** 310 (1966); Chem. Abs. **66,** 94923 (1967).

348. KUZNETSOVA, G. A., and L. M. BELENOVSKAYA: A Chemical Study of the Roots of *Prangos ornata* Kuzm. Khim. prirod. Soedinenii **1,** 430 (1965); Chem. Abs. **64,** 13088 (1965).

349. — — Isolation of Oxypeucedanin Hydrate (Prangol) from the Roots of *Prangos fedtschenkoi* (Rgl. et Schmal.) Eug. Kor. Zhur. priklad. Khim. **38,** 1146 (1965); Chem. Abs. **63,** 6021 (1965).

350. — — Prantschimgin, a New Coumarin from the Roots of *Prangos tschimganica.* Khim. prirod. Soedinenii **2,** 235 (1966); Chem. Abs. **66,** 2438 (1967).

351. — — Additional Information on the Coumarin Composition of *Prangos fedtschenkoi* (Regel et Schmalh). Korob., *P. tschimganica* B. Fedtsch and *P. isphairamica* B. Fedtsch. (Umbelliferae). Zhur. priklad. Khim. **42,** 471 (1969); Chem. Abs. **71,** 811 (1969).

352. KUZNETSOVA, G. A., and V. N. FLORYA: Coumarins from the Roots, Aboveground Parts, and Fruits of *Seseli campestre* Growing in Moldavia. Zhur. priklad. Khim. **43,** 1412 (1970); Chem. Abs. **73,** 117214 (1970).

353. KUZNETSOVA, G. A., and N. F. GASHIMOV: The Structure of a New Coumarin from *Haplophyllum pedicellatum.* Khim. prirod. Soedinenii **8,** 666 (1972); Chem. Abs. **78,** 108171 (1973).

354. — — A New Coumarin, Pedicellone from *Haplophyllum pedicellatum.* Khim. prirod. Soedinenii **9,** 113 (1973); Chem. Abs. **79,** 5223 (1973).

355. KUZNETSOVA, G. A., and G. V. PIGULEVSKII: The Structure of Prangenine. Zhur. obshchei Khim. **31,** 323 (1961); Chem. Abs. **55,** 22282 (1961).

356. KUZNETSOVA, G. A., and L. M. SOKOLOVA: Coumarins from the Roots of *Prangos*

fedtschenkoi (Rgl. et Schmalh.) Eug. Kor. Zhur. priklad. Khim. **37,** 1105 (1964); Chem. Abs. **61,** 4630 (1964).

357. KUZNETSOVA, G. A., and A. D. ZORINA: Coumarins from the Roots of *Prangos serawschanica* (Rgl. et Schmalh.) Korov. Zhur. priklad. Khim. **39,** 482 (1966); Chem. Abs. **65,** 1043 (1966).

358. LAHEY, F. N., and J. K. MACLEOD: The Coumarins of *Geijera parviflora* Lindl. Austral. J. Chem. **20,** 1943 (1967).

359. — — Halfordin and Isohalfordin — Revised Structures. Tetrahedron Letters **1968,** 447.

360. LAHEY, F. N., and D. J. WLUKA: Geijerin, a New Coumarin from the Bark of *Geijera salicifolia* SCHOTT. Austral. J. Chem. **8,** 125 (1955).

361. LAMBERTON, J. A., J. R. PRICE, and A. H. REDCLIFFE: Micromelin, a New Coumarin from *Micromelum minutum* (FORST. f.) SEEM (Family Rutaceae). Austral. J. Chem. **20,** 973 (1967).

362. LAPPER, R. D.: The Carbon-13 Nuclear Magnetic Resonance Spectrum of Siderin. Tetrahedron Letters **1974,** 4293.

363. LARSEN, P. K., and F. SANDBERG: Constituents of Umbelliferous Plants. XV. Coumarins from *Thapsia garganica* L. The Structure of a New Coumarin. Acta Chem. Scand. **24,** 1113 (1970).

364. LASSAK, E. V., and J. T. PINHEY: The Constituents of *Eriostemon trachyphyllus.* The Structure of Trachyphyllin, a New Coumarin. Austral. J. Chem. **22,** 2175 (1969).

365. LASSAK, E. V., and I. A. SOUTHWELL: The Coumarins from the Resin of *Evodia vitiflora.* Austral. J. Chem. **25,** 2491 (1972).

366. LAXMI, M. V., C. V. RATNAM, and N. V. SUBBA RAO: 7-Methoxy-8-(3-butenyl-3-methyl-2-oxo)-coumarin, a New Coumarin from *Murraya exotica* Linn. Indian J. Chem. **10,** 564 (1972).

367. LEE, K.-H., and T. O. SOINE: Coumarins VII: The Coumarins of *Lomatium nuttalli.* J. Pharm. Sci. **57,** 865 (1968).

368. — — Coumarins IX: Coumarins of *Sphenosciadium capitellatum* (A. Gray). J. Pharm. Sci. **58,** 675 (1969).

369. — — Coumarins X: Spectral Studies on Some Linear Furocoumarins. J. Pharm. Sci. **58,** 681 (1969).

370. LEMMICH, E., J. LEMMICH, and B. E. NIELSEN: Constituents of Umbelliferous Plants XIV. Coumarins of *Peucedanum oreoselinum* (L.) Moench. Acta Chem. Scand. **24,** 2893 (1970).

371. LEMMICH, J., E. LEMMICH, and B. E. NIELSEN: Constituents of Umbelliferous Plants VIII. Coumarins from the Root of *Seseli libanotis* (L.) Koch. The Structure of Three New Coumarins. Acta Chem. Scand. **20,** 2497 (1966).

372. LEMMICH, J., and B. E. NIELSEN: Constituents of Umbelliferous Plants (X). Stereochemistry of Natural Coumarins Containing the 3-Hydroxy-2,2-dimethylchroman System. Tetrahedron Letters **1969,** 3.

373. LEMMICH, J., P. A. PEDERSEN, and B. E. NIELSEN: Revised Absolute Configurations of Natural Khellactone Esters (1). Tetrahedron Letters **1969,** 3365.

374. — — — Constituents of Umbelliferous Plants XVIII. Terpenoids and Coumarins of the Root of *Ligusticum seguieri* Koch. Acta Chem. Scand. **25,** 344 (1971).

375. — — — Coumarins and Terpenoids of the Fruits of *Ligusticum seguieri.* Phytochemistry **10,** 3333 (1971).

376. MCCABE, P. H., R. MCCRINDLE, and R. D. H. MURRAY: Constituents of Sneezewood, *Ptaeroxylon obliquum* (Thunb.) Radlk. Part I. Chromones. J. Chem. Soc. (C) **1967,** 145.

377. MCLEOD, J. K.: Halfordinin, a 3,4,5-Oxygenated Furocoumarin with a Novel α,α-Dimethylallyl Ether Substituent. Tetrahedron Letters **1970,** 1319.

378. — Halfordinin — A Revised Structure. Tetrahedron Letters **1970,** 3611.

379. MAHEY, S., T. R. SESHADRI, and S. K. MUKERJEE: Synthesis of Poncitrin. Indian J. Chem. **11**, 983 (1973).

380. — — — Nuclear Prenylation of 5,7-Dihydroxycoumarin. Synthesis of Coumurrayin and Toddaculin. Indian J. Chem. **12**, 29 (1974).

381. MARTINEZ, E. A., R. E. REYES, A. G. GONZÁLEZ, and F. R. LUIS: Nuevas fuentes de cumarinas naturales VIII. Cumarinas de las raíces de la *Pimpinella rupicola* Svent. (Umbelliferae); estructura de una nueva cumarina. Anales de Quím. **63**, 205 (1967).

382. MATSUMOTO, T., K. FUKUI, and M. NANBU: Synthesis and Absolute Configuration of *trans*-Meranzinic Acid. Chem. Letters **1974**, 603.

383. MENDEZ, J., and M. I. LOJO: Spectral Characterization of Coumarins and Cinnamic Acids. Microchem. J. **14**, 567 (1969).

384. MITSUHASHI, H., and T. ITOH: Studies on the Constituents of Umbelliferous Plants. V. Studies on the Constituents of *Angelica edulis* Mayabe (1). Isolation of Edultin. Chem. and Pharm. Bull. (Japan) **10**, 511 (1962).

385. — — Studies on the Constituents of Umbelliferous Plants. VI. Studies on the Constituents of *Angelica edulis* Muyabe. (2). Chem. and Pharm. Bull. (Japan) **10**, 514 (1962).

386. MOLHO, D. P., P. JÖSSANG, and J. CARBONNIER: Structure de la sprengelianine, dihydrofurannocoumarine extraite de deux *Heracleum* indiens: *H. ceylanicum* Gardn. et *H. sprengelianum* W. et A. Bull. Soc. chim. France **1972**, 208.

387. MOLHO, D., P. JÖSSANG, M.-C. JARREAU, and J. CARBONNIER: Derivés furannocoumariniques du genre *Heracleum* et plus specialement de *Heracleum sprengelianum* Wight et Arn. et *Heracleum ceylanicum* Gardn. ex C. B. Clarke. Etude Phylogénique. In: The Biology and Chemistry of the Umbelliferae (V. H. HEYWOOD, Ed.) 337. London: Academic Press. 1971.

388. MONDON, A., and H. CALLSEN: Inhaltsstoffe der Cneoraceen, IV. Chromone und Cumarine aus *Cneorum pulverulentum*. Chem. Ber. **108**, 2005 (1975).

389. MONDON, A., H. CALLSEN, and P. HARTMANN: Inhaltsstoffe der Cneoraceen, III. Trennverfahren für *Cneorum pulverulentum* und Untersuchung der Wachsfraktion und Phytosterole. Chem. Ber. **108**, 1989 (1975).

390. MORRIS, M. P., and C. PAGÁN: The Isolation of the Toxic Principles of Mamey. J. Amer. Chem. Soc. **75**, 1489 (1953).

391. MOWAT, D., and R. D. H. MURRAY: Claisen Rearrangements — V. Synthesis of the Coumarins Dentatin (Ponicitrin) and Glabralactone (Angelicone). Tetrahedron **29**, 2943 (1973).

392. MUJUMDAR, A. S., and R. S. USGAONKAR: Benzopyrans. Part V. Synthesis of a Linear Dihydrobenzodipyrandione Isolated from Dill and of Other Similar Benzodipyrandiones. J. C. S. Perkin I **1974**, 2236.

393. MUJUMDAR, R. B., A. V. R. RAO, S. S. RATHI, and K. VENKATARAMAN: Swietenone, the First Natural *t*-Butyl Ketone, from *Chloroxylon swietenia*. Tetrahedron Letters **1975**, 867.

394. MUKHAMEDOVA, K. S., S. T. AKRAMOV, and S. Y. YUNUSOV: Structure of Prangosine. Khim. prirod. Soedinenii **3**, 117 (1967); Chem. Abs. **67**, 54284 (1967).

395. — — — Mass-Spectrometric Study of Prangosine. Khim. prirod. Soedinenii **3**, 357 (1967); Chem. Abs. **68**, 105397 (1968).

396. MURRAY, R. D. H., and M. M. BALLANTYNE: Nieshoutol, a Sternutatory Hydroxycoumarin from Sneezewood. Tetrahedron Letters **1969**, 4031.

397. — — Constituents of Sneezewood, *Ptaeroxylon obliquum* — II. Coumarins — The Structure of Nieshoutol. Tetrahedron **26**, 4473 (1970).

398. — — Claisen Rearrangements — I. Synthesis of the Coumarin, Pinnarin. Tetrahedron **26**, 4667 (1970).

399. MURRAY, R. D. H., M. M. BALLANTYNE, T. C. HOGG, and P. H. MCCABE: Claisen

Rearrangements — VI. Synthesis of the Coumarins, Sesibiricin and Toddaculin. Tetrahedron **31,** 2960 (1975).

400. Murray, R. D. H., M. M. Ballantyne, and K. P. Mathai: A Method to Introduce a 3,3-Dimethylallyl Unit Ortho to a Phenol. Tetrahedron Letters **1970,** 243.

401. — — — Claisen Rearrangements — III. Convenient Syntheses of the Coumarins, Osthenol, Demethylsuberosin and Coumurrayin. Tetrahedron **27,** 1247 (1971).

402. Murray, R. D. H., and I. T. Forbes: Syntheses of the Coumarins, Avicennol, Dipetaline and Dipetalolactone. Tetrahedron Letters **1976,** 953.

403. — — Synthesis of 8-(3-Methylbut-2-enoyl)-7-methoxycoumarin, a Novel Coumarin from *Ligusticum hultenii.* Tetrahedron Letters **1977,** 3077.

404. Murray, R. D. H., and T. C. Hogg: Synthesis of the Coumarin, Toddaculin. Tetrahedron Letters **1972,** 185.

405. Murray, R. D. H., T. C. Hogg, M. M. Ballantyne, and P. H. McCabe: Synthesis of the Coumarin, Sesibiricin. Tetrahedron Letters **1971,** 3317.

406. Murray, R. D. H., and D. Mowat: Synthesis of the Coumarin, Glabralactone. Tetrahedron Letters **1972,** 2171.

407. Murray, R. D. H., M. Sutcliffe, and M. Hasegawa: Claisen Rearrangements — VII. Novel Reactions of the Coumarin, Tomentin. Tetrahedron **31,** 2966 (1975).

408. Murray, R. D. H., M. Sutcliffe, and P. H. McCabe: Claisen Rearrangements — IV. Oxidative Cyclization of Two Coumarin *o*-Isoprenylphenols. Tetrahedron **27,** 2901 (1971).

409. Musajo, L., and G. Rodighiero: Mode of Photosensitising Action of Furocoumarins. Photophysiology **7,** 115 (1972).

410. Nakabayashi, T.: Coumarin Derivatives in *Angelica saxicola* — The Constitution of Calcicolin. Nippon Kagaku Zasshi **83,** 182 (1962); Chem. Abs. **59,** 1578 (1963).

411. Nakahara, H.: The Constituents of the Inner Bark of *Hydrangea paniculata.* I. A New Glycoside, Neohydrangin J. Fac. Sci., Niigata Univ. Ser. I **2,** 63 (1955); Chem. Abs. **50,** 4926 (1956).

412. Nakajima, M., J. Oda, and H. Fukami: Total Synthesis of Nodakenetin and Marmesin. Agric. and Biol. Chem. (Japan) **27,** 695 (1963).

413. Nakayama, M.: NMR Solvent Shifts Induced by Benzene in Coumarins. I. Methoxyl Resonances. J. Sci. Hiroshima Univ. Ser. A-II **33,** 205 (1969); Chem. Abs. **73,** 24475 (1970).

414. — NMR Solvent Shifts Induced by Benzene in Coumarins. II. The Structure of Isohalfordin. J. Sci. Hiroshima Univ. Ser. A-II **33,** 213 (1969); Chem. Abs. **73,** 25339 (1970).

415. Nakayama, M., S. Fujimoto, K. Fukui, and J. K. MacLeod: The Isolation, Structure and Synthesis of Halkendin. Austral. J. Chem. **24,** 209 (1971).

416. Nakazaki, M., Y. Hirose, and K. Ikematsu: The Absolute Configurations of Athamantin and Edultin. Tetrahedron Letters **1966,** 4735.

417. Nayar, M. N. S., and M. K. Bhan: Coumarins and Other Constituents of *Hesperathusa crenulata.* Phytochemistry **11,** 3331 (1972).

418. Nayar, M. N. S., M. K. Bhan, and V. George: A New Coumarin in *Boenninghausenia albiflora.* Phytochemistry **12,** 2073 (1973).

419. Nielsen, B. E.: Coumarins of Umbelliferous Plants. Dan. Tidsskr. Farm. **44,** 111 (1970); Chem. Abs. **74,** 20314 (1971).

420. — Coumarin Patterns in the Umbelliferae. In: The Biology and Chemistry of the Umbelliferae (V. H. Heywood, Ed.), 325. London: Academic Press. 1971.

421. Nielsen, B. E., and E. Jensen: The Structure of Two New Coumarins from the Roots of *Lomatium columbianum.* Phytochemistry **15,** 1049 (1976).

422. Nielsen, B. E., P. K. Larsen, and J. Lemmich: Constituents of Umbelliferous Plants XIII. Coumarins from *Seseli gummiferum* Pall. The Structure of Three New Coumarins. Acta Chem. Scand. **24,** 2863 (1970).

423. NIELSEN, B. E., P. K. LARSEN, and J. LEMMICH: Constituents of Umbelliferous Plants XVII. Coumarins from *Seseli gummiferum* Pall. The Structure of Two New Coumarins. Acta Chem. Scand. **25,** 529 (1971).

424. NIELSEN, B. E., and L. LEMMICH: Constituents of Umbelliferous Plants III. The Structure of Archangelicin, a Coumarin from *Angelica archangelica* L. subsp. *litoralis* Thell. Acta Chem. Scand. **18,** 932 (1964).

425. — — Constituents of Umbelliferous Plants IV. The Coumarins of *Peucedanum palustre* L. The Structure of a New Coumarin. Acta Chem. Scand. **18,** 1379 (1964).

426. — — Constituents of Umbelliferous Plants V. On the Configuration of Archangelicin and Related Coumarins. Acta Chem. Scand. **18,** 2111 (1964).

427. — — Constituents of Umbelliferous Plants VI. The Structure of Peulustrin, a New Coumarin from *Peucedanum palustre* L. Acta Chem. Scand. **19** 601 (1965).

428. — — Constituents of Umbelliferous Plants VII. Coumarins from the Fruits of *Peucedanum palustre* L. The Structure of Two New Coumarins. Acta Chem. Scand. **19,** 1810 (1965).

429. — — Constituents of Umbelliferous Plants IX. The Configuration of (+)-Oxypeucedanin Hydrate and Related Coumarins. Acta Chem. Scand. **23,** 962 (1969).

430. NIELSEN, B. E., and T. O. SOINE: Coumarins IV. Coumarins of *Pteryxia terebinthina.* Structures of Two New Coumarins, Isopteryxin and Calipteryxin. J. Pharm. Sci. **56,** 184 (1967).

431. NIGAM, S. K., C. R. MITRA, G. KUNESCH, B. C. DAS, and J. POLONSKY: Constituents of *Calophyllum tomentosum* and *Calophyllum apetalum* Nuts: Structure of a New 4-Alkyl- and of Two 4-Phenyl-Coumarins. Tetrahedron Letters **1967,** 2633.

432. NIKONOV, G. K.: Deltoin — a New Natural Furocoumarin. Doklady Akad. Nauk S.S.S.R. **156,** 1210 (1964); Chem. Abs. **61,** 7000 (1964).

433. — Investigation of the Lactones of Monnier's *Cnidium* (L.) CUSS. Zhur. obshchei Khim. **34,** 1350 (1964); Chem. Abs. **61,** 3352 (1964).

434. — Xanthogalin, a New Coumarin from Purple Xanthogalum (*Xanthogalum purpurascens* LALLEM. Zhur. obshchei Khim. **34,** 3853 (1964); Chem. Abs. **62,** 6454 (1965).

435. — A Study of Cnidimin and Its Identification with Libanotin. Khim. prirod. Soedinenii **4,** 48 (1968); Chem. Abs. **69,** 77062 (1968).

436. — Structure of the Coumarin Sachalinin. Khim. prirod. Soedinenii **6,** 623 (1970); Chem. Abs. 76262 (1971).

437. — The Structure of Mogoltin, a Coumarin from the Roots of *Peucedanum mogoltavicum.* Khim. prirod. Soedinenii **7,** 572 (1971); Chem. Abs. **76,** 59780 (1972).

438. — The Structure of Mogoltavin, a Coumarin from the Roots of *Peucedanum mogoltavicum.* Khim. prirod. Soedinenii **8,** 43 (1972); Chem. Abs. **77,** 48653 (1972).

439. — The Structure of Mogoltavinin, a Coumarin from the Roots of *Peucedanum mogoltavicum.* Khim. prirod. Soedinenii **8,** 54 (1972); Chem. Abs. **77,** 48647 (1972).

440. NIKONOV, G. K., and D. I. BARANAUSKAITE: Coumarins of Wormwood-Zosimia *Zosimia absinthifolia* (Vent.) LINK. Zhur. obshchei Khim. **34,** 3854 (1964); Chem. Abs. **62,** 6465 (1965).

441. — — Xanthogalin — a New Coumarin from *Xanthogalum purpurascens* LALLEM. Khim. prirod. Soedinenii **1,** 139 (1965); Chem. Abs. **63,** 8303 (1965).

442. — — Lactones of *Zosimia absinthifolia* (VENT) LINK. Khim. prirod. Soedinenii **1,** 220 (1965); Chem. Abs. **63,** 13701 (1965).

443. NIKONOV, G. K., and A. A. IVASHENKO: The Chemical Structure of Morison's Mountain Parsley *Peucedanum morisonii* Bess. Zhur. obshchei Khim. **33,** 2740 (1963); Chem. Abs. **60,** 1725 (1964).

444. NIKONOV, G. K., and V. B. KUVAEV: The Lactone of *Peucedanum mogoltavicum* KOROV. Zhur. obshchei Khim. **34,** 1020 (1964); Chem. Abs. **60,** 15820 (1964).

445. NIKONOV, G. K., G. Y. PEK, and V. V. VANDYSHEV: The Structure of Agasyllin. Khim. prirod. Soedinenii **5,** 119 (1969); Chem. Abs. **71,** 61256 (1969).
446. NIKONOV, G. K., and M. G. PIMENOV: A Chemical Study of *Angelica sachalinensis* Maxim. Khim. prirod. Soedinenii **1,** 73 (1965); Chem. Abs. **62,** 16630 (1965).
447. NIKONOV, G. K., and A. I. SAIDKHODZHAEV: The Structure of Pranferin, a New Coumarin from the Roots of *Prangos ferulacea.* Khim. prirod. Soedinenii **7,** 255 (1971); Chem. Abs. **75,** 110142 (1971).
448. NIKONOV, G. K., R. K. VEREMEI, and M. G. PIMENOV: (+)-5-(3′-Methyl-2′,3′-dihydroxy)-butoxypsoralen — a Natural Component of *Angelica gmelini* (D.C.) Wormsk. Ex Fisch. Zhur. obshchei Khim. **34,** 1353 (1964); Chem. Abs. **61,** 1821 (1964).
449. NOVÁK, I., Z. ROZSA, Y. W. MIRRHOM, K. SZENDREI, J. REISCH, and E. MINKER: Isolation of Active Substances from *Ruta* Roots III. Marmesin, Marmesinin, Rutacultin and a Rutamarin Alcohol. Herba Hung. **11,** 13 (1972); Chem. Abs. **79,** 102812 (1973).
450. OGNYANOV, I., and D. BOCHEVA: Natural Coumarins II. Felamidin (Benzoylmarmesin). Z. Naturforsch. **22b,** 1231 (1967).
451. — — Pangeline and Angeloylpangeline. Two New Furocoumarins in the Roots of *Angelica pancici.* Doklady Bolg. Akad. Nauk **24,** 315 (1971); Chem. Abs. **75,** 45611 (1971).
452. — — Acid-Catalyzed Cleavage of Oxypeucedanin. Doklady Bolg. Akad. Nauk **25,** 1061 (1972); Chem. Abs. **78,** 58270 (1973).
453. OVODOV, Y. S., G. M. FROLOVA, M. Y. NEFEDOVA, and G. B. ELYAKOV: Glycosides of *Eleutherococcus senticocus* II. The Structure of Eleutherosides A, B_1, C and D. Khim. prirod. Soedinenii **3,** 63 (1967); Chem. Abs. **67,** 54394 (1967).
454. PAKNIKAR, S. K., and J. K. KIRTANY: The Terpenoid Parts of Kamolone and Kamolol of *Ferula penninervis.* Experientia **30,** 224 (1974).
455. PEREL'SON, M. E.: The Configuration of Polyanthin and Polyanthinin. Khim. prirod. Soedinenii **11,** 249 (1975); Chem. Abs. **84,** 5174 (1976).
456. PEREL'SON, M. E., A. I. BAN'KOVSKII, and N. E. ERMATOV: A Study of the Structure and Configuration of the Terpenoid Coumarins Kamolol and Kamolone by PMR Spectroscopy with the addition of $Eu(DPM)_3$. Khim. prirod. Soedinenii **11,** 703 (1975); Chem. Abs. **84,** 150766 (1976).
457. PEREL'SON, M. E., N. P. KIR'YALOV, and A. I. BAN'KOVSKII: The Question of the Configurations of Farnesiferol A, Gummosin, Badrakemin, and Colladonin. Khim. prirod. Soedinenii **11,** 244 (1975); Chem. Abs. **84,** 5172 (1976).
458. PEREL'SON, M. E., A. A. KIR'YANOV, A. I. BANKOVSKII, N. P. KIR'YALOV, and T. V. BUKREEVA: Studies of Stereochemistry of Terpenoidal Coumarins of the Iresane Series, Using the Method of Paramagnetic Resonance Spectroscopy with $Eu(DPM)_3$ Additions. Khim. prirod. Soedinenii **12,** 442 (1976); Chem. Abs. **86,** 29944 (1977).
459. PEREL'SON, M. E., A. A. KIR'YANOV, Y. E. SKLYAR, and V. V. VANDYSHEV: Use of the Paramagnetic Shift Agent $Eu(DPM)_3$ for Studying the Structure and Stereochemistry of Conferol and Conferone. Khim. prirod. Soedinenii **9,** 726 (1973); Chem. Abs. **82,** 16954 (1975).
460. PEREL'SON, M. E., Y. N. SHEINKER, A. A. SAVINA, and G. P. SYROVA: NMR Spectra of Natural Coumarin Derivatives III. Dihydrofurocoumarins and Dihydropyranocoumarins. Khim. prirod. Soedinenii **7,** 712 (1971); Chem. Abs. **76,** 98583 (1972).
461. PEREL'SON, M. E., Y. N. SHEINKER, and G. P. SYROVA: NMR Spectra of Natural Coumarin Derivatives II. Furocoumarins. Khim. prirod. Soedinenii **7,** 576 (1971); Chem. Abs. **76,** 45335 (1972).
462. PEREL'SON, M. E., Y. N. SHEINKER, G. P. SYROVA, G. K. NIKONOV, and A. P. PROKOPENKO: Use of NMR Spectroscopy in the Chemistry of Natural Coumarins. Trudy vses. nauch.-issled. Inst. Lek. Aromat. Rast. **15,** 60 (1969); Chem. Abs. **75,** 35617 (1971).

463. PEREL'SON, M. E., Y. N. SHEINKER, G. P. SYROVA, and K. F. TURCHIN: NMR Spectra of Natural Coumarin Derivatives I. Coumarins. Khim. prirod. Soedinenii **6**, 6 (1970); Chem. Abs. **73**, 34402 (1970).

464. PEREL'SON, M. E., V. V. VANDYSHEV, and Y. E. SKLYAR: The Structure and Stereochemistry of Conferin. Khim. prirod. Soedinenii **11**, 248 (1975); Chem. Abs. **84**, 5173 (1976).

465. PERONE, V. B.: The Natural Occurrence and Uses of the Toxic Coumarins. Microbial Toxins **8**, 67 (1972).

466. PFAU, A. S.: Etudes sur les matières végétales volatiles, IX. Sur quelque constituants inédits de l'huile essentielle de rue. Helv. Chim. Acta **22**, 382 (1939).

467. PIGULEVSKII, G. V., and G. A. KUZNETSOVA: Structure of a New Furocoumarin, Prangenine. Zhur. obshchei Khim. **23**, 1237 (1953); Chem. Abs. **47**, 12341 (1953).

468. PIGULEVSKII, G. V., and T. N. NAUGOL'NAYA: Resin from the Roots of *Ferula gummosa*. Trudy bot. Inst. Akad. Nauk S.S.R. **5**, 80 (1955); Chem. Abs. **50**, 15743 (1956).

469. PLOUVIER, V.: Recherche d'hétérosides coumariniques: le magnolioside du *Magnolia macrophylla*, le calycanthoside et le cichorioside des *Fraxinus*. Compt. rend. **266**, D, 1526 (1968).

470. — Sur quelque constituants des écorces de *Prunus mahaleb* L. (Rosacées). Compt. rend. **250**, 594 (1960).

471. — Sur deux hétérosides nouveaux, l'isofraxoside isolé du *Diervilla lonicera* Mill. (Caprifoliacées) et le poliothyroside isolé du *Poliothyrsis sinensis* Oliv. (Flacourtiacées). Compt. rend. **268**, D, 1982 (1969).

472. POUSSET, J. L., M. DEBRAY, and R. R. PARIS: Sur le *Baissea leonensis*, présence d'un nouvel hétéroside coumarinique, le baisseoside. Compt. rend. **271**, D, 2320 (1970).

473. POZZI, H., E. SÁNCHEZ, and J. COMIN: Studies on Argentine Plants — XXII. Heliettin, a New Furocoumarin from *Helietta longifoliata* Britt. Tetrahedron **23**, 1129 (1967).

474. PROKOPENKO, A. P.: Peucenidin — a New Furocoumarin Isolated from the Fruit of Mountain *Peucedanum oreoselinum* (L.) MOENCH. Zhur. obshchei Khim. **34**, 4111 (1964); Chem. Abs. **62**, 9117 (1965).

475. — Libanotin, a New Furocoumarin from *Libanotis transcaucasia* SCHISCHK. Khim. prirod. Soedinenii **1**, 215 (1965); Chem. Abs. **63**, 14638 (1965).

476. — Coumarins of *Libanotis intermedia*. Rast Resursy **2**, 201 (1966); Chem. Abs. **65**, 12561 (1966).

477. PURUSHOTHAMAN, K. K., S. VASANTH, J. D. CONNOLLY, and C. LABBÉ: 4,6,7-Trimethoxyl-5-methylchromen-2-one, a New Coumarin from *Leonotis nepetaefolia*. J. C. S. Perkin I **1976**, 2594.

478. RAJ, K., P. P. JOSHI, D. S. BHAKUNI, R. S. KAPIL, and S. P. POPLI: Synthesis of (±)-Auraptenol and Murrayone. Indian J. Chem. **14B**, 332 (1976).

479. RAJ, K., R. S. KAPIL, and S. P. POPLI: Synthesis of Subernol and Coumarin Rr_8. Indian J. Chem. **13**, 1404 (1975).

480. RAJENDRAN, K., C. K. MESTA, S. K. PAKNIKAR, and S. C. BHATTACHARYYA: Vaginol, a New Coumarin Related to Archangelicin. Indian J. Chem. **8**, 200 (1970).

481. RAMSTAD, E., W. C. LIN, T. LIN, and W. KOO: Coumurrayin, A New Coumarin from *Murraya paniculata* (L.) JACK. Tetrahedron Letters **1968**, 811.

482. REISCH, J., S. A. KHALED, K. SZENDREI, and I. NOVÁK: 5-Alkoxyfuranocoumarins from *Peucedanum ostruthium*. Phytochemistry **14**, 1889 (1975).

483. REISCH, J., I. NOVÁK, K. SZENDREI, and E. MINKER: Rutamarin, ein Coumarin-Derivat aus *Ruta graveolens*. Acta Pharm. Suecica **4**, 179 (1967); Chem. Abs. **67**, 54057 (1967).

484. — — — — Die Anwendung der KMR-Spektroskopie zur Strukturermittlung von Cumarin-Derivaten. Die Pharmazie **22**, 205 (1967).

485. Reisch, J., and K. Szendrei: Photoisomerisierung des Chalepensin zum Rutolid (Clausindin). Arch. Pharm. (Weinheim) **308,** 983 (1975).

486. Reisch, J., K. Szendrei, E. Minker, and I. Novák: 3-(1′,1′-Dimethylallyl)-6-(3′,3′-dimethylallyl) umbelliferone (gravelliferon) aus den Wurzeln von *Ruta graveolens.* Experientia **24,** 992 (1968).

487. — — — — Chalepensin, Gravelliferon-Methyläther und 3-(1′,1′-Dimethylallyl)-herniarin aus den Wurzeln von *Ruta graveolens.* Tetrahedron Letters **1968,** 4395.

488. — — — — Neue C_3-substituierte Cumarin-Derivate aus *Ruta graveolens* L.: 3-(1′,1′-Dimethylallyl)-daphnetindimethyläther, 8-Methoxygravelliferon, 3-(1′,1′-Dimethylallyl)-8-(3″,3″-dimethylallyl)-xanthyletin. Tetrahedron Letters **1970,** 4305.

489. Reisch, J., K. Szendrei, I. Novák, and E. Minker: Suberenon, a New Coumarin Derivative from the Root of *Ruta graveolens* L. Magyar Kem. Folyóirat **78,** 6 (1972); Chem. Abs. **76,** 138149 (1972).

490. Reisch, J., K. Szendrei, Z. Rózsa, I. Novák, and E. Minker: Rutacultin und Rutamarin-Alkohol aus den Wurzeln von *Ruta graveolens.* Phytochemistry **11,** 1529 (1972).

491. Reyes, R. E., and A. G. Gonzáles: Quimica de las Rutaceas IV. Cumarinas de las hojas de la *Ruta pinnata* L. Fil: estructura de una nueva furocumarina. Anales de Quím. **61,** 803 (1965).

492. — — Structure of Pinnarin and Furopinnarin, Two New Coumarins from Roots of *Ruta pinnata.* Phytochemistry **9,** 833 (1970).

493. — — Nuevas fuentes de cumarinas naturales XV. Estructura de la sabandinina, una nueva cumarina de la *Ruta pinnata* L. fil. Anales de Quím. **67,** 207 (1971).

494. — — Química de las Rutaceas IV. Cumarinas de las hojas de la *Ruta pinnata* L. fil; estructura de una nueva furocumarina. Farmacia Nueva **37,** 729 (1972); Chem. Abs. **79,** 9814 (1973).

495. Rodighiero, P., A. Guiotto, P. Manzini, and G. Pastorini: Synthesis of Poncimarin, a New Coumarin from *Poncirus trifoliata* L. Z. Naturforsch. **30B,** 987 (1975).

496. Rondest, J., B. C. Das, M.-N. Ricroch, C. Kan-Fan, R. Potier, and J. Polonsky: Plantes Malgaches-III. Etude des constituants de *Evodia belahe* Baillon (Rutacées). Phytochemistry **7,** 1019 (1968).

497. Rybalko, K. S., O. A. Konovalova, V. I. Sheichenko, and P. I. Zakharov: Armin a New Coumarin from *Artemisia armeniaca.* Khim. prirod. Soedinenii **12,** 294 (1976).

498. Saha, S. K., and A. Chatterjee: Isolation of Allo-imperatorin and β-sitosterol from the Fruits of *Aegle marmelos* Correa. J. Indian Chem. Soc. **34,** 228 (1957).

499. Saidkhodzhaev, A. I., and G. K. Nikonov: The Configuration of Badrakemin and Gummosin, and the Identity of Isobadrakemin, Colladonin and Farnesiferol A. Khim. prirod. Soedinenii **9,** 490 (1973); Chem. Abs. **80,** 60048 (1974).

500. — — Confirmation of the Configurations of Badrakemin and Gummosin. Khim. prirod. Soedinenii **10,** 15 (1974); Chem. Abs. **80,** 121140 (1974).

501. Saiki, Y., K. Morinaga, O. Okegawa, S. Sakai, Y. Amaya, A. Ueno, and S. Fukushima: On the Coumarins of the Roots of *Angelica dahurica* Benth. et Hook. Yakugaku Zasshi **91,** 1313 (1971); Chem. Abs. **76,** 56621 (1972).

502. Saiki, Y., M. Uchida, O. Okegawa, and S. Fukushima: On the Mass Spectra of Several Furanocoumarins Having Various Isoprenoidal Residues. Chem. and Pharm. Bull. (Japan) **22,** 1227 (1974).

503. Sano, K., I. Yosioka, and I. Kitagawa: Stereostructures of Decursin, Decursidin and a New Coumarin Isolated from *Angelica decursiva.* Chem. and Pharm. Bull. (Japan) **21,** 2095 (1973).

504. — — — Stereostructures of Decursin, Decursidin and Other New Pyranocoumarin Derivatives from the Root of *Angelica decursiva.* Chem. and Pharm. Bull. (Japan) **23,** 20 (1975).

505. SAVINA, A. A., G. K. NIKONOV, and A. I. BAN'KOVSKII: Seseliflorin — a New Coumarin from *Seseli sessiliflorum*. Khim. prirod. Soedinenii **6**, 522 (1970); Chem. Abs. **74**, 50519 (1971).

506. SAVINA, A. A., G. K. NIKONOV, and M. E. PEREL'SON: Smirniorin, a New Coumarin from the Roots of *Smirniopsis aucheri*. Khim. prirod. Soedinenii **5**, 592 (1969); Chem. Abs. **73**, 25337 (1970).

507. SAVINA, A. A., and M. E. PEREL'SON: The Configuration of the Acid Residue in the Molecules of Floroselin, Seselirin and Seseliflorin. Khim. prirod. Soedinenii **7**, 831 (1971); Chem. Abs. **76**, 71910 (1972).

508. SAVINA, A. A., M. E. PEREL'SON, and G. K. NIKONOV: The Structure of Smirnioridin. Khim. prirod. Soedinenii **6**, 185 (1970); Chem. Abs. **73**, 87813 (1970).

509. SAVINA, A. A., M. E. PEREL'SON, G. K. NIKONOV, and A. I. BAN'KOVSKII: Floroselin — a New Coumarin from the Roots of *Seseli sessiliflorum*. Khim. prirod. Soedinenii **6**, 517 (1970); Chem. Abs. **74**, 84002 (1971).

510. SAVINA, A. A., and M. G. PIMENOV: *Seseli sessiliflorum*, a New Source of Coumarins and Chromones. Rast. Resursy **8**, 361 (1972); Chem. Abs. **77**, 149719 (1972).

511. SAVINA, A. A., V. V. VANDYSHEV, M. E. PEREL'SON, and M. G. PIMENOV: Edultin from *Libanotis seseloides*. Khim. prirod. Soedinenii **7**, 116 (1971); Chem. Abs. **75**, 16129 (1971).

512. SCHEEL, L. D., V. B. PERONE, R. L. LARKIN, and R. E. KUPEL: The Isolation and Characterization of Two Phototoxic Furanocoumarins (Psoralens) from Diseased Celery. Biochemistry **2**, 127 (1963).

513. SCHNEIDER, G., and H. MÜLLER: Über das Furocumaringlucosid, Rutarin, aus der Weinraute. Arch. Pharm. **300**, 913 (1967).

514. SCHNEIDER, G., H. MÜLLER, and P. PFAENDER: Rutaretin, ein neues Furocumarin aus *Ruta graveolens* L. Arch. Pharm. **300**, 73 (1967).

515. SCHÖNBERG, A., and G. AZIZ: Furo-chromones and -coumarins (X): On the Constitution of Prangenin. J. Amer. Chem. Soc. **77**, 2563 (1955).

516. SCHROEDER, H. D., W. BENCZE, O. HALPERN, and H. SCHMID: Struktur der Visnagane; Synthese von (±)-*trans*-Samidin. Chem. Ber. **92**, 2338 (1959).

517. SESHADRI, T. R., and M. S. SOOD: Synthesis of Sphondin. J. Indian Chem. Soc. **39**, 539 (1962).

518. — — New Synthesis of Linear Furanocoumarins: Psoralen, Xanthotoxin and 8-Acetylpsoralen. Indian J. Chem. **1**, 291 (1963).

519. — — Chemical Comparison of the Roots of *Selinum vaginatum* and *Nardostachys jatamansi*. Phytochemistry **6**, 445 (1967).

520. — — Absolute Configuration of Selinidin, Vaginidin and Related Compounds. Current Sci. **36**, 563 (1967).

521. SESHADRI, T. R., M. S. SOOD, K. L. HANDA, and VISHWAPAUL: Constitution of Selinidin: a New Coumarin from *Selinum vaginatum* Clarke. Tetrahedron Letters **1964**, 3367.

522. — — — — Chemical Components of the Roots of *Selinum vaginatum* — I. Coumarins of the Petroleum Ether Extract. Tetrahedron **23**, 1883 (1967).

523. SESHADRI, T. R., and VISHWAPAUL: Sesibiricin, a New Coumarin from *Seseli sibiricum*. Indian J. Chem. **8**, 202 (1970).

524. — — Chemical Components of the Roots of *Selinum vaginatum* — II. Chemistry of Vaginidin and Vaginidiol. Indian J. Chem. **9**, 418 (1971).

525. — — Recent Advances in Naturally Occurring Coumarins. J. Sci. Ind. Res., India **32**, 227 (1973).

526. SHAGOVA, L. I., V. N. FLORYA, G. A. KUZNETSOVA, and M. E. PEREL'SON: Diesters of Khellactone and Rutarin (Campesenin) from *Seseli campestre* growing in Moldavia. Khim. prirod. Soedinenii **9**, 665 (1973).

527. SHANBHAG, S. N., C. K. MESTA, M. L. MAHESHWARI, and S. C. BHATTACHARYYA: Terpenoids — LXXV. Constituents of *Nardostachys jatamansi* and Synthesis of (±)-Dihydrosamidin and Visnadin from Jatamansin. Tetrahedron **21**, 3591 (1965).
528. SHANBHAG, S. N., C. K. MESTA, M. L. MAHESHWARI, S. K. PAKNIKAR, and S. C. BHATTACHARYYA: Terpenoids — LII. Jatamansin, a New Terpenic Coumarin from *Nardostachys jatamansi*. Tetrahedron **20**, 2605 (1964).
529. SHARMA, Y. N., R. C. SHARMA, A. ZAMAN, and A. R. KIDWAI: Chemical Examination of *Heracleum candicans* II. Isolation and Structure of a New Furocoumarin, Heraclenol. Naturwiss. **51**, 537 (1964).
530. SHARMA, Y. N., A. ZAMAN, and A. R. KIDWAI: Chemical Examination of *Heracleum candicans* — I. Isolation and Structure of a New Furocoumarin — Heraclenin. Tetrahedron **20**, 87 (1964).
531. SHARMA, Y. N., A. ZAMAN, A. R. KIDWAI, R. B. BATES, and V. P. THALACKER: Coumarin Constituents of *Heracleum candicans* — III. Tetrahedron **22**, 3221 (1966).
532. SHEICHENKO, V. I., and V. V. VANDYSHEV: The Composition of Visnadin. Khim. prirod. Soedinenii **7**, 368 (1971); Chem. Abs. **75**, 110204 (1971).
533. SHEINKER, Y. N., G. K. NIKONOV, M. E. PEREL'SON, G. P. SYROVA, G. Y. PEK, N. S. VUL'FSON, V. I. ZARETSKII, and V. G. ZAIKIN: Structure of Peucenol (Peumorisin) a New Hydroxycoumarin. Khim. prirod. Soedninenii **5**, 361 (1969); Chem. Abs. **72**, 66744 (1970).
534. SHIPCHANDLER, M., and T. O. SOINE: A Revised Structure of Columbianin and Pertinent Mass Spectral Studies. J. Pharm. Sci. **57**, 747 (1968).
535. SHIPCHANDLER, M., T. O. SOINE, and P. K. GUPTA: Coumarins XI: A Total Synthesis of (±)-Columbianetin. J. Pharm. Sci. **59**, 67 (1970).
536. SKLYAR, Y. E., L. G. AVRAMENKO, and L. I. DUKHOVLINOVA: A Method of Obtaining Felamidin. Khim. prirod. Soedinenii **9**, 427 (1973); Chem. Abs. **79**, 78647 (1973).
537. SKLYAR, Y. E., M. E. PEREL'SON, and M. G. PIMENOV: Moschatol, a New Coumarin from the Roots of *Ferula moschata*. Khim. prirod. Soedinenii **9**, 428 (1973); Chem. Abs. **79**, 102791 (1973).
538. SMITH, E., N. HOSANSKY, and W. G. BYWATER: Constitution of Samidin, Dihydrosamidin and Visnadin. Chem. and Ind. **1956**, 718.
539. SMITH, E., N. HOSANSKY, W. G. BYWATER, and E. E. VAN TAMELEN: Constitution of Samidin, Dihydrosamidin and Visnadin. J. Amer. Chem. Soc. **79**, 3534 (1957).
540. SMITH, E., L. A. PUCCHI, and W. G. BYWATER: Crystalline Visnagan. Science **115**, 520 (1952).
541. SOINE, T.: Naturally Occurring Coumarins and Related Physiological Activities. J. Pharm. Sci. **53**, 231 (1964).
542. SOINE, T. O., and F. H. JAWAD: Structure of Lomatin, a New Coumarin. J. Pharm. Sci. **53**, 990 (1964).
543. SOINE, T. O., A. ZHELEVA, M. M. MAHANDRU, P. ERHARDT, and L. DUBEVA-IVANOVA: Natural Coumarins VII: Isolation and Structure of a New Coumarin, Peuruthenicin, from *Peucedanum ruthenicum* M. B. J. Pharm. Sci. **62**, 1879 (1973).
544. SOKOLOVA, A. I., and G. K. NIKONOV: Structure of Xanthalin. Khim. prirod. Soedinenii **6**, 14 (1970); Chem. Abs. **73**, 35157 (1970).
545. SOKOLOVA, A. I., G. K. NIKONOV, M. E. PEREL'SON, G. P. SYROVA, and Y. N. SHEINKER: The Structure of Xanthaline. Khim. prirod. Soedinenii **4**, 280 (1968); Chem. Abs. **70**, 77713 (1969).
546. SOKOLOVA, A. I., M. E. PEREL'SON, and G. K. NIKONOV: Tomasin, a New Coumarin from *Xanthogalum purpurascens*. Khim. prirod. Soedinenii **5**, 359 (1969); Chem. Abs. **72**, 97296 (1970).
547. SOKOLOVA, A. I., Y. E. SKLYAR, M. E. PEREL'SON, and M. G. PIMENOV: Furo-

coumarins of *Komarovia anisospermum*. Khim. prirod. Soedinenii **12,** 166 (1976); Chem. Abs. **85,** 59598 (1976).

548. STANLEY, W. L., and L. JURD: Citrus Coumarins. Agric. Food Chem. **19,** 1106 (1971).

549. STANLEY, W. L., and S. H. VANNIER: Chemical Composition of Lemon Oil. I. Isolation of a Series of Substituted Coumarins. J. Amer. Chem. Soc. **79,** 3488 (1957).

550. STANLEY, W. L., A. C. WAISS, R. E. LUNDIN, and S. H. VANNIER: Auraptenol, a Coumarin Compound in Bitter (Seville) Orange Oil. Tetrahedron **21,** 89 (1965).

551. STARKOWSKY, N. A., and N. BADRAN: Ammajin, a New Constituent of *Ammi majus* L. J. Org. Chem. **23,** 1818 (1958).

552. STECK, W.: New Synthesis of Columbianetin and Related Coumarins. Canad. J. Chem. **49,** 1197 (1971).

553. — New Syntheses of Demethylsuberosin, Xanthyletin, (±)-Decursinol, (+)-Marmesin, (−)-Nodakenetin, (±)-Decursin, and (±)-Prantschimgin. Canad. J. Chem. **49,** 2297 (1971).

554. — Paniculatin, a New Coumarin from *Murraya paniculata* (L.) Jack. Canad. J. Chem. **50,** 443 (1972).

555. — Coumarins and Chromones from *Lomatium macrocarpum*. Phytochemistry **12,** 2283 (1973).

556. STECK, W., and B. K. BAILEY: Leaf Coumarins of *Angelica archangelica*. Canad. J. Chem. **47,** 2425 (1969).

557. — — Characterization of Plant Coumarins by Combined Gas Chromatography, Ultraviolet Absorption Spectroscopy, and Nuclear Magnetic Resonance Analysis. Canad. J. Chem. **47,** 3577 (1969).

558. STECK, W., B. K. BAILEY, J. P. SHYLUK, and O. L. GAMBORG: Coumarins and Alkaloids from Cell Cultures of *Ruta graveolens*. Phytochemistry **10,** 191 (1971).

559. STECK, W., and S. A. BROWN: Biosynthesis of Angular Furanocoumarins. Canad. J. Biochem. **48,** 872 (1970).

560. — — Comparison of (+)- and (−)-Marmesin as Intermediates in the Biosynthesis of Linear Furanocoumarins. Canad. J. Biochem. **49,** 1213 (1971).

561. STECK, W., and M. MAZUREK: Identification of Natural Coumarins by Nmr Spectroscopy. Lloydia **35,** 418 (1972).

562. STECK, W., and L. R. WETTER: Apterin, an Unusual Glucoside of *Zizia aptera*. Phytochemistry **13,** 1925 (1974).

563. STEMPLE, N. F., and W. H. WATSON: The Crystal and Molecular Structure of Xanthotoxin, $C_{12}H_{18}O_4$. Acta Cryst. **28B,** 2485 (1972).

564. STERMITZ, F. R., and R. D. THOMAS: Separation of Furocoumarins by High-Pressure Liquid Chromatography. J. Chromat. **77,** 431 (1973).

565. STOUT, G. H., and K. L. STEVENS: The Structure of Costatolide. J. Org. Chem. **29,** 3604 (1964).

566. SVENDSEN, A. B.: Zur Chemie norwegischer Umbelliferen. Diss. University Oslo **1954;** Chem. Abs. **52,** 2173 (1958).

567. SZENDREI, K., E. MINKER, I. NOVÁK, and J. REISCH: Biosynthesis of Rutaceae Components. Herba Hung. **9,** 33 (1970); Chem. Abs. **75,** 95445 (1971).

568. SZENDREI, K., M. PETZ, I. NOVÁK, J. REISCH, H. E. BAILEY, and V. L. BAILEY: Isolation and Structure Analysis of Quinoline Alkaloids and Coumarins from *Ptelea trifoliata* subspecies *pallida* var. *confinis*. Herb. Hung. **13,** 49 (1974); Chem. Abs. **83,** 40169 (1975).

569. TALAPATRA, B., S. K. MUKHOPADHYAY, M. K. CHAUDHURI, and S. K. TALAPATRA: Synthesis of 6,7-Dihydro-8,8-dimethyl-2H,8H-benzo[1,2-*b*:5,4-*b*′]-dipyran-2,6-dione (Graveolone): An Unusual Bromination of Dihydroxanthyletin by N-Bromosuccinimide in Presence of Lead Tetraacetate. Indian J. Chem. **14B,** 129 (1976).

570. TALAPATRA, S. K., M. BHATTACHARYYA, B. TALAPATRA, and B. C. DAS: Xylotenin, a New Furocoumarin from *Chloroxylon swietenia* DC. J. Indian Chem. Soc. **45,** 861 (1968).

571. TALAPATRA, S. K., L. N. DUTTA, and B. TALAPATRA: The Structure and Stereochemistry of Murrangatin. A New Monomeric Coumarin from *Murraya elongata* ALPH. DC. Tetrahedron **29,** 2811 (1973).

572. — — — Structure of Murralongin, a Novel Monomeric Coumarin from *Murraya elongata:* Stereochemistry and Preferred Conformation of its Unique Side Chain. Tetrahedron Letters **1973,** 5005.

573. TALAPATRA, S. K., S. K. MUKHOPADHYAY, and B. TALAPATRA: Nodakenetin Acetate: A New Coumarin from *Boenninghausenia albiflora.* Phytochemistry **12,** 2312 (1973).

574. — — — Minor Coumarins of *Boenninghausenia albiflora.* Phytochemistry **14,** 836 (1975).

575. TANINO, H., and S. INOUE: Synthesis of Coumurrayin. Chem. and Pharm. Bull. (Japan) **17,** 1071 (1969).

576. TERAZAWA, M., and T. SASAYA: Extractives of Yachidamo, *Fraxinus mandshurica* var. *japonica.* II. Glucosides in bark. Mokuzai Gakkaishi **16,** 192 (1970); Chem. Abs. **74,** 108116 (1971).

577. TOMIMATSU, T., H. HASEGAWA, and K. TORI: Studies on the Chemical Components of the Rutaceae Plants — VII. Components of the Roots of *Poncirus trifoliata* Rafinesque (5). Poncitrin, a New Coumarin: Structure and Chemical Degradation. Tetrahedron **30,** 939 (1974).

578. TOMIMATSU, T., M. HASHIMOTO, T. SHINGU, and K. TORI: Poncitrin, a New Coumarin: Structure and Nuclear Overhauser Effects. J. Chem. Soc. (D) **1969,** 168.

579. — — — — Studies on the Chemical Components of Rutaceae Plants — VI. Components of the Root of *Poncirus trifoliata* Rafinesque (4). Poncitrin, a New Coumarin: Structure and Nuclear Overhauser Effects. Tetrahedron **28,** 2003 (1972).

580. TURABELIDZE, D. G., G. K. NIKONOV, and E. P. KEMGRTELIDZE: Coumarins of the Roots of *Libanotis transcaucasica.* Khim. prirod. Soedinenii **10,** 402 (1974); Chem. Abs. **81,** 117102 (1974).

581. VANDYSHEV, V. V., G. K. NIKONOV, and M. G. PIMENOV: Coumarins of *Agasyllis latifolia.* Rast Resursy **4,** 330 (1968); Chem. Abs. **70,** 17563 (1969).

582. VANDYSHEV, V. V., M. E. PEREL'SON, Y. E. SKLYAR, and M. D. MOROZ: Conferin — New Coumarin from the Roots of *Ferula conocaula.* Khim. prirod. Soedinenii **10,** 660 (1974); Chem. Abs. **82,** 73218 (1975).

583. VANDYSHEV, V. V., Y. E. SKLYAR, M. E. PEREL'SON, and M. D. MOROZ: Conferdione — a New Coumarin from *Ferula conocaula.* Khim. prirod. Soedinenii **10,** 658 (1974); Chem. Abs. **82,** 73217 (1975).

584. VANDYSHEV, V. V., Y. E. SKLYAR, M. E. PEREL'SON, M. D. MOROZ, and M. G. PIMENOV: Conferone, a New Terpenoid Coumarin from the Fruit of *Ferula conocaula.* Khim. prirod. Soedinenii **8,** 669 (1972); Chem. Abs. **78,** 108266 (1973).

585. — — — — — Conferol, a New Coumarin from the Roots of *Ferula conocaula* and *F. moschata.* Khim. prirod. Soedinenii **8,** 670 (1972); Chem. Abs. **78,** 108265 (1973).

586. VAN SUMERE, C. F., G. WOLF, H. TEUCHY, and J. KINT: A New Thin-Layer Method for Phenolic Substances and Coumarins. J. Chromat. **20,** 48 (1965).

587. VAN TAMELEN, E. E., and R. M. COATES: Biogenetic-Type Synthesis of (±)-Farnesiferol A and (±)-Farnesiferol C. Chem. Comm. **1966,** 413.

588. VARGA, E., K. SZENDREI, I. NOVÁK, and J. REISCH: Isolation of Coumarin Glycosides from Tissue Cultures of *Ruta graveolens.* Acta Pharm. Hung. **44,** Suppl. 36 (1974); Chem. Abs. **82,** 13996 (1975).

589. VENTURELLA, P., A. BELLINO, and F. PIOZZI: Coumarin from *Ekebergia senegalensis.* Revision. Heterocycles **2,** 345 (1974).

590. VENTURELLA, P., A. BELLINO, and F. PIOZZI: Revised Structure for Siderin. Tetrahedron Letters **1974,** 979.

591. VON BROCKE, W., E. REINHARD, G. NICHOLSON, and W. A. KÖNIG: Über das Vorkommen von 3-(1′,1′-Dimethylallyl)-scopoletin in Gewebekulturen von *Ruta graveolens.* Z. Naturforsch. **26B,** 1252 (1971).

592. WAGNER, H., and S. BLADT: Cumarine aus südafrikanischen *Pelargonium*-Arten. Phytochemistry **14,** 2061 (1975).

593. WAGNER, H., S. BLADT, D. J. ABRAHAM, and H. LOTTER: Neue Cumarine aus *Pelargonium reniforme* CURT-Wurzel. Tetrahedron Letters **1974,** 3807.

594. WESSELY, F., and E. DEMMER: Konstitution und Eigenschaften des Fraxins. Ber. **62,** 120 (1929).

595. WILLETTE, R. E., and T. O. SOINE: Coumarins I. Isolation, Purification and Structure Determination of Pteryxin and Suksdorfin. J. Pharm. Sci. **51,** 149 (1962).

596. — — Coumarins II. Structure of Columbianadin and Columbianin. J. Pharm. Sci. **53,** 275 (1964).

597. WORDEN, L. R., K. D. KAUFMAN, J. A. WEIS, and T. K. SCHAAF: Synthetic Furocoumarins. IX. A New Synthetic Route to Psoralen. J. Org. Chem. **34,** 2311 (1969).

598. YAMADA, S., N. OH-HASHI, and K. ACHIWA: Synthetic Studies on Optically Active Epoxyterpenes from L-Glutamic Acid — I. Syntheses of R-(+)-Epoxygeraniol, R-(+)-Marmin and R-(+)-Epoxyaurapten. Tetrahedron Letters **1976,** 2557.

599. YAMADA, Y., C.-S. HSU, K. IGUCHI, M. SUZUKI, H.-Y. HSU, and Y.-P. CHEN: Two New Khellactone Esters from *Peucedanum japonicum* Thunb. Tetrahedron Letters **1974,** 2513.

600. YAMAMOTO, K., H. IRIE, and S. UYEO: Studies on the Constituents of the Rhizomes of *Glaucidium palmatum* II. The Total Synthesis of Glaupalol. Yakugaku Zasshi **91,** 257 (1971); Chem. Abs. **74,** 125509 (1971).

601. YUSUPOV, M. I., and G. P. SIDYAKIN: A Hydroxycoumarin Arscotin from *Artemisia scotina.* Khim. prirod. Soedinenii **9,** 430 (1973).

602. — — Fraxidin and Isofraxidin from *Artemisia scotina.* Khim. prirod. Soedinenii **11,** 91 (1975); Chem. Abs. **83,** 128637 (1975).

603. ZAKHAROV, P. I., V. S. KABANOV, A. I. BAN'KOVSKII, G. K. NIKONOV, and N. E. ERMATOV: A Mass Spectrometric Study of Some Dihydropyranocoumarins. Khim. prirod. Soedinenii **7,** 398 (1971); Chem. Abs. **75,** 139893 (1971).

604. ZAKHAROV, P. I., V. S. KABANOV, M. E. PEREL'SON, A. I. BAN'KOVSKII, and N. E. ERMATOV: Study of Natural Coumarins by Mass Spectroscopy I. Mass Spectra of Kamolone and Kamolol. Khim. prirod. Soedinenii **6,** 296 (1970); Chem. Abs. **73,** 76374 (1970).

605. ZAKHAROV, P. I., P. B. TERENT'EV, G. K. NIKONOV, and A. I. BAN'KOVSKII: A Mass-Spectrometric Study of 3′,4′-Dihydroxy- and 3′,4′-Diacycloxydihydropyranocoumarins. Khim. prirod. Soedinenii **7,** 704 (1971); Chem. Abs. **76,** 125962 (1972).

606. — — — — A Mass-Spectrometric Study of Angular and Linear Diacycloxydihydrofurocoumarins. Khim. prirod. Soedinenii **8,** 275 (1972); Chem. Abs. **77,** 151143 (1972).

607. — — — — A Mass-Spectrometric Study of Linear Monohydroxy- and Monoacyloxydihydrofurocoumarins. Khim. prirod. Soednineii **8,** 431 (1972); Chem. Abs. **78,** 15097 (1973).

608. — — — — Mass-Spectrometric Study of the Coumarins Mogoltin, Mogoltavin, and Mogoltavinin. Khim. prirod. Soedinenii **10,** 18 (1974); Chem. Abs. **80,** 121139 (1974).

609. ZAKHAROV, P. I., P. B. TERENT'EV, G. K. NIKONOV, A. I. BAN'KOVSKII, N. D. ANTONOVA, and A. P. PROKOPENKO: The Structure of Peucenidin. Khim. prirod. Soedinenii **8,** 271 (1972); Chem. Abs. **77,** 164558 (1972).

610. ZARETSKII, V. I., N. S. VUL'FSON, L. S. CHETVERIKOVA, and V. G. ZAITKIN: Mass-

Spectrometric Investigation of Heterocyclic Compounds. Structure of Peumorisin — a New Natural Hydroxycoumarin Isolated from *Peucedanum morisonii* (Bess.) Zhur. obshchei Khim. **34,** 3655 (1964); Chem. Abs. **62,** 9097 (1965).

611. Zheleva, A. B., L. Bubeva-Ivanova, and S. L. Spasov: On the Structure of Peuarenarine, a New Coumarin from *Peucedanum arenarium*. Z. Naturforsch. **26B,** 113 (1971).

612. Zheleva, A. B., M. M. Mahandru, and L. Bubeva-Ivanova: Four New Coumarins from the Roots of *Peucedanum arenarium*. Phytochemistry **15,** 209 (1976).

613. — — — Stereochemistry of Isokhellactone Derived Coumarins from *Peucedanum arenarium*. Phytochemistry **15,** 1293 (1976).

614. Zheleva, A. B., T. O. Soine, and L. Bubeva-Ivanova: Natural Coumarins V. Isolation of Xanthalin and a New Pyranocoumarin Peuarenine from *Peucedanum arenarium*. J. Pharm. Sci. **61,** 1643 (1972).

(Received May 21, 1977)

Recent Developments in the Field of Naturally-Occurring Aroma Components

By G. Ohloff, Firmenich SA, Research Laboratories, Geneva, Switzerland

Contents

Acknowledgement. The author is indebted to Drs. I. Flament, B. Maurer, W. Pickenhagen and A. F. Thomas for corrections of the text, stimulating discussions and valuable suggestions. The partial translation by Mrs. Edith Wenger is gratefully acknowledged. Miss Gertrud Lingesleben is responsible for the lay-out of this paper, which she did with great skill and patience.

I. Occurrence and Implication of Flavors

Nutritious foods have to contain trace constituents such as minerals, vitamins and flavors. The role of flavors is critical; they are partially responsible for the digestion of food in man and control food recognition, selection and acceptance. Flavor can be considered to be the neurophysiological answer to the question simultaneously asked of the chemical senses of taste and odor by the stimulus, food (*447*).

This review deals exclusively with the chemistry of the odoriferous constituents of the volatile part of the flavor, namely the aroma. Aroma components can be considered to be direct metabolites produced in vegetable or animal organisms by intracellular biogenetic pathways and their quality and quantity therefore depend on genetic factors and are influenced by ripening and storage. Food flavors, however, are mainly produced by extracellular events, *e.g.* by enzymatic or chemical transformations of the foodstuff upon processing, by degradation of high molecular weight precursors. Proteins, carbohydrates and lipids of the food material are the three main sources of food flavors, although some less frequently occurring precursors, such as polyphenols, nucleotides, pigments, vitamins or polyterpenes, also contribute to a certain extent to the overall flavor. Fermentation, curing, dehydration, boiling, baking, frying and roasting are the most common post-harvest procedures.

Table 1. *Amount of Volatiles in Food and Number of Aroma Components Therein*

Kind of Food	Volatiles in ppm	Number of Components identified	unknown
beef	50 (*153*)	270 (*153*)	250 (*153*)
cheddar cheese	100 (?) (*639*)	150 (*639*)	130 (*639*)
potatoes (baked)	1 (*263, 440*)	110 (*694*)	?
onions/garlic	900 (*565*)	96 (*694*)	?
boletus edulis	100 (*644*)	100 (*644*)	>200 (*639*)
coffee	1* (*496*)	468 (*673*)	>500 (*710*)
"Ceylon" tea	60 (*491*)	325 (*412*)	?
"Burley" tobacco	300** (*111, 502*)	309 (*114*)	≥200 (*108*)
"Virginia" tobacco	2200 (*119*)	346 (*119, 125, 347*)	≥200 (*108*)
strawberry	10 (*714*)	324 (*694*)	?
pineapple	11,3 (*227*)	59 (*227*)	?
raspberry	1,7 (*708*)	95 (*708*)	~120 (*474*)
bell pepper	20 (*73*)	43 (*73*)	63 (*73*)
rye crispbread	15 (*613*)	92 (*613*)	30 (*613*)
bilberries	2,5 (*614*)	109 (*614*)	90 (*614*)
yellow passion fruit	12 (*711*)	194 (*711*)	~50 (*711*)

* Concentration of volatiles in brewed coffee.

** This rather low value obtained at RJR's laboratories does not include nicotine nor similar bases. For the whole volatile material including bases, a value of at least 400 would seem reasonable. At FIRMENICH (Virginia tobacco), the steam-distillation of tobacco was carried out more exhaustively, affording better yield of volatiles.

The pathways for the formation of flavoring substances are characterized by normal chemical reactions such as oxidations, rearrangements, fragmentation and recombination reactions. Fundamental knowledge of reaction mechanisms in organic chemistry can facilitate the structure elucidation of aroma compounds. Review articles of general and specific nature deal with this fascinating field of activity (*132, 199, 234, 269, 355, 439, 494, 662*).

Flavors are never attributed to a single aroma compound; rather, foods are flavored by multitudes of odoriferous components frequently present in trace concentrations. Thus, for instance, the volatile components of fruits only rarely exceed a quantity of 10 ppm (*441*) (Table 1). The isolation of the original aroma without any substantial modification requires special methods which have to be adapted with great care to the specific food (*472*). Some cases are known where a natural flavor impression is dominated by one single substance (character impact compound) (*270*) (Table 2) or a small number of compounds mostly of

Table 2. *Natural Occurrence of "Character Impact" Aroma Compounds in Food*

Structure	Chemical or Common Name	Flavor Description	Natural Occurrence
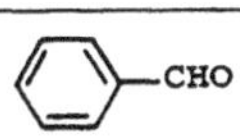	benzaldehyde	almond-like	almond
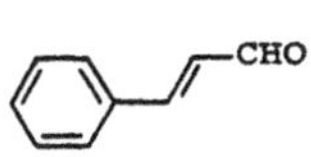	cinnamaldehyde	cinnamon-like	oil of cinnamon
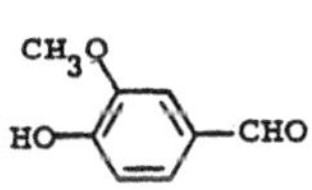	vanillin	vanilla-like	Bourbon vanilla
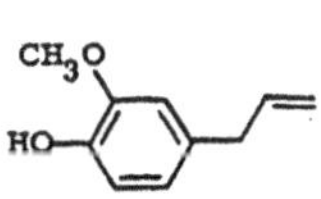	eugenol	clove-like	oil of clove buds
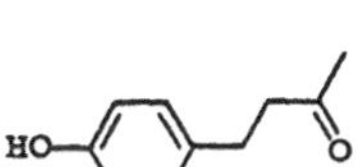	raspberry ketone	raspberry	raspberries
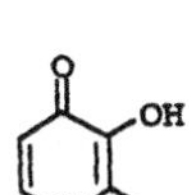	maltol	burnt sugar	malt caramel coffee

Table 2 (continued)

Structure	Chemical or Common Name	Flavor Description	Natural Occurrence
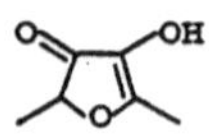	furaneol	burnt pineapple; caramel	strawberry pineapple
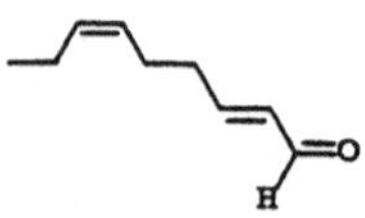	violet leaf aldehyde	cucumber	cucumber
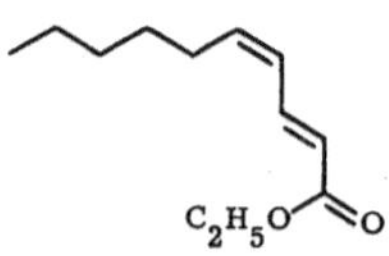	pear ester	Bartlett pear	pear
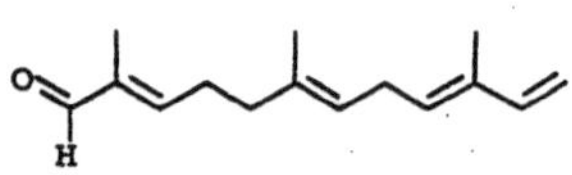	α-sinensal	mandarin	bitter orange
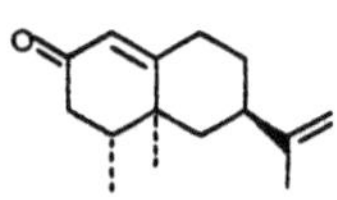	(+)-nootkatone	grapefruity	grapefruit
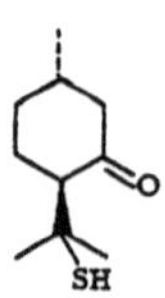	(+)-*trans*-8-mercapto-*p*-methan-3-one	catty	buchu leaves

differing chemical natures (organoleptic principle). Thus, the combination of (*E*)-2-hexenal, hexanal and ethyl 2-methylbutyrate is of critical importance for the characteristic flavor note of the Delicious apple (*155*). Although sinensal constitutes the character impact compound of mandarin flavor, a mixture of thymol and methyl anthranilate in suitable proportions is also essential for the specific response to mandarin oil (*319*).

One and the same compound can give rise to several receptor events which result in the perception of different odor qualities. This phenomenon depends on the concentration. The extent to which volatile components contribute to the total flavor is dependent on their potency and concentration. The threshold value constitutes a measure

Table 3. *Threshold Values (T) in ppb of Water*

Structure	Chemical Name	T	Ref.
	acetone	450 000	(*162*)
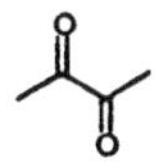	diacetyl	4	(*506*)
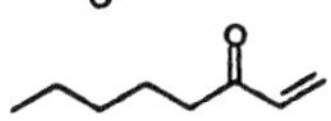	1-octen-3-one	0.1	(*162*)
	butanol	10 000	(*506*)
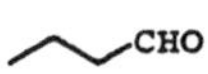	butanal	60	(*506*)
	butyric acid	3 000	(*506*)
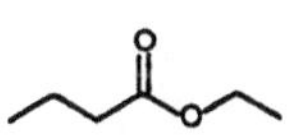	ethyl butyrate	0.1	(*506*)
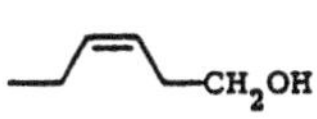	(*Z*)-3-hexenol	70	(*74*)
	(*Z*)-3-hexenal	0.25	(*74*)
	hexanal	4.5	(*74*)
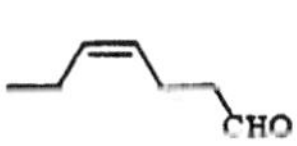	(*Z*)-4-heptenal	0.8	(*368*)
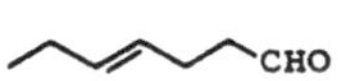	(*E*)-4-heptenal	10	(*368*)
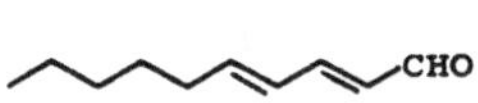	(*E,E*)-2,4-decadienal	0.07	(*74*)
	decanal	0.9	(*506*)
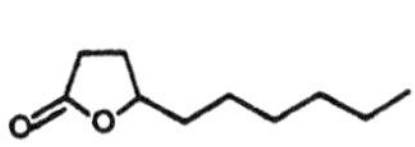	γ-decalactone	88	(*572*)
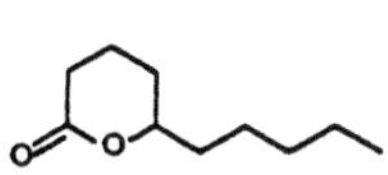	δ-decalactone	160	(*572*)

Table 3 (continued)

Structure	Chemical Name	T	Ref.
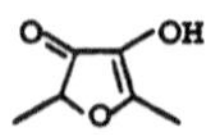	2,5-dimethyl-4-hydroxy-3(2*H*)-furanone (FURANEOL®)	0.04	(*485*)
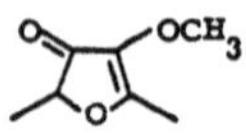	2,5-dimethyl-4-methoxy-3(2*H*)-furanone	0.03	(*485*)
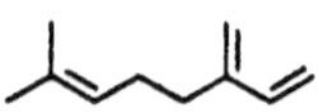	myrcene	15	(*630*)
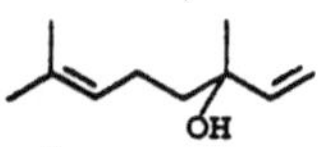	linalool	6	(*74*)
	limonene	10	(*74*)
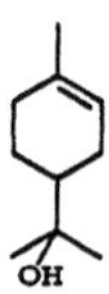	α-terpineol	350	(*74*)
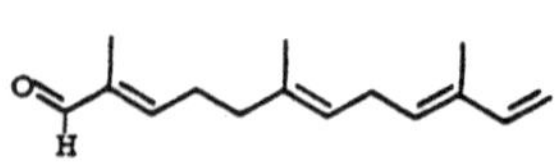	α-sinensal	0.05	(*630*)
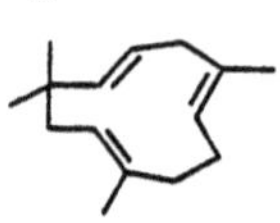	humulene	160	(*630*)
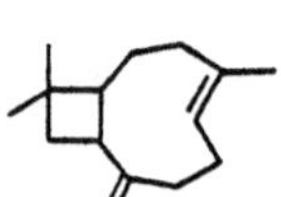	caryophyllene	64	(*202*)
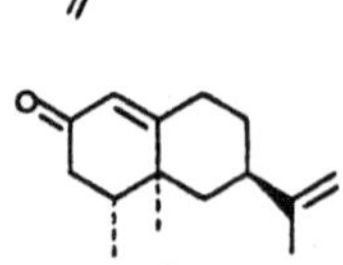	(+)-nootkatone	1	(*38*)
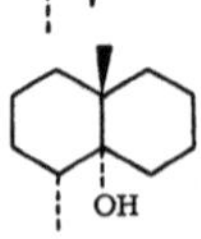	geosmin	0.021	(*70*)

Table 3 (continued)

Structure	Chemical Name	T	Ref.
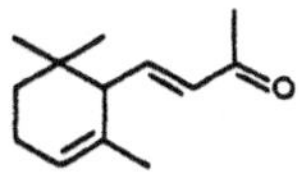	α-ionone	0.4	(*292*)
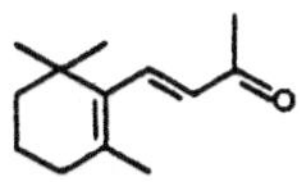	β-ionone	0.007	(*74*)
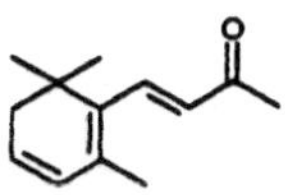	dehydro-β-ionone	1	(*378*)
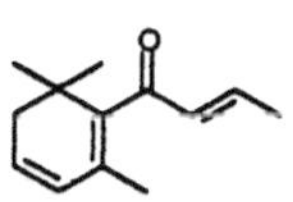	β-damascenone	500	(*378*)
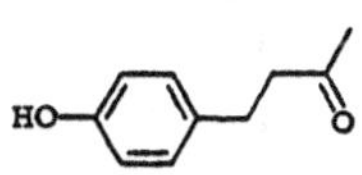	4-(*p*-hydroxyphenyl)-2-butanone ("raspberry ketone")	100	(*537*)
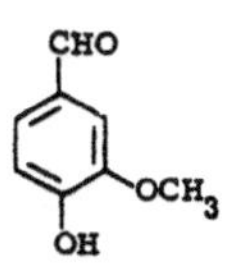	vanillin	20	(*506*)
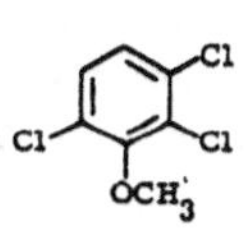	2,3,6-trichloroanisole	0.0000003	(*96*)

for the odor intensity of an aroma compound. It is defined as that quantity of a substance which will only just be perceived by the human olfactory system (Table 3), and 1,210 odor threshold values have been listed (*586*). Intensity and quality of an aroma are not necessarily determined by the components occurring in the highest proportion. Attempts have been made to determine the relative contribution of the individual components to the overall odor intensity of the total flavor (*202*). The odoriferous contribution of a single compound to the total flavor can be expressed in terms of odor units (*202*) or in aroma values (*399, 505, 506, 523*). Thus, the aroma value of methyl mercaptan in coffee is about 45,000 times higher than that of acetaldehyde and 1,500 times higher than that of H_2S in comparable concentrations (*496*).

These relative values demonstrate the organoleptic importance of trace components of high potency in an aroma complex. The analyst frequently reaches the limit of detectability for volatile substances. The most sensitive method known today consists in the direct coupling of a gas chromatograph with a mass spectrometer. This method allows the analytical detection of 10^{-12} mole of a single component in a mixture (*43, 544*).

The phenomenon of flavor interaction is not yet understood. Treshold values can vary considerably in different solvents (*267*). Synergism and antagonism between aroma components exist (*523*) even at sub-threshold concentrations (*104, 329, 489*) although it is a minor phenomenon (*380*). Furthermore, flavors in unprocessed foodstuffs give rise to further problems whose scientific investigation on corresponding model systems has recently been started (*165, 356, 464, 581*).

Up to now it was not possible to establish general rules as to correlations between odor quality and chemical structure. Some valuable concepts have, however, been established by approaching the molecular basis of stimulant interaction at the olfactory receptor level (*16, 34, 35, 36, 45, 46, 313, 346, 446, 448, 452, 602, 630, 631*). The attempts to create a code of odor impressions for flavors are still rudimentary (*213, 447, 612*). Psychologists interested in sensory measurements try to use physico-chemical data of stimuli to establish empirical sensory correlations (*131, 238, 324, 365, 392, 393, 396, 477, 533*). Several techniques for the characterization and measurement of odor qualities for practical purposes have been described (*328, 392, 477*).

At present, about 150 common flavor types are known, fifty of which have been investigated to a certain extent. Most of these flavors are composed of between 300 and 800 individual components (*501*) (Table 1). There have been 2,600 individual chemicals recognized as aroma components (*694*), and about 5,000 to 10,000 aroma components may exist (*501*).

Our knowledge of the chemical composition of natural flavors can serve many purposes (*147, 580, 696*). An essential task of applied research consists in creating true-to-nature flavors by means of the basic building blocks. The target is to formulate simple compositions with as few compounds as possible so that the very complex natural mixtures can be replaced in a biologically clear manner. Nature-identical flavors are used for flavoring foodstuffs, the flavor of which is partially or totally lost during post-harvest processing or storage. Finally, flavors are used for converting foodstuffs merely having a nutritional value into foodstuffs of highest acceptance, thus creating the possibility of developing novel food products. In view of the precarious food situation in the world — about 2 billion people are reported to be short of food — this possibility is of utmost importance. According to rough estimates, industrially produced flavors are used, at present, in the foodstuff industry as follows (*97*): soft drinks 25%, bakery 20%, candies and chewing-gum 18%, ice-cream 12%, sweets 12% and miscellaneous 13%. Due to the rapidly progressing development of novel foodstuffs, based on proteins for instance, flavor production will rise by 10 to 20% in 1980 (*97*).

Flavoring materials are considered as a special type of food additive and are, therefore, subjected to legal control which will, at some time in the future, be based on international regulations (*674*). In the USA, for instance, the synthetic equivalent of a naturally occurring aroma component can be used only if it is generally recognized as safe. Compounds of this type are put on the so-called GRAS lists of the U.S. Flavor and Extract Manufacturer's Association (*665*). Furthermore, there exist the so-called White list and the list of synthetic flavoring substances which may be safely used in foods, established by the U.S. Food and Drug Administration (*88*). Consequently, the present review deals exclusively with nature-identical compounds, the sensory properties of which are described in the literature, but does not take into consideration aroma compounds which are isolated from so-called "reaction flavors", for instance compounds which are used in imitation meat flavors (*466, 557, 558, 704*). By describing a selection of the most important classes of compounds, an attempt is made to familiarize the organic chemist with the diversity of flavor research.

II. Oxygen-Containing Aliphatic Aroma Compounds

1. Alcohols and Carbonyl Compounds

Functionalized n-alkanes belong to the most widely distributed aroma components, their proportion of the volatiles of foodstuffs varying from traces to major amounts. This class of compounds has a dominating position as aroma-forming elements in fruit aromas (*441*) in the broadest sense, fat-containing products of all kinds (*162*), and to some extent in foodstuffs produced by fermentation, such as cheese (*318*), wine (*689*), beer (*381*) or even black tea (*491, 527*). Of the 220 flavor substances so far detected in whisky, over 140 belong to the oxygen-containing aliphatic group (*523*). Aminoacids and higher unsaturated fatty acids are among the most important precursors for this class of flavors, the active compounds of which arise by enzymatic pathways. Suggestions about the mechanism of formation (*22, 159, 161, 162, 201, 450*) and model reactions (*136, 175, 176, 564*) have been published in recent years.

Among the acids, alcohols, esters and carbonyl compounds, the aldehydes with the lowest odor threshold (Table 3) have a particularly high aroma value. The thresholds of homologous series of saturated aliphatic aldehydes decrease progressively as the chain length increases from C-5 to C-10 (12 to 0.1 ppb) (*203*). The introduction of double bonds into the functionalized compounds generally causes a decrease of the odor threshold. It is generally recognized that, as a rule, the odor of *Z*-unsaturated compounds is not only much stronger but also more aroma-like and pleasant than that of the corresponding *E*-isomers. Relevant information about the sensory interaction of aliphatic alcohols, esters, acids and carbonyl compounds has been referred to in the preceding chapter. For practical reasons a selection must be made among the large number of known straight-chain oxygen-containing aroma compounds. Particularly interesting flavor properties of selected typical representatives of this class of compounds will be described.

Short-chain saturated aliphatic aldehydes constitute an essential part of the natural tomato flavor (*544*). As the time of storing of fruit preserves is prolonged, the carbonyl content decreases due to formation of the corresponding alcohols (*432*). This conversion is equivalent to a decrease of the flavor value since the odor threshold of the alcohols is considerably higher than that of the aldehydes (Table 3). 1-Hexanol was identified as the main component (22.5%) of the volatile constituents of yellow passion fruit, the flavor character of which is, however, not determined by this alcohol but by trace amounts of sulfur-containing compounds (*709, 711*). Unsaturated aliphatic aldehydes play a much more important role as aroma principles than their saturated analogues. It is very likely that a number of 2-alkenals, together with a series of alkanals, contribute to the overall flavor of dehydrated potato products (*203*). Whereas nonanal is quantitatively the main constituent (7.1%) of the flavor of bilberries (*Vaccinium myrtillus* L.), the α,β-unsaturated

(1) 

(2) 

(3) 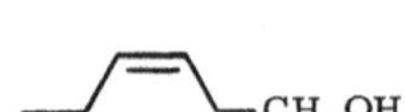

(4) 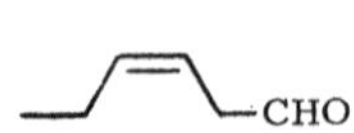

aldehydes with 6 to 11 carbon atoms as well as the C-10 dienals are much more important for the overall flavor impression (*614*).

(*E*)-2-Hexenal (**2**) ("leaf aldehyde") and (*Z*)-3-hexenol (**3**) ("leaf alcohol") are characterized by a pronounced green odor of leaves. (*Z*)-3-Hexenol (**3**), first discovered in fermented tea leaves in 1895, was later reported to occur in a wide range of higher plants. In tea leaves the alcohol is produced as a result of mechanical injury (*512*). Aliphatic C-6 compounds which decisively contribute to the "green" odor of fruit flavors (*441*) are formed from unsaturated aliphatic C-18 fatty acids by enzymatic action (*133*) after destruction of the cellular structure and its subsequent contact with molecular oxygen (*714*). Enzymatic model experiments using ^{14}C-labeled linoleic acid suggest the participation of lipoxygenase during the formation of (*E*)-2-hexenal (**2**) and (*Z*)-3-hexenal (**4**) in apples and grapes after homogenization. Aldehyde (**4**) is formed in tea leaves from linolenic acid *via* 13-L-hydroperoxylinolenic acid and serves as the common precursor for (**2**) and (**3**) (*564*). Aldehydes (**2**) and (**4**), in turn, serve as precursors of the corresponding alcohols (**1**) and (**3**) (*136, 564*). Differences in varieties of apples are due to quantitative variations in the formation of (*E*)-2-hexenal (**2**) (Jonathan, Cox's Orange, golden pearmain) and (*Z*)-3-hexenal (**4**) (bell apple, Champagne rennet) as well as of the corresponding alcohols (**1**) and (**3**) (*136*). (*Z*)-3-Hexenal (**4**) was found to be responsible for the "green" flavor note in soybean oil (*235*).

At a level of 1 ppb (*161*) (*Z*)-4-heptenal (**5**) confers a creamy flavor (*220*) to deodorized butter oil. At higher concentrations (from 2 ppm) this aldehyde causes rancidity in butter (*22*) and soybean oil (*560*).

(**5**) 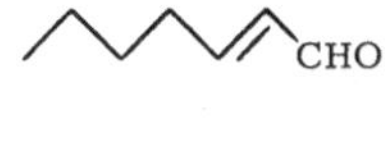

(**6**) 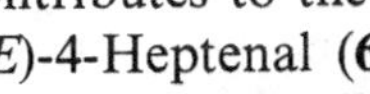

(**7**) 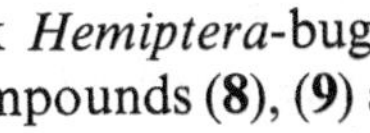

Furthermore, aldehyde (**5**) contributes to the off-flavor in frozen stored cod (*Gadus morlina*) (*368*). (*E*)-4-Heptenal (**6**) which has an aggressive putty and green odor constitutes the offensive pheromone of the tropical bedbug species [black *Hemiptera*-bug (*Pentatomidea*)] (*398*).

The three aliphatic C-8 compounds (**8**), (**9**) and (**11**) constitute the main components of the raw potato flavor (*76*). The two alcohols (**8**) and (**11**) are also present in flavor concentrates of raw mushrooms of the genus

(8) 

(12) 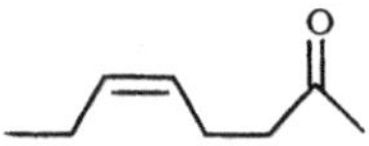

(9) 

(13) 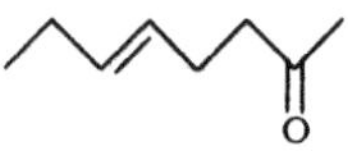

(10) 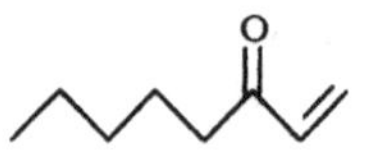

(14) 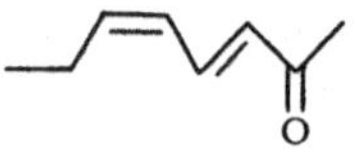

(11) 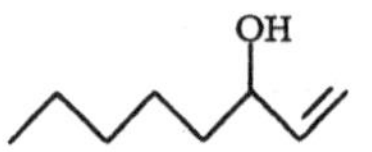

(15) 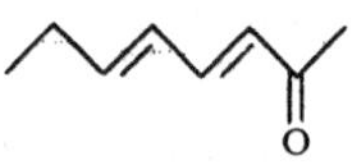

Agaricus bisporus and *Boletus edulis* (*682*). In addition, ketone **(10)** is found in the volatile parts of cooked mushrooms (*473*).

Compound **(11)**, called "mushroom alcohol" on account of its typical odor, has the lowest threshold (0.001 ppm in water) among the known alkenols (*588*). The racemic alcohol **(11)** was one substance isolated from soybeans (*21, 699*), decomposed corn oil (*317*), oxidized butter (*77, 588*) and dry red beans where the corresponding alcohol of ketone **(12)** was also found. It was suggested that alcohol **(11)** is formed from autoxidized arachidonate as well as from linoleate (*588*). Soaking of soybeans in water results in the production of significant quantities of optically active 1-octen-3-ol **(11)** (*21*). (−)-1-Octen-3-ol **(11)**, the most important flavor compound in *Agaricus bisporus*, imparts a stronger flavor response than its enantiomer (*126*). In lavender oil **(11)** constributes to the top-note. 1-Octen-3-one **(10)** has an unpleasant odor and was shown to be the compound responsible for the "metallic" flavor in oxidized butter (*587*). However, ketone **(10)** is also a flavor component of clover (*Trifolium* species) (*240*). 3,5-Octadien-2-one of unknown stereochemistry, found in oxidized butter, has a fruity odor and a threshold value of 0.3 ppm (*22*). (*E,Z*)-Stereoisomers **(14)** were found in black tea (*491*). The odor of the (*E,E*)-compound **(15)** is described as woody and pleasant, and this substance was reported to give freshness to green tea flavor (*722*). The (*Z*)- and (*E*)-5-octen-3-ones **(12)** and **(13)** are important constituents of bergamot oil (*606*). 1-Nonen-4-one **(28)**, (*E*)-2-nonen-4-one **(29)** and (*E,E*)-2,5-nonadien-4-one **(30)** as well as (*E*)-3-hepten-2-one **(7)** were found to be present in green bell peppers (*73*).

The importance of 2-alkanones of different chain length in the flavor of dairy food has been fully recognized (*103, 159, 160, 161*). Oil of rue (*Ruta graveolens* L.) consists almost completely of methyl ketones, the corresponding alcohols and esters (*386*). 2-Nonanone constitutes up to 90% of this oil (*196*). A strong blue cheese flavor is imparted to dairy products, *e.g.* butter, cheese, cream and milk, by the addition of 8-nonen-2-one (**27**) (*718*). Among the long-chain methyl ketones (*E,E*)-6,9-pentadecadien-2-one (odor threshold of about 1 ppb) and 6,9,13-pentadecatrien-2-one of unknown stereochemistry seem to play a particular role in hop oil (*71*). The presence of methyl ketones of the series ranging from acetone to 2-tridecanone in the essential oils of various plants (*163*) seems related primarily to their role as insect attractants or repellents (*157*). An interesting review deals mainly with the biosynthesis, biodegradation and biological role of methyl ketones (*157*).

(**16**) 

(**24**) 

(**17**) 

(**25**) 

(**18**) 

(**26**) 

(**19**) 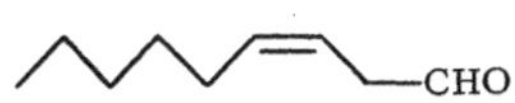

(**27**) 

(**20**) 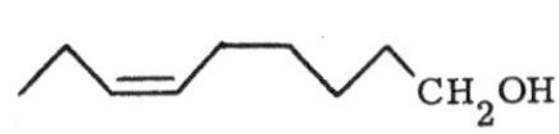

(**28**) 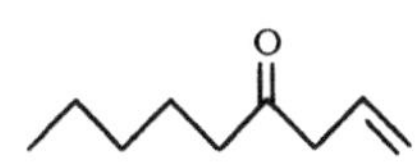

(**21**) 

(**29**) 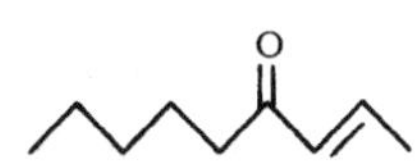

(**22**) 

(**30**) 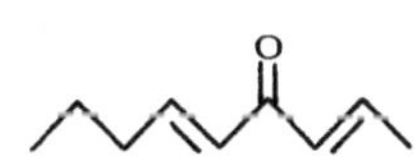

(**23**) 

Because of their structural diversity, the C-9 compounds occupy first place among the aliphatic alcohols and carbonyl compounds. Cucumber (*164, 296, 619*) and various melon fruits (*293, 294, 297, 298*) seem to be particularly rich in derivatives having this number of carbon atoms. In addition to nonanal and the corresponding alcohol, their unsaturated analogs (**16**) to (**25**) were also found in these *Cucurbitaceae* fruits. (*E,Z*)-2,6-Nonadienal (**23**) and (*E*)-2-nonenal (**17**) constitute the organoleptic principle for cucumber flavor (*Cucumis sativus* L.) (**164, 296**). Although it is present in a concentration of only 0.9 ppb (about 45 times its threshold value) (*295*), (*Z*)-6-nonenal (**21**) is the essential constituent of the flavor of musk melon (*Cucumis melo* L. var. *reticulatus Naudin*) (*293, 297, 298*), whereas (*Z,Z*)-3,6-nonadien-1-ol (**25**) embodies the aroma of water melon (*Citrullus vulgaris* Schrad) (*294, 297*). An enzymatic pathway for the biogenesis of (**17**), (**19**) and (**23**) in cucumber from linoleic acid has been proposed (*175, 176*). Aldehyde (**23**), found among the degradation products of autoxidized trilinolenin (*210*), and the corresponding alcohol (**22**) contribute decisively to the odoriferous principle of violet leaf oil (*536*) which is appreciated as a fine fragrance. On the other hand, (*E*)-6-nonenal (**21**) was found to be the off-flavor component of partially hydrogenated vegetable oils and dried milk (threshold concentration in milk 0.7 ppb) (*300, 469*). This typical "hardening flavor" also occurs in other linoleic and oleic acid-containing oils, such as peanut, cottonseed and sunflower oils, after hardening and refining (*145*). In the stereoisomeric nonenal series, the odor threshold values fall with increasing distance of the double bond from conjugation, reaching a clear minimum in 6-nonenal. Here too the (*Z*)-derivatives have a greater odor intensity than (*E*)-derivatives (*376*). The odor of (*E,E*)-2,6-nonadienal (**24**) has been associated with the tallowy cucumber flavor in beef and mutton tallow (*237*). (*E*)-6-Nonenyl acetate [ester of (**20**)] is attractive to the female melon fly (*Dacus cucurbitae* Coquillet) in laboratory tests (*264*).

A careful organoleptic evaluation of the α,β-unsaturated aldehyde (**17**) which was detected in various food products such as carrot root oil (*72*), tomato (*74*), beef (*342*) and cranberry (*17*) has shown that this aldehyde has some remarkable flavor properties. Above the threshold concentration of about 0.1 ppb (*72*), for example at 0.4 to 2 ppb in water, (*E*)-2-nonenal (**17**) possesses a woody character (*470*). Above 8 ppb the sensory impression turns into a fatty one which becomes unpleasant above 30 ppb. Finally, an aqueous solution of 1,000 ppb of aldehyde (**17**) has a strong taste of cucumber. The fresh-brew woody note of roasted and ground coffee as well as the woody effect in bell peppers is due to the presence of this compound (*470*). Addition of (*E*)-2-nonenal (**17**) to cranberry juice at a level of 1 ppb causes a considerable reduction in the normal astringent character of the juice (*470*) without any change of the original odor impression. An antagonistic effect of (**17**) is observed on the flavor of (*Z*)-3-hexenal (**5**) (*375*). Above 2 flavor units (*E*)-2-nonenal (**17**) causes the stale flavor of spoiled beers (*381*).

Aliphatic aldehydes having 8 to 10 carbon atoms, particularly decanal, belong to the odoriferous principle of citrus oils, although they are present only at low concentrations (*190*). Besides aldehyde (**19**), (*Z*)-4-decenal (**31**) was identified in cooked chicken (*212*). In the C-10 series the two 2,4-decadienals (**33**) and (**34**) are important as flavor components and have been found in a wide variety of foods (*389*), to

(31) 

(32) 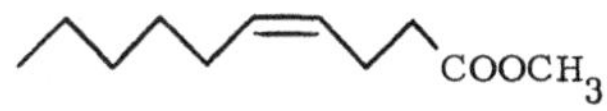

(33) 

(34) 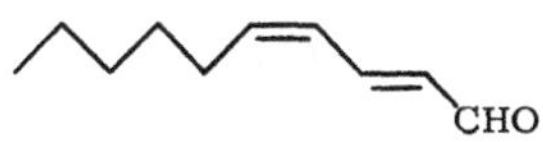

(35) 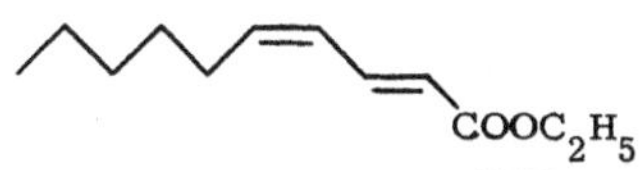

(36) 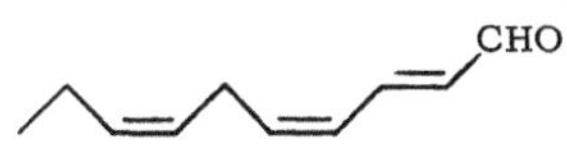

(37) 

which they are believed to impart a desirable deep fat flavor. These dienals were found especially in cottonseed oil (*611*), soybean oil (*236*), peanut oil (*272*), milk fat (*559*), beef fat (*342, 687*), cooked chicken (*475*), tomatoes (*74*) and the aroma of rye crisp bread (*613*).

Among the higher aldehydes (*E*)-2-tridecenal is responsible for the bug-like odor note in coriander seed oil (*Coriandrum sativum* L.) (*545*). In addition, (*E*)-2-dodecenal and (*E*)-2-decenal, together with the saturated members with 9 to 12 carbon atoms, were found in the same essential oil (*626*). It should be pointed out that the oil of coriander leaves, widely employed as a flavoring commodity in certain Eastern dishes consists of nearly 90% of aliphatic aldehydes (*370*). As is well known, quite a number of simple long-chain unsaturated aldehydes, alcohols and esters act as pheromones in various insects (*41*).

Whereas alkadienals are known to be widely spread in nature, alkatrienals have only recently been detected with certainty in food flavors. (*E,E,E*)-2,4,6-Nonatrienal (**26**) was found in the vacuum steam-

(38) 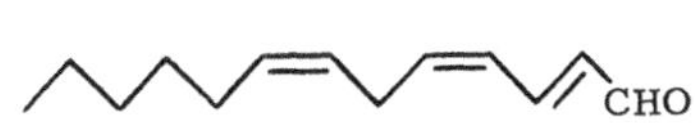

(39) 

(40) 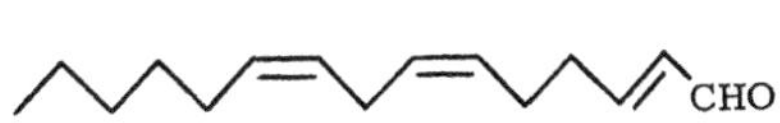

distilled volatile oil of blended dry red and white beans (*Phaseolus vulgaris* L.) (*66*). 2,4,6-Decatrienal was recently reported to be present in the aroma of black tea (*79, 80*), and its 2,4,7-isomer (both of unknown stereochemistry) was recognized as a flavor constituent of fish oil (*377*). (*E,Z,Z*)-2,4,7-Decatrienal (**36**), (*E,Z,Z*)-2,4,7-tridecatrienal (**38**), (*E,Z,Z*)-2,5,8-tetradecatrienal (**39**) and (*E,Z,Z*)-2,6,9-pentadecatrienal (**40**) were isolated for the first time from cooked chicken (*212*). (*E,Z,Z*)-2,4,7-Decatrienal (**36**) and (*E,E,Z*)-2,4,7-decatrienal (**37**), considered to be products formed by specific oxidation of linolenic acid (*377*), were found to be responsible for the formation of the "fishy" off-flavor in rancid mackerel oil (*290*). Volatile aliphatic acids can play an important role in the development of the total flavor in fruits, whereby the gustatory receptors are stimulated rather than the olfactory receptors. Thus, up to now 48 acids of very diverse structures were identified as components of fresh strawberries (*404*).

There is evidence that among the fatty acids only butyric and caproic acids (*42*), at sub-threshold concentrations, interact to contribute to the desirable butter flavor (*367*). A whole set of fatty acids provide a background flavor in the case of Cheddar cheese. Propionic acid is an essential flavor component in Swiss cheese (*330*). The characteristic flavor of blue cheese, however, is due to the presence of aliphatic methylketones (*161, 308*).

Many unsaturated fatty acids occur in nature. 2- and 3-Hexenoic acids are of a certain importance in raspberries (*441*). 2- and 3-Alkenoic acids are formed when animal and vegetable oils are heated in the presence of air. (*E*)-2-Alkenoic acids (C-6 to C-10), (*E*)-alkenoic acids (C-6 to C-9) and (*Z*)-3-decenoic acid were isolated from heated pork fat (*685*). 5-Hexenoic, 6-heptenoic and 10-undecenoic acids were found in several fried vegetable oils (*364*). 9-Decenoic acid, in addition to other acids, was observed at a relatively high concentration in wine of the Riesling variety (*134*).

Several alkyl-branched fatty acids were found to be powerful and characteristic food flavors. 2-Methylbutyric acid is an important constituent of the aroma of cranberry (*Vaccinium vitis-idaea* L.) (*17*). 3-Methylvaleric acid and isovaleric acid were identified in tobacco leaves and found to contribute to the distinct sensory properties of Turkish tobacco smoke (*590*). Isovaleric acid, which has the lowest flavor threshold (0.7 ppm) (*522*) of all saturated fatty acids, is regarded as an essential flavor component of Limburger cheese (*575*). (*E*)-3-Methyl-2-hexenoic acid was recognized as the malodorous component of schizophrenics' sweat (*579*). The undesirable odor of mutton can be attributed to branched-chain and unsaturated fatty acids having 8 to 10 carbon atoms. The 4-methyl-branched acids and in particular 4-methyloctanoic and 4-methylnonanoic acids contribute significantly to the "sweaty-sour" odor note described in Chinese as "soo" odor of mutton (*717*). (*Z*)-2-Methyl-3-pentenoic acid is used in aroma compositions for imparting a sweet, green, sharp strawberry character (*207*). (*E*)-3-Methyl-3-decenoic acid, (*Z*)-3-methyl-3-decenoic acid and the corresponding alcohol play an important role in the development of the typical odor of cassia oil (*Acacia farnesiana* Willd.) (*121*). These compounds have been synthesized in a stereospecific manner (*116*).

Aliphatic esters in various combinations play a major part in many flavors, particularly fruit flavors (*441*). Hexyl acetate develops the strongest and most typical odor in Cox's Orange Pippin apples (*219*). Hexyl 2-methylbutyrate contributes highly to the flavor of Golden Delicious apples (*265*). Methyl and ethyl esters of (*E*)-3-hexenoic, (*Z*)-4-decenoic (**32**) and (*E*)- and (*Z*)-4-octenoic acids have some importance in the flavor of pineapple (*408*). Methyl (*Z*)-4-decenoate (**32**) and methyl thiohexanoate account for about 56% of the odor provided by the components of the oxygenated fraction of hop oil (*202*). Ethyl (*E,Z*)-2,4-decadienoate (**35**) is considered as the character impact compound in the aroma of Bartlett pears (*227*) (Table 2). The application of this "pear ester" in reconstitution work has been facilitated by the development of highly stereospecific syntheses (*405, 407, 457*). Isobutyl angelate is considered as an essential constituent of Roman camomile oil (*85*).

2. Lactones

The importance of aliphatic lactones as aroma components in foods is based on their characteristic organoleptic properties. Due to their mostly low odor thresholds averaging about 0.1 ppm (*572*), lactones often have a high flavor value. Detailed reports (*128, 162, 425, 688*) exist on the formation of lactones, especially from glycerides, higher fatty acids and short-chain functionalized aliphatic compounds as their precursors. Whereas γ-lactones preferentially occur in plants, δ-lactones are mainly found in animal products. Thus, considerable quantities of δ-lactones

(41) 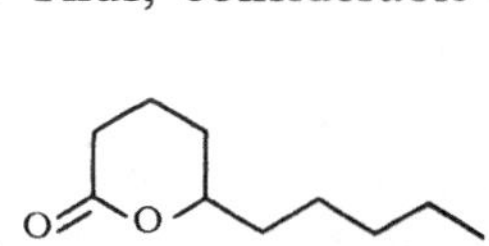

with an even number of carbon atoms from 8 to 16 were found in milk fat, while the corresponding compounds with odd numbers of carbon atoms occur only in minor amounts, together with γ-lactones in milk fat (*50*). The most butter-like compound is δ-decalactone (**41**) the organoleptic properties of which were described as being creamy, sweet, coconut-like, peach-like and milky (*162*).

Precise odor descriptions for other δ-lactones exist in the literature (*162*). δ-Lactones play an outstanding role in many dairy products (*128, 162*). They were also isolated from depot animal fat (*50, 127, 342, 683, 684, 686*), and furthermore found in certain products of vegetable origin such as coconut (*14, 344*), pineapple (*94, 408*), passion fruit (*711*), apricot (*623*), peach (*129, 268*), strawberry (*135, 661*), raspberry (*708*), roasted filberts (*307*), black tea (*79, 80*), and in beverages produced by alcoholic fermentation such as beer (*584*), rum (*343*) and whisky (*277*).

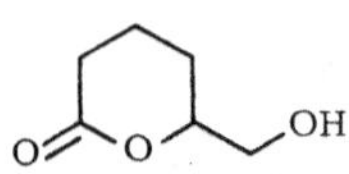

(**42**)

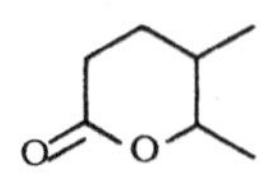

(**43**)

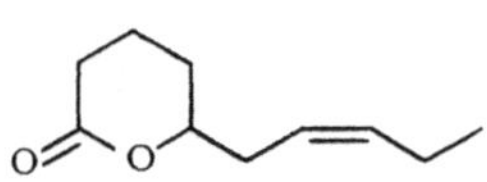

(**44**)

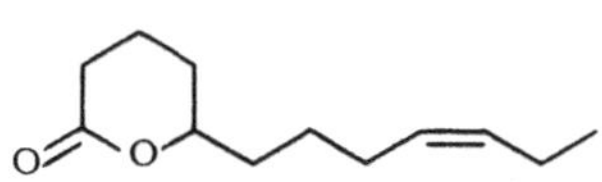

(**45**)

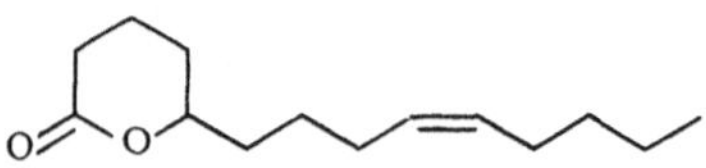

(**46**)

In addition to the δ-lactones of structure (**41**), other lactonic aroma components with a saturated ring are known, the structure of which, however, is somewhat different from that of the n-series represented by (**41**). These are 6-hydroxy-5-hexanolide (**42**), identified in tomato (*561*), and three cyclic compounds carrying an unsaturated chain in position 5, *viz.* (*Z*)-7-decen-5-olide (jasmine lactone) (**44**) (*712*), identified in the flavor of black tea (*80*) together with methyl jasmonate (*719*). (*Z*)-9-dodecen-5-olide (**45**) and (*Z*)-9-tetradecen-5-olide (**46**), the latter two lactones having been isolated from butter flavor (*50, 728*). On the other hand, lactone (**43**) was isolated from milk where it is present in a quantity of 10 μg/kg (0.001 ppb) (*241*); this quantity, however, is probably much lower than the threshold value.

The simplest and also the first known α,β-unsaturated δ-lactone, parasorbic acid (**47**), was isolated in 1894 as the main volatile consti-

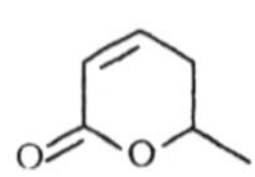

(**47**)

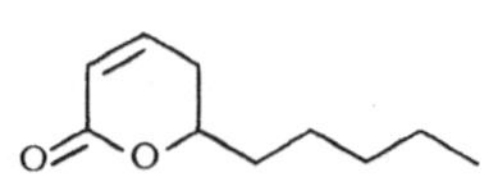

(**48**)

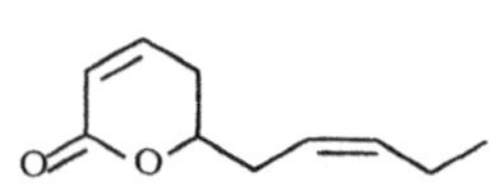

(**49**)

tuent from the fruits of *Sorbus aucuparia* L. (*Rosaceae*) (*130, 320*). It (**47**) was found to be a cytotoxic agent which has a blocking effect on mitosis at the metaphase (*495*). Massoialactone (**48**), found for the first time in bark oil of *Massoia aromatica* (*95, 382*), was also detected in wine (*30*) and recognized as an organoleptically important component of sugar molasses (*216*). Tuberolactone (**49**) [(*Z,Z*)-2,7-decadien-5-olide] is the first δ-decadienolactone found in nature. It occurs in addition to 5-decanolide (**41**), jasmine lactone (**44**), massoialactone (**48**) and a series of saturated γ- and δ-lactones in the characteristically smelling portion of the absolute of tuberose flowers (*282*).

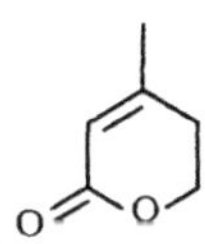
(**50**)

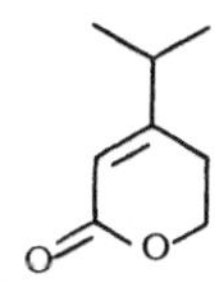
(**51**)

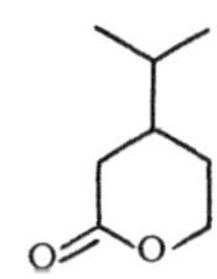
(**52**)

Two other α,β-unsaturated δ-lactones (**50**) and (**51**), substituted in position 3, were identified in the flavor of Burley tobacco (*111*). 3-Isopropyl-2-penten-5-olide (**51**) and its saturated derivative (**52**) impart a coconut-like note (*555*) to tobacco (*111*). Two doubly unsaturated δ-lactones were identified in peach flavor (*567, 568*). These are 2,4-pentadien-5-olide (or α-pyrone) (**53**) and 2,4-decadien-5-olide (**54**). The latter, which was also identified in mushrooms (*89*) has a powerful buttery note. Detailed descriptions of the odors of α-pyrone derivatives are to be found in the literature (*476*).

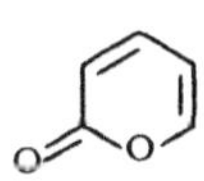
(**53**)

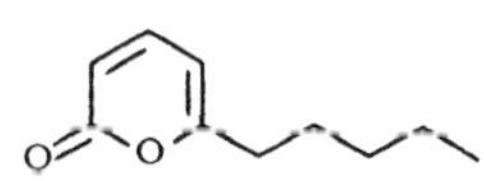
(**54**)

The γ-lactones most frequently found in flavors have structure (**55**). γ-Butyrolactone (**55a**) is the most widely spread γ-lactone; it has been identified in about 20 foodstuffs. The higher straight-chain homologs [(**55**): $R_5 = CH_3$, $C_2H_5 \ldots C_{12}H_{25}$] are found less and less frequently as the number of carbon atoms increases, and the lactones having an even number of carbon atoms predominate. They were identified in the following products: chicken meat (*705*), beef meat (*228*), beef fat (*687*), pork fat (*686*), shrimp paste (*388*), fruits such as apricot (*94*), passion fruit (*711*), peach (*271*), raspberry (*708*), strawberry (*135*), raisins (*488*), plums (*158*), tomatoes (*668*), pineapple (*573*), citrus fruits (*291*) and bilberries (*614*), cooked, grilled or roasted products such as caramel (*620*), coffee (*596, 669*), black tea (*51*), roasted onions (*338*), popcorn

(*677*), roasted filberts (*307, 569*), roasted peanuts (*678*), cocoa (*154, 359*), white bread (*399, 578*), rye bread (*613*), roasted barley (*679*), roasted pecans (*680*), or beverages such as beer (*68, 584*), wine (*692*), sherry (*690*), rum (*343*), and whisky (*277*). They are also found in vinegar (*276*), mushrooms (*644*), tobacco (*111*) and soybeans (*698*).

Other γ-lactones which are mono- or poly-substituted in other positions are represented by formula (**55**).

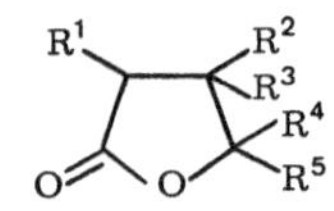

(**55**)

	R^1	R^2	R^3	R^4	R^5
(**55a**)	H	H	H	H	H
(**55b**)	CH_3	H	H	H	H
(**55c**)	$_iC_3H_7$	H	H	H	H
(**55d**)	H	H	H	CH_3	C_2H_5
(**55e**)	H	H	H	CH_3	$CH{=}CH_2$
(**55f**)	OH	CH_3	CH_3	H	H
(**55g**)	H	CH_3	H	C_4H_9	H
(**55h**)	H	H	H	$CHOHCH_3$	H
(**55i**)	H	H	H	$C(=O)CH_3$	H
(**55j**)	H	H	H	OCH_2CH_3	H
(**55k**)	H	H	H	$COOC_2H_5$	H

2-Methyl-4-butanolide (**55b**) was identified in cocoa (*358*), coffee (*594*), passion fruit (*711*), mushrooms (*644*) and wine (*547*), while 2-isopropyl-4-butanolide (**55c**) was found in Burley tobacco (*111*).

4-Methyl-5-hexen-4-olide (**55e**), presumably a metabolite of linalool oxide (furanoid), was found among the ten lactones (*655*) of lavender oil (*310*) as well as in wine (*547*) and Japanese hop (*427*). Its hydrogenation product (**55d**) was isolated from Burley tobacco (*111*). 2-Hydroxy-3,3-dimethyl-4-butanolide (**55f**) is present in sherry (*691*). 3-Methyl-4-octanolide (**55g**) (*437, 608*), also known under the name "Whisky lactone", exists in two diastereoisomeric forms. The aroma of the *cis*-lactone, which has a higher odor threshold value (0.79 ppm), is more fragrant than that of the *trans*-lactone (odor threshold 0.067 ppm)

(*465*). The γ-lactones (**55h, i, j,** and **k**) which carry oxygen-containing substituents in position 4 are found exclusively in alcoholic beverages, except for hydroxylactone (**55h**) which also occurs in tomatoes (*561*). Owing to its distinct vinous smell, solerone (**55i**) is a particularly typical representative of this series (*607*). The occurrence and the mode of formation of lactones in various wines and in other alcoholic beverages are described in several review articles (*400, 547, 609*).

Among the unsaturated γ-lactones, 5-methyl-5-hexen-4-olide (**56**), identified in the essential oil of hop (*427*), and (*Z*)-6-dodecen-4-olide (**57**), found in butter (*50*), mutton meat (*468*) and mushrooms (*617*), can be mentioned. The odor of butter fat of cows or of the meat of lambs fed a special lipid diet, high in linoleic acid, is mainly

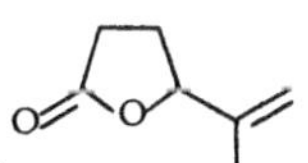

(56)

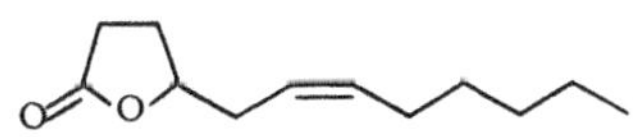

(57)

attributed to lactone (**57**) (*468, 589*). Furthermore, (**57**) was reported to be the major pheromonal component of the tarsal tuft scent of blacktail deer (*60*).

Eighteen γ-lactones having unsaturated rings have been identified in flavors. These compounds are principally of three general structural types: α,β-unsaturated (**58**), β,γ-unsaturated (**59**) and α,β,γ,δ-unsaturated (**60**).

The α,β-unsaturated γ-lactones are mainly present in tobacco (*111, 287, 570*) (**58a, c, d, e, h, i, k, l**), wood smoke (*152, 304*) (**58a, b, d, f, g, h, j**), roasted coffee (*184, 594*) (**58a, f, g, j**), fried onions (*338*) (**58c, d, h**) and milk caramel (*620*) (**58m**). Only one of these lactones is found in white bread (*399*) (**58a**), saffron (*726*) (**58a**), roasted filberts (*307*) (**58a**), roasted peanuts (*678*) (**58b**), raisins (*488*) (**58b**), burnt sugar (*144, 273*) (**58b**), soybeans (*357*) (**58b**), hop (*430*) (**58e**), Ceylon tea (*719*) (**58l**), Japanese peppermint (*520*) (**58l**), raspberry (*708*) (**58i**) and soy protein hydrolysates (*603*) (**58n**). As can be seen, with rare exceptions, these α,β-unsaturated γ-lactones are principally present in products which have been subjected to thermal treatment. However, 4-methyl-2-penten-4-olide (**58c**) was found in lavender oil (*655*) as well as hop oil (*660*).

Only two members of the series of β,γ-unsaturated γ-lactones are known today: 3-buten-4-olide (**59a**), identified in cocoa (*478*) and fried onions (*338*), and 3-penten-4-olide (**59b**) (α-angelicalactone), identified in white bread (*399*), soybeans (*357*) and raisins (*488*).

Only two α,β,γ,δ-unsaturated γ-lactones are known: protoanemonin (**60a**), identified in fried onions (*338*), and bovolide (**60b**), present in butter (*50, 331*) and tobacco (*111*). Bovolide (**60b**), due to its celery-like

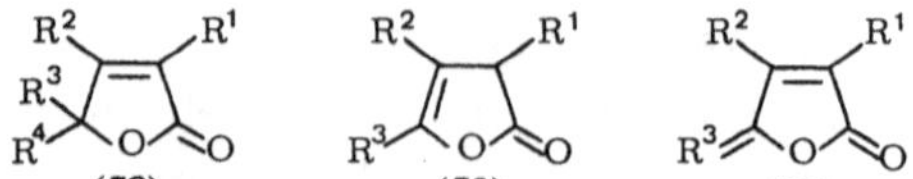

(58) **(59)** **(60)**

	R^1	R^2	R^3	R^4
(58a)	H	H	H	H
(58b)	H	H	CH_3	H
(58c)	H	CH_3	H	H
(58d)	CH_3	H	H	H
(58e)	H	H	CH_3	CH_3
(58f)	H	CH_3	CH_3	H
(58g)	CH_3	CH_3	H	H
(58h)	CH_3	H	CH_3	H
(58i)	H	H	C_2H_5	H
(58j)	CH_3	CH_3	CH_3	H
(58k)	H	$_iC_3H_7$	H	H
(58l)	CH_3	CH_3	C_5H_{11}	H
(58m)	H	H	C_8H_{17}	H
(58n)	OH	CH_3	C_2H_5	H
(59a)	H	H	H	
(59b)	H	H	CH_3	
(60a)	H	H	CH_2	
(60b)	H	H	$CH-C_4H_9$	

odor, contributes to the flavor of butter (*331*). Spiroxabovolide (**61**), which is related to lactone (**60b**), was identified in Burley tobacco (*111*).

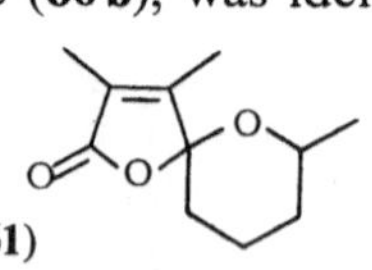

(61)

Both diastereoisomers are used as flavorings in foodstuffs, beverages and especially tobacco products, but also as perfume ingredients (*109*). Lactones (**60a**) and (**60b**) are particularly toxic; they have vesicant and carcinogenic activity (*495*).

Among the bicyclic γ-lactones having an annelated cyclohexane ring, compounds of the general formula (**62**) constitute outstanding aroma

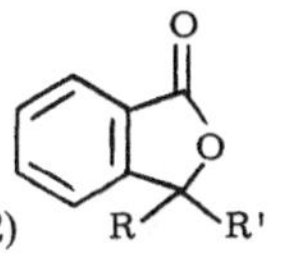

(**62**)

components. The aroma-producing phthalides (**62**) include about 15 individual compounds which occur in spice plants of certain *Umbelliferae* species and which accumulate in the essential oils of their fruits, seeds and rhizomes. These compounds are derivatives of (**62**) as well as dihydro and perhydro derivatives substituted with alkyl groups (R = H) or alkylidene groups (R and R′).

(**63**) 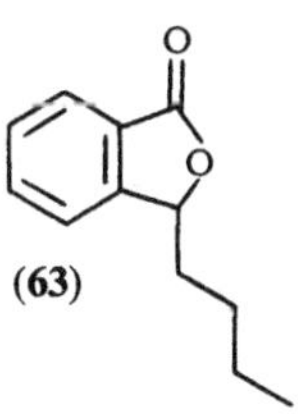 (**64**) 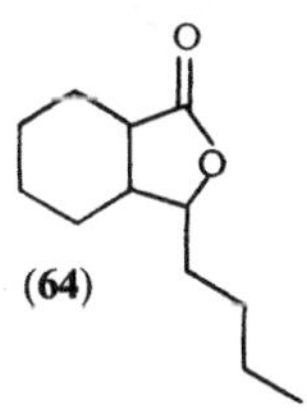 (**65**) 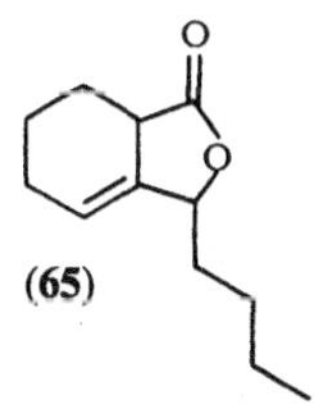

The essential oil of celery, for example, contains 3-butyl-phthalide (**63**), 3-butyl-hexahydro-phthalide (**64**) as well as sedanolide (**65**) (*703*). The latter two lactones (**64**) and (**65**) have the characteristic odor and flavor of celery (*703*). Reports on the occurrence and importance of this class of aroma compounds are to be found in specialized review articles (*193, 354*).

A final group of important aroma compounds consists of the macrocyclic lactones. Angelica root oil finds extensive use in flavor formulations for liquors such as vermouths, bitters, benedictines etc. (*149*). The musk-like odor of this essential oil is generally attributed to 15-pentadecanolide (**67**) (EXALTOLIDE®) (*301*). Recently three additional macrocyclic lactones were discovered in this oil, 13-tridecanolide (**66**), 17-heptadecanolide (**68**) and 12-methyl-13-tridecanolide (**69**) (*625, 627*) which, together, account for about 1% of the oil.

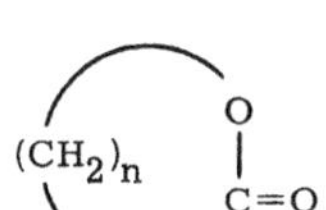

(**66**): n = 12
(**67**): n = 14
(**68**): n = 16

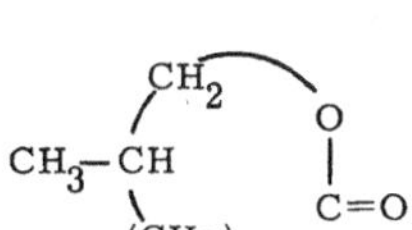

(**69**): n = 10
(**70**) :n = 12

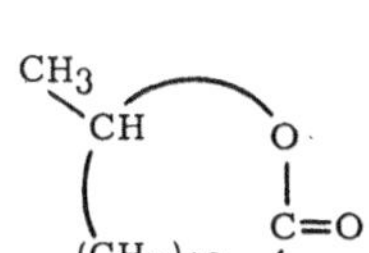

(**71**)

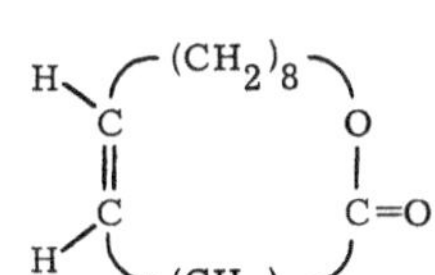

(**72**)

The homolog of (**69**), known as muscolide (**70**), is accessible by synthesis (*32*). 15-Hexadecanolide (**71**), a methyllactone related to (**69**) and (**70**), was found in optically active form for the first time in galbanum oil (*421*). Ambrettolide (**72**), originally discovered in ambrette seed oil (*301*), is recommended for flavoring ice cream, candy, puddings and chewing gum in concentrations of 0.007 to 0.7 ppm (*151*).

III. Isoprenoid Aroma Compounds

1. Monoterpenoids

Myrcene (**1**) as well as the stereoisomeric ocimenes (**2**) and (**3**) (*186*) which were detected in more than 200 different flavors and fragrances are present in particularly high quantities in the essential oils of spice belonging to the families *Compositae* and *Labiatae*. Myrcene (**1**) and *trans*-ocimene (**2**) were identified in basil oil (*Ocimum basilicum* L.) (*337*). The essential oil of a plant growing in Taiwan, *Ocimum gratissimum* L., contains, in addition to 8% of myrcene (**1**), up to 30% of ocimenes (**2**)/(**3**) (*724*). Myrcene (**1**) constitutes about 63% of the components of hop oils and thus accounts for about 58% of the total odor intensity of the whole oil and 85% of the odor of the hydrocarbon fraction (*202*). In a Thai species, locally known as "horapha", (**2**) is

(**1**) 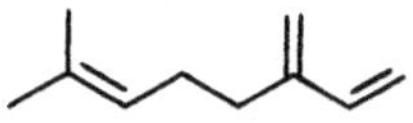

(**2**) 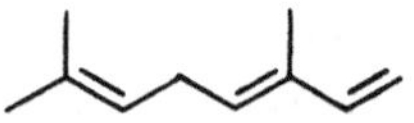

(**3**) 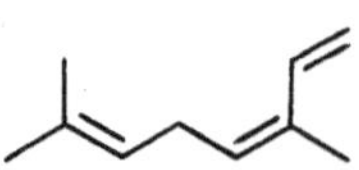

the most abundant monoterpene (*336*). On the other hand, *cis*-ocimene (**3**) is the main monoterpenoid component (9.3%) of sage oil (*Salvia officinalis* L.) (*335*). As a curiosity, the two monoterpenoid nor-terpenes (**4**) and (**5**) may be mentioned. They were detected in the same essential

(4) 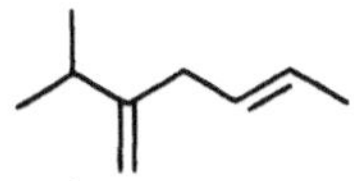

(5) 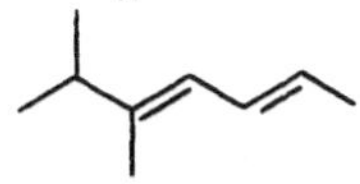

oil (*335*) as trace components. 2-Isopropyl-1,4-hexadiene (**4**), called "salvane", was fiirst isolated from various *Salvia* species (*58*). The bicyclic thujone (**113**) may be considered as the precursor from which 2-isopropyl-1,4-hexadiene (**4**) is formed by photochemical decarbonylation (*138*).

The three monoterpenes (**1**) to (**3**) are reported to contribute to the flavor of Japanese black tea (*511*). The oil of ripe fruits of *Phellodendron amurense* Rupr. contains 92% of myrcene (**1**) (*431*), and the herb oil of *Boronia dentigeroides* Cheel is reported to contain up to 80% of ocimene of unknown stereochemistry (*471*).

On an industrial scale, myrcene (**1**) and *cis*-ocimene (**3**) are produced by pyrolysis of β-pinene (**118**) (*197*) and α-pinene (**117**) (*221*), respectively.

Linalool (**6**), geraniol (**7**), nerol (**8**), and citronellol (**9**) are the oxygen-containing derivatives of the monoterpene series which occur

(6) 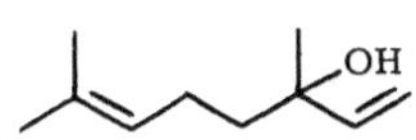

(7) 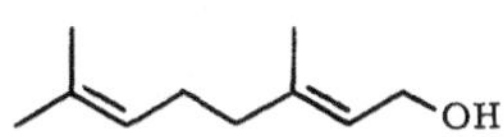

(8) 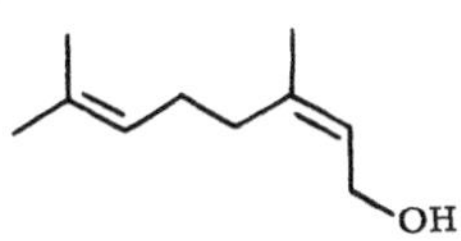

(9) 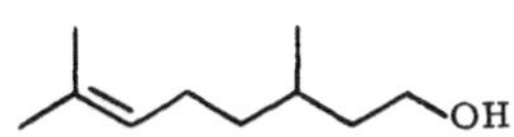

most abundantly in nature. Next come their esters and the aldehydes citral (mixture of geranial and neral) and citronellal derived from (**7**) to (**9**). All the natural compounds derived from (**6**) and (**9**) exist in optically active forms of both enantiomeric series. These optically active forms differ in their odoriferous properties (*449*). (−)-Linalool constitutes the main proportion (about 50%) of American basil oil (*433*), whereas its enantiomer accounts for 70% of the essential oil of coriander fruit (*566*). The monoterpene alcohols (**6**) to (**9**), in most cases together with the corresponding derivatives, constitute the base of many natural flavors and essential oils. Geranic acid [derivative of (**7**)] in the free form is rarely found in nature. However, this acid was found to contribute to a large extent to the overall aroma of the German wines "Traminer" and "Scheurebe" (*134*) and of various grapes (*549*).

A series of compounds derived from (**6**) to (**9**) and having the same carbon skeleton may be considered as more or less exotic compounds which, however, can impart a certain odoriferous impact to their

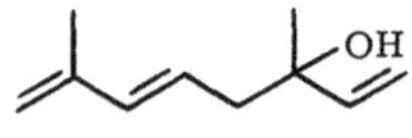
(10)

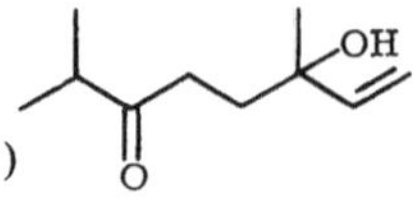
(11)

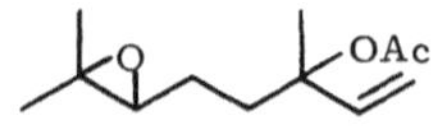
(12)

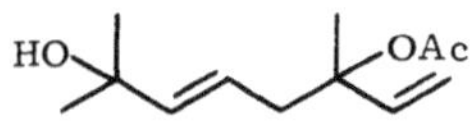
(13)

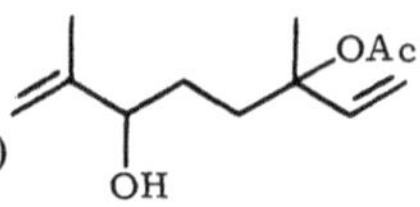
(14)

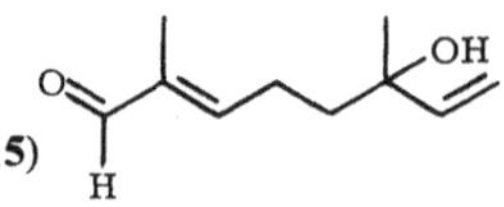
(15)

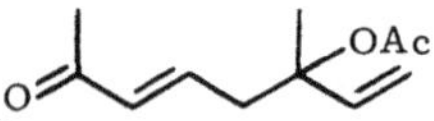
(16)

natural parent substances and thus contribute to the great diversity of natural flavors. Among the derivatives of linalool (**6**) some compounds with a higher degree of oxidation have acquired a certain renown. (−)-3,7-Dimethyl-1,5,7-octatrien-3-ol (**10**), also known as hotrienol, was found, together with the new hydroxy ketone (+)-tagetonol (**11**), in the leaf oil of the Japanese Ho tree (*725*). Hotrienol (**10**), the odor of which is similar to that of lime-tree blossoms (*31*) was detected in various grape and wine flavors (*548*), muscatel grape (*497*) and tea aroma (*413, 434, 722*). Epoxyacetate (**12**) was isolated from Indian linaloe oil (*8*). The bifunctional compounds of the linalyl type (**12**) to (**16**) and of the

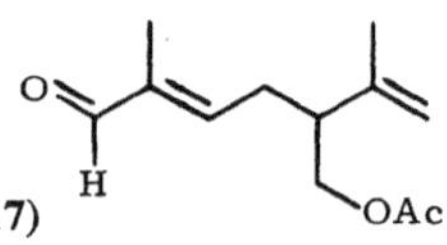
(17)

lavandulyl type (**17**) were reported for the first time to be constituents of a natural product (*Lavandula hybrida* Reverchon) (*390*).

Ambrosial, a component of the volatile oil of *Ambrosia confertiflora* D.C. is assumed to be dehydrogeranial (**18**) (*323*). A stereospecific synthesis of dehydronerol (**19**) isovalerate (*78*), isolated from the roots of *Anthemis montana* L. (*47*), has been described.

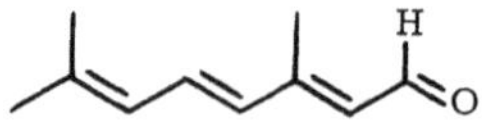
(18)

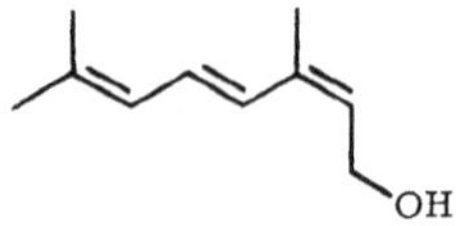
(19)

Myrcenol (**20**), which has been available synthetically for a long time by hydration of (**1**), was found in hop oil (*188*). Together with the corresponding acetate (*243*) and its acrolein adduct (LYRAL®) (*312, 628*), alcohol (**20**) counts among the modern perfumery base materials which are used in substantial quantities. The three oxygen-containing

(**20**) 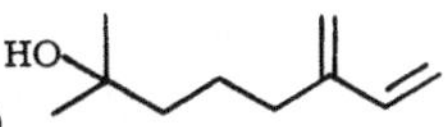 (**21**) 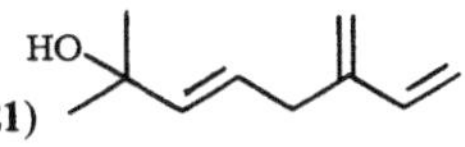 (**22**) 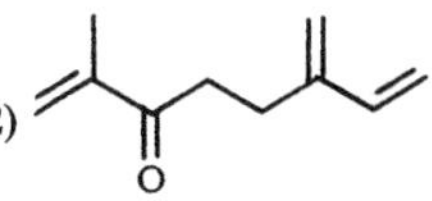

trienes (**21**), (**22**) and (**32**) were isolated from *Ledum palustre* (*530*), one of them (**21**) having previously been isolated from *Pinus ponderosa* (*574*). It is conceivable that (**21**) and (**22**) are formed from myrcene (**1**) by a photooxygenation process (*198, 362*).

Ipsdienol (**23**) and the dihydro derivative ipsenol (**24**) [formerly tagetol (*574*)] together with (+)-*cis*-verbenol, are constituents of the active principle of the sex attractant produced by some bark beetles (*214, 366, 672*). Besides myrcene (**1**), as one of the main components,

(**23**) 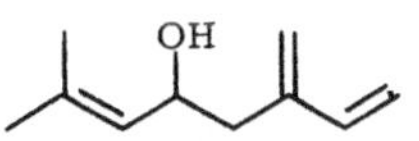 (**24**) 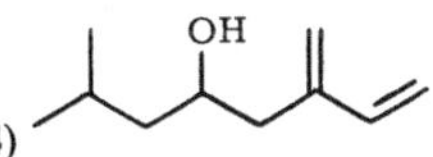

(**25**) 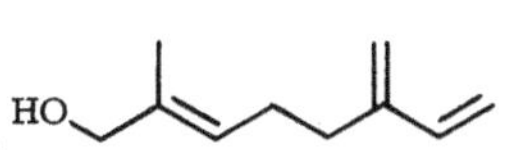 (**26**) 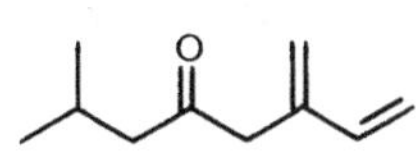

the primary alcohol (**25**) and its acetate were detected in the essential oil of *Thymus vulgaris* L. (*200*). The corresponding *E*-stereoisomer has been prepared by a multistep synthesis (*671*).

(**27**) 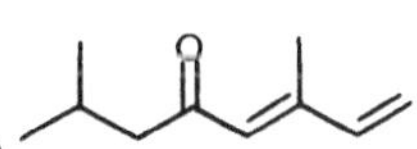 (**28**) 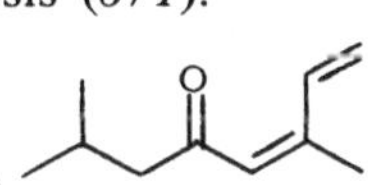

(**29**) 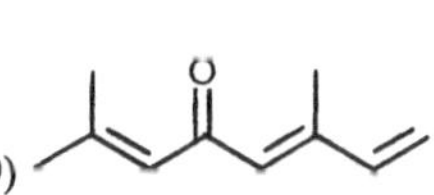 (**30**) 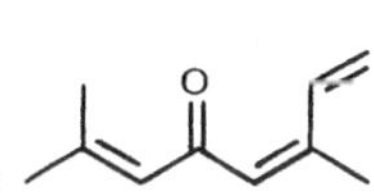

(**31**) 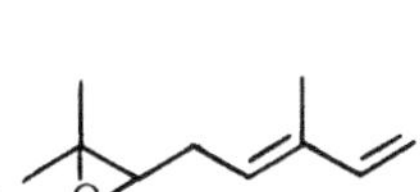 (**32**) 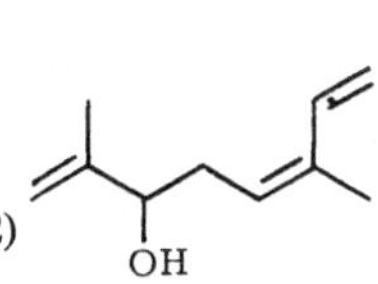

(**33**) 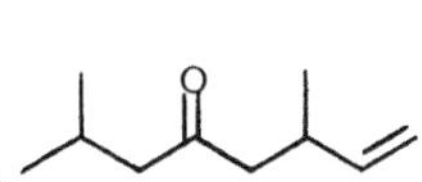

The stereoisomeric ketones of the ocimene series, (*Z*)- and (*E*)-tagetone **(27)** and **(28)**, in a proportion of 7 : 1, are the main constituents (50 to 60%) of the essential oil of *Tagetes glandulifera* Shrank (*44, 275*). The essential oil of *Tagetes minuta* L., at the stage of seed shedding, contains 26.5% of ocimenones **(29)** and **(30)** (dehydrotagetones), 9% of tagetones **(27)** and **(28)**, 13.5% of dihydrotagetone **(33)** and 41% of (*Z*)-ocimene **(3)** (*123*). The leaf oil of Australian *Rutaceae* species contains up to 97% (+)-2,6-dimethyloct-7-en-4-one **(33)** (*333*). Lewis acid-catalyzed acylation of isoprene with 3-methylcrotonyl chloride and subsequent dehydrochlorination led in 90% yield to a mixture of ocimenones **(29)** and **(30)** (9 : 1) (*10, 123*). The same reaction sequence, performed in the presence of isovaleryl chloride, yielded a mixture of *cis*- and *trans*-tagetones (40 : 60) **(27)** and **(28)** (*10*). The dehydro derivatives **(26)**, **(27)** and/or **(28)** are reported to constitute, together with 5% of ocimenes **(2)**/**(3)**, by far the major portion (80%) of the characteristic odoriferous components of *Lippia asperifolia* Rich. (*419*). For the first time, the monoepoxide of *trans*-ocimene **(31)** was found in nature, *viz.* in basil oil, though it is a minor constituent (0.1%) (*336*).

Naturally occurring monoterpenes having a carbon skeleton other than head-to-tail, *e.g.* derivatives of the santolinyl, artemisyl and lavandulyl type, have formally been considered to be fission products of natural *trans*-chrysanthemic acid (*28, 143*). Compounds of the first two

(34)

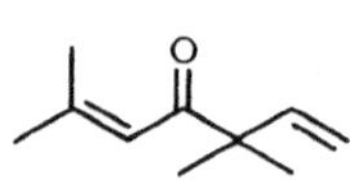

(35)

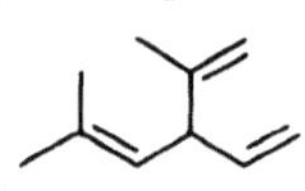

(36)

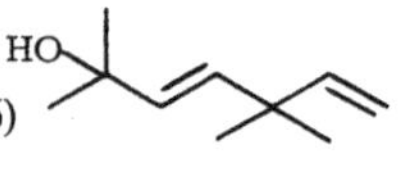

(37)

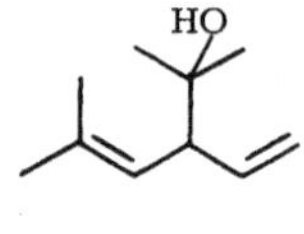

types occur in some *Compositae* and have a specific rather than a general value as aroma components. Thus, it was found that the plant extract of *Santolina chamaecyparissus* L. contained 45% of artemisia ketone **(34)** (*654*), and that the flower oil contained santolinatriene **(35)**. Yomogi alcohol **(36)** was identified in *Artemisia feddei* Lév. et Van (*700*), and the tertiary alcohol **(37)** was found to be one of the main components of Eastern Moroccan camomile oil from *Ormenis multicaulis* Braun-Blanquet and Maire (*84*).

As the most typical representative of monocyclic monoterpenes **(38)** to **(44)**, limonene **(38)** is certainly the most widely distributed aroma component (*187*). In the essential oils of most of the citrus species the (+)-form of **(38)** accounts for more than 90% of all constituents, whereas the (−)-

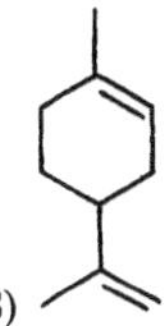
(38)

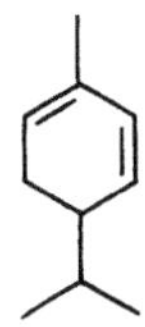
(39)

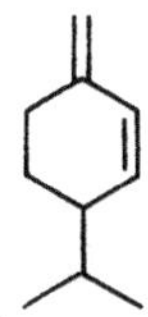
(40)

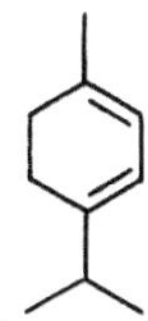
(41)

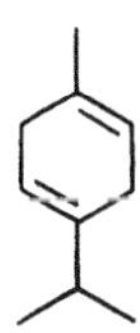
(42)

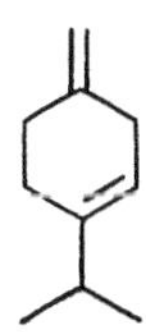
(43)

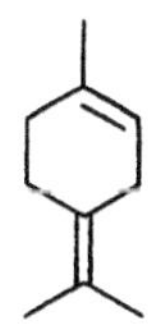
(44)

enantiomer is more common in *Mentha* and *Pinus* species (*187*). The double bond isomeric hydrocarbons (**39**) and (**44**) are considerably rarer in essential oils. In flavors α-phellandrene (**39**) ranks in the first position as far as importance and rate of distribution are concerned. γ-Terpinene (**42**) was identified as the main constituent (20%) in steam distilled oil of sweet marjoram [*Majorana hortensis* Moench (*Origanum majorana* L.)] (*350*) and in cumin seed oil (30%) (*666*).

1,3,8-Menthatriene (**45**) (*166, 177, 288, 648*) and 1-methyl-4-isopropenylbenzene (**46**) (*166*) were found to be associated with the strong and typical odor of parsley *Petroselinum crispum* (Mill.) Nyman syn.

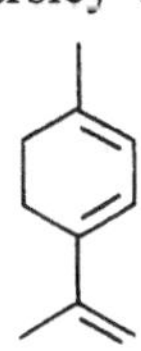
(45)

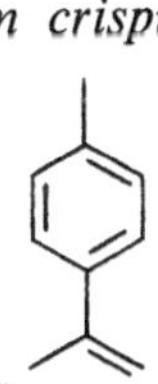
(46)

Petroselinum sativum Hoffm. β-Phellandrene (**40**) and terpinolene (**44**) are also supposed to contribute to this particular aroma note (*166*). The formation of (**46**) in commercial blackberry essence is attributed to the presence of *p*-cymen-8-ol (*242*).

The occurrence and importance of monocyclic oxygen-containing monoterpene derivatives in nature have been known for a long time as is evidenced by a voluminous literature. The alcohols found in nature include mainly the diastereoisomers of menthol (**47**), piperitol (**48**), pulegol (**49**), isopulegol (**50**), dihydrocarveol (**51**) and carveol (**52**).

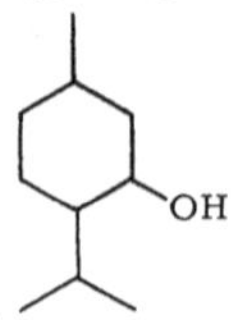

(**47**)

(**48**)

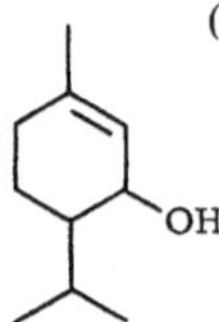

(**49**)

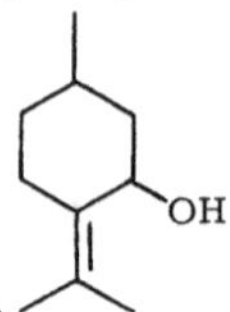

(**50**)

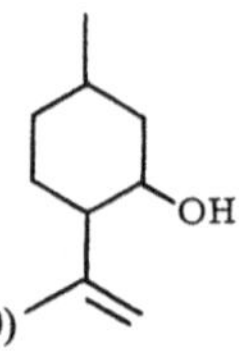

(**51**)

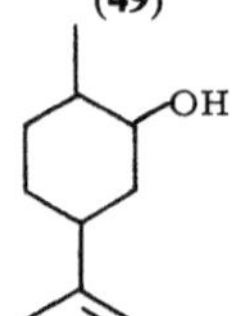

(**52**)

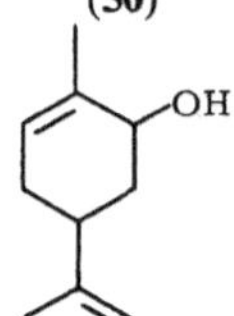

(**53**)

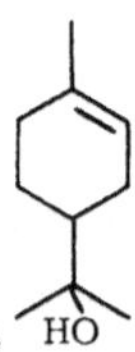

(**54**)

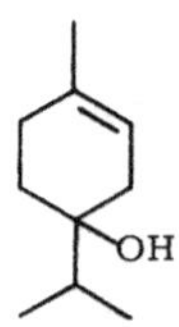

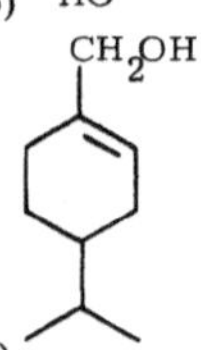

(**55**)

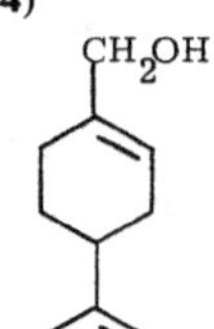

(**56**)

α-Terpineol (**53**) is more widely spread in flavors than 4-terpineol (**54**) or even phellandrol (**55**) and perilla alcohol (**56**).

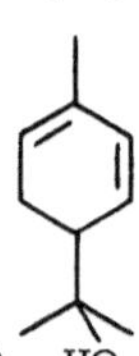

(**57**)

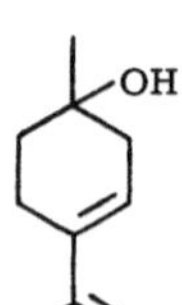

(**58**)

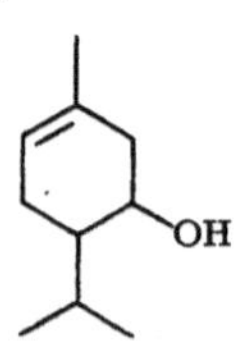

(**59**)

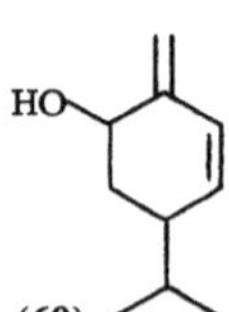

(**60**)

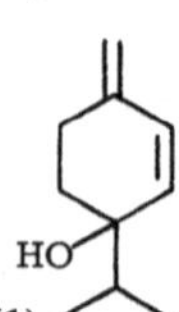

(**61**)

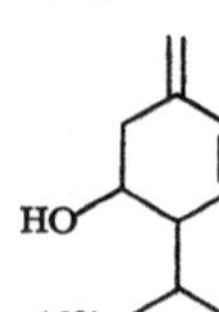

(**62**)

References, pp. 493—527

Recently, new compounds having particular sensory properties have been discovered as trace components in flavors. Four new alcohols (**58**) to (**61**) were reported to be components of the oil of black pepper, *Piper nigrum* L.; however, the true principles of pepper flavor still remain unknown (*106*). The isopropenyl derivative of (**59**) [1,8-*p*-menthadien-5-ol (*445*)] has also been found in nature (*232*). The phellandrene derivative (**57**) was found in angelica root oil for the first time in nature (*627*). The structures (**60**) and (**62**) were assigned tentatively to two further constituents of the same oil (*627*). Both diastereoisomers of (**60**) had previously been reported to occur in *Cinnamomum japonicum* Sieb. (*173*).

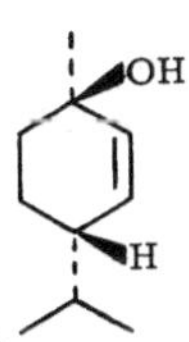

(**63**)

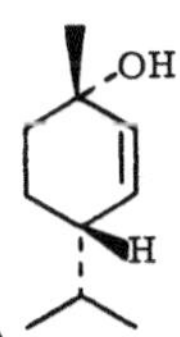

(**64**)

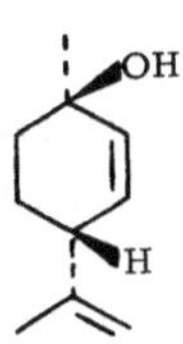

(**65**)

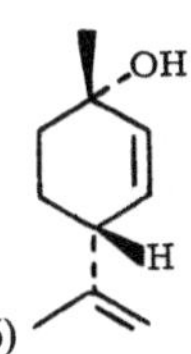

(**66**)

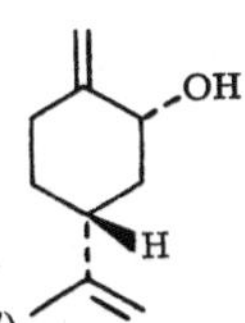

(**67**)

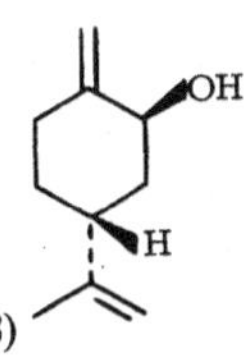

(**68**)

The compounds (**63**) and (**64**) were identified (≈ 1%) in "Phlai" (*Zingiber cassumunar* Roxb.) (*334*). Furthermore, (**63**) was isolated from nutmeg oil (*Myristica fragrans* Houttyn) (*24*), and (**64**) was also identified in Hinoki oil (*640*) and raspberries (*707*). The three allyl alcohols (**65**), (**66**) and (**68**) were detected for the first time in gingergrass oil (*423, 532*).

The alcohols (**65**) (*232, 246*), (**66**) (*232, 246*), (**67**) (*232*), (**68**) (*232*), (**69**) (*246, 309*), (**70**) (*246, 309, 391, 461, 571*) and (**73**) (*571*) were found in particular in the oils of different Citrus varieties, often in the form of the corresponding acetates. Allyl alcohols (**63**) and (**64**) can be formed by dye-sensitized photooxygenation of (+)-carvomenthene, whereas (**52**) and (**65**) to (**68**) arise from the same reaction with (+)-limonene (**38**) (*531*); indeed, the allylhydroperoxides corresponding to alcohols (**65**) and (**66**), of which they can be considered as precursors of the alcohols, were isolated from a citrus oil (*232*).

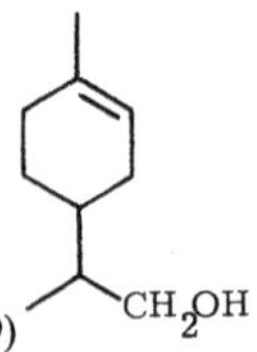
(69)
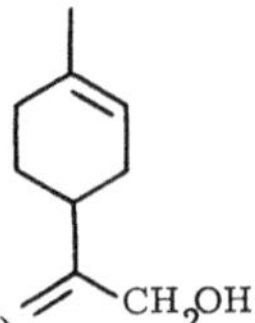
(70)

(71)
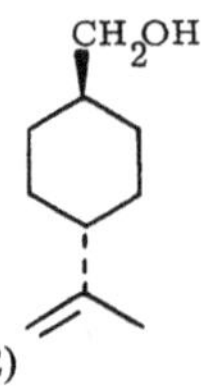
(72)

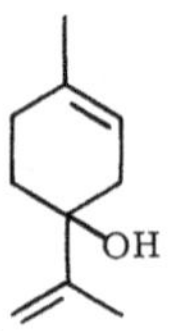
(73)
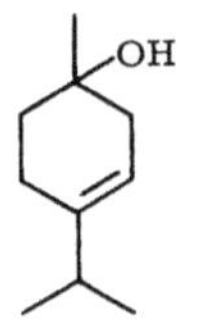
(74)
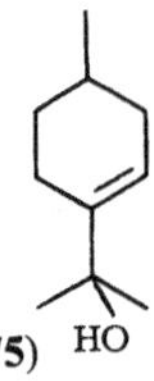
(75)

Compound (**73**) is present in Japanese pepper oil (*514*), and (**74**) occurs in *Origanum vulgare* (*57*). The two primary alcohols (**71**) and (**72**) (9%) (shisools) were found for the first time in the essential oil of *Perilla arguta* Benth. Nakai (*172*) and later in Buchu leaf oil (*283*) with the *trans*-isomer (**72**) in *Juniperus communis* (*638*). (−)-Perilla alcohol (**56**) is the main component (55%) of perilla oil, whereas the corresponding aldehyde (**78**) constitutes only up to 4.5% of this oil, although 90% of the essential oil of the intensely smelling fruits of *Laserpitium siler* L. (*Umbelliferae*) consists of (+)-perillaldehyde (**78**) (*394*). The pleasant fruity odor of the essential oil of a *Salvia* species has been attributed to a mixture of methyl perillate and perillyl acetate, derived from (**56**) (*206*). (+)-8-Hydroxy-3-*p*-menthene (**75**) and (−)-*cis*-pulegol (**49**) are two new terpene alcohols which were isolated for the first time from the essential oil of "Seto-mint" (*411*). The alcoholic distillate of sweet marjoram (*Marjorana hortensis* Moens = *Origanum majorana* L.), which is widely used for flavoring liqueurs, contains 15% of the ethyl ethers of (**54**), (**48**) (*cis* and *trans*), (**63**) and (**64**), respectively, and of *cis*-

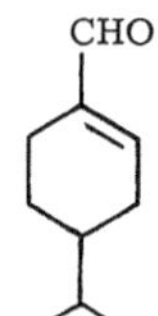
(76)
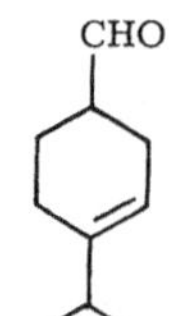
(77)
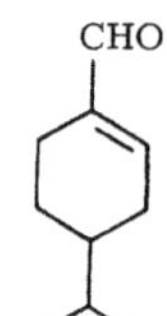
(78)
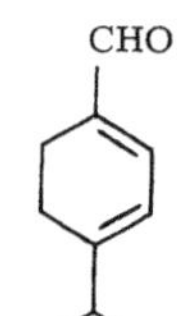
(79)

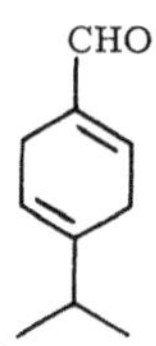
(80)
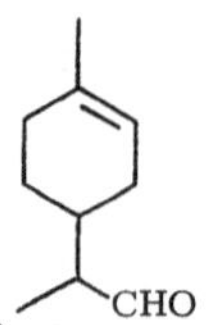
(81)
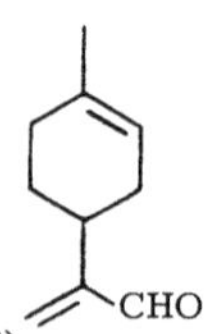
(82)

and *trans*-4-hydroxythujane. The mixture of these ethers is supposed to contribute to a large extent to the overall aroma of this particular essential oil (*624*). Furthermore, *cis*-sabinene hydrate (**115**) is considered to play an important role in the formation of the flavor of sweet marjoram (*350*). A mixture of ethyl ethers, similar to the one of the sweet marjoram distillate but in addition containing the diastereoisomeric verbenyl ethers [derived from (**116**)] as major components, proved to be a characteristic feature of the alcoholic angelica distillate (*627*). These ethers are considered to be artifacts, and are not found when ethanol is not used in the extraction (*624*). Ethyl ethers have also been detected in ethanolic extracts of *Juniperus communis* (*639*).

Due to their low threshold concentration, carbonyl compounds have a high flavor value. In addition to the known compounds of this class, which are derived essentially from alcohols (**47**) to (**52**) and (**56**), a series of isomeric derivatives of the *p*-menthane series has become known in the last few years. Apart from cuminaldehyde, which is the major constituent of pre-ground oil from cumin seeds (*Cuminum cyminum* L.), dihydroaldehydes (**76**), (**77**) and (**79**) were isolated as minor constituents (*666, 667*). Another cumin oil, distilled from freshly ground whole seeds, contained a mixture of 1,3-*p*-menthadien-7-al (**79**) and 1,4-*p*-menthadien-7-al (**80**) as main products (31.5%) (*667*). A certain importance is attributed to the two diastereoisomers of 1-*p*-menthen-9-al (**81**) in the development of the odor of Bulgarian rose oil, although this aldehyde is merely a trace component (*454*). It is also a constituent (0.1%) of the essential oil of cotton buds (*Gossypium hirsutum* L. var. Detapine Smothleaf) (*225*) as well as of Buchu leaf oil (*283*). The latter oil also contains aldehyde (**82**) in addition to the corresponding alcohol (**70**) (*283*). Aldehyde (**82**) is present as the main component (67%) together with alcohol (**70**) in the essential oil of *Dracocephalum foetidum* Bunge (*Labiatae*) (*137*).

Together with (+)-pulegone (78%) [derived from alcohol (**49**)], two new *p*-menthene ketones, *viz.* (+)-4-hydroxypiperitone (**92**) (*410*) and (−)-8-hydroxy-4-*p*-menthen-3-one (**93**) (*411*), were isolated from "Setomint" grown in the Tokai area. Both hydroxyketones (**92**) and (**93**) develop a strong minty odor impression (*443*). A detailed report on the natural occurrence, the methods of preparation and the chemistry of piperitone [carbonyl derivative of (**48**)] and related compounds was published recently (*422*).

The essential oil of Buchu leaves is a particularly rich source of bifunctional compounds of the *p*-menthane series. Among more than 120 constituents identified, 31 compounds were found to be oxygen- and sulfur-containing derivatives of (+)-menthone (**95**) and (−)-isomenthone (**96**), including (**83**) to (**91**) (*283*). In addition to the known

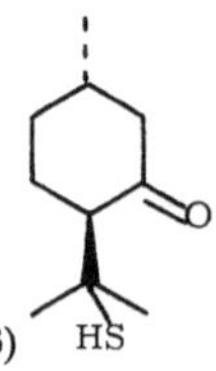
(83)
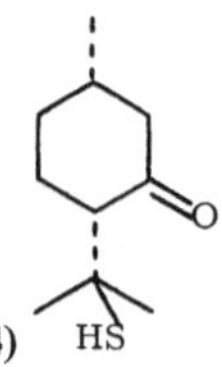
(84)

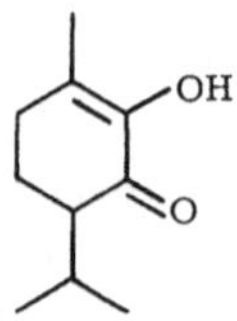
(85)
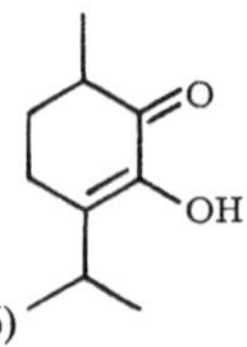
(86)
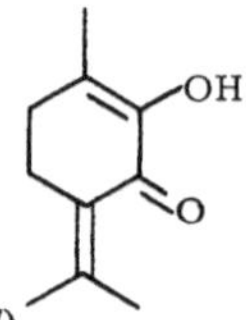
(87)
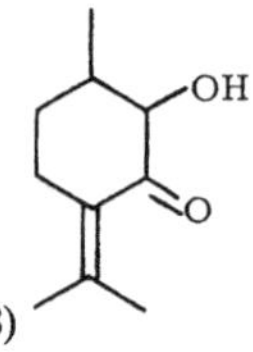
(88)
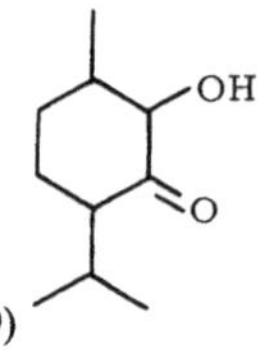
(89)

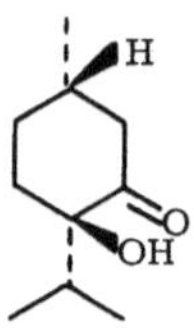
(90)
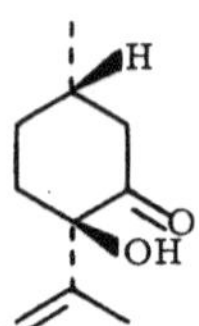
(91)
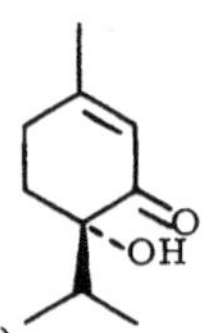
(92)
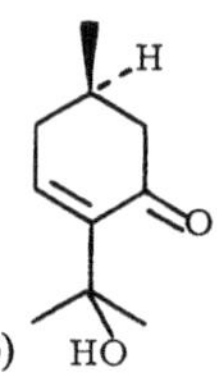
(93)
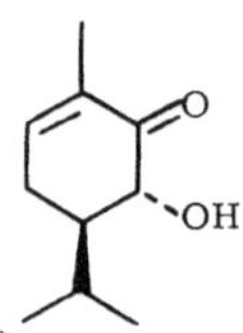
(94)

compounds diosphenol (**85**) and ψ-diosphenol (**86**), new compounds such as the diosphenol (**87**) as well as the diastereoisomers of 2-hydroxy-pulegone (**88**) and 2-hydroxymenthone (**89**) were isolated. Further-

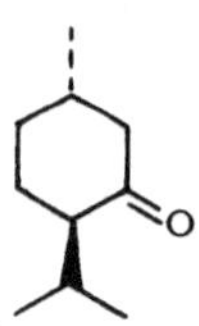
(95)
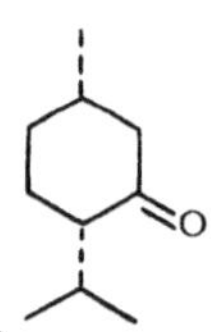
(96)

more, 4-hydroxyisomenthone (**90**) and 4-hydroxyisopulegone (**91**) were also found in the same oil. The essential oil of Buchu leaves has been known for a long time to be an important ingredient for the preparation of imitation blackcurrant flavors. Buchu leaf oil was also found to contain up to 0.7% of the diastereoisomers of 8-mercapto-*p*-menthan-3-one (**83**) and (**84**) as character impact components (*326, 605*), the threshold concentration of which was determined to be about 1 ppb (*379*). Due to its full fruity sensory impression which is typical of blackcurrants, (+)-8-mercapto-*p*-menthan-3-one (**83**) is preferred to (−)-8-mercapto-isomenthan-3-one (**84**) (*605*). The corresponding acetates and methyl ethers of the thio compounds (**83**) and (**84**), which were also isolated from Buchu leaf oil, do not possess as high a flavor impact as the unsubstituted compounds (*283*). The remaining bifunctional *p*-menthane derivatives (**85**) to (**91**) also exist in the form of their acetates, more rarely of their methyl ethers (*283*). It is worthwhile to mention the odor characteristics of 2-acetoxypulegone [ester of (**88**)] which imparts a particular minty, hay-like note to the odor complex of Buchu leaf oil (*283*).

Mercaptoketones (**83**) and (**84**) can be prepared in a simple manner by adding H_2S to pulegone [carbonyl derivative of (**49**)] which was also found to be a component of Buchu leaf oil (*311*). Hydroxyketones (**91**) and (**93**), which belong to different steric series, are both formed by photooxygenation of the corresponding pulegones (*552*). Furthermore, 3-hydroxy-6-menthen-2-one (**94**) in the form of its angelate was initially detected in the root oil of a *Pluchea* species (*48*).

It is now generally admitted that enantiomers can have differing odor qualities (*449, 637*). Carvone (**97**)/(**98**) constitutes a particularly

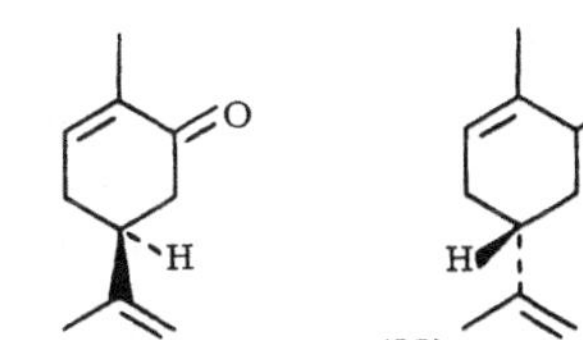

(**97**) (**98**)

impressive example in this respect. Whereas (+)-carvone (**97**) imparts its odor to caraway oil, the dominating odor impression of spearmint oil is caused by (−)-carvone (**98**). In both oils the carvones amount to about 60% of the total (*191*). While this odor discrimination had been

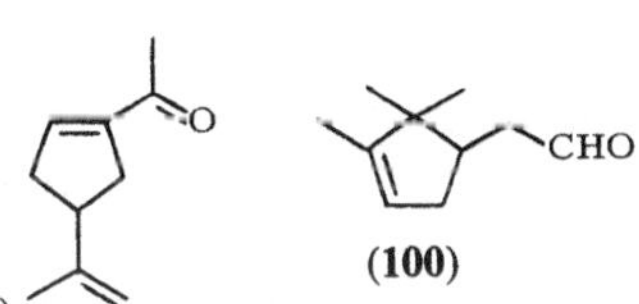

(**99**) (**100**)

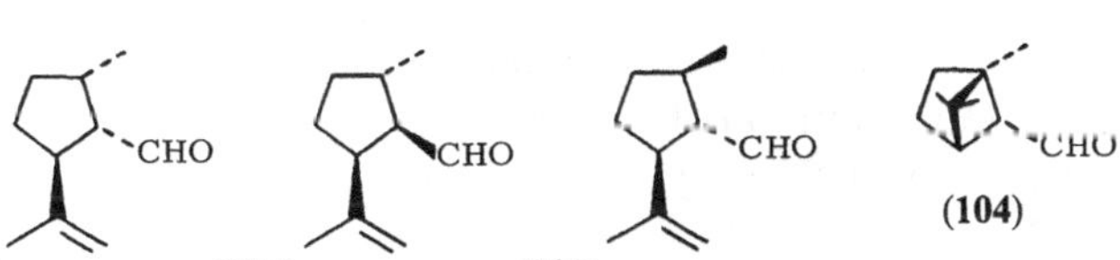

(**101**) (**102**) (**103**) (**104**)

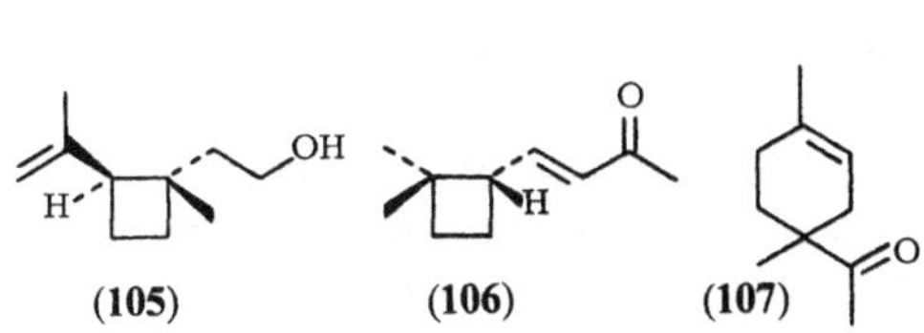

(**105**) (**106**) (**107**)

discussed for a long time, the solution of this problem was made possible only by the development of modern sensory measuring methods (*340, 508*) as well as by an unequivocal way of preparing the two enantiomeric carvones (**97**) and (**98**) by chemical interconversion (*167*).

A series of important aroma substances has become known which do not follow the normal isoprene rule and which are either metabolites or artifacts of isoprenoids or are formed *via* a hitherto unknown biogenetic pathway (besides the substances, described above, related to chrysanthemic acid). The methyl ketone (**99**), smelling like anise and caraway, was found in Spanish eucalyptus oil (*538, 539, 715*). Other cyclopentene derivatives, *i.e.* campholenic aldehyde (**100**) and its epoxide together with the corresponding alcohol and its acetate, were identified in the fruit oil of *Juniperus communis* L. (*642*). The occurrence of aldehyde (**100**) in oil of lavender was also reported recently (*390*). Junionone (**106**) was also found in juniper oil (*651*), its technical importance being recognized at the same time (*646*). In the same essential oil there was found 1,4-dimethylcyclohex-3-enyl methyl ketone (**107**) which is another compound having an "anomalous" carbon skeleton (*645*). Fraganol (**105**) is a further monoterpenoid cyclobutane derivative which was originally isolated in the form of its lower esters from the roots of *Artemisia fragrans* Willd. *(Compositae)* (*49*).

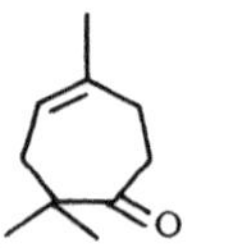
(**108**)

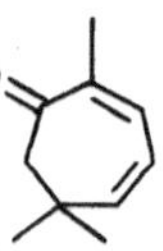
(**109**)

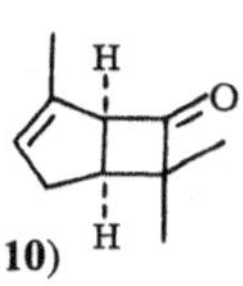
(**110**)

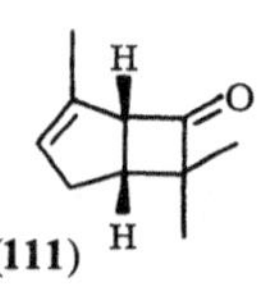
(**111**)

Karahanaenone (**108**), isolated from Japanese hop "Shinsh-wase" (*426*), is an unusual monoterpenoid of the eucarvone type (**109**). The essential oil of Arizona sage (*Artemisia filifolia* Torrey) contains 11% of (−)-filifolene (**110**), the first monoterpenoid reported to have the bicyclo[3.2.0]heptane ring system. Its enantiomer (**111**) was found in the Australian plant *Zieria Smithii* Andrews (*27*).

The diastereoisomeric 2-isopropenyl-5-methylcyclopentane carbaldehydes (**101**), (**102**) and (**103**) as well as 1,6,6-trimethylbicyclo[2.1.1]-hexane-5-carbaldehyde (**104**), which result from the photocyclization of citral (*92*), were described as important constituents of verbena oil (*280*). Photocitral A (*101*), the scent of which is described as fresh, camphor-like, green and herbal, is responsible for the typical top note of verbena oil and is, therefore, the most important member among the four photocitrals present in the oil (*279*). The odor qualities of *epi*-photocitral A (**102**) are similar, though milder, whereas *trans,trans*-photocitral (**103**) is characterized by a more fruity, resinous and woody tonality (*325*). The latter odor description also applies to aldehyde (**104**). Together, the four photocitrals (**101**) to (**104**) are stated to produce the typical fresh note of the citrus type (*279*). The base tonality of photonerols A and B [alcohols of compounds (**101**) and (**104**)] is described as rose-like, whereas the odor of *epi*-photonerol [derived from (**102**)] is characterized as sweet floral (*325*).

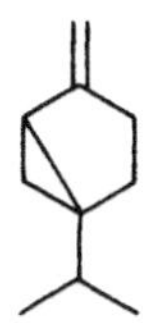
(**112**)

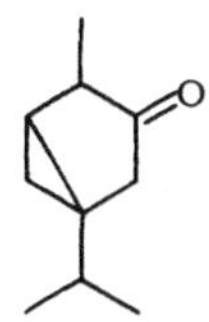
(**113**)

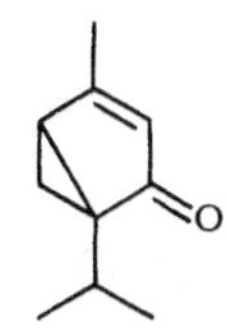
(**114**)

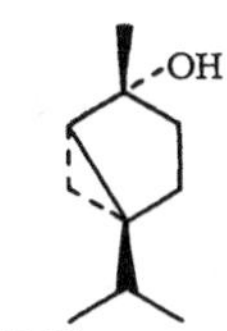

(**115**)

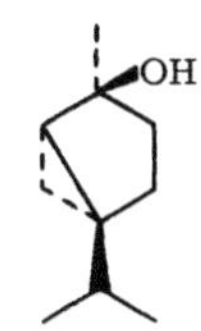

(**116**)

Several thujane derivatives were found to be important components in a series of flavor-producing plants. Sabinene (**112**) was identified as the major constituent of "Phlai" (*Zingiber cassumunar* Roxb.) (33.4%) (*334*) and of nutmeg oil (*Myristica fragrans* Houttyn) (*24, 266*), especially in the "West Indian" spice of Granada origin in which it is present in amounts of up to 50%.

Wormwood oil (*Artemisia absinthium* L.), sage oil (*Salvia officinalis* L.) and milfoil oil (*Achillea millefolium* L.) contain the diastereoisomeric thujones (**113**) as their main components (*192*). The biogenesis of these ketones was investigated in certain *Thuja, Tanacetum* and *Juniperus* species (*25*). (−)-Isothujone (*cis*-**113**) possesses an antinociceptive activity which is comparable with that of (−)-Δ^9-tetrahydrocannabinol and codeine (*498*). The acute toxicity of California bay oil (*Umbellularia californica* Nutt.) is due primarily to umbellulone (**114**) (39%), a substance, which is not present in Mediterranean or West Indian oils (*67, 369*). A similar neuropharmacological effect is reported for (−)-thujone (*trans*-**113**) (*208*).

The most important alcohols of this series are the diastereoisomeric sabinene hydrates (**115**)/(**116**). (+)-*trans*-Sabinene hydrate (**116**) was found in nature for the first time in American peppermint oil (0.8%) (*98*). The content of (+)-*cis*-sabinene hydrate (**115**) in marjoram oils is as high as 40% (*351*). Alcohol (**115**), purified by sublimation, is reported to have a strong spicy marjoram-like odor (*349*). In other spices such as black pepper (*509*), basil (*337*) and nutmeg (*24*), the mixture of the two diastereoisomeric alcohols (**115**)/(**116**) seems to have a flavor-forming function.

The occurrence of other bicyclic monoterpenes such as α-pinene (**117**), β-pinene (**118**), camphene (**119**) and 3-carene (**120**) in flavors is not only extraordinarily diverse, but also considerable from a quantitative point of view. The same is true for the corresponding oxygen-

 (117)  (118) 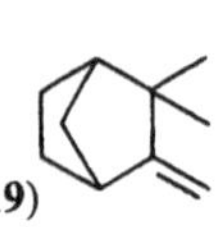 (119) 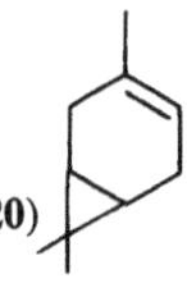 (120)

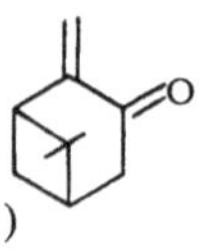 (121) 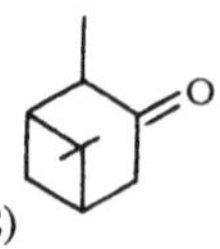 (122) 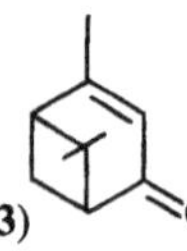 (123) 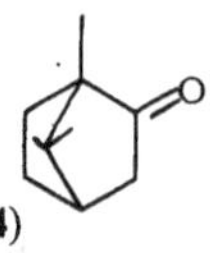 (124)

containing derivatives such as pinocarvone (**121**), pinocamphone (**122**), verbenone (**123**) and camphor (**124**) and the derived alcohols. For the organoleptic properties of these compounds see the voluminous literature (*150*).

2. Monoterpenoid Ethers and Lactones

The importance of cyclic monoterpene ethers as aroma compounds was recognized only during the last decade and in particular after the discovery of rose oxide (**134**) (*562, 563*) and the elucidation of the structure of linalool oxides (**128**) and (**129**) (*146*), although the two members of the cineole series (**125**) and (**126**) were known long before (*194*) (**126**) having been detected in most plant families (*194*). Due to its eucalyptus-like odor and its importance as an organoleptic component in eucalyptus oil (*194*), 1,8-cineole (**126**) is also known as eucalyptol. The bicyclic ether (**126**) is said to be responsible for the characteristic odor note of sudachi oil obtained from a Japanese citrus variety (*353*). From lime oil (**125**) and (**126**) are formed during processing only (*577*). Odor threshold studies indicated that 1,8-cineole (**126**), present in California bay oil (18%), was by far the major contributor to the odor (95% odor units of the whole oil), although the major constituent (39%) was umbellulone (**114**) (which contributes only 0.3% to the total odor units) (*67*). Eucalyptol (**126**) was also recognized as a relatively important substance in the secretion of the rove beetle (*Stenus comma*) where it probably serves as an antiseptic (*535*).

The diastereoisomers of dihydropinol (**127**) (*315, 541*), which also have a cineole-like odor, as well as their probable precursors (+)-dihydrocarveol and (−)-isodihydrocarveol [diastereoisomers (**51**)] (*541, 542*) are components of caraway oil. The diastereoisomeric linalool oxides (**128**) (furanoid) and (**129**) (pyranoid) have been found during the last few years in a large number of essential oils as important aroma-forming substances. In most cases these hydroxy ethers are accompanied by linalool (**6**) from which they may be formed in a suc-

(125)

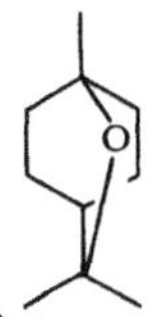

(126)

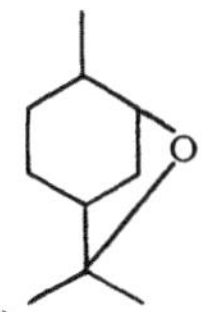

(127)

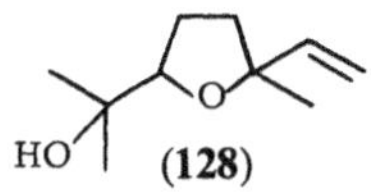

(128)

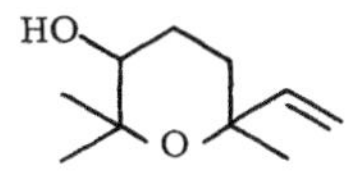

(129)

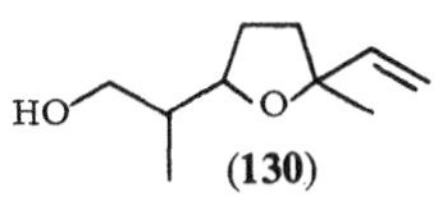

(130)

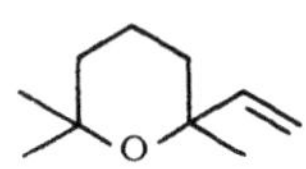

(131)

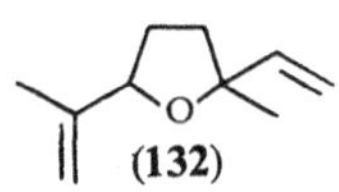

(132)

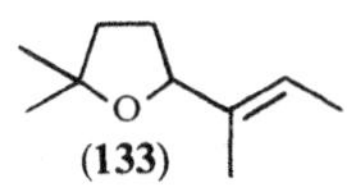

(133)

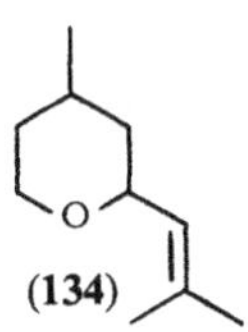

(134)

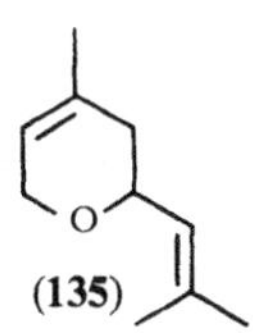

(135)

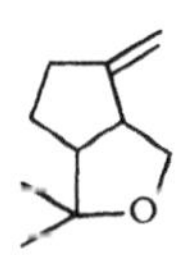

(136)

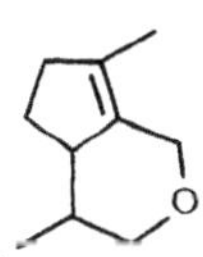

(137)

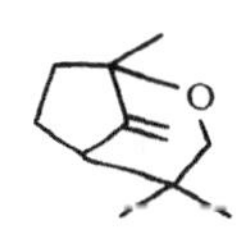

(138)

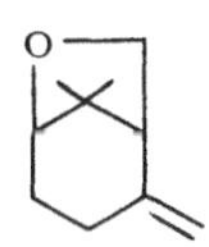

(139)

cession of reactions *via* an enzymatically or autoxidatively initiated epoxidation reaction. Thus, these compounds were detected in various citrus oils (*247*), and in the flavors of apricot (*623*), pineapple (*94*) and muscatel grape (*634*). In the flavor of black tea, linalool oxide (**128**) together with (**129**) even are members of the group of main constituents (*493*). Furthermore, several authors have independently reported that the diastereoisomers of the cyclic ethers (**128**) and (**129**), especially together with linalool (**6**), exert a favorable effect on tea aroma (*183, 316, 721*). This mixture, which contributes to the sweet-woody note, can be used to improve the flavor of instant black tea (*629*). The deoxyether (**132**) is found less often, and the tetrahydropyran derivative (**131**) was detected only in two cases (*283, 599*). Ether (**133**) is to be considered as an artifact of (**131**) (*599*). The diastereoisomeric lilac alcohols (**130**) (*676*) and the corresponding aldehydes (*675*) should probably be considered as potential flavor components in spite of the fact that their presence in a flavor has not yet been observed. The diastereoisomeric rose oxides (**134**) of the (4*R*)-series, first discovered in rose oil (*563*) and geranium oil (*424*), were also detected in tea oil (*720*) and *Ribes nigrum* (*615*). The rose oxides (**134**) of the same stereochemical series were also found in the animal kingdom, notably in the secretion of the thoracic glands of *Aromia moschata* L. (*Cerambycidae, Coleoptera*) (*670*). Long after nerol oxide (**135**) had been introduced as a synthetic odorant (*459*), it was identified in the flavor of various varieties of grape (*546*), and also in Bulgarian rose oil (*141*). The juice of "Scheurebe" grape reaches the highest content of (**135**) with 30 μg/liter. Rose oxide (**134**), nerol oxide (**135**), linalool oxide (**128**), and (**129**) were recognized as important aroma components in grape and wine varieties showing intense aroma properties (*549, 550*).

The dye-sensitized photooxygenation reaction has to be considered as a biomimetic process (*451*) involved in the formation of deoxylinalool oxide (**132**) (*459*), rose oxide (**134**) (*456*) and nerol oxide (**135**) (*459*) from linalool (**6**), citronellol (**9**) and nerol (**8**), respectively.

Verbena oil contains, besides (**134**) and (**135**), the two furan derivatives (**140**) and (**141**) as trace components (*280*).

Hop ether (**136**) and karahana ether (**139**) are monoterpenoid compounds which occur in Japanese hop (Shinshu-wase) (*426*), while the occurrence of (+)-matatabi ether (**138**) and of its isomer (**137**) is limited to the essential oil of *Actinidia polygama* Franch. et Sav. (*Ternstroemiaceae*) (*261, 716*).

Monoterpenoid furan derivatives occur in a limited number of flavor-yielding plant species. Menthofuran (**140**), as the oldest example (*195*), is present in all the oils obtained from *Mentha piperita* L. in which its pronounced odoriferous character contributes to the pleasant flavor

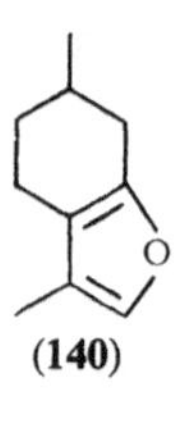

(**140**)

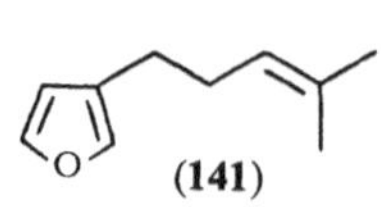

(**141**)

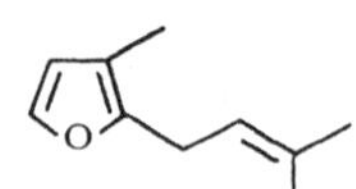

(**142**)

properties of these oils. In the less appreciated oils of *Mentha arvensis* Holmes, (**140**) was not detected (*540*).

Perillene (**141**) was isolated from the essential oil of the leaves of *Perilla citriodora* Makino (*314*) as well as from the ant *Lasius (Dendrolasius) fuliginosus* Latr. (*487*). On the other hand, rose furan (**142**) was isolated for the first time from rose oil (*61*).

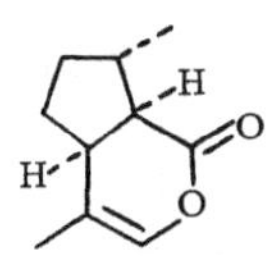
(**143**)

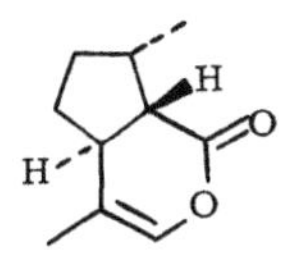
(**144**)

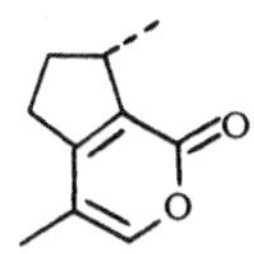
(**145**)

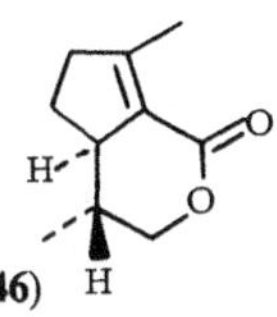
(**146**)

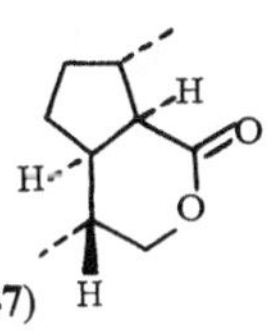
(**147**)

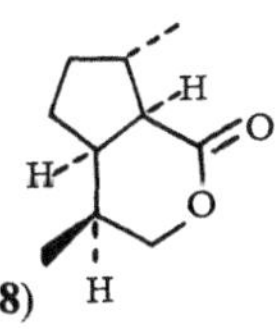
(**148**)

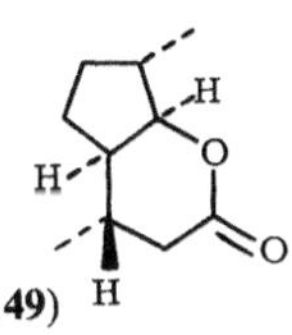
(**149**)

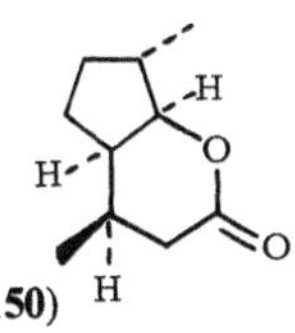
(**150**)

Among the monoterpenoid lactones, derivatives of the iridane series (*600*) have acquired limited importance as aroma substances. Nepetalactone (**143**) occurs in quantities of up to 78% in American oil of catnip (*Nepeta cataria* L.) (*490*), while epinepetalactone (**144**) is the major constituent (70%) of the essential oil of *Nepeta mussini* L. (*490*). The essential oil of this species was also found to contain neonepetalactone (**146**) as well as the two saturated compounds α- and δ-dihydronepetalactones (**147**) and (**148**) (*490, 518*) and 5,9-dehydronepetalactone (**145**) (*528*). Both dihydro compounds (**147**) and (**148**) are structurally closely related to the iridomyrmecins from *Actinidia polygama* (*515, 516*). Lactones (**143**) to (**150**) are aroma compounds which have a strong odor and which exert an attracting effect on *Felidae* (*518, 519*).

3. Sesquiterpenes

The sesquiterpenes and their derivatives comprise about 2,000 naturally occurring representatives, and are the major sub-group of isoprenoid natural products*. These compounds derive from more than

* Up to 1971 the number of naturally occurring sesquiterpenes was estimated at 1,000 individual compounds (*507*). However, if we consider the fact that up to the beginning of 1975 as many as 600 sesquiterpene lactones [at the end of 1976 their number was estimated to be 700 (*352*)] were isolated (mainly from *Compositae*) (*229*) — 300 of these compounds were known by 1971 — the estimated number of about 2,000 sesquiterpene derivatives discovered in nature by the end of 1976 is hardly exaggerated. Sesquiterpene lactones have not yet been reported as flavor compounds.

100 different carbon skeletons (*507, 124*). Although 500 sesquiterpene derivatives have been found in flavors (*501*), only about 20 seem of importance for the formation of flavors, if subjected to severe criteria. The aroma compounds of the sesquiterpene series seem to be associated with certain carbon skeletons. In any event, we still have no indication as to the occurrence in natural flavor complexes of compounds such as cedranes, santalanes, vetivanes, spirovetivanes (*360*) and germacranes which, on the other hand, constitute very important chemical classes in the chemistry of fragrances. Compared with the monoterpene derivatives, the flavor value of sesquiterpene compounds is small, although some of their representatives have been found to be character impact compounds.

While nerolidol (**155**), farnesol (**154**) and farnesal, the aldehyde corresponding to (**154**), have been known for a long time as natural flavor components, the importance of the farnesenes (**151**) to (**153**), was only recognized recently. α-Farnesene — first discovered in hop oil (*71, 583*) — has meanwhile been isolated from a wide variety of flavors

(**151**) 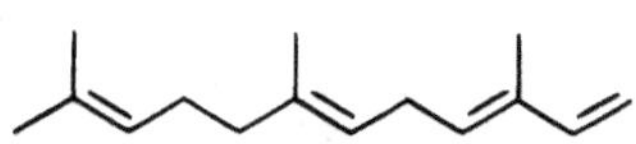

(**152**) 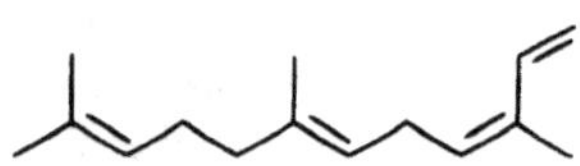

(**153**) 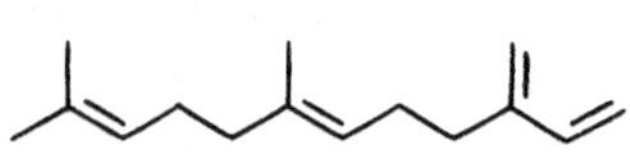

(**154**) 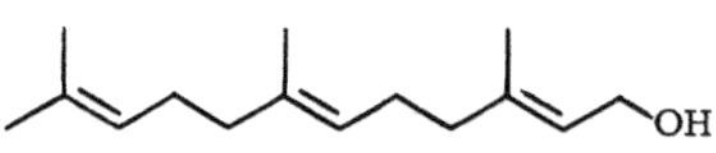

(**155**) 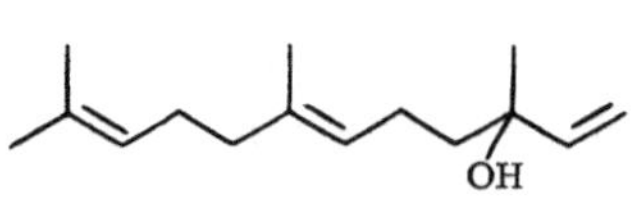

(*337, 571, 667*). The stereoisomeric α-farnesenes (**151**) and (**152**) appear to play a particular role as aroma compounds in apples (*402, 442*). (**151**) was found as the sole sesquiterpene hydrocarbon in the natural coating of apples of the variety "Granny Smith" (*245, 402*) and is

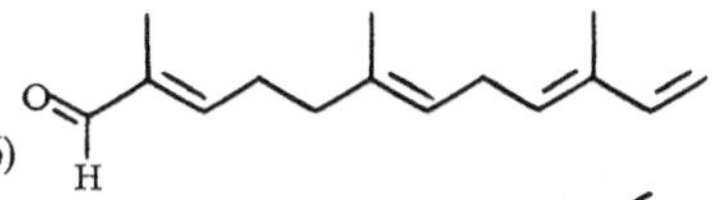

(156)

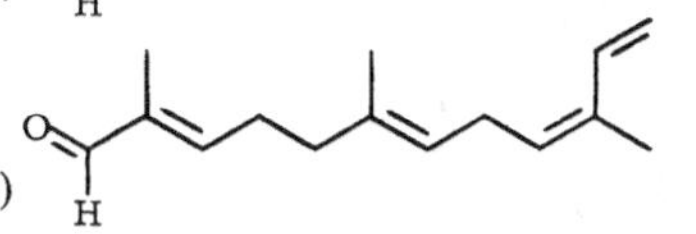

(157)

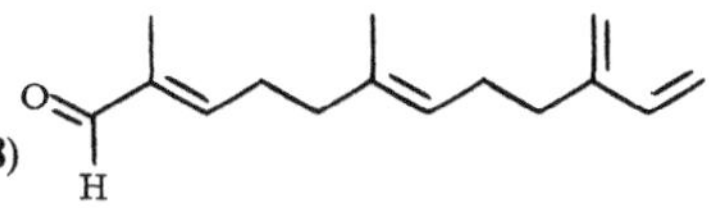

(158)

believed to be responsible for the browning of the skin surface of stored fruit (*244*). Moreover, these hydrocarbons have highly attractive properties to newly hatched codling moth (*Laspeyresia pomonella*) larvae (*610*). (*E*)-β-Farnesene (**153**), however, was identified as the repellent odor principle in the cornicle secretion of several economically important species of aphids (*53, 139, 417*). An *allo*-farnesene of unknown stereochemistry is present, besides (**151**) (up to 6.5%), in a concentration up to 4% in the essential oil of *Perilla acuta* Thunb. (*172*). Farnesol (**154**) can be converted specifically into β-farnesene (**153**) (*56*) or even selectively into the stereoisomeric α-farnesenes (**151**) and (**152**) (*622*). Alcohol (**154**) contributes considerably to the odoriferous principle in the "bouquet" of French or Georgian champagne (*499*). This compound has given rise to a considerable amount of research since its juvenile hormone activity was discovered (*54*). The orange flavor is mainly developed by a mixture of sinensals (**156**) to (**158**) (*156, 391, 593*) the nomenclature of which was established in correlation with the farnesenes (**151**) to (**153**) (*632*). Interest in these aldehydes has given rise to intensive preparative efforts which resulted in the development of a series of original syntheses (*39, 63, 64, 65, 641*). (*all-E*)-α-Sinensal (**156**) has an orange-like note, whereas (*E,E*)-β-sinensal (**158**) is dominated by a strong metallic-fishy undertone and causes, at high concentrations, an unpleasant odor sensation (*444*). (*E,E,Z*)-α-Sinensal (**157**) has the same odor character as (**156**) but is the weakest of the three aroma compounds (*443*).

Whereas the bicyclic sesquiterpene (−)-caryophyllene (**159**) [odor threshold in water solution 64 ppb (*71*)] is widely distributed in flavors, its stereoisomer (**160**) is found only rarely in nature (*206, 281, 336, 401*). Apparently (**160**) is an artifact, the formation of which from (**159**) by the action of photochemically produced radicals could be shown (*551*). Black pepper oil rich in hydrocarbon (**159**) smells sweet and flowery (*341, 500*). On the other hand, the hydrocarbon fraction of

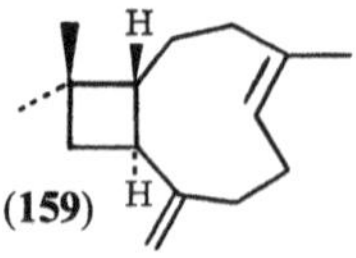
(159)

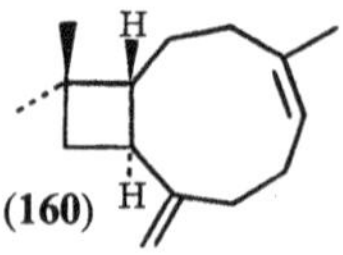
(160)

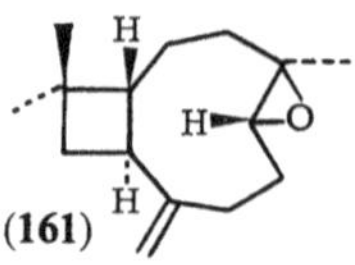
(161)

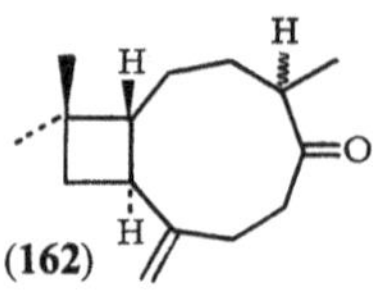
(162)

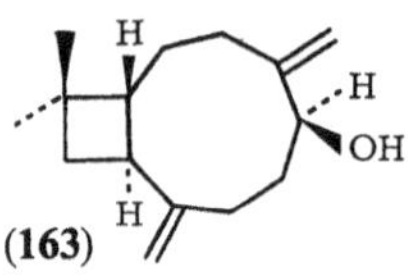
(163)

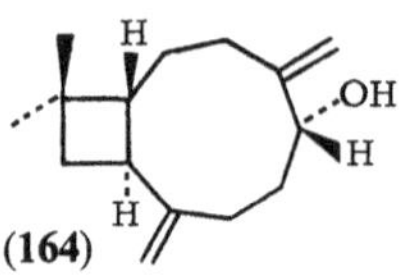
(164)

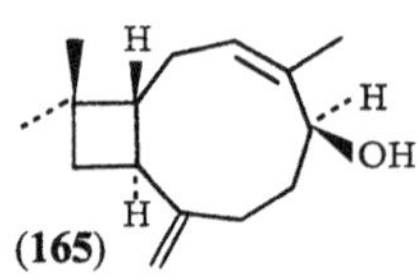
(165)

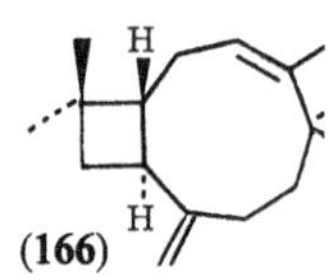
(166)

black pepper oil has at the same time a turpentine smell completely remote from the odor of pepper as it is perceived in daily life (*105*).

Monoepoxide (**161**), which was first discovered in clove bud oil (*Eugenia caryophyllata* Thunb., *Myrtaceae*) (*420, 658*), is present in most cases together with hydrocarbon (**159**). Epoxide (**161**) has a marked woody note with the celluloid-like undertone (*443*) and seems to play a relatively important role in various flavors (*100, 217, 218, 335, 336, 504, 636*). Epoxide (**161**) is present in a proportion of 23% in the essential oil of an Indian *Alpinia* species and constitutes the main component of this oil (*484*), an Indian guajava leaf oil even contains 37.5% of this epoxide (*81*). Recently, (**161**) has also been found in the essential oils from clary sage (*339*), *Daucus carota* (*601*), *Elsholtzia pilosa* (*635*) and *Eucalyptus saligna* (*11*). Caryophyllene oxide (**161**) is also a component of marihuana essential oil (*435*) and exerts a strong attraction on the cocker spaniel. Dogs of this species, who are trained to find the drug, are able to detect 0.000 001 g (1 μg) of the epoxide (**161**). These animals thus attain the limit of detectability, approximately that of the lower fatty acids, which lies at about 10^6 lower than in man (*585*). The observation that (**161**) is responsible for the recognition of *Cannabis sativa* by dogs has since been confirmed (*467*). The three alcohols (**164**), (**165**) and (**166**) which are formed during the photo-oxygenation of (−)-caryophyllene (**159**) (*551*) were found to be present in the oxygen-containing part of the essential oil of black pepper (*106*) and to have a faint woody-spicy note (*443*). One of the diastereoisomeric alcohols (**163**) or (**164**), called α-multijugenol, was isolated from the essential oil of *Copaifera multijuga* Hayne (*107*).

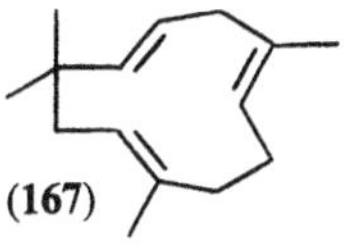
(167)

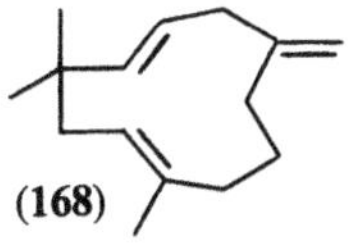
(168)

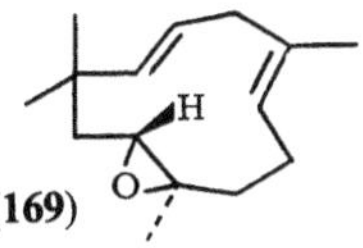

(169)

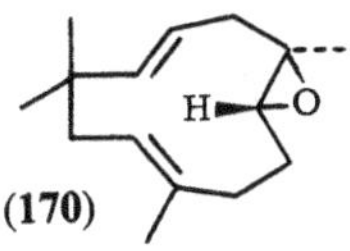

(170)

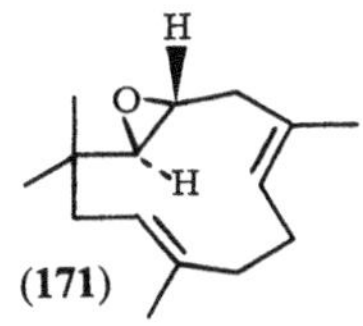

(171)

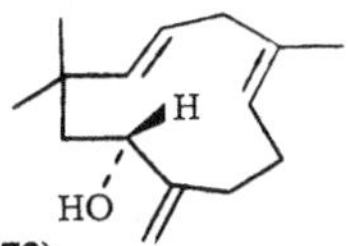

(172)

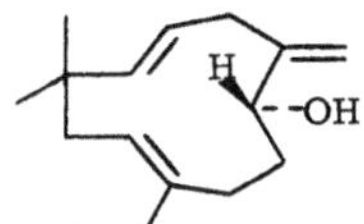

(173)

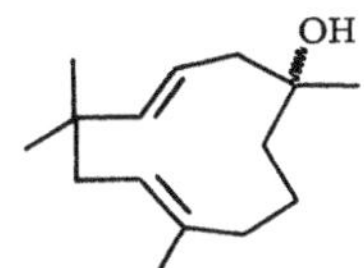

(174)

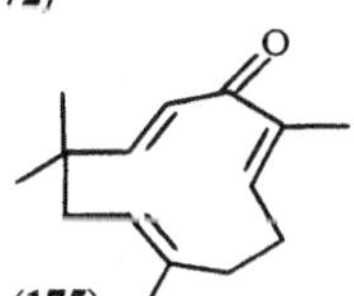

(175)

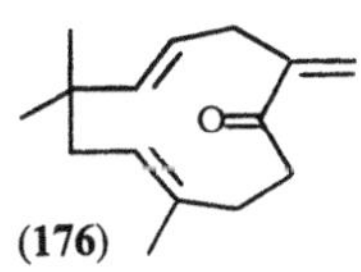

(176)

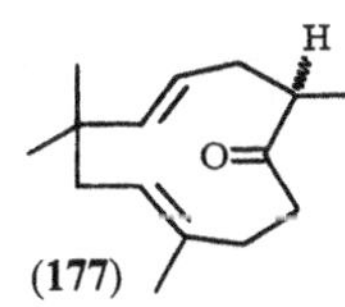

(177)

The oleoresin of *Dipterocarpus* species is composed essentially of (−)-caryophyllene and humulene derivatives **(159)**, **(161)**, **(167)**, **(169)**, **(170)**, and **(171)**, **(165)** and **(166)** constituting more than 1% of it (*205*). Allyl alcohols having a caryophyllene structure are present in the bud oil of *Betula alba, Betula lenta* and also in peppermint oil (*189*).

Pepper oil (*500*), after irradiation and aeration, as well as the untreated oil of cinnamon leaf, stem and root (*697*) contain ketone **(162)** which has approximately the same organoleptic properties as epoxide **(161)** (*443*). This ketone, **(162)**, was found for the first time in clove oil, and its pleasant and tenacious odor was described as being camphor-like and reminiscent of ionone (*659*).

Humulenes **(167)** and **(168)**, which were detected in various oils, presumably do not play an essential role in hop oil [odor threshold in water solution 120 ppb (*71*)]. Oxygen-containing humulene derivatives, among which epoxides **(169)** and **(170)** (*71*), humulol **(174)** (*428*) and ketones **(176)** and **(177)** (*429*) were identified, might contribute much to the hop character imparted to hop beverages (*71*). All these substances **(167)** to **(173)** as well as zerumbone **(175)** were first isolated from the essential oil of wild ginger (*99, 100, 101*).

The series of monocyclic sesquiterpene derivatives of the isoprenologous *p*-menthane type, includes some particularly interesting aroma components. (−)-α-Bisabolol **(178)** contributes substantially to the antiphlogistic and spasmolytic activity of camomile (*239, 255, 345*).

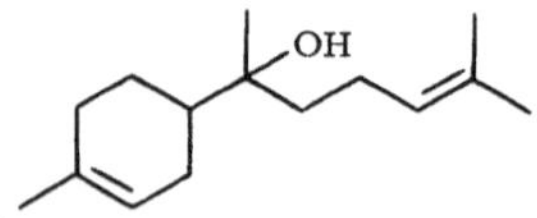

(178)

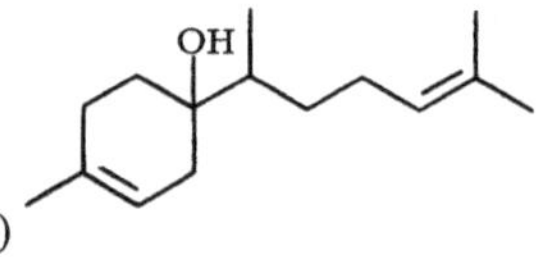

(179)

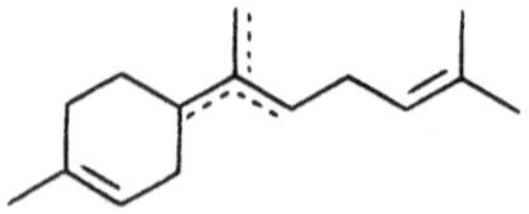

(180)

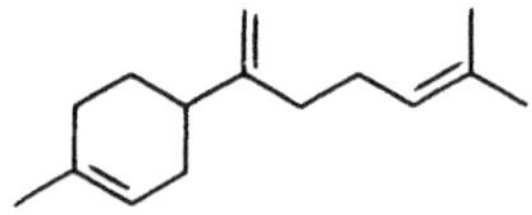

(181)

(−)-α-Bisabolol (**178**) was isolated from the essential oil of poplar buds (*Populus* species) (*239, 255*). (+)-β-Bisabolol (**179**) (*387*) and bisabolene oxide (**206**) (*226*) which were isolated from the essential oil of cotton buds have an apple blossom-like odor; (**179**) turned out to be moderately attractive to the boll weevil, a major cotton pest (*387*). The hydrocarbons [the bisabolenes (**180**)] from which this alcohol derives, together with the cadinenes and caryophyllenes (**159**), are among the sesquiterpenes most widely distributed in nature. The double bond isomer mixture (**180**) is described as herby, spicy by the manufacturer (*26*) and as pleasant, balsamic by others (*148*).

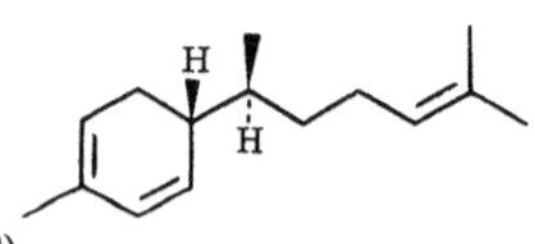

(182)

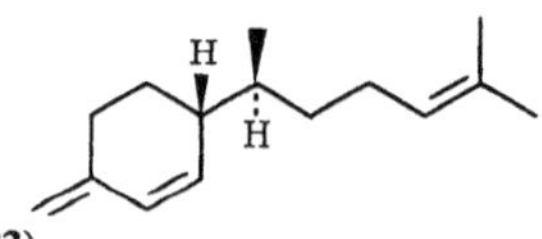

(183)

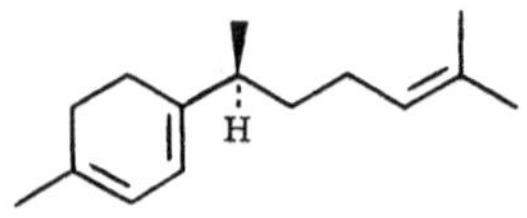

(184)

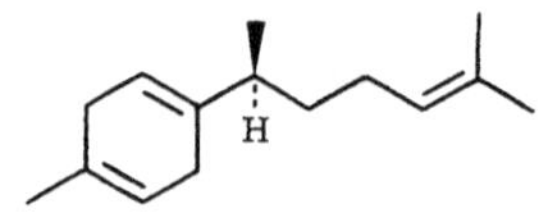

(185)

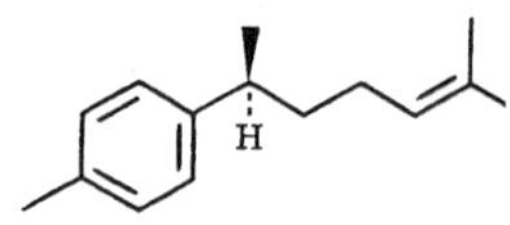

(186)

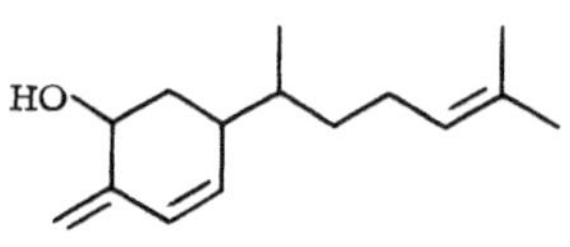

(187)

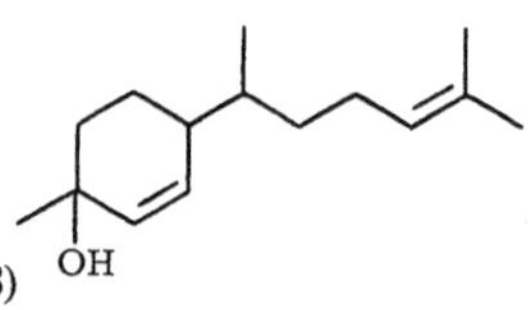

(188)

A fairly wide variety of compounds having the bisabolane skeleton is present in ginger oil. (−)-Zingiberene (**182**) (<30%), (−)-sesquiphellandrene (**183**) (<12%), (+)-*ar*-curcumene (**186**) (<19%) and (−)-β-bisabolene (**181**) (<12%) constitute as much as 50 to 70% of the total oil (*90*). According to an expert organoleptic description, a mixture of hydrocarbons (**182**) and (**183**) has a warm, woody-spicy tenacious odor with a peculiar deep sweetness and seems to be of particular value in perfumery (*19*). Remarkably, ginger oil contains up to 10% of an unknown farnesene (**151**), (**152**) or (**153**). The diastereoisomeric mixture of sesquiphellandrols (**187**) whose absolute configuration is not known was shown to have possible significance in ginger flavor (*33*). The organoleptic properties of zingiberenol (**188**) (*633*) have not yet been described.

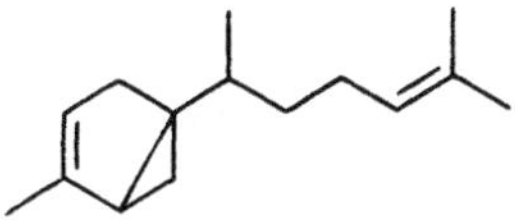
89)
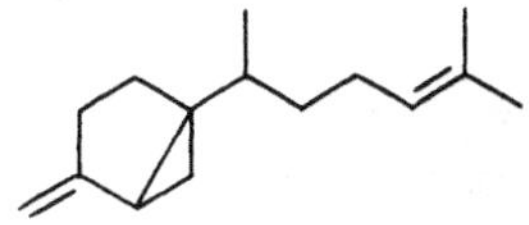
(**190**)
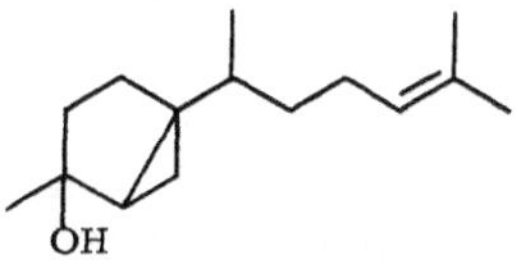
(**191**)

In addition, sesquithujene (**189**) and *cis*-sesquisabinene hydrate (**191**) were found in *Zingiber officinale*, whereas sesquisabinene (**190**) is a component of *Piper nigrum* L. (*633*). The biogenetic relation between the monocyclic hydrocarbons (**182**) to (**186**) and the bicyclic compounds (**189**) to (**191**) is obvious.

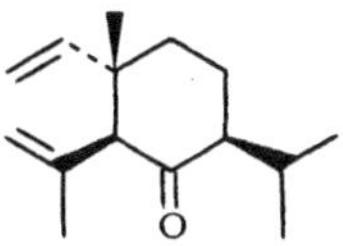
(**192**)
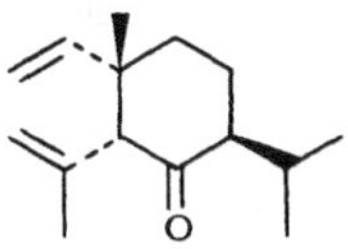
(**193**)
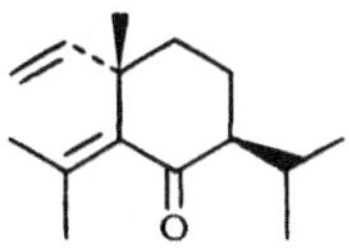
(**194**)

The three ketones (**192**), (**193**) and (**194**) having the monocyclic elemane skeleton, which together constitute about 7% of sweet flag oil (*Acorus calamus* L.), are said to reproduce the true odor of the original wild plant (*248*). The use of the oil of calamus has been banned in the United States by federal regulations (*87*). In certain European countries it is still used (*209*) as a minor flavoring agent (in concentrations below 30 ppm), especially in bitters, liqueurs and vermouths (*384*).

The eremophilane derivative (+)-nootkatone (**195**) has been valued for its most important contribution to the distinctive flavor of many citrus oils, in particular of grapefruit peel oil (*Citrus paradisi*, Macfayden) (*371, 372*). Qualitative (*211, 313, 453, 591*) and quantitative (*38, 211, 313, 591, 630*) aspects of this aroma compound have been treated. The naturally occurring (+)-nootkatone (**195**) is detectable at 1 ppb in water (*38*). The both enantiomeric forms of ketone (**195**) differ not only

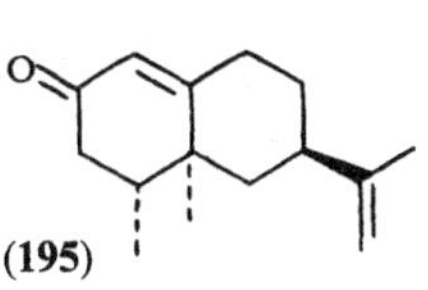
(**195**)
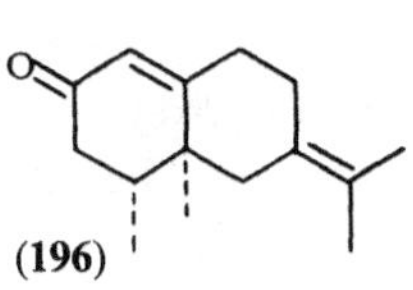
(**196**)

in odor profiles but significantly in odor strength, the threshold concentration of (+)-nootkatone (**195**) differing from that of (−)-nootkatone by a factor of 1,000 in water and by as much as 2,000 in the vapor phase (*211*). This grapefruit-like compound is also interesting because it illustrates the fact that in most cases even a small modification in the structure results in a change of the odor quality (*211, 453, 630*). This effect is particularly pronounced in the case of the double bond isomer of (**195**), (+)-α-vetivone (**196**), which occurs in the essential oil of vetiver and has a woody, floral and vetiver-like note (*313, 453*). As a result of this minor structural modification, the very specific aroma component (**195**) is converted into a typical perfume ingredient (**196**). (+)-Nootkatone (**195**) can be obtained by dye-sensitized photooxygenation of (+)-valencene (*451*), a major sesquiterpene of orange juice and peel oil (*246*).

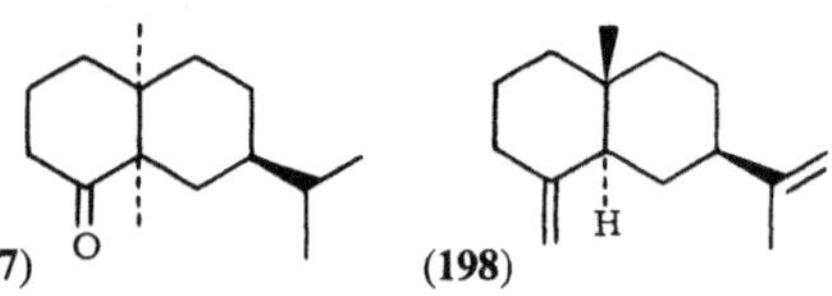

(**197**) (**198**)

(−)-Valeranone (**197**) which is present in valerian oil and has a non-isoprenoid carbon skeleton has been described as an odorless compound (*595*). It has been reported that the sesquiterpene β-selinene (**198**), a constituent of the essential oil from fresh celery, reproduces the characteristic odor of the latter (*702*).

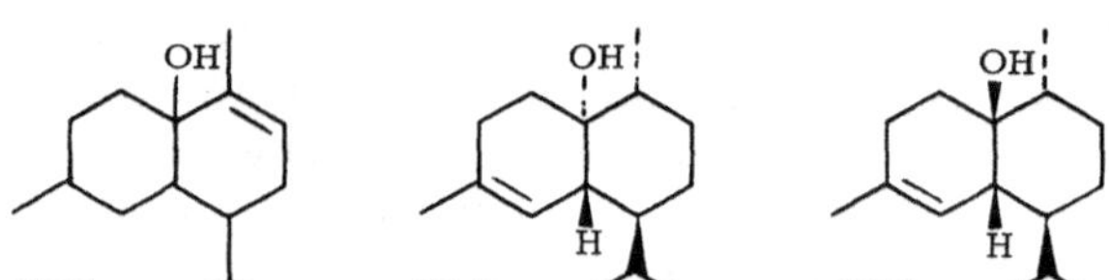

(**199**) (**200**) (**201**)

Although numerous oxygen-containing compounds having the skeleton of cadinane have been found in nature, only very few of them were described as aroma compounds. A cadin-9-en-1-ol (**199**) of unknown stereochemistry imparts to green tea flavor the heavy flowery aroma which somewhat resembles that of nerolidol (*722*). The diastereoisomeric cadin-4-en-1-ols (**200**) and (**201**) were isolated from commercial cubeb oil (*462*), *Cedrela toona* Roxb. timber (*409*) and green tea aroma (*438*). (−)-Cubenol (**200**) is described as having an earthy, woody odor character, whereas (−)-*epi*-cubenol (**201**) has a sweet-spicy note and seems to have a more intense odor (*181*). The enantiomer of (**201**) was obtained from Streptomyces sp. B-7 (*181*).

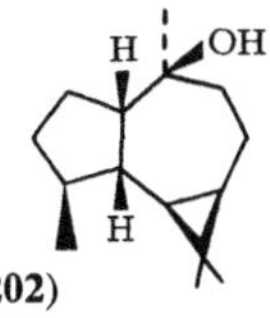

(202)

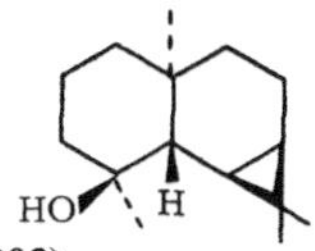

(203)

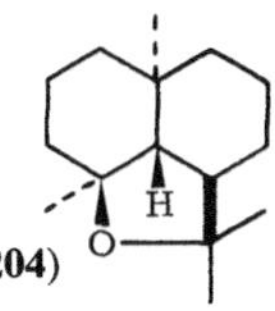

(204)

Among the tricyclic aromadendrols viridiflorol (**202**), which occurs, for instance, in peppermint oil from Piedmont (*Mentha piperita* L. Italo-Mitcham), is supposed to confer to this essential oil its characteristic note. When added to normal peppermint oils, (**202**) enhances their sweetish scent without modifying their taste and aroma (*415*). A diastereoisomer of (**202**), globulol, is present in eucalyptus oil, and is described as odorless (*204*). Maaliol (**203**), an alcohol which was first isolated from maali resin (*62*), was found in fresh roots of *Valeriana officinalis* L. (*595*) and also in the Indian variety *Valeriana Wallichii* D.C. (*416*). Maaliol-containing oils from *Valeriana Wallichii* D.C. are said to have a finer odor (*582*). In addition, maalioxide (**204**) (*62*) was found in the same oil (*416*), presumably as an artifact of alcohol (**203**). The cyclic ether is characterized by its fine woody odor quality (*443*).

The characteristic odor of carrot seeds (*Daucus carota* L.), which closely resembles that of the roots, has been attributed to the presence

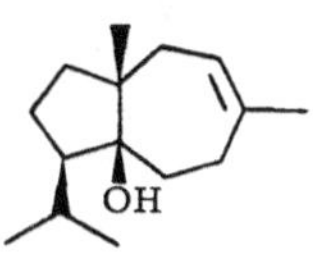

(205)

of carotol (**205**) (24 to 63%) and daucol (**220**) (2%). Alcohol (**205**) is described as having a pleasant mouldy odor with a woody note (*322*).

4. Sesquiterpenoid Ethers

Whereas sesquiterpenoid hydrocarbons, with few exceptions, have inferior characteristic organoleptic properties, the ethers of this series — as far as they are known — are distinguished by outstanding odor qualities. Once the general value of these compounds as aroma components is recognized, as it is the case with the monoterpenoid ethers, the importance of this group of compounds will rapidly grow.

Four ethers of the bisabolane series have so far been found in nature. The sole member of this series whose odoriferous properties

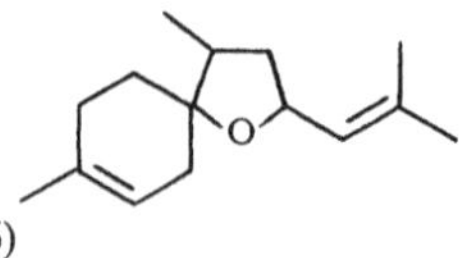

(206)

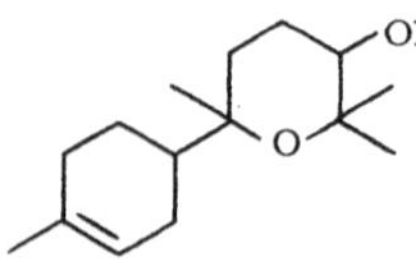

(207)

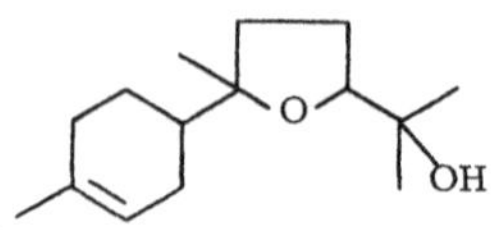

(208)

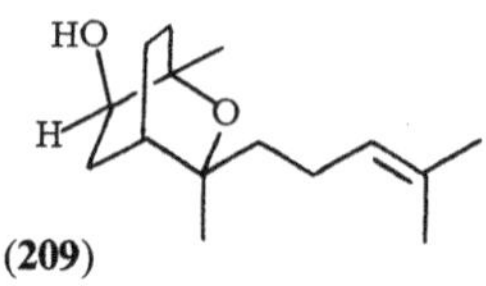

(209)

are known is the tetrahydrofuran derivative (**206**) derived from β-bisabolol (**179**). No odor description is available for α-bisabolol oxides A, B (*534*) and C (*524*) (**207**) to (**209**) among which (**207**) and (**208**) show the structural elements of the monoterpenoid oxides, derived from linalool.

Caparrapi balsam is a fragant oil from Colombia which is stated to be of some therapeutic value (*18*). A bicyclic ether (**210**) (*18, 59, 93*) called caparrapi oxide was isolated as the main component (up to 43%) from this oil. Whereas the natural product (**210**) is virtually odorless, its diastereoisomer (**211**) has pronounced amber-like odor properties (*455*).

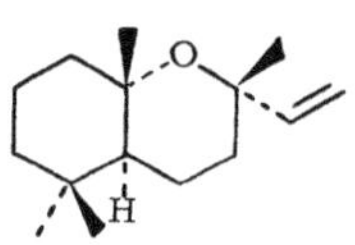

(210)

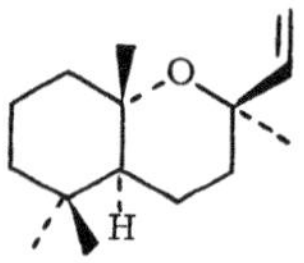

(211)

Apart from a few compounds isolated from aquatic plants (*543, 604, 652*), sesquiterpene derivatives with a reduced carotenoid structure [monocyclofarnesane (*385*)] have not been detected in terrestrial plants. Thus, 9-vinyl-7,8-dihydro-α-ionol is still a synthetic product which, however, shows a biogenetic relationship to the ethers of the caparrapi series (**210**)/(**211**) (*348*).

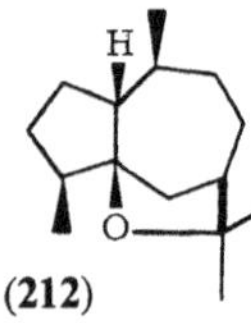

(212)

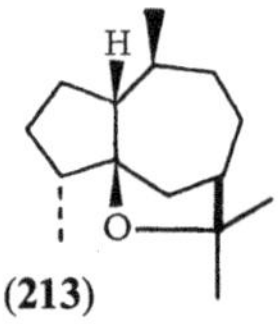

(213)

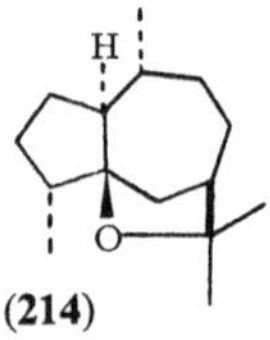

(214)

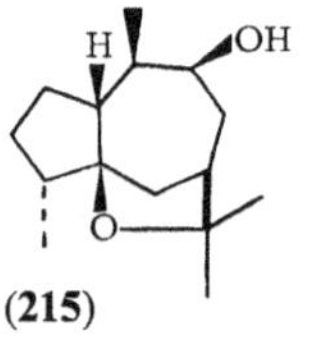

(215)

Guaioxide (**212**) which was first separated from guaiac wood oil (*Bulnesia sarmienti* Lor.) (*29, 140*) has a distinct woody, musty odor character (*443*). The aerial parts and the roots of the Chinese herb

"San-Shion" of the *Ligularia* genus *(Compositae)* contain various sesquiterpenoids having the guaiane skeleton which include liguloxide (**213**) (*256, 257*), liguloxidol (**215**) and its acetate (*256, 257*) and ogarukaya ether (**214**) (*233*). (**213**) and (**214**) were also found in the essential oil of *Cymbopogon Georingii* (Steud., A. Camus) (*171*).

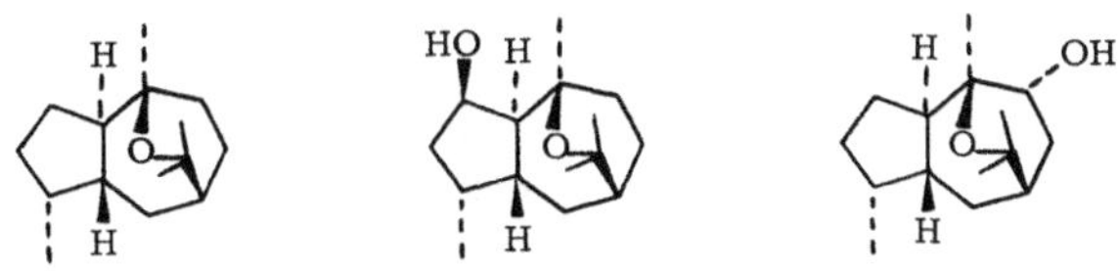

(**216**) (**217**) (**218**)

Japanese kesso root oil contains a considerable amount of tricyclic ethers among which kessane (**216**) (*231*) is the major component, together with its hydroxy derivatives (**217**) (*262*) and (**218**) (*230*). α-Kessyl alcohol (**217**) was described as being odorless (*185*).

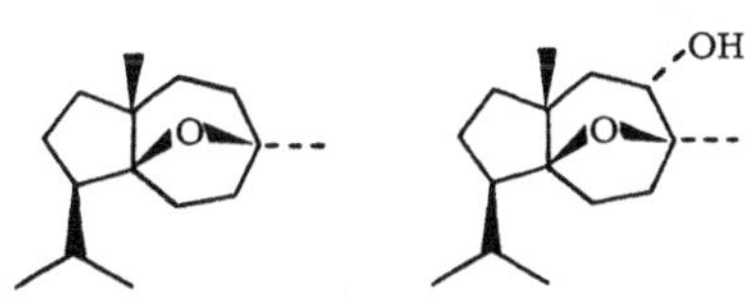

(**219**) (**220**)

On the other hand, the tricyclic hydroxy ether (**220**) having the carotane skeleton is described as having a sweetish odor reminiscent of calamus oil with an amber-like note (*322*), no corresponding odor description having been given for the parent compound (**219**) which was also found in carrot oil (*82*).

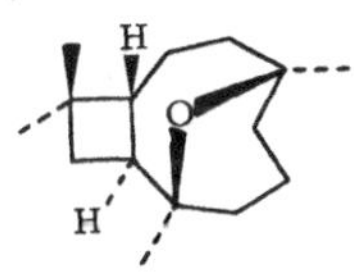

(**221**)

(−)-Caryophyllan-2,6-β-oxide (**221**) which has a woody-balsamic odor (*278*) contributes to the pleasant body of verbena oil (*281*).

Among the monocyclic ethers of the sesquiterpene series the furan derivatives have an outstanding position. Dendrolasin (**222**), first isolated from the mandibular glands of the ant *Lasius (Dendrolasius) fuliginosus* Latr. and described as having a lemon grass-like odor (*487*), was discovered together with its derivatives (**223**) and (**224**) in a Japanese plant (*513*). Upon progressive hydrogenation the dendrolasin odor disappears. Whereas the tetrahydrofuran derivative derived from **(222)** still has a faint odor, perhydrodendrolasin is described as being odorless (*487*). Sesquirose furan **(225)** with an offensive odor was found in a Lauraceae species called "Baribari" in Japan (*222*). No odor descriptions are available for the difuranoid sesquiterpenes lasiospermane **(227)** (*52*), dehydrolasiospermane **(226)** (*180*) and longifolin

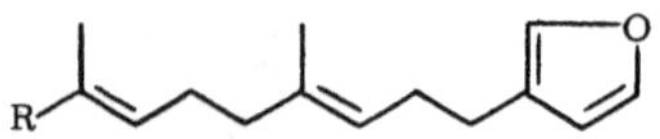

(222) R = CH_3
(223) R = CH_2OH
(224) R = CHO

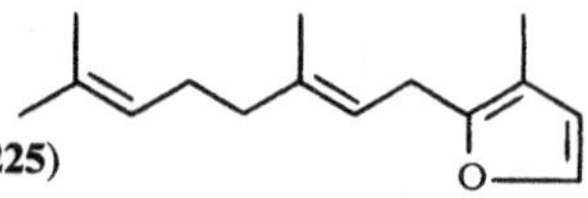

(225)

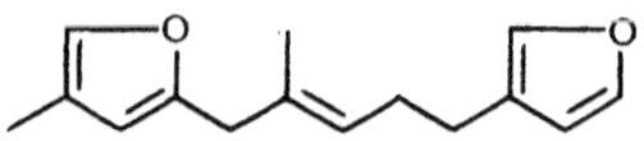

(226)

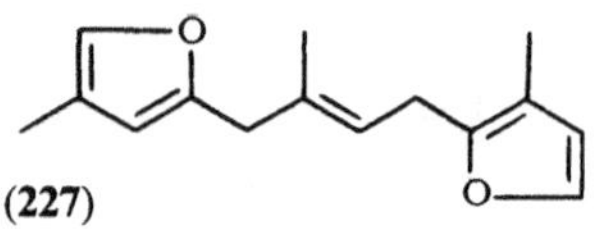

(227)

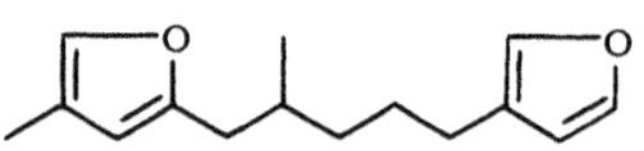

(228)

(228) (*222*). Sesquiterpene furans **(222)** to **(228)** can all be regarded as potential aroma components. Their presence can be expected in all cases where farnesenes **(151)**, **(152)** and **(153)**, farnesal [aldehyde of **(154)**] or sinensals **(156)** to **(158)** occur in flavors. Indeed, sesquirose furan **(225)** was detected in tangerine oil (*178*).

(229)

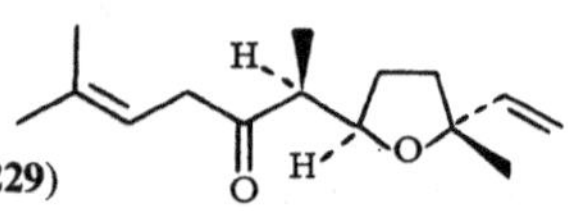

(230)

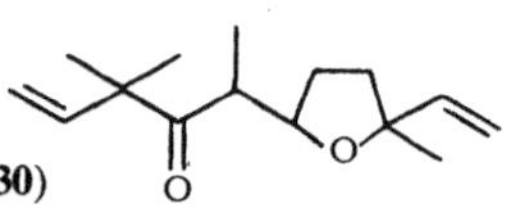

(231)

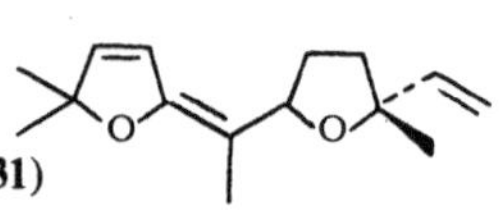

(232)

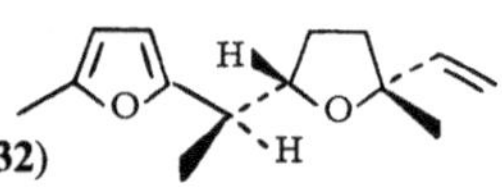

(233)

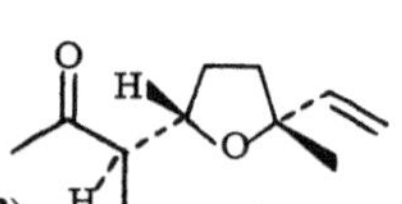

(234)

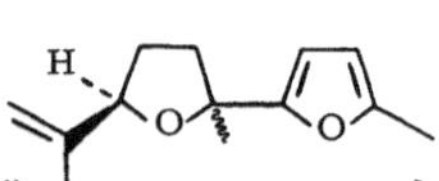

Davanone **(229)**, the main component of davana oil, and its isomer artemone **(230)**, when thoroughly purified, turned out to be odorless (*653*). The compounds which contribute to the characteristic odor of the natural oil include davana ether **(231)** (*650, 653*), the *cis,threo*-isomer of the nor-sesquiterpene davana furan **(232)** (*649*) and a lower homologue of davanone (**233**) (*652*). These compounds can be used for flavoring tobacco, and (**232**) is claimed to have a fresh green fruit-like note generally reminiscent of black-currants or limonene (*647*). Finally, we should mention the diastereoisomeric diethers (**234**) and their dihydro derivatives (isopropyl in place of isopropenyl), these four compounds having been found in Greek tobacco (*15*).

5. Norsesquiterpenoids

Due to their low vapor pressure, oxygen-containing sesquiterpene derivatives are in many cases beyond olfactive perceptibility. Since reports about synergistic effects as well as influences on the gustative receptor system are lacking, the flavor value of these compounds has to be considered as relatively small. The situation is different for the nor-compounds containing 1 to 3 carbon atoms less which occur as metabolites of normal sesquiterpenes and are generally characterized by a strong odor.

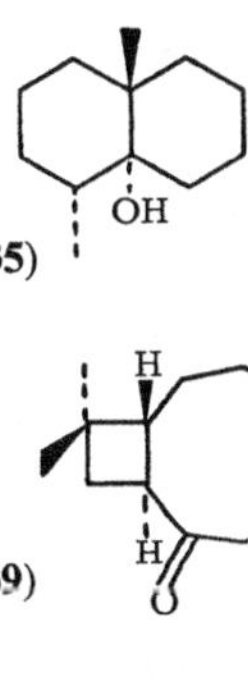
(**235**)

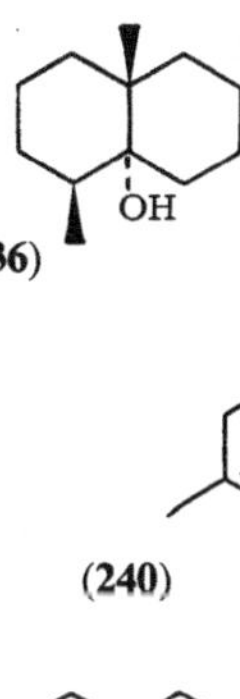
(**236**)

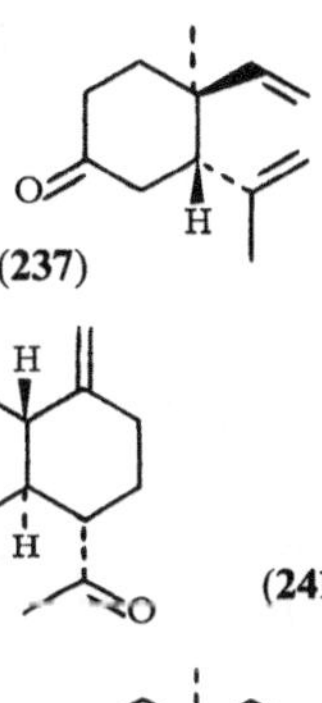
(**237**)

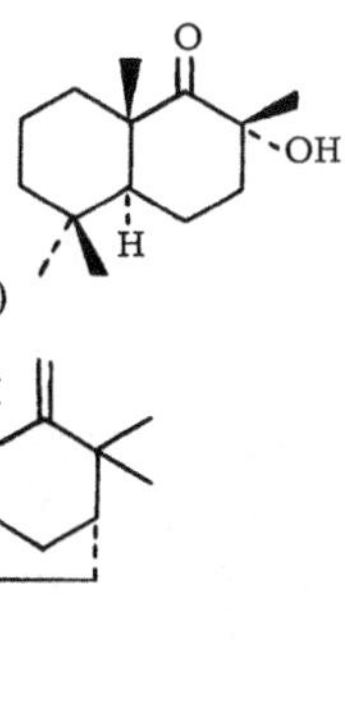
(**238**)

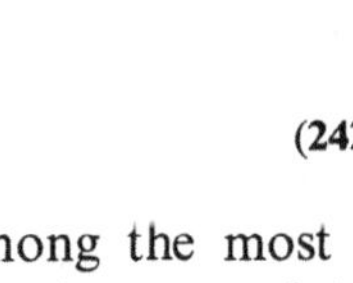
(**239**)

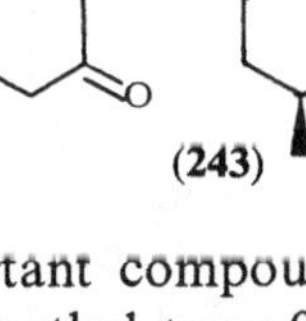
(**240**)

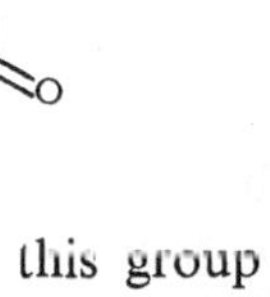
(**241**)

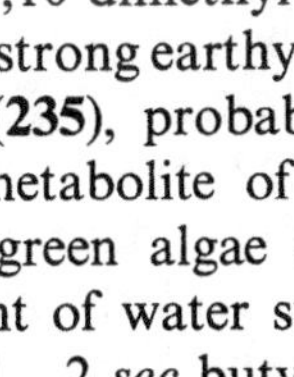
(**242**)

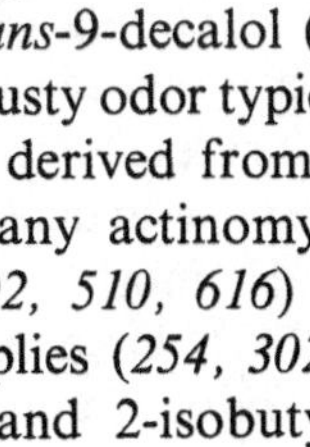
(**243**)

Among the most important compounds of this group are the two diastereoisomers of 1,10-dimethyl-*trans*-9-decalol (**235**) and (**236**) called geosmin which emit a strong earthy-musty odor typical of freshly ploughed soil (*180*). Geosmin (**235**), probably derived from a eudesmane sesquiterpene (*361*), is a metabolite of many actinomycetes (*182, 303, 373*) and of several blue-green algae (*302, 510, 616*) and is known as an off-flavor contaminant of water supplies (*254, 302, 303, 510*). Together with the 2-isopropyl-, 2-*sec*-butyl, and 2-isobutyl-3-methoxypyrazines (*403*), (**235**) is considered to make an important contribution to the aroma of raw beetroot (*7, 403*). Furthermore, geosmin was found in whiskey (*91*), in off-flavored dry white navy beans (*70*), and finally in ordinary garden soil (*69*).

The cedar wood-like amber note in the flavors of Turkish (*556*) and Greek tobacco (*2*) emanates from (+)-8α-hydroxy-11-nordriman-9-one (**238**). The corresponding 8β-compound was described as being odorless (*455*). The two diastereoisomers constitute essential examples supporting the "triaxial rule of odor sensation" (*455*).

A warm woody, amber-like note of relatively low strength is also said to be characteristic of kobusone (**239**) (*325*). This epoxyketone was one of the substances found in verbena oil (*280*).

Geijerone (**237**), which was found in the fruit oil of *Juniperus communis* L., is probably a metabolite of an elemene (*643*). The biogenetic formation of (−)-geijerone (**237**) can be simulated by dye-sensitized photooxygenation of (+)-γ-elemene and subsequent fragmentation of the corresponding allyl hydroperoxides (*460*). Nor-compounds (**240**) (*663*) and (**241**) (*664*) as well as the two C-12 derivatives (**242**) and (**243**) (*363*) were isolated from vetiver oil which is an important perfumery material.

Finally, the norisoprenoid cyclic ethers (**232**) to (**234**), discussed in the preceding chapter should be mentioned.

6. Metabolites From Higher Terpene Precursors

A class of unusually potent isoprenoid aroma components comprises substances which are formed by biodegradation of certain odor- and tasteless high molecular weight terpenoid compounds occurring in foodstuffs, such as carotenoids, diterpenes and polyterpenoids of different kinds. Nor-isoprenoid breakdown products are found as a rule in certain vegetable and fruit varieties as well as, in remarkable quantities, in tea and tobacco. Of the 269 aroma constituents identified in black tea (*525*) 10% of those having 9 to 13 carbon atoms can safely be considered as being derived from carotenoid compounds (*527*) from the same source. This hypothesis is indirectly supported by the loss of carotenoids from tea leaf tissues during the manufacture of black tea (*526, 656*). Among the 200 nor-isoprenoids (*3*) found in tobacco leaf condensates 83 are assumed to be carotenoid metabolites (*108*). The explosive development which has taken place in this field during the last ten years is illustrated by two review articles (*142, 590*).

The nor-isoprenoids most frequently found in flavors, α- and β-ionone (**244**) and (**245**), have a particularly high aroma value due to their characteristic odor qualities and their low threshold concentration (Table 3). Though both ketones had, for a long time, been

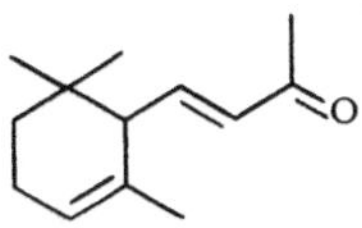

(244)

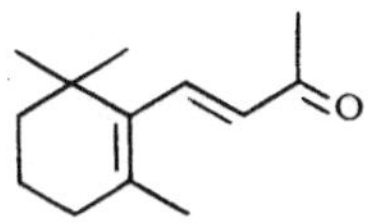

(245)

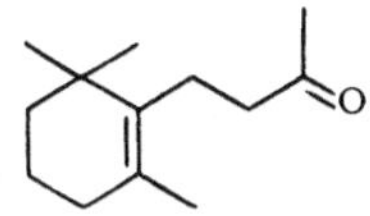

(246)

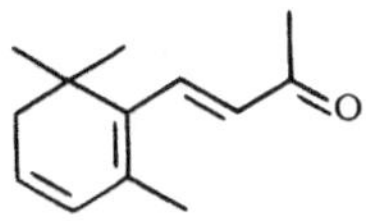

(247)

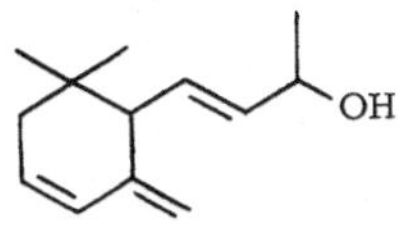

(248)

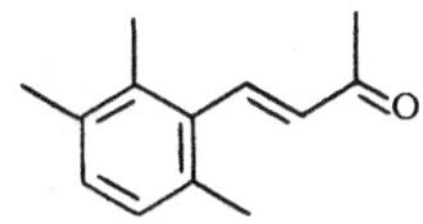

(249)

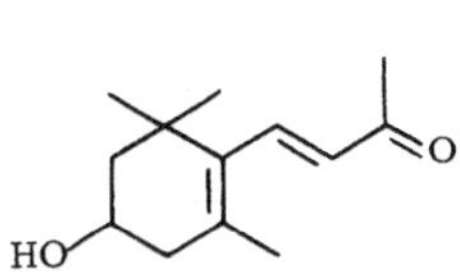

(250)

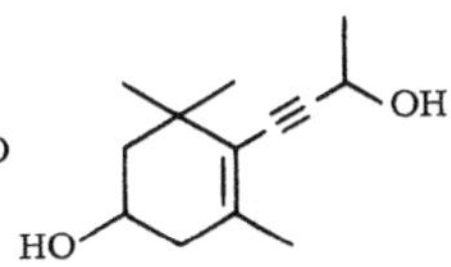

(251)

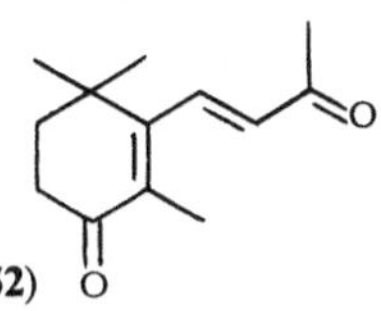

(252)

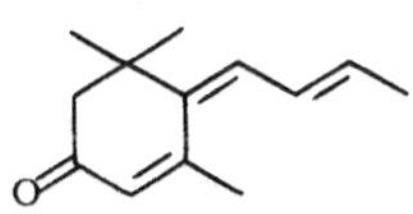

(253)

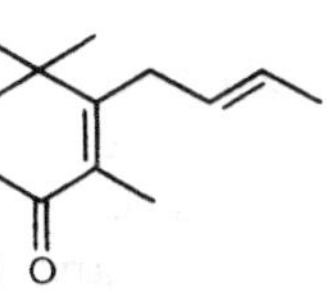

(254)

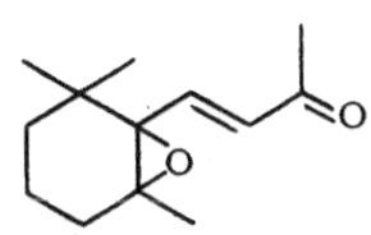

(255)

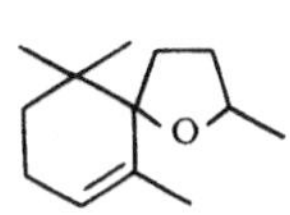

(256)

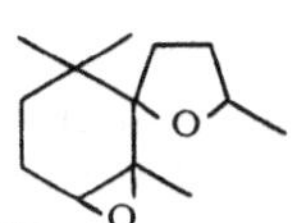

(257)

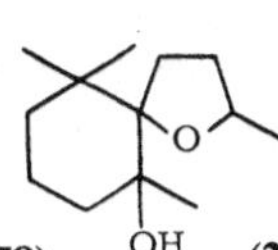

(258)

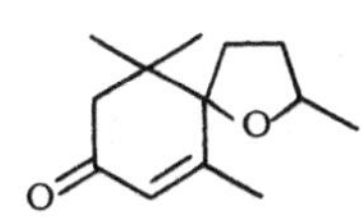

(259)

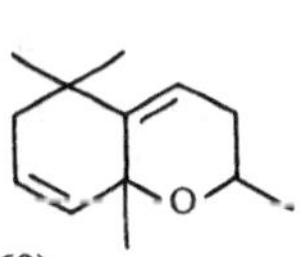

(260)

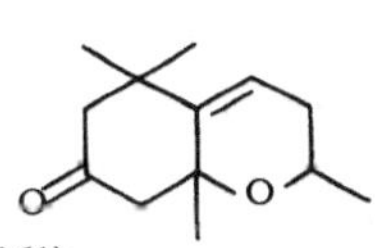

(261)

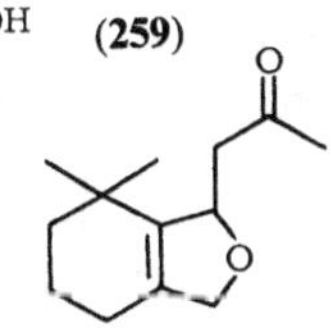

(262)

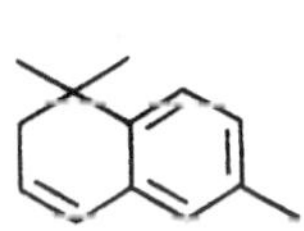

(263)

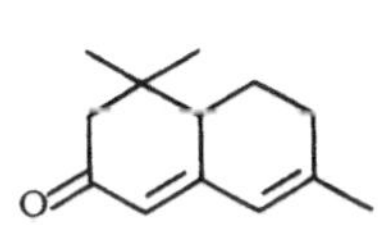

(264)

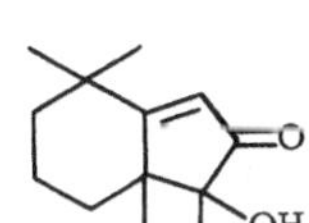

(265)

used empirically in compounding, especially of fruit flavors (*418*), it was only in 1947 that β-ionone (**245**) was first discovered in raspberry flavor (*418*). In addition to α-ionone (**244**) (*713*), derivatives of β-ionone (**245**) such as its mono-epoxide (**255**) and its dihydro derivative (**246**) were later found in the same flavor (*708*). Both ketones (**244**) and (**245**) are assumed to contribute to the flavor of tobacco (*556, 110*), black tea (*397, 722*), yellow passion fruit (*711*) and blackberry (*486*). They were also detected in distilled alcoholic beverages such as various brands of whiskey, grape brandy and grain fusel oils (*332*). In the essential oils of the blossoms of *Osmanthus fragrans* Lour. used for flavoring Chinese tea the three ionones (**244**), (**245**) and (**246**) account for more than 50%, β-ionone (**245**) and its dihydro derivative (**246**) constituting the main components present in proportions of 33.8% and 15.7%, respectively (*576*). β-Ionone (**244**) is present in orange oil (*374*) as well as in the essential oil of *Ligusticum elatum* (*621*) and at a level of 1% in celery oil (*703*). When present at abnormally high concentrations, the ionones (**244**), (**245**) and (**255**) are perceived as off-flavors, for instance in carrots (*20, 327*).

The first observation relating to the formation of ionone from carotene under conditions of autoxidation was made as early as 1910 (*701*). Later on this observation was confirmed experimentally (*259*). The action of $^1O_2$ on β-carotene also results in the formation of β-ionone (**245**) and other breakdown products (*259, 395*). Finally it was shown that (**245**) can also be formed enzymatically from β-carotene (*201, 526, 657*). The last comprehensive report on the chemistry of the ionones was published in 1951 (*536*), apart from a full list of publications each year in the Specialist Report on Terpenoids and Steroids (*83*).

Ionone compounds are found in flavors in virtually all oxidation states. Among them are the dehydro derivatives (**247**) (*111*) and (**248**) (*114*), (**250**) (*169, 170*), which is presumably their common precursor, as well as the aromatic analog (**249**) (*347*) having the terminal group of φ,φ-carotene. All of these compounds have been identified in tobacco. Bifunctional ionone derivatives, such as (**252**), have an odor reminiscent of that of the corresponding parent compounds, but they produce a substantially weaker sensory impression (*249*). Nevertheless, these compounds, when considered under the conditions of their practical utilization in flavors, undoubtedly contribute to the overall impression of a flavor (*502, 556*). The three bifunctional ionone derivatives (**250**) to (**252**) have excellent aroma qualities and are useful for enhancing the flavor quality of cigarettes (*169*). Zeaxanthin is assumed to be the precursor of (**250**) (*169*). The four stereoisomers of (**253**), which were found in various tobacco brands (*5, 111, 502, 556*), not only constitute 10% of the Burley tobacco condensate, but are also considered to be key flavoring components (*117*). The ketone (**254**) was recently isolated from tobacco flavor (*108*) and passion fruit flavor (*706*). The tetrahydro derivative of (**250**), the aromatic compound (**249**) (*479, 482*) and the bicyclic ketone (**264**) (*481*) were isolated for the first time from mares' urine. Diketone (**252**) was found in rabbit urine (*480*) and, together with the smaller carotenoid fragments (**275**) and (**281**), in the anal gland secretion of the red fox (*13*). 4-Oxo-β-ionone (**252**) was shown to be a product of the photooxygenation of α-ionone (**244**) (*251*).

The four spiro ethers of the ionone series (**256**) to (**259**) are characteristic aroma constituents of black tea. The first compound discovered in it was theaspirone (**259**) which is present in a proportion of 15 mg per ton of tea leaves (*252*). This compound is reported to have a camphor-like odor (*253*) and, according to other authors, a peach-like odor (*618*). Despite the faint aroma of (**259**), it is reported to develop a fine, mild tea aroma when mixed with dihydroactinidiolide (**275**) (*253*), although this observation has been contested (*527*). According to others authors theaspirone (**259**) constitutes a flavor and aroma improver for black tea (*521*) and, furthermore, enhances the flavor of foodstuffs in general (*223*). Theaspiran (**256**) was found as a naturally occurring substance for the first time in raspberry (*708*), then, together with ketone (**259**), in yellow passion fruit (*711*) and finally in tea itself (*491*). The use of the diastereoisomers of (**256**), (**257**) and (**258**), which have particularly interesting organoleptic properties, is described in the patent literature (*492*). The discovery of the spirocyclic ethers (**256**) to (**259**) had a highly stimulating effect on the imagination of chemists specialized in natural products insofar as it gave rise to an important series of synthetic developments (*37, 224, 253, 414, 529, 693*). The diastereoisomeric edulans (**260**) (*9, 695*) as well as two *cis*-annellated dihydro compounds (*483*) were isolated from passion fruit. The carbonyl derivative (**261**) (*111, 554*) appears to be an important tobacco component and is reported to have an oriental tobacco-type note (*556*). The ketone (**262**) is a new natural substance isolated from passion fruit (*406*).

The high reactivity of ionone derivatives can explain the formation of the bicyclic compounds (**263**) (*305*), (**264**) (*502*) and (**265**) (*111, 502*). The odor of ketone (**264**) is described as pleasant, green and woody,

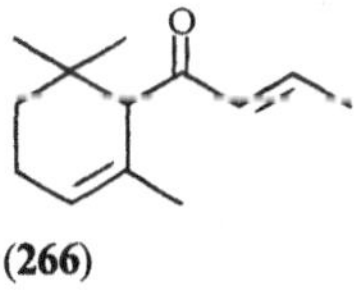
(**266**)

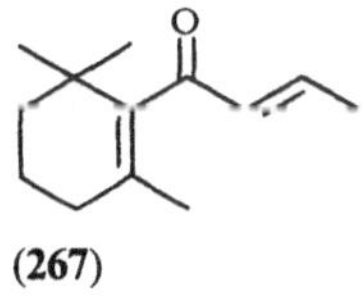
(**267**)

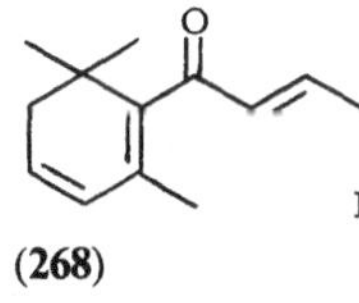
(**268**)

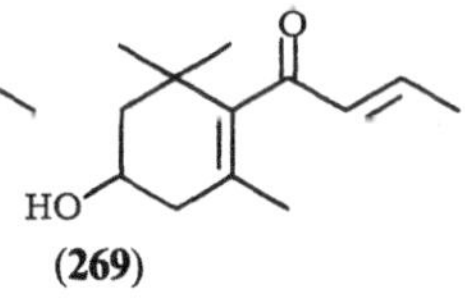

(**269**)

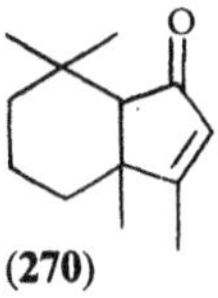
(**270**)

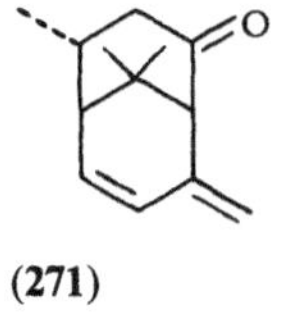
(**271**)

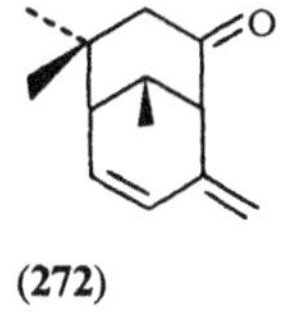
(**272**)

whereas the hydroxy ketone is said to develop a sweet flavor note in tobacco (*502*). 1,2-Dihydro-1,1,6-trimethyl-naphthalene (**263**), which apparently is formed from β-ionone epoxide (**255**) (*592*) or 3-oxo-β-ionone [the ketone corresponding to (**250**)] (*102*), was also detected in peach (*299*) and strawberry oils (*597*) as well as in the "bouquet" of Italian red wine (*40*). Ketone (**264**) is a flavor component of passion fruit (*711*).

One of the most important discoveries recently made in the field of carotenoid aroma compounds relates to the class of damascones. Seven naturally occurring members (**266**) to (**272**) of this class are known today. The discovery of these new compounds started with the isolation of damascenone (**268**) from Bulgarian rose oil in 1970 (*120*). Later, the same compound was detected in various tobacco brands (*5, 110, 306*), tea (*491*), raspberry oil (*708*), cooked apples (*442*), various grape and wine varieties (*546, 549*), beer (*598*), coffee oil (*179*), buchu leaf oil (*283*) and Roman camomile (*638*). 3-Hydroxy-β-damascone (**269**) was shown to be an important aroma component of tobacco (*108, 169, 285*) in which it develops a green, rose-floral and sweet Burley-like note (*556*). β-Damascone (**267**) was found in Burley tobacco (*110*) and rose oil (*108*), whereas α-damascone (**266**), accompanied by β-compound (**267**) and damascenone (**268**), was exclusively observed in tea oil (*491*). Among the known cyclization products (**270**) to (**272**), the last two compounds were found in tobacco (*119*), while ketone (**270**) was detected in tea (*491*).

The biosynthesis of the damascones, like that of the ionones, is related to carotene metabolism (*260, 458*). Grasshopper ketone, which contains the allenic end group of neoxanthin, is considered to be a central intermediate for the formation of β-damascenone (**268**) *in vivo*. A biosynthetic-type synthesis starting from a corresponding deoxy derivative analogously leads to β-damascone (**267**). 3-Hydroxy-β-damascone (**269**) is formed in 75% yield together with 20% of β-damascone (**267**) when 3,6-dihydroxy-7,8-dehydro-ionol is treated with 30% aqueous sulfuric acid, whereby 3-hydroxy-7,8-dehydro-β-ionol (**251**) constitutes a logical intermediate in the reaction pathway (*458*). The biomimetic relevance of this conversion is further supported by the discovery of the acetylene compound (**251**) in Burley tobacco (*168*). It would be beyond the scope of this article to discuss in detail the flood of scientific papers and patent literature dealing with this subject.

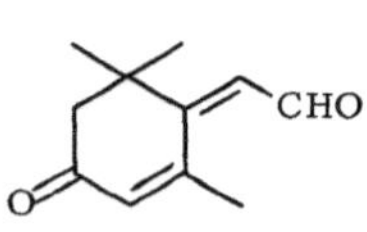

(**273**)

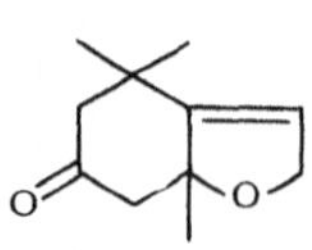

(**274**)

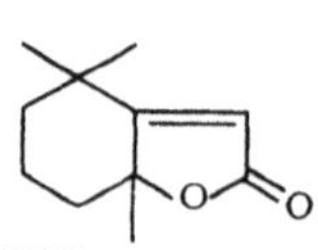

(**275**)

Among the next lower carotenoid fragments the C-11 derivatives (**273**) to (**275**) are worth mentioning. The compounds (**273**) and (**274**) demonstrate the wide structural variety of tobacco metabolites (*502*). Although by itself (**273**) seems to be odorless it adds body and enhances the flavor of Burley tobacco (*502*); (**274**) is described as a pleasant, mild aroma component (*502*). Dihydroactinidiolide (**275**) is present in various tobacco brands (*23, 110, 286*), black tea (*55, 250, 252*), tomato (*668*), oil of cassia (*115, 121*), and also occurs as a physiologically active component in *Actinidia polygama* which exerts an attracting effect on *Felidae* (*517*). It is also a component of the volatiles from the anal gland secretion of the red fox (*13*). This lactone (**275**) exerts a slightly cooling effect on the mucous membranes of the mouth cavity (*556*); it is presumably an enzymatic oxidation product of β-carotene or its metabolites and can be prepared by biomimetic processes. It is formed directly from β-carotene by dye-sensitized photooxygenation in an alkaline medium (*259*), or by peracid oxidation of β-ionone (**245**) (*215*), photooxygenation of β-ionone (**245**) (*395*) and *trans*-β-ionol (*258, 259*), or treatment of *cis*-β-ionol with Collin's reagent (*681*).

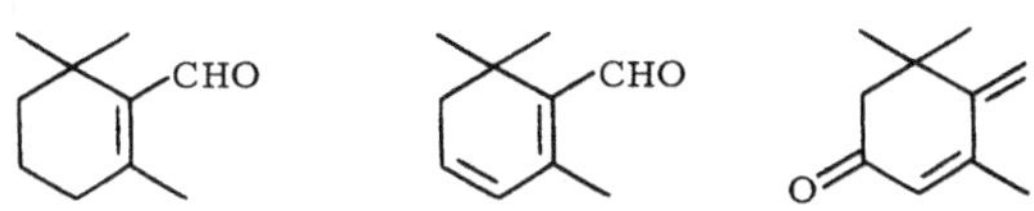

(**276**) (**277**) (**278**)

Although the three C-10 aroma components (**276**) to (**278**) are monoterpenoids from a formal point of view, they should probably be considered as degradation products of corresponding carotenoid compounds. At any rate, this assumption was made for the formation of safranal (**277**) (*321*), the main component of saffron oil (*12, 727*). The ketone (**278**) has only been detected in Burley tobacco (*111, 502*). β-Cyclocitral (**276**) and safranal (**277**) were also shown to be present in tea (*491*).

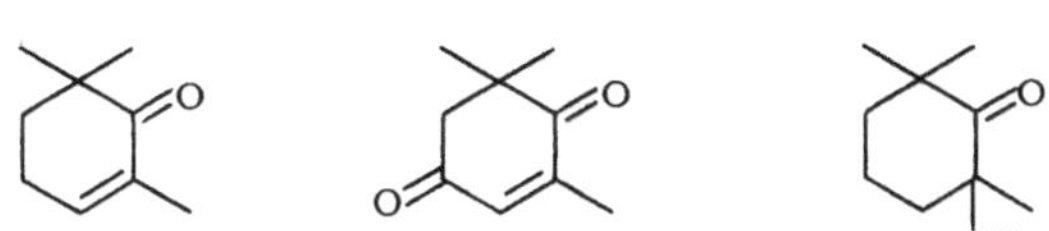

(**279**) (**280**) (**281**)

The smallest monocyclic aroma compounds of the carotenoid series are the C-9 fragments (**279**) to (**281**). They occur in various oxidation states in Burley tobacco (*111, 502*) and tea (*397, 491*). 2,2,6-Trimethylcyclohex-2-ene-1,4-dione (**280**), which also occurs in saffron (*727*), produces an odor impression similar to that of dried leaves or straw and is suitable, at levels of 20 to 200 ppm, for improving the flavor of infusions and fermented drinks (*553*).

Although the number of aroma compounds derived from acyclic carotenoids is much inferior to that of the mono- and bicyclic compounds, some of them can also be considered as breakdown products from genuine mono-, sesqui- and diterpenoids. The importance of the aliphatic isoprenoids (**282**) to (**291**) in the formation of total flavors of certain foodstuffs is not less than that of the cyclic compounds. the three methyl ketones (**282**), (**287**) and (**290**) which are related to the main tomato pigment lycopene were observed in tomato flavor (*75*). The hexahydro derivative (**291**) from coffee (*596*), jasmine oil (*122*) and green tea (*438*) is perceived as flowery and warm and can be considered as an oxidative biodegradation product of phytol and phytadiene. 6-Methyl-3,5-heptadien-2-one (**283**), with a grassy and cinnamon-like aroma (*438*) [detected in tomato (*668*), the essential oil of *Hamamelis* leaves (*383*), Ceylon tea (*722*) and passion fruit (*711*)], and pseudo-ionone (**288**) [also isolated from passion fruit (*711*)] are believed to be formed from two different dehydrolycopenes. Compounds other than carotenoids, such as solanesol or squalene, can also be considered

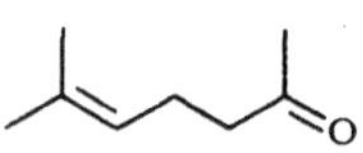
(**282**)

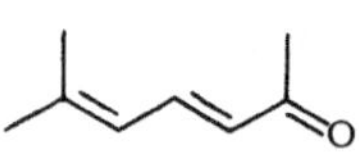
(**283**)

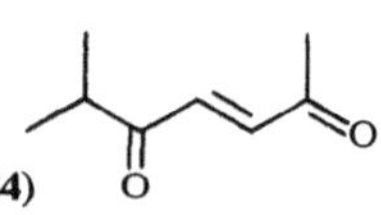
(**284**)

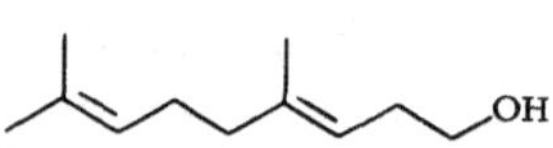
(**285**)

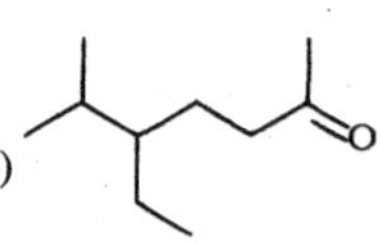
(**286**)

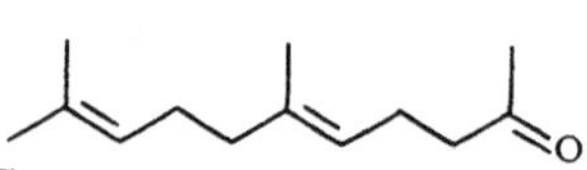
(**287**)

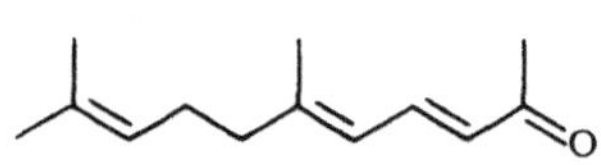
(**288**)

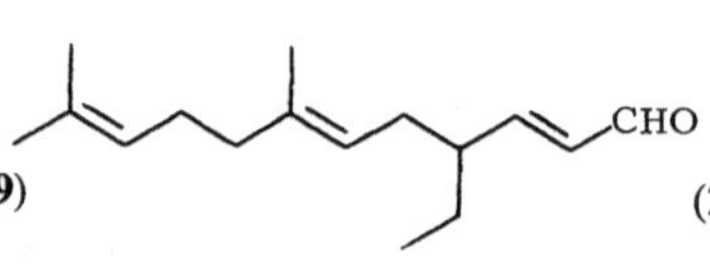
(**289**)

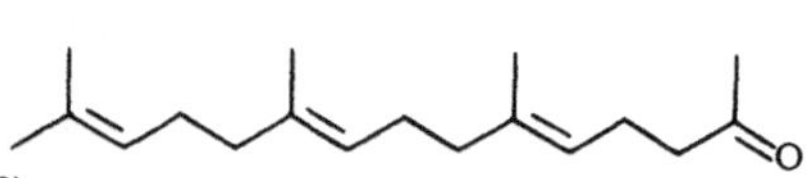
(**290**)

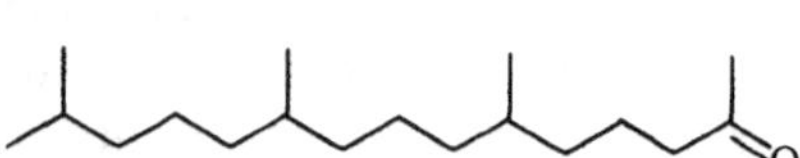
(**291**)

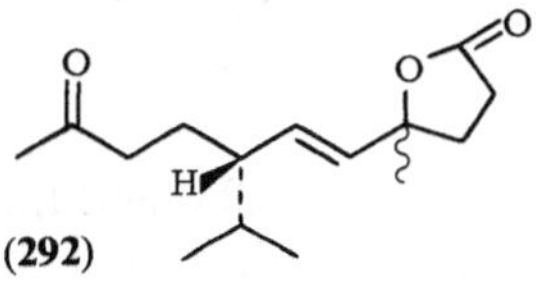
(292)

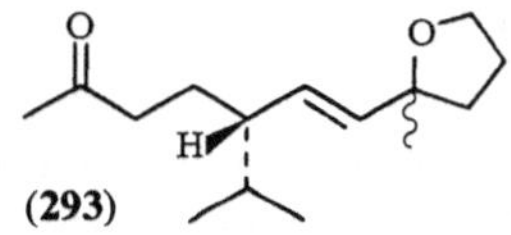
(293)

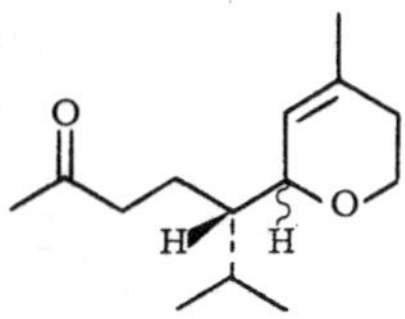
(294)

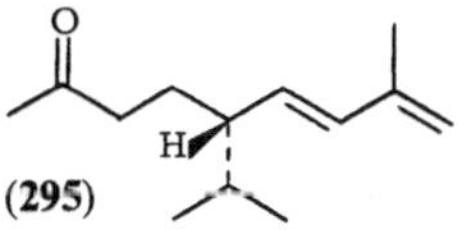
(295)

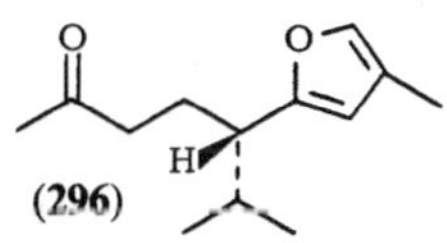
(296)

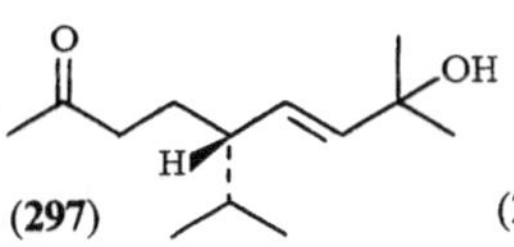
(297)

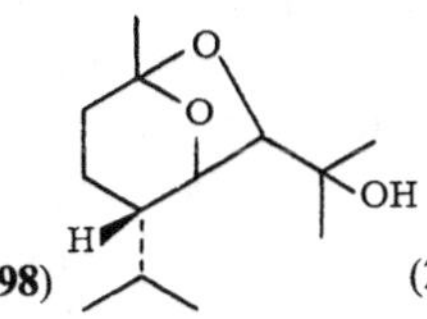
(298)

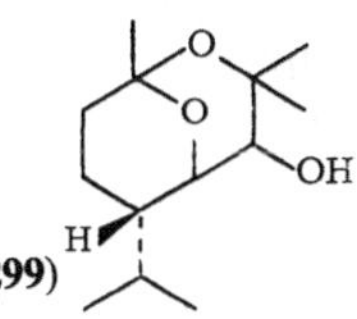
(299)

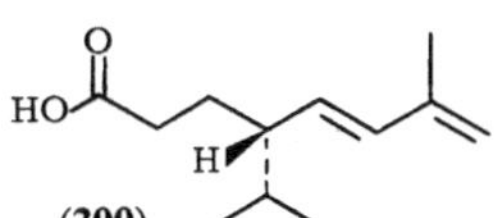
(300)

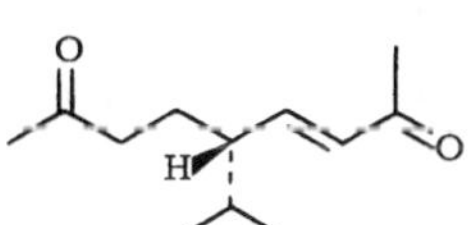
(301)

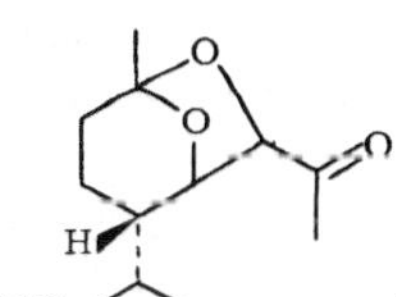
(302)

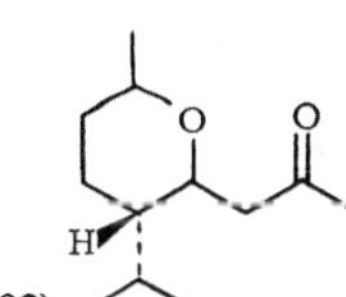
(303)

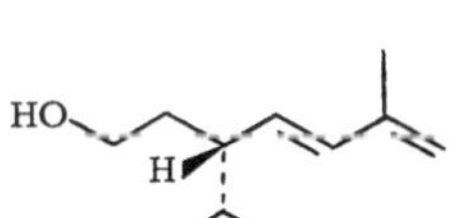
(304)

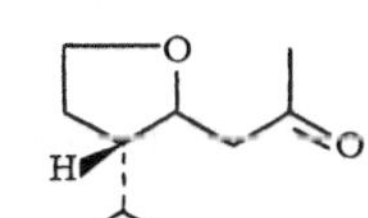
(305)

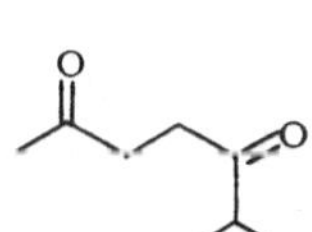
(306)

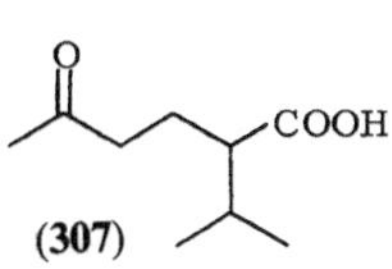
(307)

as precursors of isoprenoid acyclic aroma compounds. The development in this field has begun with two highly promising investigations (*289, 723*).

The two ethyl-homologous carbonyl compounds (**286**) and (**289**) have so far only been found in tea oil (*491*). The possibility that 5-ethyl-6-methylheptan-2-one (**286**) originates from the metabolism of a monocyclic monoterpene cannot be excluded (*491*), although this ketone seems more likely to be formed by oxidative biodegradation of the sitosterol side chain. Thus, for instance, all the aroma compounds (**282**) to (**291**) listed here except (**286**) and (**289**) were detected in tobacco oil (*111, 142, 306, 463, 590*).

In addition to the acyclic compounds, corresponding cyclic compounds are also invariably found, especially in flavors formed by fermentation processes.

A class of aroma compounds (**292**) to (**307**) having considerable organoleptic importance, which were found exclusively in tobacco condensates until now and which have an irregular terpene skeleton, must be considered as biodegradation products of thunberganoids (*1, 3, 4, 6*). Indeed, this type of macrocyclic diterpenoid has been found in no flavor-forming product other than in tobacco (*503*). Of the 309 volatiles of Burley tobacco, 43 are reported to originate from metabolites of thunberganoids (*114*). A selection of these fragrant substances is described below. The C-15 breakdown aroma compounds include lactone (**292**) (*3, 118*) and the monocyclic keto ether (**293**) (*118*). Dihydropyran derivative (**294**) contains one carbon atom less (*114*). All C-13 nor-thunberganes are derived from (+)-(*R*)-solanone (**295**) (*4*) ehich is present in Burley tobacco at a level of 0.0036% and which is reported to have a faint aroma reminiscent of carrots (*274*). Solanofuran (**296**), which displays interesting organoleptic properties, was prepared by a biogenetic-type reaction *via* the dye-sensitized photooxygenation of solanone (**295**) (*112*). The occurrence of "solanone hydrate" (**297**) and of its corresponding α-hydroxy-epoxide among the volatiles of Burley tobacco provides a ready explanation for the formation of the two isolated bicyclic internal acetals (**298**) and (**299**) (*113*). The same relationship exists between the C-12 tobacco constituent nor-solandione which can be obtained by autoxidation of (**295**), and acetal (**302**) (*113*). The acetonyltetrahydropyran derivative (**303**), described as having a sweet, phenolic and woody odor (*502*), is chemically related to a mono-alcohol from (**301**). Finally, among the C-10 solanone-related constituents the primary alcohol (**304**) (*113*), the acetonyltetrahydrofuran derivative (**305**) and diketone (**306**) (*111*) might be mentioned.

The flavor of Turkish tobacco and other Oriental tobaccos is mainly derived from two classes of compounds, the low molecular weight fatty acids and the metabolites from bicyclic diterpenes (*556*). A series of acids related to solanone (**295**) are known, for instance the carboxylic acid (**300**) (*1*), its lower homolog (*86*) and the degradation product (**307**) (*174*).

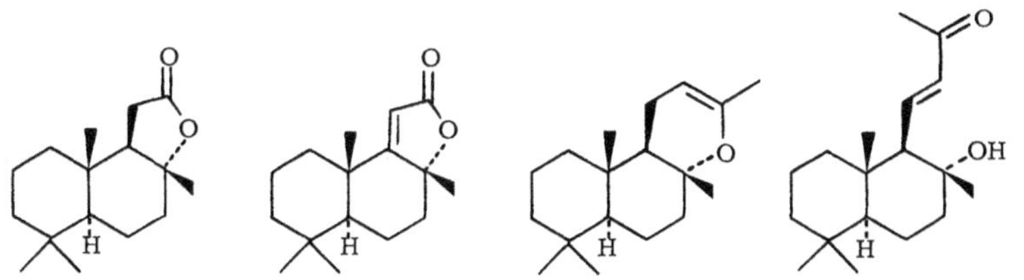

(**308**) (**309**) (**310**) (**311**)

Compounds (**308**) to (**311**), which are considered to be breakdown flavors from labdane diterpenoids, were also exclusively found in Oriental-type tobaccos. Nor-ambreinolide (**308**), obtained for the first time from Manila cigar-type tobacco (*284*), its hemi-acetal and dehydro derivative (**309**), sclareol oxide (**310**), the unsaturated hydroxyketone (**311**) (*15*) and the drimane metabolite (**238**) (*556*) "impart a pleasant cedar-like odor which is detectable to some extent in the cigarette package but is particularly detectable and pleasing in the mainstream and sidestream smokes when the cigarette is smoked" (*556*).

References

1. Aasen, A. J., T. Chuman, and C. R. Enzell: (6*S*)-6-Isopropyl-3-methyl-9-oxo-2*E*,4*E*-decadienoic Acid from Turkish *Nicotiana tabacum* L., Assignment of Absolute Configuration. Agric. Biol. Chem. **39**, 2085 (1975).

2. Aasen, A. J., and C. R. Enzell: Tobacco Chemistry. 30. The Absolute Configuration of 11-nor-8-Hydroxy-9-drimanone, a Constituent of Greek *Nicotiana tabacum* L. Acta Chem. Scand. B **28**, 1239 (1974).

3. Aasen, A. J., J. R. Hlubucek, and C. R. Enzell: Tobacco Chemistry. 29. (7*S*)-10-Oxo-4ξ-methyl-7-isopropyl-5*E*,undecen-4-olide, a New Thunbergan-Type nor-Isoprenoid Isolated from Greek *Nicotiana tabacum* L. Acta Chem. Scand. B **29**, 677 (1975).

4. Aasen, A. J., N. Junker, C. R. Enzell, J. E. Berg, and A. M. Pilotti: Tobacco Chemistry. 36. Absolute Configuration of Tobacco Thunberganoids. Tetrahedron Letters **1975**, 2607.

5. Aasen, A. J., B. Kimland, S.-O. Almqvist, and C. R. Enzell: Tobacco Chemistry. 15. New Tobacco Constituents — the Structures of Five Isomeric Megastigmatrienones. Acta Chem. Scand. **26**, 2573 (1972).

6. Aasen, A. J., T. Nishida, C. R. Enzell, and M. Devreux: Tobacco Chemistry. 37. The Absolute Configuration of Prenylsolanone. (9*S*)-6,12-Dimethyl-9-isopropyltrideca-5*E*,10*E*,12-trien-2-one, a Northunberganoid of *Nicotiana tabacum* L. Acta Chem. Scand. B **30**, 178 (1976).

7. ACREE, T. E., C. Y. LEE, R. M. BUTTS, and J. BARNARD: Geosmin, the Earthy Component of Table Beet Odor. J. Agric. Food Chem. **24,** 430 (1976).

8. ADAMS, D. R., and S. P. BHATNAGAR: Analysis of the Volatile Constituents of the Essential Oil of Indian Linaloe (*Bursera delpechiana* Poisson). Int. Flavours Food Addit. **6,** 185 (1975); THOMAS, A. F., and W. RENOLD (Firmenich SA, Geneva, Private Communication) believe the substance to be a mixture of *cis*- and *trans*-2,6,6-trimethyl-2-vinyl-5-acetoxytetrahydropyrans.

9. ADAMS, D. R., S. P. BHATNAGAR, and R. C. COOKSON: Synthesis and Structure of Edulans I and II. J. Chem. Soc. Perkin I **1975,** 1736.

10. ADAMS, D. R., S. P. BHATNAGAR, R. C. COOKSON, and R. M. TUDDENHAM: Simple Synthesis of Atlantones, the Ocimenones, the Tagetones and Filifolone from Isoprene. J. Chem. Soc. Perkin I **1975,** 1741.

11. AGARWAL, S. G., R. K. THAPPA, V. N. VASHIST, and C. K. ATAL: Essential Oil of *Eucalyptus saligna* Raised in Jammu. Indian J. Chem. (B) **14 B,** 388 (1976); Chem. Abstr. **85,** 99 021x (1976).

12. AKHMEDOV, A. I., M. I. GORYAEV, SH. K. CHOGOVADZE, and A. D. DEMBITSKII: Substances Present in Essential Oils. 55. Essential Oil of *Crocus sativus*. Izv. Akad. Nauk Kaz. SSR, Ser. Khim. **22,** 56 (1972) (Russ.); Chem. Abstr. **78,** 33 775v (1973).

13. ALBONE, E. S.: Dihydroactinidiolide in the Supracaudal Scent Gland Secretion of the Red Fox. Nature **256,** 575 (1975).

14. ALLEN, R. R.: Volatile Flavour Constituents in Coconut Oil. Chem. and Ind. **1965,** 1560.

15. ALMQVIST, S.-O., A. J. AASEN, J. R. HLUBUCEK, B. KIMLAND, and C. R. ENZELL: Tobacco Chemistry. 23. Structure and Synthesis of Four New Norisoprenoid Furans from Greek *Nicotiana tabacum* L. Acta Chem. Scand. B **28,** 528 (1974).

16. AMOORE, J. E.: In: Molecular Basis of Odor (I. NEWTON KUGELMASS, ed.). Springfield: Ch. C Thomas. 1970.

17. ANJOU, K., and E. v. SYDOW: The Aroma of Cranberries. I. *Vaccinium vitis-idaea* L. Acta Chem. Scand. **21,** 945 (1967).

18. APPEL, H. H., C. J. W. BROOKS, and M. M. CAMPBELL: Comparative Studies of Caparrapi Oil: The Essential Oil of *Ocotea caparrapi* (Nates) Dugand. Perf. Essent. Oil Rec. **58,** 776 (1967).

19. ARCTANDER, S.: In: Perfume and Flavor Chemicals (Aroma Chemicals), Vol. II, Compounds No 3102. Publ. by the Author, Montclair, N. J. (U.S.A.). 1969.

20. AYERS, J. E., M. J. FISHWICK, D. G. LAND, and T. SWAIN: Off-Flavour of Dehydrated Carrot Stored in Oxygen. Nature **203,** 81 (1964).

21. BADENHOP, A. F., and W. F. WILKENS: Formation of 1-Octen-3-ol in Soybeans During Soaking. J. Amer. Oil Chem. **46,** 179 (1969).

22. BADINGS, H. T.: Cold-Storage Defects in Butter and Their Relation to the Autoxidation of Unsaturated Fatty Acids. Ned. Melk-Zuiveltijdschr. **24,** 147 (1970).

23. BAILEY, W. C., JR., A. K. BOSE, R. M. IKEDA, R. H. NEWMAN, H. V. SECOR, and C. VARSEL: The Isolation from Tobacco of 2-Hydroxy-2,6,6-trimethylcyclohexylideneacetic Acid. γ-Lactone and its Synthesis. J. Organ. Chem. (U.S.A.) **33,** 2819 (1968).

24. BALDRY, J., J. DOUGAN, W. S. MATTHEWS, J. NABNEY, G. R. PICKERING, and F. V. ROBINSON: Composition and Flavor of Nutmeg Oils. Int. Flavours Food Addit. **1976,** 28.

25. BANTHORPE, D. V., J. MANN, and K. W. TURNBULL: Terpene Biosynthesis. II. Biosynthesis of Thujane Derivatives in *Thuja, Tanacetum* and *Juniperus* Species. J. Chem. Soc. (C) (London) **1970,** 2689.

26. BASF Riechstoffkatalog. 1976.

27. BATES, R. B., M. J. ONORE, S. K. PAKNIKAR, and C. STEELINK: D- and L-Filifolones, Monoterpenoid Cyclobutanones with the Bicyclo[3,2,0]heptane Ring System. Chem. Commun. **1967,** 1037.

28. Bates, R. B., and S. K. Paknikar: Terpenoids IX. Biogenesis of Some Monoterpenoids not Derived from a Geranyl Precursor. Tetrahedron Letters **1965**, 1453.
29. Bates, R. B., and R. C. Slagel: Terpenoids VI. β-Bulnesene, α-Guaiene, β-Patchoulene and Guaioxide in Essential Oils. Chem. and Ind. **1962**, 1715.
30. Bayer, E.: Quality and Flavor by Gas Chromatography. J. Gas Chromatogr. **4**, 67 (1966).
31. Bayonove, C., H. Richard et R. Cordonnier: Isolement et Identification du Diméthyl-3,7-octatriène-1,5,7-ol-3 on *trans*-Hotriénol à Odeur de *Tilleul de l'Huile Essentielle de Muscat de Frontignan.* C. R. hebd. séances Acad. Sci. C **1976** 549.
32. Becker, J., und G. Ohloff: Eine neuartige Fragmentierung bicyclischer Enoläther. Verfahren zur Darstellung macrocyclischer Lactone. Helv. Chim. Acta **54**, 2889 (1971).
33. Bednarczyk, A. A., W. G. Galetto, and A. Kramer: *Cis*- and *trans*-β-Sesquiphellandrol. Two New Sesquiterpene Alcohols from Oil of Ginger (*Zingiber officinale*, Roscoe). J. Agric. Food Chem. **23**, 499 (1975).
34. Beets, M. G. J.: Odor and Molecular Structure. Olfactologia **1**, 77 (1968).
35. — The Molecular Parameters of Olfactory Response. Pharmac. Rev. **22**, 1 (1970).
36. — Olfactory Reponse and Molecular Structure. In: Handbook of Sensory Physiology (L. M. Beidler, ed.), Vol. IV, p. 257. Berlin-Heidelberg-New York: Springer. 1971.
37. Bellas, T. E., R. G. Brownlee, and R. M. Silverstein: Isolation, Tentative Identification, and Synthesis Studies of the Volatile Components of the Hairpencil Secretion of the Monarch Butterfly. Tetrahedron **30**, 2267 (1974).
38. Berry, R. E., C. J. Wagner, Jr., and M. G. Moshonas: Flavor Studies of Nootkatone in Grapefruit Juice. J. Food Sci. **32**, 75 (1967).
39. Bertele, E., und P. Schudel: Synthese von α- und β-Sinensal. Helv. Chim. Acta **50**, 2445 (1967).
40. Bertuccioli, M., and R. Viani: Red Wine Aroma: Identification of Headspace Constituents. J. Sci. Food Agric. **27**, 1035 (1976).
41. Bestmann, H. J., W. Stransky, O. Vostrowsky und P. Range: Synthese von 1-substituierten (*Z*)-9-Alkenen. Chem. Ber. **108**, 3582 (1975).
42. Bills, D. D., R. A. Scanlan, R. C. Lindsay, and L. A. Sather: Free Fatty Acids and the Flavor of Dairy Products. J. Dairy Sci. **52**, 1340 (1969).
43. Blumer, M.: Organische Verbindungen in der Natur: Die Grenze unseres Wissens. Angew. Chem. **87**, 527 (1975); ibid. internat. Edit. **14**, 507 (1975).
44. Boehm, E. E., V. Thaller, and M. C. Whiting: Synthetical Studies on Terpenoids. II. The Structure of "Tagetone". J. Chem. Soc. (London) **1963**, 2535.
45. Boelens, H.: Relation Between the Chemical Structure of Compounds and their Olfactive Properties. Cosmet. Perfum. **89**, 70 (1974); Chem. Abstr. **82**, 47 606k (1975).
46. Boelens, H., und J. Heydel: Chemische Konstitution und Geruch. Eine Studie über strukturelle Eigenschaften chemischer Verbindungen mit verschiedenen Geruchscharakteristiken. Chem. Ztg. **97**, 8 (1973); Chem. Abstr. **78**, 84 544b (1973).
47. Bohlmann, F., und H. Kapteyn: Isolierung von Dehydronerolisovalerat aus *Authemis montana* L. Tetrahedron Letters **1973**, 2065.
48. Bohlmann, F., und C. Zdero: Über neue Inhaltsstoffe aus *Pluchea odorata* Cass. Chem. Ber. **109**, 2653 (1976).
49. Bohlmann, F., C. Zdero und U. Faass: Über die Inhaltsstoffe von *Artemisia fragrans* Willd. Ber. dtsch. chem. Ges. **106**, 2904 (1973).
50. Boldingh, J., and R. J. Taylor: Trace Constituents of Butterfat. IV. Isolation and Identification of Unsaturated Aliphatic Lactones. Nature **194**, 909 (1962).
51. Bondarovich, H. A., A. S. Giammarino, J. A. Renner, F. W. Shephard, A. J. Shingler, and M. A. Gianturco: Some Aspects of the Chemistry of Tea. A Contri-

bution to the Knowledge of the Volatile Constituents. J. Agric. Food Chem. **15,** 36 (1967).

52. Bornowski, H.: Die Struktur des Lasiospermans, ein neuer Typ von Furanosesquiterpenen. Tetrahedron **27,** 4101 (1971).

53. Bowers, W. S., L. R. Nault, R. E. Webb, and S. R. Dutky: Aphid Alarm Pheromone. Isolation, Identification, Synthesis. Science **177,** 1121 (1972).

54. Bowers, W. S., T. Ohta, J. S. Cleere, and P. A. Marsella: Discovery of Insect Anti-Juvenile Hormones in Plants. Plants Yield a Potential Fourth-Generation Insecticide. Science **193,** 542 (1976).

55. Bricout, J., R. Viani, F. Müggler-Chavan, J. P. Marion, D. Reymond et R. H. Egli: Sur la Composition de l'Arôme de Thé Noir. II. Helv. Chim. Acta **50,** 1517 (1967).

56. Brieger, G.: A Convenient Preparation of *trans*-β-Farnesene. J. Org. Chem. (U.S.A.) **32,** 3720 (1967).

57. Brieskorn, C. H., and H. Brunner: Terpenoids of Various Origanum Oils. Planta Med. Suppl. **1967,** 96; Chem. Abstr. **68,** 19 565p (1968).

58. Brieskorn, C. H., und S. Dalferth: Über das Vorkommen von 2-Methyl-3-Methylen-Hepten-(5) im ätherischen Öl von *Salvia officinalis* L. Liebigs Ann. Chem. **676,** 171 (1964).

59. Brooks, Ch. J. W., and M. M. Campbell: Caparrapi Oxide, a Sesquiterpenoid from Caparrapi Oil. Phytochem. **8,** 215 (1969).

60. Brownlee, R. G., R. M. Silverstein, D. Müller-Schwartze, and A. G. Singer: Isolation, Identification, and Function of the Chief Component of the Male Tarsal Scent in Black-Tailed Deer. Nature **221,** 284 (1969).

61. Büchi, G., E. sz. Kováts, P. Enggist, and G. Uhde: Syntheses of Rosefuran and Dehydroelsholtzione. J. Organ. Chem. (U.S.A.) **33,** 1227 (1968).

62. Büchi, G., M. Schach v. Wittenau, and D. M. White: Terpenes. X. The Constitution of Maaliol. J. Amer. Chem. Soc. **81,** 1968 (1959).

63. Büchi, G., und H. Wüest: Eine Synthese des β-Sinensals. Helv. Chim. Acta **50,** 2440 (1967).

64. — — A Stereoselective Synthesis of a α-Sinensal. J. Amer. Chem. Soc. **96,** 7573 (1974).

65. Büchi, G., H. Wüest, G. Ohloff, and H. Strickler (Firmenich SA, Geneva): 4-Methyl-8-vinyl-4,7- and -4,8-nonadienals. U.S. Pat. 3.963.783 (15. 6. 1976).

66. Buttery, R. G.: Nona-2,4,6-trienal, an Unusual Component of Blended Dry Beans. J. Agric. Food Chem. **23,** 1003 (1975).

67. Buttery, R. G., D. R. Black, D. G. Guadagni, L. C. Ling, G. Connolly, and R. Teranishi: California Bay Oil. I. Constituents, Odor Properties. J. Agric. Food Chem. **22,** 773 (1974).

68. Buttery, R. G., D. R. Black, M. J. Lewis, and L. Ling: A Study of the Fate of Volatile Hop Constituents in Beer. J. Food Sci. **32,** 414 (1967).

69. Buttery, R. G., and J. A. Garibaldi: Geosmin and Methylisoborneol in Garden Soil. J. Agric. Food Chem. **24,** 1246 (1976).

70. Buttery, R. G., D. G. Guadagni, and L. C. Ling: Geosmin, a Musty Off-Flavor of Dry Beans. J. Agric. Food Chem. **24,** 419 (1976).

71. Buttery, R. G., R. E. Lundin, and L. Ling: Characterization of Some C_{15} Constituents of Hop Oil. J. Agric. Food Chem. **15,** 58 (1967).

72. Buttery, R. G., R. M. Seifert, D. G. Guadagni, D. R. Black, and L. C. Ling: Characterization of Some Volatile Constituents of Carrots. J. Agric. Food Chem. **16,** 1009 (1968).

73. Buttery, R. G., R. M. Seifert, D. G. Guadagni, and L. C. Ling: Characterization of Some Volatile Constituents of Bell Peppers. J. Agric. Food Chem. **17,** 1322 (1969).

74. ———— Characterization of Additional Volatile Components of Tomato. J. Agric. Food Chem. **19,** 524 (1971).
75. BUTTERY, R. G., R. M. SEIFERT, and L. C. LING: Volatile Tomato Components: Characterization of 6,10,14-Trimethylpentadec-5,9,13-trien-2-one. Chem. and Ind. **1969,** 238.
76. ——— Characterization of Some Volatile Potato Components. J. Agric. Food Chem. **18,** 538 (1970).
77. ——— Characterization of Some Volatile Constituents of Dry Red Beans. J. Agric. Food Chem. **23,** 516 (1975).
78. CARDILLO, G., M. ORENA, and S. SANDRI: Synthesis of Compounds Containing the Isoprene Unit. A Stereospecific Synthesis of Dehydronerol Isovolerate, a Compound of *Anthemis montana* L. Tetrahedron **32,** 107 (1976).
79. CAZENAVE, P., I. HORMANN, F. MÜGGLER-CHAVAN et R. VIANI: Sur la Composition de l'Arôme de Thé Noir. IV. Helv. Chim. Acta **57,** 206 (1974).
80. CAZENAVE, P., et I. HORMANN: Sur la Composition de l'Arôme de Thé Noir. V. Helv. Chim. Acta **57,** 209 (1974).
81. CHANDRA, V., M. L. SHARMA, R. M. RAINA, and A. SINGH: Untersuchung des Guajavenblätteröles der *Psidium guajava* L.-Varietät. Riechst. Aromen Körperpflegem. **1976,** 156.
82. CHEEMA, A. S., R. S. DHILLON, B. C. GUPTA, B. R. CHHABRA und P. S. KALSI: Chemische Untersuchung der von *Daucus carota* L. stammenden Terpenoide. Riechst. Aromen Körperpflegem. **25,** 138 (1975).
83. Chemical Society: Specialist Periodical Report on Terpenoids and Steroids (K. OVERTON, ed.), London.
84. CHRÉTIEN-BESSIERE, Y., L. PEYRON, L. BENÉZET et J. GARNERO: Sur un Nouvel, Non Isoprénique, $C_{10}H_{18}O$, Présent dans *Ormenis multicaulis.* Bull. soc. chim. France **1968,** 2018.
85. ———— Contribution à la Connaissance de la Constitution de l'Huile Essentielle de Camomille Romaine (*Anthemis nobilis* L.). Bull. soc. chim. France **1970,** 381.
86. CHUMAN, T., H. KANEKO, T. FUKUZUMI, and M. NOGUCHI: Isolation of Two Terpenoid Acids. 4-Isopropyl-7-methyl-5*E*, 7-oct-adienoic Acid and 3-Isopropyl-6-methyl-4*E*, 6-heptadienoic Acid from Turkish Tobacco. Agric. Biol. Chem. **38,** 2295 (1974).
87. Code of US-Federal Regulations, Title 21: Food and Drugs, § 3.65a, b, c. Washington: U.S. Government Printing Office. 1974.
88. — § 121.101g and 121.1164. 1976.
89. COLLINS, R. P., and A. F. HALIM: Characterization of the Major Aroma Constituent of the Fungus *Trichoderma viride.* J. Agric. Food Chem. **20,** 437 (1972).
90. CONNELL, D. W.: The Chemistry of the Essential Oil and Oleoresin of Ginger (*Zingiber officinale,* Roscoe). Flavour Ind. **1970,** 677.
91. CONNER, H. A., P. A. SHIPLEY, E. T. BLESSINGER, J. H. KAHN, and E. G. LAROE: Whiskey Composition: Identification and Possible Origin of Components in Aged and Unaged Whiskey. Abstr. papers, 158th Meeting Amer. chem. Soc. **1969,** MICR-27.
92. COOKSON, R. C., J. JUDEC, S. A. KNIGHT, and B. R. D. WHITEAR: The Photochemistry of Citral. Tetrahedron **19,** 1995 (1963).
93. COOKSON, R. C., and P. LOMBARDI: Synthesis of (±)-Caparrapi Oxide and (±)-8-Epicaparrapi Oxide. Gazz. chim. ital. **105,** 621 (1975); Chem. Abstr. **83,** 193 531 u (1975).
94. CREVELING, R. K., R. M. SILVERSTEIN, and W. G. JENNINGS: Volatile Components of Pineapple. J. Food Sci. **33,** 284 (1968).
95. CROMBIE, L.: Synthesis of (±)-Massoialactone. J. Chem. Soc. (London) **1955,** 2535.
96. CURTIS, R. F., D. G. LAND, N. M. GRIFFITHS, M. GEE, D. ROBINSON, J. L. PEEL,

C. DENNIS, and J. M. GEE: 2,3,4,6-Tetrachloroanisole Association with Musty Taint in Chickens and Microbiological Formation. Nature **235,** 223 (1972).

97. DAENIKER, H. U.: Lebensmittelaromen. Bedeutung, Herstellung und Anwendung. Neue Zürcher Zeitung **1975,** Nr. 29, 45.

98. DALY, J. W., F. C. GREEN, and R. H. EASTMAN: Sabinene Hydrate, a Constituent of American Peppermint Oil. J. Amer. Chem. Soc. **80,** 6330 (1958).

99. DAMODARAN, N. P., and S. DEV: New Humulene-Based Sesquiterpenoids. Tetrahedron Letters **1963,** 1941.

100. — — Studies in Sesquiterpenes. XXXVII. Sesquiterpenoids from the Essential Oil of *Zingiber zerumbet* Smith. Tetrahedron **24,** 4113 (1968).

101. — — Studies in Sesquiterpenes. XXXIX. Structure of Humulenols. Tetrahedron **24,** 4133 (1968).

102. DAVIS, D. L., K. L. STEVENS, and L. JURD: Chemistry of Tobacco Constituents. Oxidation of α-Ionone and the Acid-Catalyzed Rearrangement of 5-Keto-α-ionone. J. Agric. Food Chem. **24,** 187 (1976).

103. DAY, E. A.: Flavor Chemistry. Advanc. Chem. Ser. **56,** 94—120 (R. F. GOULD, ed.), Washington: Amer. Chem. Soc. Publ. 1966.

104. DAY, E. A., D. A. LILLARD, and M. W. MONTGOMERY: Autoxidation of Milk Lipids. III. Effect on Flavor of the Additive Interactions of Carbonyl Compounds at Subthreshold Concentrations. J. Dairy Sci. **46,** 291 (1963).

105. DEBRAUWERE, J., and M. VERZELE: Constituents of Peppers. IV. The Hydrocarbons of Pepper Essential Oil. J. Chrom. Sci. **14,** 296 (1976).

106. DEBRAUWERE, J., and M. VERZELE: Constituents of Peppers. VI. The Oxygenated Fraction of Pepper Essential Oil. Bull. soc. chim. Belges **84,** 167 (1975).

107. DELLE MONACHE, G., J. LEONCIO D'ALBUQUERQUE, F. DELLE MONACHE, G. B. MARINI BETTOLO, and G. M. NANO: α-Multijugenol, a New Sesquiterpene Alcohol with Caryophyllane Carbon Skeleton. Tetrahedron Letters **1971,** 659.

108. DEMOLE, E. (Firmenich SA, Geneva): Personal Communication.

109. — (Firmenich et Cie): 3,4,7-Trimethyl-2-oxa-1,6-dioxa-spiro(4,5)dec-3-ene. U.S. Pat. 3.880.888 (29. 4. 1975).

110. DEMOLE, E., et D. BERTHET: Identification de la Damascénone et de la β-Damascone dans le Tabac Burley. Helv. Chim. Acta **54,** 681 (1971).

111. — — A Chemical Study of Burley Tobacco Flavour (*Nicotiana tabacum* L.). I. Volatile to Medium-Volatile Constituents (b.p. ≦84°/0.001 Torr). Helv. Chim. Acta **55,** 1866 (1972).

112. DEMOLE, E., C. DEMOLE, and D. BERTHET: A Chemical Study of Burley Tobacco Flavour (*Nicotiana tabacum* L.). III. Structure Determination and Synthesis of 5-(4-Methyl-2-furyl)-6-methylheptan-2-one *("Solanofuran")* and of 3,4,7-Trimethyl-1,6-dioxa-spiro[4.5]dec-3-en-2-one *("Spiroxabovolide")*, Two New Flavour Components of Burley Tobacco. Helv. Chim. Acta **56,** 265 (1973).

113. DEMOLE, E., and C. DEMOLE: A Chemical Study of Burley Tobacco Flavour (*Nicotiana tabacum* L.). VII. Identification and Synthesis of Twelve Irregular Terpenoids Related to Solanone Including 7,8-Dioxabicyclo[3.2.1]octane and 4,9-Dioxabicyclo[3.3.1]nonane Derivatives Helv. Chim. Acta **58,** 1867 (1975).

114. DEMOLE, E., and P. DIETRICH: The Chemistry of Burley Tobacco Flavour. ACS National Meeting, New Orleans, March 1977.

115. DEMOLE, E., et P. ENGGIST: Synthèse des (±)-Loliolide, (±)-Actinidiolide, (±)-Dihydroactinidiolide et Dérivés à Partir d'Acide Homosafranique, Photooxygénation Sensibilisée de l'Homosafranate de Méthyle. Helv. Chim. Acta **51,** 481 (1968).

116. — — Synthèses dans la Série des Constituants Odorants Caractéristiques de l'Essence Absolue de Cassie (*Acacia farnesiana* Willd.): Les Acides Méthyl-3-décène-3-oïques *cis* et *trans,* Méthyl-3-décène-4-oïques *cis* et *trans,* et les Quatre Méthyl-3-décénols-1 Correspondants. Helv. Chim. Acta **52,** 933 (1969).

117. — — Novel Synthesis of 3,5,5-Trimethyl-4-(2-butenylidene)-cyclohex-2-en-1-one, a Major Constituent of Burley Tobacco Flavour. Helv. Chim. Acta **57,** 2087 (1974).

118. — — A Chemical Study of Burley Tobacco Flavour (*Nicotiana tabacum* L.) VI. Identification and Synthesis of Four Irregular Terpenoids Related to Solanone, Including a "Prenylsolanone". Helv. Chim. Acta **58,** 1602 (1975).

119. — — A Chemical Study of Virginia Tobacco Flavour (*Nicotiana tabacum* L.). I. Isolation and Synthesis of Two Bicyclodamascenones. Helv. Chim. Acta **59,** 1938 (1976).

120. DEMOLE, E., P. ENGGIST, U. SÄUBERLI, M. STOLL et E. sz. KOVÁTS: Structure et Synthèse de la Damascénone (Triméthyl-2,6,6-*trans*-crotonoyl-1-cyclohexadiène-1,3), Constituant Odorant de l'Essence de Rose Bulgare. Helv. Chim. Acta **53,** 541 (1970).

121. DEMOLE, E., P. ENGGIST et M. STOLL: Sur les Constituants Odorants de l'Essence Absolue de Cassie (*Acacia farnesiana* Willd.). Helv. Chim. Acta **52,** 24 (1969).

122. DEMOLE, E., B. WILLHALM et M. STOLL: Propriétés et Structure de la Cétolactone $C_{12}H_{16}O_3$ de l'Essence de Jasmin (*Jasminum grandiflorum* L.). Helv. Chim. Acta **47,** 1152 (1964).

123. DE VILLIERS, D. J. J., C. F. GARBERS, and R. N. LAURIE: Synthesis of Tagetenones and Their Occurrence in Oil of *Tagetes minuta.* Phytochem. **10,** 1359 (1971).

124. DEVON, T. K., and A. I. SCOTT: Handbook of Naturally Occurring Compounds. Vol. II. Terpenes. New York-London: Academic Press. 1972.

125. DICKERSON, J. P., D. L. ROBERTS, C. W. MILLER, R. A. LLOYD, and C. E. RIX: Flue-Cured Tobacco Flavour. II. Constituents Arising from Amino Acid-Sugar Reactions. Tobacco Sci. **20,** 71 (1976).

126. DIJKSTRA, F. Y., and T. O. WILKÉN: Studies on Mushroom Flavours. 1. Organoleptic Significance of Constituents of the Cultivated Mushroom, *Agaricus bisporus.* Z. Lebensm.-Unters. Forsch. **160,** 255 (1976); Chemorec. Abstr. **4,** 46 (1976).

127. DIMICK, P. S., S. PATTON, J. E. KINSELLA, and N. J. WALKER: The Prevalence of Aliphatic δ-Lactones or Their Precursors in Animal Fats. Lipids **1,** 387 (1966).

128. DIMICK, P. S., N. J. WALKER, and S. PATTON: Occurrence and Biochemical Origin of Aliphatic Lactones in Milk Fat — A Review. J. Agric. Food Chem. **17,** 649 (1969).

129. DO, J. Y., D. K. SALUNKHE, and L. E. OLSON: Isolation, Identification, and Comparison of the Volatile of Peach Fruit as Related to Harvest Maturity and Artificial Ripening. J. Food Sci. **34,** 618 (1969).

130. DOEBNER, O.: Über das flüchtige Öl der Vogelbeeren, die Parasorbinsäure und ihre Isomerie mit der Sorbinsäure. Ber. dtsch. chem. Ges. **27,** 344 (1894); Synthese der Sorbinsäure, ibid. **33,** 2140 (1900).

131. DRAVNIEKS, A.: A Building-Block Model for the Characterization of Odorant Molecules and Their Odors. Ann. New York Acad. Sci. **237,** 144 (1974).

132. DRAWERT, F.: Biochemical Formation of Aroma Components. Proc. int. Symp. Aroma Research, Zeist., p. 13. Wageningen: Pudoc. 1975.

133. DRAWERT, F., W. HEIMANN, R. EMBERGER und R. TRESSL: Enzymatische Bildung von Hexen-(2)-al-(1), Hexanal und deren Vorstufen. Liebigs Ann. Chem. **694,** 200 (1966).

134. DRAWERT, F., P. SCHREIER und W. SCHERER: Gaschromatographisch-massenspektrometrische Untersuchung flüchtiger Inhaltsstoffe des Weines. III. Säuren des Weinaromas. Z. Lebensm.-Unters. Forsch. **155,** 342 (1974).

135. DRAWERT, F., und R. TRESSL: Moderne physikalisch-chemische Methoden und ihre Anwendung auf Probleme der Biogenese von Aromastoffen. I. Moderne physikalisch-chemische Methoden in der Aromaforschung. Ernährungsumschau **17,** 392 (1970).

136. DRAWERT, F., R. TRESSL, W. HEIMANN, R. EMBERGER und M. SPECK: Über die Biogenese von Aromastoffen und Früchten. XV. Enzymatisch-oxidative Bildung von

C_6-Aldehyden und Alkoholen und deren Vorstufen bei Äpfeln und Trauben. Chem. Mikrobiol. Technol. Lebensm. **2,** 10 (1973).

137. DUBOVENKO, Z. V., J. N. GUSKOVA, and V. A. PENTEGOVA: Essential Oils of Different *Ruta* species. Khim. Prir. Soedin. **9,** 120 (1973); Chem. Abstr. **79,** 9747 (1973).

138. EASTMAN, R. H., J. E. STARR, R. S. MARTIN, and M. K. SAKATA: The Photolysis of Thujone. J. Organ. Chem. (U.S.A.) **28,** 2162 (1963).

139. EDWARDS, L. J., J. B. SIDDALL, L. L. DUNHAM, P. UDEN, and C. J. KISLOW: *trans*-β-Farnesene, Alarm Pheromone of the Green Peach Aphid, *Myzus persicae* (Sulzer). Nature **241,** 126 (1973).

140. EHRET, CH., et G. OURISSON: Oxydation du Gaïol par le Tétraacétate de Plomb: Structure du Gaïoxyde et de l'Épi-Gaïoxyde. Bull. soc. chim. France **1968,** 2629.

141. EHRET, C., and P. TEISSEIRE: Contribution to the Knowledge of Bulgarian Rose Oil. Presence of Neroloxide. Recherches **1974,** 287.

142. ENZELL, C. R., I. WAHLBERG, and A. J. AASEN: Isoprenoids and Alkaloids of Tobacco. In: Fortschritte der Chemie organischer Naturstoffe **34,** 1—79 (W. HERZ, H. GRISEBACH, G. W. KIRBY, ed.). Wien-New York: Springer. 1977.

143. EPSTEIN, W. W., and C. D. POULTER: A Survey of Some Irregular Monoterpenes and Their Biogenetic Analogies to Presqualene Alcohol. Review Article. Phytochemistry **12,** 737 (1973).

144. FAGERSON, I. S.: Thermal Degradation of Carbohydrates. J. Agric. Food Chem. **17,** 747 (1969).

145. FEENSTRA, W. H., and P. W. MEIJBOOM: 2-*trans*-Nonenal, the Hardened Flavor Present in Hydrogenated Peanut Oil. J. Amer. Oil Chem. Soc. **48,** 684 (1971).

146. FELIX, D., A. MELERA, J. SEIBL und E. sz. KOVÁTS: Zur Kenntnis ätherischer Öle. 2. Mitt. Die Struktur der sogenannten „Linalooloxide". Helv. Chim. Acta **46,** 1513 (1963).

147. FENAROLI's Handbook of Flavor Ingredients (T. E. FURIA and N. BELLANCA, eds.), 2nd ed., Vol. I. Cleveland, Ohio: CRC Press Inc. 1975.

148. FENAROLI's Handbook of Flavor Ingredients (T. E. FURIA and N. BELLANCA, eds.), 2nd ed., Vol. I, p. 55. Cleveland, Ohio: CRC Press Inc. 1975.

149. FENAROLI's Handbook of Flavor Ingredients (T. E. FURIA and N. BELLANCA, eds.), 2nd ed., Vol. I, p. 275. Cleveland, Ohio: CRC Press Inc. 1975.

150. FENAROLI's Handbook of Flavor Ingredients (T. E. FURIA and N. BELLANCA, eds.), 2nd ed., Vol. II. Cleveland, Ohio: CRC Press Inc. 1975.

151. FENAROLI's Handbook of Flavor Ingredients (T. E. FURIA and N. BELLANCA, eds.), 2nd ed., Vol. II, p. 242. Cleveland, Ohio: CRC Press Inc. 1975.

152. FIDDLER, W., R. C. DOERR, and A. E. WASSERMAN: Composition of an Ether-Soluble Fraction of a Liquid Smoke Solution. J. Agric. Food Chem. **18,** 310 (1970).

153. FLAMENT, I. (Firmenich SA, Geneva): Unpublished Results.

154. FLAMENT, I., B. WILLHALM et M. STOLL: Recherches sur les Arômes. 16. Sur l'Arôme du Cacao III. Helv. Chim. Acta **50,** 2233 (1967).

155. FLATH, R. A., D. R. BLACK, D. G. GUADAGNI, W. H. MCFADDEN, and T. H. SCHULTZ: Identification and Organoleptic Evaluation of Compounds in Delicious Apple Essence. J. Agric. Food Chem. **15,** 29 (1967).

156. FLATH, R. A., R. E. LUNDIN, and R. TERANISHI: The Structure of β-Sinensal. Tetrahedron Letters **1966,** 295.

157. FORNEY, F. W., and A. J. MARKOVETZ: Biology of Methyl Ketones. J. Lipid Res. **12,** 383 (1971).

158. FORREY, R. R., and R. A. FLATH: Volatile Components of *Prunus salicina* var. Santa Rosa. J. Agric. Food Chem. **22,** 496 (1974).

159. FORSS, D. A.: Origin of Flavors in Lipids. In: Symposium on Foods. The Chemistry and Physiology of Flavors. The Fourth in a Series of Symposia on Foods

Held at Oregon State University, p. 492. Westport, Connecticut: The Avi Publ. Comp., Inc. 1967.

160. — Flavor of Dairy Products: A Review of Recent Advances. J. Dairy Science **52,** 832 (1969).

161. — Flavors of Dairy Fats. A Review. J. Amer. Oil Chem. Soc. **48,** 702 (1971).

162. — Odor and Flavor Compounds from Lipids. In: Progress in the Chemistry of Fats and Other Lipids, Vol. XIII, Part 4, p. 177 (R. T. HOLMANN, ed.). Oxford: Pergamon Press. 1972.

163. — Odor and Flavor Compounds from Lipids. In: Progress in the Chemistry of Fats and Other Lipids, Vol. XIII, Part 4, listed in Table 6, p. 230 (R. T. HOLMANN, ed.). Oxford: Pergamon Press. 1972.

164. FORSS, D. A., E. A. DUNSTONE, E. H. RAMSHAW, and W. STARK: The Flavor of Cucumbers. J. Food Sci. **27,** 90 (1962).

165. FRANZEN, K. L., and J. E. KINSELLA: Physiocochemical Aspects of Food Flavouring. Chem. and Ind. **1975,** 505.

166. FREEMAN, G. G., R. J. WHENHAM, R. SELF, and J. EAGLES: Volatile Flavor Components of Parsley Leaves (*Petroselinum crispum* [Mill.] Nyman). J. Sci. Food Agric. **26,** 465 (1975).

167. FRIEMAN, L., and J. G. MILLER: Odor Incongruity and Chirality. Science **172,** 1044 (1971).

168. FUJIMORI, T., R. KASUGA, H. KANEKO, and M. NOGUCHI: A New Acetylenic Diol, 3-Hydroxy-7,8-Dehydro-β-ionol, from Burley *nicotiana tabacum.* Phytochemistry **14,** 2095 (1975).

169. FUJIMORI, T., R. KASUGA, H. MATSUSHITA, H. KANEKO, and M. NOGUCHI: Neutral Aroma Constituents in Burley Tobacco. Agric. Biol. Chem. **40,** 303 (1976).

170. FUJIMORI, T., R. KASUGA, M. NOGUCHI, and H. KANEKO: Isolation at *R*-(−)-3-Hydroxy-β-ionone from Burley Tobacco. Agric. Biol. Chem. **38,** 891 (1974).

171. FUJITA, S., and Y. FUJITA: Essential Oils of Plants from Various Territories. XXX. Components of the Essential Oils of *Cymbopogon goeringii.* Yakugaku Zasshi **92,** 1285 (1972); Chem. Abstr. **78,** 33 791x (1973).

172. FUJITA, Y., S. FUJITA, and Y. HAYAMA: Miscellaneous Contributions to the Essential Oils of the Plants from Various Territories. XXIV. Essential Oils of *Perilla acuta* and *P. acuta* var. *viridis.* Nippon Nogei Kagaku Kaishi **44,** 428 (1970); Chem. Abst. **74,** 57 198c (1971).

173. FUJITA, Y., S. FUJITA, and H. YOSHIKAWA: *trans-* and *cis*-Yabunikkeol, new Monoterpene Alcohols Isolated from the Essential Oils of *Cinnamomum japonicum.* Bull. Chem. Soc. Japan **43,** 1599 (1970).

174. FUKIZUMI, T., H. KANEKO, and H. TAKAHARA: Studies on the Chemical Constituents of Tobacco Leaves. III. Isolation of (−)-2-Isopropyl-5-oxahexanoic Acid from Turkish Tobacco Leaves and Absolute Configuration of Solanone. Agric. Biol. Chem. **31,** 607 (1967).

175. GALLIARD, T., and D. R. PHILLIPS: The Enzymatic Cleavage of Linoleic Acid to C_9 Carbonyl Fragments in Extracts of Cucumber *(Cucumis sativus)* Fruit and the Possible Role of Lipoxygenase. Biochem. Biophys. Acta **431,** 278 (1976).

176. GALLIARD, T., D. R. PHILLIPS, and J. REYNOLDS: The Formation of *cis*-3-Nonenal, *trans*-2-Nonenal and Hexanal from Linoleic and Hydroperoxide Isomers by a Hydroperoxide Cleavage Enzyme System in Cucumber *(Cucumis sativus)* fruits. Biochim. Biophys. Acta **441,** 181 (1976).

177. GARNERO, J., L. BENÉZET, L. PEYRON et Y. CHRÉTIEN-BESSIÈRE: Sur la Présence du *p*-Menthatriène-1,3,8 dans l'Huile Essentielle de Feuilles de Persil de Yougoslavie. Bull. soc. chim. France **1967,** 4679.

178. GAUTSCHI, F. (Firmenich SA, Geneva): Personal Communications.

179. GAUTSCHI, F., and B. WILLHALM (Firmenich SA, Geneva): Personal Communications.
180. GERBER, N. N.: Geosmin, from Microorganisms, is *trans*-1,10-Dimethyl-*trans*-9-decalol. Tetrahedron Letters **1968,** 2971.
181. — Sesquiterpenoids from Actinomycetes: Cadin-4-ene-1-ol. Phytochem. **10,** 185 (1971).
182. GERBER, N. N., and H. A. LECHEVALIER: Geosmin, an Earthy-Smelling Substance Isolated from Actinomycetes. Appl. Microbiol. **13,** 935 (1965); Chem. Abstr. **63,** 16 658 d (1965).
183. GIANTURCO, M. A., R. E. BIGGERS, and B. H. RIDLING: Seasonal Variations in the Composition of the Volatile Constituents of Black Tea. A Numerical Approach to the Correlation Between Composition and Quality of Tea Aroma. J. Agric. Food Chem. **22,** 758 (1974).
184. GIANTURCO, M. A., A. S. GIAMMARINO, and P. FRIEDEL: Volatile Constituents of Coffee. Nature **210,** 1358 (1966).
185. GILDEMEISTER, E., und FR. HOFFMANN: Die ätherischen Oele, 3. Aufl., Bd. III, S. 942. Miltitz bei Leipzig: Verlag der Schimmel & Co. A.-G. 1931.
186. — — Die ätherischen Oele, 4. Aufl., Bd. IIIa, S. 30 und 36 (W. TREIBS und D. MERKEL, ed.). Berlin: Akademie-Verlag. 1960.
187. — — Die ätherischen Oele, 4. Aufl., Bd. IIIa, S. 45 (W. TREIBS und D. MERKEL, ed.). Berlin: Akademie-Verlag. 1960.
188. — — Die ätherischen Oele, 4. Aufl., Bd. IIIa, S. 578; see there literature on the subject (W. TREIBS und D. MERKEL, ed.). Berlin-Akademie-Verlag. 1960.
189. — — Die ätherischen Oele, 4. Aufl., Bd. IIIb, S. 302—308 (W. TREIBS und D. MERKEL, ed.). Berlin: Akademie-Verlag. 1962.
190. — — Die ätherischen Oele, 4. Aufl., Bd. IIIc, S. 23, 25 und 28 (W. TREIBS und D. MERKEL, ed.). Berlin: Akademie-Verlag. 1963.
191. — — Die ätherischen Oele, 4. Aufl., Bd. IIIc, S. 250 (W. TREIBS und D. MERKEL, ed.). Berlin: Akademie-Verlag. 1963.
192. — — Die ätherischen Oele, 4. Aufl., Bd. IIIc, S. 270 (W. TREIBS und D. MERKEL, ed.). Berlin: Akademie-Verlag. 1963.
193. — — Die ätherischen Oele, 4. Aufl., Bd. IIId, S. 622 (W. TREIBS und D. MERKEL, ed.). Berlin: Akademie-Verlag. 1966.
194. — — Die ätherischen Oele, 4. Aufl., Bd. IIId, S. 680 (W. TREIBS und D. MERKEL, ed.). Berlin: Akademie-Verlag. 1966.
195. — — Die ätherischen Oele, 4. Aufl., Bd. IIId, S. 758 (W. TREIBS und D. MERKEL, ed.). Berlin: Akademie-Verlag. 1966.
196. — — Die ätherischen Oele, 4. Aufl. Bd. V, S. 419 (W. TREIBS und K. BOURNOT, ed.). Berlin: Akademie-Verlag. 1959.
197. GOLDBLATT, L. A., and S. PALKIN: Vapor Phase Thermal Isomerization of α- and β-Pinene. J. Amer. Chem. Soc. **63,** 3517 (1941).
198. GOLLNICK, K., und G. O. SCHENCK: 1,4-Cycloaddition Reactions. In: The Diels Alder Reaction in Heterocyclic Syntheses (J. HAMER, ed.), p. 255. (Organic Chemistry, a Series of Monographs, Vol. 8.) New York: Academic Press. 1967.
199. GOLOVNYA, R. V., G. A. MIRONOV, and S. D. SOKOLOV: Flavor of Food Products. Russ. chem. Rev. **33,** 816 (1964); Chem. Abstr. **61,** 8814h (1964).
200. GRANGER, R., J. PASSET et J. P. GIRARD: Méthyl-2 méthylène-6 octadiène-2,7 ol isolé de *Thymus vulgaris*. Phytochem. **11,** 2301 (1972).
201. GROSCH, W., G. LASKAWY, and K.-H. FISCHER: Aroma Compounds Formed by Enzymic Co-Oxidation. Proc. int. Symp. Aroma Research, Zeist, p. 75. Wageningen: Pudoc. 1975.
202. GUADAGNI, D. G., R. G. BUTTERY, and J. HARRIS: Odour Intensities of Hop Oil Components. J. Sci. Food Agric. **17,** 142 (1966).

203. GUADAGNI, D. G., R. G. BUTTERY, and S. OKANO: Odour Thresholds of Some Organic Compounds Associated with Food Flavours. J. Sci. Food Agric. **14,** 761 (1963).

204. GUENTHER, E., and D. ALTHANSEN: The Essential Oils, Vol. II. The Constituents of Essential Oils, p. 286. Princeton-New Jersey-Toronto-London-New York: D. van Nostrand Comp., Inc. 1949.

205. GUPTA, A. S., and S. DEV: Studies in Sesquiterpenes. XLVI. Sesquiterpenes from the Oleoresin of *Dicterocarpus pilosus:* Humulene Epoxide-III, Caryophyllenol-I and Caryophyllenol-II. Tetrahedron **27,** 635 (1971).

206. HALIM, A. F., and R. P. COLLINS: Essential Oil of *Salvia dorisiana* (Standley). J. Agric. Food Chem. **23,** 506 (1975).

207. HALL, J. B., M. H. VOCK, J. F. VINALS und E. J. SHUSTER (IFF Incorp., New York): Verfahren zur Herstellung von Aroma und Duftzusammensetzung. DOS 2 446 826 (30. 4. 1975).

208. HALL, R. L.: Toxicants Occurring Naturally in Foods, p. 164—171. Washington, D. C.: National Academy of Sciences, National Research Council. 1966.

209. HALL, R. L.: Toxicants Occurring in Spices and Flavors. In: Toxicants Occurring Naturally in Foods (Committee on Food Protection — Food and Nutrition Board — National Council), 2nd ed., chap. 20, p. 448. Washington, D. C.: National Academy of Sciences. 1973.

210. HAMMOND, E. G., and F. D. HILL: The Oxidized-Metallic and Grassy Flavor Components of Autoxidized Fat. J. Amer. Oil. Chem. Soc. **41,** 180 (1964).

211. HARING, H. G., F. RIJKENS, H. BOELENS, and A. VAN DER GEN: Olfactory Studies on Enantiomeric Eremophilane Sesquiterpenoids. J. Agric. Food Chem. **20,** 1018 (1972).

212. HARKES, P. D., and W. J. BEGEMANN: Identification of Some Previously Unknown Aldehydes in Cooked Chicken. J. Amer. Oil. Chem. Soc. **51,** 356 (1974).

213. HARPER, R., E. C. BATE SMITH, and D. G. LAND: Odour Description and Odour Classification — A Multidisciplinary Examination. London: J. and A. Churchill Ltd. 1968.

214. HARRING, C. M., J. P. VITÉ und P. R. HUGHES: „Ipsenol", der Populationslockstoff des krummzähnigen Tannenborkenkäfers. Naturwiss. **62,** 488 (1975).

215. HASEGAWA Perf. Co.: Tri- and tetra-Methyl-2-hydroxy-cyclohexylidene-acetic lactones. Japan. Kokai 75 069 062 (9. 6. 1975).

216. HASHIZUME, T., N. KICKUCHI, Y. SASAKI, and I. SAKATA: Constituents of Cane Molasses. III. Isolation and Identification of (−)-2-Deceno-5-lactone (Massoialactone). Agric. Biol. Chem. **32,** 1306 (1968).

217. HASHIZUME, T., and I. SAKATA: Constituents of Japanese Peppermint Oil. II. Isolation and Identification of (−)-β-Caryophyllene Epoxide from "Okako No. 6". Agric. Biol. Chem. **34,** 638 (1970).

218. HASSELSTROM, T. E., E. J. HEWITT, K. S. KONIGSBACHER, and J. J. RITTER: Composition of Volatile Oil of Black Pepper, *Piper nigrum.* J. Agric. Food Chem. **5,** 53 (1957).

219. HATFIELD, S. A.: A Symposium on Flavour with Special Reference to Fruit and Vegetables. ECRO Information Bull. **1975,** No 7, p. 61.

220. HAVERKAMP-BEGEMANN, P., and J. C. KOSTER: 4-*cis*-Heptenal: A Creamy-Flavored Component of Butter. Nature **202,** 552 (1964).

221. HAWKINS, J. E., and H. G. HUNT: The Preparation of Ocimene from α-Pinene. J. Amer. Chem. Soc. **73,** 5379 (1951).

222. HAYASHI, N., H. KOMAE, S. EGUCHI, M. NAKAYAMA, S. HAYASHI, and T. SAKAO: Two New Furanosesquiterpenes from *Actinodaphne longifolia* (Blume) Nakai. Chem. and Ind. **1972,** 572.

223. HECKMANN, R. A. (R. J. Reynolds Tobacco Co): Enhancement of the Flavor of Foods with 2,6,10,10-Tetramethyl-1-oxaspiro[4.5]dec-6-en-8-one. U.S. Pat. 3.637.403 (25. 1. 1972); Chem. Abstr. **76,** 139 286k (1972).

224. HECKMANN, R. A., and D. L. ROBERTS: Synthesis of Theaspirone, a Component of Essential Oil. Tetrahedron Letters **1969,** 2701.

225. HEDIN, P. A., A. C. THOMPSON, and R. C. GUELDNER: Constituents of Cotton Bud Essential Oil. Phytochem. **14,** 2087 (1975).

226. HEDIN, P. A., A. C. THOMPSON, R. C. GUELDNER, and J. M. RUTH: Isolation of Bisabolene Oxide from the Cotton Bud. Phytochem. **11,** 2118 (1972).

227. HEINZ, D. E., and W. G. JENNINGS: Volatile Components of Bartlett Pear. V. J. Food Sci. **31,** 69 (1966).

228. HERZ, K. O.: A Study of the Nature of Boiled Beef Flavor. Thesis, Rutgers — The State University. Order No 68-14,248, University Microfilms, Ann Arbor/Mich. 1968.

229. HERZ, W.: Sesquiterpene Lactones in the Compositae: In: The Biology and Chemistry of The Compositae (J. B. HARBORNE, V. HEYWOOD, and B. L. TURNER, eds.). London: Academic Press. 1977.

230. HIKINO, H., Y. HIKINO, Y. TAKESHITA, Y. ISURUGI, and T. TAKEMOTO: Constituents of Some Japanese Valerian Roots. Yakugaku Zasshi **83,** 555 (1963); Chem. Abstr. **59,** 4971a (1963).

231. HIKINO, H., Y. HIKINO, Y. TAKESHITA, K. SHIRATA, and T. TAKEMOTO: Structure of Kessane. Chem. Pharm. Bull (Japan) **11,** 547 (1963); Chem. Abstr. **59,** 6445 (1963).

232. HIROI, M., and D. TAKAOKA: Peel Oil of *Citrus iyo.* Nippon Kagaku Kaishi **1973,** 1339, Chem. Abstr. **79,** 70 085 p (1973).

233. HIROTA, H., Y. TANAHASHI, and T. TAKAHASHI: Structure of a Guaiane-Typ Sesquiterpene Ether from "San-Shion". Tetrahedron Letters **1975,** 4579.

234. HODGE, J. E., F. D. MILLS, and B. E. FISHER: Compounds of Browned Flavor Derived from Sugar-Amine Reactions. Cereal Sci. Today **17,** 34 (1972).

235. HOFFMANN, G.: 3-*cis*-Hexenal, the "Green" Reversion Flavor of Soybean Oil. J. Amer. Oil Chem. Soc. **38,** 1 (1961).

236. — Isolation of Two Pairs of Isomeric 2,4-Alkadienals from Soybean Oil-Reversion Flavor Concentrate. J. Amer. Oil Chem. Soc. **38,** 30 (1961).

237. HOFFMANN, G., and P. W. MEIJBOOM: Isolation of Two Isomeric 2,6-Nonadienals and Two Isomeric 4-Heptenals from Beef and Mutton Tallow. J. Amer. Oil Chem. Soc. **45,** 468 (1968).

238. HOLLEY, A.: La Perception des Odeurs. La Recherche **6,** 629 (1975).

239. HOLUB, M., V. HEROUT und F. ŠORM: Synthese des α-Bisabols, eines spasmolytisch wirksamen Sesquiterpenalkohols. Ceskoslov. Farmac. **4,** 129 (1955); C. **128,** 11 308 (1957).

240. HONKANEN, E., T. MOISIO, and P. KARVONEN: Volatile Flavor Substances in Some Clover Species. Suom. Kemistilehti B **42,** 448 (1969).

241. HONKANEN, E., T. MOISIO, P. KARVONEN, A. J. VIRTANEN, and J. PAASIVIRTA: On the Occurrence of a New Lactone Compound, *trans*-4-Methyl-5-hydroxy-hexanoic Acid Lactone, in Milk. Acta Chem. Scand. **22,** 2041 (1968).

242. HOUCHEN, M., R. A. SCALAN, L. M. LIBBEY, and D. D. BILLS: Possible Precursor for 1-Methyl-4-isopropenylbenzene in Commercial Blackberry Flavor Essence. J. Agric. Food Chem. **20,** 170 (1972).

243. HOULIHAHN, W. J., J. LEVY, and J. MAYER: The Structure of the Alcohol Obtained by the Hydration of Myrcene. J. Amer. Chem. Soc. **81,** 4692 (1959).

244. HUELIN, F. E., and I. M. COGGIOLA: Superfical Scald, a Functional Disorder of Stored Apples. VI. Evaporation of a α-Farnesane from the Fruit. J. Sci. Food Agric. **21,** 82 (1970).

245. HUELIN, F. E., and K. E. MURRAY: α-Farnesene in the Natural Coating of Apples. Nature **210,** 1260 (1966).
246. HUNTER, G. L. K., and W. B. BRODGEN, JR.: Terpenes and Sesquiterpenes in Cold-Pressed Orange Oil. J. Food Sci. **30,** 1 (1965).
247. HUNTER, G. L. K., and M. G. MOSHONAS: Analysis of Alcohols in Essential Oils of Grapefruit, Lemon, Lime, and Tangerine. J. Food Sci. **31,** 167 (1966).
248. IGUCHI, M., A. NISHIYAMA, H. KOYAMA, S. YAMAMURA, and Y. HIRATA: Isolation and Structures of Three Sesquiterpenes. Tetrahedron Letters **1968,** 5315; YAMAMURA, S., M. IGUCHI, A. NISHIYAMA, M. NIWA, H. KOYAMA, and Y. HIRATA: Sesquiterpenes from *Acorus calamus* L., Tetrahedron **27,** 5419 (1971).
249. INA, K., and H. ETÔ: 3-Keto-β-ionone in the Essential Oil from Black Tea. Agric. Biol. Chem. (Tokyo) **35,** 962 (1971).
250. — — High Boiling Compounds of Neutral Essential Oil from Tea. Agric. Biol. Chem. (Tokyo) **36,** 1027 (1972).
251. — — Photo-Oxygenation of α-Ionone. Agric. Biol. Chem. (Tokyo) **36,** 1091 (1972).
252. INA, K., Y. SAKAMOTO, and H. FUKAMI: Isolation and Structure Elucidation of Theaspirone, a Component of Tea Essential Oil. Tetrahedron Letters **1968,** 2777.
253. INA, K., T. TAKANO, Y. IMAI, and Y. SAKATO: Synthesis of Theaspirone. Agric. Biol. Chem. (Tokyo) **36,** 1033 (1972).
254. IRIE, S., N. KATO, and S. KINOSHITA: Gamma-Rays Treatment of Odorous Compound in Water. Bull. Chem. Soc. Japan **49,** 2737 (1976).
255. ISAAC, O., H. SCHNEIDER und H. EGGENSCHWILLER: Beitrag zur Kenntnis des Kamillen-Bisabolols. Dtsch. Apotheker-Ztg. **108,** 293 (1968).
256. ISHII, H., T. TOZYO, and H. MINATO: Structure of Liguloxide, Liguloxidol, and Liguloxidol Acetate, Sesquiterpenes of Antipodal Guaiane Type. Chem. Commun. **1968,** 106.
257. ISHII, H., T. TOZYO, M. NAKAMURA, and H. MINATO: Studies on Sesquiterpenoids. XIX. Structure and Absolute Configuration of Liguloxide, Liguloxidol, and Liguloxidol Acetate. Tetrahedron **26,** 2911 (1970).
258. ISOE, S., S. BE HYEON, H. ICHIKAWA, S. KATSUMURA, and T. SAKAN: The Synthesis of Actinidiolide, Dihydroactinidiolide and Actinidol. Tetrahedron Letters **1968,** 5561.
259. ISOE, S., S. BE HYEON, and T. SAKAN: Photo-Oxygenation of Carotenoids. I. The Formation of Dihydroactinidiolide and β-Ionone from β-Carotene. Tetrahedron Letters **1969,** 279.
260. ISOE, S., S. KATSUMURA, and T. SAKAN: The Synthesis of Damascenone and β-Damascone and the Possible Mechanism of Their Formation from Carotenoids. Helv. Chim. Acta **56,** 1514 (1973).
261. ISOE, S., T. ONO, S. BE HYEON, and T. SAKAN: The Structure of Matatabiether. Tetrahedron Letters **1968,** 5319.
262. ITO, S., M. KODAMA, T. NOZOE, H. HIKINO, Y. HIKINO, Y. TAKESHITA, and T. TAKEMOTO: Structures of α-Kessyl Alcohol and Kessyl Glycol. Tetrahedron Letters **1963,** 1787.
263. JACKSON, H. W.: Flavor Research on Cheese. Perf. Essent. Oil Rec. **49,** 256 (1958).
264. JACOBSON, M., J. KEISER, D. L. CHAMBERS, D. H. MIYASHITA, and CH. HARDING: Synthetic Nonenyl Acetates as Attractants for Female Melon Flies. J. med. Chem. **14,** 236 (1971).
265. JAKOB, M. A., R. HIPPLER und H. R. LÜTHI: Über das Vorkommen von Hexyl-2-methylbutyrat im Apfelaroma. Mitt. Geb. Lebensm. Hyg. **60,** 223 (1969); ref. from Z. Lebensm.-Unters. Forsch. **145,** 58 (1971).
266. JAURÉGUIBERRY, G., et R. E. WOLFF: Sur la Présence de (+)-Sabinène dans l'Essence de Noix de Muscade et Quelques Propriétés de cet Hydrocarbure. Bull. soc. chim. France **1962,** 1985.

267. JELLINEK, J. S.: The Effect of Intermolecular Forces on Preceived Odors. Ann. N. Y. Acad. Sci. **116,** 725 (1964).

268. JENNINGS, W. G.: Peaches and Pears. 4th Symp. Foods, p. 419, Oregan State University 1967.

269. — Chemistry of Flavor. Lebensm.-Wiss. Technol. **2,** 75 (1969).

270. JENNINGS, W. G., and M. R. SEVENANTS: Volatile Esters of Bartlett Pear. III. J. Food Sci. **29,** 158 (1964).

271. — — Volatile Components of Peach. I. J. Food Sci. **29,** 796 (1964).

272. JOHNSON, A. E., H. E. NURSTEN, and A. A. WILLIAMS: Vegetable Volatiles: A Survey of Components Identified. I. Chem. and Ind. **1971,** 556.

273. JOHNSON, R. R., E. D. ALFORD, and G. W. KINZER: Formation of Sucrose Pyrolysis Products. J. Agric. Food Chem. **17,** 22 (1969).

274. JOHNSON, R. R., and J. A. NICHOLSON: The Structure, Chemistry, and Synthesis of Solanone. A New Anomalous Terpenoid Ketone from Tobacco. J. Organ. Chem. (U.S.A.) **30,** 2918 (1965).

275. JONES, T. G. H.: Reactions of Tagetone. I. Proc. Roy. Soc. Queensland **45,** 45 (1934).

276. KAHN, J. H., G. B. NICKOL, and H. A. CONNER: Identification of Volatile Components in Vinegars by GC-MS. J. Agric. Food Chem. **20,** 214 (1972).

277. KAHN, J. H., P. A. SHIPLEY, E. G. LAROE, and H. A. CONNER: Whiskey Composition: Identification of Additional Components by Gas Chromatography-Mass Spectrometry. J. Food Sci. **34,** 587 (1969).

278. KAISER, R., and D. LAMPARSKY (Givaudan, L. et Cie SA, Geneva): Odorous 12-norcaryophyll-6-en-3-one and 3,7-epoxy caryophyllane. DOS 2.335.046 (31. 1. 1974); Chem. Abstr. **80,** 121 147p (1974).

279. — — Verfahren zur Herstellung von Riech- und Geschmacksstoffkompositionen. DOS 2.347.454 (25. 4. 1974); Chem. Abstr. **83,** 15 508j (1975).

280. — — Inhaltsstoffe des Verbenaöls. 1. Mitt. Natürliches Vorkommen der Photocitrale und einiger ihrer Derivate. Helv. Chim. Acta **59,** 1797 (1976).

281. — — Inhaltsstoffe des Verbenaöls. 2. Mitt. Caryophyllan-2,6-β-oxid, eine neue Sesquiterpenverbindung aus dem Öl von *Lippia citridora* KUNTH. Helv. Chim. Acta **59,** 1803 (1976).

282. — — Das Lacton der 5-Hydroxy-*cis*-2,*cis*-7-decadiensäure und weitere Lactone aus dem Absolue der Blüten von *Polianthes Tuberosa* L. Tetrahedron Letters **1976,** 1659.

283. KAISER, R., D. LAMPARSKY, and P. SCHUDEL: Analysis of Buchu Leaf Oil. J. Agric. Food Chem. **23,** 943 (1975).

284. KANEKO, H.: The Aroma of Cigar Tobacco. II. Isolation of norambreinolide from Cigar Tobacco. Agric. Biol. Chem. **35,** 1461 (1971).

285. KANEKO, H., and M. HARADA: 4-Hydroxy-β-damascone and 4-Hydroxy-dihydro-β-damascone from Cigar Tobacco. Agric. Biol. Chem. **36,** 168 (1972).

286. KANEKO, H., and K. IJICHI: Aroma of Cigar Tobacco. I. Isolation of 2-Hydroxy-2,6,6-trimethyl-cyclohexylidene-1-acetic Acid Lactone (Dihydroactinidiolide) from Ether Extract of Cigar Leaves. Agric. Biol. Chem. **32,** 1337 (1968).

287. KANEKO, H., and M. MITA: Isolation from Cigar Tobacco Leaves of 2,3-Dimethyl-4-hydroxy-2-nonenoic Acid Lactone. Agric. Biol. Chem. **33,** 1525 (1969).

288. KASTING, R., J. ANDERSSON, and E. VON SYDOW: Volatile Constituents in Leaves of Parsley. Phytochem. **11,** 2277 (1972).

289. KATAYAMA, M., and S. MARUMO: 11,11-Oxidosqualene from *Sclerotinia fructicola.* Tetrahedron Letters **1976,** 1293.

290. KE, P. J., R. G. ACKMAN, and B. A. LINKE: Autoxydation of Polyunsaturated Fatty Compounds in Mackerel Oil: Formation of 2,4,7-Decatrienals. J. Amer. Oil Chem. Soc. **52,** 349 (1975).

291. KEFFORD, J. F., and B. V. CHANDLER: The Chemical Constituents of Citrus Fruits. Adv. Food Res., Suppl. 2, Academic Press 1970; Chem. Abstr. **75,** 18993j (1971).

292. KEITH, E. S., and J. J. POWERS: Determination of Flavor Threshold Levels and Sub-Threshold, Additive, and Concentration Effects. J. Food Sci. **33,** 213 (1968).

293. KEMP, T. R.: Identification of Some Volatile Compounds from *Citrullus vulgaris*. Phytochem. **14,** 2637 (1975).

294. KEMP, T. R., D. E. KNAVEL, and L. P. STOLTZ: Characterization of Some Volatile Components of Muskmelon Fruit. Phytochem. **10,** 1925 (1971).

295. — — — *cis*-6-Nonenal: A Flavor Component of Muskmelon Fruit. Phytochem. **11,** 3321 (1972).

296. — — — Identification of Some Volatile Compounds from Cucumber. J. Agric. Food Chem. **22,** 717 (1974).

297. KEMP, T. R., D. E. KNAVEL, L. P. STOLTZ, and R. L. LUNDIN: 3,6-Nonadien-1-ol from *Citrullus vulgaris* and *Cucumis melo*. Phytochem. **13,** 1167 (1974).

298. KEMP, T. R., L. P. STOLTZ, and D. E. KNAVEL: Volatile Components of Muskmelon Fruit. J. Agric. Food Chem. **20,** 196 (1972).

299. KEMP, T. R., L. P. STOLTZ, and L. V. PACKETT: Rosaceae. Aromatic Hydrocarbons: Examination of Peach Fruit and Foliage Volatiles. Phytochem. **10,** 478 (1971).

300. KEPPLER, J. G., J. A. SCHOLS, W. H. FEENSTRA, and P. W. MEIJBOOM: Components of the Hardening Flavor Present in Hardened Linseed Oil and Soybean Oil. J. Amer. Oil Chem. Soc. **42,** 246 (1965).

301. KERSCHBAUM, M.: Über Lactone mit großen Ringen — die Träger des vegetabilischen Moschus-Duftes. Ber. dtsch. chem. Ges. **60,** 902 (1927).

302. KIKUCHI, T., T. MIMURA, K. MARIMAYA, H. YANO, T. ARIMOTO, Y. MASADA, and T. INOUE: Odorous Metabolic of Blue-Green Alga, *Schizothrix muelleri* Collected in the Southern Basin of Lake Biwa. Identification of Geosmin. Chem. Pharm. Bull. (Japan) **21,** 2342 (1973); Chem. Abstr. **80,** 117938e (1974).

303. KIKUCHI, T., T. MIMURA, Y. ITOH, Y. MORIWAKI, Y. NEGORO, K. MASADA, and T. INOUE: Odorous Metabolites of Actinomyces Biwako-C and -D Strain Isolated from the Bottom Deposits of Lake Biwa. Identification of Geosmin, 2-Methylisoborneol and Furfural. Chem. Pharm. Bull. (Japan) **21,** 2339 (1973); Chem. Abstr. **80,** 117937d (1974).

304. KIM, K., T. KURATA, and M. FUJIMAKI: Identification of Flavor Constituents in Carbonyl, Non-Carbonyl, Neutral and Basic Fractions of Aqueous Smoke Condensates. Agric. Biol. Chem. **38,** 53 (1974).

305. KIMLAND, B., A. J. AASEN, and C. R. ENZELL: Tobacco Chemistry. 10. Volatile Neutral Constituents of Greek Tobacco. Acta Chem. Scand. **26,** 2177 (1972).

306. KIMLAND, B., R. A. APPLETON, A. J. AASEN, J. ROERAADE, and C. R. ENZELL: Tobacco Chemistry. 6. Neutral Oxygen-Containing Volatile Constituents of Greek Tobacco. Phytochem. **11,** 309 (1972).

307. KINLIN, T. E., R. MURALIDHARA, A. O. PITTET, A. SANDERSON, and J. P. WALRADT: Volatile Components of Roasted Filberts. J. Agric. Food Chem. **20,** 1021 (1972).

308. KINSELLA, J. E., and D. H. HWANG: Enzymes of *Penicillium roqueforti* Involved in the Biosyntheses of Cheese Flavor. Crit. Rev. Food Sci. Nutr. **8,** 191 (1977).

309. KITA, Y., Y. NAKATANI, A. KOBAYASHI, and T. YAMANISHI: Composition of Peel Oil from *Citrus Unshiu*. Agric. Biol. Chem. **33,** 1559 (1969).

310. KLEIN, E., and W. ROJAHN: (+)-γ-Methyl-γ-vinylbutyrolactone, a New Component of Lavender and Lavandin Oils. Dragoco Rep., Ger. Ed. **14,** 3 (1967); Chem. Abstr. **67,** 5640b (1967).

311. — — The More Important Components of Buchu Oil. Dragoco Rep., Ger. Ed. **9,** 183 (1967); Chem. Abstr. **68,** 16059r (1968).

312. KOGAMI, K., O. TAKAHASHI, and J. KUMANOTANI: A New Synthesis of 4-(4-Methyl-4-hydroxyamyl)-Δ^3-cyclohexenecarboxaldehyde. Canad. J. Chem. **52,** 125 (1974).

313. KÖNST, W. M. B., L. M. VAN DER LINDE, and J. G. WITTEVEEN: Recent Developments in the Chemistry of Eremophilanes. Flavours **6,** 121 (1975).

314. KONDO, H., und H. SUZUKI: Über die Konstitution des Perillens. Ber. dtsch. chem. Ges. **69,** 2459 (1936).

315. KOVÁTS, E. sz.: Zur Kenntnis ätherischer Öle. 4. Mitt. Zur Kenntnis des sog. „destillierten" Limetten-Öls (*Citrus medica* L., var. *acida*, Brandis; *Citrus aurantifolia*, Swingle). Helv. Chim. Acta **46,** 2705 (1963).

316. KOZHIN, S. A., N. D. TREIZER, N. M. MOSKALEVA, and G. I. PRIKHOD'KO: Mathematical Treatment of Quantitative Gas-Liquid Chromatographic Analysis Data to Determine Objective Differences in Tea Specimens by Their Aroma. Vestn. Leningrad. Univ., Fiz., Khim. **1972,** 147; Chem. Abstr. **77,** 86 754y (1972).

317. KRISHNAMURTHY, R. G., and ST. S. CHANG: Chemical Reactions Involved in the Deep Fat Frying of Foods. III. Identification of Nonacidic Volatile Decomposition Products of Corn Oil. J. Amer. Oil Chem. Soc. **44,** 136 (1967).

318. KRISTOFFERSEN, T.: Biogenesis of Cheese Flavor. J. Agric. Food Chem. **21,** 573 (1973).

319. KUGLER, E., und E. sz. KOVÁTS: Zur Kenntnis ätherischer Öle. 1. Zur Kenntnis des Mandarinenschalen-Öls (*Citrus reticulata* Blanco bzw. *Citrus nobilis*, var. *deliciosa* Swingle "Mandarin"). Helv. Chim. Acta **46,** 1480 (1963).

320. KUHN, R., und K. KUM: Über die absolute Konfiguration des Sorbinöls. Ber. dtsch. chem. Ges. **95,** 2009 (1962).

321. KUHN, R., und A. WINTERSTEIN: Über die Konstitution des Pikrocrocins und seine Beziehung zu den Carotin-Farbstoffen des Safrans. Ber. dtsch. chem. Ges. **67,** 346 (1934).

322. KULESZA, J., J. KULA und W. KWIATKOWSKI: Anwendung des Carotols bei der Synthese der Riechstoffe. Riechstoffe, Aromen, Körperpflegemittel **23,** 34 (1973); Chem. Abstr. **78,** 155 120s (1973).

323. KUMAMOTO, J., R. W. SCORA, and W. W. PAYNE: Ambrosial a Dehydrocitral Isolated from the Leaf Oil of *Ambrosia confertiflora* D. C. (Ragweed). J. Agric. Food Chem. **23,** 123 (1975).

324. LAFFORT, P., F. PATTE, and M. ETCHETO: Olfactory Coding on the Basis of Physicochemical Properties. Ann. New York Acad. Sci. **237,** 193 (1974).

325. LAMPARSKY, D. (Givaudan & Cie, SA, Geneva): Private Communication.

326. LAMPARSKY, D., and P. SCHUDEL: *p*-Menthane-8-thiol-3-one, a New Component of Buchu Leaf Oil. Tetrahedron Letters **1971,** 3323.

327. LAND, D. G.: Volatile Oxidation Products of β-Carotene and β-Ionone and Their Relation of the Off-Flavor Developed by Dehydrated Carrot Exposed to Oxygen. Food Sci. Technol., Proc. Int. Congr. 1st 1962 (J. MUIL LEITCH, ed.), p. 353. New York: Gordon and Breach Sci. Publ. 1969; Chem. Abstr. **73,** 45 633g (1970).

328. — Techniques for Assessing Odour: Uses and Limitations. Proc. int. Symp. Aroma Research, Zeist, p. 131. Wageningen: Pudoc. 1975.

329. LANGLER, J. E., and E. A. DAY: Development of Flavor Properties of Methyl Ketones in Milk Fat. J. Dairy Sci. **47,** 1291 (1964).

330. — — Quantitative Analysis of the Major Free Fatty Acid in Swiss Cheese. J. Dairy Sci. **49,** 91 (1966).

331. LARDELLI, G., G. DIJKSTRA, P. D. HARKES, and J. BOLDINGH: New γ-Lactone Found in Butter. Rec. trav. chim. Pays-Bas **85,** 43 (1966).

332. LAROE, E. G., and P. A. SHIPLEY: Whiskey Composition: Formation of alpha- and beta-Ionone by the Thermal Decomposition of beta-Carotene. J. Agric. Food Chem. **18,** 174 (1970).

333. LASSAK, E. V., and I. A. SOUTHWELL: Occurrence of Some Unusual Compounds in the Leaf Oils of *Eriostemon obovalis* and *Phebalium glandulosum* subsp. *glandulosum*. Austral. J. Chem. **27,** 2703 (1974).

334. LAWRENCE, B. M., J. W. HOGG und S. J. TERHUNE: Das ätherische Öl von *Zingiber cassumunar* Roxb. Riechst. Aromen Körperpflegem. **20,** 261 (1970); Chem. Abstr. **73,** 106 337n (1970).

335. — — — Les Huiles Essentielles et Leurs Constituants. IV. Quelques Nouveaux Constituants à l'Etat de Traces dans l'Huile Essentielle de *Salvia officinalis* L. Parf. Cosm. Savons France **1,** 256 (1971).

336. LAWRENCE, B. M., J. W. HOGG, S. J. TERHUNE, and N. PICHITAKUL: Essential Oils and Their Constituents. IX. The Oils of *Ocimum sanctum* and *Ocimum basilicum* from Thailand. Flavour Ind. **3,** 47 (1972).

337. LAWRENCE, B. M., S. J. TERHUNE, and J. W. HOGG: Essential Oils and Their Constituents. VI. So-Called Exotic-Oil of *Ocimum basilicum*. Flavour Ind. **2,** 173 (1971).

338. LEDEL, F.: Untersuchung des Röstzwiebelaromas. Z. Lebensm.-Unters. Forsch. **157,** 229 (1975).

339. LEFFINGWELL, J. C., J. W. STALLINGS, F. O. SELLER, R. A. LLOYD, and F. C. KANE, JR.: Clary Sage Production in the Southeastern United States. Intern. Congr. Essent. Oils (Pap.), 6th **1974,** 3, 11 pp; Chem. Abstr. **84,** 49 723m (1976).

340. LEITEREG, T. J., D. G. GUADAGNI, J. HARRIS, T. R. MON, and R. TERANISHI: Chemical and Sensory Data Supporting the Difference Between the Odor of the Enantiomeric Carvones. J. Agric. Food Chem. **19,** 785 (1971).

341. LEWIS, Y. S., E. S. NAMBUDIRI, and N. KRISHNAMURTHY: Composition of Pepper Oil. Perf. Essent. Oil Rec. **60,** 259 (1969).

342. LIEBICH, H. M., D. R. DOUGLAS, A. ZLATKIS, F. MÜGGLER-CHAVAN, and A. DONZEL: Volatile Components in Roast Beef. J. Agric. Food Chem. **20,** 96 (1972).

343. LIEBIG, H. M., W. A. KOENIG, and E. BAYER: Analysis of the Flavor of Rum by GLC and MS. J. Chrom. Science **8,** 527 (1970).

344. LIN, F. M., and W. F. WILKENS: Volatile Flavor Components of Coconut Meat. J. Food Sci. **35,** 538 (1970).

345. LINDE, H., und G. CRAMER: (−)-α-Bisabolol und *cis*-2-[Hexadien(2,4)-yliden]-1,6-dioxaspiro-[4.4]-nonen-(3) in handelsüblichen Kamillen-Extrakten. Arzneim.-Forsch. **22,** 583 (1972); Chem. Abstr. **77,** 52 392p (1972).

346. LINDER, P., and O. MARTENSSON: Electronic Structure of Some Odorant Molecules of the Benzaldehyde Type. Int. J. Quantum Chem. **6,** 363 (1972).

347. LLOYD, R. A., C. W. MILLER, D. L. ROBERTS, J. A. GILES, J. P. DICKERSON, N. H. NELSON, C. E. RIX, and P. H. AYERS: Flue-Cured Tobacco Flavor. 1. Essence and Essential Oil Components. Tob. Sci. **20,** 43 (1976).

348. LOMBARDI, P., R. C. COOKSON, H. P. WEBER, W. RENOLD, A. HAUSER, K. H. SCHULTE-ELTE, B. WILLHALM, W. THOMMEN und G. OHLOFF: Synthesen der diastereoisomeren Caparrapioxide. Helv. Chim. Acta **59,** 1158 (1976).

349. LOSSNER, G.: Constituents of *Majorana hortensis*. II. *Cis*- and *trans*-Sabinene Hydrates as Constituents of the Essential Oil. Pharmazie **22,** 51 (1967); Chem. Abstr. **66,** 118 799h (1967).

350. — Constituents of *Majorana hortensis*. III. Essential Oil. Pharmazie **22,** 324 (1967); Chem. Abstr. **67,** 111 328r (1967).

351. — mentioned in: Aromastoffe in Gewürzen und Würzkräutern. IV. (R. OBERDIEK, ed.), Riechst. Aromen Körperpflegem. **23,** 3 (1973); Chem. Abstr. **79,** 77 047j (1973).

352. MABRY, T. J. (The University of Texas at Austin, Dpt. of Botany): Private Communication.

353. MAEKAWA, K., M. KODAMA, M. KUSHII, and M. MITAMURA: Essential Oils of Some Orange Peels. Agric. Biol. Chem. **31,** 373 (1967).

354. MAGA, J. A.: Lactones in Food. Crit. Rev. Food Sci. Nutr. **8**, 1 (1976).
355. MAIER, H. G.: Flüchtige Aromastoffe in Lebensmitteln. Angew. Chem. **82**, 965 (1970).
356. — Binding of Volatile Aroma Substances to Nutrients and Foodstuffs. Proc. int. Symp. Aroma Research, Zeist, p. 143. Wageningen: Pudoc. 1975.
357. MANLEY, C. H., and I. S. FAGERSON: Major Volatile Neutral and Acid Compounds of Hydrolyzed Soy Protein. J. Food Sci. **35**, 286 (1970).
358. MARION, J. P., F. MÜGGLER-CHAVAN, R. VIANI, J. BRICOUT, D. REYMOND, et R. H. EGLI: Sur la Composition de l'Arôme de Cacao. Helv. Chim. Acta **50**, 1509 (1967).
359. — — — — — —, ibid. 1509, 1516 (1967).
360. MARSHALL, J. A., S. F. BRADY, and N. H. ANDERSEN: The Chemistry of Spiro[4.5]-decane Sesquiterpenes. In: Progress in the Chemistry of Organic Natural Products (W. HERZ, H. GRISEBACH, and G. W. KIRBY, eds., founded by L. ZECHMEISTER), Vol. 31, p. 283. Wien-New York: Springer. 1974.
361. MARSHALL, J. A., and A. R. HOCHSTETLER: The Synthesis of (±)-Geosmin and the Other 1,10-Dimethyl-9-decalol Isomers. J. Organ. Chem. (U.S.A.) **33**, 2593 (1968).
362. MATSUMOTO, M., and K. KONDO: Sensitized Photooxygenation of Linear Monoterpenes Bearing Conjugated Double Bonds. J. Organ. Chem. (U.S.A.) **40**, 2259 (1975).
363. MAURER, B., M. FRACHEBOUD, A. GRIEDER und G. OHLOFF: Zur Kenntnis der sesquiterpenoiden C_{12}-Ketone des ätherischen Oeles von *Vetiveria zizanioides* L., Nash. Helv. Chim. Acta **55**, 2371 (1972).
364. MAY, W. A.: Identification of the Decomposition Products Produced by Triolein under Simulated Deep Fat Frying Conditions and a Preliminary Study of Sensory Characteristics and Physiological Effects of Unsaturated Lactones. Thesis, Rutgers State University, New Brunswick, N. J. 1971. Univ. Microfilms Ann Arbor, Mich., No 72-9649 from Diss. Abstr. Int. B **32**, 5240 (1972).
365. MAZZIOTTI, A.: Intermolecular Basis of Odor. Nature **250**, 645 (1974).
366. MCCONNELL, J. G., und R. M. SILVERSTEIN: Neue Ergebnisse der Chemie von Insektenpheromonen. Angew. Chem. **85**, 647 (1973); see there lit. on this subject.
367. MCDANIEL, M. R., L. A. SATHER, and R. C. LINDSAY: Influence of Free Fatty Acids on Sweet Cream Butter Flavor. J. Food Sci. **34**, 251 (1969).
368. MCGILL, A. S., R. HARDY, and J. R. BURT: Hept-*cis*-4-enal and its Contribution to the Off-Flavour in Cold Stored Cod. J. Sci. Food Agric. **25**, 1477 (1974).
369. MCGREGOR, J. T., L. L. LAYTON, and R. G. BUTTERY: California Bay Oil. II. Biological Effects of Constituents. J. Agric. Food Chem. **22**, 777 (1974).
370. MCLEOD, A. J., and R. ISLAM: Volatile Flavour Components of Coriander Leaf. J. Sci. Food Agric. **27**, 721 (1976).
371. MCLEOD, W. D., JR.: The Constitution of Nootkatone, Nootkatene and Valencene. Tetrahedron Letters **1965**, 4779.
372. MCLEOD, W. D., JR., and N. M. BUIGUES: Sesquiterpenes. I. Nootkatone, a New Grapefruit Flavor Constituent. J. Food Sci. **29**, 565 (1964).
373. MEDSKER, L. L., D. JENKINS, and J. F. THOMAS: Odorous Compounds in Natural Waters. An Earthy-Smelling Compound Associated with Blue-Green Algae and Actinomycetes. Environ. Sci. Technol. **2**, 461 (1968); Chem. Abstr. **69**, 5094t (1968).
374. MEHLITZ, A., und T. MINAS: Über Pomeranzenschalenöl. 2. Mitt.: Terpenoide (sauerstoffhaltige Verbindungen). Riechst. Aromen Körperpflegem. **15**, 204 (1965).
375. MEIJBOOM, P. W.: Relation Between Molecular Structure and Flavor Perceptibility of Aliphatic Aldehydes. J. Amer. Oil Chem. Soc. **41**, 326 (1964).
376. MEIJBOOM, P. W., and G. A. JONGENOTTER: Flavor Thresholds of Unbranched Isomeric Nonenals. J. Amer. Oil Chem. Soc. **48**, 143 (1971).
377. MEIJBOOM, P. W., and J. B. A. STROINK: *trans,cis,cis*-2,4,7-Decatrienal, the Fishy

Off-Flavor Occurring in Strongly Autoxidized Oils Containing Linolenic Acid or ω-3,6,9, etc. Fatty Acids. J. Amer. Oil Chem. Soc. **49,** 555 (1972).

378. MEILGAARD, M. C. (The Stroh Brewery Comp., Detroit, Michigan): Personal Communication.

379. — kindly carried out these measurements on our products (*605*).

380. — Flavor Chemistry of Beer. I. Flavor Interaction Between Principal Volatiles. Tech. Q. Master Brew. Assoc. Am. **12,** 105 (1975).

381. — Flavor Chemistry of Beer. II. Flavor Threshold of 329 Aroma Volatiles. Tech. Q. Master Brew. Assoc. Am. **12,** 151 (1975).

382. MEJER, T. M.: Essential Oil of Massoy Bark. Rec. trav. chim. Pays-Bas **59,** 191 (1940).

383. MESSERSCHMIDT, W.: Steam Distillate from *Hamamelis virginiana* Leaves. II. Planta Med. suppl. **1967,** 102 (Ger.); Chem. Abstr. **68,** 36 759 x (1968).

384. MILLER, J. A.: Naturally Occurring Substances that can Induce Tumors. In: Toxicants Occurring Naturally in Foods (Committee on Food Protection — Food and Nutrition Board — National Council), 2nd ed., chapt. 23, p. 508. Washington, D. C.: National Academy of Sciences. 1973.

385. MILLS, R. W., and T. MONEY: Sesquiterpenoids. In: A Specialist Periodical Report. Terpenoids and Steroids (K. H. OVERTON, ed.), Vol. 4, p. 77. London: The Chemical Society, Burlington House. 1974.

386. MILTITZER Berichte über ätherische Oele, Riechstoffe usw., p. 38. Miltitz: VEB Chemisches Kombinat. 1969. Compare here the literature on this subject.

387. MINYARD, J. P., A. C. THOMPSON, and P. A. HEDIN: Constituents of the Cotton Bud. VIII. β-Bisabolol, a New Sesquiterpene Alcohol. J. Organ Chem. (U.S.A.) **33,** 909 (1968).

388. MOELJOHARDJO, D. S.: On the Volatile Flavour Compounds of Cooked Trassi, a Cured Shrimp Paste Condiment of the Far East. Thesis. Wageningen: Centre for Agricultural Publishing and Documentation. 1972.

389. MOOKHERJEE, B. D., R. E. DECK, and S. S. CHANG: Relationship Between Monocarbonyl Compounds and Flavor of Potato Chips. J. Agric. Food Chem. **13,** 131 (1965).

390. MOOKHERJEE, B. D., and R. W. TRENKLE: Isolation, Identification, and Biogenesis of Bifunctional Compounds in Lavandin Oil. J. Agric. Food Chem. **21,** 298 (1973).

391. MOSHONAS, M. G., and E. D. LUND: Aldehydes, Ketones and Esters in Valencia Orange Peel Oil. J. Food Sci. **34,** 502 (1969).

392. MOSKOWITZ, H. R.: Towards a Resolution of Dual Approaches in Flavor Analysis. Lebensm.-Wiss. Technol. **8,** 237 (1975).

393. MOSKOWITZ, H. R., and C. D. BARBE: Psychometric Analysis of Food Aromas by Profiling and Multidimensional Scaling. J. Food Sci. **41,** 567 (1976).

394. MOTL, O.: Plant Substances. XXXVI. Components of *Laserpitium siler* Fruits. Collect. Czech. Chem. Comm. **38,** 3737 (1973).

395. MOUSSERON-CANET, M., J.-C. MANI et J. P. DALLE: Photoxydation Sensibilisée dans la Série de la β-Ionone. Bull. soc. chim. France **1967,** 608.

396. MOZELL, M. M., and M. JAGODOWICZ: Chromatographic Separation of Odorants by the Nose: Retention Times Measured Across *in vivo* Olfactory Mucosa. Science **181,** 1247 (1973).

397. MÜGGLER-CHAVAN, F., R. VIANI, J. BRICOUT, J. P. MARION, H. MECHTLER, D. REYMOND et R. H. EGLI: Sur la Composition de l'Arôme de Thé. III. Identification de Deux Cétones Apparentées aux Ionones. Helv. Chim. Acta **52,** 549 (1969).

398. MUKERJI, S. K., and H. L. SHARMA: Investigation on the Offensive Odour of Hemiptera Bugs. Tetrahedron Letters **1966,** 2479.

399. MULDERS, E. J.: The Odour of White Bread. IV. Quantitative Determination of Constituents in the Vapor and Their Odour Values. Z. Lebensm.-Unters. Forsch. **151,** 310 (1973).

400. MÜLLER, C. J., R. E. KEPNER, and A. D. WEBB: Lactone in Wines. Review. Amer. J. Enol. Viticult. **24,** 5 (1973); Chem. Abstr. **79,** 16821k (1973).
401. MULLER, C. J., R. K. CREVELING, and W. G. JENNINGS: Some Minor Sesquiterpene Hydrocarbons of Black Pepper. J. Agric. Food Chem. **16,** 113 (1968).
402. MURRAY, K. E.: α-Farnesene: Isolation from the Natural Coating of Apples. Austral. J. Chem. **22,** 197 (1969).
403. MURRAY, K. E., P. A. BANNISTER, and R. G. BUTTERY: Geosmin: An Important Volatile Constituent of Beetroot *(Beta vulgaris)*. Chem. and Ind. **1975,** 973.
404. MUSSINAN, C. J., and J. P. WALRADT: Organic Acids from Fresh California Strawberries. J. Agric. Food Chem. **23,** 482 (1975).
405. NÄF, F., and R. DECORZANT: A Stereospecific Synthesis of (*E,Z*)-α,β-γ,δ-Diunsaturated Aldehydes, Ketones and Esters Using the Benary Reaction. Helv. Chim. Acta **57,** 1309 (1974).
406. NÄF, F., R. DECORZANT, B. WILLHALM, A. VELLUZ, and M. WINTER: Structure and Synthesis of Two Novel Ionones Identified in the Purple Passionfruit (*Passiflora edulis* Sims.). Tetrahedron Letters **1977,** 1413.
407. NÄF, F., and P. DEGEN: Stereospecific Olefin Synthesis *via* Lithium Vinylcuprates. Helv. Chim. Acta **54,** 1939 (1971).
408. NÄF-MÜLLER, R., und B. WILLHALM: Über die flüchtigen Anteile der Ananas. Helv. Chim. Acta **54,** 1880 (1971).
409. NAGASAMPAGI, B. A., L. YANKOV, and S. DEV: Sesquiterpenoids from the Wood of *Cedrela toona* Roxb.; Partial Synthesis of T-Muurolol, T-Cadinol and Cubenol; Structures of δ-Cadinene and δ-Cadinol. Tetrahedron Letters **1968,** 1913.
410. NAGASAWA, T., K. UMEMOTO, T. TSUNEYA, and M. SHIGA: Wild Mints of Tokai Districts. II. Absolute Configuration and Conformation of a New Terpenic Alcohol Isolated from Essential Oil of *Mentha gentilis*. Nippon Nogei Kagaku Kaishi **48,** 467 (1974); Chem. Abstr. **82,** 57958r (1975).
411. — — — — Studies on the Wild Mints of Tolai Districts. IV. New Terpene Alcohols in Essential Oil of *Mentha gentilis* L., Containing (+)-Pulegone as a Main Compound. J. Agric. Chem. Soc. Japan **49,** 217 (1975).
412. NAIMIE, H., Z. SAMEK, L. DOLEJS, and Z. REHACEK: Volatile Components of Asafoetida. Two Novel Naturally Occurring Sulfur Compounds with Pesticidal Activity. Collect. Czech. Chem. Commun. **37,** 1166 (1972).
413. NAKATANI, Y., S. SATO, and T. YAMANISHI: 3*S*-(+)-3,7-Dimethyl-1,5,7-octatrien-3-ol in the Essential Oil of Black Tea. Agric. Biol. Chem. **33,** 967 (1969).
414. NAKATANI, Y., and T. YAMANISHI: Synthèse totale des *cis*- et *trans*-1-oxa-8-oxo-2,6,10,10-tetraméthylspiro(4 : 5)-6-décènes. Tetrahedron Letters **1969,** 1995.
415. NANO, G. M., A. FUNDARO, R. CALVINO, and P. CABELLA: Sesquiterpene Components of *Mentha piperita* Italo-Mitcham Cultivated in the Piedmont. Essenze Deriv. Agrumari **42,** 325 (1972); Chem. Abstr. **78,** 546u (1973).
416. NARAYANAN, C. S., K. S. KULKARNI, A. S. VAIDYA, S. KAUTHAMANI, G. LAKSHMI KUMARI, B. V. BAPAT, S. K. PAKNIKAR, S. N. KULKARNI, G. R. KELKAR, and S. C. BHATTACHARYYA: Terpenoids. XLVI. Components of Indian Valerian Root Oil. Tetrahedron **20,** 963 (1964).
417. NAULT, L. R., M. E. MONTGOMERY, and W. S. BOWERS: Ant-Aphid Association: Role of Aphid Alarm Pheromone. Science **192,** 1349 (1976).
418. NAVES, Y. R.: Sur la Présence d'Ionones dans les Produits Végétaux. Helv. Chim. Acta **30,** 956 (1947).
419. — Etudes sur les Matières Végétales Volatiles. LVI. Sur de Nouvelles Cétones, les Tagéténones, Isolées de l'Huile Essentielle de *Lippia asperifolia* Rich. Helv. Chim. Acta **31,** 29 (1948).
420. — Études sur les Matières Végétales Volatiles. LX. Présence d'Époxydihydrocaryo-

phyllène (Caryophyllenoxyd de Treibs) dans l'Extrait Benzénique de Bourgeons Floraux du Giroflier. Helv. Chim. Acta **31,** 378 (1948).

421. — Sur Quelques Constituants Inédits de la Gomme-Résine de Galbanum. Parf. Cosm. Savons **12,** 586 (1969).

422. — Pipériténone, Isopipériténone et Dérivés. Riv. Ital. É.P.P.O.S. **58,** 136 (1976).

423. NAVES, Y. R., et A. V. GRAMPOLOFF: Études sur les Matières Végétales Volatiles CLX. Sur les *p*-Menthadiénols des Huiles Essentielles de *Cymbopogon martini* Stapf var. *sofia* (Essence de Gingergrass) et de *C. densiflorus* Stapf. Bull. Soc. Chim. France **1960,** 37.

424. NAVES, Y. R., D. LAMPARSKY et P. OCHSNER: Etudes sur les Matières Végétales Volatiles. CLXXIV (1). Présence de Tétrahydropyrannes dans l'Huile Essentielle de Géranium. Bull. soc. chim. France **1961,** 645.

425. NAWAR, W. W.: Thermal Degradation of Lipids — A Review. J. Agric. Food Chem. **17,** 18 (1969).

426. NAYA, Y., and M. KOTAKE: New Monoterpenoids from Hop Oil. Tetrahedron Letters **1968,** 1645.

427. — — Volatile Constituents of Hops. II. Nippon Kagaku Zasshi **89,** 1113 (1968); Chem. Abstr. **70,** 80786u (1969).

428. — — Natural Occurrence of Humulol and Tricyclohumuladiol. Bull. Chem. Soc. Japan **42,** 2405 (1969); Chem. Abstr. **71,** 91678g (1969).

429. — — Natural Occurrence of Humuladienone, Humulenone-II and α-Corocalene, New Constituents of Hop Oil. Bull. Chem. Soc. Japan **42,** 2088 (1969); Chem. Abstr. **71,** 91677f (1969).

430. — — Constituents of Hops. IV. Volatile Composition of *Humulus lupulus* var. *cordifolius.* Bull. Chem. Soc., Japan **43,** 2956 (1970); Chem. Abstr. **73,** 127629q (1970).

431. NELSON, E. K.: The Volatile Oil of Amur Cork Tree Fruit. J. Amer. Chem. Soc. **60,** 920 (1938).

432. NELSON, P. E., and J. E. HOFF: Tomato Volatiles: Effect of Variety, Processing and Storage Time. J. Food Sci. **34,** 53 (1969).

433. NELSON, E. K., and M. S. LOWMAN: Chemical Composition of Sweet Basil Oil from Virginia. Amer. Perf. essent. Oil Rev. **31** (Dec.), 69 (1935).

434. NGUYEN, THI-THANH, and T. YAMANISHI: Flavor Components in Vietnamese Green Tea and Lotus Tea. Agric. Biol. Chem. **39,** 1263 (1975).

435. NIGAM, M. C., K. L. HANDA, I. C. NIGAM, and L. LEVI: Essential Oils and Their Constituents. XXIX. The Essential Oil of Marihuana: Composition of Genuine Indian *Cannabis sativa* L. Canad. J. Chem. **43,** 3372 (1965).

436. NIGAM, M. C., I. C. NIGAM, L. LEVI, and K. L. HANDA: Essential Oils and Their Constituents. XXII. Detection of New Trace Components in Oil of Ginger. Canad. J. Chem. **42,** 2610 (1964).

437. NISHIMURA, K., and M. MASUDA: Minor Constituents of Whisky Fusel Oils. 1. Basis, Phenolic and Lactonic Compounds. J. Food Sci. **36,** 819 (1971).

438. NOSE, M., Y. NAKATANI, and T. YAMANISHI: Studies in the Flavor of Green Tea. IX. Identification and Composition of Intermediate and High Boiling Constituents in Green Tea Flavor. Agric. Biol. Chem. **35,** 261 (1971).

439. NURSTEN, H. E.: Chemistry of Flavours. Past, Present, and Future. Int. Flavours Food Addit. **6,** 75 (1975).

440. NURSTEN, H. E., and M. R. SHEEN: Volatile Flavour Components of Cooked Potato. J. Sci. Food Agric. **25,** 643 (1974).

441. NURSTEN, H. E., and A. A. WILLIAMS: Fruit Aromas. A Survey of Components Identified. Chem. and Ind. **1967,** 486.

442. NURSTEN, H. E., and M. L. WOOLFE: An Examination of the Volatile Compounds

Present in Cooked Bramley's Seedling Apples and the Changes They Undergo on Processing. J. Sci. Food Agric. **23**, 803 (1972).

443. OHLOFF, G. (Firmenich SA, Geneva): Personal Observation.

444. — Personal Observations with the Pure Compounds **156, 157** and **158.**

445. — Zur thermischen Isomerisation von Citral. Tetrahedron Letters **1960,** No 11, 10.

446. — Chemie der Geruchs- und Geschmacksstoffe. In: Fortschritte der chemischen Forschung (A. DAVISON, M. J. S. DEWAR, K. HAFNER et al., eds.), Vol. 12, No. 2, p. 185. Berlin-Heidelberg-New York: Springer. 1969.

447. — Classification and Genesis of Food Flavors. 3rd International Congress of Food Science and Technology, SOS, Washington 1970. Proceedings, p. 368. Chicago, Ill. 60 601: Inst. Food Technol. 1971; reprinted from Flavour Ind. **3,** 501 (1972).

448. — Relationship Between Odor Sensation and Stereochemistry of Decalin Ring Compounds. In: Gustation and Olfaction (G. OHLOFF and A. F. THOMAS, eds.), p. 178, London-New York: Academic Press. 1971.

449. — Odorous Properties of Enantiomeric Compounds. In: Olfaction and Taste (D. SCHNEIDER, ed.), Vol. IV, p. 156. Stuttgart: Wissenschaftliche Verlagsgesellschaft mbH. 1972.

450. — Fette als Aromavorstufen. In: Functional Properties of Fats in Foods (J. SOLMS, ed.), p. 119. Zürich: Forster-Verlag AG. 1973.

451. — Singlet Oxygen: A Reagent in Organic Synthesis. Pure and Appl. Chem. **43,** 481 (1975).

452. OHLOFF, G., and W. GIERSCH: The Sensory Properties of (+)-Nootkatone and Related Compounds. In: Gustation and Olfaction (G. OHLOFF and A. F. THOMAS, eds.). London-New York: Academic Press. 1971.

453. — — The Sensory Properties of (+)-Nootkatone and Related Compounds. In: Gustation and Olfaction (G. OHLOFF and A. F. THOMAS, eds.), p. 184. London-New York: Academic Press. 1971.

454. OHLOFF, G., W. GIERSCH, K. H. SCHULTE-ELTE und E. sz. KOVÁTS: Δ^1-*p*-Menthen-9-al, ein Bestandteil des bulgarischen Rosenöls. Helv. Chim. Acta **52,** 1531 (1969).

455. OHLOFF, G., W. GIERSCH, K. H. SCHULTE-ELTE und C. VIAL: Zur Stereochemie der Geruchswahrnehmung von 1-Dekalon-Derivaten und ihren oxaanalogen Verbindungen. Helv. Chim. Acta **59,** 1140 (1976).

456. OHLOFF, G., E. KLEIN und G. O. SCHENCK: Darstellung von "Rosenoxyden" und anderen Hydropyran-Derivaten über Photohydroperoxyde. Angew. Chem. **73,** 578 (1961).

457. OHLOFF, G., and M. PAWLAK: Synthesis of "Pear Ester", a Novel Synthesis of 2,4-Diolefinic Aldehydes and Esters. Helv. Chim. Acta **56,** 1176 (1973).

458. OHLOFF, G., V. RAUTENSTRAUCH und K. H. SCHULTE-ELTE: Modellreaktionen zur Biosynthese von Verbindungen der Damascon-Reihe und ihre präparative Anwendung. Helv. Chim. Acta **56,** 1503 (1973).

459. OHLOFF, G., K. H. SCHULTE-ELTE und B. WILLHALM: Zur Darstellung von Tetrahydropyran- und Tetrahydrofuranderivaten aus 1,7- bzw. 1,6-Allyldiolen durch Dehydration in der Allylstellung. Helv. Chim. Acta **47,** 602 (1964).

460. OHLOFF, G., C. VIAL, and A. F. THOMAS: Published in 451.

461. OHTA, Y., and Y. HIROSE: Constituents of Cold-Pressed Peel Oil of *Citrus natsudaidai* Hayata. Agric. Biol. Chem. **30,** 1196 (1966).

462. — — The Structure of Cubenol and epi-Cubenol. Tetrahedron Letters **1967,** 2073.

463. OSMAN, S., and J. BARSON: The Isolation of Farnesyl Acetone from Cigar Smoke Condensate. Tobacco Sci. **10,** 85 (1966); published in Tobacco **163,** 21 (1966).

464. OSMAN-ISMAIL, F., and J. SOLMS: The Formation of Inclusion Compounds of Starches with Flavor Substances. Lebensm.-Wiss. Technol. **6,** 147 (1973).

465. OTSUKA, K., Y. ZENIBAYASHI, M. ITOH, and A. TOTSUKA: Presence and Significance of Two Diasteromers of β-Methyl-γ-octalactone in Aged Distilled Liquors. Agric. Biol. Chem. **38**, 485 (1974).

466. OUWELAND, G. A. M. VAN DEN, and H. G. PEER: Components Contributing to Beef Flavor. Volatile Compounds Produced by the Reaction of 4-Hydroxy-5-methyl-3(2*H*)-furanone and Its Thio Analog with Hydrogen Sulfide. J. Agric. Food Chem. **23**, 501 (1975).

467. PARIS, M.: Cannabis Essential Oil. Mysterious Perfume. Riv. Ital. Essenze, Profumi, Piante Off., Aromi, Saponi, Cosmet., Aerosol **57**, 83 (1975); Chem. Abstr. **83**, 168 298r (1975).

468. PARK, R. J., K. E. MURRAY, and G. STANLEY: 4-Hydroxydodec-*cis*-6-enoic Acid Lactone: An Important Component of Lamb Flavour from Animals Fed a Lipid-Protected Dietary Supplement. Chem. and Ind. **1974**, 380.

469. PARKS, O. W., N. P. WONG, C. A. ALLEN, and D. P. SCHWARTZ: 6-*trans*-Nonenal: An Off-Flavor Component of Foam Spray-Dried Milks. J. Dairy Sci. **52**, 953 (1969).

470. PARLIMENT, T. H., W. CLINTON, and R. SCARPELLINO: *trans*-2-Nonenal: Coffee Compound with Novel Organoleptic Properties. J. Agric. Food Chem. **21**, 485 (1973).

471. PENFOLD, A. R.: The Essential Oil of a New Species of Anemone Leaf *Boronia* Rich in Ocimene. J. Proc. Roy. Soc. New South Wales **62**, 263 (1928/29).

472. PEYRON, L.: Les Flaveurs sur le Plan de l'Alimentation Humaine — Isolement des Flaveurs. Riv. ital. Essenze Profumi, Piante off., Aromi, Saponi, Cosmet., Aerosol **55**, 617 (1973); Chem. Abstr. **81**, 24 205u (1974).

473. PICARDI, S. M., and P. ISSENBERG: Investigation of Some Volatile Constituents of Mushrooms *(Agaricus bisporus)*: Changes which Occur During Heating. J. Agric. Food Chem. **21**, 959 (1973).

474. PICKENHAGEN, W., and A. FURRER (Firmenich SA, Geneva): Unpublished Results.

475. PIPPEN, E. L., and M. NONAKA: Volatile Carbonyl Compounds of Cooked Chicken. II. Compounds Volatilized with Steam during Cooking. Food Res. **25**, 764 (1960).

476. PITTET, O., and E. M. KLAIBER: Synthesis and Flavor Properties of Some Alkyl-Substituted α-Pyrone Derivatives. J. Agric. Food Chem. **23**, 1189 (1975).

477. POWERS, J. J., and M. C. QUINLAN: Subjective-Objective Evaluation of Model Odor Systems. Lebensm.-Wiss. Technol. **6**, 209 (1973).

478. PRAAG, M. VAN, H. S. STEIN, and M. S. TIBBETTS: Steam Volatile Aroma Constituent of Roasted Cocoa Beans. J. Agric. Food Chem. **16**, 1005 (1968).

479. PRELOG, V., J. FÜHRER, R. HAGENBACH und R. SCHNEIDER: Untersuchungen über Organextrakte. 13. Über die Isolierung von Jonon-Derivaten aus dem Harn trächtiger Stuten. Helv. Chim. Acta **31**, 1799 (1948).

480. PRELOG, V., und H. L. MEIER: Untersuchungen über Organextrakte. 18. Über die biochemische Oxydation von β-Jonon im Tierkörper. Helv. Chim. Acta **33**, 1276 (1950).

481. PRELOG, V., und B. VATERLAUS: Untersuchungen über Organextrakte und Harn. 16. Über die Konstitution des ungesättigten Ketons $C_{13}H_{18}O$ aus dem Harn trächtiger Stuten. Helv. Chim. Acta **32**, 2082 (1949).

482. — — Untersuchungen über Organextrakte und Harn. 19. Über die Konstitution der Oxy-Ketone E und G aus dem Harn trächtiger Stuten. Helv. Chim. Acta **33**, 1725 (1950).

483. PRESTWICH, G. D., F. B. WHITFIELD, and G. STANLEY: Synthesis and Structures of Dihydroedulan I and II, Trace Components from the Juice of *Passiflora edulis* Sims. Tetrahedron **32**, 2945 (1976).

484. PUROHIT, R. M., und K. DEVI: Chromatographische Analyse des aus den Rhizomen von *Alpina allughas* Rosc. gewonnenen ätherischen Oeles. Riechst. Aromen, Körperpflegem. **26**, 139 (1976).

485. PYYSALO, T.: Studies on the Volatile Compounds of Some Berries in Genus *Rubus*, especially Cloudberry (*Rubus chamaemorus* L.) and Hybrids Between Raspberry (*Rubus idaeus* L.) and Arctic Bramble (*Rubus arcticus* L.). Materials and Processing Technology, Technical Research Centre of Finland, Publication 14. Helsinki: 1976.
486. — Nachweis flüchtiger Verbindungen in Hybriden zwischen Himbeere (*Rubus ideaus* L.) und arktischer Brombeere (*Rubus arcticus* L.). Z. Lebensm.-Unters. Forsch. **162**, 263 (1976).
487. QUILICO, A., F. PIOZZI, and M. PAVAN: The Structure of Dendrolasin. Tetrahedron **1**, 177 (1957).
488. RAMSHAW, E. H., and P. J. HARDY: Volatile Compounds in Dried Grapes. J. Sci. Food Agric. **20**, 619 (1969).
489. REDDY, M. C., R. C. LINDSAY, and D. D. BILLS: Ester Production by *Pseudomonas fragi*. III. Synergistic Flavor Interaction of Esters at Subthreshold Concentrations. J. Dairy Sci. **52**, 1198 (1969).
490. REGNIER, F. E., G. R. WALLER, and E. J. EISENBRAUN: Studies on the Composition of Essential Oils of Three Nepeta Species. Phytochem. **6**, 1281 (1967).
491. RENOLD, W., R. NÄF-MÜLLER, U. KELLER, B. WILLHALM, and G. OHLOFF: An Investigation of Tea Aroma. I. New Volatile Black Tea Constituents. Helv. Chim. Acta **57**, 1301 (1974).
492. RENOLD, W., W. SKORIANETZ, K. H. SCHULTE-ELTE und G. OHLOFF (Firmenich SA): Spiranderivate. Verfahren zu deren Herstellung und deren Verwendung. Ger. Offenl. 2.504.618 (14. 8. 1975).
493. REYMOND, D., F. MÜGGLER-CHAVAN, R. VIANI, and R. H. EGLI: Characterization of the Aroma of Black Tea. Mitt. Geb. Lebensmittelunters. Hyg. (Bern) **57**, 484 (1966); C. **1968**, 25-3458.
494. REYNOLDS, T. H.: Flavors from Nonenzymatic Browning Reactions. Food Technol. Austr. **22**, 610 (1970).
495. REYNOLDS, T.: Aggressive Chemicals in Plants. Chem. and Ind. **1975**, 603.
496. RHOADES, J. W.: Analysis of the Volatile Constituents of Coffee. J. Agric. Food Chem. **8**, 136 (1960).
497. RIBÉRAU-GAYON, P., J. N. BOIDRON, and A. TERRIER: Aroma of Muscat Grape Varieties. J. Agric. Food Chem. **23**, 1042 (1975).
498. RICE, K. C., and R. S. WILSON: (−)-Isothujane, a Small Nonnitrogenous Molecule with Antinociceptive Activity in Mice. J. medizin. Chem. **19**, 1054 (1976).
499. RICH, V.: Soviet Dairy. Nature **257**, 353 (1975).
500. RICHARD, H. M., and W. G. JENNINGS: Volatile Composition of Black Pepper. J. Food Sci. **36**, 584 (1971).
501. RIJKENS, F., and H. BOELENS: The Future of Aroma Research. Proc. int. Symp. Aroma Research, Zeist, p. 203. Wageningen: Pudoc. 1975.
502. ROBERTS, D. L., and W. A. RHODE: Isolation and Identification of Flavor Components of Burley Tobacco. Tobacco Sci. **16**, 107 (1972).
503. ROBERTS, D. L., and R. L. ROWLAND: Macrocyclic Diterpenes. α- and β-4,8,13-Duvatriene-1,3-diols from Tobacco. J. Organ. Chem. (U.S.A.) **27**, 3989 (1962).
504. ROBERTS, J. B.: Hop Oil. II. Major Constituents of the Nonsaponifiable Fraction. J. Inst. Brewing **69**, 343 (1963).
505. ROTHE, M.: Aroma Values — a Useful Concept? Proc. int. Symp. Aroma Research, Zeist, p. 111. Wageningen: Pudoc. 1975.
506. ROTHE, M., G. WÖLM, L. TUNGER und H. J. SIEBERT: Schwellenkonzentrationen von Aromastoffen und ihre Nutzung zur Auswertung von Aromaanalysen. Die Nahrung **16**, 483 (1972); Chem. Abstr. **78**, 15 568c (1973).
507. RÜCKER, G.: Sesquiterpene. Angew. Chem. **85**, 895 (1973).

508. Russell, G. F., and J. I. Hills: Odor Differences Between Enantiomeric Isomers. Science **172,** 1043 (1971).

509. Russell, G. F., and W. G. Jennings: Occurrence of *cis*- and *trans*-Sabinenehydrate in Oil of Black Pepper. Synthesis by Oxy Mercuration — Demercuration Procedure. J. Agric. Food Chem. **18,** 733 (1970).

510. Safferman, R. S., A. A. Rosen, C. I. Mashni and M. E. Morris: Earthy-Smelling Substance from a Blue-Green Alga. Envir. Sci. Technol. **1,** 25 (1967); Chem. Abstr. **67,** 18 819s (1967).

511. Saijo, R.: Volatile Flavor of Black Tea. II. Examination of the First Fraction of Effluents in Gas Chromatography. Agric. Biol. Chem. **31,** 1265 (1967); Chem. Abstr. **68,** 38 353j (1968).

512. Saijo, R., and T. Takeo: Increase of *cis*-3-Hexen-1-ol Content in Tea Leaves Following Mechanical Injury. Phytochem. **14,** 181 (1975).

513. Sakai, T., K. Nishimura, and Y. Hirose: Structure and Stereochemistry of Four New Sesquiterpenes Isolated from the Wood Oil of *Torreya nucifera.* Bull. Chem. Soc. Japan **38,** 381 (1965); Chem. Abstr. **63,** 3 003d (1965).

514. Sakai, T., K. Yoshihara, and Y. Hirose: Constituents of Fruit Oil from Japanese Pepper. Bull. Chem. Soc. Japan **41,** 1945 (1968).

515. Sakan, T., A. Fujino, and F. Murai: Chemical Components in Matatabi. I. The Isolation of the Physiologically Active Principles, Matatabilactone and Actinide. Nippon Kagaku Zasshi **81,** 1320 (1960); Chem. Abstr. **56,** 11 644a (1962).

516. Sakan, T., A. Fujino, F. Murai, Y. Butsugan, and A. Suzui: The Synthesis of Actinidine. Bull. Chem. Soc. Japan **32,** 1155 (1959); Chem. Abstr. **54,** 15 419a (1960).

517. Sakan, T., S. Isoe, and S. Be Hyeon: The Structure of Actinidiolide, Dihydroactinidiolide and Actinidol. Tetrahedron Letters **1967,** 1623.

518. Sakan, T., S. Isoe, S. Be Hyeon, R. Katsumura, T. Maeda, J. Wolinsky, D. Dickerson, M. Slabaugh, and D. Nelson: The Exact Nature of Matatabilactone and the Terpenes of *Nepeta cataria.* Tetrahedron Letters **1965,** 4097.

519. Sakan, T., F. Murai, S. Isoe, S. Be Hyeon, and Y. Hayashi: Biologically Active C_9-, C_{10}- and C_{11}-Terpenes from *Actinidia polygama, Boschniakia rossica,* and *Menyanthes.* J. Chem. Soc. Japan **90,** 507 (1969) (Jap.); Chem. Abstr. **71,** 53 479k (1969).

520. Sakata, I., and T. Hashizume: Isolation and Identification of 2,3-Dimethyl-4-hydroxy-2-nonenoic Lactone from Shubi (= Japanese Peppermint). Agric. Biol. Chem. **37,** 2441 (1973).

521. Sakato, Y., K. Ina, Y. Yamamoto et al. (Kaken Industry Co., Ltd.): 1-Oxa-8-oxo-2,6,10,10-tetramethylspiro[4.5]deca-6-ene. U.S. Pat. 3.734.932 (22. 5. 1973); Brit. Pat. 1.196.897 (1. 7. 1970); Chem. Abstr. **73,** 66 410x (1970).

522. Salo, P.: Determining the Odor Thresholds for Some Compounds in Alcoholic Beverages. J. Food Sci. **35,** 95 (1970).

523. — Use of Odour Thresholds in Sensorial Testing and Comparisons with Instrumental Analysis. Proc. int. Symp. Aroma Research, Zeist, p. 121. Wageningen: Pudoc. 1975.

524. Sampath, V., M. R. Thakar, S. K. Paknikar, B. K. Sabata, and S. Bhattacharyya: Structure of D- and L-Isomers of β-Bisabolol Oxide-B. Indian J. Chem. **7,** 1060 (1969); Chem. Abstr. **72,** 19 068m (1970).

525. Sanderson, G. W.: Black Tea Aroma and its Formation. Paper delivered at Int. Symp. on Odour and Flavour Substances, Bad Pyrmont (Germany), Oct. 2—4, 1974.

526. Sanderson, G. W., H. Co, and J. G. Gonzalez: Biochemistry of Tea Fermentation: The Role of Carotenes in Black Tea Aroma Formation. J. Food Sci. **36,** 231 (1971).

527. Sanderson, G. W., and H. N. Graham: On the Formation of Black Tea Aroma. J. Agric. Food Chem. **21,** 576 (1973).

528. SASTRY, S. D., W. R. SPRINGSTUBE, and G. R. WALLER: Identification of 5,9-Dehydronepetalactone, a New Monoterpene from *Nepeta cataria*. Phytochem. **11,** 453 (1972).

529. SATO, A., and H. MISHIMA: Synthesis of Theaspirone. Tetrahedron Letters **1969,** 1803.

530. SCHANTZ, M. VON, K. G. WIDÉN, and R. HILTUNEN: Structures of Some Aliphatic Monoterpenoids Isolated from the Essential Oil of *Ledum palustre* L. Acta Chem. Scand. **27,** 551 (1973).

531. SCHENCK, G. O., K. GOLLNICK, G. BUCHWALD, S. SCHROETER und G. OHLOFF: Zur chemischen und sterischen Selektivität der photosensibilisierten O_2-Übertragung auf (+)-Limonen und (+)-Carvomenthen. Liebigs Ann. Chem. **674,** 93 (1964).

532. — — — — —, ibid.: The correct structure of gingergrass alcohol **65** is given there.

533. SCHIFFMAN, S. S.: Physicochemical Correlates of Olfactory Quality. Science **185,** 112 (1974) and Ann. New York Acad. Sci. **237,** 164 (1974).

534. SCHILCHER, H., L. NOVOTNY, K. UBIK, O. MOTL und V. HEROUT: Strukturaufklärung eines dritten Bisaboloxides aus *Matricaria camomilla* L. und massenspektrometrischer Vergleich der Bisaboloxide A, B und C. Arch. Pharm. **309** (3), 189 (1976).

535. SCHILDKNECHT, H.: Chemische Oekologie — Ein Kapitel moderner Naturstoffchemie. Angew. Chem. **88,** 235 (1976).

536. SCHINZ, H.: Die Veilchenriechstoffe. In: Progress in the Chemistry of Organic Natural Products (L. ZECHMEISTER, ed.). Vol. VIII, p. 146. Wien: Springer. 1951.

537. SCHMIDLIN-MESZAROS, J.: Probleme im Gebiet der Lebensmittelzusatzstoffe. Ermittlung des *p*-Hydroxyphenylbutanon-(3)s, des Himbeerketons. Alimenta **10,** 39 (1971).

538. SCHMIDT, H.: Untersuchung der hochsiedenden Anteile des spanischen Eucalyptusöls von *Eucalyptus globulus*. Chem. Ber. **80,** 528 (1947).

539. — 3-Isopropyliden-1-acetyl-cyclopenten-(5), ein neuer Bestandteil des spanischen Eucalyptusöls von *Eucalyptus globulus*. Chem. Ber. **80,** 533 (1947).

540. — Zur Kenntnis des Pfefferminzöls. Vorkommen von Jasmon im ätherischen Oel von *Mentha piperita* L. Ber. dtsch. chem. Ges. **80,** 538 (1947).

541. — Über Inhaltsstoffe des Kümmelöls; ein Beitrag zur Konfiguration der Dihydrocarveole. Chem. Ber. **83,** 193 (1950); ibid. **88,** 455 (1955), Fußnote 7.

542. — Zur Dehydratation der raumisomeren Dihydrosobrerole (Dihydrocarveolhydrate). Chem. Ber. **88,** 459 (1955).

543. SCHMITZ, F. J., and F. J. MCDONALD: Marine Natural Products: Dactyloxene-B a Sesquiterpene Ether from the Sea Hare, *Aplysia dactylomela*. Tetrahedron Letters **1974,** 2541.

544. SCHORMÜLLER, J., und H. J. KOCHMANN: Gaschromatographisch-massenspektrometrische Analyse leichtflüchtiger Aromastoffe der Tomate. Z. Lebensm.-Unters. Forsch. **141,** 1 (1969).

545. SCHRATZ, E., and S. M. J. S. QADRY: The Synthesis of the Etheral Oils in the Fruit and the Living Plant of *Coriandrum sativum*. I. Investigation of the Fruit Oils. Planta Med. **14,** 310 (1966).

546. SCHREIER, P., und F. DRAWERT: Gaschromatographisch-massenspektrometrische Untersuchung flüchtiger Inhaltsstoffe des Weines. I. Unpolare Verbindungen des Weinaromas. Z. Lebensm.-Unters. Forsch. **154,** 273 (1974).

547. — — Gaschromatographisch-massenspektrometrische Untersuchung flüchtiger Inhaltsstoffe des Weines. V. Alkohole, Hydroxyester, Lactone und andere polare Komponenten des Weinaromas. Chem. Mikrobiol., Technol. Lebensm. **3,** 154 (1974); Chem. Abstr. **82,** 153 714d (1945).

548. SCHREIER, P., F. DRAWERT und A. JUNKER: Identifizierung von 3,7-Dimethyl-1.5.7-octatrien-3-ol als flüchtige Komponenten des Trauben- und Weinaromas. Z. Lebensm.-Unters. Forsch. **155,** 98 (1974).

549. — — — Identification of Volatile Constituents from Grapes. J. Agric. Food Chem. **24,** 331 (1976).

550. SCHREIER, P., F. DRAWERT, Z. KERÈNYI und A. JUNKER: Gaschromatographisch-massenspektrometrische Untersuchung flüchtiger Inhaltsstoffe des Weines. VI. Aromastoffe in Tokajer-Trockenbeerauslese-(Aszu-)Weinen. a) Neutralstoffe. Z. Lebensm.-Unters. Forsch. **161,** 249 (1976).

551. SCHULTE-ELTE, K. H., und G. OHLOFF: Über En-Synthesen mit Singulett-Sauerstoff. Die Farbstoff-sensibilisierte Photooxygenierung von (−)-Caryophyllen und (−)-iso-Caryophyllen und die Stereochemie der dargestellten Caryophylladienole. Helv. Chim. Acta **51,** 494 (1968).

552. SCHULTE-ELTE, K. H., M. GADOLA und B. L. MÜLLER: Farbstoffsensibilisierte Photo-Oxygenierung von (−)-*cis*-Pulegol und (+)-Pulegon. Zur Addition von O_2 ($^1\Delta_g$) an Doppelbindungen mit allylständiger Hydroxyl- oder Carbonyl-Gruppe. Helv. Chim. Acta **54,** 1870 (1971).

553. SCHULTE-ELTE, K. H. (Firmenich SA): Use of 2,2,5-Trimethyl-cyclohex-5-ene-1,4-dione as a Flavouring Agent. Swiss Pat. 563.127 (30. 6. 1975).

554. SCHUMACHER, J. N. (R. J. Reynolds Tobacco Co.): Flavoring Tobacco with a Benzopyran. U.S. Pat. 3.389.706 (25. 6. 1968); Chem. Abstr. **69,** 41 817 (1968).

555. SCHUMACHER, J. N., W. A. ROHDE, and D. L. ROBERTS (R. J. Reynolds Tobacco Co.): Tobacco Additives. U.S. Pat. 3.380.457 (30. 4. 1968); Chem. Abstr. **70,** 35 165e (1969).

556. SCHUMACHER, J. N., and L. VESTAL: Isolation and Identification of Some Components of Turkish Tobacco. Tobacco Sci. **18,** 43 (1974).

557. SCHUTTE, L.: Precursors of Sulfur-Containing Flavor Components. CRC Critical Rev. Food Techn. **4,** 457 (1974).

558. — Flavor Precursors in Food Systems. Abstracts of Papers. 170th ACS National Meeting, Chicago, Aug. 1975, AGFD 14.

559. SCHWARTZ, D. P., and A. I. VIRTANEN: Volatile Carbonyl Compounds in the Milk Fat of Normally- and Synthetically-fed Cows. Acta Chem. Scand. **22,** 1717 (1968).

560. SEALS, R. G., and E. G. HAMMOND: Carbonyl Flavor Compounds of Oxidized Soybean and Linseed Oils. J. Amer. Oil Chem. Soc. **47,** 278 (1970).

561. SECK, S., et J. CROUZET: Constituants Volatils de *Lycopersicon esculentum.* Phytochem. **12,** 2925 (1973).

562. SEIDEL, C. F., D. FELIX, A. ESCHENMOSER, K. BIEMANN, E. PALLUY und M. STOLL: Zur Kenntnis des Rosenöls. 2. Mitt. Die Konstitution des Oxyds $C_{10}H_{18}O$ aus bulgarischem Rosenöl. Helv. Chim. Acta **44,** 598 (1961).

563. SEIDEL, C. F., and M. STOLL: Über die flüchtigen Bestandteile der grauen Ambra. 3. Mitt. Isolierung und Identifizierung von γ-Cyclohomogeraniol im grauen Ambra. Synthese von 1,1-Dimethyl-2-(β-hydroxyäthyl)-3-methylen-cyclohexan (γ-Cyclohomo-geraniol) und 1,1-Dimethyl-2-(γ-hydroxypropyl)-3-methylen-cyclohexan. Helv. Chim. Acta **40,** 1990 (1957).

564. SEKIYA, J., S. NUMA, T. KAJIWARA, and A. HATANAKA: Biosynthesis of Leaf Alcohol. Formation of 3*Z*-Hexenal from Linoleic Acid in Chloroplasts of *Thea sinensis* leaves. Agric. Biol. Chem. **40,** 185 (1976).

565. SEMMLER, F. W.: Über das ätherische Oel des Knoblauchs *(Allium sativum).* Arch. Pharm. **230,** 434 (1892).

566. — Über olefinische Bestandteile ätherischer Oele. Ber. dtsch. chem. Ges. **24,** 201 (1891).

567. SEVENANTS, M. R., and W. G. JENNINGS: Volatile Components of Peach. II. J. Food Sci. **31,** 81 (1966).

568. — — Occurrence of 6-Pentyl-α-pyrone in Peach Essence. J. Food Sci. **36,** 536 (1971).

569. SHELDON, R. M., R. C. LINDSAY, and L. M. LIBBEY: Identification of Volatile Flavor Compounds from Roasted Filberts. J. Food Sci. **37,** 313 (1972).

570. SHIGEMATSU, H., R. ONO, Y. YAMASHITA, and Y. KABURAKI: Studies on Composition of Tobacco Smoke. XVI. Volatile Compounds from the Neutral Fraction of Tobacco Smoke Condensates. Agric. Biol. Chem. **35,** 1751 (1971).

571. SHINODA, N., M. SHIGA, and K. NISHIMURA: Constituents of Yuzu *(Citrus junos)* Oil. Agric. Biol. Chem. **34,** 234 (1970).

572. SIEK, T. J., J. A. ALBIN, L. A. SATHER, and R. C. LINDSAY: Comparison of Flavor Thresholds of Aliphatic Lactones with Those of Fatty Acids, Esters, Aldehydes, Alcohols and Ketones. J. Dairy Sci. **54,** 1 (1971).

573. SILVERSTEIN, R. M., J. O. RODIN, C. M. HIMEL, and R. W. LEEPER: Volatile Flavor and Aroma Components of Pineapple. II. Isolation and Identification of Charicol and γ-Caprolactone. J. Food Sci. **30,** 668 (1965).

574. SILVERSTEIN, R. M., J. O. RODIN, D. L. WOOD, and L. E. BROWNE: Identification of Two New Terpene Alcohols from Frass Produced by *Ips confusus* in Ponderosa Pine. Tetrahedron **22,** 1929 (1966).

575. SINGH, S., and S. L. TUCKEY: Chemical Changes in Fat and Protein in Limburger Cheese During Ripening. J. Dairy Sci. **51,** 942 (1968).

576. SISIDO, K., S. KUROZUMI, K. UTIMOTO, and T. ISIDA: Fragrant Flower Constituents of *Osmanthus fragrans.* II. Perf. Essent. Oil Rec. **58,** 212 (1967).

577. SLATER, C. A.: Composition of Natural Lime Oil. Chem. and Ind. **1961,** 833.

578. SMITH, D. E., and J. R. COFFMAN: Analysis of Food Flavors by GLC. Separation and Identification of the Neutral Components from Bread Pre-Ferment Liquid. Analyt. Chemistry **32,** 1733 (1960).

579. SMITH, K., G. F. THOMPSON, and H. D. KOSTER: Sweat in Schizophrenic Patients: Identification of the Odourous Substance. Science **166,** 398 (1969).

580. SOLMS, J.: Die Bedeutung von Lebensmittelaromen. Chem. Rundschau **26** (Nr. 15), 19 (1973).

581. SOLMS, J., F. OSMAN-ISMAIL, and M. BEYELER: Interaction of Volatiles with Food Components. Can. Inst. Food Sci. Technol. **6,** A 10 (1973).

582. SOOD, V. K.: Isolation of India Valerian Oil from *Valeriana wallichii* Roots and Rhizomes. Perf. Essent. Oil Rec. **56,** 656 (1965); Chem. Abstr. **64,** 4 857h (1966).

583. ŠORM, F., J. MLEZIVA, Z. ARNOLD, and J. PLIVA: Terpenes. XIII. Sesquiterpenes from the Essential Oil of Hops. Collect. Czech. Chem. Commun. **14,** 699 (1949).

584. SPENCE, L. R., S. R. PALAMAND, and W. A. HARDWICK: Identification of C_4 and C_5 Lactones in Beer. Tech. Quart. **10,** 127 (1973); Chem. Abstr. **79,** 113 879t (1973).

585. STAHL, E.: Die Leitsubstanzen der Haschisch-Suchhunde. Kriminalistik (Hamburg) **1973,** 385.

586. STAHL, W. H.: Compilation of Odor and Taste Threshold Values Data. ASTM Data Series DS 48. Baltimore: American Society for Testing and Materials. 1973.

587. STARK, W., and D. A. FORSS: A Compound Responsible for Metallic Flavor in Dairy Products. Isolation and Identification. J. Dairy Res. **29,** 173 (1962).

588. — — A Compound Responsible for Mushroom Flavour in Dairy Products. J. Dairy Res. **31,** 253 (1964).

589. STARK, W., and G. URBACH: The Level of Saturated and Unsaturated γ-Dodecalatones in the Butter Fat from Cows on Various Rations. Chem. and Ind. **1974,** 413.

590. STEDMAN, R. L.: The Chemical Composition of Tobacco and Tobacco Smoke. Chem. Rev. **68,** 153 (1968).

591. STEVENS, K. L., D. G. GUADAGNI, and D. J. STERN: Odor Character and Threshold Values of Nootkatone and Related Compounds. J. Sci. Food Agric. **21,** 590 (1970).

592. STEVENS, K. L., R. LUNDIN, and D. L. DAVIS: Acid Catalyzed Rearrangement of β-Ionone Epoxide. Tetrahedron **31,** 2749 (1975).

593. STEVENS, K. L., R. E. LUNDIN, and R. TERANISHI: Volatiles from Oranges. III. The Structure of Sinensal. J. Organ. Chem. (U.S.A.) **30,** 1690 (1965).

594. STOFFELSMA, J., G. SIPMA, D. K. KETTENES, and J. PYPKER: New Volatile Components of Roasted Coffee. J. Agric. Food Chem. **16**, 1000 (1968).

595. STOLL, A., E. SEEBECK und D. STAUFFACHER: Isolierung und Charakterisierung von bisher unbekannten Inhaltsstoffen aus dem Neutralteil des frischen Baldrians. Helv. Chim. Acta **40**, 1203 (1957).

596. STOLL, M., M. WINTER, F. GAUTSCHI, I. FLAMENT et B. WILLHALM: Recherches sur les Arômes. 13. Sur l'Arôme de Café. I. Helv. Chim. Acta **50**, 628 (1967).

597. STOLTZ, L. P., T. R. KEMP, W. O. SMITH, JR., W. T. SMITH, JR., and C. E. CHAPLIN: 1,2-Dihydro-1,1,6-trimethylnaphthalene from Strawberry Oil. Phytochem. **9**, 1157 (1970).

598. STRATING, J., and P. VAN EERDE: The Staling of Beer. J. Inst. Brewing **79**, 414 (1973).

599. STRICKLER, H., und E. sz. KOVÁTS: Zur Kenntnis ätherischer Oele. 5. Mitt. Zwei Monoterpenoxide aus dem sog. „destillierten" Limetten-Oel (*Citrus medica* L., var. *acida* Brandis; *Citrus aurantifolia* Swingle). Helv. Chim. Acta **49**, 2055 (1966).

600. — — Zur Kenntnis ätherischer Oele. 6. Mitt. Die thermische Cyclisation des (−)-(*R*)-Linalools. Die Struktur der Plinole und einiger Derivate mit Iridan-Gerüst. Helv. Chim. Acta **50**, 759 (1967).

601. STRZELECKA, H., and T. SOROCZYNSKA: Polish Essential Oils. IX. Seed Oils of Different Varieties of *Daucus carota*. Farm. Pol. **30**, 13 (1974); Chem. Abstr. **81**, 68 331m (1974).

602. STURM, W.: Die gezielte Riechstoffsynthese — geruchliche Verwandtschaft der Isobutenyl- und Phenylgruppe. Parfüm, Kosmet. **55**, 351 (1974); Chem. Abstr. **82**, 160 100d (1975).

603. SULSER, H., J. DEPIZZOL, and W. BÜCHI: A Probable Flavoring Principle in Vegetable-Protein Hydrolysates. J. Food Sci. **32**, 611 (1967).

604. SUN, H. H., S. M. WARASZKIEWICZ, and K. L. ERICKSON: Sesquiterpenoid Alcohols from the Hawaiian Marine Alga *Laurencia nidifica* — III. Tetrahedron Letters **1976**, 585.

605. SUNDT, E., B. WILLHALM, R. CHAPPAZ und G. OHLOFF: Das organoleptische Prinzip von Cassis-Flavor im Buccublätteröl. Helv. Chim. Acta **54**, 1801 (1971).

606. SUNDT, E., B. WILLHALM et M. STOLL: Analyse des Parties Acides de l'Essence de Bergamote Saponifiée. Helv. Chim. Acta **47**, 408 (1964).

607. SUOMALAINEN, H., and M. LEKTONEN: Some Features of the Aroma Compounds of Alcoholic Beverages. Kemia-Kemi (Helsinki) **3**, 69 (1976).

608. SUOMALAINEN, H., and L. NYKANEN: Investigation on the Aroma of Alcoholic Beverages. Naeringsmiddelindustrien **23**, 15 (1970).

609. SUOMALAINEN, H., L. NYKANEN, and K. ERIKSSON: Composition and Consumption of Alcoholic Beverages — Review. Amer. J. Enol. Viticult. **25**, 179 (1974); Chem. Abstr. **82**, 110 249p (1975).

610. SUTHERLAND, O. R. W., and R. F. N. HUTCHINS: α-Farnesene, a Natural Attractant for Codling Moth Larvae. Nature **239**, 170 (1972).

611. SWIFT, C. E., F. G. DOLLEAR, L. E. BROWN, and R. T. O'CONNOR: Decomposition of Methyl Hydroperoxid oleate. J. Amer. Oil Chem. Soc. **25**, 39 (1948).

612. SYDOW, E. von: Flavor — a Chemical or Psychophysical Concept. Part 1 of the Underwood-Prescott Memorial Symposium on Food Quality and Flavor Research. Food Technol. **25**, 40 (1971); see also Z. Lebensm.-Unters. Forsch. **147**, 185 (1971).

613. SYDOW, E., von, and K. ANJOU: The Aroma of Rye Crispbread. Lebensm.-Wiss. Technol. **2**, 15 (1969).

614. — — The Aroma of Bilberries (*Vaccinium myrtillus* L.). I. Identification of Volatile Compounds. Lebensm.-Wiss. Technol. **2**, 78 (1969).

615. SYDOW, E. VON, and G. KARLSSON: The Aroma of Black Currants. IV. The Influence of Heat Measured by Instrumental Methods. Lebensm.-Wiss. Technol. **4**, 54 (1971).

616. TABACHEK, J. A. L., and M. YURKOWSKI: Isolation and Identification of Blue-Green Algae Producing Muddy Odor Metabolites, Geosmin, and 2-Methylisoborneol, in Saline Lakes in Manitoba. J. Fish. Res. Board Can. **33,** 25 (1976); Chem. Abstr. **84,** 101 985q (1976).

617. TAHARA, S., K. FUJIWARA, and J. MIZUTANI: Neutral Constituents of Volatiles in Cultured Broth of *Sporobolomyces odorus.* Agric. Biol. Chem. **37,** 2855 (1973).

618. TAKANO, T., and H. IKEDA (Kaken Industry Co., Ltd.): 1-Oxa-8-oxo-2,6,10,10-tetramethylspiro[4.5]dec-6-ene. Jap. Pat. 71.40.640 (1. 12. 1971); Chem. Abstr. **78,** 33810c (1973).

619. TAKEI, S., and M. ONO: Leaf Alcohol. III. Fragrance of Cucumber. J. Agric. Chem. Soc. Japan **15,** 193 (1939).

620. TALAPATRA, K.: Studies on the Volatile Compounds in the Aroma Fraction of Milk Caramel. Thesis, Pennsylvania State University, 1974 (Univ. Microfilm 74-21-031, Ann Arbor).

621. TALMAR, Y. P., M. C. NIGAM, and K. L. HANDA: *Ligusticum Elatum.* The Essential Oil and its Constituents. Soap, Perf. Cosm. **38,** 345 (1965).

622. TANAKA, S., A. YASUDA, H. YAMAMOTO, and H. NOZAKI: A General Method for the Synthesis of 1,3-Dienes. Simple Syntheses of β- and *trans*-α-Farnesene from Farnesol. J. Amer. Chem. Soc. **97,** 3252 (1975).

623. TANG, C. S., and W. G. JENNINGS: Lactonic Compounds of Apricot. J. Agric. Food Chem. **16,** 252 (1968).

624. TASKINEN, J.: Composition of the Essential Oil of Sweet Marjoram Obtained by Distillation with Steam and by Extraction and Distillation with Alcohol-Water Mixture. Acta Chem. Scand. B **28,** 1121 (1974).

625. TASKINEN, J.: 12-Methyl-ω-tridecanolide, a New Macrocyclic Lactone from Angelica Root Oil. Acta Chem. Scand. B **29,** 637 (1975).

626. TASKINEN, J., and L. NYKÄNEN: Volatile Constituents Obtained by the Extraction with Alcohol-Water Mixture and by Steam Distillation of Coriander Fruits. Acta Chem. Scand. B **29,** 425 (1975).

627. — —Chemical Composition of Angelica Root Oil. Acta Chem. Scand. B **29,** 757 (1975); *cis*- and *trans*-**(60)** have recently been confirmed to be constituents of angelica root oil (S. ESCHER and U. KELLER, Firmenich SA, unpublished results).

628. TEEGARDEN, R. W., and L. STEINBACH (International Flavors and Fragrances, Inc.): Cyclohexene and Hexadiene Carboxaldehydes and Their Derivatives. U.S. Pat. 2.947.780 (2. 8. 1960); Chem. Abstr. **55,** 2522d (1961).

629. TENCO BROOKE BOND Ltd.: Tea Flavour Enhancers. Brit. Pat. 1.306.017 (7. 2. 1973); Chem. Abstr. **78,** 122 965z (1973).

630. TERANISHI, R.: Odor and Molecular Structure. In: Gustation and Olfaction (G. OHLOFF and A. F. THOMAS, eds.), p. 165. London: Academic Press. 1971.

631. TERANISHI, R., R. G. BUTTERY, and D. G. GUADAGNI: Odor Quality and Chemical Structure in Fruit and Vegetable Flavors. Ann. New York Acad. Sci. **237,** 109 (1974).

632. TERANISHI, R., A. F. THOMAS, P. SCHUDEL, and G. BÜCHI: The Nomenclature of the Sinensals. Chem. Commun. **1968,** 928.

633. TERHUNE, S. J., J. W. HOGG, A. C. BROMSTEIN, and B. M. LAWRENCE: Four New Sesquiterpene Analogs of Common Monoterpenes. Canad. J. Chem. **53,** 3285 (1975).

634. TERRIER, A., J. N. BOIDRON, and P. RIBEREAU-GAYON: Identification of Terpenes in *Vitis vinifera* Grapes. C. R. hebd. séances Acad. Sci., Ser. D **275,** 495 (1972).

635. THAPPA, R. K., S. G. AGARWAL, V. N. VASHIST, C. K. ATAL, B. C. GULATI, and S. P. S. DUHAN: Essential Oil of *Elsholtzia pilosa.* Indian J. Chem. (B) **14 B,** 387 (1976); Chem. Abstr. **85,** 99 020w (1976).

636. THAPPA, R. K., V. N. VASHIST, J. SINGH, and R. K. SHARMA: Caryophyllene Epoxide from the Oil of *Artemisia scoparia, Elsholtzia polystacha, Piper hookeri,* and *Piper brachystachyum*. Current Sci. (India) **39**, 182 (1970); Chem. Abstr. **73**, 25 684w (1970).

637. THEIMER, E. T., and M. R. MCDANIEL: Odor and Optical Activity. J. Soc. Cosmet. Chem. **22**, 15 (1971).

638. THOMAS, A. F. (Firmenich SA, Geneva): Personal Communication.

639. — Unpublished Results.

640. — An Analysis of the Volatile Fraction of Hinoki Leaf Oil. Perf. Essent. Oil **56**, 301 (1965).

641. — Addition of a Funtionalized Isoprene Unit to an Allyl Alcohol. I. The Synthesis of β-Sinensal and Related Topics. J. Amer. Chem. Soc. **91**, 3281 (1969).

642. — The National Occurrence of the Campholenyl Skeleton, Another "Non-Isoprenoid" Monoterpene System. Helv. Chim. Acta **55**, 815 (1972).

643. — Geijerone (3-Isopropenyl-4-methyl-4-vinylcyclohexanone), a New Naturally Occurring C_{12} Terpenoid. Helv. Chim. Acta **55**, 2429 (1972).

644. — An Analysis of the Flavor of the Dried Mushroom, *Boletus edulis*. J. Agric. Food Chem. **21**, 955 (1973).

645. — 1,4-Dimethylcyclohex-3-enyl Methyl Ketone, a Monoterpenoid with a Novel Skeleton. Helv. Chim. Acta **56**, 1800 (1973).

646. — (Firmenich SA, Geneva): Utilisation d'un Dérivé du Cyclobutane comme Ingrédient Parfumant et/ou Aromatisant. [4-(2,2-Dimethyl-cyclobutyl)but-3-en-2-one as Ingredient in Perfumes and Foodstuffs for Imparting or Modifying Flavor]. Swiss Pat. 559 517 (14. 3. 1975).

647. — Heterocyclic Compound as Odor- and Flavor-Modifying Agent. U.S. Pat. 3.931.250 (6. 1. 1976).

648. THOMAS, A. F., and W. BUCHER: Menthatrienes and the Oxidation of Limonene. Helv. Chim. Acta **53**, 770 (1970).

649. THOMAS, A. F., and R. DUBINI: Terpenoids Derived from Linalyl Oxide. 3. The Isolation, Structure, Absolute Configuration and Synthesis of the Davanafurans, Nor-Sesquiterpenes Isolated from *Artemisia pallens*. Helv. Chim. Acta **57**, 2006 (1974).

650. — — Terpenoids Derived from Linalyl Oxide. 4. The Oxidation of Davanone. Isolation and Synthesis of the Davana Ethers, Sesquiterpenes of *Artemisia pallens*. Helv. Chim. Acta **57**, 2076 (1974).

651. THOMAS, A. F., and M. OZAINNE: "Junionone" [1-(2,2-Dimethylcyclobutyl)but-1-en-3-one] the First Vegetable Monocyclic Cyclobutane Monoterpenoid. Chem. Commun. **1973**, 746.

652. — — Terpenoids Derived from Linalyl Oxide. 2. The Isolation and Synthesis of Nordavanone, a C_{11}-Terpenoid from *Artemisia pallens*. Helv. Chim. Acta **57**, 2062 (1974).

653. THOMAS, A. F., and G. PITTON: The Isolation, Structure and Synthesis of Davana Ether, an Odoriferous Compound of the Oil of *Artemisia pallens* Wall. Helv. Chim. Acta **54**, 1890 (1971).

654. THOMAS, A. F., and B. WILLHALM: The Oil of *Santolina chamaecyparissus* L. A New, Non-Isoprenoid C_{10} Hydrocarbon. Tetrahedron Letters **1964**, 3775.

655. TIMMER, R., R. TER HEIDE, P. J. DE VALOIS, and H. J. WOBBEN: Analysis of the Lactone Fraction of Lavender Oil (*Lavendula vera* D.C.). J. Agric. Food Chem. **23**, 53 (1975).

656. TIRIMANA, A. S. L., and R. L. WICKREMASHINGHE: The Quality and Flavor of Tea. II. The Carotenoids. Tea Quart. **36**, 115 (1965); Chem. Abstr. **64**, 7284d (1966).

657. TISCHER, J.: Über die Carotinoide und die Bildung von Jonon in Trentepohlia nebst Bemerkungen über den Gehalt dieser Alge an Erythrit. (Carotinoide der Süßwasseralgen, II.) Z. physiol. Chem. **243**, 103 (1936).

658. TREIBS, W.: Über das Caryophyllenoxyd, seine Darstellung durch Autoxydation des Caryophyllens und sein Vorkommen in Pflanzenölen. Chem. Ber. **80**, 56 (1947).

659. TREIBS, W.: Über bi- und polycyclische Azulene. XIII. Das bicyclische Caryophyllen als Azulenbildner. Liebigs Ann. Chem. **576,** 125 (1952).

660. TRESSL, R. (Technische Universität, Berlin): Private Communication.

661. TRESSL, R., F. DRAWERT und W. HEIMANN: Gaschromatographisch-massenspektrometrische Bestandsaufnahme von Erdbeer-Aromastoffen. Z. Naturforsch. **24b,** 1201 (1969); Chem. Abstr. **72,** 28 877b (1970).

662. TRESSL, R., und R. RENNER: Bildung von Aromastoffen in Lebensmitteln. Deutsche Lebensmittel-Rundschau **72,** 37 (1976).

663. TRIVEDI, G. K., P. S. KALSI, and K. K. CHAKRAVARTI: Terpenoids. LIV. Structure and Absolute Configuration of Khusitone. Tetrahedron **20,** 2631 (1964).

664. UMRANI, D. C., R. SESHADRI, K. G. GORE, and K. K. CHAKRAVARTI: Terpenoids. CXL. Khusimone — a New C_{14}-Ketone from Vetiver Oil. Flavour Ind. **1,** 623 (1970).

665. U.S. Flavoring Extract Manufacturer's Association (FEMA): GRAS substances. Recent Progress in the Consideration of Flavoring Ingredients under the Food Additives Amendment (R. L. HALL, B. L. OSER et al., eds.). In: Food Technology **19** (2, part 2), 151 (1965); **24** (5), 25 (1970); **26** (5), 35 (1972); **27** (1), 64 (1973); (**27**) (11), 56 (1973); **28** (9), 76 (1974); **29** (8), 70 (1975); **31** (1), 65 (1977).

666. VARO, P. T., and D. E. HEIZ: Volatile Components of Cumin Seed Oil. J. Agric. Food Chem. **18,** 234 (1970).

667. — — Identification and Characterization of 1,4-*p*-menthadien-7-al Isolated from Cumin Seeds. J. Agric. Food Chem. **18,** 239 (1970).

668. VIANI, R., J. BRICOUT, J. P. MARION, F. MÜGGLER-CHAVAN, D. REYMOND et R. H. EGLI: Sur la Composition de l'Arôme de Tomate. Helv. Chim. Acta **52,** 887 (1969).

669. VIANI, R., F. MÜGGLER-CHAVAN, D. REYMOND et R. H. EGLI: Sur la Composition de l'Arôme de Café. Helv. Chim. Acta **48,** 1809 (1965).

670. VIDARA, G., M. DE BERNARDI, M. PAVAN, and L. RAGOZZINO: Rose Oxide and Iridodial from *Aromia moschata* L. (Coleoptera: Cerambycidae). Tetrahedron Letters **1973,** 4065.

671. VIG, O. P., O. P. CHUGH, V. K. HANDA, and A. K. VIG: Terpenoids. 83. Synthesis of 2-Methyl-6-methylene-2,7-octadiene-1-ol. J. Indian Chem. Soc. **52,** 161 (1975).

672. VITÉ, J. P., A. BAKKE und P. R. HUGHES: Ein Populationslockstoff des zwölfzähnigen Kiefernborkenkäfers *Ips sexdentatus.* Naturwiss. **61,** 365 (1974). Behavior Studies with Ipsenol and Ipsdienol.

673. VITZTHUM, O. G., und P. WERKHOFF: Neuentdeckte Stickstoffheterocyclen im Kaffee-Aroma. Z. Lebensm.-Unters. Forsch. **156,** 300 (1974).

674. VODOZ, C.-A.: Flavour Legislation. New Developments and Trends. Internation. Flavours **1975,** 219.

675. WAKAYAMA, S., and S. NAMBA: Lilac Aldehydes. Bull. Chem. Soc. Japan **47,** 1293 (1974); Chem. Abstr. **81,** 77 755e (1974).

676. WAKAYAMA, S., S. NAMBA, and M. OHNO: Lilac Alcohol-a and -b, New Naturally Occurring Odor Ingredients. Bull. Chem. Soc. Japan **43,** 3319 (1970); Chem. Abstr. **74,** 91 089c (1971).

676a. — — — Odorous Constituents of Lilac Flower Oil. Nippon Kagaku Zasshi **92,** 256 (1971); Chem. Abstr. **76,** 37 341g (1972).

677. WALRADT, J. P., R. C. LINDSAY, and L. M. LIBBEY: Popcorn Flavor: Identification of Volatile Compounds. J. Agric. Food Chem. **18,** 926 (1970).

678. WALRADT, J. P., A. O. PITTET, T. E. KINLIN, R. MURALIDHARA, and A. SANDERSON: Volatile Components of Roasted Peanuts. J. Agric. Food Chem. **19,** 972 (1971).

679. WANG, P. S., H. KATO, and M. FUJIMAKI: Studies on Flavor Components of Roasted Barley. IV. The Volatile Sulfur Compounds, Fatty Acids and Neutral Non-Carbonyl Oxygenated Compounds. Agric. Biol. Chem. **34,** 561 (1970).

680. WANG, P. S., and G. V. ODELL: Characterization of Some Volatile Constituents of Roasted Pecans. J. Agric. Food Chem. **20,** 206 (1972).

681. WARTBURG, B. R. VON, und H. R. WOLF: Ein neuer, ergiebiger Zugang zu Dihydroactinodiolid und α-substituierten Furanderivaten. Helv. Chem. Acta **57,** 916 (1974).

682. WASOWICZ, E., and E. KAMINSKI: Aroma Compounds of the Mushroom *Boletus edulis.* Przem. Spozyw. **28,** 269 (1974); Chem. Abstr. **82,** 108 847p (1975).

683. WATANABE, K., and Y. SATO: Studies on the Changes of Meat Fats by Various Processing. II. Gas Chromatographic Identification of Aliphatic γ- and δ-Lactones Obtained from Beef Fats. Agric. Biol. Chem. **32,** 191 (1968).

684. — — Changes of Meat Fats by Various Processings. III. Aliphatic γ- and δ-Lactones in Meat Fats. Agric. Biol. Chem. **32,** 1318 (1968).

685. — — Volatile Acidic Compounds in Heatdegradated Pork Fat. Agric. Biol. Chem. **33,** 1411 (1969).

686. — — Lactones in the Flavor of Heated Pork Fat. Agric. Biol. Chem. **33,** 242 (1969).

687. — — Gas Chromatographic and Mass Spectral Analyses of Heated Flavor Compounds of Beef Fats. Agric. Biol. Chem. **35,** 756 (1971).

688. — — Lactones Produced through Thermal Oxidation of Higher Fatty Acids. Agric. Biol. Chem. **35,** 278 (1971).

689. WEBB, A. D.: Volatile Aroma Compounds of Wines. In: Symposium on Foods: The Chemistry and Physiology of Flavors. The Fourth in a Series of Symposia on Foods Held at Oregon State University, p. 203. Westport, Conn.: The Avi Publ. Comp., Inc. 1967.

690. WEBB, A. D., and R. E. KEPNER: The Aroma of Flor Sherry. Amer. J. Enol. Viticult. **13,** 1 (1962); Chem. Abstr. **57,** 3873i (1962).

691. WEBB, A. D., R. E. KEPNER, and L. MAGGIORA: Sherry Aroma. VI. Some Volatile Components of Flor Sherry of Spanish Origin — Neutral Substances. Amer. J. Enol. Viticult. **18,** 190 (1967); Chem. Abstr. **68,** 76915b (1968).

692. WEBB, A. D., P. RIBEREAU-GAYON et J. N. BOIDRON: Composition d'une Essence Extraite d'un Vin de *V. vinifera* (Variété Cabernet-Sauvignon). Bull. soc. chim. France **1964,** 1415.

693. WEISS, G., M. KOREEDA, and K. NAKANISHI: Stereochemistry of Theaspirone and the Blumenols. Chem. Commun. **1973,** 565.

694. WEURMANN, C., S. VAN STRATEN, and F. de VRIJER: Lists of Volatile Compounds in Food, 3rd edit., June 1973, with suppl. 1 to 7, 1976. Zeist, Holland: Central Institut for Nutrition and Food Research TNO.

695. WHITFIELD, F. B., G. STANLEY, and K. E. MURRAY: Concerning the Structures of Edulan I and II. Tetrahedron Letters **1973,** 95.

696. WIGGERS DE VRIES, J. A.: Modern Trends in Flavor Development. Amer. Cosm. Perf. **87,** 37 (1972).

697. WIJESEKERA, R. O. B., A. L. JAYEWARDENE, and L. S. RAJAPAKSE: Volatile Constituents of Leaf, Stem and Root Oils of Cinnamon (*Cinnamomum zeylanicum*). J. Sci. Food Agric. **25,** 1211 (1974).

698. WILKENS, W. F., and F. M. LIN: Gas Chromatographic and Mass Spectral Analyses of Soybean Milk Volatiles. J Agric. Food Chem. **18,** 333 (1970).

699. — — Volatile Flavor Components of Deep Fat-Fried Soybeans. J. Agric. Food Chem. **18,** 337 (1970).

700. WILLHALM, B., and A. F. THOMAS: Synthesis and Structure of Yomogi Alcohol, and Alcohol Related to Artemisia Alcohol. Chem. Commun. **1969,** 1380.

701. WILLSTÄTTER, R., und H. H. ESCHER: Über den Farbstoff der Tomate. Z. physiol. Chem. **64,** 47 (1910).

702. WILSON III, C. W.: Terpene and Sesquiterpene Hydrocarbons in the Essential Oil from Fresh Celery. J. Food Sci. **34,** 521 (1969).

703. WILSON III, C. W.: Relative Recovery and Identification of Carbonyl Compounds from Celery Essential Oil. J. Food Sci. **35,** 766 (1970).
704. WILSON, R. A.: A Review of Thermally Produced Imitation Meat Flavors. J. Agric. Food Chem. **23,** 1032 (1975).
705. WILSON, R. A., and I. KATZ: Review of Literature on Chicken Flavor and Report of Isolation of Several New Chicken Flavor Components from Aqueous Cooked Chicken Broth. J. Agric. Food Chem. **20,** 741 (1972).
706. WINTER, M. et al. (Firmenich SA, Geneva): To be Published.
707. WINTER, M., and P. ENGGIST (Firmenich SA, Geneva): Private Communication.
708. — — Recherches sur les Arômes. 17. comm. Sur l'Arôme de Framboise, IV. Helv. Chim. Acta **54,** 1891 (1971).
709. WINTER, M., A. FURRER, B. WILLHALM, and W. THOMMEN: Identification and Synthesis of Two New Organic Sulfur Compounds from the Yellow Passion Fruit *(Passiflora edulis* f. *flavicarpa).* Helv. Chim. Acta **59,** 1631 (1976).
710. WINTER, M., F. GAUTSCHI, I. FLAMENT, B. WILLHALM und M. STOLL: Die flüchtigen Aromastoffe des Kaffees. In: Aroma- und Geschmacksstoffe in Lebensmitteln, p. 165. Zürich: Forster-Verlag AG. 1967.
711. WINTER, M., und R. KLÖTI: Über das Aroma der gelben Passionsfrucht *(Passiflora edulis* f. *flavicarpa).* Helv. Chim. Acta **55,** 1916 (1972).
712. WINTER, M., G. MALET, M. PFEIFFER et E. DEMOLE: Structure d'une Lactone Odorante Présente dans l'Essence de Jasmin (*Jasminum grandiflorum* L.). Helv. Chim. Acta **45,** 1250 (1962);
DEMOLE, E., et M. WINTER: Synthèse du *cis*-(Pentène-2-yl)-5-pentanolide-(5,1), Composant Odorant de l'Essence de Jasmin (*Jasminum grandiflorum* L.), ibid. 1256.
713. WINTER, M., et E. SUNDT: Recherches sur les Arômes. 5. comm. Analyse de l'Arôme des Framboises, I. Les Constituants Carbonylés Volatils. Helv. Chim. Acta **45,** 2195 (1962).
714. WINTER, M., and B. WILLHALM: Sur l'Arôme des Fraises Fraîches. Analyse des Composés Carbonylés, Esters et Alcools Volatils. Helv. Chim. Acta **47,** 1215 (1964).
715. WOLINSKY, J., and W. BARKER: The Synthesis of 1-Acetyl-4-isopropenyl-1-cyclopentene. J. Amer. Chem. Soc. **82,** 636 (1960).
716. WOLINSKY, J., and D. NELSON: The Synthesis of (+)-Matatabiether and Related Methylcyclopentane Monoterpenes. Tetrahedron **25,** 3767 (1969).
717. WONG, E., L. N. NIXON, and C. B. JOHNSON: Volatile Medium Chain Fatty Acids and Mutton Flavor. J. Agric. Food Chem. **23,** 495 (1975).
718. YAMAMOTO, R., I. YAMAMOTO, and T. YOSHIZAWA (Suntory Ltd. and Shiono Koryo Kaishi Ltd., Osaka): Method for Giving an Increasing Flavor Resembling Dairy Products. U.S. Pat. 3.904.780 (9. 9. 1975).
719. YAMANISHI, T., M. KAWATSU, T. YOKOYAMA, and Y. NAKATANI: Methyl Jasmonate and Lactones Including Jasmine Lactone in Ceylon Tea. Agric. Bull. Chem. **37,** 1075 (1973).
720. YAMANISHI, T., T. KIRIBUCHI, Y. MIKUMO, H. SATO, and A. OMURA: The Flavor of Green Tea. VI. Neutral Fraction of the Essential Oil of Tea Leaves. Agric. Biol. Chem. **29,** 300 (1965).
721. YAMANISHI, T., A. KOBAYASHI, H. NAKAMURA, A. UCHIDA, S. MORI, K. OHSAWA, and S. SASAKURA: Flavor of Black Tea. V. A Comparison of Aroma of Various Types of Black Tea. Agric. Biol. Chem. **32,** 379 (1968).
722. YAMANISHI, T., M. NOSE, and Y. NAKATANI: Studies on the Flavor of Green Tea. VIII. Further Investigation of Flavor Constituents in Manufactured Green Tea. Agric. Biol. Chem. **34,** 599 (1970).
723. YASUMATSU, N., M. EDA, Y. TSUJINO, and M. NOGUCHI: Isolation of Oxygenated Derivatives of Solanesol from Burley Tobacco. Agric. Biol. Chem. **40,** 1757 (1976).

724. Yeh, P. H.: The Essential Oil of *Ocimum gratissimum* var. *snavis* Hook. Perf. Ess. Oil Rec. **51,** 611 (1960); Chem. Abstr. **55,** 4889b (1961).

725. Yoshida, T., S. Muraki, H. Kawamura, and A. Komatsu: Minor Constituents of Japanese Ho-Leaf Oil. The Structures of (+)- Tagetonol and (−)-*trans*-Hotrienol. Agric. Biol. Chem. **33,** 343 (1969).

726. Zarghami, N. S., and D. E. Heinz: The Volatile Constituents of Saffron. Lebensm.-Wiss. Technol. **4,** 43 (1971).

727. Zarghami, N. S., and D. E. Heinz: Monoterpene Aldehydes and Isophorone-Related Compounds of Saffron. Phytochem. **10,** 2755 (1971).

728. Zijden, A. S. M. van der, K. de Jong, D. Sloot, J. Clifford et R. J. Taylor: Constituants de la Graisse de Beurre Présents a l'Etat de Traces. IV. Isolement et Identification de Lactones Aliphatiques Insaturées. Rev. Fr. Corps Gras **13,** 731 (1966); Chem. Abstr. **66,** 75 049u (1967).

(Received April 30, 1977)

Namenverzeichnis. Author Index

Kursiv gedruckte Seitenzahlen beziehen sich auf Literaturverzeichnisse

Page numbers printed in *italics* refer to References

Sachverzeichnis. Subject Index

Von · By

A. SIEGEL, Wien

Fortschritte der Chemie organischer Naturstoffe
Progress in the Chemistry of Organic Natural Products

All Volumes and Cumulative Index 1—20 available / Alle Bände und Generalregister 1—20 lieferbar.

Price reduction for subscribers / Preisermäßigung für Subskribenten: 10%.

Special price reduction (20% of the list price) for the Vols. 1—20 plus Cumulative Index. / Vorzugspreis (20% Nachlaß) bei Bezug der Bände 1—20 inklusive Generalregister.

Volume 32: 31 figures. VIII, 560 pages. 1975. ISBN 3-211-81255-5.

Contents: W. K. SEIFERT, Carboxylic Acids in Petroleum and Sediments – P. G. SAMMES, Naturally Occurring 2,5-Dioxopiperazines and Related Compounds – R. J. HIGHET and E. A. SOKOLOSKI, Structural Investigations of Natural Products by Newer Methods of NMR Spectroscopy – P. M. SCOPES, Applications of the Chiroptical Techniques to the Study of Natural Products – H. C. VAN HUMMEL, Chemistry and Biosynthesis of Plant Galactolipids – H. KÖSSEL and H. SELIGER, Recent Advances in Polynucleotide Synthesis – Author Index – Subject Index.

Volume 33: 48 figures. VIII, 581 pages. 1976. ISBN 3-211-81357-8.

Contents: L. MINALE, G. CIMINO, S. DE STEFANO, and G. SODANO, Natural Products from Porifera – R. M. COATES, Biogenetic-Type Rearrangements of Terpenes – K. L. RINEHART, JR., and L. S. SHIELD, Chemistry of the Ansamycin Antibiotics – A. FONTANA and C. TONIOLO, The Chemistry of Tryptophan in Peptides and Proteins – P. HEMMERICH, The Present Status of Flavin and Flavocoenzyme Chemistry – Author Index – Subject Index.

Volume 34: 63 figures. X, 620 pages. 1977. ISBN 3-211-81415-9.

Contents: C. R. ENZELL, I. WAHLBERG, and A. J. AASEN, Isoprenoids and Alkaloids of Tobacco – A. R. PINDER, The Chemistry of the Eremophilane and Related Sesquiterpenes – D. GROSS, Phytoalexine und verwandte Pflanzenstoffe – K. H. OVERTON and D. J. PICKEN, Studies in Secondary Metabolism with Plant Tissue Cultures – D. P. CHAKRABORTY, Carbazole Alkaloids – J. JACOB, Bürzeldrüsenlipide – W. VOELTER, Hypothalamus-Regulationshormone – Author Index – Subject Index.

Springer-Verlag Wien · New York

GPSR Compliance
The European Union's (EU) General Product Safety Regulation (GPSR) is a set of rules that requires consumer products to be safe and our obligations to ensure this.

If you have any concerns about our products, you can contact us on

ProductSafety@springernature.com

In case Publisher is established outside the EU, the EU authorized representative is:

Springer Nature Customer Service Center GmbH
Europaplatz 3
69115 Heidelberg, Germany

www.ingramcontent.com/pod-product-compliance
Lightning Source LLC
Chambersburg PA
CBHW071406230926
51846CB00001B/21
9783709185070